Lower Previsions

A complete list of the titles in this series appears at the end of this volume.

Lower Previsions

Matthias C.M. Troffaes
Department of Mathematical Sciences, Durham University, UK

Gert de Cooman
SYSTeMS Research Group, Ghent University, Belgium

WILEY

This edition first published 2014

Registered office
John Wiley & Sons Ltd, The Atrium, Southern Gate, Chichester, West Sussex, PO19 8SQ, United Kingdom

For details of our global editorial offices, for customer services and for information about how to apply for permission to reuse the copyright material in this book please see our website at www.wiley.com.

Library of Congress Cataloging-in-Publication Data

Troffaes, Matthias C. M., 1977 –
Lower previsions / Matthias C.M. Troffaes. Gert de Cooman.
pages cm
Includes bibliographical references and index.
ISBN 978-0-470-72377-7 (cloth)
1. Probabilities. 2. Statistical decision. 3. Games of chance (Mathematics) I. Cooman, Gert de. II. Title.
QA273.T755 2014
519.2 – dc23

2014002217

A catalogue record for this book is available from the British Library.

ISBN: 978-0-470-72377-7

Set in 10/12pt TimesLTStd by Laserwords Private Limited, Chennai, India
Printed and bound in Singapore by Markono Print Media Pte Ltd

1 2014

Matthias dedicates this book to Sabine, Nausicaa, and Linus;
Gert to Wietse, Linde and Reine.

Contents

Preface

Lower Previsions is an overview of, and reference guide to, the mathematics of lower previsions. Starting from first principles – acceptability – we derive their mathematical properties and relate them to a wide range of other uncertainty models – belief functions, Choquet capacities, possibility measures – and mathematical concepts – including filters, limits, propositional logic, integration and many other constructs from functional and convex analysis.

The material of the book is advanced and aimed at researchers, postgraduate students and lecturers. It will be of interest to statisticians, probabilists, mathematicians and anyone whose field of interest includes some form of uncertainty modelling, both from a practical and a theoretical point of view.

Work on this book started about 8 years ago. The idea was, at that time, to turn the most important results in Matthias's PhD thesis, supervised by Gert, into a coherent and more or less self-contained research monograph. Our initial plan was to focus on two things: first, the relationship between natural extension and integration and second, the discussion of lower previsions defined on unbounded gambles.

It soon became clear that, in order to make the book more self-contained, we needed to include much more material on lower previsions themselves. At the same time, we gathered from conversations with close colleagues that there was a definite interest in – given the perceived lack of – a comprehensive treatment of the existing theory of lower previsions. And so we decided to include, besides our own, a number of contributions from other people, amongst whom in particular are Peter Williams, Peter Walley, Sebastian Maaß, Dieter Denneberg and Enrique Miranda.

The present book, therefore, differs significantly from the one we started out with. While initially, the book was mostly focused on Matthias's PhD work, in its present form, it contains much more material and both authors have contributed to it on an equal footing.

In the first part of this book, we expose and expand on the main ideas behind the theory that deals exclusively with bounded gambles. We also discuss a wide variety of special cases that may be of interest when implementing these ideas in practical problems. In doing so, we demonstrate the unifying power behind the concept of coherent lower previsions, for uncertainty modelling as well as for functional analysis. In the second part of this book, we extend the scope of the theory of lower previsions by allowing it to deal with real gambles that are not necessarily bounded. In that part, we also deal with conditioning and provide practical constructions for extending lower previsions to unbounded gambles. For more in-depth overviews of each part, refer to Chapter 2 (p. 23) and Chapter 12 (p. 233). We have tried to make this book as self-contained as possible. This means, amongst other things, that we have tried to at least provide an explicit formulation – if not an actual proof – of most results that we use. We have relegated to a number of appendices supporting material that did not fit nicely into the main storyline.

If you are used to a measure-theoretic approach to probability, you may initially feel somewhat lost in this book, because we do not start out with measurability at all. Indeed, the foundations of lower previsions, for arbitrary spaces, do not rely on any notion of measurability. This may come as a surprise to some people who think that using measurability is natural and should come first. Instead, our discussion of lower previsions is founded on a notion of acceptability of gambles, which has a direct behavioural interpretation. In other words, rather than posing laws of probability, we pose laws of acceptability, from which laws of probability are derived.

As often as possible, we give detailed accounts of most steps in the proofs, with explicit references to other results that are being used. This may appear to be pedantic – or even worse, condescending – to some, but we thought it better to be too specific rather than incur the risk of explaining too little: as this is not a small book, we cannot expect any reader to remember every little result we have proved or mentioned earlier.

Because this is a rather large book, we have used a number of typographical and other devices to help readers find their way through it. Newly defined terms are set **in bold** when they appear for the first time. Things we want to emphasise are set *in italics*. Sections, theorems, definitions, examples and so on are numbered consecutively throughout each chapter. So are the more important equations. Because we need to refer continuously to material elsewhere in the book, such as chapters, sections, theorems and statements therein, equations, properties and so on, we tried to find a way to make it easy for a reader to find it. After a lot of experimentation, we chose the simple device of explicitly mentioning the page number, but doing so in as unobtrusive a manner as possible, by turning it into a subscript. To give a few examples, Section 3.4_{29} starts on p. 29, Equation $(11.15)_{225}$ can be found on p. 225, and Theorem $4.13(\text{xiv})_{53}$ refers to statement (xiv) of Theorem 4.13, and this theorem can be found on p. 53. This page number is omitted when we refer to a result on the same page, and previous and following pages are referenced by special symbols ↶ and ↷.

We hope that you will enjoy reading and working with this book as much as we have enjoyed researching and writing it.

Matthias Troffaes and Gert de Cooman
Durham and Ghent

Acknowledgements

Work on this book has been supported by various grants, institutions and people.

A substantial part of the research reported in this book is the result of Matthias's PhD work (2000–2005), which was supported by project G.0139.01 of the Fund for Scientific Research, Flanders (Belgium), and by the Belgian Programme on Interuniversity Poles of Attraction initiated by the Belgian state, Prime Minister's Office for Science, Technology and Culture. As always, the scientific responsibility rests with the authors. From July 2005 until July 2006, Matthias's work on the book was supported by the Belgian American Educational Foundation during his visit to the Carnegie Mellon University. Between 2006 and 2013, Grey College (Durham University) and the London Mathematical Society supported Gert during various visits to Durham.

We thank Richard Davies from Wiley for his friendly support, eternal patience, kind understanding and ever gentle shepherding. From Grey College, we are particularly indebted to the unforgettably kind, generous and exacting Henry Dyson. We are grateful to Teddy Seidenfeld from Carnegie Mellon University for stimulating much of the research reported in this book, with a never ending stream of counterexamples and deep foundational arguments, ranging from zero probability to infinite utility. We also thank Minou van de Kerkhove for her amazing design of the first version of the cover, which was the one that we really wanted. Her work was the starting point for the design people at Wiley, who tried their best to reconcile her ideas with the format imposed by the series.

We could not have written this book without the support of many of our colleagues and friends at work, who create the stimulating environment of companionable challenge that research can thrive in. We also thank our families, partners and children, who have been so generous in understanding and fostering the passion that kept us engaged in this work for such a long time.

Finally, we owe a great intellectual debt to Peter Walley. Although he has not been directly involved in this book, his work has been a continuous source of inspiration and will doubtlessly remain so for many a generation to come.

1

Preliminary notions and definitions

Before we set out on our exploration of lower previsions, we start with a brief overview of the basic notation used throughout this book. More specific notation will be introduced in further chapters, as we go along. Refer to Appendix E_{396} for a comprehensive list of symbols and notations used in the book.

We also recall and prove a few basic definitions and results from elementary measure and functional theory, as well as topology. We need to rely on these quite often throughout the book, so it seems convenient to herd them all into this easily accessible corner. Of course, the reader may wish to skip this chapter and simply refer back to it when necessary.

Somewhat more detailed introductions to specific topics, such as linear spaces, topology, Choquet integration, and extended real numbers, can be found in the appendices.

We assume that the reader is familiar with the basics of set theory and logic (proposition, set, union, intersection, complementation, inclusion, relation, function, map), as well as calculus (derivative, Riemann integral, sequence).

1.1 Sets of numbers

$\mathbb{N}$ denotes the set of natural numbers including zero. Sometimes, we need infinity as well, or we do not need zero; in these cases, we write

$$\mathbb{N}^* := \mathbb{N} \cup \{\infty\} \text{ and } \mathbb{N}_{>0} := \mathbb{N} \setminus \{0\} \text{ and } \mathbb{N}^*_{>0} := \mathbb{N}^* \setminus \{0\}.$$

As usual, $\mathbb{Z}$ is the set of integers, $\mathbb{Q}$ is the set of rational numbers and $\mathbb{R}$ is the set of real numbers. The set of extended real numbers is denoted by

$$\mathbb{R}^* := \mathbb{R} \cup \{-\infty, +\infty\}.$$

Lower Previsions, First Edition. Matthias C.M. Troffaes and Gert de Cooman.

We say that a real number a is **positive** when $a > 0$, **negative** when $a < 0$, **non-positive** when $a \leq 0$, and **non-negative** when $a \geq 0$. The set of non-negative real numbers is denoted by $\mathbb{R}_{\geq 0}$, and the set of strictly positive real numbers is denoted by $\mathbb{R}_{>0}$:

$$\mathbb{R}_{\geq 0} := \{x \in \mathbb{R} : x \geq 0\} \text{ and } \mathbb{R}_{>0} := \{x \in \mathbb{R} : x > 0\} = \mathbb{R}_{\geq 0} \setminus \{0\}.$$

For any subset B of $\mathbb{R}^*$, we define its **supremum** $\sup B$ and **infimum** $\inf B$ as its smallest upper and greatest lower bound, respectively:

$$\sup B := \min \left\{a \in \mathbb{R}^* : a \geq b \text{ for all } b \in B\right\}$$
$$\inf B := \max \left\{a \in \mathbb{R}^* : a \leq b \text{ for all } b \in B\right\}.$$

Addition and multiplication can be extended from the real to the extended real numbers in a straightforward manner; refer to the extensive discussion in Appendix D_{391} for more details. The only case that deserves some care is the sum of $-\infty$ and $+\infty$, which is undefined. We call a sum $a + b$ of two extended real numbers **well defined** whenever it cannot be reduced to the cases $-\infty + (+\infty)$ or $+\infty + (-\infty)$.

1.2 Gambles

In this section, we are mainly concerned with real-valued maps defined on a non-empty set, or **space**, $\mathscr{X}$. Because such maps play such an essential role in the theory of lower previsions, they are given a special name: following Walley (1991), a real map is called a **gamble** – however, in contrast to Walley's (1991) use of the term gamble, in this book, gambles need not be bounded. When we consider bounded gambles, we always explicitly say so. We denote the set of all gambles by $\mathbb{G}(\mathscr{X})$.

In the course of this book, we shall come across many other sets of this type that contain special types of gambles, defined on sets such as $\mathscr{X}$. When no confusion can arise as to what underlying set $\mathscr{X}$, we are talking about, we often omit explicit reference to $\mathscr{X}$. For instance, we simply write $\mathbb{G}$ instead of $\mathbb{G}(\mathscr{X})$, when the underlying space $\mathscr{X}$ is clear from the context.

If $*$ is a binary operation on the set of real numbers $\mathbb{R}$, then we extend $*$ point-wise to a binary operation on gambles as follows: for any two gambles f and g on $\mathscr{X}$, $f * g$ is the gamble on $\mathscr{X}$ defined by

$$(f * g)(x) := f(x) * g(x) \text{ for all } x \in \mathscr{X}.$$

This allows us, for instance, to consider the gambles $f + g, f - g$ and fg.

Gambles can be ordered point-wise: for any pair of gambles f and g on $\mathscr{X}$, we say that

$$f \leq g \text{ whenever } (\forall x \in \mathscr{X})(f(x) \leq g(x)), \tag{1.1}$$

where of course the '$\leq$' on the right-hand side is the usual ordering of real numbers. We also use the notation '$f < g$' for '$f \leq g$ and not $f = g$'. The binary relation $\leq$ on $\mathbb{G}$ is a **partial order**: it is reflexive, antisymmetrical and transitive.[1] The **supremum**, or smallest upper bound, $f \vee g$ of two gambles f and g on $\mathcal{X}$ with respect to this partial order is given by their point-wise maximum, and their **infimum**, or greatest lower bound; $f \wedge g$ is given by their point-wise minimum, so

$$(f \vee g)(x) := \max\{f(x), g(x)\} \text{ and } (f \wedge g)(x) := \min\{f(x), g(x)\} \text{ for all } x \text{ in } \mathcal{X}.$$

These operators should not be confused with the **supremum** $\sup f$ and **infimum** $\inf f$ of a gamble f on $\mathcal{X}$, which are defined as the extended real numbers:

$$\sup f := \sup\{f(x) : x \in \mathcal{X}\} = \min\left\{a \in \mathbb{R}^* : a \geq f\right\}$$
$$\inf f := \inf\{f(x) : x \in \mathcal{X}\} = \max\left\{a \in \mathbb{R}^* : a \leq f\right\}.$$

If $\sup f$ is finite, then we say that the gamble f is **bounded above**, and similarly, f is **bounded below** if $\inf f$ is finite. We say that f is **bounded** when it is bounded above and below. We denote the set of all bounded gambles on $\mathcal{X}$ by $\mathbb{B}(\mathcal{X})$, or more simply by $\mathbb{B}$.

We also define the **absolute value** $|f|$ of a gamble f by $|f|(x) := |f(x)|$ for all x in $\mathcal{X}$, and the **negation** $-f$ of f as $(-f)(x) := -f(x)$ for all x in $\mathcal{X}$. Clearly, $f + (-g)$ is the same thing as $f - g$.

It is convenient to identify a real number $a \in \mathbb{R}$ with the constant gamble $a(x) := a$ for all $x \in \mathcal{X}$.[2] For instance, the expression $a \geq f$, where f is a gamble on $\mathcal{X}$ and a is a real number, means that $a \geq f(x)$ for all x in $\mathcal{X}$; we have already used this earlier in our definition of the supremum and infimum of a gamble f. As another example, $f \geq 0$ means that the gamble f is nowhere negative (or equivalently, everywhere non-negative): we simply say that f is **non-negative**. Similarly, we say that f is **non-positive** when $f \leq 0$. For another example, we define the **scalar product** λf of a real number λ and a gamble f as the point-wise product to the constant gamble λ and f:

$$(\lambda f)(x) := \lambda f(x) \text{ for all } x \text{ in } \mathcal{X}.$$

As indicated earlier, we make one important exception to the rule of interpreting binary operators in a point-wise manner: we take $f < 0$ to mean that the gamble f is nowhere positive (or equivalently, everywhere non-positive), and actually negative in at least one element of its domain:

$$f < 0 \text{ whenever } f \leq 0 \text{ and } f \neq 0; \tag{1.2}$$

we then say that f is **negative**. Similarly, we call f **positive** and write $f > 0$ when $-f < 0$, or equivalently, $f \geq 0$ and $f \neq 0$.

[1] For a detailed discussion of partially ordered sets, refer to Davey and Priestley (1990) or Birkhoff (1995). Basic definitions can also be found in Appendix A_{368}.

[2] As it would be silly and even pedantic to distinguish notationally between constant gambles and real numbers, we also use the notation $\mathbb{R}$ for the set of all constant gambles on $\mathcal{X}$.

Given a gamble f, we can consider its **non-negative part** $f^+ := 0 \vee f$ and its **non-positive part** $f^- := -(0 \wedge f)$. These names are inspired by the fact that $f^+ \geq 0$, $-f^- \leq 0$, $f = f^+ - f^-$ and $|f| = f^+ + f^-$. Observe that also $f^+ \wedge f^- = 0$ and $f^+ \vee f^- = |f|$.

Definition 1.1 (Special sets of gambles) *We call a subset $\mathscr{K}$ of $\mathbb{G}$,*

(a) ***negation-invariant*** *if it is closed under negation, meaning that $-\mathscr{K} := \{-f : f \in \mathscr{K}\} \subseteq \mathscr{K}$;*[3]

(b) *a **cone** if it is closed under non-negative scalar multiplication,*[4] *meaning that $\lambda\mathscr{K} := \{\lambda f : f \in \mathscr{K}\} \subseteq \mathscr{K}$ for all $\lambda \in \mathbb{R}_{\geq 0}$;*

(c) ***convex*** *if it is closed under convex combinations, which means that $\lambda\mathscr{K} + (1-\lambda)\mathscr{K} := \{\lambda f + (1-\lambda)g : f, g \in \mathscr{K}\} \subseteq \mathscr{K}$ for all $\lambda \in (0, 1)$;*

(d) *a **convex cone** if it is closed under addition and non-negative scalar multiplication, or, in other words, under non-negative linear combinations:* $\operatorname{nonneg}(\mathscr{K}) \subseteq \mathscr{K}$, *where*

$$\operatorname{nonneg}(\mathscr{K}) := \left\{ \sum_{k=1}^{n} \lambda_k f_k : n \in \mathbb{N}, f_k \in \mathscr{K}, \lambda_k \in \mathbb{R}_{\geq 0} \right\};$$

(e) *a **linear space** if it is closed under addition and scalar multiplication, or, in other words, under linear combinations:* $\operatorname{span}(\mathscr{K}) \subseteq \mathscr{K}$, *where*

$$\operatorname{span}(\mathscr{K}) := \left\{ \sum_{k=1}^{n} \lambda_k f_k : n \in \mathbb{N}, f_k \in \mathscr{K}, \lambda_k \in \mathbb{R} \right\};$$

(f) *a **$\wedge$-semilattice** if it is closed under point-wise minimum and a **$\vee$-semilattice** if it is closed under point-wise maximum;*

(g) *a **lattice** if it is closed under point-wise minimum and maximum, that is, at the same time a $\wedge$-semilattice and a $\vee$-semilattice;*

(h) *a **linear lattice** if it is both a linear space and a lattice.*[5]

A linear space can also be defined as a negation-invariant convex cone.

The set $\{f \in \mathbb{G}(\mathscr{X}) : f \geq 0\}$ of all non-negative gambles on $\mathscr{X}$ is denoted by $\mathbb{G}_{\geq 0}(\mathscr{X})$, or simply by $\mathbb{G}_{\geq 0}$. Similarly, the set $\{f \in \mathbb{B}(\mathscr{X}) : f \geq 0\}$ of all non-negative bounded gambles on $\mathscr{X}$ is denoted by $\mathbb{B}_{\geq 0}(\mathscr{X})$, or simply by $\mathbb{B}_{\geq 0}$. Both $\mathbb{G}_{\geq 0}$ and $\mathbb{B}_{\geq 0}$ are convex cones.

It is clear that $\mathbb{G}$, $\mathbb{B}$ and $\mathbb{R}$ are linear lattices. In the course of this book, we shall come across quite a few other interesting subsets of $\mathbb{G}$ that are linear lattices too.

---

[3] In this and some of the subsequent conditions, the inclusion can actually be replaced by an equality.

[4] The mathematical literature is particularly messy when it comes to defining cones. A cone is sometimes defined as being closed under *positive* scalar multiplication (see, for instance, Rockafellar, 1970, p. 13), or as what we call a *convex cone* (Wong and Ng, 1973, p. 1), which in its turn is called a *wedge* by others (Holmes, 1975, p. 17). We follow Boyd and Vandenberghe (2004, Section 2.1.5).

[5] In Appendix A$_{368}$ we give a more general definition of linear spaces and linear lattices. A linear lattice in the present sense is also a linear lattice if the sense of Definition A.6$_{370}$, with respect to the point-wise order $\leq$ on gambles defined in Equation $(1.1)_2$.

1.3 Subsets and their indicators

There are important special gambles that correspond to subsets of $\mathcal{X}$. With a subset A of $\mathcal{X}$, also called an **event**, we can associate a $\{0, 1\}$-valued gamble I_A given by

$$I_A(x) := \begin{cases} 1 & \text{if } x \in A \\ 0 & \text{otherwise} \end{cases} \quad \text{for all } x \text{ in } \mathcal{X}.$$

This bounded gamble I_A is called the **indicator** of A. For a collection $\mathcal{A}$ of subsets of $\mathcal{X}$, we sometimes look at the corresponding collection of indicators:

$$I_{\mathcal{A}} := \{I_A : A \in \mathcal{A}\}.$$

The set of all subsets of $\mathcal{X}$ is the **power set** of $\mathcal{X}$, and it is denoted by $\mathcal{P}(\mathcal{X})$, or simply $\mathcal{P}$, when it is clear from the context which space $\mathcal{X}$ we are talking about. We call a subset (or event) **proper** if it is neither empty nor equal to $\mathcal{X}$.

Union, intersection and difference of sets are denoted as usual as $A \cup B$, $A \cap B$ and $A \setminus B$, respectively. For the complement $\mathcal{X} \setminus A$ of a set A, we also write A^c.

With a gamble f on $\mathcal{X}$ and a real number α, we associate the following **level sets**:

$$\begin{aligned} \{f \geq \alpha\} &:= \{x \in \mathcal{X} : f(x) \geq \alpha\} \\ \{f > \alpha\} &:= \{x \in \mathcal{X} : f(x) > \alpha\} \\ \{f = \alpha\} &:= \{x \in \mathcal{X} : f(x) = \alpha\}. \end{aligned}$$

We emphasise that, say, the '$>$' in $\{f > \alpha\}$ is interpreted here in a point-wise manner; the exception discussed near Equation $(1.2)_3$ does not apply here.

1.4 Collections of events

We now turn to special collections of subsets of $\mathcal{X}$.

Definition 1.2 (Fixed vs free) *A subset $\mathcal{C}$ of $\mathcal{P}$ is called **fixed** if its intersection is non-empty: $\bigcap \mathcal{C} \neq \emptyset$, so some element of $\mathcal{X}$ belongs to all elements of $\mathcal{C}$. Otherwise, $\mathcal{C}$ is called **free**.*

Definition 1.3 (Filters and proper filters) *A subset $\mathcal{F}$ of $\mathcal{P}$ is called a **filter** if*

(i) *$\mathcal{F}$ is **increasing**: if $A \in \mathcal{F}$ and $A \subseteq B$ then also $B \in \mathcal{F}$;*

(ii) *$\mathcal{F}$ is **closed under finite intersections**: if $A \in \mathcal{F}$ and $B \in \mathcal{F}$ then $A \cap B \in \mathcal{F}$.*

*A filter $\mathcal{F}$ is called a **proper** filter if in addition $\mathcal{X} \in \mathcal{F}$ and $\emptyset \notin \mathcal{F}$, or equivalently, if $\mathcal{F}$ is a proper subset of $\mathcal{P}$. We denote set of all proper filters by $\mathbb{F}(\mathcal{X})$, or simply by $\mathbb{F}$ if it is clear from the context which space $\mathcal{X}$ we are talking about.*

Definition 1.4 (Filter bases and sub-bases) *A set $\mathscr{C}$ is called a* ***filter base*** *for a filter $\mathscr{F}$ if $\mathscr{F} = \{A \subseteq \mathscr{X} : (\exists C \in \mathscr{C})(C \subseteq A)\}$. A set $\mathscr{C}'$ is called a* ***filter sub-base*** *for a filter $\mathscr{F}$ if the collection of finite intersections of elements of $\mathscr{C}'$ constitutes a filter base for $\mathscr{F}$.*

Definition 1.5 (Ultrafilters) *A proper filter $\mathscr{U}$ is called an* ***ultrafilter*** *if additionally either $A \in \mathscr{U}$ or $A^c \in \mathscr{U}$ for all A in $\mathscr{P}$, or equivalently, if there is no other proper filter that includes $\mathscr{U}$. We denote set of all ultrafilters by $\mathbb{U}(\mathscr{X})$, or simply by $\mathbb{U}$ if it is clear from the context which space $\mathscr{X}$ we are talking about.*

An ultrafilter $\mathscr{U}$ is *fixed* if and only if it contains a singleton, that is, if $\mathscr{U} = \{A \subseteq \mathscr{X} : x \in A\}$ for some $x \in \mathscr{X}$.

Sometimes, instead of filters, the dual notion of **ideals** is used: a decreasing collection $\mathscr{I}$ of events that is closed under finite unions. A **proper ideal** contains $\emptyset$ but not $\mathscr{X}$. In fact, from an ideal $\mathscr{I}$, we can always construct a filter $\{A^c : A \in \mathscr{I}\}$ by element-wise complementation, and vice versa.

If a lattice of gambles contains only (indicators of) events, we call it a **lattice of events**. A lattice of events can also be seen as a collection of subsets of $\mathscr{X}$ that is closed under (finite) intersection and union. If it is also closed under set complementation and contains the empty set $\emptyset$, it is a *field*.

Definition 1.6 (Fields and σ-fields) *A* ***field*** *$\mathscr{F}$* ***on*** *$\mathscr{X}$ is a collection of subsets of $\mathscr{X}$ that contains the empty set $\emptyset$ and is closed under finite unions and complementation. A* ***σ-field*** *$\mathscr{F}$* ***on*** *$\mathscr{X}$ is a field on $\mathscr{X}$ that is also closed under countable unions.*

Next, we discuss two standard ways to provide topological spaces[6] with σ-fields. So, assume that $\mathscr{X}$ is a topological space.

Definition 1.7 (Borel σ-field) *The smallest σ-field on $\mathscr{X}$ that contains all open sets is called the* ***Borel σ-field*** *on $\mathscr{X}$ and is denoted by $\mathscr{B}(\mathscr{X})$, or simply by $\mathscr{B}$. Its members are called* ***Borel sets****.*

For instance, $\mathscr{B}(\mathbb{R})$ is the smallest σ-field that contains all open intervals (a, b) (where $a, b \in \mathbb{R}$ and $a < b$). In measure theory, this is the standard way to equip $\mathbb{R}$ with a σ-field.

Another very useful way to equip a topological space $\mathscr{X}$ with a σ-field goes via the so-called G_δ sets. A set is G_δ if it is a countable intersection of open sets – remember that, generally, only finite intersections of open sets are open, so a G_δ set need not be open.

Definition 1.8 (Baire σ-field) *The smallest σ-field that contains all compact G_δ sets is called the* ***Baire σ-field*** *and is denoted by $\mathscr{B}_0(\mathscr{X})$, or simply by $\mathscr{B}_0$. Its members are called* ***Baire sets****.*

A gamble on $\mathscr{X}$ has **compact support** if it is zero outside some compact subset of $\mathscr{X}$. A very useful characterisation of $\mathscr{B}_0(\mathscr{X})$ goes as follows (Schechter, 1997, Section 20.34).

[6] For a brief overview of relevant notions from topology, refer to Appendix B_{371}.

Theorem 1.9 *If $\mathcal{X}$ is a locally compact Hausdorff space, then $\mathcal{B}_0(\mathcal{X})$ is the smallest σ-field that makes all continuous bounded gambles on $\mathcal{X}$ with compact support measurable.*

In fact, all continuous gambles with compact support are bounded – we emphasise here that they are bounded because of our focus on bounded gambles when we need to apply this result. Measurability is introduced a bit further on in Definition 1.17$_{12}$. For measurability with respect to σ-fields, such as the Baire σ-field, see in particular Proposition 1.19$_{15}$.

It can be shown that every Baire set is also a Borel set: $\mathcal{B}_0(\mathcal{X}) \subseteq \mathcal{B}(\mathcal{X})$. Equality holds for compact metric spaces $\mathcal{X}$ such as compact subsets of $\mathbb{R}$ but not for arbitrary locally compact Hausdorff spaces (Schechter, 1997, Section 20.35).

1.5 Directed sets and Moore–Smith limits

Consider a non-empty set $\mathcal{A}$ provided with a binary relation $\preceq$ that satisfies the following properties:

    (i) $\preceq$ is **reflexive**: $\alpha \preceq \alpha$ for all α in $\mathcal{A}$;

    (ii) $\preceq$ is **transitive**: if $\alpha \preceq \beta$ and $\beta \preceq \gamma$, then also $\alpha \preceq \gamma$ for all α, β, and γ in $\mathcal{A}$;

    (iii) $\preceq$ satisfies the **composition property**: for all α and β in $\mathcal{A}$, there is some γ in $\mathcal{A}$ such that both $\alpha \preceq \gamma$ and $\beta \preceq \gamma$.

Any such set equipped with a relation that is transitive, reflexive, and that satisfies the composition property, is called a **directed set** and is said to have the **Moore–Smith property**.

A **net** f in a set $\mathcal{Y}$ on a directed set $\mathcal{A}$ is a map $f : \mathcal{A} \to \mathcal{Y}$. If the set $\mathcal{Y}$ is provided with a topology[7] (of open sets) $\mathcal{T}$, then we say that the net $f : \mathcal{A} \to \mathcal{Y}$ **converges** to an element y of $\mathcal{Y}$ if for any open set $O \in \mathcal{T}$ containing y – or, in other words, for any open neighbourhood O of y – there is some $\alpha_O \in \mathcal{A}$ such that $f(\alpha) \in O$ for all $\alpha_O \preceq \alpha$. Instead of $f(\alpha)$, we also write f_α.[8]

Alternatively, if the topology on $\mathcal{Y}$ is determined by a (semi)norm $\|\bullet\|$, then the net f converges to y if, for all real $\varepsilon > 0$, there is some $\alpha_\varepsilon \in \mathcal{A}$ such that $\|f(\alpha) - y\| < \varepsilon$ for all $\alpha_\varepsilon \preceq \alpha$.

A net of propositions p is said to hold **eventually** if there is some $\alpha^* \in \mathcal{A}$ such that $p(\alpha)$ holds for all $\alpha^* \preceq \alpha$ (see, for instance, Schechter (1997, Section 7.7) for the terminology). This allows us to say that a net $f : \mathcal{A} \to \mathcal{Y}$ converges to y if f eventually belongs to every open neighbourhood of y.

In particular, an **extended real net** f on a directed set $\mathcal{A}$ is a map $f : \mathcal{A} \to \mathbb{R}^*$. If f assumes only values in $\mathbb{R}$, it is called a **real net**.

---

[7] See Appendix B$_{371}$ for more details about the various topological notions discussed in this section.

[8] We usually write f_α when the elements f_α of $\mathcal{Y}$ are themselves functions defined on some set $\mathcal{X}$, whose values $f_\alpha(x)$ in $x \in \mathcal{X}$ we then have to consider. This allows us to avoid writing $f(\alpha)(x)$ ...

An extended real net f is said to be **non-increasing** if, for all $\alpha \preceq \beta$, we have that $f(\alpha) \geq f(\beta)$. Similarly, f is said to be **non-decreasing** if, for all $\alpha \preceq \beta$, we have that $f(\alpha) \leq f(\beta)$.

An extended real net f converges to a real number if there is some real number a such that, for every $\varepsilon > 0$, there is some $\alpha_\varepsilon \in \mathscr{A}$ such that $|f(\alpha) - a| < \varepsilon$ for all $\alpha_\varepsilon \preceq \alpha$ (see Moore and Smith, 1922, Section I, p. 103). The net f converges to $+\infty$ if it is eventually greater than any real number: for all $R \in \mathbb{R}$, there is some $\alpha_R \in \mathscr{A}$ such that $f(\alpha) > R$ for all $\alpha_R \preceq \alpha$. Similarly, f converges to $-\infty$ if it is eventually smaller than any real number: for all $R \in \mathbb{R}$, there is some $\alpha_R \in \mathscr{A}$ such that $f(\alpha) < R$ for all $\alpha_R \preceq \alpha$.

If an extended real net f converges to some extended real number a, then this a is unique, and it is called the **Moore–Smith limit**, or simply the **limit**, of f. We denote this limit by $\lim_{\alpha\in\mathscr{A}} f(\alpha)$ or $\lim_\alpha f(\alpha)$ or $\lim_\alpha f_\alpha$, or simply $\lim f$. If $a = \lim f$, then we also write $f_\alpha \to a$ and say that f_α **converges to** a. If the net f has a limit $\lim f$, we say that it **converges**.

The Moore–Smith limit is a natural generalisation of the limit of sequences, $\mathbb{N}$, provided with the natural order of natural numbers is a directed set. The following examples show that nets appear in other interesting situations as well. All of them will be used at some point in this book. For a discussion of nets and their fundamental rôle in topology, refer, for instance, to Willard (1970).

Example 1.10 Consider the set $\mathbb{R}^2$, partially ordered by the component-wise ordering: $(a, b) \preceq (c, d) \Leftrightarrow a \leq c$ and $b \leq d$. This ordering is clearly reflexive and transitive and also satisfies the composition property: for any (a, b) and (c, d) in $\mathbb{R}^2$, both $(a, b) \preceq (\max\{a, c\}, \max\{b, d\})$ and $(c, d) \preceq (\max\{a, c\}, \max\{b, d\})$. An extended real net f on $\mathbb{R}^2$ is then a map $f : \mathbb{R}^2 \to \mathbb{R}^*$. It follows from the definition given earlier that this net converges to a real number f^* if and only if, for all $\varepsilon > 0$, there is some real number R such that $|f(a, b) - f^*| < \varepsilon$ for all $a, b \geq R$. We then write that $\lim_{a,b\to+\infty} f(a, b) = f^*$.

We can of course also impose other partial orders on $\mathbb{R}^2$, which lead to other limits. For instance, if we now let $(a, b) \preceq (c, d) \Leftrightarrow a \geq c$ and $b \leq d$, we again get a directed set, and the net $f : \mathbb{R}^2 \to \mathbb{R}$ converges to a real number f^* if and only if, for all $\varepsilon > 0$, there is some real number R such that $|f(a, b) - f^*| < \varepsilon$ for all $-a, b \geq R$. We then write that $\lim_{(a,b)\to(-\infty,+\infty)} f(a, b) = f^*$.

Nets on $\mathbb{N}^2$ and their convergence are similarly defined, and in this case, we also use the notation $f_{m,n}$ instead of $f(m, n)$. ♦

Example 1.11 If we partially order a proper filter $\mathscr{F}$ with the 'includes' relation $\supseteq$, then we see that $\mathscr{F}$ is a *directed set:* $\supseteq$ is clearly reflexive and transitive, and for any A and B in $\mathscr{F}$, we know that $A \cap B \in \mathscr{F}$, $A \supseteq A \cap B$ and $B \supseteq A \cap B$, so $\supseteq$ satisfies the composition property on $\mathscr{F}$. This implies that we can use proper filters to define nets, and we can associate limits with proper filters. A net p of propositions defined on $\mathscr{F}$ is said to hold **eventually** if there is some $F^* \in \mathscr{F}$ such that $p(F)$ holds for all $F^* \supseteq F$. ♦

Example 1.12 Let $\mathfrak{P}_\mathscr{F}$ denote the set of all finite partitions of $\mathscr{X}$ whose elements belong to some given field $\mathscr{F}$ on $\mathscr{X}$. We define a binary relation $\preceq$ on $\mathfrak{P}_\mathscr{F}$: say that $\mathscr{A} \preceq \mathscr{B}$ whenever $\mathscr{B}$ is a **refinement** of $\mathscr{A}$, that is, whenever every element of $\mathscr{B}$ is a subset of some element of $\mathscr{A}$. Then $\preceq$ is *reflexive*: every finite partition is a refinement of itself. $\preceq$ is also *transitive*: if a finite partition $\mathscr{A}$ refines a finite partition $\mathscr{B}$, and $\mathscr{B}$ refines a finite partition $\mathscr{C}$, then $\mathscr{A}$ also refines $\mathscr{C}$. And $\preceq$ satisfies the *composition property*: because $\mathscr{F}$ is a field, every two finite partitions in $\mathscr{F}$ have a common finite refinement in $\mathscr{F}$, that is, for every $\mathscr{A}$ and $\mathscr{B}$ in $\mathfrak{P}_{\mathscr{F}}$, there is a $\mathscr{C}$ in $\mathfrak{P}_{\mathscr{F}}$ such that $\mathscr{A} \preceq \mathscr{C}$ and $\mathscr{B} \preceq \mathscr{C}$. So $\mathfrak{P}_{\mathscr{F}}$ is a directed set with respect to $\preceq$, and as a consequence, we can take the Moore–Smith limit of nets on $\mathfrak{P}_{\mathscr{F}}$. ♦

Moore–Smith limits of nets have more or less the same important properties as limits of sequences. We list a few that will be important for the discussion in this book. Their proofs are obvious.

Proposition 1.13 *Consider extended real nets f and g, and a real number $\lambda \neq 0$.*

(i) *If f converges, then* $\inf f \leq \lim f \leq \sup f$.

(ii) *If f is non-increasing, then f converges and* $\lim f = \inf f$*; similarly, if f is non-decreasing, then f converges and* $\lim f = \sup f$.

(iii) *If any two of the nets f, g and $f + g$ converge, then the third also converges and* $\lim(f + g) = \lim f + \lim g$*, provided that the sum on the right-hand side is well defined.*

(iv) *If any one of the nets f and λf converges, then the other also converges, and* $\lim(\lambda f) = \lambda \lim f$.

(v) *If $f \geq g$ and both f and g converge, then* $\lim f \geq \lim g$.

(vi) *If the nets f and g converge, then so do the nets $f \wedge g$ and $f \vee g$, and* $\lim(f \wedge g) = \lim f \wedge \lim g$ *and* $\lim(f \vee g) = \lim f \vee \lim g$.

We can also define the **limit inferior** $\liminf f$ and the **limit superior** $\limsup f$ of an extended real net f as follows:

$$\liminf f = \liminf_{\alpha} f_\alpha := \sup_{\alpha\in\mathscr{A}} \inf_{\beta\geq\alpha} f_\beta \text{ and } \limsup f = \limsup_{\alpha} f_\alpha := \inf_{\alpha\in\mathscr{A}} \sup_{\beta\geq\alpha} f_\beta. \quad (1.3)$$

Clearly, the net $g_\alpha := \inf_{\beta\geq\alpha} f_\beta$ is non-decreasing and therefore converges, with $\liminf f = \lim g = \lim_\alpha \inf_{\beta\geq\alpha} f_\beta$. Similarly, $\limsup f = \lim_\alpha \sup_{\beta\geq\alpha} f_\beta$.

1.6 Uniform convergence of bounded gambles

A nice illustration of the convergence of nets is provided by the notion of uniform convergence of bounded gambles, which we shall need several times in the rest of this book.

We can provide the set $\mathbb{B}$ of bounded gambles on $\mathscr{X}$ with the so-called **supremum norm** $\|\bullet\|_{\inf}$ given by[9]

$$\|f\|_{\inf} := \sup |f| \text{ for all bounded gambles } f \text{ on } \mathscr{X}.$$

It is clear that $\|f\|_{\inf}$ is a (finite) real number for any bounded gamble f.

[9] The reason for this perhaps surprising notation will become clear further on, in Definition 4.25_{64}.

With the notations used in Section 1.5, we now let $\mathscr{Y} := \mathbb{B}$ and consider a net $f: \mathscr{A} \to \mathbb{B}$ in $\mathbb{B}$, whose values f_α are bounded gambles on $\mathscr{X}$. This net converges to a bounded gamble $g \in \mathbb{B}$ if

$$(\forall \varepsilon > 0)(\exists \alpha_\varepsilon)(\forall \alpha_\varepsilon \preceq \alpha) \sup |g - f_\alpha| < \varepsilon, \tag{1.4}$$

and then we say that the net f_α converges to g **in supremum norm** or that it converges to g **uniformly**. The net f_α is then also called **uniformly convergent**. Interestingly, we can consider $\sup |g - f_\alpha|$ as a real net, and we see that the net of bounded gambles f_α converges uniformly to g if and only if this real net converges to zero: $\lim_\alpha \sup |g - f_\alpha| = 0$, or in other words, $\sup |g - f_\alpha| \to 0$.

For any set $\mathscr{K} \subseteq \mathbb{B}$ of bounded gambles on $\mathscr{X}$, its **uniform closure**[10] $\mathrm{cl}(\mathscr{K})$ is defined as the set of limits of all uniformly convergent nets (or, what is equivalent in this case, sequences, see, for instance, Willard (1970, Chapter 4 and Section 11.7)) of elements of $\mathscr{K}$:

$$\begin{aligned}\mathrm{cl}(\mathscr{K}) &:= \left\{g \in \mathbb{B} : f_\alpha \to g \text{ uniformly for some net } f_\alpha \text{ in } \mathscr{K}\right\}\\ &= \left\{g \in \mathbb{B} : f_n \to g \text{ uniformly for some sequence } f_n \text{ in } \mathscr{K}\right\}.\end{aligned}$$

The set $\mathscr{K}$ is **uniformly closed** if it contains the limits of all uniformly convergent nets (or sequences) in $\mathscr{K}$, or, in other words, if $\mathrm{cl}(\mathscr{K}) = \mathscr{K}$.

1.7 Set functions, charges and measures

The following definition describes well-known extensions for set functions. By a **set function** μ, we mean a map from a collection of events $\mathscr{A}$ to the set $\mathbb{R}_{\geq 0}$ of non-negative real numbers, such that (i) $\emptyset \in \mathscr{A}$ and $\mu(\emptyset) = 0$ and (ii) μ is **monotone**: if $A \subseteq B$, then $\mu(A) \subseteq \mu(B)$ for all A and B in $\mathscr{A}$.

Definition 1.14 *Let μ be a set function defined on a lattice $\mathscr{A}$ of subsets of $\mathscr{X}$ containing the empty set. The* **inner set function** *μ_* and the* **outer set function** *μ^* of μ, or induced by μ, are the maps defined for all $B \subseteq \mathscr{X}$ by*

$$\begin{aligned}\mu_*(B) &:= \sup\{\mu(A) : A \in \mathscr{A} \text{ and } A \subseteq B\},\\ \mu^*(B) &:= \inf\{\mu(A) : A \in \mathscr{A} \text{ and } A \supseteq B\}.\end{aligned}$$

Clearly, μ_* and μ^* are monotone as well, and they coincide with μ on its domain $\mathscr{A}$, implying that $\mu_*(\emptyset) = \mu^*(\emptyset) = 0$. But μ_* and μ^* are not necessarily real valued; however, they are real valued when ($\emptyset$ and) $\mathscr{X}$ belongs to $\mathscr{A}$. In that case, they really deserve the name '(inner and outer) *set function*'.

To define probability charges, we first identify a sufficiently large collection of events of interest. For convenience, we assume that this collection has the minimal structure of a field, see Definition 1.6_6.

[10] For a brief discussion of the notion of topological closure, refer to Appendix B_{371}.

Next, to each event A in a field $\mathscr{F}$, we attach a(n extended) real number $\mu(A)$ that measures something related to A: belief in the occurrence of A, mass, charge, capacity of A and so on. We borrow the following definition from Bhaskara Rao and Bhaskara Rao (1983).

Definition 1.15 *Let $\mathscr{F}$ be a field on $\mathscr{X}$. A* ***charge*** μ ***on*** $\mathscr{F}$*, also called* ***finitely additive measure****, is an $\mathbb{R}^*$-valued map on $\mathscr{F}$ that assumes at most one of the values $+\infty$ and $-\infty$ and satisfies*

(i) $\mu(\emptyset) = 0$, *and*

(ii) $\mu(A \cup B) = \mu(A) + \mu(B)$ *whenever* $A, B \in \mathscr{F}$ *and* $A \cap B = \emptyset$.

A charge μ on $\mathscr{F}$ is called a ***probability charge*** *if additionally it is* ***positive****,*

(iii) $\mu(A) \geq 0$ *for any* $A \in \mathscr{F}$,

and ***normalised****,*

(iv) $\mu(\mathscr{X}) = 1$.

The set of all probability charges on $\mathscr{F}$ is denoted by $\mathbb{P}(\mathscr{F})$. Finally, a charge μ is said to be ***σ-additive*** *if additionally $\mathscr{F}$ is a σ-field and*

(v) *for any sequence A_n of pairwise disjoint sets in $\mathscr{F}$, the limit $\lim_{n\to+\infty} \sum_{k=0}^{n} \mu(A_k)$ exists in $\mathbb{R}^*$ and is equal to $\mu\left(\bigcup_{n\in\mathbb{N}} A_n\right)$ (and hence, the limit is independent of the order of the sequence).*

A σ-additive charge is simply called a ***measure*** *and a σ-additive probability charge a* ***probability measure****.*

A probability charge is a special set function. Perhaps the most commonly used charge is the **Lebesgue measure** λ on $\mathscr{B}(\mathbb{R})$, which is defined as the unique σ-additive measure on $\mathscr{B}(\mathbb{R})$ such that

$$\lambda((a,b)) = \lambda([a,b)) = \lambda((a,b]) = \lambda([a,b]) = b - a$$

for any $a, b \in \mathbb{R}$ ($a \leq b$): it measures the length of intervals of $\mathbb{R}$. Vitali (1905) proved that there is no σ-additive measure on all of $\mathscr{P}(\mathbb{R})$ that has this property – if we accept the Axiom of Choice (see, for instance, Schechter (1997, Section 21.22)). This is one of the reasons for introducing the Borel σ-field $\mathscr{B}(\mathbb{R})$. Sometimes the Lebesgue measure is defined on larger σ-fields (see, for instance, Halmos (1974, Section 15)). The Lebesgue measure can of course also be defined, by taking appropriate restrictions, on any real interval, rather than on the space $\mathbb{R}$ itself.

As another example, consider the **total variation** $|\mu|$ of a charge μ on a field $\mathscr{F}$. It can be defined as follows, using the set $\mathfrak{P}_{\mathscr{F}}$ of all finite partitions of $\mathscr{X}$ whose elements belong to the field $\mathscr{F}$, discussed in Example 1.12_8. This set is directed by the refinement relation $\preceq$. With any $A \in \mathscr{F}$, we can consider the extended real net ν^A

on $\mathfrak{P}_{\mathcal{F}}$ that maps any finite partition $\mathscr{C}$ in $\mathfrak{P}_{\mathcal{F}}$ to the extended real number

$$\nu^A(\mathscr{C}) := \sum_{C \in \mathscr{C}} |\mu(C \cap A)|.$$

Using the properties of the charge μ, it is not difficult to show that the net ν^A is non-decreasing, and therefore converges, by Proposition 1.13_9, to some extended real number

$$|\mu|(A) := \lim_{\mathscr{C}} \sum_{C \in \mathscr{C}} \nu^A(C) = \sup_{\mathscr{C} \in \mathfrak{P}_{\mathcal{F}}} \sum_{C \in \mathscr{C}} |\mu(C \cap A)| \text{ for any } A \in \mathcal{F}.$$

It is easy to check that $|\mu|$ is a charge as well. When μ is positive – meaning that $\mu(A) \geq 0$ for all $A \in \mathcal{F}$, which holds, for instance, for probability charges and also for the Lebesgue measure – the total variation $|\mu|$ is equal to μ.

1.8 Measurability and simple gambles

In this book, we use charges and fields much more often than we do measures and σ-fields: we seem to have little need for σ-additivity. Measurability of functions is a notion that is commonly associated with σ-fields in the context of measures (see, for instance, Kallenberg, 2002), so it will be useful to come up with a definition for measurability that works for fields $\mathcal{F}$ in the context of charges as well.

In the context of this book, we shall only have to rely on a notion of measurability for bounded gambles, so that is what we concentrate on here. We begin with a definition of $\mathcal{F}$-simple gambles, which are necessarily bounded.

Definition 1.16 (Simple gambles) *Let $\mathcal{F}$ be a field on $\mathcal{X}$. A gamble f on $\mathcal{X}$ is called **$\mathcal{F}$-simple** if it belongs to the linear span of $I_{\mathcal{F}}$, or, in other words, if $f = \sum_{i=1}^n a_i I_{A_i}$ for some $n \in \mathbb{N}$, $a_1, \dots, a_n$ in $\mathbb{R}$ and $A_1, \dots, A_n$ in $\mathcal{F}$. The sum $\sum_{i=1}^n a_i I_{A_i}$ is called a **representation** of f. The set of $\mathcal{F}$-simple gambles is denoted by* $\operatorname{span}(\mathcal{F})$.

Note that $\operatorname{span}(\mathcal{F})$ is a simplified notation for $\operatorname{span}(I_{\mathcal{F}})$, the linear span of all indicators of elements of $\mathcal{F}$.

Next, we define $\mathcal{F}$-measurability. The characterisation (A) in Definition 1.17 is due to Greco (1981), and the characterisation (B) is due to Janssen (1999), who also established equivalence with Greco's definition. We give a proof that is shorter than the one by Janssen.

Definition 1.17 *Let $\mathcal{F}$ be a field on $\mathcal{X}$ and let f be a bounded gamble on $\mathcal{X}$. Then the following conditions are equivalent; if any (and hence all) of them are satisfied, we say that f is **$\mathcal{F}$-measurable**.*

(A) *For any $a \in \mathbb{R}$ and any $\varepsilon > 0$, there is an $A \in \mathcal{F}$ such that*

$$\{f \geq a\} \supseteq A \supseteq \{f \geq a + \varepsilon\}.$$

(B) *There is a sequence f_n of $\mathcal{F}$-simple gambles that converges uniformly to f, that is, $\lim_{n \to +\infty} \sup |f - f_n| = 0$, or, in other words, $f \in \operatorname{cl}(\operatorname{span}(\mathcal{F}))$.*

The set of all $\mathscr{F}$-measurable bounded gambles is denoted by $\mathbb{B}_{\mathscr{F}}(\mathscr{X})$ and given by

$$\mathbb{B}_{\mathscr{F}}(\mathscr{X}) := \mathrm{cl}(\mathrm{span}(\mathscr{F})).$$

If it is clear from the context which space $\mathscr{X}$ we are talking about, we also use the simpler notation $\mathbb{B}_{\mathscr{F}}$.

Obviously, $\mathbb{B}_{\mathscr{F}}$ is a uniformly closed linear lattice that contains all constant gambles. Let us prove that the conditions are indeed equivalent.

Proof. (A)⇒(B). Assume that condition (A) is satisfied. Let $\varepsilon > 0$. Then (B) is established if we can find an $\mathscr{F}$-simple gamble g such that $\sup|f - g| \leq \varepsilon$.

Let $a_0, \ldots, a_n$ be any finite sequence of real numbers such that $a_0 < \inf f$, $0 < a_{i+1} - a_i < \frac{\varepsilon}{3}$ for $i \in \{0, \ldots, n-1\}$ and $\sup f < a_n$. We can always find such a sequence by making n large enough. Define $A_i := \{f \geq a_i\}$ for $i \in \{0, \ldots, n\}$. We infer from (A) that there is some sequence $B_0, \ldots, B_{n-1}$ of members of $\mathscr{F}$ such that

$$A_0 \supseteq B_0 \supseteq A_1 \supseteq B_1 \cdots \supseteq A_{n-1} \supseteq B_{n-1} \supseteq A_n.$$

With $a_{-1} := 0$, define the $\mathscr{F}$-simple gamble $g := \sum_{i=0}^{n-1}(a_i - a_{i-1})I_{B_i}$. This g has the desired property $\sup|f - g| \leq \varepsilon$: choose any $x \in \mathscr{X}$, then we show that $|f(x) - g(x)| < \varepsilon$. First, observe that, by construction of the $a_0, \ldots, a_n$, there is a unique $j_x \in \{1, \ldots, n\}$ such that $a_{j_x-1} \leq f(x) < a_{j_x}$. So $x \in A_i$ for $i \leq j_x - 1$ and $x \notin A_i$ for $i \geq j_x$. We then infer from the construction of the sequence $B_0, \ldots, B_{n-1}$ that $I_{B_i}(x) = 1$ for $i \leq j_x - 2$ and $I_{B_i}(x) = 0$ for $i \geq j_x$; for $i = j_x - 1$, both the values 0 and 1 are possible. Hence, indeed

$$\begin{aligned}
|f(x) - g(x)| &= \left| f(x) - \sum_{i=0}^{n-1}(a_i - a_{i-1})I_{B_i}(x) \right| \\
&\leq \left| f(x) - \sum_{i=0}^{j_x-2}(a_i - a_{i-1}) \right| + \left| a_{j_x-1} - a_{j_x-2} \right| \\
&= \left| f(x) - a_{j_x-2} \right| + \left| a_{j_x-1} - a_{j_x-2} \right| \\
&\leq \left| f(x) - a_{j_x-1} \right| + 2\left| a_{j_x-1} - a_{j_x-2} \right| \\
&< \frac{\varepsilon}{3} + 2\frac{\varepsilon}{3} = \varepsilon,
\end{aligned}$$

which establishes the first part of the proof.

(B)⇒(A). Consider any $a \in \mathbb{R}$ and $\varepsilon > 0$, and suppose there is a sequence of $\mathscr{F}$-simple gambles such that $\sup|f - f_n|$ converges to zero. Then there is some $n_\varepsilon \in \mathbb{N}$ such that $\sup\left|f - f_{n_\varepsilon}\right| < \frac{\varepsilon}{2}$ and therefore $f(x) - \varepsilon < f_{n_\varepsilon}(x) - \varepsilon/2 < f(x)$ for all $x \in \mathscr{X}$. This guarantees that

$$\{f \geq a\} \supseteq \left\{f_{n_\varepsilon} \geq a + \tfrac{\varepsilon}{2}\right\} \supseteq \{f \geq a + \varepsilon\}.$$

Now simply observe that $\left\{f_{n_\varepsilon} \geq a + \varepsilon/2\right\}$ belongs to $\mathscr{F}$, because f_{n_ε} is $\mathscr{F}$-simple. □

Either of these equivalent approaches to defining measurability with respect to a field has been taken by a number of authors (see, for instance, Greco (1981), Denneberg (1994), Janssen (1999) and Bhaskara Rao and Bhaskara Rao (1983); Bhaskara Rao and Bhaskara Rao speak of $\mathcal{F}$-continuity rather than $\mathcal{F}$-measurability).

Hildebrandt (1934, Section 1(f)) and Walley (1991, Section 3.2.1) use definitions for $\mathcal{F}$-measurability that are stronger than ours, unless $\mathcal{F}$ is a σ-field. However, when $\mathcal{F}$ is a field but not a σ-field, then Hildebrandt's (1934) and Walley's (1991) sets of $\mathcal{F}$-measurable bounded gambles are not even linear spaces, which makes us prefer our weaker definition given earlier.

As a special case, a $\mathcal{P}$-simple gamble is any gamble with a finite number of values, and it is simply called a **simple gamble**. For simple gambles, $\mathcal{F}$-measurability is easier to characterise than for bounded gambles.

Proposition 1.18 *Let $\mathcal{F}$ be a field, and let f be a simple gamble. Then the following statements are equivalent:*

(i) *f is $\mathcal{F}$-simple;*

(ii) *f is $\mathcal{F}$-measurable;*

(iii) *for any $x \in \mathbb{R}$, the set $\{f \geq x\}$ belongs to $\mathcal{F}$;*

(iv) *for any $x \in \mathbb{R}$, the set $\{f > x\}$ belongs to $\mathcal{F}$.*

Proof. A bounded gamble f is clearly simple if and only if it assumes a finite number of values, say $f_1 < f_2 < \cdots < f_n$, and then (see also Corollary C.4$_{381}$):

$$f = \sum_{k=1}^{n} f_k I_{\{f=f_k\}} = f_1 + \sum_{k=2}^{n} (f_k - f_{k-1}) I_{\{f \geq f_k\}}. \tag{1.5}$$

We prove that (i)⇒(ii)⇒(iii)⇒(i). Proving (i)⇒(ii)⇒(iv)⇒(i) is similar.

(i)⇒(ii). If f is $\mathcal{F}$-simple, then there are $m \geq 1$, $a_\ell \in \mathbb{R}$ and $F_\ell \in \mathcal{F}$ such that $f = \sum_{\ell=1}^{m} a_\ell I_{F_\ell}$. This implies that the f_k are of the form $\sum_{\ell \in L_k} a_\ell$ and the corresponding $\{f = f_k\}$ of the form $\bigcap_{\ell \in L_k} F_\ell \in \mathcal{F}$ for appropriately chosen $L_k \subseteq \{1, \dots, m\}$. Hence, each $\{f \geq f_k\} \in \mathcal{F}$, and this implies that f is $\mathcal{F}$-measurable: simply take $A := \{f \geq f_k\}$ in Definition 1.17(A)$_{12}$.

(ii)⇒(iii). Consider any $x \in \mathbb{R}$. It is clear that $\{f \geq x\}$ is either empty or of the form $\{f \geq f_k\}$. We may therefore assume without loss of generality that $\{f \geq x\}$ has the form $\{f \geq f_k\}$. Moreover, we can always choose a and ε in Definition 1.17(A)$_{12}$ in such a way that $\{f \geq a\} = \{f \geq f_k\} = \{f \geq a + \varepsilon\}$, so $\{f \geq f_k\} \in \mathcal{F}$.

(iii)⇒(i). Immediately from Equation (1.5) and the fact that by assumption all $\{f \geq f_k\} \in \mathcal{F}$. □

When $\mathcal{F}$ is a σ-field, this simple characterisation carries over to all $\mathcal{F}$-measurable bounded gambles, which means that our measurability definition reduces to the classical definition of $\mathcal{F}$-measurability for bounded gambles (see, for instance, also

Kallenberg, 2002, Lemma 1.11). As the proof is short, we repeat it here.

Proposition 1.19 *Let $\mathscr{F}$ be a σ-field. A bounded gamble f is $\mathscr{F}$-measurable if and only if, for any $a \in \mathbb{R}$, the set $\{f > a\}$ belongs to $\mathscr{F}$, or equivalently, if and only if, for any $a \in \mathbb{R}$, the set $\{f \geq a\}$ belongs to $\mathscr{F}$.*

Proof. 'if'. Simply take $A := \{f > a\}$ or $A := \{f \geq a\}$ in Definition 1.17(A)$_{12}$.

'only if'. Suppose f is $\mathscr{F}$-measurable. By Definition 1.17(A)$_{12}$, there is some sequence A_n in $\mathscr{F}$ such that

$$\left\{f \geq a + \tfrac{1}{n+1}\right\} \supseteq A_n \supseteq \left\{f \geq a + \tfrac{2}{n+1}\right\}.$$

Taking the countable union over $n \in \mathbb{N}$ on all sides, this leads to

$$\{f > a\} \supseteq \bigcup_{n\in\mathbb{N}} A_n \supseteq \{f > a\},$$

which means that $\{f > a\} = \bigcup_{n\in\mathbb{N}} A_n$. As $\mathscr{F}$ is a σ-field, it is closed under countable unions, and hence, $\bigcup_{n\in\mathbb{N}} A_n$ belongs to $\mathscr{F}$. This establishes the proposition.

For the other part, construct the sequence A_n in $\mathscr{F}$ such that

$$\left\{f \geq a - \tfrac{2}{n+1}\right\} \supseteq A_n \supseteq \left\{f \geq a - \tfrac{1}{n+1}\right\}$$

and take countable intersections to arrive at the desired result. □

When $\mathscr{F} = \mathscr{P}$, measurability is no longer an issue, as all bounded gambles are trivially $\mathscr{P}$-measurable.

Proposition 1.20 *The set of simple gambles* span($\mathscr{P}$) *is uniformly dense in* $\mathbb{B}$, *meaning that any bounded gamble is a uniform limit of a sequence of simple gambles:* $\mathbb{B} = \mathrm{cl}(\mathrm{span}(\mathscr{P}))$.

Proof. Immediately from the equivalence proof for Definition 1.17$_{12}$. □

The following surprisingly simple lemma, due to Denneberg (1994, Lemma 6.2, pp. 73–74) – but we believe part (iv)$_\curvearrowright$ is new – shows that *all* gambles f, and not just the bounded ones, can be uniformly approximated in a very convenient and systematic manner by a sequence of 'primitive gambles' $u_n \circ f$, which turn out to be simple whenever f is bounded. Here $u_n : \mathbb{R} \to \mathbb{R}$ is defined by

$$u_n(t) := \inf\left\{\frac{k}{n} : k \in \mathbb{Z} \text{ and } \frac{k}{n} \geq t\right\} \text{ for all } t \in \mathbb{R}, \tag{1.6}$$

where $n \in \mathbb{N}_{>0}$. It is an increasing step function with equidistant steps of width and height $1/n$ such that (see the following graph, which is also helpful in understanding a number of steps in the proof of Lemma 1.21$_\curvearrowright$ further on)

$$t \leq u_n(t) \leq t + \frac{1}{n} \text{ for all } t \in \mathbb{R}. \tag{1.7}$$

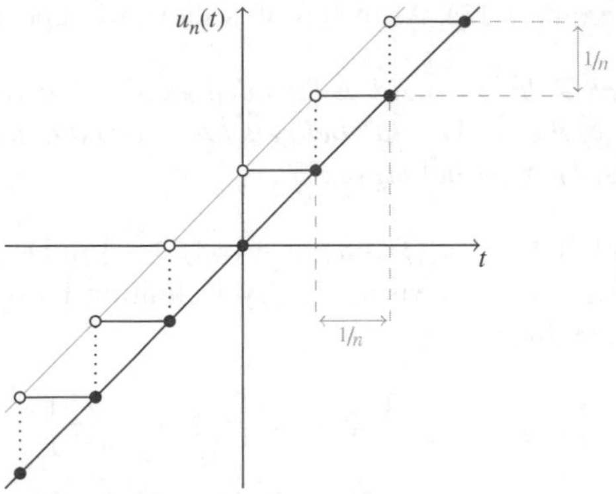

Lemma 1.21 *For any gamble f on $\mathcal{X}$ and all $n \in \mathbb{N}$, we have that $f \le u_n \circ f \le f + 1/n$, so the sequence of gambles $u_n \circ f$ converges uniformly to f. Moreover, the following statements hold for all gambles f and g on $\mathcal{X}$:*

(i) *The subsequence $u_{2^n} \circ f$ of $u_n \circ f$ is non-increasing.*

(ii) *If f is bounded, then $u_n \circ f$ is simple for all $n \in \mathbb{N}$.*

(iii) *$u_n \circ f + u_n \circ g - 2/n \le u_n \circ (f + g) \le u_n \circ f + u_n \circ g$ for all $n \in \mathbb{N}$, so the sequence $u_n \circ f + u_n \circ g$ converges uniformly to $f + g$.*

(iv) *if f and g are comonotone in the sense of Definition C.2$_{378}$, then so are $u_n \circ f$ and $u_n \circ g$.*

Proof. That $f \le u_n \circ f \le f + 1/n$ follows immediately from Equation (1.7)$_{\frown}$.

(i). This follows at once from $u_{2n}(t) \le u_n(t)$ for all $n \in \mathbb{N}$ and all $t \in \mathbb{R}$ (see also the earlier graph for more geometrical insight).

(ii). If f is bounded, then we see that $\inf f \le f \le u_n \circ f \le f + 1/n \le \sup f + 1/n$, and therefore $u_n \circ f$ is bounded as well. As it then necessarily only assumes a finite number of values, it is a simple gamble.

(iii). It suffices to prove that $u_n(t_1) + u_n(t_2) - 2/n < u_n(t_1 + t_2) \le u_n(t_1) + u_n(t_2)$ for all real t_1 and t_2. On the one hand, we see that

$$\begin{aligned}
u_n(t_1 + t_2) &= \inf\left\{\frac{\ell}{n} : \ell \in \mathbb{Z} \text{ and } \frac{\ell}{n} \ge t_1 + t_2\right\} \\
&= \inf\left\{\frac{\ell_1}{n} + \frac{\ell_2}{n} : \ell_1, \ell_2 \in \mathbb{Z} \text{ and } \frac{\ell_1}{n} + \frac{\ell_1}{n} \ge t_1 + t_2\right\} \\
&\le \inf\left\{\frac{\ell_1}{n} + \frac{\ell_2}{n} : \ell_1, \ell_2 \in \mathbb{Z} \text{ and } \frac{\ell_1}{n} \ge t_1 \text{ and } \frac{\ell_1}{n} \ge t_2\right\} \\
&= u_n(t_1) + u_n(t_2);
\end{aligned}$$

on the other hand, we infer from Equation $(1.7)_{15}$ that

$$u_n(t_1) + u_n(t_2) - u_n(t_1 + t_2) = [u_n(t_1) - t_1] + [u_n(t_2) - t_2] - [u_n(t_1 + t_2) - (t_1 + t_2)]$$
$$< \frac{1}{n} + \frac{1}{n} + 0.$$

(iv). Consider arbitrary x_1 and x_2 in $\mathcal{X}$ and assume that $u_n(f(x_1)) < u_n(f(x_2))$. Then $f(x_1) < f(x_2)$ because u_n is non-decreasing, and therefore $g(x_1) \leq g(x_2)$, because f and g are comonotone. We conclude that $u_n(g(x_1)) \leq u_n(g(x_2))$, again because u_n is non-decreasing. □

1.9 Real functionals

A real-valued function Γ defined on some subset $\mathcal{K}$ of $\mathbb{G}$ will be called a (real) **functional** on $\mathcal{K}$. It is called **monotone** if

$$f \leq g \Rightarrow \Gamma(f) \leq \Gamma(g) \text{ for all } f \text{ and } g \text{ in } \mathcal{K}.$$

In very much the same way as monotone set functions give rise to inner set functions, monotone real functionals may lend themselves to inner extension.

Definition 1.22 (Inner extension) *If Γ is a monotone real functional defined on a lattice of bounded gambles*[11] *$\mathcal{K}$ that contains all constant gambles, then its* ***inner extension*** *Γ_* is the real functional defined on all bounded gambles f on $\mathcal{X}$ by*

$$\Gamma_*(f) := \sup \{\Gamma(g) : g \in \mathcal{K} \text{ and } g \leq f\}. \tag{1.8}$$

This inner extension Γ_* is obviously monotone as well, and it coincides with Γ on its domain $\mathcal{K}$.

We now turn to a number of other possible properties of real functionals, for which it is easiest to assume that they are defined on a convex cone $\mathcal{K}$ of gambles.

Definition 1.23 *A real functional Γ defined on a convex cone of gambles $\mathcal{K}$ is called*

(a) ***positively homogeneous*** *if $\Gamma(\lambda f) = \lambda\Gamma(f)$ for all $f \in \mathcal{K}$ and all real $\lambda > 0$;*

(b) ***non-negatively homogeneous*** *if $\Gamma(\lambda f) = \lambda\Gamma(f)$ for all $f \in \mathcal{K}$ and all real $\lambda \geq 0$;*

(c) ***homogeneous*** *if $\Gamma(\lambda f) = \lambda\Gamma(f)$ for all $f \in \mathcal{K}$ and all real λ;*[12]

---

[11] Of course, we could consider inner extensions for functionals defined on gambles that are not bounded; our sole reason for not doing so here is because we do not need to in the context of this book.

[12] Sometimes, in the literature (see, for instance, Schechter, 1997, Definition 12.24), a real functional Γ is called *homogeneous* if $\Gamma(\lambda f) = |\lambda|\Gamma(f)$ for all $f \in \mathcal{K}$ and all real λ.

(d) ***convex*** *if* $\Gamma(\lambda f + (1-\lambda)g) \leq \lambda\Gamma(f) + (1-\lambda)\Gamma(g)$ *for all* $f, g \in \mathcal{K}$ *and all real* $\lambda \in (0, 1)$*;*

(e) ***concave*** *if* $\Gamma(\lambda f + (1-\lambda)g) \geq \lambda\Gamma(f) + (1-\lambda)\Gamma(g)$ *for all* $f, g \in \mathcal{K}$ *and all real* $\lambda \in (0, 1)$.

If in addition $\mathcal{K}$ *is a linear space, then we call* Γ

(a) ***super-additive*** *if* $\Gamma(f+g) \geq \Gamma(f) + \Gamma(g)$ *for all* $f, g \in \mathcal{K}$*;*

(b) ***sub-additive*** *if* $\Gamma(f+g) \leq \Gamma(f) + \Gamma(g)$ *for all* $f, g \in \mathcal{K}$*;*

(c) ***additive*** *if* $\Gamma(f+g) = \Gamma(f) + \Gamma(g)$ *for all* $f, g \in \mathcal{K}$*;*

(d) ***linear*** *if* $\Gamma(\lambda f + \mu g) = \lambda\Gamma(f) + \mu\Gamma(g)$ *for all* $f, g \in \mathcal{K}$ *and all real* λ, μ.

There is a very straightforward characterisation of real functionals that can be extended linearly, whose formulation and simple proof we borrow from Schechter (1997, Proposition 11.10).

Proposition 1.24 *A real functional* Γ *defined on a set* $\mathcal{K}$ *of gambles on* $\mathcal{X}$ *can be extended to a linear functional on* span($\mathcal{K}$) *if and only if*

$$\sum_{k=1}^{n} \lambda_k f_k = 0 \Rightarrow \sum_{k=1}^{n} \lambda_k \Gamma(f_k) = 0 \text{ for all } n \in \mathbb{N}_{>0},\ \lambda_k \in \mathbb{R} \text{ and } f_k \in \mathcal{K} \tag{1.9}$$

and in that case, this linear extension is unique.

Proof. If there is a linear extension Γ' to span $\mathcal{K}$, it must be unique, because linearity requires that it should satisfy

$$\Gamma'\left(\sum_{k=1}^{n} \lambda_k f_k\right) = \sum_{k=1}^{n} \lambda_k \Gamma(f_k) \text{ for all } n \in \mathbb{N}_{>0},\ \lambda_k \in \mathbb{R} \text{ and } f_k \in \mathcal{K}.$$

To show that there is a linear extension, we only need to show that this definition of Γ' is consistent. Assume that $\sum_{k=1}^{n} \lambda_k f_k = \sum_{\ell=1}^{m} \mu_\ell g_\ell$, where $m, n \in \mathbb{N}_{>0}$ and $\lambda_k, \mu_\ell \in \mathbb{R}$ and $f_k, g_\ell \in \mathcal{K}$. Then it follows from Equation (1.9) that $\sum_{k=1}^{n} \lambda_k \Gamma(f_k) = \sum_{\ell=1}^{m} \mu_\ell \Gamma(g_\ell)$. □

To conclude, we mention a convenient continuity property.

Definition 1.25 (Monotone convergence) *A real functional* Γ *defined on a set* $\mathcal{K}$ *of gambles on* $\mathcal{X}$ *satisfies* ***downward monotone convergence*** *if for every non-increasing sequence* f_n *in* $\mathcal{K}$ *such that*

(i) $a \geq f_1 \geq f_2 \geq \ldots$ *for some* $a \in \mathbb{R}$, *and*

(ii) *the point-wise limit* $\lim_{n\to+\infty} f_n$ *of the sequence* f_n *also belongs to* $\mathcal{K}$,

it holds that $\Gamma(\lim_{n\to+\infty} f_n) = \lim_{n\to+\infty} \Gamma(f_n)$. *It satisfies* ***upward monotone convergence*** *if for every non-decreasing sequence* g_n *in* $\mathcal{K}$ *such that*

(i) $b \leq g_1 \leq g_2 \leq \dots$ *for some* $b \in \mathbb{R}$, *and*

(ii) *the point-wise limit* $\lim_{n\to+\infty} g_n$ *of the sequence* g_n *also belongs to* $\mathscr{K}$,

it holds that $\Gamma(\lim_{n\to+\infty} g_n) = \lim_{n\to+\infty} \Gamma(g_n)$. *We say that* Γ *satisfies* ***monotone convergence*** *if it satisfies both upward and downward monotone convergence.*

In the literature, monotone convergence is more usually defined in terms of *non-negative non-decreasing* sequences (see, for instance, Schechter, 1997, Sections 21.38 and 24.29). For real linear functionals that satisfy constant additivity (meaning that $\Gamma(f + a) = a\Gamma(1) + \Gamma(f)$ for all f in $\mathscr{K}$ and $a \in \mathbb{R}$), such as the usual types of integrals, our definition is evidently equivalent with the usual formulation.

Proposition 1.26 *If* Γ *is a real linear functional defined on a linear space* $\mathscr{K}$, *then the following statements are equivalent:*

(i) Γ *satisfies upward monotone convergence;*

(ii) Γ *satisfies downward monotone convergence;*

(iii) Γ *satisfies monotone convergence.*

Proof. Immediate, once we realise that $\mathscr{K}$ is negation-invariant, $\Gamma(-f) = -\Gamma(f)$ for all $f \in \mathscr{K}$, and $\lim_{n\to+\infty} -f_n = -\lim_{n\to+\infty} f_n$. □

We differentiate between upward and downward monotone convergence here because we will mainly be working outside the ambit of lincar functionals and will come across examples of both types (see, for instance, Proposition 5.12_{91} and its proof and Proposition 5.21_{100}).

1.10 A useful lemma

We shall need the following simple lemma a few times.

Lemma 1.27 *Let a, b and* ε *be real numbers and assume that* $\varepsilon \geq 0$. *If* $a \leq b \leq a + \varepsilon$, *then, for every real number c, we have that* $ac - \varepsilon|c| \leq bc \leq ac + \varepsilon|c|$.

Proof. If $c \geq 0$, then we have that $ac \leq bc \leq (a + \varepsilon)c$, which implies that

$$ac - \varepsilon|c| \leq ac \leq bc \leq (a + \varepsilon)c \leq ac + \varepsilon|c|.$$

On the other hand, if $c < 0$, then we have that $ac \geq bc \geq (a + \varepsilon)c$, which implies that

$$ac + \varepsilon|c| \geq ac \geq bc \geq (a + \varepsilon)c \geq ac - \varepsilon|c|.$$

In both cases, we find that $ac - \varepsilon|c| \leq bc \leq ac + \varepsilon|c|$. □

Part I

LOWER PREVISIONS ON BOUNDED GAMBLES

2

Introduction

In this first part, we lay the foundations, and expose the main ideas, of the theory of lower previsions for bounded gambles.

This theory constitutes a generalisation of de Finetti's theory of previsions (de Finetti, 1937, 1970, 1974–1975). The idea of generalising previsions to lower previsions is mainly due to Williams (1975b, 1976) and, in a more developed form, to Walley (1991), based on earlier work of, among others, Boole (1854), Keynes (1921), Ramsey (1931), de Finetti (1970), Good (1950), Smith (1961) and Shafer (1976). With some imagination, lower previsions can even be traced back to Pascal's work in the late seventeenth century (Hacking, 1975, Chapter 8).

The theory of lower previsions is part of a family of so-called imprecise probability theories. They are concerned with modelling the type of uncertainty that cannot be modelled by a probability measure, because, roughly speaking, there may not be enough information to identify such a single probability. Amongst these theories, the theory of lower previsions is one of the most interesting and prominent ones, mainly because of its mathematically unifying character. Indeed, all of the following models can be treated as lower previsions from a mathematical point of view: probability charges (Bhaskara Rao and Bhaskara Rao, 1983), n-monotone set functions (Choquet, 1953–1954), belief functions (Dempster, 1967b; Shafer, 1976), possibility measures (De Cooman, 1997; Dubois and Prade, 1988; Zadeh, 1978), p-boxes (Ferson et al., 2003) and clouds (Destercke and Dubois, 2006; Neumaier, 2004). Lower previsions have also been linked to various theories of integration, such as Choquet integration (Walley, 1981, p. 53), Lebesgue integration (Walley, 1991, p. 132) and many others (De Cooman et al., 2008b; Denneberg, 1994; Troffaes, 2005). They can also be viewed as lower expectations with respect to closed convex sets of probability measures, also called credal sets (Levi, 1983).

In Chapter 3_{25}, we introduce Williams' (1975b) notion of acceptability and rationality criteria for sets of acceptable bounded gambles. We suggest and motivate

Lower Previsions, First Edition. Matthias C.M. Troffaes and Gert de Cooman.

conditions for coherence, consistency and natural extension for sets of acceptable bounded gambles. These notions provide a foundation for the remainder of the material in this book.

Starting from acceptability, we essentially derive Walley's (1991) unconditional theory of lower previsions, in Chapter 4_{37}. Here too we look at coherence, consistency and natural extension, but now for lower previsions directly. We study their properties in great detail and also pay attention to the well-known lower envelope theorem for coherent lower previsions in terms of linear previsions. We conclude the chapter with some topological considerations, leading to an extreme point theorem for coherent lower previsions.

In Chapters 5_{76}–9_{181}, we discuss a wide range of theoretical and practical special cases of coherent lower previsions. Simple standard examples are discussed in Chapter 5_{76}. In Chapter 6_{101}, we focus on n-monotone lower previsions, with more concrete examples of these discussed in Chapter 7_{122}. Amongst these, we draw attention to belief functions, possibility measures and probability boxes, all of which are widely used in the literature. In Chapter 8_{151}, we investigate the link between coherent lower previsions and integration, to arrive at a computationally more practical lower envelope result in terms of integration of probability charges. Chapter 9_{181} deals with further specific issues and applications of integration.

In Chapter 10_{191}, we investigate lower previsions that are symmetric, in the sense of being invariant under a monoid of transformations, based on the work of Walley (1991, Section 3.5) and De Cooman and Miranda (2007). We consider a weaker and a stronger type of invariance and discuss their meaning.

Finally, in Chapter 11_{214}, following the work of Maaß (2002, 2003), we use the Bishop–De Leeuw Theorem to arrive at representation theorems for various classes of coherent lower previsions, generalising Choquet's (1953–1954) Representation Theorem for completely monotone capacities.

3

Sets of acceptable bounded gambles

The term **imprecise probability** (model) relates to any mathematical description or representation of uncertainty that allows for indecision, incomplete preferences or intervals to represent chances or probabilities. We use the term **precise probability** (model) to contrast this with uncertainty representations from the classical, or usual, theory of probability, which do not allow for indecision, incomplete preferences, and where chances or probabilities are represented by real numbers, not intervals.

In modelling uncertainty, we are especially interested in representing, and making inferences based on, some subject's *beliefs*. We try to represent those beliefs by looking at a subject's *behaviour* and decisions. This makes our approach **epistemic** and **behavioural**. The main idea is that some types of behavioural model are more *reasonable* than others, and therefore, we impose certain criteria that we think rational behaviour *should* satisfy. This makes the theory both **rationalist** and **normative** and allows us to reason and make inferences based on a subject's beliefs.

Obviously, a subject's beliefs can be represented in many different ways, some of which are mutually related. For instance, *sets of acceptable bounded gambles* use risky transactions that a subject may or may not accept. Equivalently, *preference orders* consider whether a subject will accept to exchange one risky transaction for another. Similarly, *lower previsions* focus on exchanging risky transactions with fixed – or certain – rewards. Walley (1991, Section 3.8) explores relations between such representations in great detail. In this book, we are not so much concerned with various representations and equivalences between them. Although we could easily address such issues, this would mostly distract us from our main mission, which is to give an overview of the existing mathematical theory of coherent lower previsions and to extend it to unbounded gambles.

Lower Previsions, First Edition. Matthias C.M. Troffaes and Gert de Cooman.

Sets of acceptable bounded gambles are thus merely one of many imprecise probability models. We choose to start out with them for various reasons:

1. Sets of acceptable bounded gambles have a large number of other uncertainty models as special cases, including the usual theory of probability and, of course, also coherent lower previsions.

2. Reasoning (or inference) with sets of acceptable bounded gambles has a simple geometric interpretation.

3. The interpretation of a set of acceptable bounded gambles – the way it represents uncertainty – is rather straightforward.

4. When framed in terms of acceptability, the theory generalises almost trivially to unbounded gambles.

So, in summary, we start here by studying acceptability, and will then use it in the next chapter as a tool to construct a theory of lower previsions.

Specifically, after setting up the basic framework in Sections 3.1, 3.2 and 3.3_{28}, we start with sets of acceptable bounded gambles in Section 3.4_{29} and explain how to model a subject's beliefs, how to formulate rationality and how to make inferences, in terms of such sets.

In Chapter 4, we will use the sets of acceptable bounded gambles approach to introduce and motivate rationality axioms in Sections 4.2_{41} and 4.3_{46} and inference mechanisms for lower previsions in Section 4.5_{65}.

This approach to lower previsions is slightly different from Walley's (1991) construction of the theory of lower previsions. Of course, the ideas used are clearly the same, and the approaches are mathematically equivalent.

3.1 Random variables

We consider a **subject**. We call any variable whose value is possibly uncertain or unknown to the subject a **random variable**. The only restriction we impose is that the actual value of the variable should in principle be observable, or determinable. One could think of a random variable as the outcome of some experiment, not necessarily known to the subject. For example, the amount of rainfall $R(d, w)$ during a particular day d, measured at a particular weather station w, is a random variable; it takes values in the set $\mathbb{R}_{\geq 0}$ of non-negative real numbers. But also the statement 'tomorrow, it will rain in Ghent' is a random variable, with possible values 'true' and 'false'. The set of possible values of a random variable X is denoted by a calligraphic letter $\mathcal{X}$. A particular value of X is generically denoted by a lowercase letter x. It is sometimes convenient to denote the event of observing X to be x as '$X = x$'.

Let us now describe a few things that we can do with random variables. At the same time, we fix some notation that will be used further on.

Any map f defined on the set of outcomes $\mathcal{X}$ of a random variable X we can again consider as a random variable. It represents a relabelling (not necessarily one to one) of the outcomes of X. If the actual value of X is observable, then so is, in principle, the actual value of $f(X)$, and hence, the function f again constitutes a random variable. For instance, 'tomorrow, it will rain in Ghent' could be defined as Boolean function of the amount of rainfall $R(d, w)$ for d tomorrow and measured at Ghent's weather station w. We may write $f(X)$ in order to emphasise that f is a function of X. In this respect, the **identity map** $\mathrm{id}_{\mathcal{X}}$ on $\mathcal{X}$ is a special case that deserves some attention. It is defined by $\mathrm{id}_{\mathcal{X}}(x) := x$ for all $x \in \mathcal{X}$. As, obviously, $\mathrm{id}_{\mathcal{X}}(X) = X$, we often simply write X rather than $\mathrm{id}_{\mathcal{X}}(X)$ or even $\mathrm{id}_{\mathcal{X}}$, following the standard practice in many books on probability theory.

Obviously, the composition $g \circ f$ of maps f and g is a random variable if f is defined on the set of outcomes of a random variable X. As $(g \circ f)(x)$ is defined as $g(f(x))$, we can also write $(g \circ f)(X)$ as $g(f(X))$.

3.2 Belief and behaviour

Our aim in this chapter is to formulate a behavioural model for the subject's beliefs about the value of a random variable X. Before going into a detailed description of this model, we first discuss very briefly a few important hows, whys, pros and cons of behavioural belief models.

Beliefs about X can be modelled through behaviour. For instance, if we strongly believe that the outcome of the random variable 'tomorrow, it will rain in Ghent' will be 'true', then probably we will take an umbrella with us for tomorrow's city trip to Ghent – even though we might not be totally certain of rain. Thus, clearly, our behaviour is a reflection of our beliefs – possibly involving uncertainty. This has led some people, such as Ramsey (1931), de Finetti (1937), von Neumann and Morgenstern (1944), Savage (1972), Anscombe and Aumann (1963), Kahneman and Tversky (1979), Walley (1981), von Winterfeldt and Edwards (1986), Seidenfeld, Schervish and Kadane (1995) and many others, to the idea of taking behaviour – in whatever form – as the primitive notion when modelling belief, *de facto* taking some form of behaviour as a definition of belief. It is also the course we pursue here.

A simple but fairly general way to model our beliefs about the random variable X consists of considering our dispositions towards transactions whose corresponding gain depends on the outcome, or actual value, of X. To see how such dispositions relate to belief, assume for instance that we strongly believe that we will observe $X = x$. We should then be inclined to accept, before observation of X, any transaction that incurs a positive gain if $X = x$. On the other hand, if we strongly believe that $X = x$ will not be observed, we should not care too much, before observation of X, about the gains or losses connected with $X = x$. Hence, beliefs about X naturally translate into behavioural dispositions towards transactions whose value depends on the outcome of X. In what follows, we consider a very particular type of transactions: buying gains that are a *bounded* function of X. In the second part of this book, we will generalise this to possibly unbounded functions of X.

Any behavioural belief model can be made operational because measuring belief is simply obtained by measuring behaviour – at least in principle. We refer to von Winterfeldt and Edwards (1986) for an extensive discussion.

Conversely, however, not all beliefs are reflected by behaviour, and certainly not by the restricted types of behaviour that we will study. Despite its unifying character, this is a hard limit on the applicability of the theory we are about to describe. This limit is at least two-fold. Firstly, we only express belief regarding random variables. Secondly, we are only concerned with transactions expressed in terms of a linear utility scale.

Indeed, not all our beliefs concern random variables. For instance, we may believe that most of our beliefs are not concerned at all with dispositions towards transactions whose value depends on the outcome of a random variable. This belief we cannot express as dispositions towards transactions whose value depends on the outcome of a random variable. More seriously, any belief concerning an entity whose value is not observable falls beyond the scope of the behavioural belief models that are based on random variables. For instance, what about a transaction whose outcome depends on the fact whether or not the electron is a fundamental particle? Or, what about Pascal's (2001) wager, considering a transaction whose outcome depends on the fact whether God exists or not? Unless we have an experiment that allows us to determine an objective (or at least inter-subjective) answer to such questions, at least in principle, transactions whose outcome depends on these answers cannot be executed. Therefore, they fall beyond the scope of behavioural belief models.

Consequently, we ought to refrain from considering transactions whose outcome depends on quantities that can attain an infinite number of values, say, all real values representing the distance between two randomly drawn points on the blackboard. How could we possibly determine that the outcome of that distance is exactly equal to, say, $\sqrt{2}$? Nevertheless, in mainstream science, it is far beyond question that such quantities are immensely useful in mathematical and statistical modelling. Therefore, we will accept random variables that take values in infinite spaces, such as $\mathbb{R}$, under the assumption that they represent mathematical idealisations of real-life measurement processes. In fact, one could say that the second part of this book is dedicated exclusively to such idealisations.

Also, we are only concerned with transactions expressed in units of a fixed and unique linear utility scale: this means roughly that twice executing a transaction doubles its value. This is taken for granted if transactions only involve a sufficiently small amount of money (see de Finetti, 1974–1975, Volume I, Section 3.2) or an exchange of lottery tickets (see Walley, 1991, Section 2.2). However, this is no longer the case when the results of transactions involve drinking beer, eating gnocchi, sunbathing, making love or raising children. So there are behavioural dispositions, reflecting beliefs, that cannot be completely expressed using a linear utility scale. They fall beyond the scope of the behavioural belief models we now turn to.

3.3 Bounded gambles

A **bounded gamble** on $\mathcal{X}$ is a bounded real-valued map $f : \mathcal{X} \to \mathbb{R}$ and is interpreted as a gain that is a bounded function of X. This gain is assumed to be expressed in units of a linear utility scale that is assumed to be pre-determined and fixed. In particular, a bounded gamble is a real-valued random variable. We say that the subject **accepts** f, or that f is a **acceptable bounded gamble** for him, whenever he accepts, at least in principle, to engage in the following *risky transaction*, where

1. the actual value x of X is determined;
2. then the subject receives the amount of utility $f(x)$.

The latter means that, when $f(x)$ is non-negative, the total utility in his possession will be increased by $|f(x)|$ and, when $f(x)$ is non-positive, it will be decreased by $|f(x)|$. As mentioned before, we may write $f(X)$ if we want to emphasise that f is interpreted as a function of X.

As we have seen in Chapter 1_1, the set of all bounded gambles, or bounded real maps, on $\mathcal{X}$ is generally denoted by $\mathbb{B}(\mathcal{X})$ or more simply by $\mathbb{B}$ when no confusion can arise about $\mathcal{X}$. Recall that $\mathbb{B}(\mathcal{X})$ is a linear lattice with respect to the point-wise addition, the scalar multiplication and the point-wise ordering of bounded gambles.

If two bounded gambles f and g represent the outcomes associated with two uncertain transactions, then their sum $f + g$ represents the outcome of the *combined* transaction. Similarly, the scalar product λf is a *rescaled* version of the bounded gamble f. And if $f \leq g$, then we say that g **dominates** f: it leads to an outcome that is *guaranteed* to be at least as high as f.

3.4 Sets of acceptable bounded gambles

The information a subject has about X will lead him to accept or reject transactions whose reward depends on X. We formulate a model for the subject's beliefs by looking which transactions he accepts.

The basic idea is then that we consider our subject's **set of acceptable bounded gambles** $\mathcal{D}$ to be the set of bounded gambles that he accepts or is disposed to accept. It is intended to represent at least part of his behavioural dispositions, and as we have argued earlier, it can be seen as a model for his beliefs about X.

Now assume we have two subjects, one with a set of acceptable bounded gambles $\mathcal{D}_1$ and a second with a set of acceptable bounded gambles $\mathcal{D}_2$. If $\mathcal{D}_1 \subseteq \mathcal{D}_2$, then the second subject will (at least) accept all the bounded gambles that the first does. For this reason, we can interpret the set inclusion relation '$\subseteq$' between sets of acceptable bounded gambles as '*is at most as informative as*', or as '*is at most as committal as*', or even as '*is at least as conservative as*'.

3.4.1 Rationality criteria

Some behavioural dispositions are evident: for instance, an indicator does not take a negative value, and therefore, the subject should always be disposed to accept it. Also, some behavioural dispositions imply other behavioural dispositions. To give a simple example, if the subject is disposed to accept a bounded gamble, then he should also be disposed to accept any bounded gamble that dominates it. These simple rules are actually implied by the following rationality axioms that we impose on a subject's dispositions, as expressed in his set of acceptable bounded gambles $\mathcal{D}$.

Regarding BA1, remember that we take $f < 0$ to mean that '$f(x) \leq 0$ for all $x \in \mathcal{X}$ and $f(x) < 0$ for at least one $x \in \mathcal{X}$'.

Axiom 3.1 (Rationality for sets of acceptable bounded gambles) *For all bounded gambles f and g on $\mathcal{X}$, and all non-negative real numbers λ,*

BA1. *If $f < 0$, then $f \notin \mathcal{D}$: a subject should not be disposed to accept any bounded gamble that he cannot win from, and that may actually make him lose.* **(avoiding partial loss)**

BA2. *If $f \geq 0$ then $f \in \mathcal{D}$: a subject should be disposed to accept any bounded gamble he cannot lose from.* **(accepting partial gain)**

BA3. *If $f \in \mathcal{D}$, then $\lambda f \in \mathcal{D}$: a subject disposed to accept f should also be disposed to accept λf.* **(scale invariance)**

BA4. *If $f \in \mathcal{D}$ and $g \in \mathcal{D}$, then $f + g \in \mathcal{D}$: a subject disposed to accept f and g should also be disposed to accept their combination $f + g$.* **(combination)**

The first two axioms require no further motivation. Axioms BA3 and BA4 can actually be seen as consequences of the linearity of the utility scale in which bounded gambles and prices are expressed; see for instance Walley (1991, Section 2.4.4) for a detailed justification using probability currency as a linear utility scale. Also see de Finetti (1974–1975, Volume I, Section 3.2.5) for a simple solution in case bounded gambles and prices are expressed in a precise but non-linear utility scale.

There is one immediate consequence of these axioms useful bearing out at his point:

BA5. If $f \in \mathcal{D}$ and $g \geq f$, then $g \in \mathcal{D}$: any bounded gamble that dominates an acceptable bounded gamble is acceptable too. **(monotonicity)**

Proof. Simply observe that $g = f + (g - f)$ and use BA2 to see that $g - f \in \mathcal{D}$, because $g - f \geq 0$. Then use $f \in \mathcal{D}$ and BA4. □

Our search through the literature has revealed that Williams (1975a,b, 1976) was the first to introduce acceptability. Building on de Finetti's betting framework (1974–1975), he considers the acceptability of *one-sided* bets instead of *two-sided* bets. This relaxation leads to cones of bets instead of linear subspaces of them. The germ of the theory is, however, already present in Smith's work (1961, p. 15), who uses a (generally) open cone of 'exchange vectors' when talking about currency exchange. Both authors influenced Walley (1991, Section 3.7 and Appendix F). He – as well as most authors who follow his lead – uses the term 'desirability' instead of 'acceptability' and describes three variants (almost, really and strictly desirable bounded gambles). He emphasises the conceptual ease with which updated (or posterior) models can be obtained in this framework (Walley, 2000). Moral (2000; 2005) takes the next step and applies acceptability to study epistemic irrelevance, a structural assessment. He also points out how conceptually easy extension, marginalisation and conditioning are in this framework. De Cooman and Miranda (2007) study transformational symmetry assessments for acceptable bounded gambles. Recently, Couso and Moral (2009, 2011) discuss the relationship with credal sets, computer representation and maximal sets of acceptable bounded gambles. De Cooman and Quaeghebeur (2009, 2012) study exchangeability using acceptability. Quaeghebeur (2013) gives a comprehensive overview of the use of acceptability notions in the context of imprecise probabilities. The newest developments also try to incorporate a notion of indifference into the conceptual framework (Hermans, 2012; Quaeghebeur et al., 2012).

The terminology in the literature regarding acceptability is somewhat confusing: the most important difference is that some authors assume the zero gamble to be acceptable, whereas others do not. Specifically, BA2 may vary slightly. Our notion of acceptability coincides with Walley's (2000) later notion of desirability, also used by Moral (2000) and Couso and Moral (2009, 2011), and aims at capturing a weak preference to the zero gamble. Walley in his book (Walley, 1991, Appendix F) and Moral (2005) in a later paper use a slightly different notion of acceptability, which is rather aimed at representing a strict preference to the zero gamble.

That a notion of strict preference is very useful in modelling uncertainty – and even necessary for capturing some of its more subtle aspects – has been argued convincingly, and in considerable detail by Seidenfeld et al. (1995), independently of the discussion initiated by Williams and Walley, and using a different language and approach. We also mention more recent developments in this area (Hermans, 2012; Quaeghebeur et al., 2012), which try to account in full detail for the notions of weak preference, strict preference and indifference.

As far as the derivation of results in this book is concerned, the difference between strict and weak preference does not actually matter very much, if at all. Taking this into account, we have chosen to use Williams' 'acceptability' to avoid adding to the confusion and to indicate clearly that on the present approach, the zero gamble is assumed to be acceptable.

Definition 3.2 (Coherence for sets of acceptable bounded gambles) *Any set of acceptable bounded gambles $\mathcal{D}$ that satisfies the rationality criteria* BA1–BA4 *is called **coherent**. We denote the set of all coherent sets of acceptable bounded gambles on $\mathcal{X}$ by $\mathbb{D}_b(\mathcal{X})$ or simply $\mathbb{D}_b$ when no confusion can arise.*

In geometrical parlance, coherent sets of acceptable bounded gambles are convex cones in the linear space $\mathbb{B}$ that include the non-negative (first) orthant and have no point in common with the negative orthant.

Example 3.3 It is clear that the non-negative orthant of $\mathbb{B}$ or, in other words, the set of non-negative bounded gambles

$$\mathbb{B}_{\geq 0} := \{f \in \mathbb{B} : f \geq 0\}$$

is a coherent set of acceptable bounded gambles that is included in any other element of $\mathbb{D}_b$ and therefore the smallest, most conservative or least informative element of $\mathbb{D}_b$.

Because $\mathbb{B}_{\geq 0}$ is the most conservative set of acceptable bounded gambles, it is also called **vacuous**. Adopting $\mathbb{B}_{\geq 0}$, the subject is guaranteed never to incur a loss, regardless of the outcome x of X, because all bounded gambles in $\mathbb{B}_{\geq 0}$ are non-negative. Therefore, it is an appropriate model when no information about X is available. Obviously, any inferences that we may be able to make from this model will be – if anything – extremely weak. ♦

3.4.2 Inference

If we ask a subject to specify a set of acceptable bounded gambles $\mathscr{A}$, for instance, by indicating a number of bounded gambles that he accepts, then we cannot expect his **assessment** $\mathscr{A}$ to be coherent. For one thing, his assessment will typically consist of only a finite number of bounded gambles.

We can then ask whether it is possible to extend the assessment $\mathscr{A}$ to a coherent set of acceptable bounded gambles by adding bounded gambles to it. Of course, we would want to do this in a way that is as conservative as possible: we only want to add those bounded gambles that are strictly necessary in order to achieve coherence.

It will be useful to associate, with any assessment $\mathscr{A}$, the following set of bounded gambles:

$$\begin{aligned}\mathscr{E}_{\mathscr{A}} &:= \left\{ g + \sum_{k=1}^{n} \lambda_k f_k : g \geq 0, n \in \mathbb{N}, f_k \in \mathscr{A}, \lambda_k \in \mathbb{R}_{\geq 0}, k = 1, \ldots, n \right\}, \qquad (3.1)\\ &= \operatorname{nonneg}(\mathbb{B}_{\geq 0} \cup \mathscr{A}) = \mathbb{B}_{\geq 0} + \operatorname{nonneg}(\mathscr{A}),\end{aligned}$$

where $\mathbb{R}_{\geq 0}$ denotes the set of non-negative real numbers, but we can of course also replace it with the set $\mathbb{R}_{>0}$ of positive real numbers. It is not difficult to see that $\mathscr{E}_{\mathscr{A}}$ is the smallest set of bounded gambles that includes $\mathscr{A}$ and satisfies the coherence conditions $\mathrm{BA2}_{30}$–$\mathrm{BA4}_{30}$.

First of all, we want to identify those assessments $\mathscr{A}$ that *can* be extended to some coherent set of acceptable bounded gambles.

Definition 3.4 (Consistency) *A set $\mathscr{A}$ of acceptable bounded gambles is called* ***consistent****, or* ***avoids partial loss****, if one (and hence all) of the following equivalent conditions is satisfied:*

(A) *$\mathscr{A}$ is included in some coherent set of acceptable bounded gambles:*

$$\left\{ \mathscr{D} \in \mathbb{D}_b : \mathscr{A} \subseteq \mathscr{D} \right\} \neq \emptyset.$$

(B) *For all n in $\mathbb{N}$, non-negative λ_1, …, λ_n in $\mathbb{R}$, and bounded gambles f_1, …, f_n in $\mathscr{A}$:*

$$\sum_{k=1}^{n} \lambda_k f_k \nless 0.$$

Condition (B) explains why the consistency condition is also called *avoiding partial loss*: no non-negative linear combination of bounded gambles in $\mathscr{A}$ should produce a partial loss. Let us prove that these conditions are indeed equivalent.

Proof. (A)⇒(B). Assume that there is some coherent $\mathscr{D}$ that includes $\mathscr{A}$. Consider arbitrary $n \in \mathbb{N}$, real $\lambda_k \geq 0$ and f_k in $\mathscr{A}$. Then $g := \sum_{k=1}^{n} \lambda_k f_k$ belongs to $\mathscr{D}$ by coherence conditions $BA3_{30}$ and $BA4_{30}$. But then $g \nless 0$ by coherence condition $BA1_{30}$.

(B)⇒(A). Recall that $\mathscr{E}_{\mathscr{A}}$ is a set of bounded gambles that satisfies the coherence conditions $BA2_{30}$–$BA4_{30}$. We infer from (B) that $h \nless 0$ for all h in $\mathscr{E}_{\mathscr{A}}$, so $\mathscr{E}_{\mathscr{A}}$ also satisfies coherence condition $BA1_{30}$, and is therefore a coherent set of acceptable bounded gambles that includes $\mathscr{A}$. This implies that $\mathscr{A}$ is consistent. □

Once we have a consistent assessment $\mathscr{A}$, then we know it can be extended to a coherent set of acceptable bounded gambles. The next step to take, then, is to see if we can find, amongst all possible such coherent extensions, the one that is as conservative as possible. This is actually very easy to do. Indeed, what makes this possible at all is the following simple result, whose proof is straightforward and therefore omitted.

Proposition 3.5 *Consider a non-empty family $\mathscr{D}_i$, $i \in I$ of sets of acceptable bounded gambles. If all $\mathscr{D}_i$ are coherent, then so is their intersection $\bigcap_{i \in I} \mathscr{D}_i$.*

Whenever some property is preserved under arbitrary (non-empty) intersections, the idea of closure is just around the corner. Indeed, for any assessment $\mathscr{A}$, we can consider the collection $\{\mathscr{D} \in \mathbb{D}_b : \mathscr{A} \subseteq \mathscr{D}\}$ of all coherent sets of acceptable bounded gambles that include $\mathscr{A}$ and we define the intersection of this collection to be the **closure** $\mathrm{Cl}_{\mathbb{D}_b}(\mathscr{A})$ of $\mathscr{A}$:

$$\mathrm{Cl}_{\mathbb{D}_b}(\mathscr{A}) := \bigcap \{\mathscr{D} \in \mathbb{D}_b : \mathscr{A} \subseteq \mathscr{D}\}.$$

In the expression above, we take the intersection of the empty collection to be equal to the set $\mathbb{B}$ of all bounded gambles: $\bigcap \emptyset := \mathbb{B}$. The closure operator $\mathrm{Cl}_{\mathbb{D}_b}$ has the following interesting properties:

Proposition 3.6 *Let $\mathscr{A}$, $\mathscr{A}_1$ and $\mathscr{A}_2$ be sets of acceptable bounded gambles. Then the following statements hold:*

(i) $\mathscr{A} \subseteq \mathrm{Cl}_{\mathbb{D}_b}(\mathscr{A})$.

(ii) *If* $\mathscr{A}_1 \subseteq \mathscr{A}_2$, *then* $\mathrm{Cl}_{\mathbb{D}_b}(\mathscr{A}_1) \subseteq \mathrm{Cl}_{\mathbb{D}_b}(\mathscr{A}_2)$.

(iii) $\mathrm{Cl}_{\mathbb{D}_b}(\mathrm{Cl}_{\mathbb{D}_b}(\mathscr{A})) = \mathrm{Cl}_{\mathbb{D}_b}(\mathscr{A})$.

(iv) *If* $\mathscr{A} \subseteq \mathbb{B}_{\geq 0}$ *then* $\mathrm{Cl}_{\mathbb{D}_b}(\mathscr{A}) = \mathbb{B}_{\geq 0}$.

(v) $\mathscr{A}$ *is consistent if and only if* $\mathrm{Cl}_{\mathbb{D}_b}(\mathscr{A}) \neq \mathbb{B}$.

(vi) $\mathscr{A}$ *is a coherent set of acceptable bounded gambles if and only if it is consistent and* $\mathscr{A} = \mathrm{Cl}_{\mathbb{D}_b}(\mathscr{A})$.

Any $\mathscr{P}(\mathbb{B}) - \mathscr{P}(\mathbb{B})$-map that satisfies the properties (i)–(iii) of this proposition is called a **closure operator** in the literature (see, for instance, Davey and Priestley, 1990, Section 2.20), and (vi) tells us that the corresponding **closed** subsets of $\mathscr{P}(\mathbb{B})$ different from $\mathbb{B}$ are precisely the coherent sets of acceptable bounded gambles.[1]

[1] A closure operator Cl is **topological** if it also satisfies $\mathrm{Cl}(A \cup B) = \mathrm{Cl}(A) \cup \mathrm{Cl}(B)$ for all $A, B \subseteq \mathbb{B}$; see also the discussion in Appendix B_{371}.

Proof. (i)$_\curvearrowleft$. Trivial.

(ii)$_\curvearrowleft$. If $\mathscr{A}_1 \subseteq \mathscr{A}_2$, then $\left\{\mathscr{D} \in \mathbb{D}_b : \mathscr{A}_1 \subseteq \mathscr{D}\right\} \supseteq \left\{\mathscr{D} \in \mathbb{D}_b : \mathscr{A}_2 \subseteq \mathscr{D}\right\}$. It now follows at once from the definition of $\mathrm{Cl}_{\mathbb{D}_b}$ that $\mathrm{Cl}_{\mathbb{D}_b}(\mathscr{A}_1) \subseteq \mathrm{Cl}_{\mathbb{D}_b}(\mathscr{A}_2)$.

(iii)$_\curvearrowleft$. By Proposition 3.5$_\curvearrowleft$, $\mathrm{Cl}_{\mathbb{D}_b}(\mathscr{A}) \in \mathbb{D}_b$, whence

$$\bigcap \left\{\mathscr{D} \in \mathbb{D}_b : \mathrm{Cl}_{\mathbb{D}_b}(\mathscr{A}) \subseteq \mathscr{D}\right\} = \mathrm{Cl}_{\mathbb{D}_b}(\mathscr{A}).$$

Therefore, indeed, $\mathrm{Cl}_{\mathbb{D}_b}(\mathrm{Cl}_{\mathbb{D}_b}(\mathscr{A})) = \mathrm{Cl}_{\mathbb{D}_b}(\mathscr{A})$.

(iv)$_\curvearrowleft$. Clearly, if $\mathscr{A} \subseteq \mathbb{B}_{\geq 0}$, then $\left\{\mathscr{D} \in \mathbb{D}_b : \mathscr{A} \subseteq \mathscr{D}\right\} = \mathbb{D}_b$, and we know that the intersection of all sets in $\mathbb{D}_b$ is precisely $\mathbb{B}_{\geq 0}$.

(v)$_\curvearrowleft$. Suppose $\mathscr{A}$ is consistent, then $\mathrm{Cl}_{\mathbb{D}_b}(\mathscr{A})$ is the intersection of a non-empty set $\left\{\mathscr{D} \in \mathbb{D}_b : \mathscr{A} \subseteq \mathscr{D}\right\}$ of coherent sets of acceptable bounded gambles and, therefore, coherent by Proposition 3.5$_\curvearrowleft$. This implies that $\mathrm{Cl}_{\mathbb{D}_b}(\mathscr{A}) \neq \mathbb{B}$. Conversely, suppose that $\mathscr{A}$ is not consistent, then indeed $\mathrm{Cl}_{\mathbb{D}_b}(\mathscr{A}) = \bigcap \emptyset = \mathbb{B}$.

(vi)$_\curvearrowleft$. Assume that $\mathscr{A}$ is consistent and that $\mathscr{A} = \mathrm{Cl}_{\mathbb{D}_b}(\mathscr{A})$. The consistency of $\mathscr{A}$ implies that $\left\{\mathscr{D} \in \mathbb{D}_b : \mathscr{A} \subseteq \mathscr{D}\right\} \neq \emptyset$, so $\mathrm{Cl}_{\mathbb{D}_b}(\mathscr{A})$ is coherent by its definition and Proposition 3.5$_\curvearrowleft$. This implies that $\mathscr{A} = \mathrm{Cl}_{\mathbb{D}_b}(\mathscr{A})$ is coherent. On the other hand, if $\mathscr{A}$ is coherent, then $\mathscr{A} \in \mathbb{D}_b$, so obviously, $\bigcap \left\{\mathscr{D} \in \mathbb{D}_b : \mathscr{A} \subseteq \mathscr{D}\right\} = \mathscr{A}$ or, in other words, $\mathrm{Cl}_{\mathbb{D}_b}(\mathscr{A}) = \mathscr{A}$. As $\mathbb{B}$ cannot be coherent, we see that $\mathrm{Cl}_{\mathbb{D}_b}(\mathscr{A}) = \mathscr{A} \neq \mathbb{B}$, so (v)$_\curvearrowleft$ guarantees that $\mathscr{A}$ is consistent. □

Theorem 3.7 (Natural extension) *If the set $\mathscr{A}$ of acceptable bounded gambles is consistent, then there is a smallest coherent set of acceptable bounded gambles that includes $\mathscr{A}$. It is given by*

$$\begin{aligned}\mathrm{Cl}_{\mathbb{D}_b}(\mathscr{A}) = \mathscr{E}_{\mathscr{A}} &= \left\{ g + \sum_{k=1}^{n} \lambda_k f_k : g \geq 0, n \in \mathbb{N}, f_k \in \mathscr{A}, \lambda_k \in \mathbb{R}_{\geq 0}, k = 1, \dots, n \right\} \\ &= \left\{ h \in \mathbb{B} : h \geq \sum_{k=1}^{n} \lambda_k f_k \text{ for some } n \in \mathbb{N}, f_k \in \mathscr{A}, \lambda_k \in \mathbb{R}_{\geq 0} \right\},\end{aligned} \tag{3.2}$$

and it is called the ***natural extension*** *of $\mathscr{A}$.*

Proof. Assume that $\mathscr{A}$ is consistent. Then the collection $\left\{\mathscr{D} \in \mathbb{D}_b : \mathscr{D} \subseteq \mathscr{A}\right\}$ is non-empty, and therefore, its intersection $\mathrm{Cl}_{\mathbb{D}_b}(\mathscr{A})$ is a coherent set of acceptable bounded gambles by Proposition 3.5$_\curvearrowleft$. Clearly, it includes $\mathscr{A}$ and is included in any other coherent set of acceptable bounded gambles that includes $\mathscr{A}$. It is, therefore, indeed the smallest coherent set of acceptable bounded gambles that includes $\mathscr{A}$.

It remains to prove that $\mathrm{Cl}_{\mathbb{D}_b}(\mathscr{A}) = \mathscr{E}_{\mathscr{A}}$. It is obvious by considering the coherence conditions $\mathrm{BA2}_{30}$–$\mathrm{BA4}_{30}$ that $\mathscr{E}_{\mathscr{A}}$ is included in any coherent set $\mathscr{D}$ that includes $\mathscr{A}$. If we can show that $\mathscr{E}_{\mathscr{A}}$ is a coherent set of acceptable bounded gambles, then the proof is complete. The verification of the coherence conditions $\mathrm{BA2}_{30}$–$\mathrm{BA4}_{30}$ is straightforward, so we turn to condition $\mathrm{BA1}_{30}$. Consider any element $h = g + \sum_{k=1}^{n} \lambda_k f_k$ of $\mathscr{E}_{\mathscr{A}}$. As $\mathscr{A}$ avoids partial loss, we know that $\sum_{k=1}^{n} \lambda_k f_k \nless 0$ by Definition 3.4(B)$_{32}$. As $g \geq 0$, *a fortiori* $h \nless 0$. □

We see that when we apply the closure operator $\mathrm{Cl}_{\mathbb{D}_b}$, or natural extension, to a consistent assessment $\mathscr{A}$, we simply add those bounded gambles to $\mathscr{A}$ that can be obtained from bounded gambles in $\mathscr{A}$ using the **production rules** $\mathrm{BA2}_{30}$–$\mathrm{BA4}_{30}$ in Axiom 3.1_{30} and no other bounded gambles. This shows that natural extension is a conservative inference mechanism: it associates with the consistent assessment $\mathscr{A}$ the smallest coherent set of acceptable bounded gambles that these production rules lead to. Consistency simply means that the production rules will not lead to negative bounded gambles and thus to a partial loss.

That such a form of conservative inference exists guarantees that we can always assume that a subject has a coherent set of acceptable bounded gambles – simply convert the assessment $\mathscr{A}$ he has given to $\mathrm{Cl}_{\mathbb{D}_b}(\mathscr{A})$. Should $\mathscr{A}$, and therefore $\mathrm{Cl}_{\mathbb{D}_b}(\mathscr{A})$, be inconsistent, then the assessment is irrational because it leads to a partial loss. It must, therefore, be corrected by, say, removing some bounded gambles until consistency is achieved.

The inference mechanism behind natural extension and coherent sets of acceptable bounded gambles subsumes that of classical propositional logic. To see how this comes about, we can rely on the following simple arguments. Recall that an *event* is a subset of $\mathscr{X}$, and its *indicator* I_A is the bounded gamble that gives one if the actual value x of X belongs to A and zero otherwise. We restrict ourselves here to propositions p about X, which are in a one-to-one correspondence with the subsets A_p of $\mathscr{X}$. If our subject **accepts a proposition** p, we take this to mean that he accepts the bounded gambles $I_{A_p} - 1 + \varepsilon$ for all $\varepsilon > 0$. As

$$I_{A_p}(x) - 1 + \varepsilon = \begin{cases} \varepsilon & \text{if } x \in A_p \\ \varepsilon - 1 & \text{if } x \notin A_p, \end{cases}$$

our subject is then willing to bet on the occurrence of A_p at any odds $(1 - \varepsilon)/\varepsilon$ – or equivalently, at any rate strictly smaller than 1: he is *practically certain* that A_p occurs or, in other words, that X belongs to A_p.

Now suppose that our subject accepts p, and that $p \Rightarrow q$, meaning that $A_p \subseteq A_q$, whence

$$I_{A_p}(x) - 1 + \varepsilon \leq I_{A_q}(x) - 1 + \varepsilon.$$

It then follows from $\mathrm{BA5}_{30}$ that our subject should also accept q. Hence, the inference mechanism behind coherent sets of acceptable bounded gambles subsumes the *modus ponens* production rule of classical propositional logic.

Next, assume that our subject accepts both p and q, meaning that he accepts the bounded gambles $I_{A_p}(x) - 1 + \varepsilon$ and $I_{A_q}(x) - 1 + \varepsilon$, and therefore (use $\mathrm{BA4}_{30}$) also the bounded gambles $I_{A_p}(x) + I_{A_q}(x) - 2 + 2\varepsilon$, for all $\varepsilon > 0$. As for the conjunction $p \wedge q$, we have $A_{p \wedge q} = A_p \cap A_q$ and

$$I_{A_p}(x) + I_{A_q}(x) - 1 \leq I_{A_p \cap A_q} = \min\{I_{A_p}, I_{A_q}\},$$

we infer from $\mathrm{BA5}_{30}$ that our subject should also accept $p \wedge q$, meaning that the inference mechanism behind coherent sets of acceptable bounded gambles also subsumes the *conjunction* production rule of classical propositional logic.

Finally, *logical contradiction* is forbidden by the inference mechanism. If our subject were to accept both p and its negation $\neg p$, he would accept both $I_{A_p}(x) - 1 + \varepsilon$ and $I_{A_{\neg p}}(x) - 1 + \varepsilon$, and therefore also, by BA4$_{30}$, the bounded gambles

$$I_{A_p}(x) - 1 + \varepsilon + I_{A_{\neg p}}(x) - 1 + \varepsilon = -1 + 2\varepsilon$$

for all $\varepsilon > 0$, because $I_{A_p} + I_{A_{\neg p}} = 1$. This contradicts the avoiding partial loss axiom BA1$_{30}$.

We will come back to this issue in Section 5.5.2$_{88}$.

4

Lower previsions

In Chapter 3, we introduced sets of acceptable bounded gambles as a simple uncertainty model. Here, we explain how acceptability paves the way for a theory of lower previsions. Similarly to what we did in the chapter on acceptability, we first introduce and motivate rationality axioms (Sections 4.2_{41} and 4.3_{46}) and then continue with inference mechanisms (Section 4.5_{65}) for lower previsions. In doing so, we solely rely on sets of acceptable bounded gambles and borrow heavily from the results derived in Chapter 3.

The most important source for the theory of lower previsions is Walley's (1991) *magnum opus* on imprecise probabilities, but the basic ideas, also expressed in terms of acceptability, can be found already in Williams' (1975b) work. As mentioned earlier, our approach in this differs slightly from that of Walley (1991) and resembles Williams' (1975b) much more closely, although all results remain entirely equivalent mathematically. Walley starts out with rationality criteria and inference mechanisms for lower previsions directly based on the axioms of acceptability, but without using coherent sets of acceptable bounded gambles explicitly – although he does explain the link with coherent sets of acceptable bounded gambles in later chapters of his book. In contrast, we introduce sets of acceptable bounded gambles first and use these consistently to explain the theory, leading to – in our opinion – a more coherent treatment.

We then use a separating hyperplane version of the Hahn–Banach theorem to express the notions of avoiding sure loss, coherence and natural extension in terms of more familiar concepts from probability theory (Section 4.6_{70}). Again, in doing so, we follow the work of Williams and Walley – however, our method of proof is slightly different: Williams' (1975b) proof relies directly on Zorn's lemma and Walley's (1991) proof relies on a topological version of the Hahn–Banach theorem whereas ours relies on a simpler non-topological version of the Hahn–Banach theorem. Further discussion on the relationship between lower previsions and various probability concepts will be postponed until Chapter 8_{151}.

Lower Previsions, First Edition. Matthias C.M. Troffaes and Gert de Cooman.

We end the chapter with a brief technical discussion of the topological aspects of coherent lower previsions, following Walley (1991, Section 3.6).

4.1 Lower and upper previsions

4.1.1 From sets of acceptable bounded gambles to lower previsions

If our subject has some coherent set of acceptable bounded gambles $\mathscr{D}$, then we can associate with $\mathscr{D}$ two special real functionals defined on $\mathbb{B}$: the **lower prevision** (operator) $\mathrm{lpr}(\mathscr{D}) : \mathbb{B} \to \mathbb{R}$ defined by

$$\mathrm{lpr}(\mathscr{D})(f) := \sup \{ \mu \in \mathbb{R} : f - \mu \in \mathscr{D} \} \tag{4.1}$$

and the **upper prevision** (operator) $\mathrm{upr}(\mathscr{D}) : \mathbb{B} \to \mathbb{R}$ defined by

$$\mathrm{upr}(\mathscr{D})(f) := \inf \{ \mu \in \mathbb{R} : \mu - f \in \mathscr{D} \}$$

for every bounded gamble f. We see that $\mathrm{lpr}(\mathscr{D})(f)$ is the supremum price μ such that the subject accepts to **buy** f **for** μ, that is, exchange the fixed (certain) reward μ for the random reward f. In other words, the lower prevision $\mathrm{lpr}(\mathscr{D})(f)$ of f is the *supremum acceptable buying price* for f associated with the coherent set of acceptable bounded gambles $\mathscr{D}$.

Similarly, $\mathrm{upr}(\mathscr{D})(f)$ is the supremum price μ such that the subject accepts to **sell** f **for** μ, that is, exchange the random reward f for the fixed (certain) reward μ, so the upper prevision $\mathrm{upr}(\mathscr{D})(f)$ of f is the *infimum acceptable selling price* for f associated with the coherent set of acceptable bounded gambles $\mathscr{D}$.

As the real functionals $\mathrm{lpr}(\mathscr{D})$ and $\mathrm{upr}(\mathscr{D})$ satisfy the following **conjugacy relationship**:

$$\mathrm{upr}(\mathscr{D})(f) = -\,\mathrm{lpr}(\mathscr{D})(-f) \text{ for all bounded gambles } f \text{ in } \mathbb{B},$$

we can always express one type of functional in terms of the other. For this reason, we concentrate on lower previsions.

Lower prevision operators derived from coherent sets of acceptable bounded gambles have three fundamental properties, which have an important part in what follows. This becomes clear in Theorem 4.2.

Theorem 4.1 *Let* $\mathrm{lpr}(\mathscr{D})$ *be the lower prevision associated with a coherent set of acceptable bounded gambles* $\mathscr{D}$*. Then for all bounded gambles* f *and* g *and all non-negative real numbers* λ*,*

LP1. $\inf f \le \mathrm{lpr}(\mathscr{D})(f)$; **(bounds)**

LP2. $\mathrm{lpr}(\mathscr{D})(f+g) \ge \mathrm{lpr}(\mathscr{D})(f) + \mathrm{lpr}(\mathscr{D})(g)$; **(super-additivity)**

LP3. $\mathrm{lpr}(\mathscr{D})(\lambda f) = \lambda\,\mathrm{lpr}(\mathscr{D})(f)$. **(non-negative homogeneity)**

Proof. LP1. Observe that $f - \inf f \geq 0$ and therefore $f - \inf f \in \mathscr{D}$ by BA2_{30}.

LP2. Let $\alpha < \text{lpr}(\mathscr{D})(f)$ and $\beta < \text{lpr}(\mathscr{D})(g)$. By Equation (4.1), we know that $f - \alpha \in \mathscr{D}$ and $g - \beta \in \mathscr{D}$. Now use BA4_{30} to conclude that $(f + g) - (\alpha + \beta) \in \mathscr{D}$, whence $\text{lpr}(\mathscr{D})(f + g) \geq \alpha + \beta$. Taking the supremum over all $\alpha < \text{lpr}(\mathscr{D})(f)$ and $\beta < \text{lpr}(\mathscr{D})(g)$, we get the desired inequality.

LP3. First assume that $\lambda > 0$. It follows from BA3_{30} that $\lambda f - \mu \in \mathscr{D}$ if and only if $f - \mu/\lambda \in \mathscr{D}$. Hence,

$$\begin{aligned}\text{lpr}(\mathscr{D})(\lambda f) &= \sup\{\mu \in \mathbb{R} : \lambda f - \mu \in \mathscr{D}\} = \sup\left\{\mu \in \mathbb{R} : f - \frac{\mu}{\lambda} \in \mathscr{D}\right\}\\ &= \sup\{\lambda\mu \in \mathbb{R} : f - \mu \in \mathscr{D}\} = \lambda \sup\{\mu \in \mathbb{R} : f - \mu \in \mathscr{D}\}\\ &= \lambda\, \text{lpr}(\mathscr{D})(f).\end{aligned}$$

For $\lambda = 0$, consider that $\text{lpr}(\mathscr{D})(0) = \sup\{\mu \in \mathbb{R} : -\mu \in \mathscr{D}\} = 0$, where the last equality follows from BA1_{30} and BA2_{30}. □

If, on the other hand, we have a real functional $\underline{P}$ defined on $\mathbb{B}$, we can ask under what conditions this real functional can be seen as a lower prevision operator associated with some coherent set of acceptable bounded gambles.

Theorem 4.2 *Consider any real functional $\underline{P}$ defined on $\mathbb{B}$. Then there is some coherent set of acceptable bounded gambles $\mathscr{D}$ such that $\underline{P} = \text{lpr}(\mathscr{D})$ if and only if $\underline{P}$ satisfies the properties* LP1–LP3.

Proof. The 'only if' part was proved in Theorem 4.1. We prove the 'if' part. Assume that $\underline{P}$ satisfies LP1–LP3 and let $\mathscr{D} := \{f \in \mathbb{B} : \underline{P}(f) > 0\} \cup \mathbb{B}_{\geq 0}$ – we will prove that $\mathscr{D}$ is coherent later on.

Then it follows from Lemma 4.3(iii)$_{\curvearrowright}$ that for any $f \in \mathbb{B}$ and $\mu \in \mathbb{R}$,

$$\begin{aligned} f - \mu \in \mathscr{D} &\Rightarrow (\underline{P}(f - \mu) > 0 \text{ or } f - \mu \geq 0)\\ &\Rightarrow (\underline{P}(f) > \mu \text{ or } \inf f \geq \mu) \Rightarrow \underline{P}(f) \geq \mu,\end{aligned}$$

where the last implication follows from LP1. Also,

$$\underline{P}(f) < \mu \Rightarrow \underline{P}(f - \mu) < 0 \Rightarrow f - \mu \notin \mathscr{D},$$

where we again used LP1 in the last implication: $\underline{P}(f - \mu) < 0$ implies $f - \mu \notin \mathbb{B}_{\geq 0}$ because if $f - \mu \in \mathbb{B}_{\geq 0}$ then $f - \mu \geq 0$ and hence $\underline{P}(f - \mu) \geq 0$. Therefore,

$$\text{lpr}(\mathscr{D})(f) = \sup\{\mu \in \mathbb{R} : f - \mu \in \mathscr{D}\} = \underline{P}(f).$$

It remains to prove that $\mathscr{D}$ is a coherent set of acceptable bounded gambles. To prove that BA1_{30} holds, consider a bounded gamble $f < 0$. To prove that $f \notin \mathscr{D}$, it suffices to show that $\underline{P}(f) \leq 0$, because clearly $f \not\geq 0$. But this follows at once from $\sup f \leq 0$ and Lemma 4.3(i)$_{\curvearrowright}$.

It is immediate from the definition of $\mathscr{D}$ that BA2_{30} holds, so we turn to BA3_{30}. It suffices to consider $\lambda > 0$ and bounded gambles f in $\mathscr{D}$ such that $\underline{P}(f) > 0$. Then $\underline{P}(\lambda f) = \lambda \underline{P}(f) > 0$ by LP3, and therefore, $\lambda f \in \mathscr{D}$.

To prove that BA4$_{30}$ holds, it suffices to consider bounded gambles f and g in $\mathscr{D}$ such that $\underline{P}(f) > 0$. As $g \in \mathscr{D}$, it follows from LP1$_{38}$ that $\underline{P}(g) \geq 0$ and, therefore, from LP2$_{38}$ that $\underline{P}(f+g) \geq \underline{P}(f) + \underline{P}(g) > 0$, whence $f + g \in \mathscr{D}$. □

Lemma 4.3 *Consider any real functional $\underline{P}$ defined on $\mathbb{B}$ that satisfies the properties* LP1$_{38}$–LP3$_{38}$. *Then for any $f \in \mathbb{B}$ and $\mu \in \mathbb{R}$,*

(i) $\underline{P}(f) \leq \sup f$;

(ii) $\underline{P}(\mu) = \mu$;

(iii) $\underline{P}(f + \mu) = \underline{P}(f) + \mu$.

Proof. (i). By LP1$_{38}$, it follows that $\inf(-f) \leq \underline{P}(-f)$. By LP2$_{38}$, it follows that $\underline{P}(f) + \underline{P}(-f) \leq \underline{P}(0)$. Finally, by LP3$_{38}$, $\underline{P}(0) = 0$. Combining all, we find that

$$\sup f = -\inf(-f) \geq -\underline{P}(-f) \geq \underline{P}(f).$$

(ii). Immediate from LP1$_{38}$ and (i).

(iii). It follows from LP2$_{38}$ that both $\underline{P}(f+\mu) \geq \underline{P}(f) + \underline{P}(\mu)$ and $\underline{P}(f+\mu) + \underline{P}(-\mu) \leq \underline{P}(f)$. Now apply (ii). □

4.1.2 Lower and upper previsions directly

So far, we have assumed that our subject has some coherent set of acceptable bounded gambles and used that set to infer his lower and upper previsions for any bounded gamble.

Admittedly, this is a rather circuitous way of coming up with lower and upper previsions for bounded gambles, as it requires that a subject should first come up with an entire coherent set of acceptable bounded gambles. A subject's **lower prevision** $\underline{P}(f)$ for a bounded gamble f can also be defined more directly as his supremum acceptable buying price for f: $\underline{P}(f)$ is the highest price s such that for any $t < s$, he (states he) accepts to pay t before observation of X, if he is guaranteed to receive $f(x)$ when observing $X = x$. So, in this approach, he is only required to consider whether he accepts bounded gambles of the very specific type $f - \mu$. *By specifying a lower prevision $\underline{P}(f)$ for a bounded gamble f, our subject is in effect stating that he accepts the bounded gambles $f - \underline{P}(f) + \varepsilon$ for all $\varepsilon > 0$* or, equivalently, that he accepts the bounded gambles $f - \mu$ for all $\mu < \underline{P}(f)$.

It deserves to be stressed at this point that we do not give an *exhaustive interpretation* to a lower prevision (Walley, 1991, Section 2.3.1): the subject neither states that he does not accept such bounded gambles for higher μ nor states that he accepts them. He remains uncommitted and can, therefore, be corrected (without internal contradiction) towards higher values, which is what the notion of natural extension tends to do, as we shall see further on in Section 4.5$_{65}$.

Of course, our subject can do the same thing for any number of bounded gambles. Mathematically speaking, we call a **lower prevision** a real-valued map defined on some subset dom $\underline{P}$ of $\mathbb{B}$, called the **domain** of $\underline{P}$. We do not require a lower prevision to be defined on the entire set of all bounded gambles. Indeed, we do not even impose any structure on dom $\underline{P}$: it can *a priori* be any subset of $\mathbb{B}$. Further on, we describe methods for extending lower previsions to the set of all bounded gambles.

Similarly, our subject's **upper prevision** $\overline{P}(f)$ for bounded gamble f can be directly defined as his infimum acceptable selling price for f: it is the lowest price s such that for any $t > s$, he (states he) accepts to receive t before observation of X, if he is guaranteed to lose $f(x)$ when observing $X = x$. As a gain r is equivalent to a loss $-r$, we see that $\overline{P}(f) = -\underline{P}(-f)$: from any lower prevision $\underline{P}$, we can infer a so-called **conjugate upper prevision** $\overline{P}$ on $\operatorname{dom}\overline{P} = -\operatorname{dom}\underline{P} = \{-f : f \in \operatorname{dom}\underline{P}\}$ that represents the same behavioural dispositions. We can therefore restrict our attention to the study of lower previsions only, without loss of generality. Also, if further on we use the notation $\underline{P}$ for a lower prevision, $\overline{P}$ will always denote its conjugate.

It may happen that $\underline{P}$ is **self-conjugate**, meaning that $\operatorname{dom}\underline{P} = \operatorname{dom}\overline{P} = -\operatorname{dom}\underline{P}$ and $\underline{P}(f) = \overline{P}(f)$ for all bounded gambles $f \in \operatorname{dom}\underline{P}$. We then simply write P instead of $\underline{P}$ or $\overline{P}$, provided that it is clear from the context whether we are considering either buying or selling prices (or both). We call a self-conjugate lower prevision P simply a **prevision**, and $P(f)$ then represents a subject's so-called **fair price** for the bounded gamble f: he accepts to buy f for any price $s < P(f)$ and he accepts to sell f for any price $s > P(f)$. Previsions, interpreted as fair prices, were introduced and studied by de Finetti (1937), who used them to provide a behavioural definition for expectations (see also de Finetti, 1974–1975).

Recall that an *event* is simply a subset of $\mathcal{X}$. Its associated indicator I_A is a bounded gamble that gives one if the actual value x of X belongs to A and zero otherwise. It is convenient to identify indicators with their corresponding events. In particular, we denote the lower prevision $\underline{P}(I_A)$ of an indicator I_A simply as $\underline{P}(A)$. We call $\underline{P}(A)$ the *lower probability* of the event A, and if the domain of $\underline{P}$ contains only indicators, we simply call $\underline{P}$ a **lower probability**. Similarly, we denote $\overline{P}(I_A)$ as $\overline{P}(A)$ and call it the *upper probability* of A. If the domain of $\overline{P}$ contains only indicators, then we call it an **upper probability**. If P is a (self-conjugate lower) prevision, then $P(A)$ is called the *probability* of A. If the domain of a prevision P contains only indicators I_A along with their negations $-I_A$, then P is called a **probability**.

4.2 Consistency for lower previsions

4.2.1 Definition and justification

If we have a subject's assessments in terms of a lower prevision $\underline{P}$, then we know his supremum acceptable buying prices $\underline{P}(f)$ for all bounded gambles f in the set dom $\underline{P}$. We can then ask what the consequences of these assessments are: what do they tell us about the prices that a subject should pay for other bounded gambles, not in dom $\underline{P}$, or more generally, what do they tell us about which bounded gambles our subject should accept?

We answer this question in this and the next two sections. We begin our analysis by going back to our discussion of acceptability in Section 3.4_{29}. First of all, we want to find the counterpart for lower previsions of the inclusion relation for sets of acceptable bounded gambles: how can we express that a lower prevision $\underline{Q}$ is at least as informative as another lower prevision $\underline{P}$? We have to require that any behavioural disposition expressed by $\underline{P}$ should (at least) also be expressed by $\underline{Q}$. This means that any transaction implied by $\underline{P}$, such as buying a bounded gamble f for a price s, should also be implied by $\underline{Q}$. Mathematically, this means that

$$\underline{P}(f) > s \Rightarrow \underline{Q}(f) > s \text{ for all real } s,$$

or, equivalently, that $\underline{P}(f) \geq \underline{Q}(f)$ for all f in $\operatorname{dom}\underline{P}$.

Definition 4.4 *We say that a lower prevision* $\underline{Q}$ ***dominates*** *a lower prevision* $\underline{P}$ *if* $\operatorname{dom}\underline{P} \subseteq \operatorname{dom}\underline{Q}$ *and* $\underline{P}(f) \leq \underline{Q}(f)$ *for every bounded gamble* $f \in \operatorname{dom}\underline{P}$.

The following relation between the inclusion for sets of acceptable bounded gambles and the notion of dominance is immediate.

Proposition 4.5 *Let* $\mathscr{A}_1$ *and* $\mathscr{A}_2$ *be two consistent sets of acceptable bounded gambles. If* $\mathscr{A}_1 \subseteq \mathscr{A}_2$*, then* $\mathrm{Cl}_{\mathbb{D}_b}(\mathscr{A}_1) \subseteq \mathrm{Cl}_{\mathbb{D}_b}(\mathscr{A}_2)$ *and* $\mathrm{lpr}(\mathrm{Cl}_{\mathbb{D}_b}(\mathscr{A}_2))$ *dominates* $\mathrm{lpr}(\mathrm{Cl}_{\mathbb{D}_b}(\mathscr{A}_1))$.

The subject's lower prevision assessment $\underline{P}$ corresponds to his specifying a set of acceptable bounded gambles:[1]

$$\mathscr{A}_{\underline{P}} := \left\{f - \mu : f \in \operatorname{dom}\underline{P} \text{ and } \mu < \underline{P}(f)\right\}. \tag{4.2}$$

The first thing to ask, then, is whether this set of acceptable bounded gambles is *consistent*. This leads to the following definition. The essential arguments behind it apparently go back to Ramsey (1931, p. 182).

Definition 4.6 (Avoiding sure loss) *We say that a lower prevision* $\underline{P}$ *is* ***consistent****, or* ***avoids sure loss****, if any (and hence all) of the following equivalent conditions are satisfied:*

(A) *The set of acceptable bounded gambles* $\mathscr{A}_{\underline{P}}$ *is consistent.*

(B) *There is some coherent set of acceptable bounded gambles* $\mathscr{D}$ *such that* $\mathrm{lpr}(\mathscr{D})$ *dominates* $\underline{P}$.

(C) *There is some lower prevision* $\underline{Q}$ *defined on the set* $\mathbb{B}$ *of all bounded gambles that satisfies the properties* $\mathrm{LP1}_{38}$–$\mathrm{LP3}_{38}$ *and that dominates* $\underline{P}$.

(D) *For all* n *in* $\mathbb{N}$*, and bounded gambles* $f_1, \ldots, f_n$ *in* $\operatorname{dom}\underline{P}$*, it holds that*

$$\sum_{i=1}^{n} \underline{P}(f_i) \leq \sup\left(\sum_{i=1}^{n} f_i\right). \tag{4.3}$$

[1] Observe that if $\underline{Q}$ dominates $\underline{P}$, then $\mathscr{A}_{\underline{P}} \subseteq \mathscr{A}_{\underline{Q}}$.

(E) *For all n in* $\mathbb{N}$, *non-negative* $\lambda_1, \ldots, \lambda_n$ *in* $\mathbb{R}$ *and bounded gambles* $f_1, \ldots, f_n$ *in* dom $\underline{P}$, *it holds that*

$$\sum_{i=1}^{n} \lambda_i \underline{P}(f_i) \leq \sup\left(\sum_{i=1}^{n} \lambda_i f_i\right). \tag{4.4}$$

(F) *For all n in* $\mathbb{N}$, λ_0 *in* $\mathbb{R}$, *non-negative* $\lambda_1, \ldots, \lambda_n$ *in* $\mathbb{R}$, *and bounded gambles* $f_1, \ldots, f_n$ *in* dom $\underline{P}$, *it holds that*

$$\sum_{i=1}^{n} \lambda_i f_i \leq \lambda_0 \Rightarrow \sum_{i=1}^{n} \lambda_i \underline{P}(f_i) \leq \lambda_0. \tag{4.5}$$

We prove that these conditions are indeed equivalent. The proof for the equivalence of (D) and (E) is due to Walley (1991, Lemma 2.4.4). We repeat it here for the sake of completeness.

Proof. Let us prove that (A)⇒(B)⇒(C)⇒(D)⇒(E)⇒(F)⇒(A).

(A)⇒(B). By (A) and Definition 3.4(A)$_{32}$ (consistency), we find that $\mathscr{A}_{\underline{P}}$ is included in some coherent set $\mathscr{D}$ of acceptable bounded gambles. The claim is established if we can show that $\mathrm{lpr}(\mathscr{D})$ dominates $\underline{P}$. Indeed, for any $f \in \mathrm{dom}\,\underline{P}$, we have that

$$\begin{aligned}\mathrm{lpr}(\mathscr{D})(f) &= \sup\left\{\mu \in \mathbb{R} : f - \mu \in \mathscr{D}\right\} \\ &\geq \sup\left\{\mu \in \mathbb{R} : f - \mu \in \mathscr{A}_{\underline{P}}\right\} = \sup\left\{\mu \in \mathbb{R} : \mu < \underline{P}(f)\right\} = \underline{P}(f).\end{aligned}$$

(B)⇒(C). Take $\underline{Q} := \mathrm{lpr}(\mathscr{D})$ and use Theorem 4.1$_{38}$.

(C)⇒(D). Let $\underline{Q}$ be a lower prevision defined on $\mathbb{B}$ that satisfies the properties LP1$_{38}$–LP3$_{38}$ and that dominates $\underline{P}$. Then

$$\sum_{i=1}^{n} \underline{P}(f_i) \leq \sum_{i=1}^{n} \underline{Q}(f_i) \leq \underline{Q}\left(\sum_{i=1}^{n} f_i\right) \leq \sup\left(\sum_{i=1}^{n} f_i\right),$$

where the second inequality follows from LP2$_{38}$ and the third from Lemma 4.3(i)$_{40}$.

(D)⇒(E). Assume that (D) holds, and assume *ex absurdo* that

$$\sum_{i=1}^{n} \lambda_i \underline{P}(f_i) > \sup\left(\sum_{i=1}^{n} \lambda_i f_i\right)$$

for some particular choice of n in $\mathbb{N}$, non-negative $\lambda_1, \ldots, \lambda_n$ in $\mathbb{R}$ and bounded gambles $f_1, \ldots, f_n$ in dom $\underline{P}$. Let $\delta := \sum_{i=1}^{n} \lambda_i \underline{P}(f_i) - \sup\left(\sum_{i=1}^{n} \lambda_i f_i\right) > 0$, $\alpha := \sup\left(\sum_{i=1}^{n} \left|f_j\right|\right) + \sum_{i=1}^{n} \left|\underline{P}(f_j)\right| \geq 0$ and $\varepsilon := \frac{\delta}{2\alpha+1} > 0$. As $\mathbb{Q}$ is dense in $\mathbb{R}$, there are non-negative rational numbers $\rho_i \in \mathbb{Q}$ such that $\lambda_i \leq \rho_i \leq \lambda_i + \varepsilon$ for every $i \in \{1, \ldots, n\}$. Using Lemma 1.27$_{19}$, we find that

$$\rho_i f_i \leq \lambda_i f_i + \varepsilon\left|f_i\right| \text{ and } -\rho_i \underline{P}(f_i) \leq -\lambda_i \underline{P}(f_i) + \varepsilon\left|\underline{P}(f_i)\right|$$

for every $i \in \{1, \dots, n\}$. Let $k \in \mathbb{N}$ be a common denominator of $\rho_1, \dots, \rho_n$, and let $m_i := k\rho_i \in \mathbb{N}$ for every $j \in \{1, \dots, n\}$. Then

$$\begin{aligned}\sup\left(\sum_{i=1}^{n}\rho_i f_i\right) - \sum_{i=1}^{n}\rho_i \underline{P}(f_i) &\leq \sup\left(\sum_{i=1}^{n}\lambda_i f_i + \sum_{i=1}^{n}\varepsilon|f_i|\right) - \sum_{i=1}^{n}\lambda_i\underline{P}(f_i) + \sum_{i=1}^{n}\varepsilon|\underline{P}(f_i)| \\ &\leq \sup\left(\sum_{i=1}^{n}\lambda_i f_i\right) - \sum_{i=1}^{n}\lambda_i\underline{P}(f_i) + \varepsilon\left(\sup\left(\sum_{i=1}^{n}|f_i|\right) + \sum_{i=1}^{n}|\underline{P}(f_i)|\right) \\ &= -\delta + \varepsilon\alpha < -\frac{\delta}{2}.\end{aligned}$$

We conclude that

$$\sup\left(\sum_{i=1}^{n}m_i f_i\right) - \sum_{i=1}^{n}m_i\underline{P}(f_i) < -k\frac{\delta}{2} < 0,$$

which contradicts the assumption $(\mathrm{D})_{42}$.

$(\mathrm{E})_{\curvearrowleft}\Rightarrow(\mathrm{F})_{\curvearrowleft}$. Consider arbitrary n in $\mathbb{N}$, λ_0 in $\mathbb{R}$, non-negative $\lambda_1, \dots, \lambda_n$ in $\mathbb{R}$, and bounded gambles $f_1, \dots, f_n$ in $\operatorname{dom}\underline{P}$, and assume that $\sum_{i=1}^{n}\lambda_i f_i \leq \lambda_0$. Then it follows from the assumption $(\mathrm{E})_{\curvearrowleft}$ that indeed

$$\sum_{i=1}^{n}\lambda_i\underline{P}(f_i) \leq \sup\left(\sum_{i=1}^{n}\lambda_i f_i\right) \leq \lambda_0.$$

$(\mathrm{F})_{\curvearrowleft}\Rightarrow(\mathrm{A})_{42}$. Suppose *ex absurdo* that $\sum_{i=1}^{n}\lambda_i(f_i - \mu_i) < 0$ for some $n \in \mathbb{N}_{>0}$, non-negative $\lambda_1, \dots, \lambda_n$ in $\mathbb{R}$, $f_1, \dots, f_n$ in $\operatorname{dom}\underline{P}$ and $\mu_1 < \underline{P}(f_1), \dots, \mu_n < \underline{P}(f_n)$ in $\mathbb{R}$ such that $\lambda_i > 0$ for at least one i. But it follows from $(\mathrm{F})_{\curvearrowleft}$ and $\sum_{i=1}^{n}\lambda_i f_i < \sum_{i=1}^{n}\lambda_i\mu_i$ that $\sum_{i=1}^{n}\lambda_i\underline{P}(f_i) \leq \sum_{i=1}^{n}\lambda_i\mu_i$, a contradiction. □

4.2.2 A more direct justification for the avoiding sure loss condition

Because our course of reasoning in deriving the avoiding sure loss conditions is a bit indirect, let us give a more direct argument to justify the condition $(\mathrm{D})_{42}$, for instance – this argument, in a slightly simpler form, is also given by Walley (1991, Section 2.4.2, p. 68).

Suppose that Equation $(4.3)_{42}$ does not hold for some $n \in \mathbb{N}_{>0}$ and $f_1, \dots, f_n$ in $\operatorname{dom}\underline{P}$. Define the strictly positive number

$$\varepsilon = \sum_{i=1}^{n}\underline{P}(f_i) - \sup\left(\sum_{i=1}^{n}f_i\right) > 0.$$

Let $\delta = \frac{\varepsilon}{n+1}$. As $\delta > 0$, the subject is disposed to buy each f_i for $\underline{P}(f_i) - \delta$. Hence, by axiom of rationality BA4$_{30}$, he should be disposed to buy $\sum_{i=1}^{n} f_i$ for the price $\sum_{i=1}^{n} \underline{P}(f_i) - n\delta$. But, this transaction always leads to a strictly positive loss because

$$\sum_{i=1}^{n} \underline{P}(f_i) - n\delta > \sum_{i=1}^{n} \underline{P}(f_i) - \varepsilon = \sup\left(\sum_{i=1}^{n} f_i\right)$$

by the definition of ε. This violates rationality axiom BA1$_{30}$.

4.2.3 Avoiding sure loss and avoiding partial loss

A bounded gamble f **incurs a partial loss** when $f < 0$: $f(x)$ is everywhere non-positive, and actually negative for some x. f **incurs a sure loss** when $\sup f < 0$, meaning that there is some $\varepsilon > 0$ such that $f(x) \leq -\varepsilon$ everywhere: f is not only negative everywhere but also bounded away from zero. A sure loss, therefore, implies a partial loss, and avoiding partial loss is a stronger condition – more difficult to satisfy – than avoiding sure loss. To give an example, the bounded gamble $f(n) := -1/n$ on $\mathbb{N}_{>0}$ avoids sure loss because $\sup f = 0$ but incurs partial loss.

That lower prevision assessments should avoid sure loss follows from the avoiding partial loss requirement for sets of acceptable bounded gambles. But why is the requirement for lower prevision assessments weaker than that for acceptability assessments?

The reason is that a lower prevision is a *supremum* acceptable buying price: specifying a lower prevision $\underline{P}(f)$ for the bounded gamble f does not imply that the transacttion $f - \underline{P}(f)$ is acceptable – that the subject actually accepts to buy f for the price $\underline{P}(f)$. Rather, it means that $f - \underline{P}(f) + \varepsilon$ is acceptable for every $\varepsilon > 0$, no matter how small. And, clearly, for any bounded gamble g,

$$\begin{aligned}(\forall \varepsilon > 0)(g + \varepsilon \text{ avoids partial loss}) &\Leftrightarrow \neg(\exists \varepsilon > 0)(g + \varepsilon \text{ incurs partial loss})\\ &\Leftrightarrow \neg(\exists \varepsilon > 0) g + \varepsilon < 0\\ &\Leftrightarrow \sup g \geq 0\\ &\Leftrightarrow g \text{ avoids sure loss.}\end{aligned}$$

4.2.4 Illustrating the avoiding sure loss condition

Let us clarify the avoiding sure loss condition with a few more small examples.

Buying $f(X)$ for a price strictly higher than $\sup f$ makes no sense: with certainty, such a transaction incurs a strictly positive loss. In particular, suppose that for some non-empty subset A of $\mathcal{X}$, our subject's lower probability for A is equal to 1.2: $\underline{P}(A) = 1.2$. This implies that he is disposed to buy the uncertain reward I_A for a price 1.1, because 1.1 is strictly less than 1.2. But the reward associated with I_A is at most 1, so this behavioural disposition leads to a sure loss of at least $1.1 - 1 = 0.1$. This is irrational behaviour, and therefore, our subject's lower probability $\underline{P}(A)$ should not be equal to 1.2; in fact, it should never be strictly higher than 1.

More generally, a combined buy of a finite collection of bounded gambles is not acceptable if this transaction leads to a sure loss. For example, suppose that

$\underline{P}(A) = 0.7$ and $\underline{P}(B) = 0.7$, where A and B are non-empty and disjoint subsets of $\mathscr{X}$. This implies in particular that our subject is disposed to buy both I_A and I_B for 0.6. Therefore, he should be disposed to buy $I_A + I_B = I_{A\cup B}$ for $0.6 + 0.6 = 1.2$. But again, the reward associated with $I_{A\cup B}$ is at most 1, so these behavioural dispositions lead to a sure loss of at least $1.2 - 1 = 0.2$. Again, this behaviour is irrational: $\underline{P}(A) + \underline{P}(B)$ should not be equal to 1.2; it should in fact never be strictly higher than 1.

4.2.5 Consequences of avoiding sure loss

Finally, we draw attention to the following important consequence of avoiding a sure loss. It provides some justification for the qualifiers 'upper' and 'lower' that we have been using.

Proposition 4.7 *If the lower prevision $\underline{P}$ avoids sure loss, then $\underline{P}(f) \leq \overline{P}(f)$ for all $f \in \operatorname{dom} \underline{P}$ such that also $-f \in \operatorname{dom} \underline{P}$.*

Proof. Let $n := 2, f_1 := f$ and $f_2 := -f$ in Criterion $(\mathrm{D})_{42}$ of Definition 4.6_{42}, and use the conjugacy relation $\overline{P}(f) = -\underline{P}(-f)$. □

Quite a number of other properties can be derived from the avoiding sure loss conditions in Definition 4.6_{42}, many of which are listed in Walley (1991, Theorem 2.4.7). We refrain from paying more attention to them here, because we shall be concerned mostly with the stronger rationality criterion of coherence, to be introduced right away.

4.3 Coherence for lower previsions

4.3.1 Definition and justification

We now return to the problem of making inferences based on the subject's lower prevision $\underline{P}$. There is nothing that can be done if the lower prevision $\underline{P}$ is inconsistent. In that case, the subject's assessments are blatantly irrational, in that they make him subject to not just a partial but even a sure loss!

But if the lower prevision $\underline{P}$ is consistent, then the corresponding set of acceptable bounded gambles $\mathscr{A}_{\underline{P}}$ is consistent too, and we can extend this assessment to a coherent set of acceptable bounded gambles. The smallest such set, which only takes into account the assessment $\mathscr{A}_{\underline{P}}$ and the consequences of coherence, is its natural extension $\mathrm{Cl}_{\mathbb{D}_b}(\mathscr{A}_{\underline{P}})$. After a few elementary manipulations, it follows from Equations $(3.2)_{34}$ and $(4.2)_{42}$ that this natural extension is given by

$$\begin{aligned}\mathrm{Cl}_{\mathbb{D}_b}(\mathscr{A}_{\underline{P}}) &= \mathscr{E}_{\mathscr{A}_{\underline{P}}}\\ &= \left\{ h \in \mathbb{B} : h \geq \varepsilon + \sum_{k=1}^{n} \lambda_k[f_k - \underline{P}(f_k)], \varepsilon > 0, n \in \mathbb{N}, f_k \in \operatorname{dom} \underline{P}, \lambda_k \in \mathbb{R}_{\geq 0} \right\}.\end{aligned} \tag{4.6}$$

With this coherent set of acceptable bounded gambles $\mathscr{E}_{\mathscr{A}_{\underline{P}}}$, we can now associate a lower prevision (operator) $\operatorname{lpr}(\mathscr{E}_{\mathscr{A}_{\underline{P}}})$, which is a real functional defined on all bounded gambles. *It is the lower prevision that can be inferred from the subject's lower prevision assessments $\underline{P}$ using arguments of coherence alone.*

Definition 4.8 (Natural extension) *Consider a lower prevision $\underline{P}$ that avoids sure loss. Then the lower prevision $\underline{E}_{\underline{P}} := \operatorname{lpr}(\operatorname{Cl}_{\mathbb{D}_b}(\mathscr{A}_{\underline{P}})) = \operatorname{lpr}(\mathscr{E}_{\mathscr{A}_{\underline{P}}})$ is called the* ***natural extension*** *of $\underline{P}$. It is given on all bounded gambles $f \in \mathbb{B}$ by*

$$\underline{E}_{\underline{P}}(f) = \sup\left\{\alpha \in \mathbb{R} : f - \alpha \geq \sum_{k=1}^{n} \lambda_k [f_k - \underline{P}(f_k)], n \in \mathbb{N}, f_k \in \operatorname{dom} \underline{P}, \lambda_k \in \mathbb{R}_{\geq 0}\right\}. \tag{4.7}$$

The expression (4.7) for the natural extension follows immediately by combining Equations $(4.1)_{38}$ and (4.6).

If for some bounded gamble f in $\operatorname{dom} \underline{P}$, we have that $\underline{P}(f) \neq \underline{E}_{\underline{P}}(f)$, this means that in specifying the assessment $\underline{P}(f)$, the subject has not fully taken into account all the behavioural implications of his other assessments $\underline{P}(g)$, $g \in \operatorname{dom} \underline{P} \setminus \{f\}$. Indeed, the argument given earlier tells us that the supremum buying price for f that can be inferred from the assessments and arguments of coherence alone is precisely $\underline{E}_{\underline{P}}(f)$.

If a subject's lower prevision assessment does not show this kind of imperfection, then we call it *coherent*. These observations are formalised and generalised in Definition 4.10, the essence of which apparently goes back to Williams (1975b).

Definition 4.9 *Let $\underline{P}$ and $\underline{Q}$ be lower previsions. Then $\underline{P}$ is said to be a* ***restriction*** *of $\underline{Q}$, and $\underline{Q}$ is said to be an* ***extension*** *of $\underline{P}$, whenever* $\operatorname{dom} \underline{P} \subseteq \operatorname{dom} \underline{Q}$ *and* $\underline{P}(f) = \underline{Q}(f)$ *for all* $f \in \operatorname{dom} \underline{P}$.

Definition 4.10 (Coherence) *Let $\underline{P}$ be a lower prevision. Then the following conditions are equivalent. If any (and hence all) of them are satisfied, then we call $\underline{P}$* ***coherent****.*

(A) *There is some coherent set of acceptable bounded gambles $\mathscr{D}$ such that $\operatorname{lpr}(\mathscr{D})$ is an extension of $\underline{P}$.*

(B) *There is some lower prevision $\underline{Q}$ defined on $\mathbb{B}$ that satisfies the properties LP1_{38}–LP3_{38} and that is an extension of $\underline{P}$.*

(C) *$\underline{P}$ avoids sure loss and it is a restriction of its natural extension $\underline{E}_{\underline{P}}$.*

(D) *For all m, n in $\mathbb{N}$ and bounded gambles $f_0, f_1, \ldots, f_m$ in* $\operatorname{dom} \underline{P}$, *it holds that*

$$\sum_{i=1}^{n} \underline{P}(f_i) - m\underline{P}(f_0) \leq \sup\left(\sum_{i=1}^{n} f_i - mf_0\right). \tag{4.8}$$

(E) *For all n in $\mathbb{N}$, non-negative λ_0, λ_1, …, λ_n in $\mathbb{R}$ and bounded gambles f_0, f_1, …, f_n in* dom $\underline{P}$, *it holds that*

$$\sum_{i=1}^{n} \lambda_i \underline{P}(f_i) - \lambda_0 \underline{P}(f_0) \leq \sup\left(\sum_{i=1}^{n} \lambda_i f_i - \lambda_0 f_0 \right).$$

(F) *For all n in $\mathbb{N}$, λ_0 in $\mathbb{R}$, non-negative λ, λ_1, …, λ_n in $\mathbb{R}$, and bounded gambles f, f_1, …, f_n in* dom $\underline{P}$, *it holds that*

$$\lambda f \geq \lambda_0 + \sum_{i=1}^{n} \lambda_i f_i \Rightarrow \lambda \underline{P}(f) \geq \lambda_0 + \sum_{i=1}^{n} \lambda_i \underline{P}(f_i).$$

We denote the set of all coherent lower previsions with domain $\mathscr{K} \subseteq \mathbb{B}(\mathscr{X})$ by $\underline{\mathbb{P}}^{\mathscr{K}}(\mathscr{X})$, or simply by $\underline{\mathbb{P}}^{\mathscr{K}}$. The set $\underline{\mathbb{P}}^{\mathbb{B}(\mathscr{X})}(\mathscr{X})$ is also denoted by $\underline{\mathbb{P}}(\mathscr{X})$, or simply by $\underline{\mathbb{P}}$.

We recall and stress that in these expressions, $\mathbb{N}$ denotes the set of all non-negative integers (zero included). Let us prove that these conditions are indeed equivalent. The equivalence of (D)$_{\curvearrowleft}$ and (E)$_{\curvearrowleft}$ is mentioned, but not proved explicitly, by Walley (1991, Lemma 2.5.4).

Proof. We prove (A)$_{\curvearrowleft}$⇒(B)$_{\curvearrowleft}$⇒(C)$_{\curvearrowleft}$⇒(D)$_{\curvearrowleft}$⇒(E)$_{\curvearrowleft}$⇒(F)⇒(A)$_{\curvearrowleft}$.

(A)$_{\curvearrowleft}$⇒(B)$_{\curvearrowleft}$. Let $\underline{Q} := \mathrm{lpr}(\mathscr{D})$, then the lower prevision $\underline{Q}$ on $\mathbb{B}$ extends $\underline{P}$, and it satisfies LP1$_{38}$–LP3$_{38}$ by Theorem 4.1$_{38}$.

(B)$_{\curvearrowleft}$⇒(C)$_{\curvearrowleft}$. As $\underline{Q}$ dominates $\underline{P}$, we infer from condition (C)$_{42}$ in Definition 4.6$_{42}$ that $\underline{P}$ avoids sure loss. We can, therefore, consider its natural extension $\underline{E}_{\underline{P}} = \mathrm{lpr}(\mathrm{Cl}_{\mathbb{D}_b}(\mathscr{A}_{\underline{P}}))$. It is clear that $\underline{E}_{\underline{P}}$ dominates $\underline{P}$. It remains to show that $\underline{P}$ also dominates $\underline{E}_{\underline{P}}$ on dom $\underline{P}$.

It suffices to prove that $\underline{Q}$ dominates $\underline{E}_{\underline{P}}$, because this results in $\underline{E}_{\underline{P}}$ being sandwiched between $\underline{P}$ and $\underline{Q}$, and we know that $\underline{Q}$ coincides with $\underline{P}$ on dom $\underline{P}$.

Because $\underline{Q}$ coincides with $\underline{P}$ on dom $\underline{P}$, it follows that $\mathscr{A}_{\underline{P}} \subseteq \mathscr{A}_{\underline{Q}}$. By avoiding sure loss criterion (C)$_{42}$, $\underline{Q}$ avoids sure loss, and hence, $\mathscr{A}_{\underline{Q}}$ is consistent by avoiding sure loss criterion (A)$_{42}$. We then infer from Proposition 4.5$_{42}$ that

$$\underline{E}_{\underline{P}} = \mathrm{lpr}(\mathrm{Cl}_{\mathbb{D}_b}(\mathscr{A}_{\underline{P}})) \leq \mathrm{lpr}(\mathrm{Cl}_{\mathbb{D}_b}(\mathscr{A}_{\underline{Q}})) = \underline{Q},$$

where the last equality follows from Theorem 4.2$_{39}$.

(C)$_{\curvearrowleft}$⇒(D)$_{\curvearrowleft}$. Assume that (C)$_{\curvearrowleft}$ holds and consider arbitrary m, n in $\mathbb{N}$ and bounded gambles f_0, f_1, …, f_m in dom $\underline{P}$. Because $\underline{P}$ avoids sure loss, we infer from Definition 4.6$_{42}$ that we may assume without loss of generality that $m > 0$. As $\underline{E}_{\underline{P}}(f_0) = \underline{P}(f_0)$, Equation (4.7)$_{\curvearrowleft}$ tells us that for all real α,

$$f_0 - \alpha \geq \sum_{k=1}^{n} \frac{1}{m}[f_k - \underline{P}(f_k)] \Rightarrow \alpha \leq \underline{P}(f_0).$$

As the antecedent is satisfied for

$$\alpha = \frac{1}{m}\left[\sum_{k=1}^{n} \underline{P}(f_k) - \sup\left(\sum_{k=1}^{n} f_k - mf_0\right)\right],$$

it indeed follows that $\sum_{k=1}^{n} \underline{P}(f_k) - \sup\left(\sum_{k=1}^{n} f_k - mf_0\right) \leq m\underline{P}(f_0)$.

(D)$_{47}$⇒(E)$_{47}$. Assume that (D)$_{47}$ holds, and assume *ex absurdo* that

$$\sum_{i=1}^{n} \lambda_i \underline{P}(f_i) - \lambda_0 \underline{P}(f_0) > \sup\left(\sum_{i=1}^{n} \lambda_i f_i - \lambda_0 f_0\right)$$

for some particular choice of n in $\mathbb{N}$, non-negative $\lambda_0, \lambda_1, \ldots, \lambda_n$ in $\mathbb{R}$ and bounded gambles $f_0, f_1, \ldots, f_n$ in $\operatorname{dom} \underline{P}$. Let $\delta := \sum_{i=1}^{n} \lambda_i \underline{P}(f_i) - \lambda_0 \underline{P}(f_0) - \sup\left(\sum_{i=1}^{n} \lambda_i f_i - \lambda_0 f_0\right) > 0$, $\alpha := \sup\left(\sum_{i=0}^{n} \left|f_i\right|\right) + \sum_{i=0}^{n} \left|\underline{P}(f_i)\right| \geq 0$ and $\varepsilon := \frac{\delta}{2\alpha+1} > 0$. As $\mathbb{Q}$ is dense in $\mathbb{R}$, there are non-negative rational numbers $\rho_i \in \mathbb{Q}$ such that $\lambda_i \leq \rho_i \leq \lambda_i + \varepsilon$ for every $i \in \{0, \ldots, n\}$. Using Lemma 1.27$_{19}$, we find that

$$\begin{aligned}
\rho_i f_i &\leq \lambda_i f_i + \varepsilon |f_i|, \\
-\rho_0 f_0 &\leq -\lambda_0 f_0 + \varepsilon |f_0|, \\
-\rho_i \underline{P}(f_i) &\leq -\lambda_i \underline{P}(f_i) + \varepsilon \left|\underline{P}(f_i)\right|, \\
\rho_0 \underline{P}(f_0) &\leq \lambda_0 \underline{P}(f_0) + \varepsilon \left|\underline{P}(f_0)\right|
\end{aligned}$$

for every $i \in \{1, \ldots, n\}$. Let $k \in \mathbb{N}$ be a common denominator of $\rho_0, \ldots, \rho_n$ and let $m_i := k\rho_i \in \mathbb{N}$ for every $i \in \{0, \ldots, n\}$. Then

$$\begin{aligned}
&\sup\left(\sum_{i=1}^{n} \rho_i f_i - \rho_0 f_0\right) - \left(\sum_{i=1}^{n} \rho_i \underline{P}(f_i) - \rho_0 \underline{P}(f_0)\right) \\
&\leq \sup\left(\sum_{i=1}^{n} \lambda_i f_i - \lambda_0 f_0 + \sum_{i=0}^{n} \varepsilon |f_i|\right) - \left(\sum_{i=1}^{n} \lambda_i \underline{P}(f_i) - \lambda_0 \underline{P}(f_0)\right) + \sum_{i=0}^{n} \varepsilon \left|\underline{P}(f_i)\right| \\
&\leq \sup\left(\sum_{i=1}^{n} \lambda_i f_i - \lambda_0 f_0\right) - \sum_{i=1}^{n} \lambda_i \underline{P}(f_i) + \lambda_0 \underline{P}(f_0) + \varepsilon\left[\sup\left(\sum_{i=0}^{n} |f_i|\right) + \sum_{i=0}^{n} \left|\underline{P}(f_i)\right|\right] \\
&= -\delta + \varepsilon\alpha < -\frac{\delta}{2}.
\end{aligned}$$

We conclude that

$$\sup\left(\sum_{i=1}^{n} m_i f_i - m_0 f_0\right) - \left(\sum_{i=1}^{n} m_i \underline{P}(f_i) - m_0 \underline{P}(f_0)\right) < -k\frac{\delta}{2} < 0,$$

which contradicts the assumption (D)$_{47}$.

$(\mathrm{E})_{47} \Rightarrow (\mathrm{F})_{48}$. Consider arbitrary n in $\mathbb{N}$, λ_0 in $\mathbb{R}$, non-negative λ, λ_1, …, λ_n in $\mathbb{R}$, and bounded gambles $f, f_1, \ldots, f_n$ in $\operatorname{dom} \underline{P}$ and assume that $\lambda f \geq \lambda_0 + \sum_{i=1}^n \lambda_i f_i$. Then it follows from the assumption $(\mathrm{E})_{43}$ that

$$\sum_{i=1}^n \lambda_i \underline{P}(f_i) - \lambda \underline{P}(f) \leq \sup\left(\sum_{i=1}^n \lambda_i f_i - \lambda f\right) \leq -\lambda_0,$$

and, therefore, indeed, $\lambda \underline{P}(f) \geq \lambda_0 + \sum_{i=1}^n \lambda_i \underline{P}(f_i)$.

$(\mathrm{F})_{48} \Rightarrow (\mathrm{A})_{47}$. The present condition $(\mathrm{F})_{48}$ implies the consistency condition $(\mathrm{F})_{43}$ in Definition 4.6_{42} (with $\lambda := 0$), which we know is equivalent to consistency condition $(\mathrm{A})_{42}$ in that same definition. Hence, $\mathscr{A}_{\underline{P}}$ is consistent. By Theorem 3.7_{34}, the natural extension $\mathrm{Cl}_{\mathbb{D}_b}(\mathscr{A}_{\underline{P}})$ of $\mathscr{A}_{\underline{P}}$ is the smallest coherent set of acceptable bounded gambles that includes $\mathscr{A}_{\underline{P}}$. We let $\mathscr{D} := \mathrm{Cl}_{\mathbb{D}_b}(\mathscr{A}_{\underline{P}})$ and show that $\mathrm{lpr}(\mathscr{D})$ is an extension of $\underline{P}$. Consider any $f \in \operatorname{dom} \underline{P}$. As $f - \underline{P}(f_0) + \varepsilon \in \mathscr{A}_{\underline{P}} \subseteq \mathscr{D}$, we infer from Equation $(4.1)_{38}$ that $\mathrm{lpr}(\mathscr{D})(f) \geq \underline{P}(f) - \varepsilon$, for all $\varepsilon > 0$, and therefore, $\mathrm{lpr}(\mathscr{D})(f) \geq \underline{P}(f)$. Assume *ex absurdo* that $\mathrm{lpr}(\mathscr{D})(f) > \underline{P}(f)$ and recall from Definition 4.8_{47} that $\mathrm{lpr}(\mathscr{D}) = \underline{E}_{\underline{P}}$. It then follows from Equation $(4.7)_{47}$ that there are n in $\mathbb{N}$, non-negative λ_1, …, λ_n in $\mathbb{R}$ and bounded gambles $f_1, \ldots, f_n$ in $\operatorname{dom} \underline{P}$ such that $\alpha > \underline{P}(f)$ and $f - \alpha \geq \sum_{k=1}^n \lambda_k [f_k - \underline{P}(f_k)]$. But from the latter inequality and condition $(\mathrm{F})_{48}$ we infer that

$$\underline{P}(f) \geq \sum_{k=1}^n \lambda_k \underline{P}(f_k) + \left(\alpha - \sum_{k=1}^n \lambda_k \underline{P}(f_k)\right) = \alpha,$$

a contradiction. □

4.3.2 A more direct justification for the coherence condition

Because our course of reasoning in deriving the coherence conditions is a bit indirect, let us give a more direct argument to justify the condition $(\mathrm{D})_{47}$, for instance – again, this argument, in a slightly simpler form, is also given by Walley (1991, Section 2.5.2, p. 73). Suppose Equation $(4.8)_{47}$ is violated for some n, m in $\mathbb{N}$ and bounded gambles $f_0, f_1, \ldots, f_n$ in $\operatorname{dom} \underline{P}$. The case $m = 0$ means that $\underline{P}$ does not avoid sure loss. We have already argued earlier that this is irrational. So suppose that $m \neq 0$. Let

$$\sigma := \sup\left(\sum_{i=1}^n f_i - m f_0\right) \text{ and } \varepsilon := \sum_{i=1}^n \underline{P}(f_i) - m \underline{P}(f_0) - \sigma > 0,$$

and let $\delta := \frac{\varepsilon}{n+1}$. As $\delta > 0$, the subject is disposed to buy each f_i for the price $\underline{P}(f_i) - \delta$. Hence, by rationality condition $\mathrm{BA4}_{30}$, he should be disposed to buy $\sum_{i=1}^n f_i$ for the price $\sum_{i=1}^n \underline{P}(f_i) - n\delta$, or equivalently, to buy the bounded gamble $\sum_{i=1}^n f_i - \sigma$ for the price $\sum_{i=1}^n \underline{P}(f_i) - n\delta - \sigma$. But because

$$\sum_{i=1}^n f_i - \sigma = \sum_{i=1}^n f_i - \sup\left(\sum_{i=1}^n f_i - m f_0\right) \leq m f_0,$$

he should also, by rationality conditions BA2$_{30}$ and BA4$_{30}$, be disposed to buy the higher uncertain reward mf_0 for the same price $\sum_{i=1}^{n} \underline{P}(f_i) - n\delta - \sigma$ and, therefore, also, by rationality condition BA3$_{30}$, to buy f_0 for $\frac{1}{m}(\sum_{i=1}^{n} \underline{P}(f_i) - n\delta - \sigma)$. But

$$\frac{1}{m}\left(\sum_{i=1}^{n} \underline{P}(f_i) - n\delta - \sigma\right) > \frac{1}{m}\left(\sum_{i=1}^{n} \underline{P}(f_i) - \varepsilon - \sigma\right) = \underline{P}(f_0)$$

by the choice of δ, which means that $\underline{P}(f_0)$ is too low.

4.3.3 Illustrating the coherence condition

It might happen that a subject is disposed to buy a bounded gamble f for a higher price than the one implicit in $\underline{P}(f)$ after consideration of buying prices of other bounded gambles. For example, consider the lower prevision $\underline{P}$ defined by $\underline{P}(A) = 0.3$, $\underline{P}(B) = 0.4$ and $\underline{P}(C) = 0.5$, where A, B and C are non-empty subsets of $\mathcal{X}$ with $A \cap B = \emptyset$ and $A \cup B \subseteq C$. This means, for instance, that the subject is disposed to buy $I_A + I_B = I_{A\cup B}$ for $0.25 + 0.35 = 0.6$. But $I_{A\cup B} \leq I_C$ because $A \cup B \subseteq C$. This means that I_C yields a reward at least as high as $I_{A\cup B}$. Therefore, he should also be disposed to buy I_C for 0.6. This means that the supremum acceptable buying price $\underline{P}(C) = 0.5$ is too low.

4.3.4 Linear previsions

In Section 4.1.2$_{40}$, we have defined *previsions* as self-conjugate lower previsions. Indeed, because a prevision is at the same time a selling and a buying price, it is natural to define it as self-conjugate on a negation-invariant domain. If the domain is not negation-invariant, we can always naturally extend it so as to become negation-invariant, simply by imposing self-conjugacy: $P(-f) = P(f)$.

For previsions, it turns out that avoiding sure loss and coherence are equivalent. De Finetti (1974–1975, Vol. I, Section. 3.3.5) actually defines coherence for previsions as avoiding sure loss. Regarding terminology, Walley (1991, Section 2.8.1) calls a coherent prevision a linear prevision – but his definition is slightly more general than the following one: it also extends to lower previsions that are not self-conjugate. However, as just mentioned, we can always naturally extend the domain to be negation-invariant, and in this sense, our notion of a linear prevision is equivalent to Walley's.

Definition 4.11 (Linear prevision) *We say that a lower prevision is* ***linear****, or that it is a* ***linear prevision*** *or a* ***coherent prevision****, if it is self-conjugate (and therefore a prevision) and coherent. We denote the set of all linear previsions with domain $\mathcal{K} \subseteq \mathbb{B}(\mathcal{X})$ by $\mathbb{P}^{\mathcal{K}}(\mathcal{X})$, or simply by $\mathbb{P}^{\mathcal{K}}$. The set $\mathbb{P}^{\mathbb{B}(\mathcal{X})}(\mathcal{X})$ is also denoted by $\mathbb{P}(\mathcal{X})$, or simply by $\mathbb{P}$.*

Theorem 4.12 *A prevision P is linear if and only if one (and hence all) of the following equivalent conditions is satisfied:*

(A) *P is coherent (as a lower prevision).*

(B) *P avoids sure loss (as a lower prevision).*

(C) *For all n in $\mathbb{N}$, $\lambda_1, \ldots, \lambda_n$ in $\mathbb{R}$ and bounded gambles $f_1, \ldots, f_n$ in* $\operatorname{dom} P$*, it holds that*

$$\sum_{i=1}^{n} \lambda_i P(f_i) \leq \sup\left(\sum_{i=1}^{n} \lambda_i f_i\right).$$

(D) *For all n in $\mathbb{N}$, and bounded gambles $f_1, \ldots, f_n$ in* $\operatorname{dom} P$*, it holds that*

$$\sum_{i=1}^{n} P(f_i) \leq \sup\left(\sum_{i=1}^{n} f_i\right).$$

Proof. By definition, P is linear if and only if it is coherent. We now show that $(\mathrm{A})_{\frown} \Rightarrow (\mathrm{B}) \Rightarrow (\mathrm{C}) \Rightarrow (\mathrm{D}) \Rightarrow (\mathrm{A})_{\frown}$.

$(\mathrm{A})_{\frown} \Rightarrow (\mathrm{B})$. This is immediate, as coherence generally implies avoiding sure loss, see coherence criterion $(\mathrm{C})_{47}$ in Definition 4.10_{47}.

$(\mathrm{B}) \Rightarrow (\mathrm{C})$. Consider arbitrary n in $\mathbb{N}$, $\lambda_1, \ldots, \lambda_n$ in $\mathbb{R}$ and bounded gambles $f_1, \ldots, f_n$ in $\operatorname{dom} P$. We may assume without loss of generality that there is some number $n' \in \{0, 1, \ldots, n\}$ such that $\lambda_i \geq 0$ for all $1 \leq i \leq n'$ and $\lambda_i < 0$ for all $n' < i \leq n$. Let $m' := n - n' \geq 0$. Also let $f_i' := f_i$ and $\lambda_i' := \lambda_i \geq 0$ for $1 \leq i \leq n'$. Finally, let $g_j' := -f_{n'+j}$ and $\mu_j' := -\lambda_{n'+j} \geq 0$ for $1 \leq j \leq m'$. Then, indeed,

$$\begin{aligned}\sum_{i=1}^{n} \lambda_i P(f_i) &= \sum_{i=1}^{n'} \lambda_i' P(f_i') + \sum_{j=1}^{m'} \mu_j' P(g_j') \leq \sup\left(\sum_{i=1}^{n'} \lambda_i' f_i' + \sum_{j=1}^{m'} \mu_j' g_j'\right) \\ &= \sup\left(\sum_{i=1}^{n} \lambda_i f_i\right),\end{aligned}$$

where the first equality follows from the self-conjugacy of P and the inequality from the fact that the lower prevision P avoids sure loss, see criterion $(\mathrm{E})_{43}$ in Definition 4.6_{42}.

$(\mathrm{C}) \Rightarrow (\mathrm{D})$. This is immediate, by the special choice of $\lambda_1 = \ldots \lambda_n = 1$.

$(\mathrm{D}) \Rightarrow (\mathrm{A})_{\frown}$. Consider arbitrary n, m in $\mathbb{N}$ and bounded gambles $f_0, f_1, \ldots, f_n$ in $\operatorname{dom} P$. Let $n' := m + n$, $f_i' := f_i$ for $1 \leq i \leq n$ and $f_i' := -f_0$ for $n + 1 \leq i \leq n + m$. It then follows at once from (D) and self-conjugacy that

$$\sum_{i=1}^{n} P(f_i) - mP(f_0) = \sum_{i=1}^{n'} P(f_i') \leq \sup\left(\sum_{i=1}^{n'} f_i'\right) = \sup\left(\sum_{i=1}^{n} f_i - mf_0\right),$$

which means that the lower prevision P satisfies coherence criterion $(\mathrm{D})_{47}$. □

Linear previsions are effectively the classical precise probability models. For instance, when $\mathcal{X}$ is finite, we will argue in Section 5.1_{77} that P is a linear prevision on $\mathbb{B}(\mathcal{X})$ if and only if there is some probability mass function $p : \mathcal{X} \to [0, 1]$ with $\sum_{x \in \mathcal{X}} p(x) = 1$ such that

$$P(f) = \sum_{x \in \mathcal{X}} p(x) f(x) \text{ for all } f \in \mathbb{B}.$$

This shows that linear previsions are simply *expectation operators*. As we will argue in great detail in Chapter 8_{151}, this generalises to infinite $\mathcal{X}$ and bounded gambles through the mediation of the Dunford integral or, equivalently, the S-integral or the Lebesgue integral. The discussion in Section $15.10.1_{358}$ shows that this identification also extends to unbounded gambles.

In particular, a linear prevision on a field of indicators of events (and their negations) is simply a finitely additive probability measure. It is not necessarily countably additive because coherence and self-conjugacy are finitary conditions.

So the restriction of a linear prevision to a field of events is a finitely additive probability measure, or probability charge. We shall see in Corollary 8.23_{167} that a finitely additive probability measure defined on all events has a unique coherent extension to the set of all bounded gambles. So there is a one-to-one relationship between finitely additive probability measures (on all events) and linear previsions (on all bounded gambles).

This means that in probability theory, the language of events is mathematically equivalent to that of bounded gambles, and it is largely a matter of taste which of the two is preferred – but see Whittle (1992) and de Finetti (1974–1975) for interesting expositions of the viewpoint that even there, the language of bounded gambles is preferable. In fact, one of the earliest publications in probability theory (Huygens, 1657) uses expectations exclusively. We shall see, for instance in Section 5.2_{78}, that for imprecise probabilities, the language of events is not expressive enough.

4.4 Properties of coherent lower previsions

4.4.1 Interesting consequences of coherence

Coherence has a number of interesting consequences, which turn out to be rather useful when we want to reason with coherent lower previsions. They were proved by Walley (1991, Theorem 2.6.1); we sketch the proofs here for the sake of completeness.

Theorem 4.13 *Let $\underline{P}$ be a coherent lower prevision. Let f and g be bounded gambles. Let a be a constant gamble. Let λ and κ be real numbers with $\lambda \geq 0$ and $0 \leq \kappa \leq 1$. Let f_α be a net of bounded gambles. Then the following statements hold whenever every term is well defined.*

(i) $\inf f \leq \underline{P}(f) \leq \overline{P}(f) \leq \sup f$. **(bounds)**

(ii) $\underline{P}(a) = \overline{P}(a) = a$. **(normality)**

(iii) $\underline{P}(f + a) = \underline{P}(f) + a$ *and* $\overline{P}(f + a) = \overline{P}(f) + a$. **(constant additivity)**

(iv) *If* $f \leq g + a$ *then* $\underline{P}(f) \leq \underline{P}(g) + a$ *and* $\overline{P}(f) \leq \overline{P}(g) + a$. **(monotonicity)**

(v) $\underline{P}(f) + \underline{P}(g) \leq \underline{P}(f + g) \leq \underline{P}(f) + \overline{P}(g) \leq \overline{P}(f + g) \leq \overline{P}(f) + \overline{P}(g)$. **(mixed super-/sub-additivity)**

(vi) $\underline{P}(\lambda f) = \lambda \underline{P}(f)$ *and* $\overline{P}(\lambda f) = \lambda \overline{P}(f)$. **(non-negative homogeneity)**

(vii) $\kappa\underline{P}(f)+(1-\kappa)\underline{P}(g)\leq\underline{P}(\kappa f+(1-\kappa)g)\leq\kappa\underline{P}(f)+(1-\kappa)\overline{P}(g)\leq\overline{P}(\kappa f+(1-\kappa)g)\leq\kappa\overline{P}(f)+(1-\kappa)\overline{P}(g)$. **(mixed convexity/concavity)**

(viii) $\underline{P}(|f|)\geq\underline{P}(f)$ *and* $\overline{P}(|f|)\geq\overline{P}(f)$.

(ix) $|\underline{P}(f)|\leq\overline{P}(|f|)$ *and* $|\overline{P}(f)|\leq\overline{P}(|f|)$.

(x) $|\underline{P}(f)-\underline{P}(g)|\leq\overline{P}(|f-g|)$ *and* $|\overline{P}(f)-\overline{P}(g)|\leq\overline{P}(|f-g|)$.

(xi) $\underline{P}(|f+g|)\leq\underline{P}(|f|)+\overline{P}(|g|)$ *and* $\overline{P}(|f+g|)\leq\overline{P}(|f|)+\overline{P}(|g|)$. **(mixed Cauchy–Schwartz inequalities)**

(xii) $\underline{P}(f\vee g)+\underline{P}(f\wedge g)\leq\underline{P}(f)+\overline{P}(g)\leq\overline{P}(f\vee g)+\overline{P}(f\wedge g)$, $\underline{P}(f)+\underline{P}(g)\leq\underline{P}(f\vee g)+\overline{P}(f\wedge g)\leq\overline{P}(f)+\overline{P}(g)$ *and* $\underline{P}(f)+\underline{P}(g)\leq\overline{P}(f\vee g)+\underline{P}(f\wedge g)\leq\overline{P}(f)+\overline{P}(g)$.

(xiii) *$\underline{P}$ and $\overline{P}$ are uniformly continuous with respect to the topology of uniform convergence on their respective domains: for any $\varepsilon>0$ and any f and g in* dom $\underline{P}$, *if* $\sup|f-g|<\varepsilon$, *then* $|\underline{P}(f)-\underline{P}(g)|<\varepsilon$. **(uniform continuity)**

(xiv) *If* $\lim_\alpha\overline{P}(|f_\alpha-f|)=0$ *then* $\lim_\alpha\underline{P}(f_\alpha)=\underline{P}(f)$ *and* $\lim_\alpha\overline{P}(f_\alpha)=\overline{P}(f)$.

Proof. These properties involving the lower prevision $\underline{P}$ can be derived directly from the coherence criteria $(D)_{47}$ and $(E)_{47}$ in Definition 4.10_{47} by making appropriate choices for the n, m, λ_k and f_k involved. For the properties involving $\overline{P}$, we can use conjugacy.

(i)$_\curvearrowleft$. For the first inequality, let $n:=0$, $m:=1$ and $f_0:=f$ in coherence criterion $(D)_{47}$. The second inequality follows from the fact that a coherent lower prevision avoids sure loss (see Definition $4.10(C)_{47}$) and Proposition 4.7_{46}).

(ii)$_\curvearrowleft$. Follows from (i)$_\curvearrowleft$ with $f:=a$.

(iv)$_\curvearrowleft$. Let $n:=m:=1$, $f_1:=f$ and $f_0:=g$ in coherence criterion $(D)_{47}$.

(iii)$_\curvearrowleft$. Use (iv)$_\curvearrowleft$.

(v)$_\curvearrowleft$. For the first inequality, let $n:=2$, $m:=1$, $f_1:=f$, $f_2:=g$ and $f_0:=f+g$ in coherence criterion $(D)_{47}$. The other inequalities can be derived using the first inequality and conjugacy. For example, let $h:=f+g$, then $f=h-g$ and, therefore,

$$\underline{P}(f)=\underline{P}(h+(-g))\geq\underline{P}(h)+\underline{P}(-g)=\underline{P}(h)-\overline{P}(g),$$

which leads to the second inequality.

(vi)$_\curvearrowleft$. Let $n:=1$, $\lambda_0:=\lambda$, $\lambda_1:=1$, $f_0:=f$ and $f_1:=\lambda f$ in coherence criterion $(E)_{47}$ to find that $\underline{P}(\lambda f)\leq\lambda\underline{P}(f)$. For the converse inequality, choose $n:=1$, $\lambda_0:=1$, $\lambda_1:=\lambda$, $f_0:=\lambda f$ and $f_1:=f$.

(vii). Use (v)$_\curvearrowleft$ and (vi)$_\curvearrowleft$.

(viii). Recall that $|f|\geq f$ and use (iv)$_\curvearrowleft$.

(ix). First, combine (viii) and (i)$_\curvearrowleft$ to find that $\overline{P}(|f|)\geq\overline{P}(f)\geq\underline{P}(f)$. Then use $|f|=|-f|$ and conjugacy to derive that $\overline{P}(|f|)\geq-\underline{P}(f)\geq-\overline{P}(f)$.

(x). Use (viii) and $(\mathrm{v})_{53}$ to find that $\overline{P}(|f-g|) \geq \overline{P}(f-g) \geq \overline{P}(f) - \overline{P}(g)$, and similarly, $\overline{P}(|g-f|) \geq \overline{P}(g-f) \geq \overline{P}(g) - \overline{P}(f)$. Hence,

$$\overline{P}(|f-g|) \geq \max\{\overline{P}(f) - \overline{P}(g), \overline{P}(g) - \overline{P}(f)\} = \left|\overline{P}(f) - \overline{P}(g)\right|.$$

The other case follows via conjugacy.

(xi). Observe that $|f+g| \leq |f| + |g|$ and use $(\mathrm{iv})_{53}$ and $(\mathrm{v})_{53}$.

(xii). All these inequalities are proved in similar way. As an example, for the first inequality, let $n := 3$, $m := 1$, $f_0 := f$, $f_1 := f \vee g$, $f_2 := f \wedge g$ and $f_3 := -g$ in coherence criterion $(\mathrm{D})_{47}$ and recall that $f \vee g + f \wedge g = f + g$.

(xiii). From $\sup |f-g| < \varepsilon$ and $(\mathrm{i})_{53}$, we derive that $\overline{P}(|f-g|) < \varepsilon$. Now use (x).

(xiv). Again use (x). □

A closer look at the proof of Theorem 4.13(xiv) and $(\mathrm{xiii})_{53}$ reveals that these continuity properties will be satisfied not only for coherent lower previsions but actually for any real functional on bounded gambles that is *super-additive* (satisfies $(\mathrm{v})_{53}$) and satisfies the *dominance* properties $(\mathrm{i})_{53}$.[2]

If we also invoke self-conjugacy, Theorem 4.13 leads at once to the following properties of coherent (or linear) previsions.

Corollary 4.14 *Let P be a coherent (or linear) prevision. Let f and g be bounded gambles. Let a be a constant gamble. Let λ and κ be real numbers with $0 \leq \kappa \leq 1$. Let f_α be a net of bounded gambles. Then the following statements hold whenever every term is well defined.*

(i) $\inf f \leq P(f) \leq \sup f$. **(bounds)**

(ii) $P(a) = a$. **(normality, unit norm)**

(iii) $P(f + a) = P(f) + a$. **(constant additivity)**

(iv) *If* $f \leq g + a$ *then* $P(f) \leq P(g) + a$. **(monotonicity)**

(v) $P(f) + P(g) = P(f + g)$. **(additivity)**

(vi) $P(\lambda f) = \lambda P(f)$. **(homogeneity)**

(vii) $\kappa P(f) + (1-\kappa)P(g) = P(\kappa f + (1-\kappa)g)$.

(viii) $P(|f|) \geq |P(f)| \geq P(f)$.

(ix) $|P(f) - P(g)| \leq P(|f-g|)$.

(x) $P(|f+g|) \leq P(|f|) + P(|g|)$. **(Cauchy–Schwartz inequality)**

(xi) $P(f \vee g) + P(f \wedge g) = P(f) + P(g)$. **(modularity)**

---

[2] As we shall see in the discussion following Theorem 4.16_{57} further on, this can be used to argue why for linear previsions the homogeneity property follows from the dominance and additivity properties.

(xii) *P is uniformly continuous with respect to the topology of uniform convergence on their respective domains: for any* $\varepsilon > 0$ *and any f and g in* dom *P*, *if* $\sup |f - g| < \varepsilon$, *then* $|P(f) - P(g)| < \varepsilon$. **(uniform continuity)**

(xiii) *If* $\lim_\alpha P\left(|f_\alpha - f|\right) = 0$, *then* $\lim_\alpha P(f_\alpha) = P(f)$.

4.4.2 Coherence and conjugacy

If we have an upper prevision $\overline{P}$ with domain $\operatorname{dom}\overline{P}$, then conjugacy allows us to define a corresponding, conjugate, lower prevision $\underline{P}$ on $-\operatorname{dom}\overline{P}$ by letting $\underline{P}(f) = -\overline{P}(-f)$ for all $f \in -\operatorname{dom}\overline{P}$ or, in other words, $-f \in \operatorname{dom}\overline{P}$. We will say that the upper prevision $\overline{P}$ is **coherent** if its conjugate lower prevision $\underline{P}$ is, and similarly for avoiding sure loss.

Lower and upper probabilities are lower and upper previsions defined on (indicators of) events, as discussed near the end of Section 4.1.2$_{40}$. As, generally speaking, the negation $-I_A$ of the indicator of an event A is not itself an indicator, conjugacy for lower and upper probabilities needs some more attention.

We have seen near the end of Section 4.1.2$_{40}$ that when the domain of a lower prevision $\underline{P}$ contains both I_A and $-I_A$, then conjugacy leads us to the expression

$$\overline{P}(A) = \overline{P}(I_A) = -\underline{P}(-I_A)$$

for the conjugate upper probability of an event A.

If the domain of $\underline{P}$ also contains the indicator $I_{A^c} = 1 - I_A$ of A's complement A^c, then we infer from the constant additivity property (iii)$_{53}$ of coherent lower previsions, established in Theorem 4.13$_{53}$, that coherence leaves us no choice but to write

$$\underline{P}(A^c) = \underline{P}(I_{A^c}) = \underline{P}(1 - I_A) = 1 + \underline{P}(-I_A) = 1 - \overline{P}(A), \tag{4.9}$$

which we will call the **conjugacy relationship** between the lower probability of an event and the upper probability of its complement. So, in the future, whenever we consider a lower probability $\underline{P}$ *as a set function* defined on a class $\mathscr{C}$ of events, then the conjugate upper probability $\overline{P}$ will be considered *as a set function* on the class of events $\{C^c : C \in \mathscr{C}\}$ defined by $\overline{P}(C^c) := 1 - \underline{P}(C)$ for all $C \in \mathscr{C}$.

4.4.3 Easier ways to prove coherence

If the domain of a lower prevision is a linear space, then we can characterise coherence in a simpler way. Also compare this result with Definition 4.10(B)$_{47}$.

Theorem 4.15 *Let* $\underline{P}$ *be a lower prevision, and assume that* $\operatorname{dom}\underline{P}$ *is a linear subspace of* $\mathbb{B}$. *Then* $\underline{P}$ *is coherent if and only if the following conditions are met for all bounded gambles f and g in* $\operatorname{dom}\underline{P}$ *and any strictly positive real number* λ.

(i) $\underline{P}(f) \geq \inf f$. **(bounds)**

(ii) $\underline{P}(\lambda f) = \lambda \underline{P}(f)$. **(non-negative homogeneity)**

(iii) $\underline{P}(f+g) \geq \underline{P}(f) + \underline{P}(g)$. **(super-additivity)**

The following proof is due to Walley (1991, Theorem 2.5.5).

Proof. Before we start with the actual proof, observe that it follows from (i) and (iii) (let $f, g := 0$) that (ii) also holds with $\lambda = 0$: $\underline{P}(0) = 0$.

'only if'. Immediate consequence of Theorem 4.13(i), (v) and (vi)$_{53}$.

'if'. Assume that the three conditions (i)–(iii) are met. Consider arbitrary n, m in $\mathbb{N}$ and bounded gambles $f_0, f_1, \ldots, f_n$ in dom $\underline{P}$, and let $f := mf_0$, $g := \sum_{k=1}^{n} f_k$ and $h := f - g$. Then f, g and h belong to the linear space dom $\underline{P}$. Moreover,

$$\begin{aligned}\sup\left(\sum_{k=1}^{n} f_k - mf_0\right) &= \sup(g-f) = -\inf h \\ &\geq -\underline{P}(h) \geq \underline{P}(g) - \underline{P}(f) = \underline{P}(g) - m\underline{P}(f_0) \\ &\geq \sum_{k=1}^{n} \underline{P}(f_k) - m\underline{P}(f_0),\end{aligned}$$

where the first inequality follows from (i) and the second inequality from (iii): recall that $f = h + g$ and therefore $\underline{P}(f) \geq \underline{P}(h) + \underline{P}(g)$. The third equality follows from (ii) (or from $\underline{P}(0) = 0$ if $m = 0$). And the last inequality follows again from (iii). We see that coherence criterion (D)$_{47}$ is satisfied. □

For previsions that are defined on linear spaces, the characterisation of linearity (or coherence) is even simpler than that for lower previsions, as de Finetti (1974–1975, Section 3.5.1) has pointed out; see also Walley (1991, Theorem 2.8.4 and Corollary 2.8.5). We sketch the proofs for the sake of completeness.

Theorem 4.16 *Let P be a prevision, and assume that* dom P *is a linear subspace of* $\mathbb{B}$. *Then P is a linear prevision, that is, P is coherent, if and only if the following conditions are met for all bounded gambles f and g in* dom P:

(i) $P(f) \geq \inf f$. **(bounds)**

(ii) $P(f+g) = P(f) + P(g)$. **(additivity)**

Proof. 'only if'. Immediate by Corollary 4.14(i) and (v)$_{55}$.

'if'. Consider arbitrary n in $\mathbb{N}$, and bounded gambles $f_1, \ldots, f_n$ in dom P, and let $f := -\sum_{i=1}^{n} f_i$, then f belongs to the linear space dom P. Moreover

$$\sup\left(\sum_{i=1}^{n} f_i\right) = \sup(-f) = -\inf f \geq -P(f) = \sum_{i=1}^{n} P(f_i),$$

where the inequality follows from (i) and the last equality from (ii) and self-conjugacy. Hence, the prevision P satisfies the coherence criterion (D)$_{52}$. □

Interestingly, the homogeneity condition (ii)$_{\curvearrowleft}$ that is present in the characterisation of coherence for lower previsions on linear spaces, namely, Theorem 4.15$_{56}$, is no longer there for linear previsions. Indeed, it can be shown to follow from conditions (i)$_{\curvearrowleft}$ and (ii)$_{\curvearrowleft}$ in Theorem 4.16: it follows at once from (ii)$_{\curvearrowleft}$ that $P(\rho f) = \rho P(f)$ for all rational ρ, and (i)$_{\curvearrowleft}$ and (ii)$_{\curvearrowleft}$ also guarantee that P is continuous with respect to uniform convergence; see the reasoning in the proof of Theorem 4.13(xiii)$_{53}$ for the essence of the argument.[3]

Corollary 4.17 *Consider a real functional P and assume that* dom P *is a linear subspace of $\mathbb{B}$ that contains all constant gambles. Then P is a linear prevision if and only if it is a* positive linear functional with unit norm, *meaning that it satisfies, for all all bounded gambles f and g in* dom P *and all non-negative real λ,*

(i) $P(f + g) = P(f) + P(g)$. **(additivity)**

(ii) $P(\lambda f) = \lambda P(f)$. **(non-negative homogeneity)**

(iii) *if $f \geq 0$ then $P(f) \geq 0$.* **(positivity)**

(iv) $P(\mathcal{X}) = P(1) = 1$. **(unit norm)**

Proof. 'if'. Infer from (ii) that $P(0) = 0$ and combine this with (i) to show that $P(f) + P(-f) = P(0) = 0$ for all bounded gambles f, implying that P is self-conjugate, and therefore a prevision. By Theorem 4.16$_{\curvearrowleft}$, it now suffices to prove that $P(f) \geq \inf f$ for all bounded gambles f. Use (ii) and (iv) and self-conjugacy to prove that $P(\mu) = \mu$ for all real numbers μ. Then because $f - \inf f \geq 0$, infer from (iii) that $P(f - \inf f) \geq 0$ and from (i) that $P(f - \inf f) = P(f) - \inf f$.

'only if'. Assume that P is a linear prevision, then (i), (ii) and (iv) follow from Theorem 4.14$_{55}$ (use (v)$_{55}$, (vi)$_{55}$ and (ii)$_{55}$, respectively). To prove (iii), infer from $f \geq 0$ and Theorem 4.14(i)$_{55}$ that $P(f) \geq \inf f \geq 0$. □

A useful method for proving that a given lower prevision is coherent (or that a given prevision is linear) consists in showing that it is the restriction of some coherent lower prevision (or linear prevision) defined a linear subspace of $\mathbb{B}$, where coherence and linearity are much easier to check. This simple observation is important and useful enough to call it a proposition.

Proposition 4.18 *The following statements hold:*

(i) *The restriction of a lower prevision that avoids sure loss, also avoids sure loss.*

(ii) *The restriction of a coherent lower prevision is also coherent.*

(iii) *The restriction of a linear prevision to a prevision is also linear.*

Proof. Immediately from Definitions 4.6$_{42}$, 4.10$_{47}$ and 4.11$_{51}$. □

[3] As we shall see in Theorem 13.36$_{271}$, a similar argument can no longer be applied for the case for unbounded random variables, essentially because the uniform continuity part of the argument can no longer be employed.

Another way to prove that a given lower prevision is coherent is to express it as a convex combination, a lower envelope or a point-wise limit of lower previsions that are already known to be coherent. This is the essence of the following three propositions. Their respective statements (i) and (ii) are due to Walley (1991, Theorems 2.6.3–2.6.5). Again, sketches of proofs are given for the sake of completeness.

Proposition 4.19 *Suppose* $\Gamma = \{\underline{P}_1, \underline{P}_2, \ldots, \underline{P}_p\}$ *is a finite collection of lower previsions defined on a common domain* $\mathscr{K}$. *Let* $\underline{Q}$ *be a* **convex combination** *of the elements of* Γ, *meaning that*

$$\underline{Q}(f) := \sum_{i=1}^{p} \lambda_i \underline{P}_i(f) \text{ for all } f \in \mathscr{K},$$

for some $\lambda_1, \ldots, \lambda_p \geq 0$ *such that* $\sum_{i=1}^{p} \lambda_i = 1$. *Then the following statements hold:*

(i) *If all lower previsions in* Γ *avoid sure loss, then* $\underline{Q}$ *avoids sure loss as well.*

(ii) *If all lower previsions in* Γ *are coherent, then* $\underline{Q}$ *is coherent as well.*

(iii) *If all lower previsions in* Γ *are linear previsions, then* $\underline{Q}$ *is a linear prevision as well.*

Proof. We first give a proof for (ii). The proof for (i) is very similar, if slightly less involved. We show that $\underline{Q}$ satisfies the coherence criterion $(\mathrm{D})_{47}$. Consider arbitrary n, m in $\mathbb{N}$ and bounded gambles $f_0, f_1, \ldots, f_n$ in $\mathscr{K}$, then indeed

$$\sup\left(\sum_{k=1}^{n} f_k - mf_0\right) \geq \max_{i=1}^{p}\left[\sum_{k=1}^{n} \underline{P}_i(f_k) - \underline{P}_i(f_0)\right] \geq \sum_{i=1}^{p} \lambda_i \left[\sum_{k=1}^{n} \underline{P}_i(f_k) - \underline{P}_i(f_0)\right]$$
$$= \sum_{k=1}^{n} \underline{Q}(f_k) - \underline{Q}(f_0),$$

where the first inequality follows from the coherence of the $\underline{P}_i$, and the second is a general property of convex mixtures.

(iii). It suffices to check the self-conjugacy of $\underline{Q}$:

$$\underline{Q}(-f) = \sum_{i=1}^{n} \lambda_i P_i(-f) = -\sum_{i=1}^{n} \lambda_i P_i(f) = -\underline{Q}(f),$$

which completes the proof. □

Proposition 4.20 *Suppose* Γ *is a non-empty collection of lower previsions defined on a common domain* $\mathscr{K}$. *Let* $\underline{Q}$ *be the* **lower envelope** *of* Γ, *meaning that*

$$\underline{Q}(f) := \inf_{\underline{P} \in \Gamma} \underline{P}(f) \text{ for all } f \in \mathscr{K}.$$

Then the following statements hold:

(i) *If some lower prevision in* Γ *avoids sure loss and* $\underline{Q}$ *is real-valued, then the lower prevision* $\underline{Q}$ *avoids sure loss as well. Equivalently, if a lower prevision is dominated by a lower prevision that avoids sure loss, then it avoids sure loss as well.*

(ii) *If all lower previsions in* Γ *are coherent, then the lower prevision* $\underline{Q}$ *is coherent as well.*

(iii) *If all lower previsions in* Γ *are linear, then the lower prevision* $\underline{Q}$ *is coherent; it is not linear, unless* Γ *is a singleton.*

Proof. (i). This is fairly immediate if we look at the avoiding sure loss condition $(\mathrm{C})_{42}$. Suppose $\underline{P} \in \Gamma$ avoids sure loss. Then there is a coherent lower prevision $\underline{R}$ on $\mathbb{B}$ that dominates $\underline{P}$. Obviously, $\underline{R}$ will also dominate $\underline{Q}$ because $\underline{P}$ dominates $\underline{Q}$. Therefore, $\underline{Q}$ avoids sure loss as well.

(ii). First, by coherence (bounds: Theorem $4.13(\mathrm{i})_{53}$), $\underline{P}(f) \geq \inf f$ for all f in $\mathscr{K}$ and $\underline{P}$ in Γ. Hence, $\underline{Q}(f) \geq \inf f$, so $\underline{Q}$ is guaranteed to be real-valued. Consider arbitrary n, m in $\mathbb{N}$ and bounded gambles $f_0, f_1, \ldots, f_n$ in $\mathscr{K}$. Fix $\delta > 0$. Then there is some $\underline{P}$ in Γ such that $\underline{Q}(f_0) + \delta \geq \underline{P}(f_0)$, and therefore

$$\sup\left(\sum_{k=1}^{n} f_k - mf_0\right) \geq \sum_{k=1}^{n} \underline{P}(f_k) - m\underline{P}(f_0) \geq \sum_{k=1}^{n} \underline{Q}(f_k) - m\underline{Q}(f_0) - m\delta,$$

where the first inequality holds because $\underline{P}$ is coherent. As this inequality holds for all $\delta > 0$, it also holds for $\delta = 0$, and therefore $\underline{Q}$ satisfies coherence criterion $(\mathrm{D})_{47}$.

(iii). It suffices to check that $\underline{Q}$ is not self-conjugate whenever Γ contains more than one element. Choose two distinct linear previsions P_1 and P_2 in Γ. As they are distinct, there is a bounded gamble f in $\mathscr{K}$ such that, for instance, $P_1(f) < P_2(f)$. This means that $\underline{Q}(f) \leq P_1(f) < P_2(f) \leq \overline{Q}(f)$, so $\underline{Q}$ is not self-conjugate. □

The point-wise limit of a net of coherent lower previsions is a coherent lower prevision; for a discussion of nets and their limits, refer to Section 5.3_{80}.

Proposition 4.21 *Let* $\mathscr{A}$ *be a directed set, and suppose* $\underline{P}_\alpha$ *is a net of lower previsions defined on a common domain* $\mathscr{K}$. *Suppose that* $\underline{P}_\alpha$ ***converges point-wise*** *to a lower prevision* $\underline{Q}$ *defined on* $\mathscr{K}$, *meaning that for all* f *in* $\mathscr{K}$, *the real net* $\underline{P}_\alpha(f)$ *converges to the real number* $\underline{Q}(f)$. *Then the following statements hold:*

(i) *If the* $\underline{P}_\alpha$ *avoid sure loss eventually, then* $\underline{Q}$ *avoids sure loss as well.*

(ii) *If the* $\underline{P}_\alpha$ *are coherent eventually, then* $\underline{Q}$ *is coherent as well.*

(iii) *If the* $\underline{P}_\alpha$ *are linear eventually, then* $\underline{Q}$ *is linear as well.*

Proof. We give a proof for (ii). The proof for (i) is analogous but simpler. Consider n, m in $\mathbb{N}$ and bounded gambles $f_0, f_1, \ldots, f_n$ in $\mathscr{K}$ and fix $\delta > 0$. Then it follows from the assumptions that there is some α_o such that $\underline{P}_\alpha$ is coherent for all $\alpha \geq \alpha_o$ and that there is some $\alpha \geq \alpha_o$ such that $\left|\underline{P}_\alpha(f_k) - \underline{Q}(f_k)\right| < \delta$ and therefore also $\underline{P}_\alpha(f_k) > \underline{Q}(f_k) - \delta$ and $-\underline{P}_\alpha(f_k) > -\underline{Q}(f_k) - \delta$ for all $k \in \{0, 1, \ldots, n\}$. Hence, also using that this $\underline{P}_\alpha$ is coherent,

$$\sup\left(\sum_{k=1}^{n} f_k - mf_0\right) \geq \sum_{k=1}^{n} \underline{P}_\alpha(f_k) - m\underline{P}_\alpha(f_0) \geq \sum_{k=1}^{n} \underline{Q}(f_k) - m\underline{Q}(f_0) - (m+n)\delta.$$

As this inequality holds for all $\delta > 0$, it also holds for $\delta = 0$, which shows that the lower prevision $\underline{Q}$ satisfies coherence criterion $(\mathrm{D})_{47}$.

(iii). It suffices to check the self-conjugacy of $\underline{Q}$:

$$\underline{Q}(f) = \lim_\alpha \underline{P}_\alpha(f) = \lim_\alpha -\underline{P}_\alpha(-f) = -\lim_\alpha \underline{P}_\alpha(-f) = -\underline{Q}(-f) = \overline{Q}(f),$$

which completes the proof. □

If we combine Propositions 4.20_{59} and 4.21, we find a similar result for the point-wise limit inferior of a net of lower previsions. For a discussion of the limit inferior of a net, see Section 5.3_{80}.

Corollary 4.22 *Suppose $\underline{P}_\alpha$ is a net of lower previsions defined on a common domain $\mathscr{K}$. Let $\underline{Q}$, defined on $\mathscr{K}$, be the* ***point-wise limit inferior*** *of this net, meaning that*

$$\underline{Q}(f) := \liminf_\alpha \underline{P}_\alpha(f) = \sup_\alpha \inf_{\beta \geq \alpha} \underline{P}_\beta(f) \text{ for all } f \text{ in } \mathscr{K}.$$

Then the following statements hold:

(i) *If the $\underline{P}_\alpha$ avoid sure loss eventually, and $\inf_{\beta\geq\alpha} \underline{P}_\beta$ is real-valued eventually, then the lower prevision $\underline{Q}$ avoids sure loss as well.*

(ii) *If the $\underline{P}_\alpha$ are coherent eventually, then the lower prevision $\underline{Q}$ is coherent as well.*

(iii) *If the $\underline{P}_\alpha$ are linear eventually, then the lower prevision $\underline{Q}$ is coherent. Moreover, $\underline{Q}$ is linear if and only if $\underline{P}_\alpha$ converges point-wise.*

Proof. (i). By assumption, there is some α^* such that $\underline{P}_\alpha$ avoids sure loss for all $\alpha \geq \alpha^*$, and such that the functional $\underline{Q}_\alpha$ defined on $\mathscr{K}$ by $\underline{Q}_\alpha(f) := \inf_{\beta\geq\alpha} \underline{P}_\beta(f)$ is real-valued for all $\alpha \geq \alpha^*$. By Proposition 4.20_{59}, the lower prevision $\underline{Q}_\alpha$ avoids sure loss for all $\alpha \geq \alpha^*$.

For each $f \in \mathscr{K}$, the net $\underline{Q}_\alpha(f)$ is non-decreasing and bounded above by $\sup f$ and, therefore, converges to a real number.

From these two observations, it follows using Proposition 4.21 that $\underline{Q}$ must avoid sure loss as well.

(ii). Similar to the proof of (i). Note that $\underline{Q}_\alpha$ is real valued because each $\underline{P}_\alpha$ satisfies $\inf f \leq \underline{P}_\alpha(f) \leq \sup f$ by coherence (bounds). Hence, unlike in (i), no additional real-valuedness assumption regarding $\underline{Q}_\alpha$ is needed.

(iii). Similar to the proof of (i). For the second part, note that if $\underline{P}_\alpha$ converges point-wise, then $\liminf_\alpha \underline{P}_\alpha$ coincides with $\lim_\alpha \underline{P}_\alpha$ and therefore $\underline{Q}$ must be linear, by Proposition 4.21(iii)$_{60}$. Conversely, if $\underline{Q}$ is linear, then, for every $f \in \mathscr{K}$,

$$\liminf_{\alpha} \underline{P}_{\alpha}(f) = -\liminf_{\alpha} \underline{P}_{\alpha}(-f) = \limsup_{\alpha} -\underline{P}_{\alpha}(-f) = \limsup_{\alpha} \underline{P}_{\alpha}(f),$$

which means that the net $\underline{P}_{\alpha}$ converges point-wise. □

A slightly stronger version of Corollary 4.22(i)$_{\curvearrowleft}$ can be proved, based on Corollary 13.44$_{277}$ further on, where the real-valuedness assumption is weakened to assuming that the limit inferior $\underline{Q}$ must be real valued. The proof of this stronger result requires the use of (extended) lower previsions on possibly unbounded gambles, which we have not introduced yet. So for the sake of simplicity, we omit the more involved result, and leave it to the reader to fill in the details.

Finally, continuity may also help in proving that a lower prevision is coherent. The following proposition establishes a partial converse to Proposition 4.18(ii)$_{58}$.

Proposition 4.23 *Suppose $\underline{P}$ is continuous on its domain with respect to the topology of uniform convergence and coherent on a dense subset $\mathscr{K}$ of* dom $\underline{P}$. *Then $\underline{P}$ is coherent.*

Proof. Fix $\varepsilon > 0$.

First, consider any $f \in \operatorname{dom} \underline{P}$. By the continuity assumption, there is some $0 < \delta_{\varepsilon} \leq \varepsilon$ such that $|\underline{P}(f) - \underline{P}(f')| < \varepsilon$ for all $f' \in \operatorname{dom} \underline{P}$ such that $\sup|f - f'| < \delta_{\varepsilon}$. As $\mathscr{K}$ is dense in dom $\underline{P}$, there is some $f^{\varepsilon} \in \mathscr{K}$ such that $\sup|f - f^{\varepsilon}| < \delta_{\varepsilon}$, and therefore, $|\underline{P}(f) - \underline{P}(f^{\varepsilon})| < \varepsilon$.

As a result, for any $m \in \mathbb{N}$, $n \in \mathbb{N}$ and $f_0, f_1, \ldots, f_n$ in dom $\underline{P}$, we find that there are $f_0^{\varepsilon}, f_1^{\varepsilon}, \ldots, f_n^{\varepsilon}$ in $\mathscr{K}$ such that

$$\begin{aligned}\sum_{i=1}^{n} \underline{P}(f_i) - m\underline{P}(f_0) &\leq (n+m)\varepsilon + \sum_{i=1}^{n} \underline{P}(f_i^{\varepsilon}) - m\underline{P}(f_0^{\varepsilon}) \\ &\leq (n+m)\varepsilon + \sup\left(\sum_{i=1}^{n} f_i^{\varepsilon} - mf_0^{\varepsilon}\right) \\ &\leq 2(n+m)\varepsilon + \sup\left(\sum_{i=1}^{n} f_i - mf_0\right),\end{aligned}$$

where the second inequality follows from the coherence of $\underline{P}$ on $\mathscr{K}$ (use coherence criterion (D)$_{47}$). As this holds for any $\varepsilon > 0$, it must also hold for $\varepsilon = 0$, and the coherence of $\underline{P}$ follows. □

The following proposition is a variation on the same theme.

Proposition 4.24 *Let $\underline{P}$ be a coherent lower prevision. Then $\underline{P}$ has a unique extension to a coherent lower prevision defined on the uniform closure* cl(dom $\underline{P}$) *of* dom $\underline{P}$. *Moreover, if $\underline{P}$ is self-conjugate, then this unique extension is self-conjugate too.*

Proof. By coherence condition (B)$_{47}$, any coherent lower prevision $\underline{P}$ has a coherent extension to $\mathbb{B}$, and also taking into account Proposition 4.18$_{58}$, we conclude that there are coherent extensions $\underline{Q}$ of $\underline{P}$ to the uniform closure of dom $\underline{P}$. We are left to show that all these coherent extensions are unique.

Let $\underline{Q}_1$ and $\underline{Q}_2$ be two coherent lower previsions defined on the uniform closure of dom $\underline{P}$, and suppose that $\underline{Q}_1(f) = \underline{Q}_2(f) = \underline{P}(f)$ for all f in dom $\underline{P}$. We must show that $\underline{Q}_1(g) = \underline{Q}_2(g)$ for all g in dom $\underline{Q}_1$ = dom $\underline{Q}_2$. For every such g, there is a sequence f_n in dom $\underline{P}$ that converges uniformly to g. As $\underline{Q}_1$ and $\underline{Q}_2$ are coherent, it follows from their continuity with respect to uniform convergence (Theorem 4.13(xiv)$_{53}$) that

$$\underline{Q}_1(g) = \lim_{n\in\mathbb{N}} \underline{Q}_1(f_n) = \lim_{n\in\mathbb{N}} \underline{P}(f_n) = \lim_{n\in\mathbb{N}} \underline{Q}_2(f_n) = \underline{Q}_2(g),$$

and therefore, $\underline{Q}_1(g) = \underline{Q}_2(g)$ for any bounded gamble g in their domain (incidentally, the limit is independent of the choice of the sequence f_n converging uniformly to g). This proves uniqueness.

If $\underline{P}$ is self-conjugate, then, again take for every g in dom $\underline{Q}$ a sequence f_n in dom $\underline{P}$ that converges uniformly to g. As $\underline{Q}$ is coherent, its continuity with respect to uniform convergence (Theorem 4.13(xiv)$_{53}$) implies that

$$\underline{Q}(g) = \lim_{n\in\mathbb{N}} \underline{Q}(f_n) = \lim_{n\in\mathbb{N}} \underline{P}(f_n) = \lim_{n\in\mathbb{N}} \overline{P}(f_n) = \lim_{n\in\mathbb{N}} \overline{Q}(f_n) = \overline{Q}(g).$$

This establishes that $\underline{Q}$ is self-conjugate. □

4.4.4 Coherence and monotone convergence

The last two propositions above are simple consequences of Theorem 4.13(xiv)$_{53}$. Indeed, a coherent lower prevision is automatically also continuous with respect to uniform convergence: if f_α is a net of bounded gambles that **converges uniformly** to a bounded gamble f, meaning that $\lim_\alpha \sup |f_\alpha - f| = 0$, then also $\underline{P}(f_\alpha) \to \underline{P}(f)$ or, in other words, $\lim_\alpha \underline{P}(f_\alpha) = \underline{P}(f)$. This is a fairly weak sort of continuity because the condition $\lim_\alpha \sup |f_\alpha - f| = 0$ is quite strong.

There is another type of continuity, which is quite important in measure-theoretic probability theory: the monotone convergence property we have introduced in Section 1.9$_{17}$.

As we shall, for instance, see in Proposition 5.12$_{91}$, as well as in the example below, the monotone convergence properties in Definition 1.25$_{18}$ are not implied by coherence: there are coherent lower previsions that do not satisfy any of them.

But some do. Consider as an example the linear prevision on the set $\mathbb{B}(\mathbb{N})$ of all bounded real sequences defined by

$$P(f) := \sum_{n=0}^{+\infty} \frac{f(n)}{2^n} \text{ for every bounded gamble } f \text{ on } \mathbb{N}.$$

Then it is not very difficult to show that this linear prevision P satisfies (both upward and downward) monotone convergence. For instance, the non-increasing sequence of bounded gambles $f_k := I_{\{n\in\mathbb{N}\colon n\geq k\}}$ converges point-wise to $\lim_{k\to+\infty} f_k = 0$, and so does the corresponding sequence of previsions:

$$P(f_k) = \sum_{n=k}^{+\infty} \frac{1}{2^n} = \frac{1}{2^{k-1}} \to 0.$$

Contrast this with the lower prevision lim inf defined on the set $\mathbb{B}(\mathbb{N})$ of all bounded real sequences f by

$$\liminf f = \sup_{n\in\mathbb{N}} \inf_{m\geq n} f(m);$$

see the discussion in Section 1.5_7 for more details about the limit inferior operator. It is easy to see that this operator satisfies the conditions of Theorem 4.15_{56} and is therefore a coherent lower prevision. Its restriction to the linear subspace of all convergent bounded sequences is the limit operator lim, which is a linear prevision. We will come back to these lower and linear previsions in Sections 5.3_{80} and 5.5_{82}, and in particular in Example 5.13_{92}. For our present purposes, it suffices to show that lim inf satisfies neither upward nor downward monotone convergence. Indeed, we see that

$$\liminf f_k = \sup_{n\in\mathbb{N}} \inf_{m\geq n} f_k(m) = 1,$$

so $\lim_{k\to+\infty}(\liminf f_k) = 1$, whereas $\liminf(\lim_{k\to+\infty} f_k) = \liminf 0 = 0$. This shows that lim inf does not satisfy downward monotone convergence. It will satisfy upward monotone convergence if and only if the conjugate upper prevision lim sup satisfies downward monotone convergence, but clearly $\limsup f_k = 1$ too, so $\lim_{k\to+\infty}(\limsup f_k) = 1$, whereas $\limsup(\lim_{k\to+\infty} f_k) = \limsup 0 = 0$.

4.4.5 Coherence and a seminorm

We briefly mention here that the properties of a coherent lower prevision guarantee that we can use the conjugate upper prevision to define a seminorm on the set of all bounded gambles.[4] We will use this seminorm in Chapter 15_{327} to help to extend lower previsions to unbounded gambles.

Definition 4.25 ($\underline{P}$-seminorm) *Let $\underline{P}$ be a coherent lower prevision defined on the set $\mathbb{B}$ of all bounded gambles. The* ***$\underline{P}$-seminorm*** *of a bounded gamble f is defined by $\|f\|_{\underline{P}} := \overline{P}(|f|)$. A net f_α of bounded gambles* ***converges in $\underline{P}$-seminorm*** *to a bounded gamble f if the net of real numbers $\|f - f_\alpha\|_{\underline{P}}$ converges to zero.*

Let us prove that $\|\bullet\|_{\underline{P}}$ satisfies all properties of a seminorm, as defined in Definition $B.6_{374}$.

Proof. First of all, it follows from the bounds (Theorem $4.13(\text{i})_{53}$) of the coherent upper prevision $\overline{P}$ that $\overline{P}(|f|) \geq \inf|f| \geq 0$ and therefore $\|f\|_{\underline{P}} \geq 0$. Secondly, the non-negative homogeneity (Theorem $4.13(\text{vi})_{53}$) of $\overline{P}$ guarantees that $\overline{P}(|\lambda f|) = \overline{P}(|\lambda||f|) = |\lambda|\overline{P}(|f|)$ and therefore $\|\lambda f\|_{\underline{P}} = |\lambda|\|f\|_{\underline{P}}$. And finally,

[4] See Appendix B_{371} and in particular Definition $B.6_{374}$ for a definition and brief discussion of seminorms.

the monotonicity and sub-additivity (Theorem 4.13(iv) and (v)$_{53}$) of the coherent $\overline{P}$ allow us to infer from $|f+g| \leq |f|+|g|$ that $\overline{P}(|f+g|) \leq \overline{P}(|f|+|g|) \leq \overline{P}(|f|)+\overline{P}(|g|)$ and therefore $\|f+g\|_{\underline{P}} \leq \|f\|_{\underline{P}}+\|g\|_{\underline{P}}$. □

4.5 The natural extension of a lower prevision

Coherence is the minimal requirement we will impose on lower previsions. Clearly, avoiding sure loss is weaker than coherence. But we have seen earlier that as soon as a lower prevision avoids sure loss, it can be *corrected* to a coherent lower prevision, namely, its natural extension. To complete this chapter, we discuss this natural extension in some more detail.

4.5.1 Natural extension as least-committal extension

If a lower prevision $\underline{P}$ avoids sure loss, then we have seen that there are coherent lower previsions on the set of all bounded gambles $\mathbb{B}$ that dominate $\underline{P}$. The following theorem states amongst other things that the natural extension $\underline{E}_{\underline{P}}$ is the most conservative coherent lower prevision that dominates $\underline{P}$: all coherent lower previsions that dominate $\underline{P}$ also dominate $\underline{E}_{\underline{P}}$.

Theorem 4.26 (Natural extension) *Let $\underline{P}$ be a lower prevision. Then the following statements hold:*

(i) *If $\underline{P}$ avoids sure loss, then $\underline{E}_{\underline{P}}$ is the point-wise smallest coherent lower prevision on $\mathbb{B}$ that dominates $\underline{P}$.*

(ii) *If $\underline{P}$ is coherent, then $\underline{E}_{\underline{P}}$ is the point-wise smallest coherent lower prevision on $\mathbb{B}$ that coincides with $\underline{P}$ on* dom $\underline{P}$ *or, in other words, its point-wise smallest coherent extension.*

Because of this property, $\underline{E}_{\underline{P}}$ is sometimes also called the **least-committal extension** of $\underline{P}$, see, for instance, Walley (1981, p. 28).

Proof. (i). Assume that $\underline{P}$ avoids sure loss. Clearly, $\underline{E}_{\underline{P}}$ is a coherent lower prevision that dominates $\underline{P}$, by Definition 4.8$_{47}$. Let $\underline{Q}$ be any coherent lower prevision on $\mathbb{B}$ that dominates $\underline{P}$. We show that $\underline{Q}$ also dominates $\underline{E}_{\underline{P}}$.

Indeed, because $\underline{Q}$ dominates $\underline{P}$, it follows that $\mathscr{A}_{\underline{P}} \subseteq \mathscr{A}_{\underline{Q}}$ (see Equation (4.2)$_{42}$). Hence, by Proposition 4.5$_{42}$, $\operatorname{lpr}(\operatorname{Cl}_{\mathbb{D}_b}(\mathscr{A}_{\underline{Q}}))$ dominates $\operatorname{lpr}(\operatorname{Cl}_{\mathbb{D}_b}(\mathscr{A}_{\underline{P}}))$. Now simply observe that $\underline{Q} = \operatorname{lpr}(\operatorname{Cl}_{\mathbb{D}_b}(\mathscr{A}_{\underline{Q}}))$ by coherence condition (C)$_{47}$, and obviously, $\underline{E}_{\underline{P}} = \operatorname{lpr}(\operatorname{Cl}_{\mathbb{D}_b}(\mathscr{A}_{\underline{P}}))$ by definition of $\underline{E}_{\underline{P}}$.

(ii). By coherence condition (C)$_{47}$, $\underline{E}_{\underline{P}}$ is indeed a coherent extension of $\underline{P}$. By (i), it is also the point-wise smallest one. □

Proposition 4.27 *Let $\underline{P}$ and $\underline{Q}$ be lower previsions that avoid sure loss. If $\underline{Q}$ dominates $\underline{P}$, then $\underline{E}_{\underline{Q}}$ dominates $\underline{E}_{\underline{P}}$ as well: $\underline{E}_{\underline{Q}}(f) \geq \underline{E}_{\underline{P}}(f)$ for every bounded gamble f on $\mathscr{X}$.*

Proof. This is just a reformulation, in terms of lower previsions and natural extension, of Proposition 4.5_{42}. If $\underline{Q}$ dominates $\underline{P}$, then $\mathscr{A}_{\underline{P}} \subseteq \mathscr{A}_{\underline{Q}}$. Indeed, consider any g in $\mathscr{A}_{\underline{P}}$, then there are $f \in \operatorname{dom} \underline{P}$ and $\mu < \underline{P}(f)$ such that $g = f - \mu$. But because, by assumption, then also $f \in \operatorname{dom} \underline{Q}$ and $\mu < \underline{Q}(f)$, we see that $g \in \mathscr{A}_{\underline{Q}}$ too. Now apply Proposition 4.5_{42}.

Alternatively, if $\underline{Q}$ dominates $\underline{P}$, then any coherent lower prevision that dominates $\underline{Q}$ also dominates $\underline{P}$. So $\underline{E}_{\underline{Q}}$ dominates $\underline{Q}$, and therefore of $\underline{P}$, by Theorem 4.26(i)$_{\curvearrowleft}$. And by the same theorem, $\underline{E}_{\underline{P}}$ is the point-wise smallest coherent lower prevision that dominates $\underline{P}$. □

4.5.2 Natural extension and equivalence

The notion of *equivalence* for two lower previsions captures that their behavioural consequences are the same:

Definition 4.28 (Equivalence) *Two lower previsions $\underline{P}$ and $\underline{Q}$ that avoid sure loss are called* **equivalent** *if their natural extensions are equal: $\underline{E}_{\underline{P}} = \underline{E}_{\underline{Q}}$. Similarly, two consistent sets of acceptable bounded gambles $\mathscr{A}_1$ and $\mathscr{A}_2$ are called* **equivalent** *if their natural extensions are the same:* $\mathrm{Cl}_{\mathbb{D}_b}(\mathscr{A}_1) = \mathrm{Cl}_{\mathbb{D}_b}(\mathscr{A}_2)$.

The following is then immediate.

Proposition 4.29 *Equivalence on lower previsions is an equivalence relation (i.e. a reflexive, symmetric and transitive binary relation) and so is equivalence on sets of acceptable bounded gambles.*

Proposition 4.30 *Let $\underline{P}$ be a lower prevision that avoids sure loss. Let $\underline{Q}$ be any coherent lower prevision that dominates $\underline{P}$. Then $\underline{Q}$ is equivalent to $\underline{P}$ if and only if $\underline{E}_{\underline{P}}$ is an extension of $\underline{Q}$, that is, if and only if $\underline{Q}$ and $\underline{E}_{\underline{P}}$ coincide on* $\operatorname{dom} \underline{Q}$.

Proof. As $\underline{P}$ avoids sure loss and $\underline{Q}$ is coherent, both $\underline{E}_{\underline{P}}$ and $\underline{E}_{\underline{Q}}$ exist.

'if'. As $\underline{Q}$ dominates $\underline{P}$, any coherent lower prevision that dominates $\underline{Q}$ also dominates $\underline{P}$. Hence, $\underline{E}_{\underline{Q}} \geq \underline{E}_{\underline{P}}$. To prove the converse inequality, let $\underline{R}$ be any coherent lower prevision on $\mathbb{B}$ that dominates $\underline{P}$. By Theorem 4.26(ii)$_{\curvearrowleft}$, the claim is established if we can show that $\underline{R}$ dominates $\underline{Q}$. Indeed, $\underline{R} \geq \underline{E}_{\underline{P}}$ by Theorem 4.26(i)$_{\curvearrowleft}$. As $\underline{Q} = \underline{E}_{\underline{P}}$ on $\operatorname{dom} \underline{Q}$ it follows that also $\underline{R} \geq \underline{Q}$ on $\operatorname{dom} \underline{Q}$, which means that $\underline{R}$ dominates $\underline{Q}$.

'only if'. Suppose $\underline{E}_{\underline{P}} = \underline{E}_{\underline{Q}}$. As $\underline{Q}$ is coherent, it follows from Definition 4.10_{47} that $\underline{Q}$ and $\underline{E}_{\underline{Q}}$ coincide on $\operatorname{dom} \underline{Q}$, and therefore $\underline{Q}$ and $\underline{E}_{\underline{P}}$ coincide on $\operatorname{dom} \underline{Q}$. □

4.5.3 Natural extension to a specific domain

It will sometimes be useful to consider the restriction of the coherent lower prevision $\underline{E}_{\underline{P}}$ to some set of bounded gambles $\mathscr{K}$ that includes $\operatorname{dom} \underline{P}$. This leads to a coherent lower prevision whose domain is $\mathscr{K}$. We denote it by $\underline{E}_{\underline{P}}^{\mathscr{K}}$ and call it the **natural extension of $\underline{P}$ to $\mathscr{K}$.**

Theorem 4.31 *Let $\underline{P}$ be a lower prevision. Then the following statements hold:*

(i) *If $\underline{P}$ avoids sure loss, then $\underline{E}_{\underline{P}}^{\mathscr{K}}$ is the point-wise smallest coherent lower prevision on $\mathscr{K}$ that dominates $\underline{P}$.*

(ii) *If $\underline{P}$ is coherent, then $\underline{E}_{\underline{P}}^{\mathscr{K}}$ is the point-wise smallest coherent lower prevision on $\mathscr{K}$ that coincides with $\underline{P}$ on* dom $\underline{P}$ *or, in other words, its point-wise smallest coherent extension to $\mathscr{K}$.*

Proof. (i). Assume that $\underline{P}$ avoids sure loss. Clearly, $\underline{E}_{\underline{P}}^{\mathscr{K}}$ is a coherent lower prevision that dominates $\underline{P}$, by Definition 4.8_{47}. Let $\underline{Q}$ be any coherent lower prevision on $\mathscr{K}$ that dominates $\underline{P}$. Then $\underline{E}_{\underline{Q}}$ dominates $\underline{E}_{\underline{P}}$ by Proposition 4.27_{65}, and hence, $\underline{Q}$ dominates $\underline{E}_{\underline{P}}^{\mathscr{K}}$ because $\underline{Q} = \underline{E}_{\underline{Q}}^{\mathscr{K}}$ by coherence condition $(\mathrm{C})_{47}$.

(ii). By coherence condition $(\mathrm{C})_{47}$, $\underline{E}_{\underline{P}}^{\mathscr{K}}$ is indeed a coherent extension of $\underline{P}$ to $\mathscr{K}$. By (i), it is also the point-wise smallest one. □

4.5.4 Transitivity of natural extension

Corollary 4.32 *Let $\underline{P}$ be a lower prevision that avoids sure loss, and let* dom $\underline{P} \subseteq \mathscr{K} \subseteq \mathbb{B}$. *Then*

$$\underline{E}_{\underline{E}_{\underline{P}}^{\mathscr{K}}}(f) = \underline{E}_{\underline{P}}(f) \text{ for all bounded gambles } f.$$

In other words, $\underline{P}$ and $\underline{E}_{\underline{P}}^{\mathscr{K}}$ are equivalent. In particular, let dom $\underline{P} \subseteq \mathscr{J} \subseteq \mathscr{K} \subseteq \mathbb{B}$, *then $\underline{E}_{\underline{P}}^{\mathscr{K}}$ is an extension of $\underline{E}_{\underline{P}}^{\mathscr{J}}$, and*

$$\underline{E}_{\underline{E}_{\underline{P}}^{\mathscr{J}}}^{\mathscr{K}}(f) = \underline{E}_{\underline{P}}^{\mathscr{K}}(f) \text{ for all bounded gambles } f \text{ in } \mathscr{K}.$$

Proof. By definition, $\underline{E}_{\underline{P}}$ is an extension of $\underline{E}_{\underline{P}}^{\mathscr{K}}$. By Proposition 4.30, $\underline{P}$ and $\underline{E}_{\underline{P}}^{\mathscr{K}}$ are therefore equivalent. By definition, $\underline{E}_{\underline{P}}^{\mathscr{K}}$ is an extension of $\underline{E}_{\underline{P}}^{\mathscr{J}}$. The last equality follows from the first by taking $\mathscr{J}$ for $\mathscr{K}$ and restricting to $\mathscr{K}$. □

This result, the transitivity of the relation 'is the natural extension of', is sometimes loosely referred to as the '*transitivity of natural extension*': let $\underline{P}$, $\underline{Q}$ and $\underline{R}$ be three coherent lower previsions, and assume that dom $\underline{P} \subseteq$ dom $\underline{Q} \subseteq$ dom $\underline{R}$; if $\underline{Q}$ is the natural extension of $\underline{P}$ to dom $\underline{Q}$ and $\underline{R}$ is the natural extension of $\underline{Q}$ to dom $\underline{R}$, then $\underline{R}$ is the natural extension of $\underline{P}$ to dom $\underline{R}$.

4.5.5 Natural extension and avoiding sure loss

The following theorem summarises the exact details of the relationship between avoiding sue loss and natural extension. It also provides us with a number of alternative constructive expressions for calculating the natural extension and a number of alternative criteria – in addition to the ones in Definition 4.6_{42} – for checking whether a lower prevision avoids sure loss.

Recall that $\mathbb{R}^* = \mathbb{R} \cup \{-\infty, +\infty\}$ denotes the set of extended real numbers.

Theorem 4.33 *Let $\underline{P}$ be a lower prevision, and let* dom $\underline{P} \subseteq \mathscr{K} \subseteq \mathbb{B}$. *Define the* $\mathbb{B}-\mathbb{R}^*$*-map* $\underline{E}$ *by letting* $\underline{E}(f)$ *be equal to any of the following equivalent expressions:*

$$\sup\left\{\alpha \in \mathbb{R} : f - \alpha \geq \sum_{k=1}^{n} \lambda_k[f_k - \underline{P}(f_k)], n \in \mathbb{N}, f_k \in \operatorname{dom}\underline{P}, \lambda_k \in \mathbb{R}_{\geq 0}\right\} \tag{4.10a}$$

$$\sup\left\{\inf\left(f - \sum_{k=1}^{n} \lambda_k[f_k - \underline{P}(f_k)]\right) : n \in \mathbb{N}, f_k \in \operatorname{dom}\underline{P}, \lambda_k \in \mathbb{R}_{\geq 0}\right\} \tag{4.10b}$$

$$\sup\left\{\alpha + \sum_{i=1}^{n} \lambda_i \underline{P}(f_i) : \alpha + \sum_{i=1}^{n} \lambda_i f_i \leq f, n \in \mathbb{N}, \alpha \in \mathbb{R}, f_k \in \operatorname{dom}\underline{P}, \lambda_k \in \mathbb{R}_{\geq 0}\right\} \tag{4.10c}$$

for every bounded gamble $f \in \mathbb{B}$. *Then the following conditions are equivalent:*

(i) $\underline{P}$ *avoids sure loss.*

(ii) $\underline{E}$ *is a coherent lower prevision on* $\mathbb{B}$.

(iii) $\underline{E}(f) < +\infty$ *for all bounded gambles* $f \in \mathbb{B}$.

(iv) $\underline{E}(f_0) < +\infty$ *for some bounded gamble* $f_0 \in \mathbb{B}$.

(v) *There is a point-wise smallest coherent lower prevision on* $\mathscr{K}$ *that dominates* $\underline{P}$ *– its natural extension* $\underline{E}_{\underline{P}}^{\mathscr{K}}$. *This lower prevision coincides with* $\underline{E}$ *on* $\mathscr{K}$.

(vi) *There is a lower prevision on* $\mathscr{K}$ *that is coherent and dominates* $\underline{P}$.

(vii) *There is a lower prevision on* $\mathscr{K}$ *that avoids sure loss and dominates* $\underline{P}$.

Proof. It is clear that the three expressions yield identical values for $\underline{E}(f)$. We prove the equivalence of the conditions. We first prove that (i)⇒(v)⇒(vi)⇒(vii)⇒(i).

(i)⇒(v). The first part is an immediate consequence of Theorem 4.31(i)$_{\curvearrowleft}$. Compare the expressions (4.10a) and (4.6)$_{46}$ to see that $\underline{E}_{\underline{P}}$ and $\underline{E}$ coincide.

(v)⇒(vi). Immediate.

(vi)⇒(vii). Immediate, because a coherent lower prevision in particular avoids sure loss; see coherence criterion (C)$_{47}$.

(vii)⇒(i). Let $\underline{Q}$ be a lower prevision on $\mathscr{K}$ that avoids sure loss and dominates $\underline{P}$. We must show that $\underline{P}$ avoids sure loss. Consider the natural extension $\underline{E}_{\underline{Q}}$ of $\underline{Q}$. By Proposition 4.27$_{65}$, $\underline{E}_{\underline{Q}}$ dominates $\underline{P}$. Now apply avoiding sure loss criterion (C)$_{42}$.

Next, we prove that (i)⇒(ii)⇒(iii)⇒(iv)⇒(i).

(i)⇒(ii). We have already established (i)⇒(v), and hence, (v) holds for $\mathscr{K} = \mathbb{B}$. So $\underline{E}$ is a coherent lower prevision on $\mathbb{B}$.

(ii)⇒(iii). Immediate from the bounds of the coherent lower prevision $\underline{E}$ (Theorem 4.13(i)$_{53}$).

(iii)⇒(iv). Immediate.

(iv)⇒(i). We see that there is some real number B such that for all n in $\mathbb{N}$, non-negative $\lambda_1, \ldots, \lambda_n$ in $\mathbb{R}$ and bounded gambles $f_1, \ldots, f_n$ in $\operatorname{dom} \underline{P}$

$$B \geq \inf\left(f_0 - \sum_{k=1}^{n} \lambda_k[f_k - \underline{P}(f_k)]\right) \geq \inf f_0 - \sup \sum_{k=1}^{n} \lambda_k[f_k - \underline{P}(f_k)], \tag{4.11}$$

where we have used expression (4.10b) for $\underline{E}$. We see that criterion $(\mathrm{E})_{43}$ for avoiding sure loss is satisfied: otherwise there would be some n in $\mathbb{N}$, non-negative $\lambda_1, \ldots, \lambda_n$ in $\mathbb{R}$ and bounded gambles $f_1, \ldots, f_n$ in $\operatorname{dom} \underline{P}$ such that $\sup \sum_{k=1}^{n} \lambda_k[f_k - \underline{P}(f_k)] < 0$, and then the second term in the right-hand side of Equation (4.11) could be made arbitrarily large by multiplying the λ_k with an arbitrarily large non-negative number. □

4.5.6 Simpler ways of calculating the natural extension

The expression for the natural extension of a lower prevision $\underline{P}$ becomes much simpler when $\underline{P}$ is defined on a linear space and is already coherent. This is stated in the following theorem, also due to Walley (1991, Theorem 3.1.4); its proof is very short, we provide it here for the sake of completeness.

Theorem 4.34 *Let $\underline{P}$ be a coherent lower prevision defined on a linear space. Then its natural extension $\underline{E}_{\underline{P}}$ is given by*

$$\underline{E}_{\underline{P}}(f) = \sup\left\{a + \underline{P}(g) : a \in \mathbb{R},\ g \in \operatorname{dom} \underline{P} \text{ and } a + g \leq f\right\}$$
$$\text{for all bounded gambles } f \text{ on } \mathcal{X}.$$

When $\operatorname{dom} \underline{P}$ *also contains (one and hence) all constant gambles $a \in \mathbb{R}$, then the expression simplifies even further to*

$$\underline{E}_{\underline{P}}(f) = \sup\left\{\underline{P}(g) : g \in \operatorname{dom} \underline{P} \text{ and } g \leq f\right\} \text{ for all bounded gambles } f \text{ on } \mathcal{X},$$

so $\underline{E}_{\underline{P}}$ coincides on all bounded gambles with the inner extension $\underline{P}_*$ of $\underline{P}$.

Proof. Look at the expression (4.10c) for natural extension and note that, by the consequences of coherence (super-additivity and non-negative homogeneity) in Theorem 4.13(v) and $(\mathrm{vi})_{53}$,

$$\underline{P}\left(\sum_{i=1}^{n} \lambda_i f_i\right) \geq \sum_{i=1}^{n} \lambda_i \underline{P}(f_i).$$

As we are looking for the supremum, we can replace $\sum_{i=1}^{n} \lambda_i \underline{P}(f_i)$ by $\underline{P}(g)$ where $g := \sum_{i=1}^{n} \lambda_i f_i$. The rest of the proof is now immediate, if we also take into account the constant additivity of the coherent $\underline{P}$ (Theorem $4.13(\mathrm{iii})_{53}$). □

There is another special case where natural extension takes an especially simple form: when $\underline{P}$ is a coherent lower probability defined on a field $\mathscr{F}$ of subsets of $\mathscr{X}$, meaning that $\operatorname{dom}\underline{P} = \{I_F : F \in \mathscr{F}\}$.

Theorem 4.35 *Let $\underline{P}$ be a coherent lower probability defined on a field of events $\mathscr{F}$. Then its natural extension $\underline{E}_{\underline{P}}$ to (indicators of) events is given by*

$$\underline{E}_{\underline{P}}(I_A) = \sup\left\{\underline{P}(I_F) : F \in \mathscr{F} \text{ and } F \subseteq A\right\} \text{ for all } A \subseteq \mathscr{X},$$

so, roughly speaking, $\underline{E}_{\underline{P}}$ coincides on (indicators of) events with the inner set function $\underline{P}_*$ of the lower probability $\underline{P}$.

The following proof is again due to Walley (1991, Theorem 3.1.5).

Proof. Fix $A \subseteq \mathscr{X}$. The expression (4.10a)$_{68}$ for natural extension leads to

$$\underline{E}_{\underline{P}}(I_A) \\ = \sup\left\{\alpha \in \mathbb{R} : I_A - \alpha \geq \sum_{k=1}^{n} \lambda_k[I_{F_k} - \underline{P}(I_{F_k})], n \in \mathbb{N}, F_k \in \mathscr{F}, \lambda_k \in \mathbb{R}_{\geq 0}\right\}.$$

As for all $F \in \mathscr{F}$ such that $F \subseteq A$, $I_A - \underline{P}(I_F) \geq I_F - \underline{P}(I_F)$, it follows from the expression above that $\underline{E}_{\underline{P}}(I_A) \geq \sup\left\{\underline{P}(I_F) : F \in \mathscr{F} \text{ and } F \subseteq A\right\}$.

We now prove the converse inequality. Consider any $\alpha \in \mathbb{R}$, $n \in \mathbb{N}$, real $\lambda_k \geq 0$ and $F_k \in \mathscr{F}$ are such that $I_A - \alpha \geq \sum_{k=1}^{n} \lambda_k[I_{F_k} - \underline{P}(I_{F_k})]$. Define the bounded gamble $h := \alpha + \sum_{k=1}^{n} \lambda_k[I_{F_k} - \underline{P}(I_{F_k})]$, then $h \leq I_A$. If we also let $G := \{h > 0\}$ then obviously $h \leq I_G \leq I_A$. Moreover, because h is constant on the atoms of the finite partition generated by the F_k, G must be a finite union of such atoms and must therefore belong to the field $\mathscr{F}$. Finally, because $\underline{P}$ is coherent, we find in particular that

$$\begin{aligned} 0 &\leq \sup\left(\sum_{k=1}^{n} \lambda_k[I_{F_k} - \underline{P}(I_{F_k})] - [I_G - \underline{P}(I_G)]\right) \\ &= \sup\left(h - I_G\right) + \underline{P}(I_G) - \alpha \leq \underline{P}(I_G) - \alpha, \end{aligned}$$

whence $\alpha \leq \underline{P}(I_G) \leq \sup\left\{\underline{P}(I_F) : F \in \mathscr{F} \text{ and } F \subseteq A\right\}$. □

4.6 Alternative characterisations for avoiding sure loss, coherence, and natural extension

In the final section of this chapter, we use the already established properties of linear previsions to shed more light on an intriguing connection, called **duality**, between coherent lower previsions and compact convex sets of linear previsions. Original proofs for most of these results were given by Williams (1975b, Theorem 2) and Walley (1991, Sections 3.4 and 3.6). We will repeat their proofs here unless explicitly stated otherwise.

Definition 4.36 (Duality maps) *We first define a **duality map** from lower previsions to sets of linear previsions. With any lower prevision* $\underline{P}$*, we can associate the set of all linear previsions on* $\mathbb{B}$ *that dominate* $\underline{P}$*,*

$$\operatorname{lins}(\underline{P}) := \left\{Q \in \mathbb{P} : (\forall f \in \operatorname{dom} \underline{P}) \left(Q(f) \geq \underline{P}(f)\right)\right\},$$

*called the **dual model** of* $\underline{P}$*.*

*Conversely, we can also define a **duality map** from sets of linear previsions to lower previsions. With any set* $\mathcal{M}$ *of linear previsions on* $\mathbb{B}$*, we can associate a lower prevision* $\operatorname{lpr}(\mathcal{M})$ *on* $\mathbb{B}$*, defined by*

$$\operatorname{lpr}(\mathcal{M})(f) := \inf\{Q(f) : Q \in \mathcal{M}\} \text{ for any } f \in \mathbb{B},$$

*called the **dual model** of* $\mathcal{M}$*.*

A lower prevision and its natural extension have the same dual model: this is an immediate consequence of Proposition 4.27_{65}.

Proposition 4.37 *Consider a lower prevision* $\underline{P}$ *that avoids sure loss, and its natural extension* $\underline{E}_{\underline{P}}$*. Then for any linear prevision* P *in* $\mathbb{P}$*,* $\underline{P} \leq P \Leftrightarrow \underline{E}_{\underline{P}} \leq P$ *and, therefore,* $\operatorname{lins}(\underline{P}) = \operatorname{lins}(\underline{E}_{\underline{P}})$*.*

The following theorem is a direct consequence of the Hahn–Banach theorem. It allows us to characterise the notions of avoiding sure loss, coherence and natural extension of lower previsions in terms of their dual models. There are many ways to prove this important result, all of which invoke one of the equivalent forms of the Hahn–Banach Theorem. We give a different proof from the one first given by Walley (1991, Sections 3.3.3 and 3.4.1). In contradistinction with his approach, ours is based on a non-topological version of the Hahn–Banach theorem, see Theorem $A.12_{370}$.

Theorem 4.38 (Lower Envelope Theorem) *Let* $\underline{P}$ *be any lower prevision. Then the following statements hold:*

(i) $\underline{P}$ *avoids sure loss if and only if* $\operatorname{lins}(\underline{P}) \neq \emptyset$*.*

(ii) $\underline{P}$ *is coherent if and only if it avoids sure loss and*

$$\underline{P}(f) = \min\left\{Q(f) : Q \in \operatorname{lins}(\underline{P})\right\} \text{ for all } f \text{ in } \operatorname{dom} \underline{P}.$$

(iii) *If* $\underline{P}$ *avoids sure loss, then its natural extension* $\underline{E}_{\underline{P}}$ *is the lower envelope of the non-empty set* $\operatorname{lins}(\underline{P})$*, that is, it satisfies*

$$\underline{E}_{\underline{P}}(f) = \min\left\{Q(f) : Q \in \operatorname{lins}(\underline{P})\right\} \text{ for all } f \text{ in } \mathbb{B}. \tag{4.12}$$

Proof. (iii). Fix any f in $\mathbb{B}$, and define the real functional Ψ on $\mathbb{B}$ by $\Psi(g) := \underline{E}_{\underline{P}}(g+f) - \underline{E}_{\underline{P}}(f)$ for all bounded gambles g. For all $\kappa \in [0,1]$ and all bounded gambles g and h, we see that

$$\begin{aligned}\Psi(\kappa g + (1-\kappa)h) &= \underline{E}_{\underline{P}}(\kappa g + (1-\kappa)h + f) - \underline{E}_{\underline{P}}(f)\\ &\geq \kappa \underline{E}_{\underline{P}}(g+f) + (1-\kappa)\underline{E}_{\underline{P}}(h+f) - \underline{E}_{\underline{P}}(f)\\ &= \kappa\Psi(g) + (1-\kappa)\Psi(h),\end{aligned}$$

where the inequality follows from the concavity (Theorem 4.13(vii)$_{53}$) of the coherent lower prevision $\underline{E}_{\underline{P}}$. So Ψ is concave, and $\Psi(0) = 0$. It, therefore, follows from the Hahn–Banach Theorem A.12$_{370}$ (with $V = \mathbb{B}$ and $V' = \{0\}$) that there is a linear functional Λ on $\mathbb{B}$ that dominates Ψ on $\mathbb{B}$. Now for any bounded gamble g,

$$\Lambda(g) \geq \Psi(g) = \underline{E}_{\underline{P}}(g+f) - \underline{E}_{\underline{P}}(f) \geq \underline{E}_{\underline{P}}(g),$$

where the last inequality follows from the super-additivity (Theorem 4.13(v)$_{53}$) of the coherent lower prevision $\underline{E}_{\underline{P}}$. Hence, Λ dominates $\underline{E}_{\underline{P}}$, is therefore a linear prevision by Lemma 4.39 and, hence, belongs to $\mathrm{lins}(\underline{E}_{\underline{P}})$. This shows that $\mathrm{lins}(\underline{E}_{\underline{P}}) = \mathrm{lins}(\underline{P})$ is non-empty. It is moreover clear that $\underline{E}_{\underline{P}}(f) \leq \inf\left\{Q(f) : Q \in \mathrm{lins}(\underline{P})\right\}$.

To prove that the converse inequality holds and that the infimum is actually a minimum, consider that, because by coherence $\underline{E}_{\underline{P}}(0) = 0$,

$$-\Lambda(f) = \Lambda(-f) \geq \Psi(-f) = \underline{E}_{\underline{P}}(-f+f) - \underline{E}_{\underline{P}}(f) = -\underline{E}_{\underline{P}}(f),$$

and therefore $\underline{E}_{\underline{P}}(f) = \Lambda(f)$.

(i)$_{\curvearrowleft}$. If $\mathrm{lins}(\underline{P}) \neq \emptyset$, then $\underline{P}$ is dominated by some linear prevision P. As P is in particular a lower prevision that avoids sure loss (Theorem 4.12(B)$_{51}$), it follows from Proposition 4.20(i)$_{59}$ that $\underline{P}$ avoids sure loss as well. The converse implication follows from (iii)$_{\curvearrowleft}$.

(ii)$_{\curvearrowleft}$. This is an immediate consequence of (iii)$_{\curvearrowleft}$, because coherence criterion (C)$_{47}$ says that $\underline{P}$ is coherent if and only if it avoids sure loss and is a restriction of its natural extension. □

Lemma 4.39 *Let Λ be a linear functional on $\mathbb{B}$ that dominates a coherent lower prevision $\underline{P}$ on $\mathbb{B}$. Then Λ is a linear prevision.*

Proof. The linear functional Λ is additive, and it follows from Theorem 4.15$_{56}$ that $\Lambda(f) \geq \underline{P}(f) \geq \inf f$ for all f in $\mathbb{B}$. Theorem 4.16$_{57}$, therefore, tells us that Λ is a linear prevision. □

Walley's proof of Theorem 4.38$_{\curvearrowleft}$ is based on a separation lemma (Walley, 1991, Lemma 3.3.2) that is of some interest in itself and will occasionally be useful in the context of this book as well. We give a proof for this lemma that, again, is based on a non-topological version of the Hahn–Banach theorem.

Lemma 4.40 (Separation) *Consider a non-empty subset $\mathscr{D}$ of $\mathbb{B}$. Then there is a linear prevision P on $\mathbb{B}$ such that $P(g) \geq 0$ for all $g \in \mathscr{D}$ if and only if*

$$\sup\left(\sum_{k=1}^{n} \lambda_k g_k\right) \geq 0 \text{ for all } n \in \mathbb{N},\ g_k \in \mathscr{D} \text{ and non-negative real } \lambda_k. \tag{4.13}$$

Proof. For sufficiency, consider the lower prevision $\underline{P}$ defined on $\mathscr{D}$ by letting $\underline{P}(g) = 0$ for all $g \in \mathscr{D}$. Then condition (4.13) tells us that $\underline{P}$ avoids sure loss. Its natural extension $\underline{E}_{\underline{P}}$ is, therefore, a coherent lower prevision on $\mathbb{B}$ that dominates $\underline{P}$ on $\mathscr{D}$, by Theorem 4.26(i)$_{65}$. Coherence implies that $\underline{E}_{\underline{P}}$ is concave (Theorem 4.13(vii)$_{53}$) and that $\underline{E}_{\underline{P}}(0) = 0$ (Theorem 4.13(ii)$_{53}$). It, therefore, follows from the Hahn–Banach Theorem A.12$_{370}$ (with $V = \mathbb{B}$ and $V' = \{0\}$) that there is a linear functional Λ on $\mathbb{B}$ that dominates $\underline{E}_{\underline{P}}$ on $\mathbb{B}$. By Lemma 4.39, this Λ is a linear prevision. Moreover, for any bounded gamble g in $\mathscr{D}$, we see that indeed

$$\Lambda(g) \geq \underline{E}_{\underline{P}}(g) \geq \underline{P}(g) = 0.$$

For necessity, suppose there is some linear prevision P on $\mathbb{B}$ such that $P(g) \geq 0$ for all $g \in \mathscr{D}$. Consider $n \in \mathbb{N}$, $g_k \in \mathscr{D}$ and non-negative real λ_k, then indeed

$$\sup\left(\sum_{k=1}^{n} \lambda_k g_k\right) \geq P\left(\sum_{k=1}^{n} \lambda_k g_k\right) \geq \sum_{k=1}^{n} \lambda_k P(g_k) \geq 0,$$

where the first and second inequalities follow from the linearity of P (see Corollary 4.14(i), (v) and (vi)$_{55}$). □

Corollary 4.41 *A lower prevision is coherent if and only if it is the lower envelope of some set of linear previsions.*

Proof. For the 'if' part, use Proposition 4.20(iii)$_{59}$; the 'only if' part follows from Theorem 4.38(ii)$_{71}$. □

Theorem 4.42 (Linear Extension Theorem) *A prevision P is linear if and only if it can be extended to a linear prevision on all bounded gambles. Moreover,* $\mathrm{lins}(P)$ *is the set of all linear previsions on all bounded gambles that coincide with P on its domain* $\mathrm{dom}\,P$. *And the natural extension $\underline{E}_P$ is the lower envelope of all linear previsions that extend P to all bounded gambles.*

Proof. We begin with the second statement. It suffices to consider any Q in $\mathrm{lins}(P)$ and to show that Q coincides with P on $\mathrm{dom}\,P$. Consider any f in $\mathrm{dom}\,P$, then on the one hand $Q(f) \geq P(f)$. But because both P and Q are self-conjugate and $-f \in \mathrm{dom}\,P$, we also have that $-Q(f) = Q(-f) \geq P(-f) = -P(f)$ and, therefore, also $Q(f) \leq P(f)$. Hence, $Q(f) = P(f)$, and therefore, Q is indeed an extension of P.

For the first statement, it clearly suffices to prove the 'only if' part. Suppose, therefore, that the prevision P is linear. This implies that P is in particular a lower prevision that avoids sure loss (Theorem 4.12$_{51}$) and, therefore, that $\mathrm{lins}(P) \neq \emptyset$ (Theorem 4.38(i)$_{71}$). We have just proved that any element of $\mathrm{lins}(P)$ is a linear prevision on all bounded gambles that extends P.

The last statement now follows from Theorem 4.38(iii)$_{71}$. □

For coherent *lower* previsions, we have a similar result: a lower prevision is coherent if and only if it can be extended to a coherent lower prevision on all bounded gambles; see coherence criterion $(B)_{47}$ and Theorem 4.15_{56}. But the argumentation for these two results is quite different: for coherent lower previsions, we have shown that there is a coherent extension that can actually be *constructed*, namely, the natural extension; for linear previsions, the argument is based on the Hahn–Banach theorem and is, therefore, essentially *non-constructive*. Indeed, when $\mathcal{X}$ is infinite, the linear extensions of a linear prevision P are usually **intangibles**, meaning that they cannot be found in a constructive manner; see Schechter (1997, Chapter 6) for more details.

4.7 Topological considerations

Let us close this chapter with a brief discussion on some of the more topological aspects of coherent lower previsions. For a brief overview of the topological terminology used in the following discussion, refer to Appendix B_{371}.

In this book, we usually provide the set $\mathbb{B}$ of all bounded gambles on a space $\mathcal{X}$ with the topology of uniform convergence, that is, we provide this set with the *supremum norm* $\|\bullet\|_{\inf}$ given by

$$\|f\|_{\inf} := \sup |f| \text{ for all bounded gambles } f \text{ on } \mathcal{X}.$$

This supremum norm is actually the $\underline{P}$-seminorm $\|\bullet\|_{\underline{P}}$ associated with the vacuous lower prevision $\underline{P} = \inf$; see Definition 4.25_{64}. This should explain and justify our perhaps at first sight surprising notation for it.

We have already seen in Theorem $4.13(\text{xiii})_{53}$ that any coherent lower prevision is continuous with respect to this topology of uniform convergence. This is interesting, because together with Proposition 1.20_{15} – stating that any bounded gamble is a uniform limit of simple gambles – this establishes that any coherent lower prevision on $\mathbb{B}$ is completely determined by its values on simple gambles, that is, bounded gambles that take only a finite number of values.

The linear space $\mathbb{B}$, provided with this topology, is a Banach space (Schechter, 1997, Section 22.8). Its **topological dual** $\mathbb{B}^*$ is the set of all continuous real linear functionals on $\mathbb{B}$. We see that $\mathbb{P} \subseteq \mathbb{B}^*$, because linear previsions are coherent lower previsions and therefore continuous.

We will provide the linear space $\mathbb{B}^*$ with the weak* topology: this is the topology of point-wise convergence or, equivalently, the weakest topology on $\mathbb{B}^*$ such that all the so-called evaluation functionals $f^*, f \in \mathbb{B}$ are continuous. These **evaluation functionals** f^* are defined by

$$f^*(\Lambda) := \Lambda(f) \text{ for all } \Lambda \text{ in } \mathbb{B}^*.$$

$\mathbb{B}^*$ is locally convex and Hausdorff under the weak* topology (Schechter, 1997, Section 28.15).

All sets of the type $\{f^* \geq x\} = \{\Lambda \in \mathbb{B}^* : \Lambda(f) \geq x\}$ are weak*-closed, as inverse images of the closed set of real numbers $[x, +\infty)$ under the continuous map f^*. They are clearly also convex. This implies that the set

$$\mathbb{P} = \bigcap_{f \in \mathbb{B}} \{\Lambda \in \mathbb{B}^* : \Lambda(f) \geq \inf f\}$$

of all linear previsions on $\mathbb{B}$ is weak*-closed and convex too (to see why the equality holds, recall that all linear previsions are continuous and use Theorem 4.16_{57}), because convexity and closedness are preserved when taking arbitrary intersections. It then follows from the Banach–Alaoglu–Bourbaki theorem (Schechter, 1997, Section 28.29) that $\mathbb{P}$ is weak*-compact.[5] Because $\mathbb{B}^*$ is Hausdorff, all weak*-closed subsets of $\mathbb{P}$ are weak*-compact and *vice versa*. This leads at once to the following result, due to Walley (1991, Section 3.6, second paragraph).

Proposition 4.43 *Let $\underline{P}$ be a lower prevision that avoids sure loss. Then the non-empty subset* $\mathrm{lins}(\underline{P})$ *of* $\mathbb{P}$ *is convex and weak*-closed (or equivalently, weak*-compact).*

Proof. Immediate if we recall that

$$\begin{aligned}\mathrm{lins}(\underline{P}) &= \left\{Q \in \mathbb{P} : (\forall f \in \mathrm{dom}\,\underline{P})\left(Q(f) \geq \underline{P}(f)\right)\right\}\\ &= \mathbb{P} \cap \bigcap_{f \in \mathrm{dom}\,\underline{P}} \{\Lambda \in \mathbb{B}^* : \Lambda(f) \geq \underline{P}(f)\}\end{aligned}$$

and that convexity and closedness are preserved by taking arbitrary intersections. □

Note, by the way, that this is consistent with the fact that we have a minimum rather than an infimum in Theorem 4.38(ii) and $(\text{iii})_{71}$: the weak*-continuous map f^* actually *reaches* its infimum (and supremum) on the weak*-compact $\mathrm{lins}(\underline{P})$ (Schechter, 1997, Section 17.26).

If we apply the Krein–Milman Theorem $\mathrm{B.8}_{374}$, then the following result, again due to Walley (1991, Section 3.6.2), is immediate.

Theorem 4.44 (Extreme Point Theorem) *Let $\underline{P}$ be a lower prevision that avoids sure loss, and consider the set of extreme points* $\mathrm{ext}(\mathrm{lins}(\underline{P}))$ *of the convex and weak*-compact set* $\mathrm{lins}(\underline{P})$*. Then the following statements hold:*

(i) $\mathrm{ext}(\mathrm{lins}(\underline{P})) \neq \emptyset$.

(ii) $\mathrm{lins}(\underline{P})$ *is smallest convex and weak*-compact subset of* $\mathbb{P}$ *that includes* $\mathrm{ext}(\mathrm{lins}(\underline{P}))$.

(iii) *If $\underline{P}$ is coherent, then for all f in* $\mathrm{dom}\,\underline{P}$ *there is some P in* $\mathrm{ext}(\mathrm{lins}(\underline{P}))$ *such that $\underline{P}(f) = P(f)$. Hence, $\underline{P}$ is the lower envelope of* $\mathrm{ext}(\mathrm{lins}(\underline{P}))$.

Proof. Because, by Proposition 4.43, $\mathrm{lins}(\underline{P})$ is a convex and weak*-compact subset of the locally convex linear topological space $\mathbb{B}^*$, we may indeed apply the Krein–Milman Theorem $\mathrm{B.8}_{374}$. (i) and (ii) are now immediate. So is (iii) if we recall that the evaluation map f^* is weak*-continuous. □

[5] See Proposition 11.2_{216} for a sketch of the proof of this theorem.

5

Special coherent lower previsions

After the fairly general discussion on coherent lower previsions, their properties and the inference method of natural extension behind them, we now turn to a number of special cases. This will allow us to begin to show that, even though they have not always been explicitly recognised as such, there are a number of instances of coherent lower previsions and natural extension in a diversity of mathematical contexts. Indeed, as we shall have occasion to witness also in the later chapters, coherent lower previsions seem to pop up naturally whenever we rid ourselves of the limiting focus on linearity.

We begin in Section 5.1 with a discussion of probability mass functions on finite spaces and show that their natural extensions to bounded gambles are linear previsions: the usual expectation operators associated with mass functions and probabilities. Here we get a first inkling of the relation between natural extension and integration, an issue that will be taken up and studied in greater depth in Chapters 8_{151} and 9_{181}.

Section 5.2_{78} explains how we can establish a one-to-one correspondence between a coherent lower prevision on a finite space $\mathscr{X}$ and a closed convex subset of the simplex of all probability mass functions on $\mathscr{X}$.

In Section 5.3_{80}, we consider bounded real-valued nets and show that the limit operator is a (finitely additive) linear prevision on the linear space of all convergent bounded nets, whose natural extension to a coherent lower prevision on all bounded nets is the limit inferior operator.

The belief that a random variable cannot assume a value outside a given set is typically modelled using a *vacuous lower prevision*. These special coherent lower previsions are introduced in Section 5.4_{81}.

Lower Previsions, First Edition. Matthias C.M. Troffaes and Gert de Cooman.

Both limits (inferior) and vacuous lower previsions only assume the values 0 and 1 on events. In Section 5.5_{82}, we take up the general study of the very interesting and fundamentally important class of coherent lower previsions whose restrictions to events are $\{0, 1\}$-valued. We show that they are in a one-to-one relationship with proper filters (Section $5.5.1_{82}$). This tells us that classical propositional logic can be embedded into the theory of coherent lower previsions (Section $5.5.2_{88}$). Their connection with proper filters also indicates that these special lower previsions can always be interpreted as limits inferior associated with some type of convergence (Section $5.5.3_{90}$). This provides connections with topology, some aspects of which include adherent probability mass (Section $5.5.5_{93}$) and the Riesz representation theorem (Section $5.5.6_{98}$).

5.1 Linear previsions on finite spaces

Consider the case that the variable X may assume only a finite number of values, so $\mathcal{X}$ is a *finite* non-empty set. Let us consider the **unit simplex** $\Sigma_{\mathcal{X}}$ in the linear space $\mathbb{R}^{\mathcal{X}}$ of all $\mathcal{X} - \mathbb{R}$-maps, defined by

$$\Sigma_{\mathcal{X}} := \left\{ p \in \mathbb{R}^{\mathcal{X}} : p \geq 0 \text{ and } \sum_{x \in \mathcal{X}} p(x) = 1 \right\}.$$

Then it is clear from Theorem 4.16_{57} that P is a linear prevision on $\mathbb{B}$ if and only if there is some p in $\Sigma_{\mathcal{X}}$ such that

$$P(f) = \sum_{x \in \mathcal{X}} p(x) f(x) \text{ for all bounded gambles } f \text{ on } \mathcal{X}, \tag{5.1}$$

and in that case, $p(x) = P(I_{\{x\}}) = P(\{x\})$ for all x in $\mathcal{X}$. In other words, we can see $P(f)$ as a *weighted average* of the values of f, with weights p or, in a more familiar probabilistic terminology, as the *expectation* of f associated with the **(probability) mass function** p. We also call p the mass function of the linear prevision P. In general, with each element q of $\Sigma_{\mathcal{X}}$, there corresponds a unique linear prevision P_q that extends it in the sense that $P_q(I_{\{x\}}) = q(x)$ for all x in $\mathcal{X}$, and we, therefore, call the elements of $\Sigma_{\mathcal{X}}$ probability mass functions on $\mathcal{X}$.

So we see that there is a one-to-one correspondence between linear previsions defined on $\mathbb{B}(\mathcal{X})$ *and mass functions on* $\Sigma_{\mathcal{X}}$. This is related to the following proposition, which essentially tells us that the natural extension of a mass function is the associated expectation operator.

Proposition 5.1 *Let $\mathcal{X}$ be a finite non-empty set. Consider any mass function p in $\Sigma_{\mathcal{X}}$, and define the lower prevision $\underline{P}_p$ with domain $\mathcal{K} := \left\{ I_{\{x\}} : x \in \mathcal{X} \right\}$ by letting $\underline{P}_p(I_{\{x\}}) := p(x)$ for all x in $\mathcal{X}$. Then* $\operatorname{lins}(\underline{P}_p) = \{P_p\}$ *and therefore* $\underline{E}_{\underline{P}_p} = P_p$.

Proof. Consider any P in $\operatorname{lins}(\underline{P}_p)$. Let $q \in \Sigma_{\mathcal{X}}$ be its mass function: $P = P_q$. As for any x in $\mathcal{X}$, $q(x) = P(I_{\{x\}}) \geq \underline{P}_p(I_{\{x\}}) = p(x)$, it follows that $p = q$, because both p and q belong to $\Sigma_{\mathcal{X}}$. Hence, $P = P_p$, whence indeed $\operatorname{lins}(\underline{P}_p) = \{P_p\}$. Theorem 4.38(i) and $(\text{iii})_{71}$ then guarantee that $\underline{E}_{\underline{P}_p}$ is the lower envelope of the set of linear previsions $\{P_p\}$ and therefore equal to P_p. □

This one-to-one correspondence between linear previsions and their restrictions to the (indicators of) certain – elementary – events on finite spaces can be carried somewhat further. Consider a linear prevision P defined on $\mathbb{B}$. If we define a real-valued function μ on $\mathscr{P}$ by letting $\mu(A) := P(I_A)$ for all subsets A of $\mathscr{X}$, then μ has the following properties:

(i) $\mu(\emptyset) = 0$ and $\mu(\mathscr{X}) = 1$;

(ii) $\mu(A \cap B) = \mu(A) + \mu(B)$ for any disjoint subsets A and B of $\mathscr{X}$.

In fact, we derive from Equation $(5.1)_{\curvearrowleft}$ that

$$\mu(A) = \sum_{x \in \mathscr{X}} I_A(x)p(x) = \sum_{x \in A} p(x) \text{ for all } A \subseteq \mathscr{X}, \tag{5.2}$$

where p is the mass function of P. Any real-valued map on $\mathscr{P}$ with these properties is called a *probability charge* on $\mathscr{P}$; see Definition 1.15_{11} and the discussion in Chapter 8_{151} for more details. Because Equation (5.2) allows us to associate a mass function $p \in \Sigma_{\mathscr{X}}$ with a probability charge μ by letting $p(x) := \mu(\{x\})$ for all x in $\mathscr{X}$, it is clear that there is a one-to-one correspondence between mass functions and probability charges and, therefore, also between linear previsions and probability charges, defined on finite spaces. In fact, we have the following immediate counterpart of Proposition $5.1_{\curvearrowleft}$, which essentially tells us that the natural extension of a probability charge is the associated expectation operator.

Proposition 5.2 *Let $\mathscr{X}$ be a finite non-empty set. Consider any probability charge μ in $\Sigma_{\mathscr{X}}$, with mass function p, and define the lower prevision $\underline{P}_\mu$ with domain $\mathscr{K} = \{I_A : A \subseteq \mathscr{X}\}$ by letting $\underline{P}_\mu(I_A) := \mu(A)$ for all A in $\mathscr{P}$. Then* $\text{lins}(\underline{P}_\mu) = \{P_p\}$ *and therefore* $\underline{E}_{\underline{P}_\mu} = P_p$.

In Chapter 8_{151}, we will discuss the extension to infinite spaces $\mathscr{X}$ of this one-to-one correspondence between probability charges and linear previsions.

5.2 Coherent lower previsions on finite spaces

The discussion in Section 5.1 can be extended in a straightforward manner to coherent *lower* previsions on $\mathbb{B}(\mathscr{X})$ when $\mathscr{X}$ is finite. If we provide the finite-dimensional space $\mathbb{R}^{\mathscr{X}}$ with the usual Euclidean topology, then the unit simplex $\Sigma_{\mathscr{X}}$ is a closed compact subset.

Consider any coherent lower prevision $\underline{P}$ on $\mathbb{B}(\mathscr{X})$. Then any linear prevision P in its set $\text{lins}(\underline{P})$ of dominating linear previsions is completely characterised by its probability mass function p, as we have seen earlier, and the set $\text{lins}(\underline{P})$ can, therefore, be identified with a closed (and therefore compact) convex subset of $\Sigma_{\mathscr{X}}$.[1] As $\text{lins}(\underline{P})$ also completely determines $\underline{P}$ (by taking its lower envelope, see Theorem 4.38_{71}),

[1] This is because the (relativisation of the) weak* topology on the set $\mathbb{P}(\mathscr{X})$ of linear previsions can be identified with the (relativisation of the) usual Euclidean topology on $\Sigma_{\mathscr{X}}$: for finite $\mathscr{X}$, point-wise convergence coincides with convergence in Euclidean norm.

we see that *there is a natural correspondence between coherent lower previsions and convex closed subsets of the unit simplex* $\Sigma_{\mathscr{X}}$. In particular, natural extensions of finite assessments correspond to convex closed polyhedra in this simplex.

Example 5.3 Let $\mathscr{X} = \{a, b, c\}$ and consider the following mass functions on $\mathscr{X}$:

$$p_1 = (\frac{1}{10}, \frac{6}{10}, \frac{3}{10}) \quad p_2 = (\frac{3}{10}, \frac{1}{10}, \frac{6}{10}) \quad p_3 = (\frac{6}{10}, \frac{3}{10}, \frac{1}{10}) \quad p_4 = (\frac{5}{10}, \frac{4}{10}, \frac{1}{10})$$
$$p_5 = (\frac{1}{10}, \frac{3}{10}, \frac{6}{10}) \quad p_6 = (\frac{6}{10}, \frac{1}{10}, \frac{3}{10}) \quad p_7 = (\frac{3}{10}, \frac{6}{10}, \frac{1}{10}) \quad p_8 = (\frac{6}{10}, \frac{2}{10}, \frac{2}{10})$$

Denote by P_k the linear prevision with mass function p_k, $k \in \{1, \dots, 8\}$. Let the coherent lower prevision $\underline{P}$ be the lower envelope of the (convex hull of the) set $\{P_1, P_2, P_3, P_5, P_6, P_7\}$, and let the coherent lower prevision $\underline{Q}$ be the lower envelope of the set $\{P_1, P_2, P_4, P_5, P_6, P_7, P_8\}$. The convex closed sets of mass functions corresponding to $\text{lins}(\underline{P})$ and $\text{lins}(\underline{Q})$ are depicted in the following left and right figures, respectively:

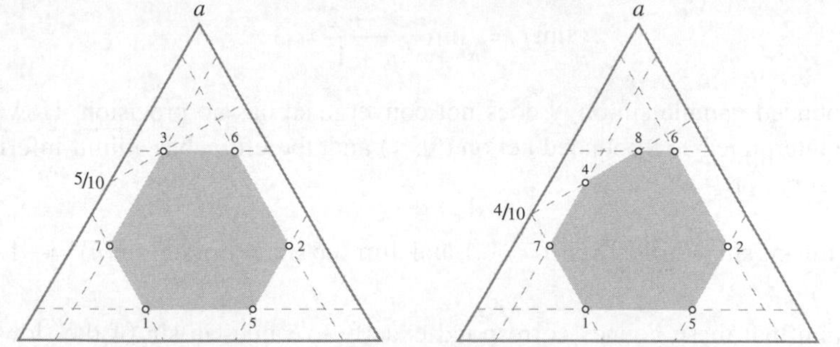

These models give the same lower and upper probabilities to all events: any singleton has lower probability $1/10$ and upper probability $6/10$, and consequently any doubleton has lower probability $4/10$ and upper probability $9/10$. But the bounded gamble $I_{\{a\}} - I_{\{c\}}$ has upper prevision $5/10$ for the model on the left and $4/10$ for the model on the right. ◆

This simple example illustrates a very important point: two lower previsions can have the same values on all events but different values in bounded gambles that are not events – meaning essentially that they assume more than two different values. This implies that, generally speaking, *a coherent lower prevision is not completely determined by the values that it assumes on events*. This should be contrasted with linear previsions: we have seen in Section 5.1 that they are completely determined by their values on singletons when $\mathscr{X}$ is finite, and we will show in Chapter 8_{151}, and in particular Corollary 8.23_{167}, that more generally, for any $\mathscr{X}$, *a linear prevision is completely determined by its behaviour on events*. To put it differently: in order to construct an interesting theory of lower previsions, the language of events is not expressive enough. This was stressed very strongly by Walley (1991, Section 2.7). It is for this reason that, following Walley, we have considered from the outset bounded gambles, rather than events, to deal with imprecise probability models. Ultimately, this also provides the motivation for the developments in Part II_{231}, and in particular Chapter 13_{235}: dealing explicitly with assessments on unbounded gambles would be decidedly less interesting, let alone exciting, if imprecise probability models too, similar to their precise counterparts, were completely determined by their behaviour on events – essentially, two-valued gambles.

5.3 Limits as linear previsions

Directed sets, nets, and Moore–Smith limits were introduced in Section 1.5_7. Interestingly, the limit operation itself can be interpreted as a linear prevision, whose natural extension corresponds to the limit inferior.

For example, consider the set of bounded gambles on $\mathbb{N}$, with the usual ordering of natural numbers. Then $\lim_{n\to+\infty} f(n)$, or briefly $\lim f$, defines a linear prevision lim on all bounded gambles f that converge; for instance, the prevision of the bounded gamble $f(n) := \frac{3n+2}{n+1}$ is

$$\lim f = \lim_{n\to+\infty} \frac{3n+2}{n+1} = 3.$$

The bounded gamble sin on $\mathbb{N}$ does not converge so has no prevision. However, it can be interpreted as a bounded net on $(\mathbb{N}, \leq)$ and, therefore, has a limit inferior and a limit superior:

$$\liminf \sin = \liminf_{n\to+\infty} \sin(n) = -1 \text{ and } \limsup \sin = \limsup_{n\to+\infty} \sin(n) = +1.$$

We claim that these bounds correspond exactly to values in sin of the (lower and upper) natural extension of the lim operator.

Generally, let $(\mathscr{A}, \leq)$ be a directed set. Bounded gambles on $\mathscr{A}$ are then simply bounded nets on $(\mathscr{A}, \leq)$. If we consider the set

$$\mathscr{I}_{\text{lim}}(\mathscr{A}) := \{f \in \mathbb{B}(\mathscr{A}) : f \text{ converges}\}$$

of all convergent bounded nets, then we infer from Theorem 4.16_{57} and the boundedness and additivity properties of the limit operator lim (Proposition 1.13(i) and $(\text{iii})_9$) that $\mathscr{I}_{\text{lim}}(\mathscr{A})$ is a linear lattice that contains all constant gambles and that lim is a *linear prevision* on this linear lattice $\mathscr{I}_{\text{lim}}(\mathscr{A})$.[2]

For any bounded gamble f, $\liminf f$ and $\limsup f$ are real numbers, so lim inf and lim sup are real functionals. As lim inf dominates the infimum operator, is super-additive and is non-negatively homogeneous, it is a coherent lower prevision by Theorem 4.15_{56}. Because $\liminf(-f) = -\limsup f$, lim sup is its conjugate upper prevision. These operators are the natural extensions of the linear prevision lim to all bounded gambles or, in other words, the lower and upper envelopes of all the linear previsions that extend lim to all bounded gambles.

[2] The reasons for this notation $\mathscr{I}_{\text{lim}}(\mathscr{A})$ will become clear in Chapter 8_{151} and, more in particular, also in Section 9.2_{182}.

Proposition 5.4 *Let $(\mathscr{A}, \leq)$ be a directed set. For all bounded gambles f on $\mathscr{A}$,*

$$\underline{E}_{\lim}(f) = \liminf f \text{ and } \overline{E}_{\lim}(f) = \limsup f.$$

Proof. It suffices to prove that $\underline{E}_{\lim} = \liminf$. As $\mathscr{I}_{\lim}(\mathscr{A})$ is a linear space that contains all constant gambles, we can use Theorem 4.34_{69} to find that

$$\underline{E}_{\lim}(f) = \sup \{\lim g \colon g \text{ converges and } g \leq f\}. \tag{5.3}$$

Consider any bounded gamble f on $\mathscr{A}$, and define the bounded gamble g by $g(\alpha) := \inf_{\beta \geq \alpha} f(\beta)$. Then g is non-decreasing and bounded above, and it, therefore, converges to $\lim g = \liminf f$, by Proposition 1.13(ii)$_9$. As moreover $g \leq f$, we gather from Equation (5.3) that $\underline{E}_{\lim}(f) \geq \liminf f$.

Suppose, *ex absurdo*, that $\underline{E}_{\lim}(f) > \liminf f$. Then we infer from Equation (5.3) that there is some convergent bounded gamble $h \leq f$ and some $\delta > 0$ such that $\lim h > \liminf f + 2\delta$. We also know that, for this $\delta > 0$, there is some α_δ such that, for all $\beta \geq \alpha_\delta$, $h(\beta) > \lim h - \delta$. Clearly, also $\liminf f \geq \inf_{\beta \geq \alpha_\delta} f(\beta)$, and therefore, there is also some $\beta_\delta \geq \alpha_\delta$ for which $\liminf f + \delta > f(\beta_\delta)$. Putting everything together, we find that

$$h(\beta_\delta) > \lim h - \delta > \liminf f + \delta > f(\beta_\delta),$$

which contradicts $h \leq f$. □

5.4 Vacuous lower previsions

Consider a non-empty event $A \subseteq \mathscr{X}$, and assume that our subject is given the information that the event A will occur or, in other words, that $X \in A$. The subject has no other relevant information about X. How can our subject's beliefs be modelled? Given this information, he is certain that this event will occur and is, therefore, willing to bet on the occurrence of this event at all odds, meaning that he will accept the bounded gamble $I_A - 1 + \varepsilon$ for all $\varepsilon > 0$. This leads to the following very simple assessment in terms of a lower probability $\underline{Q}_A$ with domain $\{I_A\}$:

$$\underline{Q}_A(I_A) = 1.$$

It is easy to see that this simple lower probability is always coherent. Let us find out what is its natural extension to all bounded gambles. We denote this natural extension by $\underline{P}_A$. We proceed by applying one of the equivalent formulae in Theorem 4.33_{68} and are thus led to $\underline{P}_A(f) = \sup_{\lambda \geq 0} \inf_{x \in \mathscr{X}} [f(x) - \lambda(I_A(x) - 1)]$. After a few elementary manipulations, we find that

$$\underline{P}_A(f) = \inf_{x \in A} f(x) \text{ for all bounded gambles } f \text{ on } \mathscr{X}. \tag{5.4}$$

We call this coherent lower prevision the **vacuous lower prevision relative to** A. It is a model for the subject's beliefs if he knows that A will occur, that is, $X \in A$, and nothing more. This is also borne out by

$$\operatorname{lins}(\underline{P}_A) = \left\{ P \in \mathbb{P}(\mathscr{X}) \colon P(I_A) = 1 \right\}.$$

If $\mathcal{X}$ is finite, then $\text{ext}(\text{lins}(\underline{P}_A)) = \{P_x : x \in A\}$, where $P_x := \underline{P}_{\{x\}}$ is the **degenerate** linear prevision representing the subject's beliefs that $X = x$. It is given by $P_x(f) := f(x)$ for all bounded gambles f on $\mathcal{X}$.

For general and possibly infinite $\mathcal{X}$, we infer from the discussion in Section 5.5, and in particular from Theorem 5.11_{86}, that $\text{ext}(\text{lins}(\underline{P}_A)) = \{P_{\mathcal{U}} : \mathcal{U} \text{ ultrafilter and } A \in \mathcal{U}\}$. The linear previsions P_x, $x \in A$ correspond to the *fixed* ultrafilters $\mathcal{U}_x := \{B \subseteq \mathcal{X} : x \in B\}$ in this set.

5.5 {0, 1}-valued lower probabilities

We now take the example discussed in Section 5.4 one step further. Consider a set of events $\mathscr{C} \subseteq \mathscr{P}$, and assume that we have a subject who is **practically certain**[3] that all events in $\mathscr{C}$ will occur, meaning that he is willing to bet on the occurrence of each of these events at all odds. For all other events, he expresses complete ignorance or, in other words, a total lack of commitment: he is only willing to bet on these other events at zero odds. We can model this assessment in terms of a lower probability $\underline{Q}_{\mathscr{C}}$ with domain $\{I_A : A \subseteq \mathcal{X}\}$, where

$$\underline{Q}_{\mathscr{C}}(I_A) := \begin{cases} 1 & \text{if } A \in \mathscr{C} \\ 0 & \text{if } A \notin \mathscr{C}. \end{cases}$$

Similarly, when we consider an assessment $\mathscr{C}$ of this type as defining a *probability*, rather than a lower probability, we have an additional self-conjugacy assessment that allows us to define the *probability* $Q_{\mathscr{C}}$ with negation-invariant domain $\bigcup_{A \subseteq \mathcal{X}} \{I_A, -I_A\}$:[4]

$$Q_{\mathscr{C}}(I_A) := -Q_{\mathscr{C}}(-I_A) := 1 \text{ if } A \in \mathscr{C} \text{ and } Q_{\mathscr{C}}(I_A) := Q_{\mathscr{C}}(-I_A) := 0 \text{ otherwise.}$$

In this case, the subject is disposed not only to *bet on* any A in $\mathscr{C}$ but also to *bet against* any B not in $\mathscr{C}$, at all odds.

5.5.1 Coherence and natural extension

Under what conditions on this $\mathscr{C}$ does the corresponding (lower) probability satisfy the rationality criteria of avoiding sure loss and coherence? To investigate this further, we make a small digression and take a closer look at the notion of a filter, as defined in Section 1.4_5. We have seen in Example 1.11_8 that if we partially order a proper filter $\mathscr{F}$ with the 'includes' relation $\supseteq$, then $\mathscr{F}$ is a directed set and we can use proper filters to define nets.

In particular, for any bounded gamble f, we can consider the real net $\underline{P}_A(f)$, $A \in \mathscr{F}$, where $\underline{P}_A$ is the vacuous lower prevision relative to A, discussed in Section $5.4_{\curvearrowleft}$. This net is bounded above by $\sup f$ and non-decreasing: if $A \supseteq B$, then $\underline{P}_A(f) \leq \underline{P}_B(f)$.

[3] Similarly, an event is **practically impossible** to a subject if he is willing to bet *against* its occurrence at all odds.

[4] See our definition for a probability near the end of Section $4.1.2_{40}$.

It, therefore, converges to a real number (Proposition 1.13(ii)$_9$) and we use the following notation for its limit:[5]

$$\underline{P}_{\mathscr{F}}(f) = \lim_{A\in\mathscr{F}} \underline{P}_A(f) = \sup_{A\in\mathscr{F}} \inf_{x\in A} f(x). \tag{5.5}$$

This leads to the definition of a lower prevision $\underline{P}_{\mathscr{F}}$ on $\mathbb{B}$, which is guaranteed to be coherent as the point-wise limit of the net of coherent lower previsions $\underline{P}_A$, $A \in \mathscr{F}$; see the discussion in Section 5.4$_{81}$ and Proposition 4.21(ii)$_{60}$.

How can we interpret this coherent lower prevision $\underline{P}_{\mathscr{F}}$? What kind of beliefs or information is it a model for? With each A in $\mathscr{F}$, we can associate the vacuous lower prevision $\underline{P}_A$ relative to A, which represents a subject's beliefs that X belongs to A. This vacuous lower prevision $\underline{P}_A$ becomes increasingly more precise as A goes 'down the directed set $\mathscr{F}$'; therefore, if we take the limit, we end up with a coherent lower prevision $\underline{P}_{\mathscr{F}}$ that represents the subject's belief that '$X \in A$ for all $A \in \mathscr{F}$'. We now intend to make this heuristic argument more formal by relating $\underline{P}_{\mathscr{F}}$ to the type of assessment $\underline{Q}_{\mathscr{C}}$ introduced earlier.

Most of the material in this section goes back to Walley (1991, Sections 2.9.8 and 3.6.3), but our treatment is somewhat more systematic and proofs may differ. In the following sections, we look at special instances of lower previsions associated with proper filters.

We begin with the following interesting result.

Proposition 5.5 *Let $\mathscr{F}$ be a proper filter. Then $\underline{P}_{\mathscr{F}}$ is a coherent lower prevision on $\mathbb{B}$. Moreover, $\underline{P}_{\mathscr{F}}$ is a linear prevision if and only if $\mathscr{F}$ is an ultrafilter. In that case, we denote this linear prevision on $\mathbb{B}$ by $P_{\mathscr{F}}$.*

Proof. The first statement follows immediately from the preceding discussion, so we proceed at once to the second statement. As we know that $\underline{P}_{\mathscr{F}}$ is a coherent lower prevision, it will be a linear prevision if and only if it is self-conjugate or, in other words, if and only if

$$\sup_{A\in\mathscr{F}} \inf_{x\in A} f(x) \geq \inf_{B\in\mathscr{F}} \sup_{y\in B} f(y) \text{ for all bounded gambles } f \text{ on } \mathscr{X}. \tag{5.6}$$

We show that this condition holds if and only if $\mathscr{F}$ is an ultrafilter.

'only if'. Consider any subset C of $\mathscr{X}$ and look at the inequality (5.6) for $f = I_C$. Then the left-hand side is equal to 1 if $A \subseteq C$ for at least one $A \in \mathscr{F}$ – this happens precisely when $C \in \mathscr{F}$ because $\mathscr{F}$ is a proper filter – and 0 otherwise:

$$\sup_{A\in\mathscr{F}} \inf_{x\in A} I_C(x) = \begin{cases} 1 & \text{if } C \in \mathscr{F} \\ 0 & \text{if } C \notin \mathscr{F}. \end{cases}$$

Similarly, the right-hand side is equal to 1 if $C \cap B \neq \emptyset$ for all $B \in \mathscr{F}$ – this happens precisely when $C^c \notin \mathscr{F}$ because $\mathscr{F}$ is a proper filter – and 0 otherwise:

$$\inf_{B\in\mathscr{F}} \sup_{y\in B} I_C(y) = \begin{cases} 1 & \text{if } C^c \notin \mathscr{F} \\ 0 & \text{if } C^c \in \mathscr{F}. \end{cases}$$

[5] Notice the formal similarity with the definition of the limit inferior of a real net in Equation (1.3)$_9$. We discuss this similarity in Section 5.5.3$_{90}$.

So we should have for all events C that $C^c \notin \mathscr{F} \Rightarrow C \in \mathscr{F}$ or, in other words, that either $C^c \in \mathscr{F}$ or $C \in \mathscr{F}$. This tells us that $\mathscr{F}$ indeed is an ultrafilter; see Definition 1.5_6.

'if'. Fix any bounded gamble f on $\mathscr{X}$, and let $\alpha < \overline{P}_{\mathscr{F}}(f)$. If we use Lemma 5.6, we see that then $\{f \leq \alpha\} \notin \mathscr{F}$. As $\mathscr{F}$ is an ultrafilter, this implies that $\{f > \alpha\} = \{f \leq \alpha\}^c \in \mathscr{F}$ and, because $\{f > \alpha\} \subseteq \{f \geq \alpha\}$, that $\{f \geq \alpha\} \in \mathscr{F}$. So $\alpha \leq \underline{P}_{\mathscr{F}}(f)$, again using Lemma 5.6. As this holds for all real α, we find that indeed $\underline{P}_{\mathscr{F}}(f) \geq \overline{P}_{\mathscr{F}}(f)$. □

Lemma 5.6 *Let $\mathscr{F}$ be a proper filter. Then for all bounded gambles f on $\mathscr{X}$,*

$$\begin{aligned}
\underline{P}_{\mathscr{F}}(f) &= \sup\{\alpha \in \mathbb{R} : \{f \geq \alpha\} \in \mathscr{F}\} = \inf\{\alpha \in \mathbb{R} : \{f \geq \alpha\} \notin \mathscr{F}\} \\
&= \sup\{\alpha \in \mathbb{R} : \{f > \alpha\} \in \mathscr{F}\} = \inf\{\alpha \in \mathbb{R} : \{f > \alpha\} \notin \mathscr{F}\}, \\
\overline{P}_{\mathscr{F}}(f) &= \inf\{\alpha \in \mathbb{R} : \{f \leq \alpha\} \in \mathscr{F}\} = \sup\{\alpha \in \mathbb{R} : \{f \leq \alpha\} \notin \mathscr{F}\} \\
&= \inf\{\alpha \in \mathbb{R} : \{f < \alpha\} \in \mathscr{F}\} = \sup\{\alpha \in \mathbb{R} : \{f < \alpha\} \notin \mathscr{F}\}.
\end{aligned}$$

Proof. If we can prove the equalities for $\underline{P}_{\mathscr{F}}$, the proof of the others will then be immediate, by conjugacy.

First, consider any $\alpha \in \mathbb{R}$ such that $\{f \geq \alpha\} \in \mathscr{F}$. Then there is some $A := \{f \geq \alpha\}$ in $\mathscr{F}$ for which $\inf_{x \in A} f(x) \geq \alpha$ and, therefore, $\sup_{A \in \mathscr{F}} \inf_{x \in A} f(x) \geq \alpha$. Hence, $\sup_{A \in \mathscr{F}} \inf_{x \in A} f(x) \geq \sup\{\alpha \in \mathbb{R} : \{f \geq \alpha\} \in \mathscr{F}\}$.

Conversely, consider any $A \in \mathscr{F}$. Then for all $\alpha < \inf_{x \in A} f(x)$, we have that $A \subseteq \{f \geq \alpha\}$ and therefore $\{f \geq \alpha\} \in \mathscr{F}$, because $\mathscr{F}$ is a filter. This implies that $\sup\{\alpha \in \mathbb{R} : \{f \geq \alpha\} \in \mathscr{F}\} \geq \inf_{x \in A} f(x)$. Because this holds for every $A \in \mathscr{F}$, it follows that $\sup\{\alpha \in \mathbb{R} : \{f \geq \alpha\} \in \mathscr{F}\} \geq \sup_{A \in \mathscr{F}} \inf_{x \in A} f(x)$.

For the second equality, observe that $\{\alpha \in \mathbb{R} : \{f \geq \alpha\} \in \mathscr{C}\}$ is a decreasing subset of $\mathbb{R}$ and, therefore, by the first equality, equal to either $(-\infty, \underline{P}_{\mathscr{P}}(f))$ or $(-\infty, \underline{P}_{\mathscr{P}}(f)]$. The second equality is now immediate.

Finally, the proofs for the equalities involving the level sets $\{f > \alpha\}$ are completely analogous. □

We are now ready to investigate the coherence of the lower probability $\underline{Q}_{\mathscr{C}}$ and the probability $Q_{\mathscr{C}}$.

Proposition 5.7 *Let $\mathscr{C}$ be a non-empty subset of $\mathscr{P}$, and consider the associated lower probability $\underline{Q}_{\mathscr{C}}$ and probability $Q_{\mathscr{C}}$.*

(i) *$\underline{Q}_{\mathscr{C}}$ avoids sure loss if and only if $\mathscr{C}$ satisfies the **finite intersection property**, meaning that $\bigcap_{k=1}^{n} A_k \neq \emptyset$ for all $n \in \mathbb{N}_{>0}$ and all $A_1, \ldots, A_n$ in $\mathscr{C}$.*

(ii) *$\underline{Q}_{\mathscr{C}}$ is a coherent lower probability if and only if $\mathscr{C}$ is a proper filter.*

(iii) *$Q_{\mathscr{C}}$ is a coherent probability if and only if $\mathscr{C}$ is an ultrafilter.*

Proof. (i). We use criterion $(\mathrm{D})_{42}$ for avoiding sure loss, which, in this particular case, turns into the requirement that, for all $n \in \mathbb{N}_{>0}$ and all $A_1, \ldots, A_n$ in $\mathscr{C}$, $\sup \sum_{k=1}^{n} I_{A_k} \geq n$, which is equivalent to $\bigcap_{k=1}^{n} A_k \neq \emptyset$.

(ii). 'only if'. Assume that $\underline{Q}_{\mathscr{C}}$ is coherent, and use the properties of coherent lower previsions established in Theorem 4.13$_{53}$. If $A \in \mathscr{C}$ and $A \subseteq B$, then $I_A \leq I_B$, and therefore, coherence property (iv)$_{53}$ (monotonicity) guarantees that $1 = \underline{Q}_{\mathscr{C}}(I_A) \leq \underline{Q}_{\mathscr{C}}(I_B)$ and also $\underline{Q}_{\mathscr{C}}(I_B) = 1$, whence $B \in \mathscr{C}$. Similarly, if $A \in \mathscr{C}$ and $B \in \mathscr{C}$, then it follows from what we have just established that $\underline{Q}_{\mathscr{C}}(I_{A\cup B}) = 1$ and, therefore, from coherence property (i)$_{53}$ (bounds) that $\overline{Q}_{\mathscr{C}}(I_{A\cup B}) = 1$. If we combine this with coherence property (xii)$_{54}$, we see that $\underline{Q}_{\mathscr{C}}(I_A) + \underline{Q}_{\mathscr{C}}(I_B) \leq \underline{Q}_{\mathscr{C}}(I_{A\cap B}) + \overline{Q}_{\mathscr{C}}(I_{A\cup B})$, whence $\underline{Q}_{\mathscr{C}}(I_{A\cap B}) \geq 1$ and therefore indeed $A \cap B \in \mathscr{C}$. As coherence also guarantees that $\underline{Q}_{\mathscr{C}}(I_{\emptyset}) = 0$, we see that $\emptyset \notin \mathscr{C}$, so $\mathscr{C}$ is indeed a proper filter.

'if'. If $\mathscr{C}$ is a proper filter, then $\underline{Q}_{\mathscr{C}}$ is the restriction of the coherent lower prevision $\underline{P}_{\mathscr{C}}$ (Proposition 5.5$_{83}$) to (indicators of) events and, therefore, a coherent lower probability.

(iii). 'only if'. Assume that $Q_{\mathscr{C}}$ is a coherent probability. Then the restriction $\underline{Q}_{\mathscr{C}}$ of $Q_{\mathscr{C}}$ to $\{I_A : A \subseteq \mathscr{X}\}$ is a coherent lower prevision, so we infer from (ii) that $\mathscr{C}$ is a proper filter. To prove that it is an ultrafilter, consider any subset C of $\mathscr{X}$ and assume that $C \notin \mathscr{C}$ or, in other words, $Q_{\mathscr{C}}(I_C) = 0$. Then we have to prove that $C^c \in \mathscr{C}$. It follows from the linearity criterion (C)$_{52}$ in Theorem 4.12$_{51}$ that for all real λ,

$$\lambda Q_{\mathscr{C}}(I_{C^c}) = \lambda Q_{\mathscr{C}}(I_C) + \lambda Q_{\mathscr{C}}(I_{C^c}) \leq \sup(\lambda I_C + \lambda I_{C^c}) = \lambda.$$

For $\lambda = -1$, this implies that $Q_{\mathscr{C}}(I_{C^c}) \geq 1$, and so by coherence, $Q_{\mathscr{C}}(I_{C^c}) = 1$. Whence, indeed, $C^c \in \mathscr{C}$.

'if'. If $\mathscr{C}$ is an ultrafilter, then $Q_{\mathscr{C}}$ is the restriction of the linear prevision $P_{\mathscr{C}}$ (Proposition 5.5$_{83}$) to the negation-invariant set $\bigcup_{A\subseteq\mathscr{X}}\{I_A, -I_A\}$ and, therefore, a coherent probability. □

Proposition 5.8 *If $\mathscr{F}$ is a proper filter on $\mathscr{X}$, then $\underline{P}_{\mathscr{F}}$ is the only coherent lower prevision that extends the coherent lower probability $\underline{Q}_{\mathscr{F}}$ to all bounded gambles, and therefore, $\underline{P}_{\mathscr{F}}$ is the natural extension of $\underline{Q}_{\mathscr{F}}$. Similarly, if $\mathscr{U}$ is an ultrafilter, then $P_{\mathscr{U}}$ is the only linear prevision that extends the coherent probability $Q_{\mathscr{U}}$ to all bounded gambles, and therefore, $P_{\mathscr{U}}$ is the natural extension of $Q_{\mathscr{U}}$.*

Proof. Consider any coherent lower prevision $\underline{P}$ that extends $\underline{Q}_{\mathscr{F}}$. Fix any bounded gamble f on $\mathscr{X}$. We show that the coherence of $\underline{P}$ implies that then necessarily $\underline{P}(f) = \underline{P}_{\mathscr{F}}(f)$. As we already know that $\underline{P}_{\mathscr{F}}$ is a coherent lower prevision, this suffices to prove the first statement.

First, consider any real α such that $\{f \geq \alpha\} \in \mathscr{F}$ and, therefore, $\underline{P}(\{f \geq \alpha\}) = 1$. Then we infer in particular from coherence criterion (E)$_{47}$ that for all real $\lambda \geq 0$,

$$\begin{aligned} 0 &\leq \sup\left(\lambda[I_{\{f\geq\alpha\}} - 1] - [f - \underline{P}(f)]\right) \\ &= \underline{P}(f) - \min\left\{\inf_{x:\,f(x)\geq\alpha} f(x),\ \inf_{x:\,f(x)<\alpha} f(x) + \lambda\right\}, \end{aligned}$$

and for large enough λ, this implies that $\underline{P}(f) \geq \inf_{x:\,f(x)\geq\alpha} f(x) \geq \alpha$. Hence, $\{f \geq \alpha\} \in \mathscr{F}$ implies that $\underline{P}(f) \geq \alpha$, and if we invoke Lemma 5.6, we conclude that $\underline{P}(f) \geq \underline{P}_{\mathscr{F}}(f)$.

Conversely, consider any β such that $\{f \geq \beta\} \notin \mathscr{F}$ and, therefore, $\underline{P}(\{f \geq \beta\}) = 0$. We infer in particular from coherence criterion $(E)_{47}$ that for all real $\lambda \geq 0$,

$$\begin{aligned} 0 &\leq \sup\left(f - \underline{P}(f) - \lambda I_{\{f\geq\beta\}}\right) \\ &= \max\left\{\sup_{x\colon f(x)\geq\beta} f(x) - \lambda,\ \sup_{x\colon f(x)<\beta} f(x)\right\} - \underline{P}(f), \end{aligned}$$

and for large enough λ, this implies that $\underline{P}(f) \leq \sup_{x\colon f(x)<\beta} f(x) \leq \beta$. Again, Lemma 5.6_{84} then guarantees that indeed also $\underline{P}(f) \leq \underline{P}_{\mathscr{F}}(f)$.

We continue with the second statement. The probability $Q_{\mathscr{U}}$ dominates the lower probability $\underline{Q}_{\mathscr{U}}$. Therefore, by Proposition 4.27_{65}, the natural extension of $Q_{\mathscr{U}}$ dominates the natural extension $P_{\mathscr{U}} = \underline{P}_{\mathscr{U}}$ (see the first statement and Proposition 5.5_{83}) of $\underline{Q}_{\mathscr{U}}$ and, therefore, coincides with it. Now apply Proposition 4.37_{71} to find that $\mathrm{lins}(Q_{\mathscr{U}}) = \mathrm{lins}(P_{\mathscr{U}}) = \{P_{\mathscr{U}}\}$. □

Corollary 5.9 *Consider a coherent lower prevision $\underline{P}$ on $\mathbb{B}$ that assumes only the values zero and one on (indicators of) events. Let $\mathscr{F} = \{A \subseteq \mathscr{X} : \underline{P}(A) = 1\}$, then $\mathscr{F}$ is a proper filter and $\underline{P} = \underline{P}_{\mathscr{F}}$. Similarly, consider a linear prevision P on $\mathbb{B}$ that assumes only the values zero and one on (indicators of) events. Let $\mathscr{U} = \{A \subseteq \mathscr{X} : P(A) = 1\}$, then $\mathscr{U}$ is an ultrafilter and $P = P_{\mathscr{U}}$.*

Proof. We only prove the first part; the second part is proved analogously. The restriction of the coherent lower prevision $\underline{P}$ to indicators of events is the lower probability $\underline{Q}_{\mathscr{F}}$, which is therefore coherent as well. Hence, $\mathscr{F}$ is a proper filter, by Proposition $5.7(\mathrm{ii})_{84}$. By Proposition $5.8_{\curvearrowleft}$, $\underline{P}_{\mathscr{F}}$ is the only coherent lower prevision that extends $\underline{Q}_{\mathscr{F}}$ to all bounded gambles, and it, therefore, coincides with the coherent extension $\underline{P}$ of $\underline{Q}_{\mathscr{F}}$. □

Corollary 5.10 *Consider a coherent lower prevision $\underline{P}$ on $\mathbb{B}$. Then there is some proper filter $\mathscr{F}$ such that $\underline{P} = \underline{P}_{\mathscr{F}}$ if and only if*

$$\underline{P}(A_1 \cap A_2) = \underline{P}(A_1)\underline{P}(A_2) \text{ for all } A_1, A_2 \in \mathscr{P}. \tag{5.7}$$

Proof. 'if'. Consider any $A \subseteq \mathscr{X}$. Then it follows from Equation (5.7) that $\underline{P}(A) = \underline{P}(A)^2$. This implies that the coherent lower prevision $\underline{P}$ assumes only the values 0 and 1 on events. Now use Corollary 5.9.

'only if'. Consider any proper filter $\mathscr{F}$ and assume that $\underline{P} = \underline{P}_{\mathscr{F}}$. Then $\underline{P}$ assumes only the values 0 and 1 on events, and it is clear that Equation (5.7) is satisfied because $\mathscr{F}$ is increasing and closed under finite intersections. □

Theorem 5.11 (Ultrafilter Theorem) *Let $\mathscr{F}$ be a proper filter. Then*

$$\mathrm{lins}(\underline{P}_{\mathscr{F}}) = \{P \in \mathbb{P} : (\forall A \in \mathscr{F})(P(A) = 1)\} \tag{5.8}$$

$$\mathrm{ext}(\mathrm{lins}(\underline{P}_{\mathscr{F}})) = \left\{P_{\mathscr{U}} : \mathscr{U} \in \mathbb{U} \text{ and } \mathscr{F} \subseteq \mathscr{U}\right\}. \tag{5.9}$$

Hence, every proper filter is included in some ultrafilter, and every proper filter is the intersection of the ultrafilters it is included in.

Proof. We begin with the proof of Equation (5.8). Proposition 5.8_{85} guarantees that $\underline{P}_{\mathscr{F}}$ is the natural extension of the coherent lower probability $\underline{Q}_{\mathscr{F}}$ and therefore, by Proposition 4.37_{71}, that $\mathrm{lins}(\underline{P}_{\mathscr{F}}) = \mathrm{lins}(\underline{Q}_{\mathscr{F}})$. Consider any P in $\mathbb{P}$. Because a linear prevision always assumes a value between 0 and 1 on events, we see that, therefore, indeed,

$$P \in \mathrm{lins}(\underline{P}_{\mathscr{F}}) \Leftrightarrow (\forall A \subseteq \mathcal{X})(P(A) \geq \underline{Q}_{\mathscr{F}}(A)) \Leftrightarrow (\forall A \in \mathscr{F})(P(A) = 1).$$

We now proceed with the proof of Equation (5.9). We first show that any element of $\mathrm{lins}(\underline{P}_{\mathscr{F}})$ is an extreme point (of this set) if and only if it only assumes the values 0 and 1 on (indicators of) events.

'if'. Assume that $P \in \mathrm{lins}(\underline{P}_{\mathscr{F}})$ only assumes the values zero and one on events. Consider any P_1 and P_2 in $\mathrm{lins}(\underline{P}_{\mathscr{F}})$ and any $\lambda \in (0, 1)$, and assume that $P = \lambda P_1 + (1 - \lambda)P_2$, then $P(A) = 1$ implies that also $P_1(A) = P_2(A) = 1$, and, similarly, $P(B) = 0$ implies that also $P_1(B) = P_2(B) = 0$. This means that the linear previsions P_1 and P_2 coincide on all events, so they also coincide on all bounded gambles, by Corollary 5.9. This shows that P is indeed an extreme point.

'only if'. Assume that P is an extreme point of $\mathrm{lins}(\underline{P}_{\mathscr{F}})$, and assume, *ex absurdo*, that there is some $A \subseteq \mathcal{X}$ such that $0 < P(A) < 1$. Consider the linear functionals Q_1 and Q_2 defined by

$$Q_1(f) := [1 - P(A)]P(f) + P(I_A f) \text{ and } Q_2(f) := [1 + P(A)]P(f) - P(I_A f)$$

for all bounded gambles f on $\mathcal{X}$. Then

$$\begin{aligned} Q_1(f) &= P([1 - P(A) + I_A]f) \\ &\geq P([1 - P(A) + I_A]\inf f) = [1 - P(A) + P(A)]\inf f = \inf f, \end{aligned}$$

and similarly,

$$\begin{aligned} Q_2(f) &= P([1 + P(A) - I_A]f) \\ &\geq P([1 + P(A) - I_A]\inf f) = [1 + P(A) - P(A)]\inf f = \inf f, \end{aligned}$$

where we have used two properties of linear previsions, namely, monotonicity (Corollary $4.14(\mathrm{iv})_{55}$) and homogeneity (Corollary $4.14(\mathrm{vi})_{55}$). By Theorem 4.16_{57}, it follows that Q_1 and Q_2 are linear previsions.

For any B in $\mathscr{F}$, $P(B) = 1$ and, therefore, $P(I_A I_B) = P(A)$; hence, $Q_1(B) = Q_2(B) = 1$. So, by Equation (5.8), both Q_1 and Q_2 belong to $\mathrm{lins}(\underline{P}_{\mathscr{F}})$.

Finally, observe that $Q_1 \neq Q_2$, because $Q_1(A) = [2 - P(A)]P(A) > P(A)^2 = Q_2(A)$. But $P = 1/2Q_1 + 1/2Q_2$, so P cannot be an extreme point of $\mathrm{lins}(\underline{P}_{\mathscr{F}})$. We have arrived at a contradiction.

So all extreme points of $\mathrm{lins}(\underline{P}_{\mathscr{F}})$ assume only the values 0 and 1 on events and are, therefore, of the form $P_{\mathscr{U}}$, for some ultrafilter $\mathscr{U}$, by Corollary 5.9. If we now invoke Corollary 5.9 and the Extreme Point Theorem 4.44_{75}, we indeed get Equation (5.9), if we also observe that $P_{\mathscr{U}} \geq \underline{P}_{\mathscr{F}}$ is equivalent to $\mathscr{F} \subseteq \mathscr{U}$.

The last two statements now follow easily. As $\underline{P}_{\mathscr{F}}$ is a coherent lower prevision, we infer from the Extreme Point Theorem 4.44_{75} that (i) $\mathrm{ext}(\mathrm{lins}(\underline{P}_{\mathscr{F}}))$ is non-empty, so there is some ultrafilter that includes $\mathscr{F}$, and (ii) $\underline{P}_{\mathscr{F}}$ is the lower envelope of $\mathrm{ext}(\mathrm{lins}(\underline{P}_{\mathscr{F}}))$, so we see that for all events A

$$A \in \mathscr{F} \Leftrightarrow \underline{P}_{\mathscr{F}}(A) = 1 \Leftrightarrow (\forall \mathscr{U} \supseteq \mathscr{F})(\underline{P}_{\mathscr{U}}(A) = 1) \Leftrightarrow (\forall \mathscr{U} \supseteq \mathscr{F})(A \in \mathscr{U})$$

and, therefore, $\mathscr{F} = \bigcap \{\mathscr{U} : \mathscr{U} \in \mathbb{U} \text{ and } \mathscr{F} \subseteq \mathscr{U}\}$. □

That every proper filter is included in some ultrafilter is also known as the *ultrafilter theorem*, or *ultrafilter principle*, which appears in many equivalent forms. Amongst these is the Banach–Alaoglu–Bourbaki theorem (Schechter, 1997, Section 28.29), which we use in Sections 4.7_{74} and 8.8_{177}. In the context of Proposition 11.2_{216}, we sketch a proof for a closely related result that is based on Tychonov's theorem for Hausdorff spaces, which is another equivalent form of the ultrafilter principle (Schechter, 1997, Section 17.22). The above-mentioned proof for the ultrafilter theorem relies on an argument by Walley (1991, Theorems 3.6.4 and 3.6.5) and hinges on the use of the Extreme Point Theorem 4.44_{75}, which in turn relies on the Krein–Milman Theorem $\mathrm{B}.8_{374}$. The ultrafilter principle is known to be weaker than the Axiom of Choice and stronger than the Hahn–Banach Theorem $\mathrm{A}.12_{370}$, see for instance Luxemburg (1962, Sec. 3), Pincus (1972, 1974) and Walley (1991, Note 12 to Chapter 3). For a detailed discussion of the Hahn–Banach theorem in its many equivalent forms, refer to Schechter (1997).

As $\mathbb{P} = \mathrm{lins}(\inf)$ and the vacuous lower prevision $\underline{P}_{\mathscr{X}} = \inf$ is also the lower prevision $\underline{P}_{\mathscr{F}}$ associated with the proper filter $\mathscr{F} = \{\mathscr{X}\}$, we see that

$$\mathrm{ext}(\mathbb{P}) = \left\{P_{\mathscr{U}} : \mathscr{U} \in \mathbb{U}\right\}, \tag{5.10}$$

meaning that the extreme linear previsions are those linear previsions that assume only the values 0 and 1 on events.

5.5.2 The link with classical propositional logic

There is a very close link between classical propositional logic and $\{0, 1\}$-valued lower probabilities: using the notions of filter and ultrafilter, *we can formally embed propositional logic into the theory of coherent lower previsions*.

To sketch how this can be done, we make things as easy as possible by limiting ourselves to 'propositions about a variable X assuming values in a set $\mathscr{X}$'. Such propositions are in a *one-to-one correspondence* with the subsets of $\mathscr{X}$: a proposition about X is simply a statement of the form:

'$X \in A$' for some subset A of $\mathscr{X}$.

It is possible, and relatively easy, to make the embedding work for any propositional system, or object language L of well-formed formulas with the usual axiomatisation, by applying the Stone representation theorem on the Lindenbaum algebra associated with L, see Davey and Priestley (1990, Chapters 7 and 10) and Schechter (1997, Chapters 13 and 14). In this more general setup, $\mathscr{X}$ will be the Stone space (or 'set of possible worlds') associated with L. This goes far beyond the scope of this section, however, and we refer to De Cooman (2005b) for a more detailed discussion.

In any case, in propositional logic, a belief model is simply a collection of propositions about X that are accepted, or held to be true. In our simplified discussion here, it is, therefore, a collection $\mathscr{C}$ of subsets of $\mathscr{X}$: those subsets that X is held to belong to or, in other words, those events that are held to occur. What are *possible* properties of such a collection of events $\mathscr{C}$?

(i) A set of propositions is **deductively closed** if it is closed under finite conjunctions and modus ponens or, equivalently, if the corresponding set $\mathscr{C}$ of events is a *filter*: closed under finite intersections and increasing.

(ii) Given a set of propositions, its **deductive closure** is the smallest deductively closed set of propositions that includes it, or equivalently in terms of sets of events:

$$\mathrm{Cl}_{\mathbb{F}}(\mathscr{C}) := \bigcap \{\mathscr{F} \in \mathbb{F} : \mathscr{C} \subseteq \mathscr{F}\}$$

is the smallest filter of events that includes $\mathscr{C}$, because being a filter is preserved under arbitrary intersections. In this expression, we let $\bigcap \emptyset := \mathscr{P}$.

(iii) A set of propositions is **consistent** if its deductive closure is a strict subset of the set of all propositions, because the latter represents inconsistency. Equivalently, a set of events $\mathscr{C}$ is consistent if and only if its deductive closure is a proper filter: $\mathrm{Cl}_{\mathbb{F}}(\mathscr{C}) \in \mathbb{F}$ or, in other words, $\mathrm{Cl}_{\mathbb{F}}(\mathscr{C}) \neq \mathscr{P}$.

This implies that the set $\mathbb{F}$ of all proper filters corresponds to the set of all deductively closed and consistent sets of events.

(iv) A set of propositions is **deductively complete** if adding any proposition to it would make it inconsistent; this means that a set of events is deductively closed and complete if and only if it is an *ultrafilter*.

To establish a formal link between sets of propositions (or events) and lower previsions, we make the following observation: if a subject believes a proposition to be true, or the corresponding event A to occur, he will believe that event to be practically certain in the sense that he will be willing to bet on it at all odds, so his lower probability for that event will be 1. In other words, there is a correspondence between an assessment that a collection of events $\mathscr{C}$ occurs and the lower probability assessment $\underline{Q}_{\mathscr{C}}$ as introduced in the introduction to this Section 5.5_{82}.

It should now be clear that there is a one-to-one correspondence between $\{0, 1\}$-valued lower probabilities and sets of events, which allows us to establish the following identification.

Sets of propositions (events)	**$\{0, 1\}$-valued lower probabilities**
consistent	avoiding sure loss
deductively closed and consistent	coherent
deductive closure	natural extension
complete	linear

It is in this particular sense that inference in classical propositional logic can be identified with inference with $\{0, 1\}$-valued lower probabilities as it has been studied in this section. As the latter is a specific special case of inference with lower previsions, we have established that classical propositional logic can be embedded in the theory of coherent lower previsions: the theory of coherent lower previsions is a generalisation of classical propositional logic. In this embedding, precise previsions (or probabilities) play the role of maximal elements and correspond to the maximal consistent deductively closed sets of propositions.

It is sometimes claimed that probability measures are the *only* reasonable extension of classical logic able to deal with partial beliefs, see for instance Lindley (1982, 1987) and Jaynes (2003). In light of what we have just established, we are clearly at odds with such a claim. For instance, how many logicians would claim that the only rational deductively closed sets of sentences are the maximal ones or that the only rational theories are complete? Interestingly, the need to model partial beliefs led Boole (1854) to one of the first formal mathematical treatments of imprecise probabilities.

The results given earlier furthermore tell us that, in a very definite sense, precise probability theory is not powerful enough to generalise all of classical propositional logic, whereas the theory of coherent lower previsions is. Of course, this theory is not the only reasonable generalisation of classical propositional logic, although it seems to be the smallest theory that fully encompasses both logic and precise probabilities. For a much more formal discussion of this embedding and the issues addressed here, we refer to De Cooman's (2005b) discussion on the order-theoretic aspects of belief models and belief change.

5.5.3 The link with limits inferior

We have already drawn attention to the formal similarity between the lower prevision associated with a proper filter in Equation $(5.5)_{83}$ and the limit inferior of a real net in Equation $(1.3)_{9}$. In fact, the expression in Equation $(5.5)_{83}$ can be used to generalise the notion of a limit inferior to convergence with respect to a proper filter; see, for instance, König (1997) and compare with Example 5.13_{92}. See, for instance, also Willard (1970, Chapter 4) for more details on the relation between filter and net convergence.

Let us uncover the systematic two-way connection that exists between the two types of lower previsions. First, consider a proper filter $\mathscr{F}$ on a set $\mathscr{X}$, and consider the set

$$\mathscr{A} := \{(x, F) : x \in F \text{ and } F \in \mathscr{F}\}$$

and the relation $\leq$ on $\mathscr{A}$ defined by

$$(x, F) \leq (y, G) \Leftrightarrow G \subseteq F.$$

Then $\mathscr{A}$ is clearly directed by $\leq$, and a bounded gamble f on $\mathscr{X}$ induces a bounded real net ϕ on $\mathscr{A}$ by $\phi(x, F) := f(x)$. Moreover, we see at once that $\underline{P}_{\mathscr{F}}(f) = \liminf_{\alpha \in \mathscr{A}} \phi(\alpha)$.

On the other hand, if we start out with a non-empty set $\mathscr{A}$ directed by a binary relation $\leq$, we can define

$$\uparrow\alpha := \{\beta \in \mathscr{A} : \alpha \leq \beta\} \text{ and } \mathscr{F} := \{F \subseteq \mathscr{A} : (\exists \alpha \in \mathscr{A}) \uparrow\alpha \subseteq F\}.$$

Let us prove that $\mathscr{F}$ is a proper filter of subsets of $\mathscr{A}$[6] and that $\underline{P}_{\mathscr{F}}(f) = \liminf f$ for all bounded nets f on $\mathscr{A}$.

It is clear that $\mathscr{A} \in \mathscr{F}$ and that $\emptyset \notin \mathscr{F}$.

To show that $\mathscr{F}$ is increasing, consider any $F \in \mathscr{F}$ and any subset G of $\mathscr{A}$ such that $F \subseteq G$. But then there is some $\alpha \in \mathscr{A}$ such that $\uparrow\alpha \subseteq F$ and, therefore, also $\uparrow\alpha \subseteq G$, which implies that indeed $G \in \mathscr{F}$.

To show that $\mathscr{F}$ is closed under finite intersections, consider any $F, G \in \mathscr{F}$. Then there are $\alpha, \beta \in \mathscr{A}$ such that $\uparrow\alpha \subseteq F$ and $\uparrow\beta \subseteq G$. As $\mathscr{A}$ is directed, there is some $\gamma \in \mathscr{A}$ such that $\alpha \leq \gamma$ and $\beta \leq \gamma$ and, therefore, $\uparrow\gamma \subseteq \uparrow\alpha \cap \uparrow\beta \subseteq F \cap G$. This tells us that indeed $F \cap G \in \mathscr{F}$.

Finally, because $\{\uparrow\alpha : \alpha \in \mathscr{A}\} \subseteq \mathscr{F}$, we see that

$$\underline{P}_{\mathscr{F}}(f) = \sup_{F \in \mathscr{F}} \inf_{\beta \in F} f(\beta) \geq \sup_{\alpha \in \mathscr{A}} \inf_{\beta \in \uparrow\alpha} f(\beta) \geq \sup_{\alpha \in \mathscr{A}} \inf_{\beta \geq \alpha} f(\beta) = \liminf f.$$

For the converse inequality, it suffices to recall that for every $F \in \mathscr{F}$, there is some $\alpha \in \mathscr{A}$ such that $\uparrow\alpha \subseteq F$ and, therefore, $\inf_{\beta \in F} f(\beta) \leq \inf_{\beta \geq \alpha} f(\beta)$.

5.5.4 Monotone convergence

Monotone convergence was introduced in Section 1.9_{17}. When does a coherent lower prevision $\underline{P}_{\mathscr{F}}$ associated with a proper filter $\mathscr{F}$ satisfy one of its versions?

Proposition 5.12 *Let $\mathscr{F}$ be a proper filter. Then the following statements are equivalent:*

(i) *The coherent lower prevision $\underline{P}_{\mathscr{F}}$ satisfies downward monotone convergence.*

(ii) *$\mathscr{F}$ is closed under limits (intersections) of non-increasing sequences: for any sequence $A_1 \supseteq A_2 \supseteq \dots$ of elements of $\mathscr{F}$, $\lim_{n \to +\infty} A_n = \bigcap_{n \in \mathbb{N}_{>0}} A_n \in \mathscr{F}$.*

(iii) *$\mathscr{F}$ is closed under countable intersections.*

The following statements are equivalent as well:

(iv) *The coherent lower prevision $\underline{P}_{\mathscr{F}}$ satisfies upward monotone convergence.*

(v) *$\mathscr{F}^c$ is closed under limits (unions) of non-decreasing sequences: for any sequence $B_1 \subseteq B_2 \subseteq \dots$ of elements of $\mathscr{F}^c$, $\lim_{n \to +\infty} B_n = \bigcup_{n \in \mathbb{N}_{>0}} B_n \in \mathscr{F}^c$.*

[6] It could be called the filter of eventually increasing subsets of $\mathscr{A}$.

Proof. We first prove that (i)$_\cap$⇔(ii)$_\cap$⇔(iii)$_\cap$ in a circular manner.

(i)$_\cap$⇒(ii)$_\cap$. Assume that $\underline{P}_{\mathscr{F}}$ satisfies downward monotone convergence, and consider a non-increasing sequence $A_1 \supseteq A_2 \supseteq \dots$ of elements of $\mathscr{F}$. Then the sequence of bounded gambles I_{A_n} is non-increasing too, is uniformly bounded above by one and converges point-wise to the bounded gamble I_A, where $A := \bigcap_{n\in\mathbb{N}_{>0}} A_n$. As all $A_n \in \mathscr{F}$, we also have that $\underline{P}_{\mathscr{F}}(A_n) = 1$, and therefore, it follows from the assumption that $\underline{P}_{\mathscr{F}}(A) = \underline{P}_{\mathscr{F}}(I_A) = \lim_{n\to+\infty} \underline{P}_{\mathscr{F}}(I_{A_n}) = \lim_{n\to+\infty} \underline{P}_{\mathscr{F}}(A_n) = 1$. Hence, indeed $\bigcap_{n\in\mathbb{N}_{>0}} A_n \in \mathscr{F}$.

(ii)$_\cap$⇒(iii)$_\cap$. Consider a countable collection C_n, $n \in \mathbb{N}_{>0}$ of elements of $\mathscr{F}$. Then the sequence of subsets $A_n := \bigcap_{k=1}^{n} C_n$ is non-increasing and $A_n \in \mathscr{F}$ because $\mathscr{F}$ is closed under finite intersections. Hence, $\bigcap_{n\in\mathbb{N}_{>0}} C_n = \bigcap_{n\in\mathbb{N}_{>0}} A_n \in \mathscr{F}$, by assumption.

(iii)$_\cap$⇒(i)$_\cap$. Assume that $\mathscr{F}$ is closed under countable intersections. Consider a non-increasing sequence of bounded gambles f_n that are uniformly bounded above, and assume that $g := \lim_{n\to+\infty} f_n = \inf_{n\in\mathbb{N}_{>0}} f_n$ is again a bounded gamble (this is equivalent to the f_n being also uniformly bounded below). As $g \le f_n$, we find that $\underline{P}_{\mathscr{F}}(g) \le \underline{P}_{\mathscr{F}}(f_n)$ by coherence (monotonicity) and, therefore, $\underline{P}_{\mathscr{F}}(g) \le \inf_{n\in\mathbb{N}_{>0}} \underline{P}_{\mathscr{F}}(f_n) = \lim_{n\to+\infty} \underline{P}_{\mathscr{F}}(f_n)$.

To prove the converse inequality, let $\beta > \underline{P}_{\mathscr{F}}(g)$. Then by Lemma 5.6$_{84}$, $\{g \ge \beta\} \notin \mathscr{F}$, and because $\{g \ge \beta\} = \bigcap_{n\in\mathbb{N}_{>0}} \{f_n \ge \beta\}$, we infer from the assumption that there must be some $m \in \mathbb{N}_{>0}$ such that $\{f_m \ge \beta\} \notin \mathscr{F}$. By Lemma 5.6$_{84}$, $\underline{P}_{\mathscr{F}}(f_m) \le \beta$ and, therefore, $\lim_{n\to+\infty} \underline{P}_{\mathscr{F}}(f_n) = \inf_{n\in\mathbb{N}_{>0}} \underline{P}_{\mathscr{F}}(f_n) \le \underline{P}_{\mathscr{F}}(f_m) \le \beta$. Because this holds for every $\beta > \underline{P}_{\mathscr{F}}(g)$, it follows that indeed $\underline{P}_{\mathscr{F}}(g) \ge \lim_{n\to+\infty} \underline{P}_{\mathscr{F}}(f_n)$.

We complete the proof by showing that (iv)$_\cap$⇔(v)$_\cap$.

(iv)$_\cap$⇒(v)$_\cap$. Assume that $\underline{P}_{\mathscr{F}}$ satisfies upward monotone convergence, and consider a non-decreasing sequence $B_1 \subseteq B_2 \subseteq \dots$ of elements of $\mathscr{F}^c$. Then the sequence of bounded gambles I_{B_n} is non-decreasing too, is uniformly bounded below by zero and converges point-wise to the bounded gamble I_B, where $B := \bigcup_{n\in\mathbb{N}_{>0}} B_n$. As each $B_n \notin \mathscr{F}$, we also have that $\underline{P}_{\mathscr{F}}(B_n) = 0$, so it follows from the assumption that $\underline{P}_{\mathscr{F}}(B) = \underline{P}_{\mathscr{F}}(I_B) = \lim_{n\to+\infty} \underline{P}_{\mathscr{F}}(I_{B_n}) = \lim_{n\to+\infty} \underline{P}_{\mathscr{F}}(B_n) = 0$. Hence, indeed, $\bigcup_{n\in\mathbb{N}} B_n \notin \mathscr{F}$.

(v)$_\cap$⇒(iv)$_\cap$. Consider a non-decreasing sequence of bounded gambles g_n that are uniformly bounded below, and assume that $f := \lim_{n\to+\infty} g_n = \sup_{n\in\mathbb{N}_{>0}} g_n$ is again a bounded gamble (this is equivalent to the g_n being also uniformly bounded above). As $f \ge g_n$, we find that $\underline{P}_{\mathscr{F}}(f) \ge \underline{P}_{\mathscr{F}}(g_n)$ by coherence (monotonicity) and, therefore, $\underline{P}_{\mathscr{F}}(f) \ge \sup_{n\in\mathbb{N}_{>0}} \underline{P}_{\mathscr{F}}(g_n) = \lim_{n\to+\infty} \underline{P}_{\mathscr{F}}(g_n)$.

To prove the converse inequality, let $\beta < \underline{P}_{\mathscr{F}}(f)$. Then by Lemma 5.6$_{84}$, $\{f > \beta\} \in \mathscr{F}$, and because $\{f > \beta\} = \bigcup_{n\in\mathbb{N}_{>0}} \{g_n > \beta\}$, we infer from the assumption that there must be some $m \in \mathbb{N}_{>0}$ such that $\{g_m > \beta\} \in \mathscr{F}$. By Lemma 5.6$_{84}$, $\underline{P}_{\mathscr{F}}(g_m) \ge \beta$ and, therefore, $\lim_{n\to+\infty} \underline{P}_{\mathscr{F}}(g_n) = \sup_{n\in\mathbb{N}_{>0}} \underline{P}_{\mathscr{F}}(g_n) \ge \underline{P}_{\mathscr{F}}(g_m) \ge \beta$. Because this holds for all $\beta < \underline{P}_{\mathscr{F}}(f)$, it follows that $\underline{P}_{\mathscr{F}}(f) \le \lim_{n\to+\infty} \underline{P}_{\mathscr{F}}(g_n)$. □

Example 5.13 Let $\mathscr{X} := \mathbb{N}_{>0}$ and consider the proper filter

$$\mathscr{F} := \left\{A \subseteq \mathbb{N}_{>0} : A^c \text{ is finite}\right\}$$

containing all **co-finite** subsets, that is, complements of finite subsets, of $\mathbb{N}_{>0}$. As

$$A \in \mathscr{F} \Leftrightarrow (\exists n \in \mathbb{N}_{>0})(\{1, \ldots, n\}^c \subseteq A),$$

the coherent lower prevision $\underline{P}_{\mathscr{F}}$ given by

$$\underline{P}_{\mathscr{F}}(f) = \sup_{A \in \mathscr{F}} \inf_{m \in A} f(m) = \sup_{n \in \mathbb{N}_{>0}} \inf_{m \geq n} f(m) = \liminf f \text{ for all } f \in \mathbb{B}(\mathbb{N}_{>0})$$

models a subject's beliefs that the random variable X assumes a value that is larger than any finite number. Because $\bigcap_{n \in \mathbb{N}_{>0}} \{1, \ldots, n\}^c = \emptyset \notin \mathscr{F}$, $\mathscr{F}$ is not closed under countable intersections and, therefore, the operator $\underline{P}_{\mathscr{F}} = \liminf$ on bounded real sequences (bounded gambles on $\mathbb{N}_{>0}$) does not satisfy downward monotone convergence. Similarly, because $\bigcup_{n \in \mathbb{N}_{>0}} \{1, \ldots, n\} = \mathbb{N}_{>0} \in \mathscr{F}$, $\mathscr{F}^c$ is not closed under non-decreasing countable unions and, therefore, $\liminf$ does not satisfy upward monotone convergence either.

If a linear prevision P is compatible with this assessment, that is, if $P \in \text{lins}(\underline{P}_{\mathscr{F}})$, then

$$P(A) = 1 \text{ if } \{1, \ldots, n\}^c \subseteq A \text{ for some } n \in \mathbb{N}_{>0},$$

or equivalently, because $P(A^c) = 1 - P(A)$,

$$P(A) = 0 \text{ if } A \subseteq \{1, \ldots, n\} \text{ for some } n \in \mathbb{N}_{>0},$$

so P is zero on all finite subsets of $\mathbb{N}_{>0}$. This means that P cannot be countably additive on events: if it were, then in particular we would have that $1 = P(\mathbb{N}_{>0}) = \sum_{n \in \mathbb{N}_{>0}} P(\{n\}) = 0$. We conclude that no countably additive probability model on $\mathbb{N}_{>0}$ is compatible with the assessment that the random variable X assumes a value that is larger than any finite number. However, there is such a countably additive probability model on $\mathbb{N}_{>0} \cup \{+\infty\}$: the one that assigns probability mass one to $+\infty$. In the language of Section 5.5.5, the finitely additive models we have described earlier have all probability mass left-adherent to $+\infty$; see in particular Example 5.17$_{97}$ for related discussion. ♦

5.5.5 Lower oscillations and neighbourhood filters

As an interesting application of the material on $\{0, 1\}$-valued lower probabilities, we consider the notion of discrete probability mass: suppose we want to model a subject's assessment that all probability mass lies 'arbitrarily close' to some element x of $\mathscr{X}$ or, in other words, that the value of the random variable X lies arbitrarily close to x.

In order to be able to capture what 'arbitrarily close' means, we provide the space $\mathscr{X}$ with a *topology* $\mathscr{T}$ of open sets. An **open neighbourhood** of x is any open set $O \in \mathscr{T}$ that contains x, and a **neighbourhood** N of x is then any subset of $\mathscr{X}$ that includes an open neighbourhood of x. The set

$$\mathscr{N}_x \coloneqq \{N \subseteq \mathscr{X} : (\exists O \in \mathscr{T})(x \in O \text{ and } O \subseteq N)\} = \left\{N \subseteq \mathscr{X} : (\exists O \in \mathscr{T}_x)(O \subseteq N)\right\}$$

of all neighbourhoods of x is then a proper filter that is called the **neighbourhood filter** of x. The filter base $\mathscr{T}_x := \{O \in \mathscr{T} : x \in O\}$ of $\mathscr{N}_x$ is called a **neighbourhood base** in x. It is an easy exercise to verify that a set is open if and only if it is a neighbourhood of each of its elements.

A subject who assesses that all probability mass lies arbitrarily close to x will be practically certain – or prepared to bet at all odds – that the value of X lies within any given neighbourhood of x:

$$\underline{P}(N_x) = 1 \text{ for all } N_x \in \mathscr{N}_x.$$

Then it follows from the arguments in this section that the natural extension of this assessment – the lower prevision on all bounded gambles that represents this assessment – is the lower prevision $\underline{P}_{\mathscr{N}_x}$, where

$$\underline{P}_{\mathscr{N}_x}(f) = \sup_{N_x \in \mathscr{N}_x} \inf_{y \in N_x} f(y) = \sup_{O_x \in \mathscr{T}_x} \inf_{y \in O_x} f(y) \text{ for all bounded gambles } f \text{ on } \mathscr{X},$$

so $\underline{P}_{\mathscr{N}_x}$ is actually the smallest coherent lower prevision that assumes the value 1 on any neighbourhood of x.

Let us introduce a new notation for the conjugate functionals $\underline{P}_{\mathscr{N}_x}$ and $\overline{P}_{\mathscr{N}_x}$, whose meaning will become clear right away.

Definition 5.14 *Define, for any $x \in \mathscr{X}$, the functionals $\underline{\mathrm{osc}}_x$ and $\overline{\mathrm{osc}}_x$ on $\mathbb{B}$ by*

$$\underline{\mathrm{osc}}_x(f) := \underline{P}_{\mathscr{N}_x}(f) = \sup_{N_x \in \mathscr{N}_x} \inf_{y \in N_x} f(y) = \sup_{O_x \in \mathscr{T}_x} \inf_{y \in O_x} f(y)$$

$$\overline{\mathrm{osc}}_x(f) := \overline{P}_{\mathscr{N}_x}(f) = \inf_{N_x \in \mathscr{N}_x} \sup_{y \in N_x} f(y) = \inf_{O_x \in \mathscr{T}_x} \sup_{y \in O_x} f(y)$$

for all bounded gambles f on $\mathscr{X}$.

The functional $\underline{\mathrm{osc}}_x$ is a coherent lower prevision on $\mathbb{B}$ and $\overline{\mathrm{osc}}_x$ is its conjugate upper prevision. For any bounded gamble f on $\mathscr{X}$, the difference

$$\overline{\mathrm{osc}}_x(f) - \underline{\mathrm{osc}}_x(f) = \inf_{N_x \in \mathscr{N}_x} \sup_{y, y' \in N_x} |f(y) - f(y')| =: \mathrm{osc}_x(f) \tag{5.11}$$

is the so-called **oscillation** of f in x, see for instance Schechter (1997, Section 18.28). Clearly, f is continuous in x if and only if $\mathrm{osc}_x(f) = 0$ or, in other words, if $\underline{\mathrm{osc}}_x(f) = \overline{\mathrm{osc}}_x(f)$. We will, therefore, call $\underline{\mathrm{osc}}_x(f) \leq f(x)$ the **lower oscillation** of f in x and $\overline{\mathrm{osc}}_x(f) \geq f(x)$ the **upper oscillation**.

The following proposition provides a nice interpretation for the bounded gamble $\underline{\mathrm{osc}}(f)$ that maps any x in $\mathscr{X}$ to $\underline{\mathrm{osc}}_x(f)$. Recall that a real map f on $\mathscr{X}$ is called **lower semi-continuous** if the sets $\{f > \alpha\}$ are open (belong to $\mathscr{T}$) for all real α; f is called **upper semi-continuous** if $-f$ is lower semi-continuous, see for instance Willard (1970, Section 7K) or Schechter (1997, Section 15.22). We denote the set of all lower semi-continuous bounded gambles on $\mathscr{X}$ by $\underline{\mathscr{C}}(\mathscr{X})$ or by $\underline{\mathscr{C}}$ if no confusion can arise. It is not difficult to see that $\underline{\mathscr{C}}$ is closed under finite point-wise minima and under arbitrary point-wise suprema.

Proposition 5.15 *Consider any bounded gambles f and g on $\mathcal{X}$. Then $\underline{\mathrm{osc}}(f)$ is lower semi-continuous, and*

(i) $\underline{\mathrm{osc}}(f) \leq f$;

(ii) *if* $f \leq g$ *then* $\underline{\mathrm{osc}}(f) \leq \underline{\mathrm{osc}}(g)$;

(iii) $\underline{\mathrm{osc}}(\underline{\mathrm{osc}}(f)) = \underline{\mathrm{osc}}(f)$;

(iv) $f = \underline{\mathrm{osc}}(f)$ *if and only if* $f \in \underline{\mathscr{C}}$.

Finally, $\underline{\mathrm{osc}}(f)$ is the point-wise greatest lower semi-continuous bounded gamble on $\mathcal{X}$ that is dominated by f.

Proof. We first prove that $\underline{\mathrm{osc}}(f)$ is always lower semi-continuous. Consider any real α, then we need to prove that the set $\{\underline{\mathrm{osc}}(f) > \alpha\}$ is open. Consider any x in this set, then it follows from Definition 5.14 that there is some $O_x \in \mathscr{T}_x$ such that $\inf_{z \in O_x} f(z) > \alpha$. Consider any y in O_x, then also $O_x \in \mathscr{T}_y$ and, therefore, $\underline{\mathrm{osc}}_y(f) \geq \inf_{z \in O_x} f(z) > \alpha$. This shows that $O_x \subseteq \{\underline{\mathrm{osc}}(f) > \alpha\}$, so $\{\underline{\mathrm{osc}}(f) > \alpha\} \in \mathscr{N}_x$. As this shows that $\{\underline{\mathrm{osc}}(f) > \alpha\}$ is a neighbourhood of each of its points, $\{\underline{\mathrm{osc}}(f) > \alpha\}$ is indeed an open set.

(i) and (ii) follow directly from the definition of the $\underline{\mathrm{osc}}$ operator.

We turn to (iv). To prove the 'only if' part, it suffices to recall the lower semi-continuous character of $\underline{\mathrm{osc}}(f)$. To prove the 'if' part, consider any $g \in \underline{\mathscr{C}}$, then for any real α, $\{g > \alpha\}$ is open, meaning that for any $x \in \{g > \alpha\}$, there is some $N_x \in \mathscr{N}_x$ such that $N_x \subseteq \{g > \alpha\}$. Consequently, $g(x) > \alpha$ implies that $\underline{\mathrm{osc}}_x(g) = \sup_{N_x \in \mathscr{N}_x} \inf_{z \in N_x} g(z) \geq \alpha$, whence $\underline{\mathrm{osc}}(g) \geq g$, and therefore $\underline{\mathrm{osc}}(g) = g$, taking into account (i).

To prove (iii), recall that $\underline{\mathrm{osc}}(f)$ is lower semi-continous and use (iv).

To prove the last statement, let $g \in \underline{\mathscr{C}}$ be such that $g \leq f$. Then it follows from (ii) that $\underline{\mathrm{osc}}(g) \leq \underline{\mathrm{osc}}(f)$, whence $g \leq \underline{\mathrm{osc}}(f) \leq f$, by (iv) and (i). As we know that $\underline{\mathrm{osc}}(f)$ is lower semi-continuous, we are done. □

For $\overline{\mathrm{osc}}(f)$, there is a similar interpretation in terms of upper semi-continuity.[7] Also, the bounded gamble $\mathrm{osc}(f)$ that maps any x in $\mathcal{X}$ to $\mathrm{osc}_x(f)$ is upper semi-continuous as the sum of two upper semi-continuous bounded gambles $\overline{\mathrm{osc}}(f)$ and $-\underline{\mathrm{osc}}(f) = \overline{\mathrm{osc}}(-f)$.

In the language of topology (see for instance Willard, 1970, Definition 12.3), a proper filter $\mathscr{F}$ is said to **converge to** an element x of the topological space $\mathcal{X}$ if $\mathscr{N}_x \subseteq \mathscr{F}$. We denote this by $\mathscr{F} \to x$. If we invoke the Extreme Point Theorem 4.44$_{75}$ and the Ultrafilter Theorem 5.11$_{86}$, we see that

$$\mathrm{ext}(\mathrm{lins}(\underline{\mathrm{osc}}_x)) = \{P_{\mathscr{U}} : \mathscr{U} \in \mathbb{U} \text{ and } \mathscr{U} \to x\} \text{ and } \underline{\mathrm{osc}}_x = \inf_{\mathscr{U} \to x} P_{\mathscr{U}}.$$

The $P_{\mathscr{U}}$ with $\mathscr{U} \to x$ are the precise probability models (linear previsions) for which 'all probability mass lies arbitrarily close to x'. In a metrisable space (see

[7] The corresponding properties of $\overline{\mathrm{osc}}$ guarantee that it is a(n order-theoretic) closure operator (Davey and Priestley, 1990), compare with Proposition 3.6$_{33}$. The corresponding dual closure operator is then $\underline{\mathrm{osc}}$.

Appendix B_{371} for a definition), these precise models can be studied somewhat more closely.

Proposition 5.16 *Let $\mathscr{X}$ be metrisable, and let $x \in \mathscr{X}$. Then for any ultrafilter $\mathscr{U} \to x$, either $\bigcap \mathscr{U} = \{x\}$ or $\bigcap \mathscr{U} = \emptyset$, and the following statements are equivalent:*

(i) *$\mathscr{U}$ is **fixed**, meaning that $\bigcap \mathscr{U} \neq \emptyset$, and therefore $\bigcap \mathscr{U} = \{x\}$ and $\mathscr{U} = \{A \in \mathscr{X} : x \in A\}$.*

(ii) *$\mathscr{U}$ is closed under countable, and even arbitrary, intersections.*

(iii) *$P_{\mathscr{U}}$ satisfies monotone convergence.*

(iv) *The restriction of $P_{\mathscr{U}}$ to events is countably additive.*

Proof. Fix $x \in \mathscr{X}$. As $\mathscr{X}$ is metrisable, $\{x\}$ is the intersection of a countable number of open neighbourhoods of x (see the discussion in Appendix B_{371} and in particular Section $B.2_{372}$), and therefore $\mathscr{N}_x \subseteq \mathscr{U}$ implies that $\{x\} = \bigcap \mathscr{N}_x \supseteq \bigcap \mathscr{U}$. Hence, indeed, either $\bigcap \mathscr{U} = \{x\}$ or $\bigcap \mathscr{U} = \emptyset$. We now prove the equivalences in a circular manner: (i)⇒(ii)⇒(iii)⇒(iv)⇒(i).

(i)⇒(ii). Immediate, given that $\mathscr{U} = \{A \in \mathscr{X} : x \in A\}$.

(ii)⇒(iii). See Propositions 5.12_{91} and 1.26_{19}.

(iii)⇒(iv). The proof is standard. Consider a countable collection of mutually disjoint events A_n, $n \in \mathbb{N}_{>0}$. Define $F_n := \bigcup_{k=1}^{n} A_k$, then $\bigcup_{n\in\mathbb{N}_{>0}} A_n = \bigcup_{n\in\mathbb{N}_{>0}} F_k$ and $P_{\mathscr{U}}(F_n) = \sum_{k=1}^{n} P_{\mathscr{U}}(A_k)$ (Corollary $4.14(\mathrm{v})_{55}$). We have to prove that $P_{\mathscr{U}}(\bigcup_{n\in\mathbb{N}_{>0}} A_n) = \sum_{k=1}^{+\infty} P_{\mathscr{U}}(A_k)$, which is equivalent to $P_{\mathscr{U}}(\bigcup_{n\in\mathbb{N}_{>0}} F_n) = \lim_{n\to+\infty} P_{\mathscr{U}}(F_n)$. But this follows from (upward) monotone convergence, by considering the uniformly bounded and non-decreasing sequence of bounded gambles I_{F_n} that converges point-wise to $I_{\bigcup_{n\in\mathbb{N}_{>0}} F_n}$.

(iv)⇒(i). It suffices to prove that $\{x\} \in \mathscr{U}$ or, in other words, that $P_{\mathscr{U}}(\{x\}) = 1$. But we know that $\{x\}$ is a countable intersection of open neighbourhoods $N_n \in \mathscr{N}_x$ and because $\mathscr{N}_x \subseteq \mathscr{U}$, $P_{\mathscr{U}}(N_n) = 1$. We can assume this sequence to be non-increasing. Define the sequence of mutually disjoint events $G_n := N_n \setminus N_{n+1}$, then $P_{\mathscr{U}}(G_n) = P_{\mathscr{U}}(N_n) - P_{\mathscr{U}}(N_{n+1}) = 0$ and $\bigcup_{k=1}^{n} G_k = N_1 \setminus N_{n+1}$. Hence, $\bigcup_{n\in\mathbb{N}_{>0}} G_n = N_1 \setminus \{x\}$, and by countable and finite additivity, we find that, indeed,

$$0 = \sum_{n=1}^{+\infty} P_{\mathscr{U}}(G_n) = P_{\mathscr{U}}(N_1 \setminus \{x\}) = P_{\mathscr{U}}(N_1) - P_{\mathscr{U}}(\{x\}) = 1 - P_{\mathscr{U}}(\{x\}),$$

which completes the proof. □

The meaning of this result should be clear. Among the ultrafilters $\mathscr{U}$ converging to x, there is only one for which $P_{\mathscr{U}}$ is σ-additive on events, namely, the fixed ultrafilter $\mathscr{U} = \{A \subseteq \mathscr{X} : x \in A\}$. This precise model corresponds to the assessment 'all probability mass lies in x'. Any other ultrafilter $\mathscr{U}$ converging to x is **free**, meaning that the intersection of all the sets in $\mathscr{U}$ is the empty set: $\bigcap \mathscr{U} = \emptyset$. The corresponding precise models $P_{\mathscr{U}}$ correspond to the assessment 'all probability mass is **adherent to** x', meaning that it lies arbitrarily close to, but not in, x. The corresponding restrictions to events $Q_{\mathscr{U}}$ are only finitely additive.

Example 5.17 Let $\mathscr{X}$ be the real line $\mathbb{R}$ provided with the Euclidean metric topology. Recall that for any $x \in \mathbb{R}$, the set of neigbourhoods of x is the set of all events that include some open interval around x:

$$\mathcal{N}_x := \{N_x \subseteq \mathbb{R} : (\exists \varepsilon > 0)((x - \varepsilon, x + \varepsilon) \subseteq N_x)\}.$$

We are going to argue anew for this special case that the linear previsions $P_{\mathscr{U}}$ associated with ultrafilters $\mathscr{U}$ can be interpreted as models for unit probability mass concentrated in, or adherent to, points in $\mathbb{R}$.

Consider any ultrafilter $\mathscr{U}$. To start the argument, note that for any $d \in \mathbb{R}$, exactly one of these three mutually exclusive possibilities holds:

(i) $(-\infty, d) \in \mathscr{U}$,

(ii) $(d, +\infty) \in \mathscr{U}$,

(iii) $\{d\} \in \mathscr{U}$.

Indeed, because $\mathscr{U}$ is a proper filter, (i), (ii) and (iii) are mutually exclusive. Because, for instance, if we had that both $(-\infty, d) \in \mathscr{U}$ and $(d, +\infty) \in \mathscr{U}$, this would imply that their intersection $\emptyset = (-\infty, d) \cap (d, +\infty) \in \mathscr{U}$, a contradiction. To see that the three possibilities are exhaustive, consider the following argument. As $\mathscr{U}$ is an ultrafilter, either $(-\infty, d)$ or $[d, +\infty)$ must belong to $\mathscr{U}$. Assume that $(-\infty, d) \notin \mathscr{U}$, so $[d, +\infty) \in \mathscr{U}$. But then, again because $\mathscr{U}$ is an ultrafilter, either $\{d\}$ or $\mathbb{R} \setminus \{d\}$ belongs to $\mathscr{U}$. So if $\{d\} \notin \mathscr{U}$, we have that both $[d, +\infty) \in \mathscr{U}$ and $\mathbb{R} \setminus \{d\} \in \mathscr{U}$. As $\mathscr{U}$ is a proper filter, this is equivalent to $(d, +\infty) = [d, +\infty) \cap (\mathbb{R} \setminus \{d\}) \in \mathscr{U}$.

To continue the argument, consider specifically

$$d := \inf\{x \in \mathbb{R} : (-\infty, x) \in \mathscr{U}\}.$$

If d is a real number (finite),[8] then it turns out that $\mathscr{U} \to d$: the ultrafilter $\mathscr{U}$ *converges to* d, meaning that $\mathcal{N}_d \subseteq \mathscr{U}$ or, equivalently, $P_{\mathscr{U}} \in \mathrm{lins}(\underline{P}_{\mathcal{N}_d})$. Indeed, it suffices to show that $(d - \varepsilon, d + \varepsilon) \in \mathscr{U}$ for all $\varepsilon > 0$. In case (iii), this is trivial, because $\{d\} \in \mathscr{U}$ and $\mathscr{U}$ is increasing. In case (ii), we see that $(d - \varepsilon, +\infty) \in \mathscr{U}$, because $(d, +\infty) \in \mathscr{U}$ and $\mathscr{U}$ is increasing. But it also follows from the definition of d and the increasing character of $\mathscr{U}$ that $(-\infty, d + \varepsilon) \in \mathscr{U}$, and therefore, indeed, $(d - \varepsilon, d + \varepsilon) \in \mathscr{U}$, since $\mathscr{U}$ is closed under finite intersections. And finally, in case (i), we see that $(-\infty, d + \varepsilon) \in \mathscr{U}$, because $(-\infty, d) \in \mathscr{U}$ and $\mathscr{U}$ is increasing. But it also follows from the definition of d that $(-\infty, d - \varepsilon/2) \notin \mathscr{U}$ and, therefore, $[d - \varepsilon/2, +\infty) \in \mathscr{U}$, because $\mathscr{F}$ is an ultrafilter. Hence, indeed, $(d - \varepsilon, d + \varepsilon) \in \mathscr{U}$, because $\mathscr{U}$ is closed under finite intersections and is increasing.

Case (iii), $\{d\} \in \mathscr{U}$, singles out the unique fixed ultrafilter, with $P_{\mathscr{U}}(A) = 1$ if and only if $d \in A$. The (cumulative) **distribution function** $F(x) := P_{\mathscr{U}}((-\infty, x])$ for this linear prevision is the rightmost simple break function.

[8] Similarly, if $d = -\infty$ or $d = +\infty$, then we can interpret $\mathscr{U}$ as a neighbourhood filter for $-\infty$ or $+\infty$, respectively.

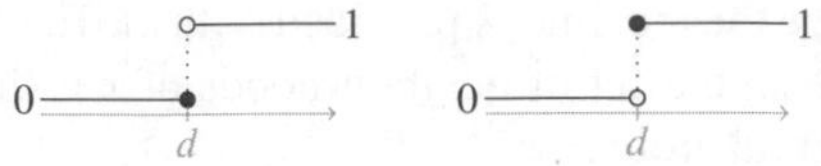

This is the only case where $P_{\mathscr{U}}$ satisfies monotone convergence or, equivalently, where $\mathscr{U}$ is closed under countable intersections (see Proposition 5.16_{96}). In all other cases, $P_{\mathscr{U}}$ is *only finitely additive*.

For case $(\mathrm{i})_{\frown}$, $(-\infty, d) \in \mathscr{U}$, we see that $P_{\mathscr{U}}((d-\varepsilon, d)) = 1$ for all $\varepsilon > 0$ and $P_{\mathscr{U}}([d, +\infty)) = 0$, and the distribution function for this linear prevision is again the rightmost simple break function. We say that $\mathscr{U}$ represents probability mass **left-adherent** to d. In the language of non-standard analysis (see, for instance, Nelson, 1987), all probability mass is concentrated in some non-standard real interval infinitesimally close to, and to the left of, d, that is, in $(d-\varepsilon, d)$ where ε is some strictly positive infinitesimal.

Similarly, case $(\mathrm{ii})_{\frown}$ describes probability mass that is **right-adherent to** d, and the distribution function for the corresponding linear prevision is now the leftmost simple break function. ♦

5.5.6 Extending a lower prevision defined on all continuous bounded gambles

To finish this section on $\{0, 1\}$-valued lower probabilities, we address a topic that will bring us to a discussion of the Riesz representation theorem in Section 9.5_{187}. The results in this section build on and extend the work of De Cooman and Miranda (2008).

Consider, as in the discussion above, a topological space $\mathscr{X}$ provided with a topology of open sets $\mathscr{T}$. We assume, in addition, that this topological space is compact. We denote the set of all continuous real functions on $\mathscr{X}$ by $\mathscr{C}(\mathscr{X})$ or simply by $\mathscr{C}$ when no confusion can arise. As any continuous function on a compact space is bounded (see, for instance, Willard, 1970, Theorem 17.13), we see that $\mathscr{C}$ is a linear subspace of $\mathbb{B}$ that contains all constant gambles.

If we, therefore, consider any coherent lower prevision $\underline{P}$ on $\mathscr{C}$, we infer from Theorem 4.34_{69} that its natural extension $\underline{E}_{\underline{P}}$ to all bounded gambles is given by its inner extension:

$$\underline{E}_{\underline{P}}(f) = \sup\left\{\underline{P}(g) \colon g \in \mathscr{C} \text{ and } g \leq f\right\}. \tag{5.12}$$

The discussion of lower oscillations above now allows us to characterise this natural extension more easily: it is completely determined by the values it assumes on lower semi-continuous bounded gambles.

Proposition 5.18 *Let $\mathscr{X}$ be a compact topological space, and let $\underline{P}$ be a coherent lower prevision on $\mathscr{C}$. Then for any bounded gamble f on $\mathscr{X}$, $\underline{E}_{\underline{P}}(f) = \underline{E}_{\underline{P}}(\underline{\mathrm{osc}}(f))$.*

Proof. For any $g \in \mathscr{C}$ such that $g \leq f$, it follows from Proposition 5.15_{95} that $g = \underline{\mathrm{osc}}(g) \leq \underline{\mathrm{osc}}(f) \leq f$. Therefore,

$$\underline{P}(g) = \underline{E}_{\underline{P}}(g) \leq \underline{E}_{\underline{P}}(\underline{\mathrm{osc}}(f)) \leq \underline{E}_{\underline{P}}(f),$$

where the equality follows from the coherence of $\underline{P}$ on $\mathscr{C}$ (Theorem 4.26(ii)$_{65}$), and the inequalities follow because $\underline{E}_{\underline{P}}$ is coherent (Theorem 4.26(ii)$_{65}$) and, therefore, monotone (Theorem 4.13(iv)$_{53}$). By Equation (5.12), we then find that indeed $\underline{E}_{\underline{P}}(f) \leq \underline{E}_{\underline{P}}(\underline{\mathrm{osc}}(f)) \leq \underline{E}_{\underline{P}}(f)$. □

As it follows from Dini's Monotone Convergence Theorem B.5$_{373}$ that any non-increasing (as well as any non-decreasing) sequence of continuous bounded gambles that converges point-wise also converges uniformly, we see that $\underline{P}$ satisfies the (downward and upward) monotone convergence property on $\mathscr{C}$, because $\underline{P}$ is coherent and therefore continuous with respect to uniform convergence (Theorem 4.13(xiii)$_{53}$).

Proposition 5.19 *Let $\mathscr{X}$ be a compact topological space, and let $\underline{P}$ be a coherent lower prevision on $\mathscr{C}$. Then $\underline{P}$ satisfies monotone convergence.*

Once we leave the ambit of continuous functions and move from $\underline{P}$ to $\underline{E}_{\underline{P}}$, it appears we will generally lose the downward monotone convergence property. But we can show that $\underline{E}_{\underline{P}}$ does not shed its upward monotone convergence property all at once.

Proposition 5.20 *Let $\mathscr{X}$ be a compact topological space, and let $\underline{P}$ be a coherent lower prevision on $\mathscr{C}$. Then $\underline{E}_{\underline{P}}$ satisfies upward monotone convergence on $\underline{\mathscr{C}}$.*

Proof. Consider a non-decreasing sequence of lower semi-continuous bounded gambles f_n that is uniformly bounded below and that converges point-wise to some bounded gamble f (which is equivalent to requiring that the sequence should also be uniformly bounded above). Then $f = \sup_{n\in\mathbb{N}_{>0}} f_n$ and therefore $\{f > \alpha\} = \bigcup_{n\in\mathbb{N}_{>0}} \{f_n > \alpha\}$ is open for all real α, so $f \in \underline{\mathscr{C}}$ too. Fix $\varepsilon > 0$. Then it follows from Equation (5.12) that there is some $f_\varepsilon \in \mathscr{C}$ such that $f_\varepsilon \leq f$ and $\underline{E}_{\underline{P}}(f) - \underline{P}(f_\varepsilon) < \varepsilon/2$. Consider, for any $n \in \mathbb{N}_{>0}$, the bounded gamble $f_{\varepsilon,n} := \min\{f_\varepsilon, f_n\}$ on $\mathscr{X}$. The sequence $f_{\varepsilon,n}$ satisfies the following properties: (i) it converges point-wise to f_ε: indeed, for any $x \in \mathscr{X}$,

$$\lim_{n\to+\infty} \min\{f_\varepsilon(x), f_n(x)\} = \min\{f_\varepsilon(x), \lim_{n\to+\infty} f_n(x)\} = \min\{f_\varepsilon(x), f(x)\} = f_\varepsilon(x);$$

(ii) $f_{\varepsilon,n} \in \underline{\mathscr{C}}$ for all $n \in \mathbb{N}_{>0}$, because it is the point-wise minimum of two elements of $\underline{\mathscr{C}}$, and (iii) the sequence $f_{\varepsilon,n}$ is non-decreasing because the sequence f_n is. By Dini's Convergence Theorem B.5$_{373}$, the sequence $f_{\varepsilon,n}$ converges uniformly to f_ε. As the lower prevision $\underline{E}_{\underline{P}}$ is coherent (Theorem 4.26(ii)$_{65}$), and therefore continuous with respect to uniform convergence and monotone (Theorem 4.13(xiv) and (iv)$_{53}$), we find that $\lim_{n\to+\infty} \underline{E}_{\underline{P}}(f_{\varepsilon,n}) = \sup_{n\in\mathbb{N}_{>0}} \underline{E}_{\underline{P}}(f_{\varepsilon,n}) = \underline{E}_{\underline{P}}(f_\varepsilon) = \underline{P}(f_\varepsilon)$, so there is some $n_\varepsilon \in \mathbb{N}_{>0}$ such that $\underline{P}(f_\varepsilon) - \underline{E}_{\underline{P}}(f_{\varepsilon,n}) < \frac{\varepsilon}{2}$ for all $n \geq n_\varepsilon$. As also $f_{\varepsilon,n} \leq f_n$ for all $n \in \mathbb{N}_{>0}$, we deduce that $\underline{E}_{\underline{P}}(f_{\varepsilon,n}) \leq \underline{E}_{\underline{P}}(f_n)$ for all $n \in \mathbb{N}_{>0}$, and so for all $n \geq n_\varepsilon$,

$$\underline{E}_{\underline{P}}(f) - \underline{E}_{\underline{P}}(f_n) = \underline{E}_{\underline{P}}(f) - \underline{P}(f_\varepsilon) + \underline{P}(f_\varepsilon) - \underline{E}_{\underline{P}}(f_{\varepsilon,n}) + \underline{E}_{\underline{P}}(f_{\varepsilon,n}) - \underline{E}_{\underline{P}}(f_n) < \varepsilon.$$

Hence, $\underline{E}_{\underline{P}}(f) = \sup_{n\in\mathbb{N}_{>0}} \underline{E}_{\underline{P}}(f_n) = \lim_{n\to+\infty} \underline{E}_{\underline{P}}(f_n)$. □

When $\mathscr{X}$ is metrisable, it turns out that this upward monotone convergence on $\underline{\mathscr{C}}$ completely characterises the natural extension of $\underline{P}$ to $\underline{\mathscr{C}}$.

Proposition 5.21 *Let $\mathscr{X}$ be a compact and metrisable topological space, and let $\underline{P}$ be a coherent lower prevision on $\mathscr{C}$. Then $\underline{P}$ has a unique extension $\hat{\underline{P}}$ to $\underline{\mathscr{C}}$ that satisfies upward monotone convergence. This unique extension $\hat{\underline{P}}$ coincides with $\underline{E}_{\underline{P}}$ on $\underline{\mathscr{C}}$ and is therefore also the point-wise smallest coherent lower prevision that extends $\underline{P}$ to $\underline{\mathscr{C}}$.*

Proof. In a compact and metrisable space, any lower semi-continuous bounded gamble f is the supremum (point-wise limit) of a non-decreasing sequence f_n of continuous bounded gambles; see Willard (1970, Section 7K.4). We may assume without loss of generality that this sequence is uniformly bounded below, for instance, by $\inf f$. Then any extension of $\underline{P}$ to $\underline{\mathscr{C}}$ that satisfies upward monotone convergence must assume the value $\sup_{n\in\mathbb{N}_{>0}} \underline{P}(f_n)$ on f and is, therefore, uniquely determined on $\underline{\mathscr{C}}$. If we call this unique extension $\hat{\underline{P}}$, then we infer from Proposition 5.20$_\frown$ that it coincides with $\underline{E}_{\underline{P}}$ on $\underline{\mathscr{C}}$. Now use Theorem 4.31(ii)$_{67}$. □

Corollary 5.22 *Let $\mathscr{X}$ be a compact and metrisable topological space, and let $\underline{P}$ be a coherent lower prevision on $\mathscr{C}$. Then for any bounded gamble f on $\mathscr{X}$, $\underline{E}_{\underline{P}}(f) = \hat{\underline{P}}(\underline{\mathrm{osc}}(f))$, where $\hat{\underline{P}}$ is the unique extension of $\underline{P}$ to $\underline{\mathscr{C}}$ that satisfies upward monotone convergence.*

Proof. We infer from Proposition 5.21 that $\hat{\underline{P}}$ is the natural extension of $\underline{P}$ to $\underline{\mathscr{C}}$. As $\underline{\mathrm{osc}}(f)$ belongs to $\underline{\mathscr{C}}$ by Proposition 5.15$_{95}$, we infer that $\underline{E}_{\underline{P}}(\underline{\mathrm{osc}}(f)) = \hat{\underline{P}}(\underline{\mathrm{osc}}(f))$ for any bounded gamble f on $\mathscr{X}$. Now use Proposition 5.18$_{98}$. □

6

n-Monotone lower previsions

In this chapter, we study the properties of a special subclass of the coherent lower previsions, namely, those that are *n-monotone*, for $n \geq 1$. We start out from Choquet's (1953–1954) original and very general definition of n-monotonicity, and we pave the way towards a representation theorem for n-monotone coherent lower previsions in terms of the Choquet integral.

Section $6.1_{\curvearrowright}$ is concerned with the precise definition of n-monotonicity for lower previsions and lower probabilities. Indeed, there is no real reason to restrict the notion of n-monotonicity to lower probabilities, as seems to be usually done in the literature. We shall see that it is fairly easy, and completely within the spirit of Choquet's original definition, to define and study this property for lower previsions. Doing this does not bring on just another generalisation of something that existed before but leads to genuinely new insights.

In Section 6.2_{107}, we establish many interesting properties and generalise a number of results from the literature for n-monotone lower probabilities defined on fields of events. In Section 6.3_{113}, we relate n-monotone coherent lower previsions to comonotone additive coherent lower previsions and prove a representation theorem for them in terms of Choquet integration. Our discussion in that section also shows that the procedure of natural extension is of particular interest for n-monotone lower previsions: not only does it provide the behaviourally most conservative (i.e. point-wise smallest) extension to all bounded gambles but it is also the only extension to be n-monotone. Any other extension expresses behavioural dispositions not implied by coherence (alone) *and, at the same time,* does not satisfy 2-monotonicity. We also deduce that, under coherence, 2-monotonicity is actually equivalent to comonotone additivity and, therefore, to being representable as a Choquet functional.

Most of the results in this chapter are based on the work by De Cooman et al. (2005a, 2005b, 2008a). We mostly follow De Cooman et al. (2005a) quite closely, with additional results taken from De Cooman et al. (2008a). In the latter paper, we have provided a more general discussion of n-monotonicity for the exact functionals (Maaß, 2002) that will be discussed briefly in Chapter 11_{214} (see Equation $(11.9)_{219}$) – and that coherent lower previsions are a special case of. But we also prove new results. In particular, as far as the domains of n-monotone lower previsions are concerned, we have generalised the discussion from lattices to $\wedge$-semilattices, where feasible.

6.1 *n*-Monotonicity

n-Monotonicity will only be defined for lower previsions whose domains have a special structure, namely, $\wedge$-*semilattices of bounded gambles* $\mathcal{S}$ in the sense of Definition $1.1(\text{f})_4$: sets of bounded gambles that are closed under point-wise minimum $\wedge$, meaning that for all f and g in $\mathcal{S}$, $f \wedge g$ also belongs to $\mathcal{S}$.

The following definition (De Cooman et al., 2005a, Definition 1) is a special case of Choquet's (1953–1954) general definition of n-monotonicity for functions from an Abelian semigroup to an Abelian group.

Definition 6.1 *Let $n \in \mathbb{N}^*_{>0}$, and let $\underline{P}$ be a lower prevision whose domain* $\operatorname{dom}\underline{P}$ *is a $\wedge$-semilattice of bounded gambles on $\mathcal{X}$. Then we call $\underline{P}$* **n-monotone** *if, for all $p \in \mathbb{N}_{>0}$, $p \leq n$ and all $f, f_1, \dots, f_p$ in* $\operatorname{dom}\underline{P}$,

$$\sum_{I \subseteq \{1,\dots,p\}} (-1)^{|I|} \underline{P}\left(f \wedge \bigwedge_{i \in I} f_i\right) \geq 0.$$

The conjugate of an n-monotone lower prevision is called **n-alternating**.[1] *An ∞-monotone lower prevision, that is, a lower prevision that is n-monotone for all $n \in \mathbb{N}_{>0}$, is also called* **completely monotone** *and its conjugate* **completely alternating**.

In this definition, and further on, we use the convention that for $I = \emptyset$, $\bigwedge_{i \in I} f_i$ simply drops out of the expressions – we could let it be equal to $+\infty$. Clearly, if a lower prevision $\underline{P}$ is n-monotone, it is also m-monotone for $1 \leq m \leq n$.

An n-monotone lower prevision on a ($\wedge$-semi)lattice of (indicators of) events is called an **n-monotone lower probability**. A **completely monotone lower probability** is one that is ∞-monotone or, equivalently, n-monotone for all $n \in \mathbb{N}_{>0}$.

In the remainder of this section, we prove basic results for n-monotonicity that will help us establish the links between coherence and n-monotonicity in later sections.

The following proposition (De Cooman et al., 2005a, Proposition 1) gives an immediate alternative characterisation for the n-monotonicity of lower previsions that are defined on a *lattice of bounded gambles*, that is, a set of bounded gambles that is closed under point-wise maxima as well as under point-wise minima. We give its proof for the sake of completeness, as it is not given elsewhere.

Proposition 6.2 *Let $n \in \mathbb{N}^*_{>0}$. Consider a lower prevision $\underline{P}$ whose domain* $\operatorname{dom}\underline{P}$ *is a lattice of bounded gambles on $\mathcal{X}$. Then $\underline{P}$ is n-monotone if and only if*

[1] It is defined on the $\vee$-semilattice $-\operatorname{dom}\underline{P}$.

(i) *$\underline{P}$ is (1-)monotone, that is, for all f and g in* dom $\underline{P}$ *such that $f \leq g$, we have $\underline{P}(f) \leq \underline{P}(g)$; and*

(ii) *for all $p \in \mathbb{N}_{>0}$, $2 \leq p \leq n$ and all $f_1, \ldots, f_p$ in* dom $\underline{P}$,

$$\underline{P}\left(\bigvee_{i=1}^{p} f_i\right) \geq \sum_{\emptyset \neq I \subseteq \{1,\ldots,p\}} (-1)^{|I|+1} \underline{P}\left(\bigwedge_{i \in I} f_i\right).$$

Proof. First, we prove that an n-monotone lower prevision satisfies (i) and (ii). If we look at the n-monotonicity condition involving $p = 1$, we find that, for all f and f_1 in dom $\underline{P}$, $\underline{P}(f) - \underline{P}(f \wedge f_1) \geq 0$. Hence, if $f_1 \leq f$, then $f_1 = f_1 \wedge f$ and therefore $\underline{P}(f_1) \leq \underline{P}(f)$, so $\underline{P}$ is (1-)monotone and (i) holds. To see that (ii) holds, consider any $2 \leq p \leq n$ and any $f_1, \ldots, f_p$ in dom $\underline{P}$. Let $f := \bigvee_{i=1}^{p} f_i$. Then the n-monotonicity condition involving p can be rewritten as

$$\underline{P}\left(\bigvee_{i=1}^{p} f_i\right) - \sum_{\emptyset \neq I \subseteq \{1,\ldots,p\}} (-1)^{|I|+1} \underline{P}\left(\bigwedge_{i \in I} f_i\right) \geq 0,$$

where we have separated off the case $I = \emptyset$ in the summation and took into account that, for this particular choice for f, $f \wedge \bigwedge_{i \in I} f_i = \bigwedge_{i \in I} f_i$ whenever $I \neq \emptyset$.

Next, we prove that a lower prevision $\underline{P}$ that satisfies (i) and (ii) is n-monotone. Consider $1 \leq p \leq n$ and $f, f_1, \ldots f_p$ in dom $\underline{P}$. For $p = 1$, we find, using (i), that $\underline{P}(f) - \underline{P}(f \wedge f_1) \geq 0$. Suppose, therefore, that $p \geq 2$. Then it follows from applying (ii) to $f \wedge f_1, \ldots, f \wedge f_p$ that

$$\underline{P}\left(f \wedge \bigvee_{i=1}^{p} f_i\right) \geq \sum_{\emptyset \neq I \subseteq \{1,\ldots,p\}} (-1)^{|I|+1} \underline{P}\left(f \wedge \bigwedge_{i \in I} f_i\right),$$

and because also $\underline{P}(f) \geq \underline{P}(f \wedge \bigvee_{i=1}^{p} f_i)$ by (i), the proof is now complete. □

In particular, a monotone lower prevision $\underline{P}$ defined on a lattice of bounded gambles is 2-monotone if and only if

$$\underline{P}(f \vee g) + \underline{P}(f \wedge g) \geq \underline{P}(f) + \underline{P}(g) \text{ for all } f, g \in \text{dom } \underline{P}. \tag{6.1}$$

This is a stronger requirement than the inequality

$$\min\left\{\overline{P}(f \vee g) + \underline{P}(f \wedge g), \underline{P}(f \vee g) + \overline{P}(f \wedge g)\right\}$$
$$\geq \underline{P}(f) + \underline{P}(g) \text{ for all } f, g \in \text{dom } \underline{P}$$

that is guaranteed by coherence (see Theorem 4.13(xii)$_{53}$). This already makes us suspect that n-monotonicity (with $n \geq 2$) is in some sense a stronger requirement than coherence, a conjecture we will come back to in detail in Section 6.2$_{107}$.

It turns out that taking the *inner extension* in the sense of Definition 1.22$_{17}$ preserves n-monotonicity. This follows from results in De Cooman et al. (2005a, Theorem 3 and Proposition 5). We provide a more direct proof, borrowed from De Cooman et al. (2008a, Theorem 12).

Theorem 6.3 *Let $n \in \mathbb{N}^*_{>0}$. Let $\underline{P}$ be a lower prevision defined on a lattice of bounded gambles that contains all constant gambles. If $\underline{P}$ is n-monotone, then its inner extension $\underline{P}_*$ is n-monotone as well.*

Proof. Let $p \in \mathbb{N}_{>0}$, $p \leq n$, and consider arbitrary bounded gambles $f, f_1, \dots, f_p$ on $\mathcal{X}$. Fix $\varepsilon > 0$. As $\operatorname{dom} \underline{P}$ is assumed to contain all constant gambles, we infer from Definition 1.22_{17} that for each $I \subseteq \{1, \dots, p\}$, there is some g_I in $\operatorname{dom} \underline{P}$ such that $g_I \leq f \wedge \bigwedge_{i \in I} f_i$ and

$$\underline{P}_*\left(f \wedge \bigwedge_{i \in I} f_i\right) - \varepsilon \leq \underline{P}(g_I) \leq \underline{P}_*\left(f \wedge \bigwedge_{i \in I} f_i\right).$$

Define, for any $I \subseteq \{1, \dots, p\}$, $h_I := \bigvee_{I \subseteq J \subseteq \{1, \dots, p\}} g_J$, then clearly $h_I \in \operatorname{dom} \underline{P}$ and also $g_I \leq h_I \leq f \wedge \bigwedge_{i \in I} f_i$.

Now consider the bounded gambles $q := h_\emptyset$ and $q_k := h_{\{k\}} \leq q$ for $k = 1, \dots, p$. Then q and all the q_k belong to $\operatorname{dom} \underline{P}$, and we have for any $K \subseteq \{1, \dots, p\}$ and any $k \in K$ that $h_K \leq h_{\{k\}} = q_k \leq f \wedge f_k$, whence $h_K \leq \bigwedge_{k \in K} q_k = q \wedge \bigwedge_{k \in K} q_k \leq f \wedge \bigwedge_{k \in K} f_k$. Summarising, we find that for every given $\varepsilon > 0$, there are q and q_k in $\operatorname{dom} \underline{P}$, such that for all $I \subseteq \{1, \dots, p\}$,

$$g_I \leq q \wedge \bigwedge_{i \in I} q_i \leq f \wedge \bigwedge_{i \in I} f_i,$$

and using the monotonicity of $\underline{P}_*$ and the fact that it coincides with $\underline{P}$ on its domain $\operatorname{dom} \underline{P}$, because $\underline{P}$ is monotone:

$$\underline{P}_*\left(f \wedge \bigwedge_{i \in I} f_i\right) - \varepsilon \leq \underline{P}\left(q \wedge \bigwedge_{i \in I} q_i\right) \leq \underline{P}_*\left(f \wedge \bigwedge_{i \in I} f_i\right).$$

Consequently, for every $\varepsilon > 0$, we find that

$$\begin{aligned}
&\sum_{I \subseteq \{1, \dots, p\}} (-1)^{|I|} \underline{P}_*\left(f \wedge \bigwedge_{i \in I} f_i\right) \\
&\qquad = \sum_{\substack{I \subseteq \{1, \dots, p\} \\ |I| \text{ even}}} \underline{P}_*\left(f \wedge \bigwedge_{i \in I} f_i\right) - \sum_{\substack{I \subseteq \{1, \dots, p\} \\ |I| \text{ odd}}} \underline{P}_*\left(f \wedge \bigwedge_{i \in I} f_i\right) \\
&\qquad \geq \sum_{\substack{I \subseteq \{1, \dots, p\} \\ |I| \text{ even}}} \underline{P}\left(q \wedge \bigwedge_{i \in I} q_i\right) - \sum_{\substack{I \subseteq \{1, \dots, p\} \\ |I| \text{ odd}}} \left[\underline{P}\left(q \wedge \bigwedge_{i \in I} q_i\right) + \varepsilon\right] \\
&\qquad = \sum_{I \subseteq \{1, \dots, p\}} (-1)^{|I|} \underline{P}\left(q \wedge \bigwedge_{i \in I} q_i\right) - N_p \varepsilon \geq -N_p \varepsilon,
\end{aligned}$$

where $N_p = 2^{p-1}$ is the number of subsets of $\{1, \dots, p\}$ with an odd number of elements, and the last inequality follows from the n-monotonicity of $\underline{P}$. As this holds for all $\varepsilon > 0$, we conclude that $\underline{P}_*$ is n-monotone on the lattice of bounded gambles $\mathbb{B}$. □

The inner extension is formally similar to the notion of an *inner set function*, introduced in Definition 1.14_{10}. Taking the inner set function also preserves n-monotonicity; this result is actually due to Choquet (1953–1954, Lemma 18.3), once it is noted that Choquet's 'interior capacity' coincides with our inner set function. The proof in Choquet's paper consists of no more than a hint (Choquet, 1953–1954, p. 186, ll. 6–9), but it is clearly possible to give a proof that is very similar to the one given earlier for Theorem 6.3. We omit the proof to avoid unnecessary repetition.

Theorem 6.4 *Let $n \in \mathbb{N}^*_{>0}$. Let $\underline{P}$ be a lower probability defined on a lattice of events containing $\emptyset$ and $\mathcal{X}$. If $\underline{P}$ is n-monotone, then its inner set function $\underline{P}_*$ is n-monotone as well.*

The next proposition tells us how to construct n-monotone lower previsions via $\wedge$-homomorphisms. It is mentioned by De Cooman et al. (2005b, Proposition 3), and its proof is borrowed from De Cooman et al. (2008a, Lemma 6). It can be useful in proving that a lower prevision is n-monotone, by writing it as a concatenation of a simpler n-monotone lower prevision and a $\wedge$-homomorphism. It generalises a similar result by Choquet (1953–1954, Sections 23.2 and 24.3) from events (using $\cap$-homomorphisms) to bounded gambles.

A **$\wedge$-homomorphism** r is a map from a $\wedge$-semilattice to a $\wedge$-semilattice that preserves the $\wedge$ operation: $r(f \wedge g) = r(f) \wedge r(g)$ for all f and g in the domain of r. A $\wedge$-homomorphism is necessarily monotone: $f \geq g$ implies $r(f) \geq r(g)$ (if $f \geq g$, then $f \wedge g = g$, so $r(g) = r(f \wedge g) = r(f) \wedge r(g)$ which can only hold if $r(f) \geq r(g)$).

Proposition 6.5 *Let $n \in \mathbb{N}^*_{>0}$, let $\underline{P}$ be an n-monotone lower prevision defined on a $\wedge$-semilattice of bounded gambles and let r be a $\wedge$-homomorphism from a $\wedge$-semilattice of bounded gambles* dom r *to the $\wedge$-semilattice of bounded gambles* dom $\underline{P}$. *Then $\underline{Q} := \underline{P} \circ r$ is an n-monotone lower prevision on* dom r.

Proof. We prove that the condition of Definition 6.1_{102} is satisfied. Consider any $p \in \mathbb{N}_{>0}$, $p \leq n$ and any $f, f_1, \ldots, f_p$ in dom $\underline{P}$, then

$$\sum_{I \subseteq \{1,\ldots,p\}} (-1)^{|I|} \underline{Q}\left(f \wedge \bigwedge_{i \in I} f_i\right) = \sum_{I \subseteq \{1,\ldots,p\}} (-1)^{|I|} \underline{P}\left(r\left(\bigwedge_{i \in I} f_i\right)\right)$$

$$= \sum_{I \subseteq \{1,\ldots,p\}} (-1)^{|I|} \underline{P}\left(\bigwedge_{i \in I} r(f_i)\right) \geq 0,$$

where the second equality holds because r is a $\wedge$-homomorphism, and the inequality holds because $\underline{P}$ is n-monotone. □

The next theorem, and its proof, are due to De Cooman et al. (2005a, Theorem 2); see also De Cooman et al. (2008a, Theorem 5).

Theorem 6.6 *A linear prevision P defined on a lattice of bounded gambles is always completely monotone and completely alternating.*

Proof. By Theorem 4.42_{73}, the linear prevision P is the restriction of some linear prevision Q on $\mathbb{B}$. Now recall that Q is a positive real-valued linear functional (by Corollary 4.17_{58}), and apply it to both sides of the following well-known identity; for indicators of events, this is known as the *sieve formula*, or *inclusion-exclusion principle* (Aigner, 1997):

$$\bigvee_{i=1}^{p} f_i = \sum_{\emptyset \neq I \subseteq \{1,\ldots,p\}} (-1)^{|I|+1} \bigwedge_{i \in I} f_i. \tag{6.2}$$

This yields

$$Q\left(\bigvee_{i=1}^{p} f_i\right) = \sum_{\emptyset \neq I \subseteq \{1,\ldots,p\}} (-1)^{|I|+1} Q\left(\bigwedge_{i \in I} f_i\right).$$

As any positive linear functional, such as Q, is also (1-)monotone, we derive from Proposition 6.2_{102} that it is completely monotone, and because in this case condition (ii)$_{103}$ in Proposition 6.2_{102} holds with equality, it is completely alternating as well. Now recall that Q and P coincide on the lattice of bounded gambles dom P, which contains all the suprema and infima in the above expression as soon as the bounded gambles f_i belong to dom P. □

The next two propositions are very easy to prove. They provide alternative ways for proving n-monotonicity and are similar in spirit to the results we proved for coherence, to wit, Propositions 4.19_{59} and 4.21_{60}, respectively.

Proposition 6.7 *Let $n \in \mathbb{N}^*_{>0}$, and let $\underline{Q}$ be a convex combination of n-monotone lower previsions $\underline{P}_k$, $k = 1, \ldots, m$ defined on a $\wedge$-semilattice of bounded gambles $\mathcal{K}$. Then $\underline{Q}$ is n-monotone as well.*

Proof. We know that $\underline{Q} = \sum_{k=1}^{m} \lambda_k \underline{P}_k$ for some $\lambda_1, \ldots, \lambda_m \geq 0$ such that $\sum_{k=1}^{m} \lambda_k = 1$. Then for any $p \in \mathbb{N}_{>0}$, $2 \leq p \leq n$, and any $f_1, \ldots, f_p \in \mathcal{K}$, it holds that

$$\begin{aligned}\sum_{I \subseteq \{1,\ldots,p\}} (-1)^{|I|} \underline{Q}\left(f \wedge \bigwedge_{i \in I} f_i\right) &= \sum_{I \subseteq \{1,\ldots,p\}} (-1)^{|I|} \sum_{k=1}^{m} \lambda_k \underline{P}_k\left(f \wedge \bigwedge_{i \in I} f_i\right) \\ &= \sum_{k=1}^{m} \lambda_k \sum_{I \subseteq \{1,\ldots,p\}} (-1)^{|I|} \underline{P}_k\left(f \wedge \bigwedge_{i \in I} f_i\right) \geq 0,\end{aligned}$$

which shows that $\underline{Q}$ is indeed n-monotone. □

A lower envelope of n-monotone lower previsions is not necessarily n-monotone. If it were, then any coherent lower prevision, that is, any lower envelope of linear previsions, which are completely monotone by Theorem $6.6_{\frown}$, would be completely monotone as well. We give examples of coherent lower previsions that are not 2-monotone in Section 6.2.1.

But taking point-wise limits does preserve n-monotonicity.

Proposition 6.8 *Let $n \in \mathbb{N}^*_{>0}$, and let $\underline{Q}$ be a point-wise limit of a net of n-monotone lower previsions $\underline{P}_\alpha$, defined on a $\wedge$-semilattice of bounded gambles $\mathscr{K}$. Then the limit lower prevision $\underline{Q}$ is n-monotone as well.*

Proof. First fix α. Then for any $p \in \mathbb{N}_{>0}$, $2 \leq p \leq n$, and any $f, f_1, \ldots, f_p \in \mathscr{K}$, it holds that

$$\sum_{I \subseteq \{1,\ldots,p\}} (-1)^{|I|} \underline{P}_\alpha\left(f \wedge \bigwedge_{i \in I} f_i\right) \geq 0.$$

As taking a limit preserves inequalities (see Proposition 1.13(v)$_9$), and because taking a limit is a linear operation (see Section 5.3$_{80}$, or alternatively, Proposition 1.13(iii) and (iv)$_9$), this leads to the desired result,

$$\begin{aligned}\sum_{I \subseteq \{1,\ldots,p\}} (-1)^{|I|} \underline{Q}\left(f \wedge \bigwedge_{i \in I} f_i\right) &= \sum_{I \subseteq \{1,\ldots,p\}} (-1)^{|I|} \lim_\alpha \underline{P}_\alpha\left(f \wedge \bigwedge_{i \in I} f_i\right) \\ &= \lim_\alpha \sum_{I \subseteq \{1,\ldots,p\}} (-1)^{|I|} \underline{P}_\alpha\left(f \wedge \bigwedge_{i \in I} f_i\right) \geq 0,\end{aligned}$$

which shows that $\underline{Q}$ is indeed n-monotone. □

The proof we have given cannot be adapted to work for limits inferior rather than limits, and indeed, because taking a lower envelope can always be seen as a trivial special case of taking a limit inferior (with respect to an appropriately chosen directed set), we see that taking a limit inferior need not preserve n-monotonicity.

6.2 *n*-Monotonicity and coherence

We now turn to the relation between n-monotonicity and coherence. Before attempting a more systematic exploration, we make a few preliminary observations.

6.2.1 A few observations

The first was mentioned by Walley (1981, p. 58).

Proposition 6.9 (Three-element space) *Any coherent lower probability on a three-element space is* 2*-monotone.*

Let us spell out the simple argument here for the sake of completeness.

Proof. Let $\mathscr{X} := \{a, b, c\}$, and let $\underline{P}$ be a coherent lower probability on $\mathscr{P}(\mathscr{X})$. We want to show that $\underline{P}$ is 2-monotone. As any coherent lower probability is monotone (or equivalently, 1-monotone), we infer from Proposition 6.2$_{102}$ that it remains to show that

$$\underline{P}(A \cup B) + \underline{P}(A \cap B) \geq \underline{P}(A) + \underline{P}(B) \text{ for all subsets } A \text{ and } B \text{ of } \mathscr{X}.$$

This inequality trivially satisfied whenever $A \subseteq B$ or $B \subseteq A$. It is an immediate consequence of coherence (sub-additivity, Theorem 4.13(v)$_{53}$) whenever $A \cap B = \emptyset$. The only remaining case is therefore, essentially and without loss of generality, the case that $A = \{a, b\}$ and $B = \{b, c\}$, so $A \cup B = \mathscr{X}$ and $A \cap B = \{b\}$. But then, because $\underline{P}(A \cup B) = 1$ by coherence (normality, Theorem 4.13(ii)$_{53}$), the inequality given earlier turns into $\overline{P}(\{a, c\}) \leq \overline{P}(\{a\}) + \overline{P}(\{c\})$, after passing to the conjugate upper probabilities using Equation (4.9)$_{56}$, and this inequality is satisfied by coherence (super-additivity, Theorem 4.13(v)$_{53}$). □

Interestingly, coherence guarantees n-monotonicity only if $n = 1$: a coherent lower prevision on a lattice of bounded gambles is monotone (or equivalently, 1-monotone) but not necessarily 2-monotone, as the following counterexample, taken from De Cooman et al. (2005a, 2008a, Counterexample 1), shows.

Example 6.10 Let $\mathscr{X} := \{a, b, c\}$, and consider the lower prevision $\underline{P}$ defined on $\{1, f\}$ by $\underline{P}(f) := \underline{P}(1) := 1$, where $f(a) := 0$, $f(b) := 1$ and $f(c) := 2$. The natural extension $\underline{E}_{\underline{P}}$ of $\underline{P}$, defined on the lattice $\mathbb{B}$, is given by

$$\underline{E}_{\underline{P}}(g) = \min\left\{g(b), g(c), \frac{g(a) + g(c)}{2}\right\} \text{ for all bounded gambles } g \text{ on } \mathscr{X}.$$

The restriction of $\underline{E}_{\underline{P}}$ to the lattice of $\{0, 1\}$-valued bounded gambles – indicators – on $\mathscr{X}$ is a 2-monotone coherent lower probability, simply because, as we have just found out in Proposition 6.9$_{\frown}$, any coherent lower probability on a three-element space is 2-monotone. However, $\underline{E}_{\underline{P}}$ itself is *not* 2-monotone:

$$\underline{E}_{\underline{P}}(f \vee 1) + \underline{E}_{\underline{P}}(f \wedge 1) = 1 + \frac{1}{2} < 1 + 1 = \underline{E}_{\underline{P}}(f) + \underline{E}_{\underline{P}}(1),$$

and this indeed violates the condition for 2-monotonicity. ♦

This example also shows that a coherent and 2-monotone lower prevision may have coherent extensions that are no longer 2-monotone on the larger domain. In addition, it tells us that there are coherent lower previsions that are not 2-monotone. Here is another example to show this, involving only lower probabilities. It was mentioned by Walley (1981, p. 51) and goes back to Suppes (1974, p. 171).

Example 6.11 Consider flipping two coins independently; the first is considered to be fair, and the second to have an unknown bias towards heads. The model for the outcome X of the first coin flip is the linear prevision P_X on $\mathbb{B}(\mathscr{X})$, given by

$$P_X(f) := \frac{f(h) + f(t)}{2} \text{ for all bounded gambles } f \text{ on } \mathscr{X},$$

where of course $\mathscr{X} := \{h, t\}$, and h stands for 'heads' and t for 'tails'. The model for the outcome Y of the second coin flip is the coherent lower prevision $\underline{P}_Y$ on $\mathbb{B}(\mathscr{X})$, given by

$$\begin{aligned}\underline{P}_Y(g) &:= \min\left\{\frac{g(h) + g(t)}{2}, g(h)\right\} \\ &= \frac{1}{2}g(h) + \frac{1}{2}\min\{g(h), g(t)\} \text{ for all bounded gambles } g \text{ on } \mathscr{Y},\end{aligned}$$

where also $\mathcal{Y} := \{h,t\}$. The appropriate model for the outcome $Z = (X,Y)$ in $\mathcal{Z} := \mathcal{X} \times \mathcal{Y} = \{(h,h),(h,t),(t,h),(t,t)\}$ is the 'independent product'[2] of P_X and $\underline{P}_Y$ that is the lower envelope of the independent products of P_X with the two extreme points of lins$(\underline{P}_Y)$: the lower prevision $\underline{P}_Z$ on $\mathbb{B}(\mathcal{Z})$ is given by

$$\underline{P}_Z(f) = \frac{1}{2}\frac{f(h,h)+f(t,h)}{2} + \frac{1}{2}\min\left\{\frac{f(h,h)+f(t,h)}{2}, \frac{f(h,t)+f(t,t)}{2}\right\}$$

for all bounded gambles f on $\mathcal{Z}$. It is coherent as a convex mixture of a linear prevision and a lower envelope of linear previsions (see Propositions 4.19_{59} and 4.20_{59}). But

$$\begin{aligned}\underline{P}_Z(\{(h,h),(h,t)\}) + \underline{P}_Z(\{(h,h),(t,t)\}) &= \frac{1}{2} + \frac{1}{2} \\ &> \frac{1}{2} + \frac{1}{4} = \underline{P}_Z(\{(h,h),(h,t),(t,t)\}) + \underline{P}_Z(\{(h,h)\}),\end{aligned}$$

so $\underline{P}_Z$ is not 2-monotone. ♦

6.2.2 Results for lower probabilities

We first concentrate on the relationship between n-monotonicity and coherence for lower probabilities.

A coherent lower probability on a lattice of events is always monotone or, in other words, 1-monotone. In Example 6.10, we have shown that a coherent lower prevision that is 2-monotone on all events need not be 2-monotone on all bounded gambles. But at the same time, a lower probability defined on a lattice of events can be coherent without necessarily being 2-monotone, as we have seen in Example 6.11. Conversely, a 2-monotone lower probability defined on a lattice of events need not be coherent: it suffices to consider any constant non-zero lower probability on $\mathcal{P}(\mathcal{X})$. In the following discussion, we give very simple necessary and sufficient conditions for the coherence of an n-monotone lower probability, we characterise its natural extension, and we prove that this natural extension is still n-monotone.

We work towards these results in a number of steps. As a first and crucial step, we recall an argument that Walley (1981, Section 6) has used for lower probabilities defined on fields of events. This result essentially seems to go back to Choquet (1953–1954, Section 54.2).[3] We give a proof for completeness. A detailed discussion of the Choquet functional (or Choquet integral) that appears in this proposition, and is used extensively in the rest of this section, is given in Appendix C_{376}.

---

[2] This is the so-called *strong*, or type-1, product (Walley, 1991, Section 9.3.3). As P_1 is a linear prevision, it coincides with other independent products, such as the *independent natural extension* (Walley, 1991, Section 9.3), and the forward and backward irrelevant products (De Cooman and Miranda, 2009); see De Cooman and Troffaes (2004), Couso et al. (2000) or De Cooman et al. (2011) for more details.

[3] We say 'seems' because we are not entirely convinced by the proof that Choquet gives for an earlier result (Choquet, 1953–1954, Theorem 54.1), which the theorem in the afore-mentioned Section 54.2 of Choquet's (1953–1954) paper is based on, namely, the equivalence of convexity and 2-monotonicity for real functionals on positive cones of a Riesz space. See also Marinacci and Montrucchio (2008) for a recent and very general discussion of this so-called *Choquet property*.

Proposition 6.12 *Consider a lower probability $\underline{P}$ defined on all events. Assume that $\underline{P}$ is non-negative and that $\underline{P}(\emptyset) = 0$ and $\underline{P}(\mathcal{X}) = 1$. Consider the associated real* ***Choquet functional*** *$C_{\underline{P}}$ on $\mathbb{B}$ defined by*

$$C_{\underline{P}}(f) := C\int f\,\mathrm{d}\underline{P} = \inf f + R\int_{\inf f}^{\sup f} \underline{P}(\{f \geq x\})\,\mathrm{d}x \text{ for all bounded gambles } f.$$

Then $C_{\underline{P}}$ extends $\underline{P}$, and the following statements are equivalent:

(i) *$\underline{P}$ is 2-monotone*

(ii) *$C_{\underline{P}}$ is super-additive: $C_{\underline{P}}(f+g) \geq C_{\underline{P}}(f) + C_{\underline{P}}(g)$ for all f and g in $\mathbb{B}$*

(iii) *$C_{\underline{P}}$ is a coherent lower prevision on $\mathbb{B}$.*

Proof. For a start, observe that for all $A \subseteq \mathcal{X}$, $C_{\underline{P}}(I_A) = \underline{P}(A)$ (see also Proposition C.5(ii)$_{382}$). We now turn to a proof of the equivalences.

(i)⇔(ii). This now follows from Proposition C.7$_{385}$.

(ii)⇒(iii). This is immediate: it is clear from Proposition C.5(i) and (iii)$_{382}$ that $C_{\underline{P}}$ is already guaranteed to satisfy conditions (i)$_{56}$ and (ii)$_{57}$ of the simplified characterisation of coherence in Theorem 4.15$_{56}$.

(iii)⇒(ii). Any coherent lower prevision is super-additive (Theorem 4.13(v)$_{53}$). □

As a second step, we prove a result that we will come back to in Chapter 8$_{151}$: a linear prevision is completely determined by its restriction to events.

Theorem 6.13 *Let P be a linear prevision on $\mathbb{B}$. Then P coincides with the Choquet functional associated with its restriction to events: $P = C_P$ or, in other words,*

$$P(f) = C_P(f) = \inf f + R\int_{\inf f}^{\sup f} P(\{f \geq t\})\,\mathrm{d}t \text{ for all bounded gambles } f \text{ on } \mathcal{X}.$$

Proof. Because both P (Corollary 4.14(xii)$_{55}$) and C_P (Proposition C.5(ix)$_{382}$) are continuous with respect to the topology of uniform convergence, and each bounded gamble is a uniform limit of simple gambles (Proposition 1.20$_{15}$), it suffices to prove the equality for simple gambles only. So assume that f has only a finite number n of values $f_1 < f_2 < \cdots < f_n$. Then we know from Corollary C.4$_{381}$ that $f = f_1 + \sum_{k=2}^{n}(f_k - f_{k-1})I_{\{f \geq f_k\}}$ and $C_P(f) = f_1 + \sum_{k=2}^{n}(f_k - f_{k-1})P(\{f \geq f_k\})$. On the other hand, it follows from the linearity (Corollary 4.14(v) and (vi)$_{55}$) of the linear prevision P that

$$P(f) = P\left(f_1 + \sum_{k=2}^{n}(f_k - f_{k-1})I_{\{f \geq f_k\}}\right) = f_1 + \sum_{k=2}^{n}(f_k - f_{k-1})P(I_{\{f \geq f_k\}}),$$

so indeed $C_P(f) = P(f)$. □

Next, we draw attention to a related result by Walley (1981, Theorem 6.1), which explores the relation between the Choquet functional and the natural extension of a lower probability.

Theorem 6.14 *Consider a lower probability* $\underline{P}$ *defined on all events, which avoids sure loss, is non-negative and monotone and satisfies* $\underline{P}(\emptyset) = 0$ *and* $\underline{P}(\mathscr{X}) = 1$. *Then for all bounded gambles* f, $\underline{E}_{\underline{P}}(f) \geq C_{\underline{P}}(f)$. *There is equality for all* $f \in \mathbb{B}$ *if and only if* $\underline{P}$ *is 2-monotone.*

Proof. Consider any linear prevision P in the set $\text{lins}(\underline{P})$ (non-empty because of Theorem 4.38(i)$_{71}$) and any bounded gamble f on $\mathscr{X}$. Then for all real t, $P(\{f \geq t\}) \geq \underline{P}(\{f \geq t\})$ and therefore $C_P(f) \geq C_{\underline{P}}(f)$. But because, by Theorem 6.13, $P = C_P$, we see that $P(f) \geq C_{\underline{P}}(f)$, and therefore indeed, using the Lower Envelope Theorem 4.38(iii)$_{71}$, $\underline{E}_{\underline{P}}(f) = \min\left\{P(f) : P \in \text{lins}(\underline{P})\right\} \geq C_{\underline{P}}(f)$.

To complete the proof, we turn to the equivalence. If $\underline{E}_{\underline{P}} = C_{\underline{P}}$, then $C_{\underline{P}}$ is a coherent lower prevision on $\mathbb{B}$ because $\underline{E}_{\underline{P}}$ is (Theorem 4.26$_{65}$), and, therefore, $\underline{P}$ is 2-monotone, by Proposition 6.12. Conversely, if $\underline{P}$ is 2-monotone, then $C_{\underline{P}}$ is a coherent lower prevision on $\mathbb{B}$ that extends $\underline{P}$ (again by Proposition 6.12), and it, therefore, dominates the point-wise smallest coherent extension $\underline{E}_{\underline{P}}$ (Theorem 4.26$_{65}$): $C_{\underline{P}} \geq \underline{E}_{\underline{P}}$. But we already know from the argument above that $C_{\underline{P}} \leq \underline{E}_{\underline{P}}$, so indeed $C_{\underline{P}} = \underline{E}_{\underline{P}}$. □

And as a fourth and final step, we show that Choquet integration preserves n-monotonicity. The proof is borrowed from De Cooman et al. (2005a, Theorem 7); also see De Cooman et al. (2008a, Theorem 9) for a more general form.

Theorem 6.15 *Let* $n \in \mathbb{N}^*_{>0}$, $n \geq 2$, *and let* $\underline{P}$ *be a lower probability defined on all events, such that* $\underline{P}(\emptyset) = 0$. *If* $\underline{P}$ *is n-monotone, then so is the associated Choquet functional* $C_{\underline{P}}$.

Proof. Let $p \in \mathbb{N}_{>0}$, $p \leq n$, and let $f, f_1, \ldots, f_p$ be arbitrary bounded gambles on $\mathscr{X}$. Let $a := \min\{\inf f, \min_{k=1}^{p} \inf f_k\}$ and $b := \max\{\sup f, \max_{k=1}^{p} \sup f_k\}$. Consider any $I \subseteq \{1, \ldots, p\}$ and let $g_I := f \wedge \bigwedge_{i \in I} f_i$. then $a \leq \inf g_I$ and $b \geq \sup g_I$. As $\underline{P}$ is a set function (it is monotone and $\underline{P}(\emptyset) = 0$), we get

$$
\begin{aligned}
C_{\underline{P}}\left(f \wedge \bigwedge_{i \in I} f_i\right) &= \underline{P}(\mathscr{X}) \inf g_I + R\int_{\inf g_I}^{\sup g_I} \underline{P}\left(\{g_I \geq x\}\right) \mathrm{d}x \\
&= \underline{P}(\mathscr{X}) \inf g_I + R\int_{a}^{b} \underline{P}\left(\{g_I \geq x\}\right) \mathrm{d}x \\
&\qquad + R\int_{\inf g_I}^{a} \underline{P}\left(\{g_I \geq x\}\right) \mathrm{d}x + R\int_{b}^{\sup g_I} \underline{P}\left(\{g_I \geq x\}\right) \mathrm{d}x \\
&= \underline{P}(\mathscr{X}) \inf g_I + R\int_{a}^{b} \underline{P}\left(\{g_I \geq x\}\right) \mathrm{d}x + (a - \inf g_I)\underline{P}(\mathscr{X}) + 0 \\
&= \underline{P}(\mathscr{X}) a + R\int_{a}^{b} \underline{P}\left(\left\{f \wedge \bigwedge_{i \in I} f_i \geq x\right\}\right) \mathrm{d}x.
\end{aligned}
$$

As it is obvious that for any x in $\mathbb{R}$

$$
\left\{f \wedge \bigwedge_{i \in I} f_i \geq x\right\} = \{f \geq x\} \cap \bigcap_{i \in I} \{f_i \geq x\},
$$

it follows from the n-monotonicity of $\underline{P}$ that for all real x

$$\sum_{I\subseteq\{1,\dots,p\}} (-1)^{|I|}\underline{P}\left(\left\{f \wedge \bigwedge_{i\in I} f_i \geq x\right\}\right) \geq 0.$$

If we take the Riemann integral over $[a, b]$ on both sides of this inequality, and recall moreover that $\sum_{I\subseteq\{1,\dots,p\}}(-1)^{|I|} = 0$, we get

$$\sum_{I\subseteq\{1,\dots,p\}} (-1)^{|I|}C_{\underline{P}}\left(f \wedge \bigwedge_{i\in I} f_i\right) \geq 0.$$

This tells us that $C_{\underline{P}}$ is n-monotone. □

We can now combine the results in these four steps to arrive at the following important conclusions. The first is due to De Cooman et al. (2005a, Proposition 4); also see De Cooman et al. (2008a, Theorem 7).

Corollary 6.16 *Let $\underline{P}$ be an n-monotone lower probability defined on a lattice of events that contains $\emptyset$ and $\mathcal{X}$, with $n \in \mathbb{N}^*_{>0}$ and $n \geq 2$. Then $\underline{P}$ is coherent if and only if $\underline{P}(\emptyset) = 0$ and $\underline{P}(\mathcal{X}) = 1$.*

Proof. Clearly, $\underline{P}(\emptyset) = 0$ and $\underline{P}(\mathcal{X}) = 1$ are necessary for coherence.

Conversely, by Theorem 6.4$_{105}$, the inner set function $\underline{P}_*$ of $\underline{P}$ to all events is also n-monotone and, hence, 2-monotone. This implies that $\underline{P}_*$ is in particular monotone, and therefore, $0 = \underline{P}_*(\emptyset) \leq \underline{P}_*(A) \leq \underline{P}_*(\mathcal{X}) = 1$ for all events A. We can now invoke Proposition 6.12$_{110}$ to find that $C_{\underline{P}_*}$ is a coherent lower prevision that extends $\underline{P}_*$. It follows from the monotonicity of $\underline{P}$ that $\underline{P}$ is also a restriction of the coherent $C_{\underline{P}_*}$ and is therefore coherent too. □

The next result is due to De Cooman et al. (2005a, Theorems 6 and 7); also see De Cooman et al. (2008a, Theorems 8 and 9).

Corollary 6.17 *Let $n \in \mathbb{N}^*_{>0}$, $n \geq 2$, and let $\underline{P}$ be an n-monotone coherent lower probability defined on a lattice of events that contains both $\emptyset$ and $\mathcal{X}$. Then its natural extension $\underline{E}_{\underline{P}}$ is equal to the Choquet functional $C_{\underline{P}_*}$ associated with its inner set function $\underline{P}_*$ and is, therefore, n-monotone as well:*

$$\underline{E}_{\underline{P}}(f) = (C)\int f\,\mathrm{d}\underline{P}_* = \inf f + R\int_{\inf f}^{\sup f} \underline{P}_*(\{f \geq x\})\,\mathrm{d}x. \tag{6.3}$$

Proof. We first consider the natural extension to events. Take $A \subseteq \mathcal{X}$. Then for any Q in $\mathrm{lins}(\underline{P})$, because Q is monotone and dominates $\underline{P}$,

$$Q(A) \geq \sup_{B\subseteq A, B\in\mathrm{dom}\,\underline{P}} Q(B) \geq \sup_{B\subseteq A, B\in\mathrm{dom}\,\underline{P}} \underline{P}(B) = \underline{P}_*(A).$$

As we know from Theorem 4.38$_{71}$ that $\underline{E}_{\underline{P}}(A) = \min\left\{Q(A) : Q \in \mathrm{lins}(\underline{P})\right\}$, we deduce that $\underline{E}_{\underline{P}}(A) \geq \underline{P}_*(A)$ for all $A \subseteq \mathcal{X}$.

Conversely, by Theorem 6.4_{105}, $\underline{P}_*$ is n-monotone because $\underline{P}_*$ is. Hence, by Corollary 6.16, $\underline{P}_*$ is a coherent extension of $\underline{P}$ to all events. Therefore, by Theorem 4.26_{65}, $\underline{P}_*$ must dominate the natural extension $\underline{E}_{\underline{P}}$ of $\underline{P}$ on all events: $\underline{E}_{\underline{P}}(A) \leq \underline{P}_*(A)$ for all $A \subseteq \mathscr{X}$. This shows that $\underline{E}_{\underline{P}}$ and $\underline{P}$ coincide on events.

Next, we consider the natural extension to bounded gambles. Because natural extension is transitive (see Corollary 4.32_{67}), we see that $\underline{E}_{\underline{P}}$ and $\underline{E}_{\underline{P}_*}$ coincide on all bounded gambles. Hence, indeed, by Theorem 6.14_{111}, $\underline{E}_{\underline{P}} = \underline{E}_{\underline{P}_*} = C_{\underline{P}_*}$.

That $\underline{E}_{\underline{P}}$ is n-monotone now follows immediately, because by Theorems 6.4_{105} and 6.15_{111}, $\underline{P}_*$ and $C_{\underline{P}_*}$ are. □

Equation (6.3) also ensures that $\underline{E}_{\underline{P}}$ is *comonotone additive* on $\mathbb{B}$, because that is a property of the Choquet functional associated with any set function (see Proposition C.5(vii)_{382}): if two bounded gambles f and g are *comonotone* in the sense of Definition C.2_{378}, meaning that

$$(\forall x_1, x_2 \in \mathscr{X})(f(x_1) < f(x_2) \Rightarrow g(x_1) \leq g(x_2)),$$

then $\underline{E}_{\underline{P}}(f + g) = \underline{E}_{\underline{P}}(f) + \underline{E}_{\underline{P}}(g)$. We will come back to this in Section 6.3.

The following is due to De Cooman et al. (2005a, Corollary 8); also see De Cooman et al. (2008a, Corollary 10).

Corollary 6.18 *Let $\underline{P}$ be any coherent lower probability defined on a lattice of events containing both $\emptyset$ and $\mathscr{X}$. Let $n \in \mathbb{N}^*_{>0}$, $n \geq 2$. Then $\underline{P}$ is n-monotone, if and only if $\underline{E}_{\underline{P}}$ is n-monotone, if and only if $C \int \bullet \, \mathrm{d}\underline{P}_*$ is n-monotone.*

Proof. If $\underline{P}$ is n-monotone, then $\underline{E}_{\underline{P}}$ is n-monotone by Corollary 6.17.

If $\underline{E}_{\underline{P}}$ is n-monotone, then $\underline{P}$ is n-monotone because $\underline{E}_{\underline{P}}$ is an extension of $\underline{P}$ (because $\underline{P}$ is coherent, see Theorem 4.26(ii)_{65}), and so, by Corollary 6.17, $\underline{E}_{\underline{P}}$ must coincide with $C \int \bullet \, \mathrm{d}\underline{P}_*$, which must therefore be n-monotone as well.

Finally, if $C \int \bullet \, \mathrm{d}\underline{P}_*$ is n-monotone, then $\underline{P}_*$ must be n-monotone because $C \int \bullet \, \mathrm{d}\underline{P}_*$ is an extension of $\underline{P}_*$ by Proposition C.5(ii)_{382} (or alternatively, by Proposition 6.12_{110}). But, $\underline{P}_*$ is also an extension of $\underline{P}$ (because $\underline{P}$ is also 1-monotone), so $\underline{P}$ is n-monotone as well. □

6.3 Representation results

Let us now focus on the notion of n-monotonicity we have given for lower previsions. We investigate whether a result akin to Corollary 6.17 holds for n-monotone coherent lower previsions: when will the natural extension of an n-monotone coherent lower prevision be n-monotone? For Corollary 6.17, we needed the domain of the lower probability to be a lattice of events containing $\emptyset$ and $\mathscr{X}$. It turns out that for our generalisation, we have to impose a similar condition on the domain: it will have to be a linear lattice containing all constant gambles. Recall that a subset of $\mathbb{B}$ is called a *linear lattice* if it is a linear space under point-wise addition and scalar multiplication with real numbers and if it is, moreover, closed under point-wise minimum $\wedge$ and point-wise maximum $\vee$.

Consider a coherent lower prevision whose domain is a linear lattice of bounded gambles that contains all constant gambles. Then its natural extension to the set of all bounded gambles $\mathbb{B}$ is its inner extension $\underline{P}_*$, by Theorem 4.34$_{69}$. With Theorem 6.3$_{104}$, this leads us to the following immediate conclusion, which is a counterpart of Corollary 6.17$_{112}$ for n-monotone coherent lower previsions; see also De Cooman et al. (2005a, Theorem 9 and Theorem 10) and De Cooman et al. (2008a, Theorem 13).

Theorem 6.19 *Let $n \in \mathbb{N}^*_{>0}$, and let $\underline{P}$ be a coherent lower prevision defined on a linear lattice of bounded gambles that contains all constant gambles. If $\underline{P}$ is n-monotone, then its natural extension $\underline{E}_{\underline{P}}$ is equal to its inner extension $\underline{P}_*$ and is, therefore, n-monotone as well.*

Example 6.10$_{108}$ tells us that this result cannot be extended to lattices of bounded gambles that are not at the same time linear spaces.

Here also, following De Cooman et al. (2005a, Theorem 11) and De Cooman et al. (2008a, Theorem 14), the natural extension can, to some extent, be related to Choquet integration. Let us explore this idea. Consider a linear lattice of bounded gambles $\mathscr{K}$ that contains all constant gambles. Then the set

$$\mathscr{F}_{\mathscr{K}} := \left\{A \subseteq \mathscr{X} : I_A \in \mathscr{K}\right\}$$

of events that belong to $\mathscr{K}$ is a field of subsets of $\mathscr{X}$. As set out in Definition 1.17$_{12}$, we denote by $\mathbb{B}_{\mathscr{F}_{\mathscr{K}}}$ the uniformly closed linear lattice

$$\mathbb{B}_{\mathscr{F}_{\mathscr{K}}} := \operatorname{cl}(\operatorname{span}(I_{\mathscr{F}_{\mathscr{K}}})) = \operatorname{cl}(\operatorname{span}(\mathscr{F}_{\mathscr{K}})),$$

where $I_{\mathscr{F}_{\mathscr{K}}} := \left\{I_A : I_A \in \mathscr{K}\right\}$, $\operatorname{cl}(\bullet)$ denotes uniform closure and $\operatorname{span}(\bullet)$ takes the linear span; see also the notions and notations established in Section 1.8$_{12}$. Observe that $\mathbb{B}_{\mathscr{F}_{\mathscr{K}}}$ contains all constant gambles as well. Its elements are the $\mathscr{F}_{\mathscr{K}}$*-measurable bounded gambles*. By definition, every such $\mathscr{F}_{\mathscr{K}}$-measurable bounded gamble is a uniform limit of $\mathscr{F}_{\mathscr{K}}$*-simple gambles*–elements of $\operatorname{span}(\mathscr{F}_{\mathscr{K}})$. Moreover, it is clear that $\mathbb{B}_{\mathscr{F}_{\mathscr{K}}} \subseteq \operatorname{cl}(\mathscr{K})$.

Theorem 6.20 *Let $\underline{P}$ be an n-monotone coherent lower prevision on a linear lattice of bounded gambles $\mathscr{K}$ that contains all constant gambles. Then $\underline{P}$ has a unique coherent extension to* $\operatorname{cl}(\mathscr{K})$*, and this extension is n-monotone as well. Denote by $\underline{Q}$ the restriction of $\underline{P}$ to $\mathscr{F}_{\mathscr{K}}$. Then for all f in $\mathbb{B}_{\mathscr{F}_{\mathscr{K}}}$,*

$$\underline{E}_{\underline{P}}(f) = \underline{E}_{\underline{Q}}(f) = C\int f \,\mathrm{d}\underline{Q}_* = \inf f + R\int_{\inf f}^{\sup f} \underline{Q}_*(\{f \geq x\})\,\mathrm{d}x.$$

Consequently, $\underline{E}_{\underline{P}}$ is both n-monotone and comonotone additive on $\mathbb{B}_{\mathscr{F}_{\mathscr{K}}}$.

Proof. We begin with a proof of the equalities. As $\underline{P}$ is n-monotone and coherent, its restriction $\underline{Q}$ to the field $\mathscr{F}_{\mathscr{K}}$ is an n-monotone coherent lower probability. By Corollary 6.17_{112}, the natural extension $\underline{E}_{\underline{Q}}$ of $\underline{Q}$ to the set $\mathbb{B}$ of all bounded gambles is the Choquet functional $C_{\underline{Q}_*}$ associated with the n-monotone inner set function $\underline{Q}_*$ of $\underline{Q}$: for any bounded gamble f on $\mathscr{X}$,

$$\underline{E}_{\underline{Q}}(f) = C\int f\,\mathrm{d}\underline{Q}_* = \inf f + R\int_{\inf f}^{\sup f} \underline{Q}_*(\{f \geq x\})\,\mathrm{d}x. \tag{6.4}$$

If we can prove that $\underline{E}_{\underline{Q}}$ and $\underline{E}_{\underline{P}}$ coincide on $\mathbb{B}_{\mathscr{F}_{\mathscr{K}}}$, then it will follow at once that $\underline{E}_{\underline{P}}$ is both n-monotone and comonotone additive on $\mathbb{B}_{\mathscr{F}_{\mathscr{K}}}$, because $\underline{E}_{\underline{Q}}$ is n-monotone (use Corollary 6.17_{112}) as well as comonotone additive (use Proposition C.5(vii)_{382}) on $\mathbb{B}$. To prove that $\underline{E}_{\underline{Q}}$ and $\underline{E}_{\underline{P}}$ indeed coincide on the subset $\mathbb{B}_{\mathscr{F}_{\mathscr{K}}}$ of $\mathrm{cl}(\mathscr{K})$, it suffices to prove that $\underline{E}_{\underline{Q}}$ and $\underline{P}$ coincide on $\mathrm{span}(\mathscr{F}_{\mathscr{K}})$, because $\underline{E}_{\underline{Q}}$ and $\underline{E}_{\underline{P}}$ are guaranteed by coherence to be continuous (Theorem 4.13(xiii)_{53}), and because $\underline{E}_{\underline{P}}$ and $\underline{P}$ coincide on $\mathrm{span}(\mathscr{F}_{\mathscr{K}}) \subseteq \mathscr{K}$ (because $\underline{P}$ is coherent on $\mathscr{K}$, see Theorem 4.26(ii)_{65}). Let, therefore, h be any element of $\mathrm{span}(\mathscr{F}_{\mathscr{K}})$, that is, let h be an $\mathscr{F}_{\mathscr{K}}$-simple gamble. This means that h assumes only a finite number n of values $h_1 < h_2 < \cdots < h_n$, and that $\{h \geq h_k\} \in \mathscr{F}$ for $k = 1, \ldots, n$, by Proposition 1.18_{14}. We then infer from Corollary C.4_{381} that $h = \mu_1 + \sum_{k=2}^{n} \mu_k I_{\{h \geq h_k\}}$ and that, also using Equation (6.4),

$$\underline{E}_{\underline{Q}}(h) = C_{\underline{Q}_*}(h) = \mu_1 \underline{Q}_*(\mathscr{X}) + \sum_{k=2}^{n} \mu_k \underline{Q}_*(\{h \geq h_k\}) = \mu_1 + \sum_{k=2}^{n} \mu_k \underline{P}(\{h \geq h_k\}), \tag{6.5}$$

where we let, for ease of notation, $\mu_1 := h_1$ and $\mu_k := h_k - h_{k-1} > 0$ for $k = 2, \ldots, n$. On the other hand, it follows from the coherence (normality, Theorem 4.13(ii)_{53}) and the 2-monotonicity of $\underline{P}$ that

$$\begin{aligned}\underline{P}(h) &= \mu_1 + \underline{P}\left(\sum_{k=2}^{n} \mu_k I_{\{h \geq h_k\}}\right) = \mu_1 - \mu_2 + \underline{P}\left(\sum_{k=2}^{n} \mu_k I_{\{h \geq h_k\}}\right) + \underline{P}(\mu_2) \\ &\leq \mu_1 - \mu_2 + \underline{P}\left(\mu_2 \vee \sum_{k=2}^{n} \mu_k I_{\{h \geq h_k\}}\right) + \underline{P}\left(\mu_2 \wedge \sum_{k=2}^{n} \mu_k I_{\{h \geq h_k\}}\right).\end{aligned}$$

Now it is easily verified that

$$\mu_2 \vee \sum_{k=2}^{n} \mu_k I_{\{h \geq h_k\}} = \mu_2 + \sum_{k=3}^{n} \mu_k I_{\{h \geq h_k\}} \text{ and } \mu_2 \wedge \sum_{k=2}^{n} \mu_k I_{\{h \geq h_k\}} = \mu_2 I_{\{h \geq h_2\}},$$

and consequently, again using the coherence (constant additivity and non-negative homogeneity, Theorem 4.13(iii) and (vi)_{53}) and the 2-monotonicity of $\underline{P}$, and continuing in the same manner,

$$\underline{P}(h) \leq \mu_1 - \mu_2 + \underline{P}\left(\mu_2 + \sum_{k=3}^{n} \mu_k I_{\{h \geq h_k\}}\right) + \underline{P}\left(\mu_2 I_{\{h \geq h_2\}}\right)$$

$$= \mu_1 + \mu_2 \underline{P}\left(\{h \geq h_2\}\right) + \underline{P}\left(\sum_{k=3}^{n} \mu_k I_{\{h \geq h_k\}}\right)$$

$$\leq \mu_1 + \mu_2 \underline{P}\left(\{h \geq h_2\}\right) + \mu_3 \underline{P}\left(\{h \geq h_3\}\right) + \underline{P}\left(\sum_{k=4}^{n} \mu_k I_{\{h \geq h_k\}}\right)$$

$$\vdots$$

$$\leq \mu_1 + \sum_{k=2}^{n} \mu_k \underline{P}\left(\{h \geq h_k\}\right).$$

Hence, $\underline{E}_{\underline{Q}}(h) \geq \underline{P}(h)$, recalling Equation $(6.5)_{\frown}$. On the other hand, because $\underline{P}$ is a coherent extension of $\underline{Q}$, and because the natural extension $\underline{E}_{\underline{Q}}$ is the point-wise smallest coherent extension of $\underline{Q}$ (Theorem 4.31(ii)$_{67}$), we also find that $\underline{E}_{\underline{Q}}(h) \leq \underline{P}(h)$. So $\underline{P}$ and $\underline{E}_{\underline{Q}}$ indeed coincide on span($\mathscr{F}_{\mathscr{K}}$).

To complete the proof, it follows from Proposition 4.24$_{62}$ that $\underline{P}$ has a unique coherent extension to cl($\mathscr{K}$), which, by Theorem 4.31(ii)$_{67}$, coincides with the natural extension $\underline{E}_{\underline{P}}$ on cl($\mathscr{K}$). By Theorem 6.19$_{114}$, $\underline{E}_{\underline{P}}$ is n-monotone. □

We have already had occasion to mention that, in general, coherent lower previsions are not determined by the values they assume on events: two coherent lower previsions may coincide on events but fail to do so on some bounded gambles; see, for instance, the discussion in Section 5.2$_{78}$ and in particular Example 5.3$_{79}$. Interestingly, the preceding theorem tells us that for coherent lower previsions that are 2-monotone and defined on a sufficiently rich domain, we can somewhat improve on this negative result: on $\mathscr{F}_{\mathscr{K}}$-measurable bounded gambles, the natural extension $\underline{E}_{\underline{P}}$ of an n-monotone coherent lower prevision $\underline{P}$ is completely determined by the values that $\underline{P}$ assumes on the events in $\mathscr{F}_{\mathscr{K}}$. Nevertheless, the following counterexample, taken from De Cooman et al. (2005a, 2008a, Counterexample 2), tells us that, in general, we cannot expect to take this result beyond the set $\mathbb{B}_{\mathscr{F}_{\mathscr{K}}}$ of $\mathscr{F}_{\mathscr{K}}$-measurable bounded gambles.

Example 6.21 Let $\mathscr{X}$ be the closed unit interval $[0, 1]$ in $\mathbb{R}$, and let $\underline{P}$ be the lower prevision on the set $\mathscr{C}([0, 1])$ of all continuous bounded gambles on $\mathscr{X}$, defined by $\underline{P}(g) := g(0)$ for any g in $\mathscr{C}([0, 1])$. This puts us squarely in the context described in Section 5.5.6$_{98}$. As $\underline{P}$ is actually a linear prevision, it must be completely monotone (see Theorem 6.6$_{105}$). Observe that $\mathscr{C}([0, 1])$ is a uniformly closed linear lattice that contains all constant gambles. Moreover, because the only continuous indicators on $[0, 1]$ are the constant gambles 0 and 1, we see that $\mathscr{F}_{\mathscr{C}([0,1])} = \{\emptyset, \mathscr{X}\}$, so $\mathbb{B}_{\mathscr{F}_{\mathscr{C}([0,1])}}$ is the set of all constant gambles, and the natural extension $\underline{E}_{\underline{Q}}$ of the restriction $\underline{Q}$ of $\underline{P}$ to $\mathscr{F}_{\mathscr{C}([0,1])}$ is the vacuous lower prevision on $\mathbb{B}([0, 1])$: $\underline{E}_{\underline{Q}}(f) = \inf f$ for all bounded gambles f on $[0, 1]$. Therefore, for any g in $\mathscr{C}([0, 1])$ such that $g(0) > \min g$, it follows that $\underline{E}_{\underline{Q}}(g) < \underline{P}(g)$: the equality in Theorem 6.20$_{114}$ holds only for those bounded gambles in $\mathscr{C}([0, 1])$ that satisfy $g(0) = \min g$ or, equivalently, $(\forall x \in (0, 1])g(0) \leq g(x)$. ♦

We conclude that, in general, an n-monotone coherent lower prevision $\underline{P}$ defined on a linear lattice of bounded gambles that contains the constant gambles cannot be written (on its entire domain) as a Choquet functional associated with its restriction $\underline{Q}$ to events.

Instead, however, we can always represent such an n-monotone coherent lower prevision by a Choquet integral with respect to *the restriction to events of its inner extension*, and this Choquet integral also immediately provides us with an alternative expression for the natural extension. *This is because 2-monotonicity and comonotone additivity turn out to be equivalent under coherence.* This result, and its proof, is taken from De Cooman et al. (2005a, Theorem 12); see also De Cooman et al. (2008a, Theorem 15).

Theorem 6.22 *Let $\underline{P}$ be a coherent lower prevision defined on a linear lattice of bounded gambles that contains all constant gambles. Then $\underline{P}$ is comonotone additive if and only if it is 2-monotone, and in both cases, we have*

$$\underline{P}(f) = C\int f \, \mathrm{d}\underline{P}_* = \inf f + R\int_{\inf f}^{\sup f} \underline{P}_*(\{f \geq x\}) \, \mathrm{d}x \text{ for all } f \text{ in } \operatorname{dom} \underline{P}.$$

Proof. 'only if'. Assume that the coherent lower prevision $\underline{P}$ is comonotone additive. Then we may apply our simplified version of Greco's Representation Theorem C.9$_{389}$ and conclude that $\underline{P}(f) = C_\alpha(f)$ for all f in $\mathscr{K}$, where α is the restriction to events of the inner extension $\underline{P}_*$ of $\underline{P}$:

$$\alpha(A) = \underline{P}_*(I_A) = \sup\left\{\underline{P}(f) : f \leq I_A \text{ and } f \in \operatorname{dom} \underline{P}\right\} \text{ for all } A \subseteq \mathcal{X}. \quad (6.6)$$

It is clear that α is a (monotone) set function on $\mathscr{P}(\mathcal{X})$ with $\alpha(\emptyset) = 0$ and $\alpha(\mathcal{X}) = 1$. By Theorem 4.34$_{69}$, α is also the restriction to events of the natural extension $\underline{E}_{\underline{P}} = \underline{P}_*$ of $\underline{P}$.

We now prove that α is 2-monotone. To do so, we first consider $A \subseteq B \subseteq \mathcal{X}$ and show that $\underline{E}_{\underline{P}}(I_A + I_B) = \underline{E}_{\underline{P}}(I_A) + \underline{E}_{\underline{P}}(I_B) = \alpha(A) + \alpha(B)$. As the coherence of $\underline{E}_{\underline{P}}$ implies that it is super-additive (Theorem 4.13(v)$_{53}$), we only need to prove that $\underline{E}_{\underline{P}}(I_A + I_B) \leq \alpha(A) + \alpha(B)$. Given $\varepsilon > 0$, we deduce from Equation (1.8)$_{17}$ that there is some f in $\operatorname{dom} \underline{P}$ such that $f \leq I_A + I_B$ and $\underline{E}_{\underline{P}}(I_A + I_B) = \underline{P}_*(I_A + I_B) \leq \underline{P}(f) + \varepsilon$. We may assume without loss of generality that f is non-negative (because $f \vee 0$ belongs to the lattice $\operatorname{dom} \underline{P}$ and also satisfies the same inequality). Let $g_1 := f \wedge 1$ and $g_2 := f - f \wedge 1$. Both these bounded gambles belong to the linear lattice $\operatorname{dom} \underline{P}$, and moreover, $g_1 + g_2 = f$. Let us show that $g_1 \leq I_B$ and $g_2 \leq I_A$:

- If $x \notin B$, we have $0 \leq f(x) \leq (I_A + I_B)(x) = 0$, whence $g_1(x) = g_2(x) = 0$.

- If $x \in A$, there are two possibilities. If $f(x) \leq 1$, then $g_2(x) = 0$ and $g_1(x) = f(x) \leq 1$. If, on the other hand, $f(x) > 1$, then $g_1(x) = 1$ and $g_2(x) = f(x) - 1 \leq 2 - 1 = 1$.

- If $x \in B \setminus A$, we have $f(x) \leq 1$, whence $g_1(x) = f(x) \leq 1$ and $g_2(x) = 0$.

Moreover, g_1 and g_2 are comonotone: consider any x_1 and x_2 in $\mathscr{X}$, and assume that $g_2(x_1) < g_2(x_2)$. Then $g_2(x_2) > 0$ and consequently $x_2 \in A$ and $f(x_2) > 1$. This implies in turn that indeed $g_1(x_2) = 1 \geq g_1(x_1)$. Hence, because $\underline{P}$ is assumed to be comonotone additive, and because $\underline{E}_{\underline{P}}$ is monotone (because it is coherent, see Theorem 4.13(iv)$_{53}$) and coincides on dom $\underline{P}$ with $\underline{P}$ (Theorem 4.26(ii)$_{65}$), we get

$$\begin{aligned}\underline{E}_{\underline{P}}(I_A + I_B) &\leq \underline{P}(f) + \varepsilon = \underline{P}(g_1 + g_2) + \varepsilon = \underline{P}(g_1) + \underline{P}(g_2) + \varepsilon \\ &\leq \underline{E}_{\underline{P}}(A) + \underline{E}_{\underline{P}}(B) + \varepsilon,\end{aligned}$$

and because this holds for all $\varepsilon > 0$, we find that indeed $\underline{E}_{\underline{P}}(I_A + I_B) \leq \underline{E}_{\underline{P}}(A) + \underline{E}_{\underline{P}}(B) = \alpha(A) + \alpha(B)$.

Next, we consider two arbitrary subsets C and D of $\mathscr{X}$. Then $C \cap D \subseteq C \cup D$, and therefore, using the reasoning above,

$$\begin{aligned}\alpha(C \cup D) + \alpha(C \cap D) &= \underline{E}_{\underline{P}}(I_{C\cup D} + I_{C\cap D}) = \underline{E}_{\underline{P}}(I_C + I_D) \\ &\geq \underline{E}_{\underline{P}}(I_C) + \underline{E}_{\underline{P}}(I_D) = \alpha(C) + \alpha(D),\end{aligned}$$

where the inequality follows from the super-additivity the coherent lower prevision $\underline{E}_{\underline{P}}$ (Theorem 4.13(v)$_{53}$). We conclude that α is indeed 2-monotone on $\mathscr{P}(\mathscr{X})$.

From Corollary 6.16$_{112}$, we conclude that α is a coherent lower probability on $\mathscr{P}(\mathscr{X})$, so, by Corollary 6.17$_{112}$, its natural extension is the Choquet functional C_α associated with α, which we have already proved to be equal to $\underline{P}$ on dom $\underline{P}$. If we now apply Corollary 6.17$_{112}$, we see that C_α is 2-monotone and, therefore, so is $\underline{P}$.

'if'. Assume that $\underline{P}$ is 2-monotone. Applying Theorems 6.3$_{104}$ and 6.19$_{114}$, its natural extension $\underline{E}_{\underline{P}} = \underline{P}_*$ to all bounded gambles is also 2-monotone, and consequently, so is its restriction α to events. Moreover, $\mathbb{B}_{\mathscr{P}(\mathscr{X})} = \mathbb{B}$, because any bounded gamble is the uniform limit of some sequence of simple gambles. If we now apply Theorem 6.20$_{114}$, we see that $\underline{E}_{\underline{P}}(f) = C_\alpha(f)$ for all f in $\mathbb{B}$. Consequently, $\underline{E}_{\underline{P}}$ is comonotone additive, because the Choquet functional associated with any monotone set function is (Proposition C.5(vii)$_{382}$). So is, therefore, $\underline{P}$. □

The natural extension of an n-monotone ($n \geq 2$) coherent lower prevision defined on a linear lattice of bounded gambles that contains the constant gambles is therefore always comonotone additive. Indeed, this natural extension is the Choquet functional associated to its restriction to events; see also De Cooman et al. (2005a, Corollary 13) and De Cooman et al. (2008a, Corollary 16) for earlier formulations of this corollary.

Corollary 6.23 *Let $n \in \mathbb{N}^*_{>0}$, $n \geq 2$, and let $\underline{P}$ be an n-monotone coherent lower prevision defined on a linear lattice that contains all constant gambles. Then $\underline{E}_{\underline{P}}$ is n-monotone, comonotone additive and equal to the Choquet integral with respect to $\underline{P}_*$ restricted to events.*

Moreover, such a coherent lower prevision is generally not uniquely determined by its restriction to events, but it is uniquely determined by the values that its natural extension $\underline{E}_{\underline{P}} = \underline{P}_*$ assumes on events. Of course, this natural extension also depends in general on the values that $\underline{P}$ assumes on bounded gambles, as is evident from Equation (6.6)$_{117}$. On the other hand, we also deduce from Theorem 6.22$_{117}$ that the procedure of natural extension preserves comonotone additivity from (indicators of) events to bounded gambles.

The following example illustrates that, while an n-monotone coherent lower prevision is generally not uniquely determined by its restriction to events, it is uniquely determined by the values that its natural extension $\underline{E}_{\underline{P}} = \underline{P}_*$ assumes on events.

Example 6.24 We continue the discussion of Example 6.21$_{116}$ and borrow the notions and notations introduced there. It follows from the discussion involving Proposition 5.21$_{100}$ and Corollary 5.22$_{100}$ in Section 5.5.6$_{98}$ that the unique extension $\hat{\underline{P}}$ of $\underline{P}$ to the set $\underline{\mathscr{C}}([0,1])$ of all lower semi-continuous bounded gambles on $[0,1]$ that satisfies upward monotone convergence is given by $\hat{\underline{P}}(h) := h(0)$ for all $h \in \underline{\mathscr{C}}([0,1])$. We then infer from Corollary 5.22$_{100}$ that the natural extension $\underline{E}_{\underline{P}} = \underline{P}_*$ of $\underline{P}$ is given by

$$\underline{E}_{\underline{P}}(f) = \underline{\operatorname{osc}}_0(f) = \sup_{x>0} \inf_{0\leq y<x} f(y) \text{ for all bounded gambles } f \text{ on } [0,1].$$

The restriction $\underline{Q}$ of $\underline{E}_{\underline{P}}$ to (indicators of) events satisfies

$$\underline{Q}(A) = \begin{cases} 1 & \text{if } (\exists x > 0)[0,x) \subseteq A \\ 0 & \text{otherwise,} \end{cases}$$

so the proper filter that is associated with this lower prevision is the neighbourhood filter $\mathscr{N}_0$ of 0, and for the Choquet functional associated with the set function $\underline{Q}$, we have that indeed

$$\begin{aligned} C_{\underline{Q}}(f) &= \inf f + R\int_{\inf f}^{\sup f} \underline{Q}(\{f \geq z\})\,\mathrm{d}z = \sup\left\{z \in \mathbb{R} : \underline{Q}(\{f \geq z\}) = 1\right\} \\ &= \sup\{z \in \mathbb{R} : (\exists x > 0)[0,x) \subseteq \{f \geq z\}\} = \sup_{x>0} \inf_{0\leq y<x} f(y) = \underline{E}_{\underline{P}}(f), \end{aligned}$$

for all bounded gambles f on $[0,1]$. ♦

As a nice side result, we deduce that an n-monotone ($n \geq 2$) coherent lower probability $\underline{P}$ on $\mathscr{P}(\mathscr{X})$, which usually has many coherent extensions to $\mathbb{B}$, has actually *only one 2-monotone* coherent extension to $\mathbb{B}$. This unique 2-monotone coherent extension coincides with the natural extension of $\underline{P}$; see also De Cooman et al. (2005a, Corollary 14) and De Cooman et al. (2008a, Corollary 17).

Corollary 6.25 *Let $n \in \mathbb{N}^*_{>0}$, $n \geq 2$. An n-monotone coherent lower probability defined on all events has a* unique *2-monotone (or equivalently, comonotone additive) coherent extension to all bounded gambles, namely, its natural extension. This natural extension is furthermore automatically also n-monotone.*

Proof. Let $\underline{P}$ be an n-monotone coherent lower probability defined on all events. By Corollary 6.17_{112}, its natural extension $\underline{E}_{\underline{P}}$ to $\mathbb{B}$ is an n-monotone and, hence, 2-monotone coherent extension of $\underline{P}$. The proof is complete if we can show that $\underline{E}_{\underline{P}}$ is the only 2-monotone coherent extension of $\underline{P}$ to all bounded gambles.

So, let $\underline{R}$ be any 2-monotone coherent extension of $\underline{P}$ to all bounded gambles. We must show that $\underline{R} = \underline{E}_{\underline{P}}$. Let f be any bounded gamble on $\mathcal{X}$. Then $\underline{R}(f) = C_{\underline{P}}(f) = \underline{E}_{\underline{P}}(f)$, where the first equality follows from Corollary 6.23_{118} and the second by applying Corollary 6.17_{112}. This establishes uniqueness. □

We can summarise some of the comments and results in this section in the following diagram, which depicts the relationships between the properties of a coherent lower prevision $\underline{P}$ on $\mathbb{B}$ and its restriction $\underline{Q}$ to events; implications are depicted using single arrows, equivalences using double arrows.

$$
\begin{array}{ccccc}
 & & \underline{P} \text{ is } n\text{-monotone} & \longrightarrow & \underline{Q} \text{ is } n\text{-monotone} \\
 & & \downarrow & & \downarrow \\
\underline{P} \text{ is comonotone additive} & \longleftrightarrow & \underline{P} \text{ is 2-monotone} & \longrightarrow & \underline{Q} \text{ is 2-monotone} \\
 & & \updownarrow & & \updownarrow \\
 & & \underline{P} = \underline{E}_{\underline{Q}} = C\int \bullet \, \mathrm{d}\underline{Q} & \longrightarrow & \underline{E}_{\underline{Q}} = C\int \bullet \, \mathrm{d}\underline{Q}
\end{array}
$$

This diagram is essentially based on De Cooman et al. (2008a, Figure 1).

Next, we come to a result that relates comonotone additivity (or, equivalently, 2-monotonicity) of coherent lower previsions to properties of their sets of dominating linear previsions; see also De Cooman et al. (2005a, Proposition 15) and De Cooman et al. (2008a, Proposition 18).

Proposition 6.26 *Let $\underline{P}$ be a coherent lower prevision on a linear lattice of bounded gambles.*

(a) *If $\underline{P}$ is comonotone additive on its domain, then, for all comonotone f and g in* dom $\underline{P}$, *there is some Q in* lins$(\underline{P})$ *such that $Q(f) = \underline{P}(f)$ and $Q(g) = \underline{P}(g)$.*

(b) *Assume in addition that* dom $\underline{P}$ *contains all constant gambles. Then $\underline{P}$ is comonotone additive (or equivalently 2-monotone) on its domain if and only if, for all comonotone f and g in* dom $\underline{P}$, *there is some Q in* lins$(\underline{P})$ *such that $Q(f) = \underline{P}(f)$ and $Q(g) = \underline{P}(g)$.*

Proof. To prove the first statement, assume that $\underline{P}$ is comonotone additive on its domain, and consider f and g in dom $\underline{P}$ that are comonotone. Then $f + g$ also belongs to dom $\underline{P}$, so we know that $\underline{P}(f + g) = \underline{P}(f) + \underline{P}(g)$. On the other hand, because $\underline{P}$ is coherent, there is some Q in lins$(\underline{P})$ such that $\underline{P}(f + g) = Q(f + g) = Q(f) + Q(g)$ (see the Lower Envelope Theorem $4.38(\text{ii})_{71}$). So $Q(f) + Q(g) = \underline{P}(f) + \underline{P}(g)$ and because we know that $\underline{P}(f) \leq Q(f)$ and $\underline{P}(g) \leq Q(g)$, this implies that $\underline{P}(f) = Q(f)$ and $\underline{P}(g) = Q(g)$.

The 'only if' part of the second statement is an immediate consequence of the first. To prove the 'if' part, consider arbitrary comonotone f and g in dom $\underline{P}$. Then it is easy to see that $f \vee g$ and $f \wedge g$ are comonotone as well and belong to dom $\underline{P}$, so, by assumption, there is a Q in lins($\underline{P}$) such that $\underline{P}(f \wedge g) = Q(f \wedge g)$ and $\underline{P}(f \vee g) = Q(f \vee g)$. Then, using Theorem 6.6_{105},

$$\underline{P}(f \vee g) + \underline{P}(f \wedge g) = Q(f \vee g) + Q(f \wedge g) = Q(f) + Q(g) \geq \underline{P}(f) + \underline{P}(g).$$

This tells us that $\underline{P}$ is 2-monotone and, by Theorem 6.22_{117}, also comonotone additive. □

As a corollary, we deduce the following, apparently first proved by Walley (1981, Corollaries 6.4 and 6.5) for coherent lower probabilities, although the ideas behind it seem to go back to Shapley (1971, Theorem 3). Our formulation again borrows from De Cooman et al. (2005a, Proposition 16) and De Cooman et al. (2008a, Corollary 19).

Corollary 6.27 *Let $\underline{P}$ be a coherent lower probability on a lattice of events. Then $\underline{P}$ is 2-monotone if and only if, for all A and B in* dom $\underline{P}$ *such that $A \subseteq B$, there is some Q in* lins($\underline{P}$) *such that $Q(A) = \underline{P}(A)$ and $Q(B) = \underline{P}(B)$.*

Proof. We show that the direct implication is a consequence of the previous results; the converse implication follows easily by applying the condition to $A \cap B \subseteq A \cup B$, for A and B in dom $\underline{P}$.

Let $\underline{P}$ be a 2-monotone coherent lower prevision defined on a lattice of events. By Corollary 6.17_{112}, the natural extension $\underline{E}_{\underline{P}}$ of $\underline{P}$ to all bounded gambles is 2-monotone and coherent. Hence, given $A, B \in \text{dom}\,\underline{P}$ with $A \subseteq B$, because I_A and I_B are comonotone, Proposition 6.26 implies the existence of a Q in lins($\underline{E}_{\underline{P}}$) = lins($\underline{P}$) (see also Proposition 4.37_{71}) such that $Q(A) = \underline{E}_{\underline{P}}(A) = \underline{P}(A)$ and $Q(B) = \underline{E}_{\underline{P}}(B) = \underline{P}(B)$. where the second equalities follow from Theorem 4.26(ii)_{65}. □

7

Special n-monotone coherent lower previsions

Most of the uncertainty models introduced in the literature related to imprecise probabilities are lower previsions or lower probabilities that are at least 2-monotone. In this chapter, we present an overview of what we believe to be relevant and interesting examples.

Upper and lower mass functions represent bounds on mass functions, and we show in Section 7.1 that they give rise to 2-monotone lower (and upper) previsions.

The simplest completely monotone lower previsions that we can think of are the minimum preserving ones, and in particular the vacuous lower previsions. We consider these in Section 7.2_{127}. They constitute the starting point for the remainder of this chapter.

As we have seen in Proposition 6.7_{106} that taking convex combinations preserves n-monotonicity, we know that any convex combination of vacuous lower previsions will be completely monotone. This brings us to the topic of belief functions in Section 7.3_{128}.

On the other hand, Proposition 6.8_{107} guarantees that point-wise limits of vacuous lower previsions are completely monotone as well. This observation leads us to continue, in Section 7.4_{129}, the discussion of lower previsions associated with proper filters, already started in Section 5.5_{82}.

Our novel approach to induced lower previsions in Section 7.5_{131} generalises and subsumes the existing treatment of random sets, belief functions and lower previsions associated with proper filters and foreshadows our account of the Choquet-like representation results in Section 11.6_{225}.

We show in Section 7.7_{142} that minimum preserving lower probabilities arise naturally as natural extensions of lower probability assessments on chains of sets. This leads to the closely related discussion of possibility and necessity measures in Section 7.8_{143}.

Lower Previsions, First Edition. Matthias C.M. Troffaes and Gert de Cooman.

Probability boxes provide a last example of imprecise probability models that can be described in terms of completely monotone lower previsions, and the discussion in Section 7.9_{147} provides a brief overview.

7.1 Lower and upper mass functions

On a finite space $\mathcal{X}$, it is straightforward to specify a precise probability model: as we have seen in Section 5.1_{77}, we simply specify a probability mass function p, which amounts to assessing, for each x in $\mathcal{X}$, that both the lower and upper probabilities of the **elementary event** $\{x\}$ are equal to $p(x)$.

Perhaps the most straightforward way of specifying an imprecise probability model on a finite space $\mathcal{X}$ is based on an obvious extension of this idea: allow the lower probabilities of such elementary events to differ from their upper probabilities. This means that we consider two [0, 1]-valued maps on $\mathcal{X}$, the **lower mass function** $\underline{p}$ and the **upper mass function** $\overline{p}$, and use these to construct a lower prevision $\underline{P}_{\underline{p},\overline{p}}$ with domain $\mathcal{K} := \bigcup_{x\in\mathcal{X}}\{I_{\{x\}}, -I_{\{x\}}\}$ from the following assessments:

$$\underline{P}_{\underline{p},\overline{p}}(I_{\{x\}}) := \underline{p}(x) \text{ and } \underline{P}_{\underline{p},\overline{p}}(-I_{\{x\}}) := -\overline{p}(x) \text{ for all } x \text{ in } \mathcal{X}.$$

Here, we intend to find out when this lower prevision $\underline{P}_{\underline{p},\overline{p}}$ avoids sure loss, when it is coherent and what is its natural extension to all bounded gambles. These results are mentioned without proof by Walley (1991, Section 4.6.1) for the apparently more general case of a lower and upper probability specification on a finite partition.[1] Some of them have also been proved by de Campos et al. (1994), but our proofs are different (and arguably easier) in some of the details.

We denote the natural extension of $\underline{P}_{\underline{p},\overline{p}}$ by $\underline{E}_{\underline{p},\overline{p}}$. It will also be useful to consider the set $\text{lins}(\underline{P}_{\underline{p},\overline{p}}) = \text{lins}(\underline{E}_{\underline{p},\overline{p}})$ of those linear previsions that dominate these lower previsions. If we recall the results and notations from Section 5.1_{77}, we find after some manipulations that

$$\text{lins}(\underline{P}_{\underline{p},\overline{p}}) = \left\{P_p : p \in \Sigma_{\mathcal{X}} \text{ and } \underline{p} \le p \le \overline{p}\right\}. \tag{7.1}$$

The following simple lemma will come in handy a few times. We explicitly point out that the lemma and its proof remain valid for the extreme choices where (at least one of) A or B is equal to $\emptyset$ or $\mathcal{X}$.

Lemma 7.1 *Consider lower and upper mass functions $\underline{p}$ and $\overline{p}$ with $\underline{p} \le \overline{p}$, and let A and B be any subsets of $\mathcal{X}$ such that $A \subseteq B$. Then there is some mass function $q \in \Sigma_{\mathcal{X}}$ such that $\underline{p} \le q \le \overline{p}$ and $P_q(A) = \sum_{x\in A}\underline{p}(A)$ and $P_q(B^c) = \sum_{x\in B^c}\overline{p}(x)$ if and only if*

$$\sum_{x\in A}\underline{p}(x) + \sum_{x\in B\setminus A}\underline{p}(x) + \sum_{x\in B^c}\overline{p}(x) \le 1 \le \sum_{x\in A}\underline{p}(x) + \sum_{x\in B\setminus A}\overline{p}(x) + \sum_{x\in B^c}\overline{p}(x). \tag{7.2}$$

[1] However, as we shall see when discussing refining the set of possible values for a random variable in Section 7.6_{138} and in particular in Theorem 7.12_{140}, there is no essential difference between the two approaches.

Proof. We begin with the 'if' part. There are two possibilities. If $\sum_{x\in B\setminus A}\underline{p}(x) = \sum_{x\in B\setminus A}\overline{p}(x)$, let q coincide with $\underline{p}$ on A, with $\overline{p}$ on B^c, and let $q(x) := \underline{p}(x) = \overline{p}(x)$ for $x \in B \setminus A$. That $q \in \Sigma_{\mathscr{X}}$ follows from Equation (7.2)$_\curvearrowleft$, where the inequalities are now equalities.

So assume that $\sum_{x\in B\setminus A}\underline{p}(x) < \sum_{x\in B\setminus A}\overline{p}(x)$. Define q by letting $q(x) := \underline{p}(x)$ for all $x \in A$, $q(x) := \overline{p}(x)$ for all $x \in B^c$ and $q(x) := (1-\varepsilon)\underline{p}(x) + \varepsilon\overline{p}(x)$ for all $x \in B \setminus A$, where we let

$$\varepsilon := \frac{1 - \sum_{x\in A}\underline{p}(x) - \sum_{x\in B\setminus A}\underline{p}(x) - \sum_{x\in B^c}\overline{p}(x)}{\sum_{x\in B\setminus A}[\overline{p}(x) - \underline{p}(x)]}.$$

The inequalities in Equation (7.2)$_\curvearrowleft$ can be rewritten as $0 \le \varepsilon \le 1$, which implies that $\underline{p} \le q \le \overline{p}$. We also find that $q \in \Sigma_{\mathscr{X}}$, because

$$\begin{aligned}\sum_{x\in\mathscr{X}} q(x) &= \sum_{x\in A}\underline{p}(x) + \sum_{x\in B\setminus A}[(1-\varepsilon)\underline{p}(x) + \varepsilon\overline{p}(x)] + \sum_{x\in B^c}\overline{p}(x)\\ &= \sum_{x\in A}\underline{p}(x) + \sum_{x\in B\setminus A}\underline{p}(x) + \sum_{x\in B^c}\overline{p}(x) + \varepsilon \sum_{x\in B\setminus A}[\overline{p}(x) - \underline{p}(x)] = 1.\end{aligned}$$

Finally, it is clear that $P_q(A) = \sum_{x\in A} q(x) = \sum_{x\in A}\underline{p}(x)$ and $P_q(B^c) = \sum_{x\in B^c} q(x) = \sum_{x\in B^c}\overline{p}(x)$.

For the 'only if' part, it follows from the assumptions that $\underline{p}$ and q coincide on A and that $\overline{p}$ and q coincide on B^c, whence

$$1 = \sum_{x\in A} q(x) + \sum_{x\in B\setminus A} q(x) + \sum_{x\in B^c} q(x) = \sum_{x\in A}\underline{p}(x) + \sum_{x\in B\setminus A} q(x) + \sum_{x\in B^c}\overline{p}(x).$$

As $\underline{p} \le q \le \overline{p}$, Equation (7.2)$_\curvearrowleft$ follows. □

Proposition 7.2 *Let $\underline{P}_{\underline{p},\overline{p}}$ be the lower prevision associated with the lower and upper mass functions $\underline{p}$ and $\overline{p}$.*

(i) *$\underline{P}_{\underline{p},\overline{p}}$ avoids sure loss if and only if $\underline{p} \le \overline{p}$ and $\sum_{x\in\mathscr{X}}\underline{p}(x) \le 1 \le \sum_{x\in\mathscr{X}}\overline{p}(x)$.*

(ii) *$\underline{P}_{\underline{p},\overline{p}}$ is coherent if and only if $\underline{p} \le \overline{p}$ and*

$$\sum_{y\in\mathscr{X}\setminus\{x\}}\underline{p}(y) + \overline{p}(x) \le 1 \le \sum_{y\in\mathscr{X}\setminus\{x\}}\overline{p}(y) + \underline{p}(x) \text{ for all } x \text{ in } \mathscr{X}.$$

Proof. (i). By Theorem 4.38(i)$_{71}$, $\underline{P}_{\underline{p},\overline{p}}$ avoids sure loss if and only if $\text{lins}(\underline{P}_{\underline{p},\overline{p}}) \neq \emptyset$. The proof is immediate if we look at Equation (7.1)$_\curvearrowleft$ and apply Lemma 7.1$_\curvearrowleft$ for $A = \emptyset$ and $B = \mathscr{X}$.

(ii). Using Theorem 4.38(ii)$_{71}$ and Equation (7.1)$_{123}$, we see that $\underline{P}_{\underline{p},\overline{p}}$ is coherent if and only if $\underline{p} \leq \overline{p}$ and

$$\left.\begin{aligned} \underline{p}(x) &= \min\left\{p(x) : p \in \Sigma_{\mathcal{X}} \text{ and } \underline{p} \leq p \leq \overline{p}\right\} \\ \overline{p}(x) &= \max\left\{p(x) : p \in \Sigma_{\mathcal{X}} \text{ and } \underline{p} \leq p \leq \overline{p}\right\} \end{aligned}\right\} \text{ for all } x \text{ in } \mathcal{X}.$$

By applying Lemma 7.1$_{123}$ for the choices of $A = \{x\}$ and $B = \mathcal{X}$, with $x \in \mathcal{X}$, as well as for the choices of $A = \emptyset$ and $B = \mathcal{X} \setminus \{x\}$, with $x \in \mathcal{X}$, we see that this last statement is equivalent to the listed condition. □

Finding the natural extension $\underline{E}_{\underline{p},\overline{p}}$ of a $\underline{P}_{\underline{p},\overline{p}}$ that avoids sure loss is a bit more involved. In Proposition 7.3, we proceed in a number of steps. To unburden the notations, we introduce the set functions L and U as

$$L(A) := \sum_{x \in A} \underline{p}(x) \text{ and } U(A) := \sum_{x \in A} \overline{p}(x) \text{ for all } A \subseteq \mathcal{X}.$$

Proposition 7.3 *Let $\underline{P}_{\underline{p},\overline{p}}$ be the lower prevision associated with the lower and upper mass functions $\underline{p}$ and $\overline{p}$. Assume that $\underline{P}_{\underline{p},\overline{p}}$ avoids sure loss. Then the following statements hold:*

(i) *For all $A \subseteq \mathcal{X}$,*

$$\underline{E}_{\underline{p},\overline{p}}(A) = \max\{L(A), 1 - U(A^c)\} \text{ and } \overline{E}_{\underline{p},\overline{p}}(A) = \min\{U(A), 1 - L(A^c)\}.$$

(ii) *$\underline{E}_{\underline{p},\overline{p}}$ is a 2-monotone coherent lower prevision, and for any bounded gamble f on $\mathcal{X}$,*

$$\begin{aligned} \underline{E}_{\underline{p},\overline{p}}(f) &= C\int f \, \mathrm{d}\underline{E}_{\underline{p},\overline{p}} \\ &= \inf f + R\int_{\inf f}^{\sup f} \max\{L(\{f \geq x\}), 1 - U(\{f < x\})\} \, \mathrm{d}x. \end{aligned}$$

Proof. (i). Fix any subset D of $\mathcal{X}$. Consider any mass function $p \in \Sigma_{\mathcal{X}}$. It follows from $\underline{p} \leq p \leq \overline{p}$ that $L(D) \leq P_p(D) \leq U(D)$ and that $L(D^c) \leq P_p(D^c) \leq U(D^c)$ or, equivalently, $1 - U(D^c) \leq P_p(D) \leq 1 - L(D^c)$. This implies that

$$\max\{L(D), 1 - U(D^c)\} \leq P_p(D) \leq \min\{U(D), 1 - L(D^c)\}.$$

If we now combine Theorem 4.38(iii)$_{71}$ with Equation (7.1)$_{123}$, we see that the proof will be complete if, for any choice of $D \subseteq \mathcal{X}$, we can find a mass function q in $\Sigma_{\mathcal{X}}$ such that $\underline{p} \leq q \leq \overline{p}$, and for which this lower bound is actually reached. Indeed, because the lower bound in D equals 1 minus the upper bound in D^c, this will automatically prove that the upper bound is also reached for some mass function q' in $\Sigma_{\mathcal{X}}$ such that $\underline{p} \leq q' \leq \overline{p}$. So consider any $D \subseteq \mathcal{X}$.

On the one hand, suppose that $L(D) \geq 1 - U(D^c)$ or, equivalently,

$$\sum_{x\in D} \underline{p}(x) + \sum_{x\in D^c} \overline{p}(x) \geq 1 \geq \sum_{x\in D} \underline{p}(x) + \sum_{x\in D^c} \underline{p}(x),$$

where the last inequality holds because $\underline{P}_{\underline{p},\overline{p}}$ avoids sure loss (Proposition 7.2(i)$_{124}$). Lemma 7.1$_{123}$ with $A = D$ and $B = \mathcal{X}$ now guarantees that we can find a mass function $q \in \Sigma_{\mathcal{X}}$ such that $\underline{p} \leq q \leq \overline{p}$ and $P_q(D) = \sum_{x\in D} \underline{p}(x) = L(D)$.

On the other hand, suppose that $L(D) \leq 1 - U(D^c)$ or, equivalently,

$$\sum_{x\in D} \underline{p}(x) + \sum_{x\in D^c} \overline{p}(x) \leq 1 \leq \sum_{x\in D} \overline{p}(x) + \sum_{x\in D^c} \overline{p}(x).$$

where, once again, the last inequality holds because $\underline{P}_{\underline{p},\overline{p}}$ avoids sure loss (Proposition (i)$_{124}$). Lemma 7.1$_{123}$ with $A = \emptyset$ and $B = D$ now guarantees that we can find a mass function $q \in \Sigma_{\mathcal{X}}$ such that $\underline{p} \leq q \leq \overline{p}$ and $P_q(D) = 1 - P_q(D^c) = 1 - \sum_{x\in D^c} \overline{p}(x) = 1 - U(D^c)$.

(ii)$_{\curvearrowright}$. We begin by proving that the restriction of the natural extension $\underline{E}_{\underline{p},\overline{p}}$ of $\underline{P}_{\underline{p},\overline{p}}$ to (indicators of) events is coherent and 2-monotone. The coherence of this lower probability follows from the coherence of the natural extension (Theorem 4.26(i)$_{65}$). To prove that this restriction to events is 2-monotone, we use Corollary 6.27$_{121}$ and the results in the proof of (i)$_{\curvearrowright}$ above. Consider any subsets C and D of $\mathcal{X}$ such that $C \subseteq D$, then, also considering Equation (7.1)$_{123}$, we must prove that there is some mass function $q \in \Sigma_{\mathcal{X}}$ with $\underline{p} \leq q \leq \overline{p}$ such that both $P_q(C) = \underline{E}_{\underline{p},\overline{p}}(C) = \max\{L(C), 1 - U(C^c)\}$ and $P_q(D) = \underline{E}_{\underline{p},\overline{p}}(D) = \max\{L(D), 1 - U(D^c)\}$. There are a number of possible cases.

The first case is that $L(C) \leq 1 - U(C^c)$, meaning that $\sum_{x\in C} \underline{p}(x) + \sum_{x\in C^c} \overline{p}(x) \leq 1$. As $\underline{p} \leq \overline{p}$, this implies that $\sum_{x\in D} \underline{p}(x) + \sum_{x\in D^c} \overline{p}(x) \leq 1$ or, equivalently, that also $L(D) \leq 1 - U(D^c)$. It then follows from Lemma 7.1$_{123}$ with $A = \emptyset$ and $B = C$ that we can always choose a mass function $q \in \Sigma_{\mathcal{X}}$ with $\underline{p} \leq q \leq \overline{p}$ that coincides on C^c, and therefore also on D^c, with $\overline{p}$. For this q, wc have, therefore, that $P_q \in \text{lins}(\underline{P}_{\underline{p},\overline{p}})$ and that $P_q(C) = 1 - P_q(C^c) = 1 - U(C^c) = \underline{E}_{\underline{p},\overline{p}}(C)$ and $P_q(D) = 1 - P_q(D^c) = 1 - U(D^c) = \underline{E}_{\underline{p},\overline{p}}(D)$.

The second case is that $L(D) \geq 1 - U(D^c)$, meaning that $\sum_{x\in D} \underline{p}(x) + \sum_{x\in D^c} \overline{p}(x) \geq 1$. As $\underline{p} \leq \overline{p}$, this implies that $\sum_{x\in C} \underline{p}(x) + \sum_{x\in C^c} \overline{p}(x) \geq 1$, or equivalently, that also $L(C) \geq 1 - U(C^c)$. It then follows from Lemma 7.1$_{123}$ with $A = D$ and $B = \mathcal{X}$ that we can always choose a mass function $q \in \Sigma_{\mathcal{X}}$ with $\underline{p} \leq q \leq \overline{p}$ that coincides on D, and therefore also on C, with $\underline{p}$. For this q, we have, therefore, that $P_q \in \text{lins}(\underline{P}_{\underline{p},\overline{p}})$ and that $P_q(C) = L(C) = \underline{E}_{\underline{p},\overline{p}}(C)$ and $P_q(D) = L(D) = \underline{E}_{\underline{p},\overline{p}}(D)$.

The only remaining case is that both $L(C) > 1 - U(C^c)$ and $L(D) < 1 - U(D^c)$, meaning that

$$\sum_{x \in C} \underline{p}(x) + \sum_{x \in D \setminus C} \underline{p}(x) + \sum_{x \in D^c} \overline{p}(x) < 1 < \sum_{x \in C} \underline{p}(x) + \sum_{x \in D \setminus C} \overline{p}(x) + \sum_{x \in D^c} \overline{p}(x).$$

It then follows from Lemma 7.1_{123} with $A = C$ and $B = D$ that we can always choose a mass function $q \in \Sigma_{\mathcal{X}}$ with $\underline{p} \leq q \leq \overline{p}$ that coincides with $\underline{p}$ on C and with $\overline{p}$ on D^c. For this q, we have, therefore, that $P_q \in \text{lins}(\underline{P}_{\underline{p},\overline{p}})$ and that $P_q(C) = L(C) = \underline{E}_{\underline{p},\overline{p}}(C)$ and $P_q(D) = 1 - P_q(D^c) = 1 - U(D^c) = \underline{E}_{\underline{p},\overline{p}}(D)$.

The rest of the proof is now easy: (i) by transitivity of natural extension (Corollary 4.32_{67}), the natural extension of $\underline{P}_{\underline{p},\overline{p}}$ to all bounded gambles is the natural extension to bounded gambles of its restriction to events, and (ii) the natural extension of a coherent and 2-monotone lower probability can be found by Choquet integration (Theorem 6.14_{111} or Corollary 6.17_{112}). □

7.2 Minimum preserving lower previsions

7.2.1 Definition and properties

A lower prevision $\underline{P}$ defined on a lattice of bounded gambles is called **minimum preserving** if $\underline{P}(f \wedge g) = \min\{\underline{P}(f), \underline{P}(g)\}$ for all f and g in dom $\underline{P}$ or, in other words, if it is a $\wedge$-homomorphism between its domain and $\mathbb{R}$.

Now, $\wedge$-homomorphisms (see Proposition 6.5_{105}) and natural extension (see Corollary 6.17_{112}) provide two ways of deducing n-monotone lower previsions from other n-monotone lower previsions. By combining them, we easily obtain that any minimum preserving lower prevision is completely monotone. This generalises a result by Nguyen et al. (1997, Theorem 1) from lower probabilities to lower previsions.

Proposition 7.4 *Any minimum preserving lower prevision, defined on a lattice of bounded gambles, is completely monotone.*

In contradistinction with Nguyen et al.'s (1997) proof, ours does not rely on combinatorics.

Proof. Let $\underline{P}$ be a minimum preserving lower prevision defined on a lattice of bounded gambles. Define the lower probability $\underline{Q}$ on $\{\emptyset, \mathcal{X}\}$ by $\underline{Q}(\emptyset) = 0$ and $\underline{Q}(\mathcal{X}) = 1$. Clearly, $\underline{Q}$ is a completely monotone coherent lower probability (it is even a probability charge). Hence, its natural extension $\underline{E}_{\underline{Q}}$ to $\mathbb{B}$ is completely monotone, by Corollary 6.17_{112}. As $\underline{Q}$ is dominated by all linear previsions on $\mathbb{B}$ (and in particular by the degenerate linear previsions P_x for all $x \in \mathcal{X}$), it is not difficult to see that $\underline{E}_{\underline{Q}}(f) = \inf f$ for all bounded gambles f on $\mathcal{X}$.

Now, define the map $r: \operatorname{dom} \underline{P} \to \mathbb{B}$ by $r(f)(x) := \underline{P}(f)$ for all f in $\operatorname{dom} \underline{P}$ and all $x \in \mathcal{X}$. As $\underline{P}$ is minimum preserving, r is a $\wedge$-homomorphism. Observe that $\underline{P} = \underline{E}_{\underline{Q}} \circ r$, and apply Proposition 6.5_{105}. □

The first part of this proof is used to show that the inf operator is a completely monotone lower prevision (see also the next section), and its complete monotonicity is used in the second part to show that any minimum preserving lower prevision is completely monotone. In the proof, we could just as well have used any other completely monotone lower prevision instead of inf, for instance, any degenerate linear prevision P_x, $x \in \mathcal{X}$.

7.2.2   Vacuous lower previsions

As an example, the vacuous lower prevision relative to a non-empty subset A of $\mathcal{X}$, given by

$$\underline{P}_A(f) := \inf_{x \in A} f(x) \text{ for all bounded gambles } f \text{ on } \mathcal{X},$$

is minimum preserving, so $\underline{P}_A$ is a completely monotone lower prevision on $\mathbb{B}$.

Corollary 7.5 *The vacuous lower prevision $\underline{P}_A$ relative to a non-empty subset A of $\mathcal{X}$ is completely monotone and coherent.*

7.3   Belief functions

Consider, for each non-empty subset H of a *finite* space $\mathcal{X}$, the vacuous lower probability $\underline{P}_H$ on $\mathscr{P}(\mathcal{X})$ relative to H, defined by

$$\underline{P}_H(A) := \begin{cases} 1 & \text{if } H \subseteq A \\ 0 & \text{otherwise} \end{cases} \text{ for all } A \subseteq \mathcal{X},$$

see also Equation $(5.4)_{81}$ and the related discussion in Section 5.4_{81}. We call any convex mixture of such vacuous lower probabilities a **belief function**. It is completely determined by the mixture coefficients $m(H)$, $H \in \mathscr{P}(\mathcal{X})$, with of course $m(\emptyset) = 0$:

$$\underline{Q}_m(A) := \sum_{H \subseteq \mathcal{X}} m(H)\underline{P}_H(A) = \sum_{H:\, H \subseteq A} m(H) \text{ for all } A \subseteq \mathcal{X}. \tag{7.3}$$

The $m(H)$ constitute a probability mass function m on $\mathscr{P}(\mathcal{X})$ with $m(\emptyset) = 0$, so $m \in \Sigma_{\mathscr{P}(\mathcal{X})}$; this m is called the **basic probability assignment** of the belief function $\underline{Q}_m$. Any event H for which $m(H) > 0$ is called a **focal element** of $\underline{Q}_m$. We can invert the relation (7.3) between $\underline{Q}_m$ and m using the **Möbius inversion formula**,

$$m(H) = \sum_{A:\, A \subseteq H} (-1)^{|H \backslash A|} \underline{Q}_m(A) \text{ for all } H \subseteq \mathcal{X}. \tag{7.4}$$

For details, we refer to Shafer (1976, 1979), whose work made the theory of belief functions very popular.

The notion of a belief function can be trivially extended to *infinite* $\mathcal{X}$ by considering as a basic probability assignment m any real-valued map on $\mathcal{P}(\mathcal{X})$ that only assumes non-zero values on a *finite* subset $\mathcal{H}$ of $\mathcal{P}(\mathcal{X}) \setminus \{\emptyset\}$, called the set of **focal elements**, such that

$$(\forall H \in \mathcal{H})m(H) > 0 \text{ and } \sum_{H \in \mathcal{H}} m(H) = 1.$$

In this case, we define the corresponding belief function $\underline{Q}_m$ by

$$\underline{Q}_m(A) := \sum_{H \in \mathcal{H}} m(H)\underline{P}_H(A) = \sum_{H \in \mathcal{H}\,:\,H \subseteq A} m(H) \text{ for all } A \subseteq \mathcal{X}. \tag{7.5}$$

The following proposition formally embeds the notion of a belief function into the theory of coherent lower previsions. We will come back to these ideas in Section 7.6_{138}.

Proposition 7.6 *Let $\mathcal{X}$ be any non-empty set, possibly infinite. Any belief function $\underline{Q}_m$ is a coherent and completely monotone lower probability, and its natural extension to all bounded gambles, denoted by $\underline{P}_m$, is the coherent and completely monotone lower prevision given by*

$$\underline{P}_m(f) := \sum_{H \in \mathcal{H}} m(H)\underline{P}_H(f) = \sum_{H \in \mathcal{H}} m(H) \inf_{x \in H} f(x) \tag{7.6}$$

for all bounded gambles f on $\mathcal{X}$.

Proof. Taking convex mixtures preserves coherence (Proposition 4.19_{59}) and n-monotonicity (Proposition 6.7_{106}). Therefore, because vacuous lower previsions $\underline{P}_F$ relative to an event H are coherent (see Section 5.4_{81}) and completely monotone (Corollary 7.5), so is therefore the lower prevision $\underline{P}_m$ defined by Equation (7.6). Its restriction to events $\underline{Q}_m$ is, therefore, a coherent and completely monotone lower probability.

It remains to prove that $\underline{P}_m$ is the natural extension of $\underline{Q}_m$. This is an immediate consequence of Corollary 6.25_{119}. Alternatively, use the fact that Choquet integration preserves convex combinations (of the lower probabilities involved). □

In summary, we find that any convex combination of vacuous lower previsions is a coherent and completely monotone lower prevision and, therefore, the natural extension to bounded gambles of a belief function. In Section $11.6.1_{225}$, we come across a converse result: on a finite space, any coherent and completely monotone lower prevision is a convex mixture of vacuous lower previsions and, therefore, the natural extension to bounded gambles of a belief function.

7.4 Lower previsions associated with proper filters

We now turn from convex combinations of vacuous lower previsions to their pointwise limits or, in other words, to the lower previsions $\underline{P}_{\mathcal{F}}$ associated with proper filters $\mathscr{F}$ on $\mathscr{X}$, which we introduced in Section $5.5.1_{82}$,

$$\underline{P}_{\mathscr{F}}(f) = \sup_{F\in\mathscr{F}} \inf_{x\in F} f(x) = \lim_{F\in\mathscr{F}} \underline{P}_F(f) \text{ for all } f \in \mathbb{B}(\mathscr{X}),$$

where the limit is associated with the directed set $\mathscr{F}$. As taking point-wise limits preserves n-monotonicity (by Proposition 6.8_{107}), we can use the n-monotonicity of vacuous lower previsions (Corollary 7.5_{128}) to strengthen the conclusion reached about $\underline{P}_{\mathscr{F}}$ in Proposition 5.5_{83}:

Proposition 7.7 *Let $\mathscr{F}$ be a proper filter. Then $\underline{P}_{\mathscr{F}}$ is a coherent and completely monotone lower prevision on $\mathbb{B}(\mathscr{X})$.*

The set $\underline{\mathbb{P}}_{\infty}(\mathscr{X})$ of all coherent and completely monotone lower previsions on $\mathbb{B}(\mathscr{X})$ is a convex set that is closed (even compact, see Section 11.2_{217}) in the topology of point-wise convergence.[2] As Choquet (1953–1954, Section 45) has shown,[3] the lower previsions $\underline{P}_{\mathscr{F}}$ associated with proper filters $\mathscr{F}$ constitute the extreme points of this convex compact set. Our proof for this is different from, and rather less involved than, Choquet's, because we can build on the development in Section $5.5.1_{82}$.

Proposition 7.8 $\operatorname{ext}(\underline{\mathbb{P}}_{\infty}(\mathscr{X})) = \left\{\underline{P}_{\mathscr{F}} : \mathscr{F} \in \mathbb{F}(\mathscr{X})\right\}$.

Proof. First of all, consider any proper filter $\mathscr{F}$ on $\mathscr{X}$, then we show that $\underline{P}_{\mathscr{F}}$ is an extreme point of $\underline{\mathbb{P}}_{\infty}(\mathscr{X})$. Consider any $\alpha \in (0,1)$, and any $\underline{P}_1$ and $\underline{P}_2$ in $\underline{\mathbb{P}}_{\infty}(\mathscr{X})$. Assume that $\underline{P}_{\mathscr{F}} = \alpha\underline{P}_1 + (1-\alpha)\underline{P}_2$. For any $A \in \mathscr{F}$, we have $\underline{P}_{\mathscr{F}}(A) = 1$ and, therefore, also $\underline{P}_1(A) = \underline{P}_2(A) = 1$. For any $A \notin \mathscr{F}$, we have $\underline{P}_{\mathscr{F}}(A) = 0$ and therefore also $\underline{P}_1(A) = \underline{P}_2(A) = 0$. Hence the coherent lower previsions $\underline{P}_1$ and $\underline{P}_2$ coincide with $\underline{P}_{\mathscr{F}}$ on all events and, therefore, by Proposition 5.8_{85}, on all bounded gambles: $\underline{P}_{\mathscr{F}} = \underline{P}_1 = \underline{P}_2$. This proves that $\underline{P}_{\mathscr{F}}$ is an extreme point.

Conversely, consider an arbitrary extreme point $\underline{P}$ of $\underline{\mathbb{P}}_{\infty}(\mathscr{X})$. Assume *ex absurdo* that there is no proper filter $\mathscr{F}$ such that $\underline{P} = \underline{P}_{\mathscr{F}}$, which implies, by Corollary 5.9_{86}, that there is some $A \subseteq \mathscr{X}$ such that $0 < \underline{P}(A) < 1$. We now define the lower probabilities $\underline{Q}_1$ and $\underline{Q}_2$ as follows:

$$\underline{Q}_1(B) := \frac{\underline{P}(A\cap B)}{\underline{P}(A)} \text{ and } \underline{Q}_2(B) := \frac{\underline{P}(B) - \underline{P}(A\cap B)}{1-\underline{P}(A)} \text{ for all } B \subseteq \mathscr{X}.$$

Then $\underline{Q}_1(\emptyset) = \underline{Q}_2(\emptyset) = 0$ and $\underline{Q}_1(\mathscr{X}) = \underline{Q}_2(\mathscr{X}) = 1$. It is obvious that $\underline{Q}_1$ is completely monotone because $\underline{P}$ is. We now show that $\underline{Q}_2$ is completely monotone as well. Consider any $p \in \mathbb{N}_{>0}$ and any subsets $B, B_1, \dots, B_p$ of $\mathscr{X}$, and let $B_{p+1} := A$. Because $\underline{P}$ is completely monotone, we see that

$$0 \leq \sum_{I\subseteq\{1,\dots,p+1\}} (-1)^{|I|}\underline{P}\left(B \cap \bigcap_{i\in I} B_i\right)$$

[2] For more details on the topological aspects related to this point-wise convergence, refer to the discussion in Section 4.7_{74} and Chapter 11_{214}.

[3] Choquet considered the case of what we are calling here coherent and completely monotone lower probabilities on $\mathscr{P}(\mathscr{X})$. It is clear from Corollary 6.25_{119} that these are in a bijective (one-to-one and onto) correspondence with the elements of $\underline{\mathbb{P}}_{\infty}(\mathscr{X})$.

$$= \sum_{I\subseteq\{1,\dots,p\}} (-1)^{|I|} \underline{P}\left(B \cap \bigcap_{i\in I} B_i\right) + \sum_{I\subseteq\{1,\dots,p\}} (-1)^{|I|+1} \underline{P}\left(B \cap A \cap \bigcap_{i\in I} B_i\right)$$

$$= \sum_{I\subseteq\{1,\dots,p\}} (-1)^{|I|} \underline{Q}_2\left(B \cap \bigcap_{i\in I} B_i\right)[1 - \underline{P}(A)],$$

so $\underline{Q}_2$ is a completely monotone lower probability as well. By Corollary 6.16_{112}, both $\underline{Q}_1$ and $\underline{Q}_2$ are coherent and completely monotone lower probabilities. By Corollary 6.25_{119}, they therefore each have a (unique) completely monotone and coherent extension to all bounded gambles (their natural extension), which we denote by $\underline{P}_1$ and $\underline{P}_2$, respectively. As $\underline{P}_1(A) = \underline{Q}_1(A) = 1$ and $\underline{P}_2(A) = \underline{Q}_2(A) = 0$, these lower previsions do not coincide: $\underline{P}_1 \neq \underline{P}_2$. Their convex combination $\underline{P}(A)\underline{P}_1 + [1 - \underline{P}(A)]\underline{P}_2$ is a completely monotone and coherent lower prevision on all bounded gambles, by Propositions $4.19(\text{ii})_{59}$ and 6.7_{106}. It is readily checked that it coincides with $\underline{P}$ on all events, and therefore on all bounded gambles, again by Corollary 6.25_{119}: indeed, for all $B \subseteq \mathcal{X}$,

$$\underline{P}(A)\underline{P}_1(B) + [1 - \underline{P}(A)]\underline{P}_2(B) = \underline{P}(A)\underline{Q}_1(B) + [1 - \underline{P}(A)]\underline{Q}_2(B)$$
$$= \underline{P}(A \cap B) + \underline{P}(B) - \underline{P}(A \cap B) = \underline{P}(B).$$

So $\underline{P}$ is a non-trivial convex combination of $\underline{P}_1$ and $\underline{P}_2$, which contradicts the assumption that it is an extreme point. □

7.5 Induced lower previsions

We now intend to generalise the ideas in the previous two sections in a number of ways at once, and see how so-called filter maps – a generalisation of multivalued maps that will be defined further on – fit into the theory of coherent lower previsions. The material in this section is a generalisation from multivalued maps to filter maps of work by Miranda et al. (2005). We begin with a slightly heuristic motivation for the new notions we are about to introduce.

7.5.1 Motivation

We consider two variables X and Y assuming values in the non-empty (but not necessarily finite) sets $\mathcal{X}$ and $\mathcal{Y}$. For simplicity, we suppose that any combination of values of X and Y is logically possible, so the set of possible values for the joint random variable (X, Y) is simply $\mathcal{X} \times \mathcal{Y}$.

Let us first look at a (single-valued) map γ between the spaces $\mathcal{Y}$ and $\mathcal{X}$. Given a linear prevision P on $\mathbb{B}(\mathcal{Y})$, such a map induces a linear prevision P_γ on $\mathbb{B}(\mathcal{X})$ by

$$P_\gamma(A) := P(\gamma^{-1}(A)) = P(\{y \in \mathcal{Y} : \gamma(y) \in A\}) \text{ for all } A \subseteq \mathcal{X}$$

or, equivalently,

$$P_\gamma(f) := P(f \circ \gamma) \text{ for all bounded gambles } f \text{ on } \mathcal{X}, \tag{7.7}$$

which is a well-known 'change of variables result' for previsions (or expectations). If we have a variable Y and the variable X is given by $X := \gamma(Y)$, then any bounded gamble $f(X)$ on the value of X can be translated back to a bounded gamble $f(\gamma(Y)) = (f \circ \gamma)(Y)$ on the value of Y, which explains where Equation (7.7)$_{\curvearrowleft}$ comes from: *If the uncertainty about Y is represented by the model P, then the uncertainty about $X = \gamma(Y)$ is represented by the model P_γ.*

There is another way of motivating the same formula, which lends itself more readily to generalisation. We can interpret the map γ as conditional information: *if we know that $Y = y$, then we know that $X = \gamma(y)$.* This conditional information can be represented by the so-called conditional linear prevision $P(\bullet|y)$ on $\mathbb{B}(\mathcal{X})$, defined by[4]

$$P(f|y) := f(\gamma(y)) = (f \circ \gamma)(y) \text{ for all } y \in \mathcal{Y} \text{ and all bounded gambles } f \text{ on } \mathcal{X}. \quad (7.8)$$

It states that conditional on $Y = y$, all probability mass for X is located in the single point $\gamma(y)$. If $P(f|Y)$ is the bounded gamble on $\mathcal{Y}$ that assumes the value $P(f|y)$ in any $y \in \mathcal{Y}$, then clearly $P(f|Y) = (f \circ \gamma)(Y)$, which allows us to rewrite Equation (7.7)$_{\curvearrowleft}$ as

$$P_\gamma(f) = P(P(f|Y)) \text{ for all bounded gambles } f \text{ on } \mathcal{X}. \quad (7.9)$$

This shows that Equation (7.7)$_{\curvearrowleft}$ is actually a special case of the law of iterated expectations – the expectation form of the law of total probability – in classical probability (see, for instance, DeGroot and Schervisch, 2011, Theorem 4.7.1).

Assume now that, more generally, the relation between X and Y is determined as follows. There is a so-called **multivalued map** $\Gamma : \mathcal{Y} \to \mathcal{P}(\mathcal{X})$ that associates with any $y \in \mathcal{Y}$ a non-empty subset $\Gamma(y)$ of $\mathcal{X}$, and *if we know that $Y = y$, then all we know about X is that it can assume any value in $\Gamma(y)$.* There is no immediately obvious way of representing this conditional information using a precise probability model. In other words, if we want to remain within the framework of precise probabilities, we must abandon the simple and powerful device of interpreting the multivalued map Γ as conditional information. But if we work with the theory of imprecise probabilities, as we are doing here, it is still perfectly possible to interpret Γ as conditional information that can be represented by a special conditional *lower* prevision $\underline{P}(\bullet|y)$ on $\mathbb{B}(\mathcal{X})$, where

$$\underline{P}(f|y) := \underline{P}_{\Gamma(y)}(f) = \inf_{x \in \Gamma(y)} f(y) \text{ for all bounded gambles } f \text{ on } \mathcal{X} \quad (7.10)$$

is the vacuous lower prevision relative to the event $\Gamma(y)$. It states that conditional on $Y = y$, the lower prevision for X is given by $\underline{P}_{\Gamma(y)}(f)$, representing the assessment that $X \in \Gamma(y)$ (see Sections 5.4_{81} and 7.2_{127} for more details and motivation). Given information about Y in the form of a coherent lower prevision $\underline{P}$ on $\mathbb{B}(\mathcal{Y})$, it follows from Walley's Marginal Extension Theorem[5] (see Walley, 1991, Section 6.7) that the

[4] See the discussion in Chapter 13_{235} and in particular Section 13.9_{288} for more technical details about conditional lower and linear previsions. Note that here, we have abbreviated $P(\bullet|\mathcal{X} \times \{y\})$ to $P(\bullet|y)$.

[5] This theorem is also discussed in Section 13.9_{288} of the second part of this book. It is an imprecise-probabilistic generalisation of the afore-mentioned law of iterated expectations.

corresponding information about X is the lower prevision $\underline{P}_\Gamma$ on $\mathbb{B}(\mathscr{X})$ defined by

$$\underline{P}_\Gamma(f) := \underline{P}(\underline{P}(f|Y)) \text{ for all bounded gambles } f \text{ on } \mathscr{X}, \tag{7.11}$$

which is an immediate generalisation of Equation (7.9). This formula provides a well-justified method for using the conditional information embodied in the multivalued map Γ to turn the uncertainty model $\underline{P}$ about Y into an uncertainty model $\underline{P}_\Gamma$ about X. This approach has been introduced and explored in great detail by Miranda et al. (2005).

What we intend to do here is take this idea of conditional information one useful step further. To motivate going beyond multivalued maps, assume, for instance, that the information about the relation between X and Y is the following: *If we know that $Y = y$, then all we know about X is that it lies arbitrarily close to $\gamma(y)$*, in the sense that X lies inside any neighbourhood of $\gamma(x)$. Here, $\gamma : \mathscr{Y} \to \mathscr{X}$ is the single-valued map also considered earlier. We are assuming that, in order to capture what 'arbitrarily close' means, we have provided $\mathscr{X}$ with a topology $\mathscr{T}$ of open sets, as in Section 5.5.5$_{93}$. The discussion in that section then tells us that we can model this type of conditional information using the conditional *lower* prevision $\underline{P}(\bullet|y)$ on $\mathbb{B}(\mathscr{X})$, where

$$\underline{P}(f|y) := \underline{P}_{\mathscr{N}_{\gamma(y)}}(f) = \sup_{N\in\mathscr{N}_{\gamma(y)}} \inf_{x\in N} f(x) = \underline{\mathrm{osc}}_{\gamma(y)}(f)$$
$$\text{for all bounded gambles } f \text{ on } \mathscr{X} \tag{7.12}$$

is the lower prevision associated with the neighbourhood filter $\mathscr{N}_{\gamma(y)}$ of $\gamma(y)$, which we associated with the lower oscillation in Section 5.5.5$_{93}$. Information about Y in the form of a coherent lower prevision $\underline{P}$ on $\mathbb{B}(\mathscr{Y})$ can now be turned into information $\underline{P}(\underline{P}(\bullet|Y))$ about X, via this conditional model, using Equation (7.12).

7.5.2 Induced lower previsions

So, given this motivation, let us try and capture these ideas in an abstract model. All the special cases mentioned earlier can be captured by considering a so-called **filter map** Φ from $\mathscr{Y}$ to $\mathscr{X}$, that is, a map $\Phi : \mathscr{Y} \to \mathbb{F}(\mathscr{X})$ that associates a proper[6] filter $\Phi(y)$ with each element y of $\mathscr{Y}$. The simple idea underlying the arguments of this section is that this filter map represents some type of conditional information and that this information can be represented by a (specific) conditional lower prevision.

Using this filter map Φ, we associate with any bounded gamble f on $\mathscr{X} \times \mathscr{Y}$ a **lower inverse** $f_\circ$ (under Φ), which is the bounded gamble on $\mathscr{Y}$ defined by

$$f_\circ(y) := \underline{P}_{\Phi(y)}(f(\bullet, y)) = \sup_{F\in\Phi(y)} \inf_{x\in F} f(x, y) \text{ for all } y \text{ in } \mathscr{Y}, \tag{7.13}$$

[6] We assume that the filter $\Phi(y)$ is proper (meaning that $\Phi(y) \neq \emptyset$ and $\emptyset \notin \Phi(y)$) mainly to make things as simple as possible. For details about how to manage without this and similar assumptions, see, for instance, Miranda et al. (2005, Technical Remarks 1 and 2).

where, of course, $\underline{P}_{\Phi(y)}$ is the lower prevision on $\mathbb{B}(\mathscr{X})$ associated with the proper filter $\Phi(y)$. Similarly, we define for any bounded gamble g on $\mathscr{X}$ its **lower inverse** $g_\bullet$ (under Φ) as the bounded gamble on $\mathscr{Y}$ defined by

$$g_\bullet(y) := \underline{P}_{\Phi(y)}(g) = \sup_{F\in\Phi(y)} \inf_{x\in F} g(x) \text{ for all } y \text{ in } \mathscr{Y}. \tag{7.14}$$

Equations $(7.13)_\curvearrowleft$ and (7.14) are obviously very closely related to, and inspired by, Equations $(7.8)_{132}$, $(7.10)_{132}$ and $(7.12)_\curvearrowleft$.

In particular, we find for any $A \subseteq \mathscr{X} \times \mathscr{Y}$ that $(I_A)_\circ = I_{A_\circ}$, where we let

$$A_\circ := \{y \in \mathscr{Y} : (\exists F \in \Phi(y))F \times \{y\} \subseteq A\}$$

denote the so-called **lower inverse** of A (under Φ). And if $B \subseteq \mathscr{X}$, then

$$B_\bullet := (B \times \mathscr{Y})_\circ = \{y \in \mathscr{Y} : (\exists F \in \Phi(y))F \subseteq B\}$$

is the set of all y for which B occurs eventually with respect to the proper filter $\Phi(y)$.

Now consider any lower prevision $\underline{P}$ on Y that avoids sure loss. Then we can consider its natural extension $\underline{E}_{\underline{P}}$ and use it together with the filter map Φ to construct an **induced lower prevision** $\underline{P}_\circ$ on $\mathbb{B}(\mathscr{X} \times \mathscr{Y})$:

$$\underline{P}_\circ(f) := \underline{E}_{\underline{P}}(f_\circ) \text{ for all bounded gambles } f \text{ on } \mathscr{X} \times \mathscr{Y}. \tag{7.15}$$

The so-called $\mathscr{X}$**-marginal** $\underline{P}_\bullet$ of this lower prevision is the lower prevision on $\mathbb{B}(\mathscr{X})$ given by

$$\underline{P}_\bullet(g) := \underline{E}_{\underline{P}}(g_\bullet) \text{ for all bounded gambles } g \text{ on } \mathscr{X}. \tag{7.16}$$

Equations (7.15) and (7.16) are very closely related to, and inspired by, Equations $(7.7)_{131}$, $(7.9)_{132}$, and $(7.11)_\curvearrowleft$. An induced lower prevision is what results if we use the conditional information embodied in the filter map to turn an uncertainty model about Y into an uncertainty model about X.

In the next section, we study a number of interesting mathematical properties of such induced lower previsions. We lay bare the formal connections with aspects of random set theory in Section 7.6_{138}.

7.5.3 Properties of induced lower previsions

Let us define a **prevision kernel from** $\mathscr{Y}$ **to** $\mathscr{X}$ as any map K from $\mathscr{Y} \times \mathbb{B}(\mathscr{X})$ to $\mathbb{R}$ such that $K(y, \bullet)$ is a linear prevision on $\mathbb{B}(\mathscr{X})$ for all y in $\mathscr{Y}$. Prevision kernels are clear generalisations of probability or Markov kernels (Kallenberg, 2002, p. 20) but without the measurability conditions.

We can extend $K(y, \bullet)$ to a linear prevision on $\mathbb{B}(\mathscr{X} \times \mathscr{Y})$ by letting $K(y,f) := K(y,f(\bullet, y))$ for all bounded gambles f on $\mathscr{X} \times \mathscr{Y}$. Using this extension, for any lower prevision $\underline{P}$ on Y that avoids sure loss, we denote by $\underline{P}K$ the lower prevision on $\mathbb{B}(\mathscr{X} \times \mathscr{Y})$ defined by

$$\underline{P}K(f) := \underline{E}_{\underline{P}}(K(\bullet,f)) \text{ for all bounded gambles } f \text{ on } \mathscr{X} \times \mathscr{Y}. \tag{7.17}$$

If $\underline{P}$ is a linear prevision on $\mathbb{B}(\mathcal{Y})$, then $\underline{P}K$ is a linear prevision on $\mathbb{B}(\mathcal{X}\times\mathcal{Y})$ (use the simple characterisation of a linear prevision on a linear space, Theorem 4.16_{57}). As an immediate consequence, $\underline{P}K$ is always a coherent lower prevision on $\mathbb{B}(\mathcal{X}\times\mathcal{Y})$, as a lower envelope of linear previsions (use Theorem $4.38(\text{iii})_{71}$ and Corollary 4.41_{73}).

We also use the following notation for any filter map $\Phi : \mathcal{Y} \to \mathbb{F}(\mathcal{X})$:

$$\mathbb{K}(\Phi) := \left\{ K \text{ prevision kernel} : (\forall y \in \mathcal{Y}) K(y, \bullet) \in \operatorname{lins}(\underline{P}_{\Phi(y)}) \right\} \tag{7.18}$$

$$= \{ K \text{ prevision kernel} : (\forall y \in \mathcal{Y})(\forall A \in \Phi(y)) K(y, A) = 1 \}, \tag{7.19}$$

where the last equality follows from Theorem 5.11_{86}.

There is an interesting relation between prevision kernels and induced lower previsions. We have seen in the introductory motivation in Section $7.5.1_{131}$ that we can interpret the lower inverse $f_\circ$ of the bounded gamble f on $\mathcal{X}\times\mathcal{Y}$ under the filter map Φ as the conditional lower prevision $\underline{P}(f|y)$ defined by

$$\underline{P}(f|y) := \underline{P}_{\Phi(y)}(f(\bullet, y)) = \sup_{F\in\Phi(y)} \inf_{x\in F} f(x, y) \text{ for all } y \text{ in } \mathcal{Y}.$$

And then $\underline{P}_\circ$ is the marginal extension $\underline{E}_{\underline{P}}(\underline{P}(f|Y))$ of the marginal lower prevision $\underline{P}$ and the conditional lower prevision $\underline{P}(\bullet|Y)$; see Walley (1991, Section 6.7), Miranda and De Cooman (2007) and the material in Section 13.9_{288} for more information about marginal extension. The second statement in the next proposition can be seen as a special case of Walley's lower envelope theorem for marginal extension (Walley, 1991, Theorem 6.7.4). Our proof closely follows Walley's original proof. All three statements extend results by Miranda et al. (2005, Theorems 1 and 3).

Proposition 7.9 *Let $n \in \mathbb{N}^*_{>0}$, and let $\underline{P}$ be a lower prevision that avoids sure loss and is defined on some subset of $\mathbb{B}(\mathcal{Y})$. Then the following statements hold:*

(i) *$\underline{P}_\circ$ is a coherent lower prevision.*

(ii) *For all $K \in \mathbb{K}(\Phi)$ and $P \in \operatorname{lins}(\underline{P})$, $PK \in \operatorname{lins}(\underline{P}_\circ)$, and for all bounded gambles f on $\mathcal{X}\times\mathcal{Y}$, there are $K \in \mathbb{K}(\Phi)$ and $P \in \operatorname{lins}(\underline{P})$ such that $\underline{P}_\circ(f) = PK(f)$.*

(iii) *If $\underline{E}_{\underline{P}}$ is n-monotone, then so is $\underline{P}_\circ$.*

Proof. We begin with the proof of (i). We use coherence condition $(\text{D})_{47}$. For arbitrary n, m in $\mathbb{N}$ and bounded gambles $f_0, f_1, \ldots, f_n$ in $\mathbb{B}(\mathcal{X}\times\mathcal{Y})$, we find that

$$\sum_{i=1}^n \underline{P}_\circ(f_i) - m\underline{P}_\circ(f_0) = \sum_{i=1}^n \underline{E}_{\underline{P}}((f_i)_\circ) - m\underline{E}_{\underline{P}}((f_0)_\circ)$$

$$\leq \sup_{y\in\mathcal{Y}} \left[\sum_{i=1}^n (f_i)_\circ(y) - m(f_0)_\circ(y) \right]$$

$$= \sup_{y\in\mathcal{Y}} \left[\sum_{i=1}^{n} \underline{P}_{\Phi(y)}(f_i(\bullet, y)) - m\underline{P}_{\Phi(y)}(f_0(\bullet, y)) \right]$$

$$\leq \sup_{y\in\mathcal{Y}} \sup_{x\in\mathcal{X}} \left[\sum_{i=1}^{n} f_i(x, y) - mf_0(x, y) \right] = \sup \left[\sum_{i=1}^{n} f_i - mf_0 \right],$$

where the first inequality follows from the coherence of the natural extension $\underline{E}_{\underline{P}}$ (Theorem 4.26(i)$_{65}$), and the second from the coherence of the lower previsions $\underline{P}_{\Phi(y)}$, $y \in \mathcal{Y}$ (Proposition 7.7$_{130}$).

(ii)$_{\curvearrowleft}$. First, fix any P in $\text{lins}(\underline{P})$ and any $K \in \mathbb{K}(\Phi)$. Consider any bounded gamble f on $\mathcal{X} \times \mathcal{Y}$. We infer from Equations (7.18)$_{\curvearrowleft}$ and (7.13)$_{133}$ that $K(y, f) \geq f_\circ(y)$ for all $y \in \mathcal{Y}$ and, therefore,

$$PK(f) = P(K(\bullet, f)) \geq P(f_\circ) \geq \underline{E}_{\underline{P}}(f_\circ) = \underline{P}_\circ(f),$$

where the first inequality follows from the monotonicity (Corollary 4.14(iv)$_{55}$) of the linear prevision P, and the second inequality follows from the fact that a lower prevision and its natural extension have the same dominating linear previsions (Proposition 4.37$_{71}$); the first equality follows from an appropriate version of Equation (7.17)$_{134}$, if we recall from the Linear Extension Theorem 4.42$_{73}$ that $\underline{E}_P = P$. This shows that $PK \in \text{lins}(\underline{P}_\circ)$.

Next, fix any bounded gamble f on $\mathbb{B}(\mathcal{X} \times \mathcal{Y})$. We infer from Equation (7.18)$_{\curvearrowleft}$ and the Lower Envelope Theorem 4.38(iii)$_{71}$ that there is some $K \in \mathbb{K}(\Phi)$ such that $f_\circ = K(\bullet, f)$. Similarly, there is some $P \in \text{lins}(\underline{P})$ such that $\underline{E}_{\underline{P}}(f_\circ) = P(f_\circ)$ and, therefore, indeed, $\underline{P}_\circ(f) = \underline{E}_{\underline{P}}(f_\circ) = P(f_\circ) = P(K(\bullet, f)) = PK(f)$.

(iii)$_{\curvearrowleft}$. For any bounded gambles f and g on $\mathcal{X} \times \mathcal{Y}$, $(f \wedge g)_\circ = f_\circ \wedge g_\circ$, using Equation (7.15)$_{134}$. This tells us that taking the lower inverse constitutes a $\wedge$-homomorphism between $\mathbb{B}(\mathcal{X} \times \mathcal{Y})$ and $\mathbb{B}(\mathcal{Y})$. Recall that a $\wedge$-homomorphism preserves n-monotonicity (Proposition 6.5$_{105}$). □

If we restrict our attention to the $\mathcal{X}$-marginal $\underline{P}_\bullet$ of the lower prevision $\underline{P}_\circ$ on $\mathbb{B}(\mathcal{X} \times \mathcal{Y})$, we can go somewhat further: the following simple proposition is a considerable generalisation of a result mentioned by Wasserman (1990a, Section 2.4);[7] see also Wasserman (1990b, Section 2) and Miranda et al. (2010, Theorem 14).

Proposition 7.10 *Let $n \in \mathbb{N}^*$, $n \geq 2$, and let $\underline{P}$ be a lower prevision that avoids sure loss and is defined on a subset of $\mathbb{B}(\mathcal{Y})$. If $\underline{E}_{\underline{P}}$ is n-monotone, then so is $\underline{P}_\bullet$ and, moreover,*

$$C\int g\,\mathrm{d}\underline{P}_\bullet = \underline{P}_\bullet(g) = \underline{E}_{\underline{P}}(g_\bullet) = C\int g_\bullet\,\mathrm{d}\underline{E}_{\underline{P}} \textit{ for all } g \in \mathbb{B}(\mathcal{X}). \tag{7.20}$$

Proof. That $\underline{P}_\bullet$ is n-monotone follows from Theorem 7.9(iii)$_{\curvearrowleft}$. To prove Equation (7.20), it suffices to prove that the first and last equalities hold, because of

[7] Wasserman considers the special case that $\underline{P}$ is a probability measure and that g satisfies appropriate measurability conditions, so the rightmost Choquet integral coincides with the usual expectation of $g_\bullet$. See also Section 9.6$_{188}$ for more details and an explicit comparison.

Equation $(7.16)_{134}$. To this end, use that both $\underline{E}_{\underline{P}}$ and $\underline{P}_{\bullet}$ are n-monotone and apply Theorem 6.22_{117}. □

If $\mathscr{K}$ is the domain of $\underline{P}$, then, as we shall witness in Section $7.6_{\curvearrowright}$, it is also useful to consider the lower prevision $\underline{P}_{\circ}^{\mathrm{r}}$, defined on the set of bounded gambles ${}_{\circ}\mathscr{K} := \{f \in \mathbb{B}(\mathscr{X} \times \mathscr{Y}) : f_{\circ} \in \mathscr{K}\}$ as follows:

$$\underline{P}_{\circ}^{\mathrm{r}}(f) := \underline{P}(f_{\circ}) \text{ for all } f \text{ such that } f_{\circ} \in \mathscr{K}. \tag{7.21}$$

If $\underline{P}$ is coherent, then of course $\underline{P}_{\circ}^{\mathrm{r}}$ is the restriction of $\underline{P}_{\circ}$ to ${}_{\circ}\mathscr{K}$, because then $\underline{E}_{\underline{P}}$ and $\underline{P}$ coincide on $\mathscr{K}$ (see the Natural Extension Theorem $4.26(\mathrm{ii})_{65}$). We now show that, interestingly and perhaps surprisingly, all the 'information' present in $\underline{P}_{\circ}$ is then already contained in the restricted model $\underline{P}_{\circ}^{\mathrm{r}}$; this generalises a result proved by Miranda et al. (2005, Theorem 7) A closer look at Section $7.6_{\curvearrowright}$ reveals that a counterpart (Proposition 7.12_{140}) of Proposition 7.11 allows us to easily connect this discussion of induced lower previsions with specific material about random sets and belief functions in the literature.

Proposition 7.11 *Let $\underline{P}$ be a coherent lower prevision, defined on a set of bounded gambles $\mathscr{K} \subseteq \mathbb{B}(\mathscr{Y})$. Then the following statements hold:*

(i) *$\underline{P}_{\circ}^{\mathrm{r}}$ is the restriction of $\underline{P}_{\circ}$ to the set of bounded gambles ${}_{\circ}\mathscr{K}$ and, therefore, a coherent lower prevision on ${}_{\circ}\mathscr{K}$.*

(ii) *The natural extension $\underline{E}_{\underline{P}_{\circ}^{\mathrm{r}}}$ of $\underline{P}_{\circ}^{\mathrm{r}}$ coincides with the induced lower prevision: $\underline{E}_{\underline{P}_{\circ}^{\mathrm{r}}} = \underline{P}_{\circ}$.*

Proof. (i). It follows from the Natural Extension Theorem $4.26(\mathrm{ii})_{65}$ that $\underline{E}_{\underline{P}}$ and $\underline{P}$ coincide on $\mathscr{K}$. For any $f \in {}_{\circ}\mathscr{K}$, we have $f_{\circ} \in \mathscr{K}$ and therefore $\underline{P}_{\circ}(f) = \underline{E}_{\underline{P}}(f_{\circ}) = \underline{P}(f_{\circ}) = \underline{P}_{\circ}^{\mathrm{r}}(f)$, also using Equations $(7.15)_{134}$ and (7.21).

(ii). As $\underline{P}_{\circ}$ is coherent (Proposition $7.9(\mathrm{i})_{135}$) and coincides with $\underline{P}_{\circ}^{\mathrm{r}}$ on ${}_{\circ}\mathscr{K}$, we infer from the Natural Extension Theorem $4.26(\mathrm{ii})_{65}$ that $\underline{E}_{\underline{P}_{\circ}^{\mathrm{r}}} \leq \underline{P}_{\circ}$. To complete the proof, we consider any bounded gamble f on $\mathscr{X} \times \mathscr{Y}$ and show that $\underline{E}_{\underline{P}_{\circ}^{\mathrm{r}}}(f) \geq \underline{P}_{\circ}(f)$.

Fix $\varepsilon > 0$. If we use the definition of natural extension (for $\underline{P}$) in Equation $(4.7)_{47}$, we see that there are n in $\mathbb{N}$, non-negative $\lambda_1, \ldots, \lambda_n$ in $\mathbb{R}$ and bounded gambles $g_1, \ldots, g_n$ in $\mathscr{K}$ such that

$$f_{\circ}(y) - \underline{P}_{\circ}(f) + \frac{\varepsilon}{2} \geq \sum_{k=1}^{n} \lambda_k[g_k(y) - \underline{P}(g_k)] \text{ for all } y \in \mathscr{Y}. \tag{7.22}$$

It also follows from the definition of the lower inverse $f_{\circ}$ that for each $y \in \mathscr{Y}$, there is some set $F(y) \in \Phi(y)$ such that

$$\inf_{x \in F(y)} f(x, y) \geq f_{\circ}(y) - \frac{\varepsilon}{2}. \tag{7.23}$$

Now define the corresponding bounded gambles h_k on $\mathscr{X} \times \mathscr{Y}$, $k = 1, \ldots, n$ by

$$h_k(x, y) := \begin{cases} g_k(y) & \text{if } y \in \mathscr{Y} \text{ and } x \in F(y) \\ L & \text{if } y \in \mathscr{Y} \text{ and } x \notin F(y), \end{cases}$$

where L is some real number strictly smaller than $\min_{k=1}^{n} \inf g_k$, to be determined shortly. Then for any $y \in \mathscr{Y}$,

$$(h_k)_\circ(y) = \sup_{F \in \Phi(y)} \inf_{x \in F} h_k(x,y) = \sup_{F \in \Phi(y)} \inf_{x \in F} \begin{cases} g_k(y) & \text{if } x \in F(y) \\ L & \text{otherwise} \end{cases}$$

$$= \sup_{F \in \Phi(y)} \begin{cases} g_k(y) & \text{if } F \subseteq F(y) \\ L & \text{otherwise} \end{cases} = g_k(y),$$

so $(h_k)_\circ = g_k \in \mathscr{K}$ and, therefore, $h_k \in {}_\circ\mathscr{K}$ and $\underline{P}^{\mathrm{r}}_\circ(h_k) = \underline{P}(g_k)$. This, together with Equations (7.22)$_\curvearrowleft$ and (7.23)$_\curvearrowleft$, allows us to infer that

$$f(x,y) - \underline{P}_\circ(f) + \varepsilon \geq \sum_{k=1}^{n} \lambda_k [h_k(x,y) - \underline{P}^{\mathrm{r}}_\circ(h_k)] \text{ for all } y \in \mathscr{Y} \text{ and } x \in F(y).$$

Moreover, by an appropriate choice of L (small enough), we can always make sure that the inequality above holds for *all* $(x,y) \in \mathscr{X} \times \mathscr{Y}$, but then the definition of natural extension (for $\underline{P}^{\mathrm{r}}_\circ$) in Equation (4.7)$_{47}$ guarantees that $\underline{E}_{\underline{P}^{\mathrm{r}}_\circ}(f) \geq \underline{P}_\circ(f) - \varepsilon$. As this inequality holds for any $\varepsilon > 0$, the proof is complete. □

7.6 Special cases of induced lower previsions

To conclude this discussion of lower previsions induced by filter maps, we consider a number of special cases that have been studied in the literature.

A multivalued map $\Gamma : \mathscr{Y} \to \mathscr{P}(\mathscr{X})$ – associating a non-empty subset $\Gamma(y)$ of $\mathscr{X}$ with any $y \in \mathscr{Y}$ – allows us to define a filter map $\Phi : \mathscr{Y} \to \mathbb{F}(\mathscr{X})$ as follows:

$$\Phi(y) := \{B \subseteq \mathscr{X} : \Gamma(y) \subseteq B\} \text{ for all } y \in \mathscr{Y}.$$

Lower inverses and induced lower previsions for this specific type of filter map were discussed by Walley (1991, Section 4.3.5) and studied in detail by Miranda et al. (2005): Section 7.5.3$_{134}$ extends their results to general filter maps. The credit for taking the first steps in this domain and for associating lower and upper probabilities and previsions – or expectations – induced using multivalued maps is commonly given to Dempster (1967a, 1967b). But, as Carl Wagner has pointed out to us in private communication, important work by Straßen (1964) pre-dates Dempster's by 3 years was published in a well-known and widely read journal and has many of the relevant notions and results. To give an example, Proposition 7.9(ii)$_{135}$ in a related form was already present in Straßen's paper: his result holds for finite (or compact) $\mathscr{X}$, multivalued maps Φ and $\underline{P}$ that are linear previsions P. In those cases, he goes even further than we do, because he proves equality of the sets $\mathrm{lins}(P_\circ)$ and $\{PK : K \in \mathbb{K}(\Phi)\}$.

Some special cases deserve particular mention, because of the attention they have received in the literature.

7.6.1 Belief functions

Consider a mass function p on the *finite* space $\mathscr{Y}$ and the linear prevision P_p on $\mathbb{B}(\mathscr{Y})$ associated with it; see the discussion in Section 5.1_{77} for notation and more details. The multivalued map Γ provides a connection between the elements of $\mathscr{Y}$ and a finite collection $\mathscr{H}$ of non-empty subsets $\Gamma(y)$ of $\mathscr{X}$.

$$\mathscr{H} := \{\Gamma(y) : y \in \mathscr{Y}\}.$$

It induces a completely monotone lower prevision $(P_p)_\bullet$ on $\mathbb{B}(\mathscr{X})$ given by

$$(P_p)_\bullet(f) = P_p(f_\bullet) = \sum_{y\in\mathscr{Y}} p(y) \inf_{x\in\Gamma(y)} f(x) \text{ for all bounded gambles } f \text{ on } \mathscr{X},$$

so Proposition 7.6_{129} tells us that the induced lower prevision $(P_p)_\bullet$ is actually the natural extension $\underline{P}_m$ of a belief function $\underline{Q}_m$ with set of focal elements $\mathscr{H} := \{\Gamma(y) : y \in \mathscr{Y} \text{ and } p(y) > 0\}$ and basic probability assignment m defined by

$$m(\Gamma(y)) := p(y) \text{ for all } y \in \mathscr{Y} \text{ with } p(y) > 0.$$

See Section 7.3_{128} for more details about belief functions. We will turn to the case of infinite $\mathscr{Y}$ in Section 9.6_{188}.

Obviously, (the natural extension of) any belief function can be seen as an induced lower probability (prevision) for an appropriately chosen finite space $\mathscr{Y}$ and multivalued map Γ. We will generalise this observation towards all completely monotone lower previsions in Section $11.6.4_{227}$.

7.6.2 Refining the set of possible values for a random variable

In his discussion of useful assessment methods for imprecise probability models, Walley (1991, Section 4.3) discusses in particular how multivalued maps can be used to refine the set of possible values for a random variable. It is instructive to delve a bit further into this matter, using the tools we have developed earlier for lower previsions induced by filter maps.

Suppose our subject has an imprecise probability model about a variable Y assuming values in a set $\mathscr{Y}$: a coherent lower prevision $\underline{P}$ on $\mathbb{B}(\mathscr{Y})$. Now suppose that he comes to the conclusion that his initial set of possible values was not fine enough: each possible value y of Y can be further refined into a non-empty set B_y of possible values, where of course the sets B_y, $y \in \mathscr{Y}$ are *pairwise disjoint*.

If we now consider the set

$$\mathscr{X} := \bigcup_{y\in\mathscr{Y}} B_y \tag{7.24}$$

of refined possibilities, and the multivalued map $\Gamma : \mathscr{Y} \to \mathscr{P}(\mathscr{X})$ defined by

$$\Gamma(y) := B_y \text{ for all } y \in \mathscr{Y}, \tag{7.25}$$

we are led to consider a new variable X, assuming values in $\mathscr{X}$. The relationship between X and Y can be described by the conditional statement: *if Y assumes the* *value y, then we know that X assumes a value in* $\Gamma(y)$, which puts us squarely within the context of Section 7.5_{131}, but in the specific case that $\mathscr{Y}$ is in a one-to-one correspondence with the *partition* $\mathscr{B} := \{B_y : y \in \mathscr{Y}\}$ of $\mathscr{X}$,

$$\bigcup_{y \in \mathscr{Y}} \Gamma(y) = \mathscr{X} \text{ and } \Gamma(y_1) \cap \Gamma(y_2) = \emptyset \text{ for all } y_1, y_2 \in \mathscr{Y}.$$

With every bounded gamble f on $\mathscr{X}$, there corresponds a bounded gamble (lower inverse) $f_\bullet$ on $\mathscr{Y}$ given by

$$f_\bullet(y) := \inf_{x \in B_y} f(x), \tag{7.26}$$

and the lower prevision $\underline{P}_\bullet$ on $\mathbb{B}(\mathscr{X})$ that can be deduced from the lower prevision $\underline{P}$ on $\mathbb{B}(\mathscr{Y})$ and the conditional information embodied in Γ is given by

$$\underline{P}_\bullet(f) = \underline{P}(f_\bullet) \text{ for all bounded gambles } f \text{ on } \mathscr{X}.$$

This leads us to the following question. Suppose we have an assessment in the form of some lower prevision $\underline{P}$ defined on some set $\operatorname{dom} \underline{P}$ of bounded gambles on the initial space $\mathscr{Y}$. We make no further assumptions about $\underline{P}$: neither does it need to be coherent nor does it need to avoid sure loss. Clearly, the corresponding assessment on the refined space $\mathscr{X}$ is the lower prevision $\underline{P}^{\mathrm{r}}_\bullet$ defined on the following set of bounded gambles on $\mathscr{X}$

$$\operatorname{dom} \underline{P}^{\mathrm{r}}_\bullet = {}_\bullet(\operatorname{dom} \underline{P}) := \{f \in \mathbb{B}(\mathscr{X}) : f_\bullet \in \operatorname{dom} \underline{P}\} \tag{7.27}$$

by

$$\underline{P}^{\mathrm{r}}_\bullet(f) := \underline{P}(f_\bullet) \text{ for all } f \in {}_\bullet(\operatorname{dom} \underline{P}). \tag{7.28}$$

Then there is the following intuitively obvious connection between inference about Y and its refined counterpart X.

Proposition 7.12 *Consider a lower prevision $\underline{P}$ and the corresponding lower prevision $\underline{P}^{\mathrm{r}}_\bullet$ defined through Equations* (7.24)$_{\curvearrowleft}$–(7.28).

(i) *$\underline{P}^{\mathrm{r}}_\bullet$ avoids sure loss if and only if $\underline{P}$ does.*

(ii) *$\underline{P}^{\mathrm{r}}_\bullet$ is coherent if and only if $\underline{P}$ is.*

(iii) *If $\underline{P}$ avoids sure loss, then the natural extension $\underline{E}_{\underline{P}^{\mathrm{r}}_\bullet}$ of $\underline{P}^{\mathrm{r}}_\bullet$ is the induced lower prevision $\underline{P}_\bullet$.*

Proof. We begin with the proof of (ii), as the proof of (i) is similar but less involved. First assume that $\underline{P}^{\mathrm{r}}_\bullet$ is coherent, then we have to prove that $\underline{P}$ is coherent too. Consider arbitrary n, m in $\mathbb{N}$ and bounded gambles $g_0, g_1, \ldots, g_n$ in $\operatorname{dom} \underline{P}$. Define the bounded gambles f_k on $\mathscr{X}$ by $f_k(x) := g_k(y)$ for all $y \in \mathscr{Y}$ and $x \in B_y$, then clearly $(f_k)_\bullet = g_k$ and $\underline{P}^{\mathrm{r}}_\bullet(f_k) = \underline{P}(g_k)$ and, therefore,

$$\sup_{y \in \mathscr{Y}} \left[\sum_{k=1}^{n} g_k(y) - m g_0(y) \right] = \sup_{x \in \mathscr{X}} \left[\sum_{k=1}^{n} f_k(x) - m f_0(x) \right]$$

$$\geq \sum_{k=1}^{n} \underline{P}^{\mathrm{r}}_{\bullet}(f_k) - m\underline{P}^{\mathrm{r}}_{\bullet}(f_0) = \sum_{k=1}^{n} \underline{P}(g_k) - m\underline{P}(g_0),$$

where the inequality follows from the coherence of $\underline{P}^{\mathrm{r}}_{\bullet}$. This tells us that $\underline{P}$ is indeed coherent as well.

Conversely, assume that $\underline{P}$ is coherent, then we have to prove that $\underline{P}^{\mathrm{r}}_{\bullet}$ is coherent too. Consider arbitrary n, m in $\mathbb{N}$ and bounded gambles $f_0, f_1, \dots, f_n$ in ${}_{\bullet}(\operatorname{dom} \underline{P})$, then

$$\begin{aligned}
\sup\left[\sum_{k=1}^{n} f_k - mf_0\right] &= \sup_{y\in\mathcal{Y}} \sup_{x\in B_y}\left(\sum_{k=1}^{n} f_k(x) - mf_0(x)\right) = \sup_{y\in\mathcal{Y}} \overline{P}_{B_y}\left(\sum_{k=1}^{n} f_k - mf_0\right)\\
&\geq \sup_{y\in\mathcal{Y}}\left[\overline{P}_{B_y}(-mf_0) + \underline{P}_{B_y}\left(\sum_{k=1}^{n} f_k\right)\right]\\
&\geq \sup_{y\in\mathcal{Y}}\left[\sum_{k=1}^{n} \underline{P}_{B_y}(f_k) - m\underline{P}_{B_y}(f_0)\right] = \sup\left[\sum_{k=1}^{n} (f_k)_{\bullet} - m(f_0)_{\bullet}\right]\\
&\geq \sum_{k=1}^{n} \underline{P}((f_k)_{\bullet}) - m\underline{P}((f_0)_{\bullet}) = \sum_{k=1}^{n} \underline{P}^{\mathrm{r}}_{\bullet}(f_k) - m\underline{P}^{\mathrm{r}}_{\bullet}(f_0),
\end{aligned}$$

which tells us that $\underline{P}^{\mathrm{r}}_{\bullet}$ is indeed coherent as well. The first two inequalities follow from the mixed super-/sub-additivity properties (Theorem 4.13(vi)$_{53}$) of the coherent vacuous lower/upper previsions $\underline{P}_{B_y}$ and $\overline{P}_{B_y}$, and the last inequality follows from the coherence of $\underline{P}$.

To conclude, we prove (iii). If $\underline{P}$ avoids sure loss, then (i) tells us that $\underline{P}^{\mathrm{r}}_{\bullet}$ avoids sure loss, so we can use the expression (4.10a)$_{68}$ in Theorem 4.33$_{68}$ to find its natural extension $\underline{E}_{\underline{P}^{\mathrm{r}}_{\bullet}}$, a coherent lower prevision. So consider any bounded gamble f on $\mathcal{X}$, then we have to prove that $\underline{E}_{\underline{P}^{\mathrm{r}}_{\bullet}}(f) = \underline{E}_{\underline{P}}(f_{\bullet})$.

Fix any real number α, and assume that there are n in $\mathbb{N}$, non-negative $\lambda_1, \dots, \lambda_n$ in $\mathbb{R}$ and bounded gambles $f_1, \dots, f_n$ in ${}_{\bullet}(\operatorname{dom} \underline{P})$ such that $f - \alpha \geq \sum_{k=1}^{n} \lambda_k[f_k - \underline{P}^{\mathrm{r}}_{\bullet}(f_k)]$. By taking infima over the B_y on both sides of this inequality and recalling that $\underline{P}^{\mathrm{r}}_{\bullet}(f_k) = \underline{P}((f_k)_{\bullet})$, this shows that there are $g_k := (f_k)_{\bullet} \in \operatorname{dom} \underline{P}$ such that $f_{\bullet} - \alpha \geq \sum_{k=1}^{n} \lambda_k[g_k - \underline{P}(g_k)]$. The expression in Equation (4.10a)$_{68}$ for the natural extension then allows us to conclude that $\alpha \leq \underline{E}_{\underline{P}}(f_{\bullet})$ and, therefore, also that $\underline{E}_{\underline{P}^{\mathrm{r}}_{\bullet}}(f) \leq \underline{E}_{\underline{P}}(f_{\bullet})$.

To prove the converse inequality, consider any real β such that $\underline{E}_{\underline{P}}(f_{\bullet}) > \beta$. Then we see that there are n in $\mathbb{N}$, non-negative $\lambda_1, \dots, \lambda_n$ in $\mathbb{R}$ and bounded gambles $g_1, \dots, g_n$ in $\operatorname{dom} \underline{P}$ such that $f_{\bullet} - \beta \geq \sum_{k=1}^{n} \lambda_k[g_k - \underline{P}(g_k)]$. Define the bounded gambles f_k on $\mathcal{X}$ by $f_k(x) := g_k(y)$ for all $y \in \mathcal{Y}$ and $x \in B_y$, then clearly $(f_k)_{\bullet} = g_k$ and $\underline{P}^{\mathrm{r}}_{\bullet}(f_k) = \underline{P}(g_k)$, and therefore, it follows that $f - \beta \geq \sum_{k=1}^{n} \lambda[f_k - \underline{P}^{\mathrm{r}}_{\bullet}(f_k)]$, implying that $\underline{E}_{\underline{P}^{\mathrm{r}}_{\bullet}}(f) \geq \beta$. Hence, indeed $\underline{E}_{\underline{P}^{\mathrm{r}}_{\bullet}}(f) \geq \underline{E}_{\underline{P}}(f_{\bullet})$ and, therefore, $\underline{E}_{\underline{P}^{\mathrm{r}}_{\bullet}}(f) = \underline{E}_{\underline{P}}(f_{\bullet}) = \underline{P}_{\bullet}(f)$. □

This shows that, to give only one example, the discussion and results about lower and upper probability mass assessments in Section 7.1$_{123}$ can be extended without further ado to lower and upper probability assessments on finite partitions, replacing singletons by atoms (or partition classes). That, incidentally, is the form in which those results are mentioned by Walley (1991, Section 4.6.1).

7.7 Assessments on chains of sets

Another interesting special case concerns a lower (or equivalently, upper) probability assessment $\underline{P}$ on a chain of events $\mathscr{C}$. That $\mathscr{C}$ is a chain means that the events C in $\mathscr{C}$ are *totally ordered* by set inclusion:

CH1. For any two elements C_1 and C_2 of $\mathscr{C}$, $C_1 \subseteq C_2$ or $C_2 \subseteq C_1$.

As this implies that $\mathscr{C}$ is closed under finite unions and intersections, $\mathscr{C}$ is in particular a lattice of events.

As any coherent lower prevision is monotone (Theorem 4.13(iv)$_{53}$), we may assume without loss of generality that the lower probability assessment $\underline{P}$ is monotone in the sense that

CH2. $C_1 \subseteq C_2 \Rightarrow \underline{P}(C_1) \leq \underline{P}(C_2)$.

As a coherent lower prevision satisfies normality (Theorem 4.13(ii)$_{53}$), it may also be assumed without loss of generality that

CH3. $\emptyset \in \mathscr{C}$ and $\underline{P}(\emptyset) = 0$, and also $\mathscr{X} \in \mathscr{C}$ and $\underline{P}(\mathscr{X}) = 1$.

It follows readily from CH1 and CH2 that $\underline{P}$ is minimum preserving on the lattice $\mathscr{C}$:

$$\underline{P}(C_1 \cap C_2) = \min\{\underline{P}(C_1), \underline{P}(C_2)\} \text{ for all } C_1, C_2 \in \mathscr{C},$$

so Proposition 7.4$_{127}$ guarantees that the assessment $\underline{P}$ is a completely monotone lower probability, that is, therefore, also coherent, taking into account CH3 and Corollary 6.16$_{112}$. In addition, we infer from Corollary 6.17$_{112}$ that its natural extension $\underline{E}_{\underline{P}}$ is completely monotone as well, and given by the Choquet functional $C_{\underline{P}_*}$ associated with the inner set function $\underline{P}_*$,

$$\underline{P}_*(A) = \sup\left\{\underline{P}(C) \colon C \in \mathscr{C} \text{ and } C \subseteq A\right\} \text{ for all } A \subseteq \mathscr{X}.$$

Interestingly, the natural extension $\underline{P}_*$ to events is not only completely monotone but also minimum preserving, as is the assessment $\underline{P}$ itself.

Proposition 7.13 *When a lower probability assessment $\underline{P}$ on $\mathscr{C}$ satisfies CH1–CH3, then its natural extension $\underline{P}_*$ to all events is minimum preserving:*

$$(\forall A, B \subseteq \mathscr{X})\underline{P}_*(A \cap B) = \min\{\underline{P}_*(A), \underline{P}_*(B)\}.$$

Proof. Consider any subsets A and B of $\mathscr{X}$. Because the coherent $\underline{P}_*$ is monotone (Theorem 4.13(iv)$_{53}$), we see at once that $\underline{P}_*(A \cap B) \leq \min\{\underline{P}_*(A), \underline{P}_*(B)\}$. Suppose *ex absurdo* that $\underline{P}_*(A \cap B) < \min\{\underline{P}_*(A), \underline{P}_*(B)\}$. Then it follows from the definition of the inner set function $\underline{P}_*$ that there are C_1 and C_2 in $\mathscr{C}$ such that $C_1 \subseteq A$

and $\underline{P}(C_1) > \underline{P}_*(A \cap B)$ and $C_2 \subseteq B$ and $\underline{P}(C_2) > \underline{P}_*(A \cap B)$. But because $C_1 \cap C_2 \subseteq A \cap B$ and $C_1 \cap C_2 \in \mathscr{C}$ (by CH1), we must have that $\underline{P}(C_1 \cap C_2) = \underline{P}_*(C_1 \cap C_2) \leq \underline{P}_*(A \cap B)$, because the inner set function $\underline{P}_*$ is monotone and coincides with $\underline{P}$ on $\mathscr{C}$. This contradicts that $C_1 \cap C_2 = C_1$ or $C_1 \cap C_2 = C_2$, because $\mathscr{C}$ is a chain (by CH1). □

These very simple ideas underly the much more involved discussions of minimum preserving lower probabilities by Dubois and Prade (1992) and De Cooman and Aeyels (1999). They have been used by Walley and De Cooman (2001) to argue that lower probabilities with such simple properties occur quite naturally in contexts where uncertainty is linguistic: the result of simple (but possibly vague) affirmative statements in natural language.

7.8 Possibility and necessity measures

The discussion in Section 7.7 centred on lower probabilities that preserve finite minima (or infima). A more restrictive class of lower probabilities will preserve arbitrary infima and their conjugate upper probabilities preserve arbitrary suprema. This leads directly to the topic of *possibility measures*. These constitute a special type of coherent upper probability that has received a lot of attention in the literature, especially after they were considered by Zadeh (1978) in the context of fuzzy set theory. They were also used before that time in a different context (and under a different name) by Shackle (1961). After Zadeh's seminal discussion, important early progress in this field was made by Dubois and Prade (1988); see also De Cooman (1997) for an interesting order-theoretic approach. Early work on the incorporation of possibility measures into the theory of coherent lower previsions was done by Dubois and Prade (1992), Walley (1997), De Cooman and Aeyels (1999) and De Cooman (2001). The discussion of the specific interpretation of possibility measures in this context by Walley and De Cooman (2001) has lead to an extension of imprecise probabilities designed to deal with vague probability assessments (De Cooman, 2005a).

A **possibility measure** Π on a non-empty set $\mathscr{X}$ is defined as a supremum preserving set function $\mathscr{P}(\mathscr{X}) \to [0,1]$: for any family A_i, $i \in I$ of subsets of $\mathscr{X}$, where I is any index set, it should hold that

$$\Pi\left(\bigcup_{i \in I} A_i\right) = \sup_{i \in I} \Pi(A_i).$$

Clearly, a set function Π is a possibility measure if and only if

$$\Pi(A) = \sup_{x \in A} \pi(x) \text{ for all } A \subseteq \mathscr{X},$$

where the map $\pi : \mathscr{X} \to [0,1]$ is given by $\pi(x) := \Pi(\{x\})$ for all x in $\mathscr{X}$ and is called the **possibility distribution** of Π. The conjugate set function N of Π defined by $N(A) := 1 - \Pi(A^c) = 1 - \sup_{x \notin A} \pi(x)$ for all $A \subseteq \mathscr{X}$ is called a **necessity measure**.[8]

[8] In all these expressions, we let $\sup_{x \in \emptyset} \pi(x) = 0$.

A possibility measure is easily seen to be sub-additive and is, therefore, naturally interpreted as an upper probability (recall Theorem 4.13(v)$_{53}$). The following proposition shows how possibility and necessity measures can be made to fit into the uncertainty framework of this book. Because it is more common to the emphasise possibility rather than necessity measures, this is one of the very few places in this book where we formulate results in terms of upper, rather than lower, probabilities and previsions.

Proposition 7.14 *Consider a map* $\pi : \mathcal{X} \to [0, 1]$ *and the corresponding possibility measure* Π *defined by* $\Pi(A) := \sup_{x \in A} \pi(x)$ *for all* $A \subseteq \mathcal{X}$. *Then the following statements are equivalent:*

(i) $\sup \pi = 1$

(ii) *the upper probability* Π *avoids sure loss*

(iii) *the upper probability* Π *is coherent.*

In that case, the upper probability Π *is completely alternating, and its natural extension to all bounded gambles, denoted by* $\overline{P}_\pi$, *is also completely alternating and given by Choquet integration: for all bounded gambles* f *on* $\mathcal{X}$,

$$\overline{P}_\pi(f) = C\int f \, \mathrm{d}\Pi = \inf f + R\int_{\inf f}^{\sup f} \sup\{\pi(x) : f(x) \geq t\} \, \mathrm{d}t \tag{7.29}$$

$$= R\int_0^1 \sup\{f(x) : \pi(x) \geq \alpha\} \, \mathrm{d}\alpha \tag{7.30}$$

$$= R\int_0^1 \sup\{f(x) : \pi(x) > \alpha\} \, \mathrm{d}\alpha. \tag{7.31}$$

The explicit formulation of the first part was given by De Cooman and Aeyels (1999), and the second part involving the equalities by De Cooman (2001). We repeat and adapt their proofs. These results were also known to Walley (1997), but his proofs for the equalities are different.

Proof. The conjugate lower probability N is minimum preserving on the lattice of events $\mathcal{P}(\mathcal{X})$ and, therefore, completely monotone by Proposition 7.4$_{127}$. As $N(\emptyset) = 1 - \sup \pi$ and $N(\mathcal{X}) = 1 - 0 = 1$, we infer from Corollary 6.16$_{112}$ that the lower probability N is coherent if and only if $\sup \pi = 1$. Moreover, if N avoids sure loss, then $N(\emptyset) \leq 0$ (use avoiding sure loss criterion (D)$_{42}$ with $n = 1$ and $f_1 = I_\emptyset = 0$) or, equivalently, $\sup \pi = 1$. This completes the proof of the equivalences.

We have also learned in the argument above that if any of these equivalent statements holds, the lower probability N is completely monotone. By Theorem 6.14$_{111}$, its natural extension $\underline{P}_\pi$ is given by Choquet integration. If we now invoke Proposition C.5(iv)$_{382}$, we see that for any bounded gamble f on $\mathcal{X}$,

$$\overline{P}_\pi(f) = -\underline{P}_\pi(-f) = -C_N(-f) = C_\Pi(f) = \inf f + R\int_{\inf f}^{\sup f} \Pi(\{f \geq t\}) \, \mathrm{d}t,$$

which proves Equation (7.29).

We now turn to Equation (7.30). Because of the constant additivity of the coherent upper prevision $\overline{P}_\pi$ (Theorem 4.13(iii)$_{53}$) and of the functional defined on the right-hand side of Equation (7.30), it suffices to consider non-negative f. We invoke Proposition C.8(ii)$_{388}$, and using the notations established there, we let ϕ be the set function determined by $\phi(A) := 1$ for all non-empty subsets A of $\mathcal{X}$. Then $\phi_\pi(A) = \sup\left\{t : A \cap \{\pi \geq t\} \neq \emptyset\right\} = \sup_{x\in A} \pi(x)$ and, similarly, $\phi_f(A) = \sup_{x\in A} f(x)$. Moreover, $\phi_\pi(f) = \phi_f(\pi)$. So, by repeatedly invoking Proposition C.3$_{379}$, we get

$$
\begin{aligned}
C_\Pi(f) &= \inf f + R\int_{\inf f}^{\sup f} \sup\{\pi(x) : f(x) \geq t\}\,\mathrm{d}t \\
&= R\int_0^{+\infty} \sup\{\pi(x) : f(x) \geq t\}\,\mathrm{d}t = R\int_0^{+\infty} \phi_\pi(\{f(x) \geq t\})\,\mathrm{d}t \\
&= \phi_\pi(f) = \phi_f(\pi) \\
&= R\int_0^{+\infty} \phi_f(\{\pi(x) \geq t\})\,\mathrm{d}t = R\int_0^{+\infty} \sup\{f(x) : \pi(x) \geq t\}\,\mathrm{d}t \\
&= R\int_0^1 \sup\{f(x) : \pi(x) \geq \alpha\}\,\mathrm{d}\alpha.
\end{aligned}
$$

To conclude, we prove Equation (7.31) by showing that

$$
R\int_0^1 \sup\{f(x) : \pi(x) > \alpha\}\,\mathrm{d}\alpha = R\int_0^1 \sup\{f(x) : \pi(x) \geq \alpha\}\,\mathrm{d}\alpha.
$$

Use the notation $\Gamma_1(f)$ for the left-hand side and $\Gamma_2(f)$ for the right-hand side. It clearly suffices to prove the equality $\Gamma_1(f) = \Gamma_2(f)$ for non-negative f (because $\Gamma_1(f + \mu) = \Gamma_1(f) + \mu$ and $\Gamma_2(f + \mu) = \Gamma_2(f) + \mu$ for any real constant μ). Consider any $\varepsilon \in (0, 1)$ and observe that $\{\pi > \alpha\} \subseteq \{\pi \geq \alpha\} \subseteq \{\pi > \alpha - \varepsilon\}$, which leads to the inequalities

$$
\Gamma_1(f) \leq \Gamma_2(f) \leq R\int_0^1 \sup\{f(x) : \pi(x) > \alpha - \varepsilon\}\,\mathrm{d}\alpha.
$$

Now observe that

$$
\begin{aligned}
&R\int_0^1 \sup\{f(x) : \pi(x) > \alpha - \varepsilon\}\,\mathrm{d}\alpha \\
&\quad = R\int_{-\varepsilon}^{1-\varepsilon} \sup\{f(x) : \pi(x) > \alpha\}\,\mathrm{d}\alpha \\
&\quad = R\int_{-\varepsilon}^{0} \sup\{f(x) : \pi(x) > \alpha\}\,\mathrm{d}\alpha + R\int_0^{1-\varepsilon} \sup\{f(x) : \pi(x) > \alpha\}\,\mathrm{d}\alpha \\
&\quad \leq \varepsilon \sup f + R\int_0^1 \sup\{f(x) : \pi(x) > \alpha\}\,\mathrm{d}\alpha \\
&\quad = \varepsilon \sup f + \Gamma_1(f),
\end{aligned}
$$

where the inequality holds because we assumed that $f \geq 0$. Letting $\varepsilon \to 0$ leads readily to the desired equality. □

Possibility measures can also be seen as special induced upper previsions and, therefore, also fit well within the framework established in Section 7.5_{131}. This idea was explored in great detail by Nguyen (Nguyen, 2006; Nguyen et al., 1997), but we limit ourselves here to outlining its essential components.

With the possibility distribution π on $\mathscr{X}$, we associate a filter map $\Phi : [0,1) \to \mathbb{F}(\mathscr{X})$ given by

$$\Phi(t) := \{A \subseteq \mathscr{X} : \{\pi > t\} \subseteq A\} \text{ for all } 0 \leq t < 1,$$

so $\Phi(t)$ is the fixed filter with smallest element $\{\pi > t\}$.[9] Then with any bounded gamble f on $\mathscr{X}$, there corresponds a lower inverse $f_\bullet$ on $[0,1)$ given by

$$f_\bullet(t) = \underline{P}_{\Phi(t)}(f) = \underline{P}_{\{\pi>t\}}(f) = \inf\{f(x) : \pi(x) > t\} \text{ for all } 0 \leq t < 1. \tag{7.32}$$

Now let P be any linear prevision on $\mathbb{B}(\mathscr{Y})$, where $\mathscr{Y} = [0,1)$, that coincides with the Lebesgue measure λ on all intervals of the type $[a,1)$, meaning that $P([a,1)) = \lambda([a,1)) = 1-a$ for all $a \in [0,1]$,[10] then we infer from Theorem 6.13_{110} that its lower inverse $P_\bullet$ is given by

$$P_\bullet(f) = P(f_\bullet) = \inf f_\bullet + R\int_{\inf f_\bullet}^{\sup f_\bullet} P(f_\bullet \geq \alpha)\,\mathrm{d}\alpha, \tag{7.33}$$

and because $\inf f_\bullet \geq \inf f$ and $\sup f_\bullet \leq \sup f$ and, moreover,

$$\begin{aligned}
f_\bullet(t) \geq \alpha &\Leftrightarrow (\forall x \in \mathscr{X})(\pi(x) > t \Rightarrow f(x) \geq \alpha)\\
&\Leftrightarrow (\forall x \in \mathscr{X})(f(x) < \alpha \Rightarrow \pi(x) \leq t)\\
&\Leftrightarrow t \geq \sup\{\pi(x) : f(x) < \alpha\} = \Pi(\{f < \alpha\}) = 1 - N(\{f \geq \alpha\}),
\end{aligned}$$

we infer from Equation (7.33) that, with obvious notations,

$$\begin{aligned}
P_\bullet(f) &= \inf f + R\int_{\inf f}^{\sup f} P([1 - N(\{f \geq \alpha\}), 1))\,\mathrm{d}\alpha\\
&= \inf f + R\int_{\inf f}^{\sup f} \lambda([1 - N(\{f \geq \alpha\}), 1))\,\mathrm{d}\alpha\\
&= \inf f + R\int_{\inf f}^{\sup f} N(\{f \geq \alpha\})\,\mathrm{d}\alpha = C_N(f) = \underline{P}_\pi(f).
\end{aligned}$$

[9] Observe that, because $\sup \pi = 1$, $\{\pi > t\} \neq \emptyset$ for all $t \in [0,1)$, so each $\Phi(t)$ is indeed a proper filter.

[10] That there are such linear previsions, follows from the Linear Extension Theorem 4.42_{73} and Theorem 8.15_{161} further on. It should also be clear that the argument that follows is supported by finite additivity alone, nothing hinges on σ-additivity.

The second equality follows from the fact that P and λ coincide on all intervals $[a, 1)$; the next to last equality follows from the definition of the Choquet functional C_N associated with the set function N, and the last equality from Equation $(7.29)_{144}$ in Proposition 7.14_{144}, after applying conjugacy.[11] We conclude that the lower prevision associated with a possibility measure is the lower inverse of the Lebesgue measure under this special filter map (essentially a multivalued map) (see also the discussion in Section 9.4_{183}).

7.9 Distribution functions and probability boxes

A probability measure, or more generally a linear prevision P, on the set $\mathbb{R}$ of real numbers is commonly represented via its (cumulative) distribution function:[12]

$$F(x) := P(X \leq x) \text{ for all } x \in \mathbb{R}. \tag{7.34}$$

A straightforward generalisation of this idea is to consider the set of all probability measures, or more generally linear previsions, whose distribution function is bounded by lower and upper distribution functions. These imprecise probability models are called *probability boxes* (or *p-boxes*, for short) and were introduced by Ferson et al. (2003), see also the more recent discussion by Ferson and Tucker (2006). Probability boxes have been related to random sets (Kriegler and Held, 2005) and generalised to arbitrary totally preordered spaces (Destercke and Dubois, 2006), see also the discussions by Destercke et al. (2008) and Miranda et al. (2008b). In the present short overview, we follow the rather general treatment of Troffaes and Destercke (2011).

7.9.1 Distribution functions

A binary relation $\preceq$ on a non-empty set $\mathcal{X}$ is called a **total preorder** (i) if $\preceq$ is a preorder – meaning that it is transitive and reflexive (see Definition A.1_{368}) – and (ii) if any two elements are comparable. The latter means that for all $x, y \in \mathcal{X}$, exactly one of the statements $x \prec y$, $x \simeq y$ or $x \succ y$ holds – we use commonly accepted notation: '$x \prec y$' means '$x \preceq y$ and $x \not\succeq y$', '$x \succ y$' means '$y \prec x$' and '$x \simeq y$' means '$x \preceq y$ and $y \preceq x$'. We also define intervals in $\mathcal{X}$ in the usual way:

$$\begin{aligned} [x, y] &:= \{z \in \mathcal{X} : x \preceq z \preceq y\} & (x, y) &:= \{z \in \mathcal{X} : x \prec z \prec y\} \\ [x, y) &:= \{z \in \mathcal{X} : x \preceq z \prec y\} & (x, y] &:= \{z \in \mathcal{X} : x \prec z \preceq y\} \end{aligned} \quad \text{for all } x, y \in \mathcal{X}.$$

For the remainder of this section, we assume that $\mathcal{X}$ is equipped with a total preorder $\preceq$ that has a smallest element $0_{\mathcal{X}}$ and a largest element $1_{\mathcal{X}}$; for example, it could be the real unit interval $[0, 1]$, or $\mathbb{R}^*$ with the usual ordering $\leq$ of extended real numbers, where $0_{\mathcal{X}} = -\infty$ and $1_{\mathcal{X}} = +\infty$.

[11] We can now also easily derive the conjugate counterpart of Equation $(7.31)_{144}$: because $f_\bullet$ is non-decreasing, it is Riemann integrable and, therefore, also Lebesgue integrable, and therefore, $P(f_\bullet) = L\int_{[0,1)} f_\bullet \, \mathrm{d}\lambda = R\int_0^1 f_\bullet(t) \, \mathrm{d}t$. The first equality follows from the discussion in Section 9.4_{183}. Now use Equation (7.32) to find that $\underline{P}_\pi(f) = R\int_0^1 \inf\{f(x) : \pi(x) > t\}$.

[12] We have already discussed a special case in Example 5.17_{97}.

A (cumulative) **distribution function** is a map $F\colon \mathcal{X} \to [0,1]$ that is **non-decreasing**, meaning that

$$x \preceq y \Rightarrow F(x) \leq F(y) \text{ for all } x, y \in \mathcal{X},$$

and that satisfies $F(1_{\mathcal{X}}) = 1$. It is interpreted similarly as in Equation (7.34)$_\curvearrowleft$: $F(x)$ represents the probability of the interval $[0_{\mathcal{X}}, x]$. No other constraints on F are imposed. In particular, we do not impose $F(0_{\mathcal{X}}) = 0$, so $\{0_{\mathcal{X}}\}$ can have non-zero probability, which is common if $\mathcal{X}$ is finite. Also, we do not impose any form of continuity on F, which is in keeping with the fact that we generally work with *finitely* additive linear previsions as precise probability models.

Let $\mathcal{H}$ be the field of events generated by the chain of intervals

$$\left\{[0_{\mathcal{X}}, x]\colon x \in \mathcal{X}\right\}$$

or, in other words, the smallest field that contains all these intervals. As is easy to see and argued explicitly by Denneberg (1994, Proposition 2.10), $\mathcal{H}$ is the collection of events of the type

$$[0_{\mathcal{X}}, x_1] \cup (x_2, x_3] \cup \cdots \cup (x_{2n}, x_{2n+1}] \text{ for } x_1 \prec x_2 \prec x_3 \prec \cdots \prec x_{2n+1} \text{ in } \mathcal{X}$$

(if n is 0 we simply take this expression to be $[0_{\mathcal{X}}, x_1]$) and

$$(x_2, x_3] \cup \cdots \cup (x_{2n}, x_{2n+1}] \text{ for } x_2 \prec x_3 \prec \cdots \prec x_{2n+1} \text{ in } \mathcal{X}.$$

These events indeed form a field: the union and intersection of any two events in $\mathcal{H}$ is again in $\mathcal{H}$, and so is the complement of any event in $\mathcal{H}$. To simplify its description, and the discussion further on, we introduce an extra element $0_{\mathcal{X}^*}$ such that

$$F(0_{\mathcal{X}^*}) := 0 \text{ and } 0_{\mathcal{X}^*} \prec x \text{ for all } x \in \mathcal{X},$$

where F is any distribution function to be considered further on. In particular, $(0_{\mathcal{X}^*}, x] = [0_{\mathcal{X}}, x]$, so if we let $\mathcal{X}^* := \mathcal{X} \cup \{0_{\mathcal{X}^*}\}$, then

$$\mathcal{H} = \left\{(x_0, x_1] \cup (x_2, x_3] \cup \cdots \cup (x_{2n}, x_{2n+1}]\colon x_0 \prec x_1 \prec \cdots \prec x_{2n+1} \text{ in } \mathcal{X}^*\right\}. \tag{7.35}$$

The distribution function F allows us to define a set function μ_F on the chain of events $\left\{(0_{\mathcal{X}^*}, x]\colon x \in \mathcal{X}^*\right\} = \{\emptyset\} \cup \left\{[0_{\mathcal{X}}, x]\colon x \in \mathcal{X}\right\}$ by letting

$$\mu_F((0_{\mathcal{X}^*}, x]) := F(x) \text{ for all } x \in \mathcal{X}^*.$$

It is fairly trivial – and proved explicitly by Denneberg (1994, Proposition 2.10) – that this set function can be extended uniquely to a charge α_F on the field $\mathcal{H}$. This charge is actually a probability charge and is given by

$$\alpha_F((x_0, x_1] \cup (x_2, x_3] \cup \cdots \cup (x_{2n}, x_{2n+1}]) := \sum_{k=0}^{n} [F(x_{2k+1}) - F(x_{2k})]$$
$$\text{for all } x_0 \prec x_1 \prec \cdots \prec x_{2n+1} \text{ in } \mathcal{X}^*.$$

If we also observe that $\alpha_F((x, 1_{\mathcal{X}}]) = F(1_{\mathcal{X}}) - F(x) = 1 - F(x)$ for all $x \in \mathcal{X}$, we come to the following simple result.

Theorem 7.15 *Let F be any distribution function, and define the lower probability $\underline{P}_F$ on the set of events*

$$\mathscr{K} := \{[0_{\mathscr{X}}, x] : x \in \mathscr{X}\} \cup \{(y, 1_{\mathscr{X}}] : y \in \mathscr{X}\}$$

by

$$\underline{P}_F([0_{\mathscr{X}}, x]) := F(x) \text{ and } \underline{P}_F((y, 1_{\mathscr{X}}]) := 1 - F(y).$$

Then $\underline{P}_F$ is coherent, and its natural extension $\underline{E}_F$ coincides with α_F on all events in $\mathscr{H}$: $\underline{E}_F(H) = \alpha_F(H)$ for all $H \in \mathscr{H}$.[13]

Of course, this means that a distribution function induces a *precise* probability model on the field of events $\mathscr{H}$.

Proof. We borrow the notations and results from Chapter 8 on linear previsions – which, it matters to say, do not depend on the present theorem. Consider the prevision P_{α_F} induced by the probability charge α_F:

$$P_{\alpha_F}(I_H) := -P_{\alpha_F}(-I_H) := \alpha_F(H) \text{ for all } H \in \mathscr{H}.$$

Theorem 8.15_{161} guarantees that P_{α_F} is a linear prevision, which means that it is coherent as a lower prevision (Theorem $4.12(\text{A})_{51}$). As $\mathscr{H}$ includes $\mathscr{K}$ and P_{α_F} and $\underline{P}_F$ coincide on $\mathscr{K}$ by construction, this guarantees that the lower prevision $\underline{P}_F$ is coherent (Theorem $4.18(\text{ii})_{58}$).

For its natural extension $\underline{E}_F$, consider any linear prevision P defined on the set $\mathbb{B}(\mathscr{X})$ of all bounded gambles that dominates $\underline{P}_F$, meaning that $P([0_{\mathscr{X}}, x]) \geq F(x)$ and $P((x, 1_{\mathscr{X}}]) \geq 1 - F(x)$ for all $x \in \mathscr{X}$. From the second inequality, we also infer that $1 - P([0_{\mathscr{X}}, x]) \geq 1 - F(x)$ and, therefore, $P([0_{\mathscr{X}}, x]) = F(x)$ and $P((x, 1_{\mathscr{X}}]) = 1 - F(x)$. Consequently, for all $x < y$, we infer from the additivity of the linear prevision P (Corollary $4.14(\text{v})_{55}$) that

$$\begin{aligned} P((x, y]) &= P([0_{\mathscr{X}}, y] \cap (x, 1_{\mathscr{X}}]) \\ &= P([0_{\mathscr{X}}, y]) + P((x, 1_{\mathscr{X}}]) - P([0_{\mathscr{X}}, y] \cup (x, 1_{\mathscr{X}}]) \\ &= F(y) + 1 - F(x) - 1 = F(y) - F(x) = \alpha_F((x, y]) \end{aligned}$$

and, therefore, also $P(H) = \alpha_F(H)$ for all $H \in \mathscr{H}$, guaranteeing that P and α_F coincide on $\mathscr{H}$. Now use the Lower Envelope Theorem $4.38(\text{iii})_{71}$. □

7.9.2 Probability boxes

We now extend this idea to allow for imprecision in the probability assessments. A **probability box** – or **p-box** for short – is simply a pair $(\underline{F}, \overline{F})$ of distribution functions satisfying $\underline{F} \leq \overline{F}$. In the framework of lower previsions, this probability box leads to a lower probability (assessment) $\underline{P}_{\underline{F},\overline{F}}$ defined on the set of events

$$\mathscr{K} := \{[0_{\mathscr{X}}, x] : x \in \mathscr{X}\} \cup \{(y, 1_{\mathscr{X}}] : y \in \mathscr{X}\}$$

[13] We refer to Section 9.1_{181} for a discussion of how this relates to integration associated with a distribution function.

by

$$\underline{P}_{\underline{F},\overline{F}}([0_{\mathcal{X}},x]) := \underline{F}(x) \text{ and } \underline{P}_{\underline{F},\overline{F}}((y,1_{\mathcal{X}}]) := 1-\overline{F}(y).$$

The following basic result (Troffaes, 2005, Theorem 3.59, p. 93) now follows almost immediately.

Theorem 7.16 *The lower probability $\underline{P}_{\underline{F},\overline{F}}$ is coherent.*

Proof. We infer from Theorem 7.15$_{\curvearrowleft}$ that the lower previsions $\underline{P}_{\underline{F}}$ and $\underline{P}_{\overline{F}}$ on $\mathcal{K}$ are coherent. It then follows from Theorem 4.20(ii)$_{59}$ that their lower envelope $\underline{P}_{\underline{F},\overline{F}}$ is coherent too. □

Consequently, $\underline{P}_{\underline{F},\overline{F}}$ has a natural extension, which we denote by $\underline{E}_{\underline{F},\overline{F}}$. It will turn out that $\underline{E}_{\underline{F},\overline{F}}$ is completely monotone. We can therefore calculate it using a Choquet integral, by invoking the results of Chapter 6_{101}. To make this work, we first find the natural extension of $\underline{P}_{\underline{F},\overline{F}}$ to the field of events $\mathcal{H}$ and prove that this extension is completely monotone. We then apply Theorem 6.17_{112}. In this way, we split the procedure of natural extension into two stages. This works because of transitivity of natural extension (Corollary 4.32_{67}).

The natural extension of $\underline{P}_{\underline{F},\overline{F}}$ to $\mathcal{H}$ is easy to calculate (Troffaes and Destercke, 2011, Proposition 4).

Proposition 7.17 *For any $A \in \mathcal{H}$, so $A = (x_0, x_1] \cup (x_2, x_3] \cup \cdots \cup (x_{2n}, x_{2n+1}]$ with $x_0 \prec x_1 \prec \cdots \prec x_{2n+1}$ in $\mathcal{X}^*$, it holds that $\underline{E}_{\underline{F},\overline{F}}(A) = \underline{P}^{\mathcal{H}}_{\underline{F},\overline{F}}(A)$, where*

$$\underline{P}^{\mathcal{H}}_{\underline{F},\overline{F}}(A) = \sum_{k=0}^{n} \max\{0, \underline{F}(x_{2k+1}) - \overline{F}(x_{2k})\}. \tag{7.36}$$

Once it is checked that $\underline{P}^{\mathcal{H}}_{\underline{F},\overline{F}}$ is completely monotone (Troffaes and Destercke, 2011, Theorem 17),[14] we immediately infer the following from Theorem 6.17_{112}.

Theorem 7.18 *The natural extension $\underline{E}_{\underline{F},\overline{F}}$ of $\underline{P}_{\underline{F},\overline{F}}$ is given by*

$$\underline{E}_{\underline{F},\overline{F}}(f) = (C)\int f \,\mathrm{d}\left(\underline{P}^{\mathcal{H}}_{\underline{F},\overline{F}}\right)_* = \inf f + \int_{\inf f}^{\sup f} \left(\underline{P}^{\mathcal{H}}_{\underline{F},\overline{F}}\right)_* (\{f \geq t\})\,\mathrm{d}t$$

for all $f \in \mathbb{B}(\mathcal{X})$.

For many more results relating to the natural extension of a probability box, we refer to Troffaes and Destercke (2011). To mention a few, Proposition 7.17 generalises to a property, called **additivity on full components**, that holds for arbitrary events (Troffaes and Destercke, 2011, Theorem 13). Also, the natural extension can be related to the lower oscillation introduced in Section $5.5.5_{93}$ (Troffaes and Destercke, 2011, Proposition 19).

[14] The meat of the proof is already contained in Destercke et al. (2008, Proposition 3.5).

8

Linear previsions, integration and duality

Amongst the coherent lower previsions, the linear previsions have the special property that they are self-conjugate. We have seen in Section 4.6_{70} that such linear previsions feature prominently in the theory of coherent lower previsions, because they can be used to provide interesting alternative characterisations of avoiding sure loss, coherence and natural extension.

As we have already mentioned, linear previsions are the *precise* probability models in this book. They do not really allow for indecision. If we have any two bounded gambles f and g and a linear prevision P defined on a sufficiently large space, then there are only three possibilities: (i) $P(f-g)>0$, which means that our subject is willing to pay some positive amount of utility to exchange g for f, so he strictly prefers f to g; (ii) $P(f-g)<0$, in which case the subject is willing to pay some positive amount of utility to exchange f for g, so he strictly prefers g to f; and (iii) $P(f-g)=0$, meaning that the subject is indifferent between f and g.

The contrast with a lower prevision $\underline{P}$ is striking, because here there are essentially four possibilities: (i) $\underline{P}(f-g)>0$ still means that our subject is willing to pay some positive amount of utility to exchange g for f, so he strictly prefers f to g; (ii) $\overline{P}(f-g)<0$ or equivalently $\underline{P}(g-f)>0$ means that the subject strictly prefers g to f; (iii) $\underline{P}(f-g)=\overline{P}(f-g)=0$ means that the subject is indifferent between f and g and (iv) in all other cases ($\underline{P}(f-g)\leq 0\leq \overline{P}(f-g)$ with at least one of the inequalities strict), the subject has no expressed preference for f or g.

If no such indecision – or lack of expressed preference – is deemed acceptable, linear previsions are the model of choice.

We have seen in Section 5.1_{77} that, on finite spaces, the restriction of a linear prevision to events is a probability charge – actually a probability measure in this case – and that the natural extension of this probability charge is again the linear

---

Lower Previsions, First Edition. Matthias C.M. Troffaes and Gert de Cooman.

prevision we set out from and can also be identified with the associated expectation operator.

The correspondence between probabilities on events and expectations (or previsions) on bounded gambles is an interesting one. In classical approaches to probability theory, this correspondence is usually brought about through an integration procedure.

For these reasons, we devote this chapter to a closer study of linear previsions and their relation to integration. We follow and extend the treatment first given by Troffaes in his PhD thesis (Troffaes, 2005, Chapter 4). We begin with a simple idea, due to Troffaes (2005, Definition 4.11): we can use the concept of natural extension to extend a linear prevision to larger domains, and if this extension is unique(ly linear) on some set, it defines an integral there. This allows us to give a unifying account of the notion of integration in Section 8.1. This abstract type of integration via natural extension can be used to establish a one-to-one relationship between linear previsions and finitely additive probability measures (or probability charges, as we usually call them). This is the subject of Sections 8.2_{159}–8.4_{166}. Moreover, we show that, for probability charges, it coincides with three known types of integration, based on the S-integral (Section 8.5_{168}), the Lebesgue integral (Section 8.6_{171}) and the Dunford integral (Section 8.7_{172}). We also mention at the end of Section 8.7_{172} that it coincides with Choquet integration but defer a full discussion of that interesting idea to the second part of the book (Section 15.7_{352}).

When extending linear functionals to larger domains, it is current practice in many mathematical fields – and most notably in probability and measure theory – to restrict attention to those domains where the functional still has a unique linear extension or, in other words, to concentrate on linear functionals only. Whilst this attitude may have perfectly respectable historical antecedents, we want to show, by working out the theoretical details here and by discussing a number of relevant examples in Chapter 9, that it is unnecessary and perhaps even unproductive. Indeed, it tends to hide very interesting mathematical structure that quite often becomes apparent only when leaving the linear ambit. Let us mention an example here, which is discussed in more detail further on. In integration theory, it is quite common to define an integral by first looking at lower and upper integrals and then to conveniently forget about them by zooming in on integrable functions. These are the functions for which the lower and upper integrals coincide and for which an integral can, therefore, be uniquely defined. Often, it is only this integral that is considered to be of interest or even to have any meaning, and much effort is then devoted to studying its properties. But, as we will argue, the lower and upper integrals are of considerable interest in themselves as well. They quite often have mathematical properties that are worthy of consideration per se and that, in addition, allow us to derive results about integrals and integrability in a straightforward manner. This is often harder to do when limiting the attention to integrals alone.

So far, we have found it useful to identify an event A and its indicator I_A. We continue to make use of this identification in the chapters that follow. But in the very specific context of the present chapter – and only here – it will be very important to make a clear distinction between events and their indicators and also between lower probabilities, defined on (sets of) indicators of events, and set functions, defined on (sets of) events; see for instance Definition 8.12_{159} and the whole of Section 8.4_{166}.

8.1 Linear extension and integration

In the following sections, we study how integration with respect to probability charges can be obtained through natural extension, and *vice versa*. To pave the way, we begin with some general considerations about integration. Many of the integrals we know are linear functionals that can be written as linear combinations of linear previsions defined on a linear space of bounded gambles. Indeed, by the Jordan decomposition theorem, any bounded charge is a linear combination of two probability charges (see Bhaskara Rao and Bhaskara Rao, 1983, Theorem 2.5.3). Therefore, as long as integrals are linear, we can reduce integration with respect to bounded charges to integration with respect to probability charges. So, it seems natural to define integration as a kind of linear natural extension for lower previsions; see Definition 8.1.

We explore its relation with some of the more common types of integrals, namely, S-integrals and Dunford integrals, further on in Section 8.5_{168} and Section 8.7_{172}, respectively. The idea of using the linear extension to define an integral has been explored much further than can be done here: there are similar connections with many other integrals, such as the Riemann–Stieltjes and the Lebesgue–Stieltjes types. But a treatment of these matters falls outside the scope of this book, and we refer to Troffaes (2005, Sections 4.3.6 and 4.3.9), De Cooman et al. (2008b), Miranda et al. (2008a) and Miranda et al. (2007).

The remainder of this section closely follows Troffaes (2005, Section 4.2).

Definition 8.1 *Let $\underline{P}$ be a lower prevision that avoids sure loss. Then the* ***linear extension*** *$E_{\underline{P}}$ of $\underline{P}$ is defined as the natural extension $\underline{E}_{\underline{P}}$ restricted to the domain where it is self-conjugate, that is, to the negation-invariant set of bounded gambles*

$$\mathcal{I}_{\underline{P}}(\mathcal{X}) = \operatorname{dom} E_{\underline{P}} := \left\{ f \in \mathbb{B}(\mathcal{X}) : \underline{E}_{\underline{P}}(f) = \overline{E}_{\underline{P}}(f) \right\}.$$

We say that f is ***$\underline{P}$-integrable*** *iff $f \in \mathcal{I}_{\underline{P}}(\mathcal{X})$, and we call $E_{\underline{P}}(f)$ the* ***integral of f with respect to $\underline{P}$*** *or simply the* ***$\underline{P}$-integral*** *of f. When no confusion can arise about $\mathcal{X}$, we will resort to the simpler notation $\mathcal{I}_{\underline{P}}$.*

As $E_{\underline{P}}$ is a restriction of a coherent lower prevision, namely, $\underline{E}_{\underline{P}}$, and because $E_{\underline{P}}$ is self-conjugate by its definition, it follows that it is a linear prevision; see Definition 4.11_{51}.

Proposition 8.2 *$E_{\underline{P}}$ is a linear prevision.*

So the $\underline{P}$-integral we introduce here is a linear functional, even though $\underline{P}$ need not be linear. Of course, not all integrals encountered in the literature are linear. The most important integral that is generally non-linear is probably the Choquet integral we have encountered in Chapter 6_{101}; see also Appendix C_{376}. A Choquet integral is usually not a $\underline{P}$-integral; it is the only generally non-linear integral that we consider in this book.

The linear extension is in some sense the *unique* coherent extension, as is made clear in the following proposition.

Proposition 8.3 *Let $\underline{P}$ be a lower prevision that avoids sure loss. Then for any coherent lower prevision $\underline{Q}$ that dominates $\underline{P}$, $E_{\underline{P}}(f) = \underline{Q}(f)$ whenever $f \in \mathscr{I}_{\underline{P}} \cap \operatorname{dom} \underline{Q}$.*

Proof. Let $f \in \mathscr{I}_{\underline{P}} \cap \operatorname{dom} \underline{Q}$. Simply note that the linear extension coincides with the natural extension on its domain, and the natural extension is the point-wise smallest coherent lower prevision on $\mathbb{B}$ that dominates $\underline{P}$. In particular,

$$E_{\underline{P}}(f) = \underline{E}_{\underline{P}}(f) \leq \underline{E}_{\underline{Q}}(f) \leq -\underline{E}_{\underline{Q}}(-f) \leq -\underline{E}_{\underline{P}}(-f) = -E_{\underline{P}}(-f) = E_{\underline{P}}(f),$$

where the first and third inequalities follow from Proposition 4.27_{65}, and the second from the coherence of $\underline{E}_{\underline{Q}}$ (Theorem 4.13(i)$_{53}$). Hence, $E_{\underline{P}}(f) = \underline{E}_{\underline{Q}}(f)$. Now, by coherence of $\underline{Q}$ and Theorem 4.26_{65}, we have that $\underline{Q}(f) = \underline{E}_{\underline{Q}}(f)$. Hence, $E_{\underline{P}}(f) = \underline{Q}(f)$. □

For linear previsions, not only do we have uniqueness, but we can even establish the equivalence of P and E_P. We prove an even more general statement; compare with Proposition 4.30_{66}.

Proposition 8.4 *Let P be a linear prevision. Then for any coherent lower prevision $\underline{Q}$ that dominates P, and satisfying $\operatorname{dom} \underline{Q} \subseteq \mathscr{I}_P$, it holds that $\underline{E}_{\underline{Q}} = \underline{E}_P$, meaning that $\underline{Q}$ is equivalent to P.*

Proof. As any coherent lower prevision that dominates $\underline{Q}$ must also dominate P, it clearly holds that $\underline{E}_{\underline{Q}} \geq \underline{E}_P$ (use Proposition 4.27_{65}). The converse inequality is established if we can show that any coherent lower prevision $\underline{R}$ on $\mathbb{B}$ that dominates P also dominates $\underline{Q}$. Indeed, by Proposition 8.3, we easily find that $E_P = \underline{R}$ on $\mathscr{I}_P$, and $E_P = \underline{Q}$ on $\operatorname{dom} \underline{Q}$. Hence, $\underline{Q} = \underline{R}$ on $\operatorname{dom} \underline{Q}$. But this implies that $\underline{R}$ certainly dominates $\underline{Q}$. □

The following is an immediate consequence of Proposition 4.27_{65}.

Proposition 8.5 *Let $\underline{P}$ and $\underline{Q}$ be lower previsions that avoid sure loss. If $\underline{Q}$ dominates $\underline{P}$, then $E_{\underline{Q}}$ is an extension of $E_{\underline{P}}$: $\mathscr{I}_{\underline{P}} \subseteq \mathscr{I}_{\underline{Q}}$ and $E_{\underline{P}}(f) = E_{\underline{Q}}(f)$ for all f in $\mathscr{I}_{\underline{P}}$.*

Proof. By Proposition 4.27_{65}, we already have that $\underline{E}_{\underline{Q}}$ dominates $\underline{E}_{\underline{P}}$. Consequently, for any bounded gamble f in $\mathscr{I}_{\underline{P}}$, it holds that

$$E_{\underline{P}}(f) = \underline{E}_{\underline{P}}(f) \leq \underline{E}_{\underline{Q}}(f) \leq \overline{E}_{\underline{Q}}(f) \leq \overline{E}_{\underline{P}}(f) = E_{\underline{P}}(f),$$

so f belongs to $\mathscr{I}_{\underline{Q}}$ and $E_{\underline{Q}}(f) = E_{\underline{P}}(f)$. □

$\underline{P}$-integrable bounded gambles interact *additively* with other bounded gambles.

Proposition 8.6 *Let $\underline{P}$ be a lower prevision that avoids sure loss. For any pair of bounded gambles f and g of which at least one is $\underline{P}$-integrable, it holds that*

$$\underline{E}_{\underline{P}}(f+g) = \underline{E}_{\underline{P}}(f) + \underline{E}_{\underline{P}}(g) \text{ and } \overline{E}_{\underline{P}}(f+g) = \overline{E}_{\underline{P}}(f) + \overline{E}_{\underline{P}}(g).$$

Proof. Suppose, for instance, that f is $\underline{P}$-integrable: $\underline{E}_{\underline{P}}(f) = \overline{E}_{\underline{P}}(f)$. Then the first equality follows from

$$\underline{E}_{\underline{P}}(f) + \underline{E}_{\underline{P}}(g) \leq \underline{E}_{\underline{P}}(f+g) \leq \overline{E}_{\underline{P}}(f) + \underline{E}_{\underline{P}}(g) = \underline{E}_{\underline{P}}(f) + \underline{E}_{\underline{P}}(g),$$

where we used the coherence of $\underline{E}_{\underline{P}}$ (Theorem 4.26(i)$_{65}$) and Theorem 4.13(v)$_{53}$. For the second equality involving upper previsions, the proof is similar. □

We shall see in Section 8.3$_{163}$ that the linear extension is a natural generalisation to bounded gambles of the Jordan extension (see Denneberg, 1994, p. 29) encountered in measure theory. Indeed, by restricting the self-conjugacy condition to indicators, we recover the condition for Jordan measurability. The following corollary shows that the additivity property of Proposition 8.6 in fact characterises the $\underline{P}$-integrable bounded gambles and, hence, the linear extension. In this sense, it also tells us that linear extension is also very similar to the construction of the Carathéodory extension (see Denneberg, 1994, p. 24) to be considered in Section 8.3$_{163}$. However, the following condition, restricted to indicators, is not directly similar to the condition for Carathéodory measurability: the equivalence is true, but it is not immediate.

Corollary 8.7 *Let $\underline{P}$ be a lower prevision that avoids sure loss. A bounded gamble f is $\underline{P}$-integrable if and only if*

$$\underline{E}_{\underline{P}}(f+g) = \underline{E}_{\underline{P}}(f) + \underline{E}_{\underline{P}}(g) \text{ for all } g \in \mathbb{B}.$$

Proof. 'only if'. Immediately from Proposition 8.6.
'if'. Take $g = -f$. Then $0 = \underline{E}_{\underline{P}}(0) = \underline{E}_{\underline{P}}(f) + \underline{E}_{\underline{P}}(-f) = \underline{E}_{\underline{P}}(f) - \overline{E}_{\underline{P}}(f)$. □

The next proposition is concerned with the domain of the linear extension. The condition $\underline{P}(f) = \sup f$ is related to the notions of null set and null bounded gamble, to be discussed much more extensively in Chapter 14$_{304}$. It will suffice for our present purposes to recall that, with respect to a probability measure μ, a strict subset A of $\mathcal{X}$ is called a null set if its outer measure (or outer set function) $\mu^*(A)$ is zero; see Definition 1.14$_{10}$. We establish in Section 8.3$_{163}$ that the outer measure μ^* corresponds to a coherent upper prevision, and, therefore, $\mu^*(A) = 0$ translates into $\overline{P}(I_A) = 0$ or, equivalently, $\underline{P}(-I_A) = \sup -I_A$. The following proposition then states that null sets belong to the domain of the linear extension, and this idea is also extended to null bounded gambles f, which satisfy $\overline{P}(|f|) = 0$.

Proposition 8.8 *Let $\underline{P}$ be a lower prevision that avoids sure loss. The following statements hold:*

(i) *If both f and $-f$ are in* dom $\underline{P}$, *and* $\underline{P}(f) = -\underline{P}(-f)$, *then* $f \in \mathscr{I}_{\underline{P}}$ *and* $\underline{P}(f) = E_{\underline{P}}(f)$.

(ii) *If* $f \in \operatorname{dom} \underline{P}$ *and* $\underline{P}(f) = \sup f$, *then* $f \in \mathscr{I}_{\underline{P}}$ *and* $\underline{P}(f) = E_{\underline{P}}(f)$.

(iii) *If* $|f| \in \operatorname{dom} \overline{P}$ *and* $\overline{P}(|f|) = 0$, *then any bounded gamble* g *such that* $|g| \leq |f|$ *belongs to* $\mathscr{I}_{\underline{P}}$, *and* $E_{\underline{P}}(g) = 0$. *In particular,* $E_{\underline{P}}(f) = E_{\underline{P}}(|f|) = E_{\underline{P}}(-|f|) = 0$.

(iv) $\mathbb{R} \subseteq \mathscr{I}_{\underline{P}}$.[1]

(v) $\mathscr{I}_{\underline{P}}$ *is a uniformly closed linear space that contains all constant gambles.*

(vi) *If* $\underline{E}_{\underline{P}}$ *is 2-monotone then* $\mathscr{I}_{\underline{P}}$ *is a uniformly closed linear lattice that contains all constant gambles. In this case, a bounded gamble is* $\underline{P}$*-integrable if and only if its level sets* $\{f \geq t\}$ *are* $\underline{P}$*-integrable for all but a countable number of* t *in* $[\inf f, \sup f]$.

Proof. (i). If both f and $-f$ are in dom $\underline{P}$ and $\underline{P}(f) = -\underline{P}(-f)$ then, because the natural extension is the point-wise smallest coherent lower prevision that dominates $\underline{P}$ (Proposition 4.27$_{65}$), $\overline{P}(f) = \underline{P}(f) \leq \underline{E}_{\underline{P}}(f) \leq \overline{E}_{\underline{P}}(f) \leq \overline{P}(f)$ (recall that $-\underline{P}(-f) = \overline{P}(f)$ by conjugacy and use the properties of coherent lower previsions from Theorem 4.13$_{53}$). Hence, $\underline{P}(f) = \underline{E}_{\underline{P}}(f) = \overline{E}_{\underline{P}}(f) = E_{\underline{P}}(f)$.

(ii). If $f \in \operatorname{dom} \underline{P}$ and $\underline{P}(f) = \sup f$, then, using similar arguments as for (i), $\sup f = \underline{P}(f) \leq \underline{E}_{\underline{P}}(f) \leq \overline{E}_{\underline{P}}(f) \leq \sup f$ and, hence, also $\underline{P}(f) = \underline{E}_{\underline{P}}(f) = \overline{E}_{\underline{P}}(f) = E_{\underline{P}}(f)$.

(iii). If $\overline{P}(|f|) = 0$, then, because $\overline{P}(|f|) \geq \inf |f| \geq 0$, it follows that $\inf |f| = 0$ or, equivalently, $\sup -|f| = \underline{P}(-|f|)$. Now, apply (ii) to find that $-|f|$ belongs to $\mathscr{I}_{\underline{P}}$, and $E_{\underline{P}}(-|f|) = 0$, and, hence, $E_{\underline{P}}(|f|) = -E_{\underline{P}}(-|f|) = 0$ too. For any bounded gamble g such that $|g| \leq |f|$, we find that

$$0 = E_{\underline{P}}(-|f|) \leq \underline{E}_{\underline{P}}(-|g|) \leq \underline{E}_{\underline{P}}(g) \leq \overline{E}_{\underline{P}}(g) \leq \overline{E}_{\underline{P}}(|g|) \leq E_{\underline{P}}(|f|) = 0,$$

where the third inequality follows from the coherence of $\underline{E}_{\underline{P}}$ (Theorem 4.13(i)$_{53}$), and the other inequalities from the monotonicity of the coherent $\underline{E}_{\underline{P}}$ (Theorem 4.13(iv)$_{53}$) and $-|f| \leq -|g| \leq g \leq |g| \leq |f|$. Therefore, $\underline{E}_{\underline{P}}(g) = \overline{E}_{\underline{P}}(g) = 0$, so g belongs to $\mathscr{I}_{\underline{P}}$, and $E_{\underline{P}}(g) = 0$.

(iv). $\underline{E}_{\underline{P}}$ is coherent, so by Theorem 4.13(ii)$_{53}$, $\underline{E}_{\underline{P}}(a) = \overline{E}_{\underline{P}}(a) = E_{\underline{P}}(a) = a$ for any $a \in \mathbb{R}$.

(v). Let $\alpha_1, \ldots, \alpha_n, \alpha'_1, \ldots, \alpha'_m$ be non-negative reals and $h_1, \ldots, h_n, h'_1, \ldots, h'_m$ bounded gambles in $\mathscr{I}_{\underline{P}}$. By the coherence of the natural extension $\underline{E}_{\underline{P}}$, it follows from Theorem 4.13(vi) and (v)$_{53}$ that

$$\underline{E}_{\underline{P}}\left(\sum_{i=1}^{n} \alpha_i h_i - \sum_{i=1}^{n} \alpha'_i h'_i\right) \geq \sum_{i=1}^{n} \alpha_i \underline{E}_{\underline{P}}(h_i) + \sum_{i=1}^{m} \alpha'_i \underline{E}_{\underline{P}}(-h'_i)$$

[1] Recall that $\mathbb{R}$ can be identified with the set of constant gambles on $\mathscr{X}$.

and because $\underline{E}_{\underline{P}}(h_i) = \overline{E}_{\underline{P}}(h_i)$ and $\underline{E}_{\underline{P}}(-h'_i) = -\overline{E}_{\underline{P}}(h'_i) = -\underline{E}_{\underline{P}}(h'_i) = \overline{E}_{\underline{P}}(-h'_i)$,

$$= \sum_{i=1}^{n} \alpha_i \overline{E}_{\underline{P}}(h_i) + \sum_{i=1}^{m} \alpha'_i \overline{E}_{\underline{P}}(-h'_i)$$

$$\geq \overline{E}_{\underline{P}}\left(\sum_{i=1}^{n} \alpha_i h_i - \sum_{i=1}^{m} \alpha'_i h'_i\right).$$

Now use $\underline{E}_{\underline{P}} \leq \overline{E}_{\underline{P}}$ to conclude that $\mathscr{I}_{\underline{P}}$ is a linear space. It is uniformly closed, because of Proposition 4.24_{62}.

(vi). To prove that $\mathscr{I}_{\underline{P}}$ is a linear lattice, it suffices to show that $|f|$ belongs to $\mathscr{I}_{\underline{P}}$ whenever f belongs to $\mathscr{I}_{\underline{P}}$ – indeed, $f \vee g = (f + g + |f - g|)/2$ and $f \wedge g = (f + g - |f - g|)/2$, and by (v), we already have that $\mathscr{I}_{\underline{P}}$ is a uniformly closed linear space. By the assumed 2-monotonicity of $\underline{E}_{\underline{P}}$ (see Equation $(6.1)_{103}$),

$$\underline{E}_{\underline{P}}(f \vee -f) + \underline{E}_{\underline{P}}(f \wedge -f) \geq \underline{E}_{\underline{P}}(f) + \underline{E}_{\underline{P}}(-f),$$

or, equivalently, because $f \vee -f = |f|$, $f \wedge -f = -|f|$ and $\underline{E}_{\underline{P}}(-|f|) = -\overline{E}_{\underline{P}}(|f|)$,

$$\overline{E}_{\underline{P}}(f) - \underline{E}_{\underline{P}}(f) \geq \overline{E}_{\underline{P}}(|f|) - \underline{E}_{\underline{P}}(|f|).$$

If f belongs to $\mathscr{I}_{\underline{P}}$, then the left-hand side is zero. By the coherence of $\underline{E}_{\underline{P}}$ (Theorem $4.13(\text{i})_{53}$), the right-hand side is non-negative, and therefore, the right-hand side must be zero too. But this means that $|f|$ indeed belongs to $\mathscr{I}_{\underline{P}}$.

As $\underline{E}_{\underline{P}}$ is both coherent and 2-monotone, we infer from Theorem 6.22_{117} that $\underline{E}_{\underline{P}} = C_{\underline{E}_{\underline{P}}}$. If we use Proposition C.5(iv)_{382} to find that also $\overline{E}_{\underline{P}} = C_{\overline{E}_{\underline{P}}}$, we conclude, using Equation $(\text{C.6})_{379}$, that for any bounded gamble f on $\mathscr{X}$:

$$\overline{E}_{\underline{P}}(f) - \underline{E}_{\underline{P}}(f) = R\int_{\inf f}^{\sup f} \left[\overline{E}_{\underline{P}}(\{f \geq t\}) - \underline{E}_{\underline{P}}(\{f \geq t\})\right] \mathrm{d}t.$$

The integrand $\overline{E}_{\underline{P}}(\{f \geq t\}) - \underline{E}_{\underline{P}}(\{f \geq t\})$ of the Riemann integral on the right-hand side is non-negative, by coherence of $\underline{E}_{\underline{P}}$ (Theorem $4.13(\text{i})_{53}$). Now the Riemann integral of a non-negative Riemann integrable function is zero if and only if this function is non-zero only in its points of discontinuity (see, for instance, Hildebrandt, 1963, p. 76). As here the integrand is a difference of two non-increasing functions, it is Riemann integrable and has at most a countable number of discontinuities. □

For just about any notion of integrability in the literature, the set of integrable bounded functions is at least uniformly closed: for instance, Darboux (1875, Théorème V (second one), p. 82) proves this result for Riemann integrability.

We now give some conditions under which the $\underline{P}$-integrals of two bounded gambles are equal.

Proposition 8.9 *Let $\underline{P}$ be a lower prevision that avoids sure loss, let f and g be bounded gambles and define $N := \{f \neq g\}$. The following statements hold:*

(i) *If f is $\underline{P}$-integrable, and $\overline{E}_{\underline{P}}(I_N) = 0$, then g is $\underline{P}$-integrable and $E_{\underline{P}}(f) = E_{\underline{P}}(g)$.*

(ii) *If f and g are $\underline{P}$-integrable, $\underline{E}_{\underline{P}}(I_N) = 0$ and $\underline{E}_{\underline{P}}$ is 2-monotone, then $E_{\underline{P}}(f) = E_{\underline{P}}(g)$.*

Proof. Let $\lambda := \sup |f - g| \geq 0$, and note that $0 \leq |f - g| \leq \lambda I_N$.
(i). By the coherence of $\underline{E}_{\underline{P}}$ and Theorem 4.13(iv) and (vi)$_{53}$, we get

$$0 \leq \overline{E}_{\underline{P}}(|f - g|) \leq \lambda \overline{E}_{\underline{P}}(I_N) = 0.$$

Applying Theorem 4.13(x)$_{53}$, it follows that $\underline{E}_{\underline{P}}(g) = E_{\underline{P}}(f) = \overline{E}_{\underline{P}}(g)$.

(ii). If $\underline{E}_{\underline{P}}$ is 2-monotone, then by Proposition 8.8(vi)$_{155}$, $\mathcal{I}_{\underline{P}}$ is a linear lattice. Therefore, if f and g are $\underline{P}$-integrable, then also $|f - g|$ is $\underline{P}$-integrable, and hence, by the coherence of $\underline{E}_{\underline{P}}$ and Theorem 4.13(iv) and (vi)$_{53}$,

$$0 \leq \overline{E}_{\underline{P}}(|f - g|) = \underline{E}_{\underline{P}}(|f - g|) \leq \lambda \underline{E}_{\underline{P}}(I_N) = 0.$$

Again applying Theorem 4.13(x)$_{53}$, it follows that $E_{\underline{P}}(g) = E_{\underline{P}}(f)$. □

Let us conclude with a few obvious but nevertheless important observations. The first one states that the linear extension of a linear prevision is actually an extension – that is moreover linear. The second one concerns the 'transitivity' of linear extension, compare with Corollary 4.32$_{67}$.

Corollary 8.10 *Let P be a linear prevision. Then E_P is an extension of P, that is,* dom $P \subseteq \mathcal{I}_P$ *and $P(f) = E_P(f)$ for all $f \in$* dom P.

Proof. Immediately from Proposition 8.8(i)$_{155}$. □

If dom $\underline{P} \subseteq \mathcal{K} \subseteq \mathcal{I}_{\underline{P}}$, then we denote $\underline{E}^{\mathcal{K}}_{\underline{P}}$ also by $E^{\mathcal{K}}_{\underline{P}}$ to emphasise that it is a linear prevision.

Corollary 8.11 *Let P be a linear prevision, and let* dom $P \subseteq \mathcal{J} \subseteq \mathcal{K} \subseteq \mathcal{I}_P$. *Then*

$$E^{\mathcal{K}}_{E^{\mathcal{J}}_P}(f) = E^{\mathcal{K}}_P(f) \text{ for all } f \in \mathcal{K}, \text{ and } E^{\mathcal{J}}_P(f) = E^{\mathcal{K}}_P(f) \text{ for all } f \in \mathcal{J}.$$

Proof. Immediately from Corollary 4.32$_{67}$, once it is observed that $E^{\mathcal{K}}_P = \underline{E}^{\mathcal{K}}_P$, $E^{\mathcal{J}}_P = \underline{E}^{\mathcal{J}}_P$ and $E^{\mathcal{K}}_{E^{\mathcal{J}}_P} = \underline{E}^{\mathcal{K}}_{E^{\mathcal{J}}_P}$. □

8.2 Integration of probability charges

Our idea of defining an integral, as explained in Section 8.1, has many interesting applications, and we refer to De Cooman et al. (2008b) for a detailed treatment. Here, we restrict the discussion to the simplest special case: $\underline{P}$-integrability where the lower prevision $\underline{P}$ is generated by a probability charge.

Probability charges, as introduced in Definition 1.15_{11}, are also called finitely additive probability measures, and they are a rather common tool for representing uncertainty. They are slightly more general than the so-called probability measures used in the classical theory of probability, in that they are only required to be finitely additive and not σ-additive. In Section 8.8_{177} further on, we establish that the notions of avoiding sure loss, coherence and natural extension as introduced and discussed in Chapter 4_{37} can be characterised completely using probability charges only. This makes probability charges play a rather important part in the theory of coherent lower previsions. In the present section, we investigate how we can use the ideas in Section 8.1 in order to associate an integral with such a probability charge. In later sections, we compare our integral with a few other integrals commonly associated with probability charges.

The programme for this section is quite simple and loosely follows Troffaes (2005, Section 4.3.2). First, we show how to associate a precise prevision P_μ with a probability charge μ. Next, we apply the ideas of Section 8.1: our integral will be the restriction E_{P_μ} of the natural extension $\underline{E}_{P_\mu}$ to the domain $\mathscr{I}_{P_\mu}$ where $\underline{E}_{P_\mu}$ is self-conjugate and, therefore, linear.

If we use the correspondence of events with indicators, we see that any real-valued map μ on a class of events corresponds to a prevision, defined on indicators.

Definition 8.12 *Let μ be any real-valued map defined on a class of subsets of $\mathcal{X}$. The* ***prevision induced by*** *μ is defined as the prevision P_μ that maps all bounded gambles I_A for $A \in \operatorname{dom}\mu$ to $\mu(A)$ and all bounded gambles $-I_A$ for $A \in \operatorname{dom}\mu$ to $-\mu(A)$:*

$$P_\mu(I_A) := -P_\mu(-I_A) := \mu(A) \text{ for all } A \in \operatorname{dom}\mu.$$

A natural question is: are these induced previsions actually linear previsions? In general, this question is hard to answer. But when the domain of μ is a field $\mathscr{F}$, we establish in Theorem 8.15_{161} that P_μ is a linear prevision if and only if μ is a probability charge. We now work our way towards proving this result.

To this end, we invoke an extension of P_μ for a probability charge μ to the linear span of the set of bounded gambles $\{I_A : A \in \mathscr{F}\}$, called the *Dunford integral*. We postpone the definition of the Dunford integral on larger spaces of bounded gambles, and a discussion of its origins, until Section 8.7_{172}.

Although we only need to associate Dunford integrals with probability charges, we shall follow Bhaskara Rao and Bhaskara Rao (1983, Section 4.4) and begin with a more general definition of the Dunford integral associated with any charge μ. The $|\mu|$

in the following definition is the total variation associated with this charge, introduced near the end of Section 1.7_{10}. If μ is in particular a probability charge, then $|\mu| = \mu$.

Definition 8.13 (Dunford integral for simple gambles) *Let $\mathcal{F}$ be a field on $\mathcal{X}$, and let μ be a charge on $\mathcal{F}$. An $\mathcal{F}$-simple gamble f is called* ***Dunford integrable*** *with respect to μ if it has a representation $f = \sum_{i=1}^n a_i I_{A_i}$ such that $|\mu|(A_i) < +\infty$ for all $i \in \{1, \dots, n\}$. The* ***Dunford integral*** *with respect to μ is then defined as the unique linear functional on the linear space of all Dunford integrable $\mathcal{F}$-simple gambles that coincides on indicators with μ. It is given by*

$$D\int f \,\mathrm{d}\mu := \sum_{i=1}^{n} a_i \mu(A_i) \text{ for all Dunford integrable } \mathcal{F}\text{-simple gambles } f, \quad (8.1)$$

where $\sum_{i=1}^n a_i I_{A_i}$ is any representation of f.

Let us first prove that this definition does not suffer from ambiguity.

Proof. Consider the linear space of bounded gambles spanned by

$$\mathcal{K} := \left\{ I_A : A \in \mathcal{F} \text{ and } |\mu|(A) < +\infty \right\},$$

then $\mathrm{span}(\mathcal{K})$ is exactly the linear space of Dunford integrable $\mathcal{F}$-simple gambles. Define the map $\psi(I_A) = \mu(A)$ on $\mathcal{K}$. This ψ satisfies the condition

$$\sum_{i=1}^{n} \alpha_i I_{A_i} = 0 \Rightarrow \sum_{i=1}^{n} \alpha_i \psi(I_{A_i}) = 0 \text{ for all } \alpha_1, \dots, \alpha_n \in \mathbb{R} \text{ and } I_{A_1}, \dots, I_{A_n} \in \mathcal{K}. \quad (8.2)$$

Indeed, suppose that $\sum_{i=1}^n \alpha_i I_{A_i} = 0$. Let $\mathcal{B} \subseteq \mathcal{F}$ be any finite partition of $\mathcal{X}$ such that each A_i, $i \in \{1, \dots, n\}$ is a union of elements of $\mathcal{B}$ (it is easy to see that such a partition always exists). Then

$$0 = \sum_{i=1}^{n} \alpha_i I_{A_i} = \sum_{i=1}^{n} \alpha_i \sum_{B \in \mathcal{B},\, B \subseteq A_i} I_B = \sum_{B \in \mathcal{B}} I_B \sum_{i \in \{1, \dots, n\},\, B \subseteq A_i} \alpha_i,$$

and because $\left\{ I_B : B \in \mathcal{B} \right\}$ is linearly independent, this implies that $\sum_{i \in \{1, \dots, n\}, B \subseteq A_i} \alpha_i = 0$ for every $B \in \mathcal{B}$. But then we also get

$$\begin{aligned}
\sum_{i=1}^{n} \alpha_i \psi(I_{A_i}) &= \sum_{i=1}^{n} \alpha_i \psi\Bigg(\sum_{B \in \mathcal{B},\, B \subseteq A_i} I_B \Bigg) = \sum_{i=1}^{n} \alpha_i \mu\Bigg(\bigcup_{B \in \mathcal{B},\, B \subseteq A_i} B \Bigg) \\
&= \sum_{i=1}^{n} \alpha_i \sum_{B \in \mathcal{B},\, B \subseteq A_i} \mu(B) = \sum_{i=1}^{n} \alpha_i \sum_{B \in \mathcal{B},\, B \subseteq A_i} \psi(I_B) \\
&= \sum_{B \in \mathcal{B}} \psi(I_B) \sum_{i \in \{1, \dots, n\},\, B \subseteq A_i} \alpha_i = 0,
\end{aligned}$$

where we have used the finite additivity of μ. Hence, Equation (8.2) holds.

By a well-known linear extension result (see Proposition 1.24_{18} or Schechter (1997, Proposition 11.10)) it follows from Equation (8.2) that ψ has a unique linear extension Ψ to span($\mathscr{K}$) given by

$$\Psi\left(\sum_{i=1}^{n} a_i I_{A_i}\right) := \sum_{i=1}^{n} a_i \psi(I_{A_i}).$$

But this means that Ψ is the Dunford integral. In other words, the Dunford integral is uniquely determined by Equation (8.1). □

For $A \in \mathscr{F}$, we say that an $\mathscr{F}$-simple gamble f is **Dunford integrable over** A with respect to μ whenever $I_A f$ is Dunford integrable with respect to μ. We then call $D\int I_A f \,\mathrm{d}\mu$ the **Dunford integral** of f **over** A with respect to μ, and we also write

$$D\int_A f \,\mathrm{d}\mu := D\int I_A f \,\mathrm{d}\mu.$$

We also consider the *Dunford functional* D_μ associated with the charge μ. For now, this functional is only defined on $\mathscr{F}$-simple gambles, but we extend it to larger domains further on in Section 8.7_{172}.

Lemma 8.14 *Let $\mathscr{F}$ be a field on $\mathscr{X}$, and let μ be a probability charge on $\mathscr{F}$. Then all $\mathscr{F}$-simple gambles are Dunford integrable with respect to μ, and the* ***Dunford functional*** *D_μ given by*

$$D_\mu(f) := D\int f \,\mathrm{d}\mu \text{ for any } \mathscr{F}\text{-simple gamble } f \text{ on } \mathscr{X}, \tag{8.3}$$

defines a linear prevision on the linear space span($\mathscr{F}$).

Proof. Dunford integrability of all $\mathscr{F}$-simple gambles follows from $|\mu| = \mu$ for probability charges μ. Note that D_μ is self-conjugate (this follows easily from its definition) and, hence, D_μ is a prevision. Now simply check the coherence conditions of Theorem 4.16_{57}. □

This brings us to the first simple but important result of this section: probability charges can be identified with special linear previsions.

Theorem 8.15 *Let $\mathscr{F}$ be a field. Let μ be any real-valued function defined on $\mathscr{F}$. Then P_μ is a linear prevision if and only if μ is a probability charge. In that case, D_μ is the unique extension of the linear prevision P_μ to a linear prevision on* span($\mathscr{F}$).

Proof. 'if'. Let μ be a probability charge defined on a field $\mathscr{F}$. Observe that the prevision P_μ is a restriction of D_μ, which is a linear prevision by Lemma 8.14. Now apply Proposition 4.18_{58} to arrive at the desired result.

'only if'. Suppose that P_μ is a linear prevision. From $\mu(A) = P_\mu(A)$ for all $A \in \mathscr{F}$, it is very easy to show that μ satisfies the properties of a probability charge simply by applying the linearity of P_μ.

The rest of the proof is now obvious, if we look at the consistency proof given earlier. □

From now on, therefore, we assume that the set function μ is a probability charge on the field $\mathscr{F}$. We are interested in the natural extension $\underline{E}_{P_\mu}$ of the linear prevision P_μ, which we denote by $\underline{E}_\mu$. The conjugate upper prevision is denoted by $\overline{E}_\mu$.

Theorem 8.16 *Let μ be a probability charge on the field $\mathscr{F}$. Then the natural extension $\underline{E}_\mu := \underline{E}_{P_\mu}$ of the linear prevision P_μ is a completely monotone coherent lower prevision on $\mathbb{B}$. It coincides with D_μ on $\mathrm{span}(\mathscr{F})$, and it, therefore, also coincides on all bounded gambles with the inner extension $(D_\mu)_*$ of D_μ.*

Proof. As a first step, we show that $\underline{E}_\mu$ and D_μ coincide on $\mathrm{span}(\mathscr{F})$. As for any bounded gamble f, both $\underline{E}_\mu(f) = \inf f + \underline{E}_\mu(f - \inf f)$ and $D_\mu(f) = \inf f + D_\mu(f - \inf f)$ (by coherence, see Theorem $4.13(\mathrm{iii})_{53}$), it suffices to prove that $\underline{E}_\mu$ and D_μ coincide on non-negative $\mathscr{F}$-simple gambles. For any such bounded gamble f, we know that there are $n \geq 1$, $A_k \in \mathscr{F}$ and $a_k \geq 0$ such that $f = \sum_{k=1}^n a_k I_{A_k}$, whence

$$\underline{E}_\mu(f) \geq \sum_{k=1}^{n} a_k P_\mu(I_{A_k}) = D_\mu(f) \geq \overline{E}_\mu(f) \geq \underline{E}_\mu(f),$$

where the first inequality follows from the super-additivity (Theorem $4.13(\mathrm{v})_{53}$) and non-negative homogeneity (Theorem $4.13(\mathrm{vi})_{53}$) of the coherent lower prevision $\underline{E}_\mu$; the second inequality is derived in a completely similar way, but now using the sub-additivity and the non-negative homogeneity of the coherent upper prevision $\overline{E}_\mu$; and the third inequality follows from Theorem $4.13(\mathrm{i})_{53}$. So $\underline{E}_\mu$ and D_μ indeed coincide on $\mathrm{span}(\mathscr{F})$.

From the transitivity of natural extension (Corollary 4.32_{67}), we now infer that $\underline{E}_\mu$ and the natural extension of D_μ coincide on all bounded gambles. Observe that $\mathrm{span}(\mathscr{F})$ is a linear lattice that contains all constant gambles. By Theorem 6.6_{105}, the linear prevision D_μ is completely monotone, and by Theorem 6.19_{114}, so is its natural extension $(D_\mu)_*$. □

We are now ready to take the next step in defining our integral associated with a probability charge μ on a field $\mathscr{F}$. As the corresponding P_μ is a linear prevision, we can consider its linear extension E_{P_μ}, which we denote as E_μ. We denote the domain $\mathrm{dom}\, E_\mu$ of this linear extension by $\mathscr{I}_\mu(\mathscr{X})$, and we call the elements of this set **μ-integrable**.

$$\mathscr{I}_\mu(\mathscr{X}) := \left\{ f \in \mathbb{B}(\mathscr{X}) : \underline{E}_\mu(f) = \overline{E}_\mu(f) \right\}$$

*is the set of those bounded gambles to which the probability charge μ can be uniquely extended as a linear prevision, and we call this linear prevision E_μ the **μ-integral** of* f. When no confusion can arise about $\mathscr{X}$, we will resort to the simpler notation $\mathscr{I}_\mu$ instead of $\mathscr{I}_\mu(\mathscr{X})$.

What can we say at this point about $\mathscr{I}_\mu$? Applying Proposition 8.8(i) and $(\mathrm{v})_{155}$ with $\underline{P}$ equal to P_μ, it follows that $\mathscr{I}_\mu$ is a linear lattice that contains all constant gambles and that includes at least the uniform closure of $\mathrm{span}(\mathscr{F})$. Bounded gambles in this uniform closure have been called *$\mathscr{F}$-measurable* in Definition $1.17(\mathrm{B})_{12}$. All $\mathscr{F}$-measurable bounded gambles are guaranteed to be μ-integrable.

Proposition 8.17 *Let $\mathscr{F}$ be a field on $\mathscr{X}$ and let μ be a probability charge on $\mathscr{F}$. Then any $\mathscr{F}$-measurable bounded gamble is μ-integrable:*

$$\mathrm{span}(\mathscr{F}) \subseteq \mathbb{B}_{\mathscr{F}} \subseteq \mathscr{I}_{\mu}.$$

Moreover, E_μ coincides with D_μ on span($\mathscr{F}$).

Proof. Look at Definition 1.17(B)$_{12}$ of $\mathscr{F}$-measurability: span($\mathscr{F}$) $\subseteq \mathbb{B}_{\mathscr{F}}$ is clear. By Proposition 8.8(i)$_{155}$, all bounded gambles in span($\mathscr{F}$) are μ-integrable (alternatively, use Theorem 8.16). By Proposition 8.8(v)$_{155}$, all bounded gambles in the uniform closure of this set are also μ-integrable. But, by Definition 1.17$_{12}$, these are exactly the $\mathscr{F}$-measurable bounded gambles. The last statement follows from Theorem 8.16. □

Clearly, the lower prevision $\underline{E}_\mu$ determines the linear extension E_μ as its restriction to $\mathscr{I}_\mu$, but conversely, E_μ also completely determines $\underline{E}_\mu$ as its natural extension. Indeed, if we recall the transitivity of natural extension (Corollary 4.32$_{67}$) and the special form that natural extension from a linear space takes (Theorem 4.34$_{69}$), it is easy to see that for all bounded gambles f on $\mathscr{X}$,

$$\underline{E}_\mu(f) = \sup\left\{D_\mu(g) \colon g \in \mathrm{span}(\mathscr{F}) \text{ and } g \leq f\right\} \tag{8.4}$$

$$= \sup\left\{E_\mu(g) \colon g \in \mathbb{B}_{\mathscr{F}} \text{ and } g \leq f\right\} \tag{8.5}$$

$$= \sup\left\{E_\mu(g) \colon g \in \mathscr{I}_\mu \text{ and } g \leq f\right\},$$

and similar expressions can be given for $\overline{E}_\mu$.

8.3 Inner and outer set function, completion and other extensions

We have seen that for a probability charge μ, there is a uniformly closed linear lattice $\mathscr{I}_\mu$ of those so-called μ-integrable bounded gambles to which μ (or rather, the associated probability, or linear prevision P_μ) has a unique extension as a linear prevision. There are in the literature various approaches to extending probability charges to larger domains, leading, for instance, to the Jordan and Carathéodory extensions, introduced in Definition 8.19$_{\curvearrowright}$. Here we investigate their relationship with the notions of linear extension and μ-integrability.

The next theorem relates the inner and outer set function of a probability charge μ to the natural extension of the associated linear prevision P_μ.

Theorem 8.18 *Let μ be a probability charge defined on a field $\mathscr{F}$, and let P_μ be the associated linear prevision, with natural extension $\underline{E}_\mu$. Then*

$$\mu_*(A) = \underline{E}_\mu(I_A) \text{ and } \mu^*(A) = \overline{E}_\mu(I_A) \text{ for all events } A \subseteq \mathscr{X}. \tag{8.6}$$

Moreover,

$$\underline{E}_{\mu}(f) = C_{\mu_*}(f) = \inf f + R\int_{\inf f}^{\sup f} \mu_*(\{f \geq t\})\,\mathrm{d}t \textit{ for all bounded gambles } f \textit{ on } \mathcal{X}, \tag{8.7}$$

and therefore, f is μ-integrable if and only if its level sets $\{f \geq t\}$ *are μ-integrable for all but a countable number of t in* $[\inf f, \sup f]$.

Proof. See Theorem 4.35$_{70}$ for Equation (8.6)$_{\curvearrowleft}$. The set function μ is completely monotone by Theorems 6.6$_{105}$ and 8.15$_{161}$. Corollary 6.17$_{112}$ and Equation (8.6)$_{\curvearrowleft}$ then lead to Equation (8.7).

For the last statement, observe that, by Proposition C.5(iv)$_{382}$, $\overline{E}_{P}(f) = -\underline{E}_{P}(-f) = -C_{\mu_*}(-f) = C_{\mu^*}(f)$, and therefore, using Proposition C.3(ii)$_{379}$,

$$\overline{E}_{\mu}(f) - \underline{E}_{\mu}(f) = R\int_{\inf f}^{\sup f} [\mu^*(\{f \geq t\}) - \mu_*(\{f \geq t\})]\,\mathrm{d}t,$$

and this Riemann integral is zero if and only if its non-negative integrand is zero in all but a countable number of elements of $[\inf f, \sup f]$; see also the related course of reasoning in the proof of Proposition 8.8(vi)$_{155}$. □

This shows that the set $\mathcal{I}_{\mu}$ of μ-integrable bounded gambles is completely determined by the set $\mathcal{I}_{\mu} \cap I_{\mathcal{P}}$ of **μ-integrable** (indicators of) **events**. Let us now take a closer look at this set and relate it to what we know from the literature, where (see for instance Denneberg, 1994, pp. 24–29) we find the following extensions of probability charges to larger domains; see also Troffaes (2005, Definition 4.38).

Definition 8.19 *Let μ be a probability charge defined on a field* $\mathcal{F}$. *The* ***completion*** *of μ is the probability charge* $\overline{\mu}$ *defined by* $\overline{\mu}(A \triangle N) := \mu(A)$ *for any* $A \in \mathcal{F}$ *and* $N \subseteq \mathcal{X}$, *whenever there is some* $M \in \mathcal{F}$ *such that* $N \subseteq M$ *and* $\mu(M) = 0$. *The* ***Carathéodory field*** *of μ is defined as*

$$\mathcal{I}_{\mu}^{C} := \left\{A \subseteq \mathcal{X} : \mu^*(B) = \mu^*(B \cap A) + \mu^*(B \setminus A) \textit{ for all } B \subseteq \mathcal{X}\right\},$$

and the ***Jordan field*** *of μ is defined as*

$$\mathcal{I}_{\mu}^{J} := \left\{A \subseteq \mathcal{X} : \mu_*(A) = \mu^*(A)\right\}.$$

The ***Carathéodory extension*** *of a probability charge μ is the unique extension of μ to a probability charge* μ_C *defined on* $\mathcal{I}_{\mu}^{C}$, *and the* ***Jordan extension*** *of a probability charge μ is the unique extension of μ to a probability charge* μ_J *defined on* $\mathcal{I}_{\mu}^{J}$.

The relationship between the linear Jordan and Carathódory extensions of a probability charge is explored in the following theorem, taken from Troffaes (2005, Theorem 4.39).

Theorem 8.20 *Let $\mathscr{F}$ be a field on $\mathscr{X}$, and let μ be a probability charge. Then, for any $A \subseteq \mathscr{X}$, $A \in \operatorname{dom}\overline{\mu}$ implies that $I_A \in \mathscr{I}_\mu$ and*

$$E_\mu(I_A) = \overline{\mu}(A) \text{ for any } A \in \operatorname{dom}\overline{\mu}.$$

The linear extension of μ restricted to events, the Carathéodory extension of μ, and the Jordan extension of μ coincide: $\mathscr{I}_\mu^J = \mathscr{I}_\mu^C = \mathscr{I}_\mu \cap I_{\mathscr{P}}$ and

$$E_\mu(I_A) = \mu_J(A) = \mu_C(A) \text{ for any } A \text{ in these (identical) sets.}$$

Proof. The completion $\overline{\mu}$ agrees on its domain with the linear extension of μ. Indeed, let $A \in \operatorname{dom}\overline{\mu}$. We must show that $\overline{\mu}(A) = \mu_*(A) = \mu^*(A)$. As $\mu_*(A) \leq \mu^*(A)$, it suffices to show that $\mu_*(A) \geq \overline{\mu}(A) \geq \mu^*(A)$. As $A \in \operatorname{dom}\overline{\mu}$, there are $B \in \mathscr{F}$, $M \in \mathscr{F}$ and $N \subseteq M$, such that $A = B \triangle N$, $\mu(M) = 0$ and $\overline{\mu}(A) = \mu(B)$; see Definition 8.19. Hence,

$$\mu_*(B \triangle N) \geq \left|\mu_*(B) - \mu_*(N)\right| = \mu_*(B) = \mu(B) = \mu^*(B) + \mu^*(N) \geq \mu^*(B \triangle N),$$

where we used $\mu_*(N) \leq \mu^*(N) \leq \mu^*(M) = \mu(M) = 0$, the coherence of the natural extension $\underline{E}_\mu$ (Theorem 4.13$_{53}$), and Theorem 8.18$_{163}$ – for the first inequality, use $I_{B\triangle N} = \left|I_B - I_N\right|$, and for the second inequality, use $I_B + I_N \geq I_{B\triangle N}$. Therefore, $\mu_*(A) \geq \overline{\mu}(A) \geq \mu^*(A)$, and hence, also $\mu_*(A) = \overline{\mu}(A) = \mu^*(A) = \mu(B)$.

It follows from Equation (8.6)$_{163}$ that $\mu_*(A) = \mu^*(A)$ if and only if $\underline{E}_\mu(I_A) = \overline{E}_\mu(I_A)$, and therefore, the Jordan extension coincides with the linear extension of μ restricted to indicators.

We finish the proof by showing that the Jordan extension coincides with the Carathéodory extension.[2] Clearly, if $A \in \mathscr{I}_\mu^C$, then $A \in \mathscr{I}_\mu^J$; simply take $B = \mathscr{X}$ in the definition of the Carathéodory field, and recall that $\mu_*(A) = 1 - \mu^*(\mathscr{X} \setminus A) = \mu^*(\mathscr{X}) - \mu^*(\mathscr{X} \setminus A)$. Hence, $\mathscr{I}_\mu^C \subseteq \mathscr{I}_\mu^J$. Conversely, assume that $A \in \mathscr{I}_\mu^J$. By Theorem 8.18$_{163}$, μ^* coincides on all events B with the coherent upper prevision $\overline{E}_\mu$, whence, using the sub-additivity of the coherent $\overline{E}_\mu$ (Theorem 4.13(v)$_{53}$),

$$\mu^*(B) = \overline{E}_\mu(I_B) = \overline{E}_\mu(I_{B\cap A} + I_{B\setminus A}) \leq \overline{E}_\mu(I_{B\cap A}) + \overline{E}_\mu(I_{B\setminus A}) = \mu^*(B \cap A) + \mu^*\left(\frac{B}{A}\right).$$

To prove the converse inequality, recall from Theorem 8.16$_{162}$ that μ^* is in particular 2-monotone. Hence,

$$\mu^*(B) \geq \mu^*(B \cap A) + \mu^*(B \cup A) - \mu^*(A).$$

Again, apply Theorem 8.18$_{163}$, and use Proposition 8.6$_{155}$, to see that

$$\mu^*(B \cup A) = \overline{E}_\mu(I_{B\cup A}) = \overline{E}_\mu(I_A + I_{B\setminus A}) = \overline{E}_\mu(I_A) + \overline{E}_\mu(I_{B\setminus A}) = \mu^*(A) + \mu^*(B \setminus A),$$

[2] Denneberg (1994, Proposition 2.9) also gives a proof for the equality of the Jordan and Carathéodory extensions. The proof we present here is shorter.

where we have used the fact that A belongs to the Jordan field $\mathscr{I}^J_\mu$, and $\underline{E}_\mu(I_A) = \mu_*(A) = \mu^*(A) = \overline{E}_\mu(I_A)$, in order to apply Proposition 8.6$_{155}$. Hence, $A \in \mathscr{I}^C_\mu$ (alternatively, recall that we have already shown that $\mathscr{I}^J_\mu = \mathscr{I}_\mu \cap I_{\mathscr{P}}$). □

8.4 Linear previsions and probability charges

Through the mediation of $\mathscr{F}$-measurability, we can now establish a fairly general correspondence between linear previsions and probability charges. Similar results were proved by Hildebrandt (1934), Dunford and Schwartz (1957, Chapter VI, p. 492 ff.) and Bhaskara Rao and Bhaskara Rao (1983, Theorem 4.7.4), amongst others. This section follows Troffaes (2005, Section 4.3.3), but the notation here is slightly different.

Definition 8.21 *We define a map taking linear previsions to probability charges: let P be a linear prevision, and let $\mathscr{F}$ be any field such that $I_{\mathscr{F}} \subseteq \operatorname{dom} P$. The restriction of P to $I_{\mathscr{F}}$ corresponds to a probability charge* $\operatorname{ch}^{\mathscr{F}}(P)$ *on $\mathscr{F}$, defined by*

$$\operatorname{ch}^{\mathscr{F}}(P)(A) := P(I_A) \text{ for all } A \in \mathscr{F}. \tag{8.8}$$

This leads to the definition of a map $\operatorname{ch}^{\mathscr{F}}$ *taking linear previsions P to the corresponding probability charges* $\operatorname{ch}^{\mathscr{F}}(P)$.

Conversely, we also define a map taking probability charges to linear previsions: let μ be a probability charge defined on the field $\mathscr{F}$, and let $\mathscr{K}$ be any negation-invariant set of μ-integrable bounded gambles: $\mathscr{K} = -\mathscr{K} \subseteq \mathscr{I}_\mu$. Then the functional $\operatorname{lin}^{\mathscr{K}}(\mu)$ *defined on $\mathscr{K}$ by*

$$\operatorname{lin}^{\mathscr{K}}(\mu)(f) := E_\mu(f) \text{ for all } f \in \mathscr{K}, \tag{8.9}$$

is a linear prevision. This leads to the definition of a map $\operatorname{lin}^{\mathscr{K}}$ *taking probability charges μ to the corresponding linear previsions* $\operatorname{lin}^{\mathscr{K}}(\mu)$.

Let us first prove that these maps are well defined.

Proof. Let P be a linear prevision, and let $\mathscr{F}$ be a field such that $I_{\mathscr{F}} \subseteq \operatorname{dom} P$. To see that $\operatorname{ch}^{\mathscr{F}}(P)$ is a probability charge, observe that the conditions of Definition 1.15$_{11}$ follow from the linearity of P.

Let μ be a probability charge on $\mathscr{F}$, and let $\mathscr{K} = -\mathscr{K} \subseteq \mathscr{I}_\mu$. To see that $\operatorname{lin}^{\mathscr{K}}(\mu)$ is a linear prevision, observe that, by its definition, it is a restriction to a negation-invariant domain of the linear extension E_μ, which is obviously a linear prevision. Therefore, by Proposition 4.18$_{58}$, $\operatorname{lin}^{\mathscr{K}}(\mu)$ is a linear prevision as well.

Also observe that, by Proposition 8.8(iv)$_{155}$ ($\mathbb{R} \subseteq \mathscr{I}_P$) and Proposition 8.3$_{154}$, any linear prevision P has a unique coherent extension to a linear prevision on $\mathbb{R} \cup \operatorname{dom} P$, through $P(a) = a$ for all $a \in \mathbb{R}$. So we can always assume that there is some field $\mathscr{F}$ such that $I_{\mathscr{F}} \subseteq \operatorname{dom} P$. □

The two types of maps introduced earlier constitute equivalences between particular sets of probability charges and sets of linear previsions. The negation-invariant set of $\mathscr{F}$-measurable bounded gambles $\mathbb{B}_{\mathscr{F}} = \mathrm{cl}(\mathrm{span}(\mathscr{F}))$, which we mention in the following text, was introduced in Definition 1.17_{12}.

Theorem 8.22 *Let $\mathscr{F}$ be a field of subsets of $\mathscr{X}$. Then $\mathrm{ch}^{\mathscr{F}}$ and $\mathrm{lin}^{\mathbb{B}_{\mathscr{F}}}$ are bijective – that is, onto and one-to-one – maps between the set of linear previsions on $\mathbb{B}_{\mathscr{F}}$ and the set of probability charges on $\mathscr{F}$. They are in fact each other's inverses.*

Proof. Let P be any linear prevision on $\mathbb{B}_{\mathscr{F}}$, and let μ be any probability charge on $\mathscr{F}$. In order to establish that the maps are onto and one-to-one, it suffices to show that both

$$\mathrm{ch}^{\mathscr{F}}(\mathrm{lin}^{\mathbb{B}_{\mathscr{F}}}(\mu)) = \mu \text{ and } \mathrm{lin}^{\mathbb{B}_{\mathscr{F}}}(\mathrm{ch}^{\mathscr{F}}(P)) = P.$$

For any $A \in \mathscr{F}$, $I_A \in \mathbb{B}_{\mathscr{F}}$, so $\mathrm{lin}^{\mathbb{B}_{\mathscr{F}}}(\mu)(I_A) = E_\mu(I_A)$, and applying Corollary 8.10_{158}, $E_\mu(I_A) = P_\mu(I_A) = \mu(A)$, so the first equality holds. To see that the second equality also holds, let f be any $\mathscr{F}$-measurable bounded gamble. By Definition 1.17_{12}, f can be uniformly approximated by a sequence of $\mathscr{F}$-simple gambles f_n. For any such $\mathscr{F}$-simple gamble $f_n = \sum_{k_n=1}^{m_n} a_{k_n} I_{A_{k_n}}$, we find that

$$\begin{aligned}\mathrm{lin}^{\mathbb{B}_{\mathscr{F}}}(\mathrm{ch}^{\mathscr{F}}(P))(f_n) &= E_{\mathrm{ch}^{\mathscr{F}}(P)}(f_n) = \sum_{k_n=1}^{m_n} a_{k_n} P_{\mathrm{ch}^{\mathscr{F}}(P)}(I_{A_{k_n}}) \\ &= \sum_{k_n=1}^{m_n} a_{k_n} \mathrm{ch}^{\mathscr{F}}(P)(A_{k_n}) = \sum_{k_n=1}^{m_n} a_{k_n} P(I_{A_{k_n}}) = P(f_n).\end{aligned}$$

Let us explain the (perhaps) less trivial equalities in this sequence. The first equality follows from Equation (8.9), if we take into account that $\mathrm{span}(\mathscr{F}) \subseteq \mathbb{B}_{\mathscr{F}}$ (use Proposition 8.17_{163}). The second one holds because $E_{\mathrm{ch}^{\mathscr{F}}(P)}$ and $D_{\mathrm{ch}^{\mathscr{F}}(P)}$ coincide on $\mathrm{span}(\mathscr{F})$ (Proposition 8.17_{163} again). The fourth equality follows from Equation (8.8) because all A_{k_n} belong to $\mathscr{F}$. For the last equality, we invoke the linearity of the linear prevision P.

As both $\mathrm{lin}^{\mathbb{B}_{\mathscr{F}}}(\mathrm{ch}^{\mathscr{F}}(P))$ and P are linear previsions, they are continuous with respect to the topology of uniform convergence (Corollary $4.14(\mathrm{xii})_{55}$), so

$$\mathrm{lin}^{\mathbb{B}_{\mathscr{F}}}(\mathrm{ch}^{\mathscr{F}}(P))(f) = \lim_{n\to+\infty} \mathrm{lin}^{\mathbb{B}_{\mathscr{F}}}(\mathrm{ch}^{\mathscr{F}}(P))(f_n) = \lim_{n\to+\infty} P(f_n) = P(f).$$

This establishes the claim. □

This allows us to prove the following important identifications between probability charges on $\mathscr{P}$ and linear previsions on $\mathbb{B}$ as their unique linear extensions.

Corollary 8.23 $\mathrm{ch}^{\mathscr{P}}$ *and* $\mathrm{lin}^{\mathbb{B}}$ *are bijective maps between probability charges on $\mathscr{P}$ and linear previsions on $\mathbb{B}$. They are in fact each other's inverses.*

Proof. See Theorem 8.22 and observe that $\mathbb{B}_{\mathscr{P}} = \mathbb{B}$. □

8.5 The S-integral

In the present section and the next section, we intend to show that the abstract integral E_μ introduced in Sections 8.1_{153} and 8.2_{159} coincides with three other well-known types of integral commonly associated with a probability charge μ: the S-integral, the Lebesgue integral and the Dunford integral. We refer to De Cooman et al. (2008b) for a more extensive discussion of connections with still other types of integrals. The following treatment borrows from Troffaes (2005, Section 4.3.5).

The S-integral is one of the simplest kinds of integrals for charges one can think of. It is also most closely related to the idea of natural extension, as we prove shortly in Theorem 8.25. The S-integral also provides us with a tool to find the natural extension itself, not only for probability charges but also for any lower prevision that avoids sure loss. This comes to the fore in Section 8.8_{177}.

Recall from Section 1.5_7 that the set $\mathfrak{P}_{\mathscr{F}}$ of all finite partitions of $\mathscr{X}$ whose elements belong to the field $\mathscr{F}$ is directed when provided with the *refinement* relation. We can, therefore, take the Moore–Smith limit of nets on $\mathfrak{P}_{\mathscr{F}}$; see Example 1.12_8. This type of limit is used in the definition of the S-integral.

Definition 8.24 (S-integrals) *Let $\mathscr{F}$ be a field on $\mathscr{X}$, and let μ be a bounded positive charge on $\mathscr{F}$. A bounded gamble f is called **S-integrable** with respect to μ if and only if the **lower** and **upper S-integrals** of f, defined respectively by*

$$S\underline{\int} f\,\mathrm{d}\mu := \lim_{\mathscr{B}\in\mathfrak{P}_{\mathscr{F}}} \sum_{B\in\mathscr{B}} \underline{P}_B(f)\mu(B) = \sup_{\mathscr{B}\in\mathfrak{P}_{\mathscr{F}}} \sum_{B\in\mathscr{B}} \underline{P}_B(f)\mu(B) \tag{8.10}$$

and

$$S\overline{\int} f\,\mathrm{d}\mu := \lim_{\mathscr{B}\in\mathfrak{P}_{\mathscr{F}}} \sum_{B\in\mathscr{B}} \overline{P}_B(f)\mu(B) = \inf_{\mathscr{B}\in\mathfrak{P}_{\mathscr{F}}} \sum_{B\in\mathscr{B}} \overline{P}_B(f)\mu(B), \tag{8.11}$$

*coincide. In that case, the **S-integral** of f with respect to μ is defined as the common value:*

$$S\int f\,\mathrm{d}\mu = S\underline{\int} f\,\mathrm{d}\mu = S\overline{\int} f\,\mathrm{d}\mu.$$

To understand the above-mentioned expressions, recall that $\underline{P}_B$ and $\overline{P}_B$ denote the *vacuous* lower and upper previsions relative to the non-empty set B, given by $\underline{P}_B(f) = \inf_{x\in B} f(x)$ and $\overline{P}_B(f) = \sup_{x\in B} f(x)$.

We first prove that these Moore–Smith limits exist and are real numbers.

Proof. The real net α on $\mathfrak{P}_{\mathscr{F}}$, defined by $\alpha(\mathscr{B}) := \sum_{B\in\mathscr{B}} \underline{P}_B(f)\mu(B)$ for all $\mathscr{B} \in \mathfrak{P}_{\mathscr{F}}$, is non-decreasing and bounded above by $\mu(\mathscr{X})\sup f$: indeed, if $\mathscr{B}_1$ refines $\mathscr{B}_2$, then

$$\begin{aligned}\alpha(\mathscr{B}_1) &= \sum_{B_1\in\mathscr{B}_1} \underline{P}_{B_1}(f)\mu(B_1) = \sum_{B_2\in\mathscr{B}_2} \sum_{B_1\in\mathscr{B}_1, B_1\subseteq B_2} \underline{P}_{B_1}(f)\mu(B_1)\\ &\geq \sum_{B_2\in\mathscr{B}_2} \underline{P}_{B_2}(f) \sum_{B_1\in\mathscr{B}_1, B_1\subseteq B_2} \mu(B_1) = \sum_{B_2\in\mathscr{B}_2} \underline{P}_{B_2}(f)\mu(B_2) = \alpha(\mathscr{B}_2),\end{aligned}$$

and, similarly,

$$\alpha(\mathscr{B}_1) = \sum_{B_1\in\mathscr{B}_1} \underline{P}_{B_1}(f)\mu(B_1) \leq \sup f \sum_{B_1\in\mathscr{B}_1} \mu(B_1) = \mu(\mathcal{X}) \sup f.$$

Hence, it converges to a real number and its Moore–Smith limit over $\mathfrak{P}_{\mathscr{F}}$ coincides with the supremum $\sup \alpha$ over $\mathfrak{P}_{\mathscr{F}}$ (recall Proposition 1.13(ii)$_9$). The proof of the existence and real character of the upper S-integral is similar. □

When the bounded positive charge μ is in particular a probability charge, the lower S-integral with respect to μ is the point-wise limit of a net of convex combinations of lower previsions $\underline{P}_B$ that are coherent and completely monotone, by Corollary 7.5$_{128}$. As taking convex combinations and point-wise limits preserves both coherence and n-monotonicty (see Propositions 4.19(ii)$_{59}$, 4.21(ii)$_{60}$, 6.7$_{106}$ and 6.8$_{107}$), we conclude that – up to some rescaling – any lower S-integral is a coherent and completely monotone lower prevision, and the upper S-integral its conjugate, coherent and completely alternating upper prevision. The following theorem yields an even stronger conclusion.

Theorem 8.25 *Let $\mathscr{F}$ be a field on $\mathcal{X}$, and let μ be a probability charge on $\mathscr{F}$. Then*

$$S\underline{\int} f \,\mathrm{d}\mu = \underline{E}_\mu(f) \text{ and } S\overline{\int} f \,\mathrm{d}\mu = \overline{E}_\mu(f) \text{ for all bounded gambles } f \text{ on } \mathcal{X}. \quad (8.12)$$

Proof. We only prove the equality for the lower S-integral. The equality for the upper S-integral follows by conjugacy. We start with a simple observation. For every finite subset $\mathscr{A} \Subset \mathscr{F}$ and $\lambda_A \in \mathbb{R}$ (for all $A \in \mathscr{A}$), there are a finite partition $\mathscr{B}_{\mathscr{A}} \in \mathfrak{P}_{\mathscr{F}}$ and a $\kappa_B \in \mathbb{R}$ (for all $B \in \mathscr{B}_{\mathscr{A}}$) such that $\sum_{A\in\mathscr{A}} \lambda_A I_A = \sum_{B\in\mathscr{B}_{\mathscr{A}}} \kappa_B I_B$ and therefore also $\sum_{A\in\mathscr{A}} \lambda_A \mu(A) = \sum_{B\in\mathscr{B}_{\mathscr{A}}} \kappa_B \mu(B)$. If $\sum_{B\in\mathscr{B}_{\mathscr{A}}} \kappa_B I_B \leq f$, then for any $B \in \mathscr{B}_{\mathscr{A}}$, it holds that $\kappa_B \leq f(x)$ for all $x \in B$, and hence, $\kappa_B \leq \inf_{x\in B} f(x) = \underline{P}_B(f)$. So $\sum_{A\in\mathscr{A}} \lambda_A I_A \leq f$ implies that

$$\sum_{A\in\mathscr{A}} \lambda_A I_A = \sum_{B\in\mathscr{B}_{\mathscr{A}}} \kappa_B I_B \leq \sum_{B\in\mathscr{B}_{\mathscr{A}}} \underline{P}_B(f) I_B \leq f \text{ and } \sum_{A\in\mathscr{A}} \lambda_A \mu(A) \leq \sum_{B\in\mathscr{B}_{\mathscr{A}}} \underline{P}_B(f)\mu(B). \quad (8.13)$$

Therefore, we get

$$\begin{aligned} S\underline{\int} f \,\mathrm{d}\mu &= \sup_{\mathscr{B}\in\mathfrak{P}_{\mathscr{F}}} \sum_{B\in\mathscr{B}} \underline{P}_B(f)\mu(B), \\ &\leq \sup_{\mathscr{A}\Subset\mathscr{F}} \left\{ \sum_{A\in\mathscr{A}} \lambda_A \mu(A) : A \in \mathscr{F},\ \lambda_A \in \mathbb{R},\ \sum_{A\in\mathscr{A}} \lambda_A I_A \leq f \right\} = \underline{E}_\mu(f) \\ &\leq \sup_{\mathscr{A}\Subset\mathscr{F}} \sum_{B\in\mathscr{B}_{\mathscr{A}}} \underline{P}_B(f)\mu(B) \leq S\underline{\int} f \,\mathrm{d}\mu. \end{aligned}$$

Let us justify the steps we have just taken in more detail. The first equality derives from the definition of the lower S-integral. The first inequality holds because we are taking the supremum over a larger set, and the second equality follows from the fact that $\mathcal{X} \in \mathcal{F}$ (so we can omit the constant in Equation (4.10c)$_{68}$) and from the self-conjugacy of P_μ (so we can allow the coefficients λ_A to be real numbers in Equation (4.10c)$_{68}$, instead of only non-negative real numbers). The second inequality follows from Equation (8.13)$_{\curvearrowleft}$, and the last one holds because $\{\mathcal{B}_{\mathcal{A}} : \mathcal{A} \in \mathcal{F}\}$ is a subset of $\mathfrak{P}_{\mathcal{F}}$. □

Theorem 8.26 *Let $\mathcal{F}$ be a field on $\mathcal{X}$, and let μ be a probability charge on $\mathcal{F}$. Then a bounded gamble f is μ-integrable if and only if it is S-integrable with respect to μ, and in that case,*

$$S\int f\,\mathrm{d}\mu = E_\mu(f).$$

Proof. Immediately from Theorem 8.25$_{\curvearrowleft}$. □

For a general bounded positive charge μ, $\mu' = {}^{\mu}/_{\mu(\mathcal{X})}$ is a probability charge, and the lower S-integral satisfies

$$S\underline{\int} f\,\mathrm{d}\mu = \mu(\mathcal{X})S\underline{\int} f\,\mathrm{d}\mu' = \mu(\mathcal{X})\underline{E}_{\mu'}(f),$$

assuming that $\mu(\mathcal{X}) > 0$. If we call the right-hand side the **natural extension of** μ, and if we say that a bounded gamble is μ**-integrable** whenever it is ${}^{\mu}/_{\mu(\mathcal{X})}$-integrable, then Theorems 8.25$_{\curvearrowleft}$ and 8.26 remain valid for all bounded positive charges.

On an historical note, the S-integral was suggested by Moore and Smith (1922, Section 5, p. 114, ll. 10–13) to provide a conceptually simpler definition of the Lebesgue integral, which we discuss in Section 8.6. It was then extended by Kolmogorov (1930, Zweites Kapitel, Section 2, p. 663, Nr. 12) to arbitrary functions and by Hildebrandt (1934, Section 1(f), p. 869) to bounded functions. Gould (1965, Definition 4.3, p. 201, and Definition 6.1 and Theorem 6.2, p. 213) extended the S-integral to unbounded functions and charges that assume values in a Banach space – incidentally, Gould aimed at a generalisation of the Dunford integral.

In this section, we have only considered the S-integral of bounded real-valued functions – bounded gambles – with respect to real-valued bounded positive charges. For this case, Bhaskara Rao and Bhaskara Rao (1983, Section 4.5) have defined the S-integral through the mediation of a lower and an upper S-integrals; equivalence with Gould's approach is immediate from Gould (1965, Theorem 4.7(c)), for real-valued positive charges. This means that Bhaskara Rao and Bhaskara Rao's construction of Hildebrandt's S-integral – restricted to positive charges and bounded functions – is similar to Darboux's (1875) construction of Riemann's (1868) integral. When associated with the Lebesgue measure (or the Lebesgue–Stieltjes measure), the S-integral is also called the Y-integral, or the Young–Stieltjes integral. This integral was introduced by Young (1905) and is extensively discussed by Hildebrandt (1963, Chapter VII, Section 3). The S-integral is a straightforward generalisation of the Young–Stieltjes integral. For bounded gambles, the Young–Stieltjes integral provides us with an alternative to, and, in our opinion, also a simpler formulation of, the Lebesgue–Stieltjes integral, see Hildebrandt (1963). In particular, it does not involve any measurability conditions. In any case, the Young–Stieltjes integral and the Lebesgue–Stieltjes integral agree on a very large class of bounded gambles; see Hildebrandt (1963, Chapter VII, Theorem 3.9).

8.6 The Lebesgue integral

One way to introduce the well-known Lebesgue integral for bounded gambles goes as follows. It relies on the (Dunford) integral for $\mathcal{F}$-simple gambles introduced in Definition 8.13_{160}.

Definition 8.27 (Lebesgue integral) *Let $\mathcal{F}$ be a field on $\mathcal{X}$, and let μ be a bounded positive charge on $\mathcal{F}$. A bounded gamble f is called **Lebesgue integrable** with respect to μ if and only if the **lower** and **upper Lebesgue integrals** of f, respectively given by*

$$L\underline{\int} f\,\mathrm{d}\mu := \sup\left\{D\int g\,\mathrm{d}\mu : g \in \operatorname{span}(\mathcal{F}) \text{ and } g \leq f\right\} \tag{8.14}$$

and

$$L\overline{\int} f\,\mathrm{d}\mu := \inf\left\{D\int g\,\mathrm{d}\mu : g \in \operatorname{span}(\mathcal{F}) \text{ and } g \geq f\right\}, \tag{8.15}$$

*coincide. In that case, the **Lebesgue integral** of f with respect to μ is defined as the common value:*

$$L_\mu(f) := L\int f\,\mathrm{d}\mu := L\underline{\int} f\,\mathrm{d}\mu = L\overline{\int} f\,\mathrm{d}\mu.$$

For an alternative definition of the Lebesgue integral that is closer in spirit to the original construction of Lebesgue (1904), we refer to De Cooman et al. (2008b, Section 3.2.3), where it is also shown that the Lebesgue integral coincides with the S-integral. With Definition 8.27, proving this equality is even easier.

Theorem 8.28 *Let $\mathcal{F}$ be a field on $\mathcal{X}$, and let μ be a bounded positive charge on $\mathcal{F}$. For any bounded gamble f,*

$$S\underline{\int} f\,\mathrm{d}\mu = L\underline{\int} f\,\mathrm{d}\mu \text{ and } S\overline{\int} f\,\mathrm{d}\mu = L\overline{\int} f\,\mathrm{d}\mu. \tag{8.16}$$

In particular, f is S-integrable if and only if it is Lebesgue integrable, and in that case, both integrals coincide.

Proof. In the definition of the lower Lebesgue integral, simply note that we may restrict our attention to bounded gambles of the form $g = \sum_{B\in\mathcal{B}} \underline{P}_B(f) I_B$ for $\mathcal{B} \in \mathfrak{P}_{\mathcal{F}}$. This is because for any $g \in \operatorname{span}(\mathcal{F})$, there is some partition $\mathcal{B} \in \mathfrak{P}_{\mathcal{F}}$ such that $g = \sum_{B\in\mathcal{B}} \kappa_B I_B$ with $\kappa_B \in \mathbb{R}$ for all $B \in \mathcal{B}$; $g \leq f$ is then equivalent to $\kappa_B \leq \underline{P}_B(f)$ for all $B \in \mathcal{B}$.

Alternatively, compare Equations $(8.4)_{163}$ and (8.14) to find that $S\underline{\int} \bullet\,\mathrm{d}\mu = \underline{E}_\mu$ and use Theorem 8.25_{169}. □

8.7 The Dunford integral

The Dunford integral was introduced by Dunford (1935, Section 3, p. 443) as an integral of vector-valued functions with respect to measures and extended to an integral of vector-valued functions with respect to charges by Dunford and Schwartz (1957, Part I, Chapter III, Definition 2.17); see also Bhaskara Rao and Bhaskara Rao (1983, Chapter 4) for a detailed study of the Dunford integral on scalar-valued functions. So far, in order to establish the coherence of probability charges in Theorem 8.15_{161}, we have only had to consider the definition of the Dunford integral on $\mathscr{F}$-simple gambles. Here we want to look at a more general definition. This section also closely follows Troffaes (2005, Section 4.3.8).

As discussed by Dunford and Schwartz (1957, Part I, Chapter III, Section 2), the Dunford integral is first defined on $\mathscr{F}$-simple gambles – which we have already done in Definition 8.13_{160} – and then extended to more general functions through Cauchy sequences. This extension method is due to Dunford (1935, Lemma 6, p. 444), and it is discussed in detail in the following text. It is the core idea behind the definition of the Dunford integral.

Interestingly, a similar method forms a possible basis for extending coherent lower previsions from bounded to unbounded functions of X. This extension will be the subject of Chapters 14_{304} and 15_{327}.

We mention in passing that on $\mathscr{F}$-simple gambles, the Dunford integral coincides with the S-integral introduced by Hildebrandt (1934, Section 1(f), p. 869) as an integral associated with charges. We have discussed a more general version of the S-integral in Section 8.5_{168}. As Hildebrandt notes, 'it is possible to define the Lebesgue integral [with respect to a charge] by the Lebesgue process' and for $\mathscr{F}$-simple gambles (or more generally, $\mathscr{F}$-measurable bounded gambles), 'obviously [the Lebesgue integral] exists', and the '[S-integral] exists also in this case and agrees with the [Lebesgue integral]'. Concluding, the integral we are about to introduce in Definition 8.13_{160} coincides with just about *any* integral found in the literature for charges and $\mathscr{F}$-simple gambles.

The following definition of the Dunford integral relies on the outer set function μ^* induced by a probability charge μ – recall Definition 1.14_{10}. It applies to any kind of gamble, not just the bounded ones.

Definition 8.29 (Dunford integral) *Let $\mathscr{F}$ be a field on $\mathscr{X}$ and let μ be a probability charge on $\mathscr{F}$. Then a gamble $f : \mathscr{X} \to \mathbb{R}$ is called* ***Dunford integrable*** *with respect to μ if there is a sequence of $\mathscr{F}$-simple gambles f_n such that*

(i) $\lim_{n,m\to+\infty} D\int |f_n - f_m|\,\mathrm{d}\mu \to 0$,

(ii) *for any $\varepsilon > 0$ it holds that* $\lim_{n\to+\infty} \mu^*\left(\left\{|f - f_n| > \varepsilon\right\}\right) \to 0$.

In that case, the ***Dunford integral*** *of f with respect to μ is defined as*

$$D\int f\,\mathrm{d}\mu := \lim_{n\to+\infty} D\int f_n\,\mathrm{d}\mu,$$

where we should note that the limit on the right-hand side is independent of the sequence f_n satisfying the above-mentioned two conditions. Any such sequence is called a ***determining sequence*** *for the gamble f.*

The second condition identifies a special case of general property called *convergence in probability*, which we define in the following text and can be associated with coherent lower previsions. It generalises the classical definition of convergence in probability associated with probability measures (see for instance Kallenberg, 2002, Chapter 4). We will study this type of convergence for gambles in some detail in Chapter 15_{327}.

Definition 8.30 (Convergence in probability) *Let $\underline{P}$ be a coherent lower prevision, and let f_n be a sequence of gambles on $\mathcal{X}$. Then f_n* ***converges in probability*** *$[\underline{P}]$ to the gamble f on $\mathcal{X}$ if for all $\varepsilon > 0$ it holds that*

$$\lim_{n\to+\infty} \overline{E}_{\underline{P}}\left(\left\{|f-f_n| > \varepsilon\right\}\right) = 0.$$

The second condition in the definition of Dunford integrability now amounts to requiring that the sequence of $\mathcal{F}$-simple gambles f_n should converge in probability (μ, or rather P_μ) to the gamble f.

In investigating on which *bounded gambles* Dunford integrals and μ-integrals coincide, the following lemma will be very useful. It immediately and trivially leads to an alternative characterisation of convergence in probability *for bounded gambles*.

Lemma 8.31 *Let $\underline{P}$ be any coherent lower prevision on $\mathbb{B}$. Let f_n be a bounded sequence of bounded gambles: there is some $\alpha \in \mathbb{R}$ such that $\sup_{n\in\mathbb{N}} \sup|f_n| \leq \alpha$. Then the following statements are equivalent:*

(i) $\lim_{n\to+\infty} \overline{P}\left(|f_n|\right) = 0$

(ii) *For any $\varepsilon > 0$, $\lim_{n\to+\infty} \overline{P}\left(\left\{|f_n| > \varepsilon\right\}\right) = 0$.*

Proof. Let, for ease of notation, $A_{n,\varepsilon} := \left\{|f_n| > \varepsilon\right\}$ for any $\varepsilon > 0$ and $n \in \mathbb{N}$.

(i)⇒(ii). Fix $\varepsilon > 0$ and consider any $0 < \delta < \varepsilon$. Then there is some $N_\delta \in \mathbb{N}$ such that $\overline{P}\left(|f_n|\right) < \delta^2$ for all $n \geq N_\delta$. As $\varepsilon I_{A_{n,\varepsilon}} \leq |f_n|$, it follows from the coherence of $\underline{P}$ that (use Theorem 4.13(iv) and $(\text{vi})_{53}$)

$$\overline{P}\left(A_{n,\varepsilon}\right) \leq \frac{\overline{P}\left(|f_n|\right)}{\varepsilon} < \frac{\delta^2}{\varepsilon} < \delta \text{ for all } n \geq N_\delta.$$

So indeed $\lim_{n\to+\infty} \overline{P}\left(A_{n,\varepsilon}\right) = 0$.

(ii)⇒(i). Fix $\varepsilon > 0$. There is some $N_\varepsilon \in \mathbb{N}$ such that $\overline{P}\left(A_{n,\varepsilon}\right) < \frac{\varepsilon}{\alpha+1}$ for all $n \geq N_\varepsilon$. It now follows from the coherence of $\underline{P}$ that

$$\overline{P}\left(|f_n|\right) \leq \overline{P}\left(|f_n| I_{A_{n,\varepsilon}}\right) + \overline{P}\left(|f_n| I_{A^c_{n,\varepsilon}}\right) < \alpha\frac{\varepsilon}{\alpha+1} + \varepsilon < 2\varepsilon \text{ for every } n \geq N_\varepsilon;$$

indeed, the first inequality follows from Theorem $4.13(\text{v})_{53}$, and the second from Theorem 4.13(iv) and $(\text{vi})_{53}$. Hence, indeed, $\lim_{n\to+\infty} \overline{P}\left(|f_n|\right) = 0$. □

Bhaskara Rao and Bhaskara Rao (1983, Theorem 4.5.7 and Proposition 4.5.8) give a rather long proof for the equivalence of Dunford integrability and S-integrability of bounded gambles and for the equality of the corresponding integrals on bounded gambles. We give a much shorter and conceptually simpler proof for a more general equivalence based on a repeated application of Lemma 8.31$_{\frown}$.

Theorem 8.32 *Let $\mathscr{F}$ be a field on $\mathscr{X}$, and let μ be a probability charge on $\mathscr{F}$. Let f be any bounded gamble. Then the following conditions are equivalent:*

(i) *f is Dunford integrable with respect to μ.*

(ii) *f is Lebesgue integrable with respect to μ.*

(iii) *f is S-integrable with respect to μ.*

(iv) *f is μ-integrable.*

If any (and hence all) of these conditions are satisfied, then

$$D\int f\,\mathrm{d}\mu = L\int f\,\mathrm{d}\mu = S\int f\,\mathrm{d}\mu = E_\mu(f).$$

Proof. The equivalence of S-integrability with respect to μ and μ-integrability, and the equality of the S-integral and the linear extension, has already been established in Theorem 8.26$_{170}$. The link between S-integrability and Lebesgue integrability was established in Theorem 8.28$_{171}$. Equality of the S-integral and the Dunford integral for $\mathscr{F}$-simple gambles is immediate from their respective definitions; see Definitions 8.13$_{160}$ and 8.24$_{168}$ and, alternatively, use Theorems 8.16$_{162}$ and 8.25$_{169}$. We are left to prove equivalence of S-integrability and the Dunford integrability and equality of the corresponding integrals.

Suppose that the bounded gamble f is S-integrable with respect to μ. Then, for any $n \in \mathbb{N}_{>0}$, there is some finite partition $\mathscr{B}_n \in \mathfrak{P}_{\mathscr{F}}$ such that

$$\sum_{B\in\mathscr{B}_n} \overline{P}_B(f)\mu(B) - \sum_{B\in\mathscr{B}_n} \underline{P}_B(f)\mu(B) < \frac{1}{n} \text{ and}$$

$$\sum_{B\in\mathscr{B}_n} \overline{P}_B(f)\mu(B) \geq S\int f\,\mathrm{d}\mu \geq \sum_{B\in\mathscr{B}_n} \underline{P}_B(f)\mu(B). \quad (8.17)$$

Define the bounded gambles $g_n := \sum_{B\in\mathscr{B}_n} \overline{P}_B(f)I_B$ and $h_n := \sum_{B\in\mathscr{B}_n} \underline{P}_B(f)I_B$. As both g_n and h_n are $\mathscr{F}$-simple gambles, the above condition can be written as $S\int(g_n - h_n)\,\mathrm{d}\mu < 1/n$ or, because $g_n \geq h_n$, also as $S\int |g_n - h_n|\,\mathrm{d}\mu < 1/n$. But $g_n \geq f \geq h_n$, and hence, $|g_n - h_n| \geq |g_n - f|$. Recalling that the upper S-integral is a coherent upper prevision on all bounded gambles (see Theorem 8.25$_{169}$), and invoking Theorem 4.13(iv)$_{53}$, we get

$$S\overline{\int} |g_n - f|\,\mathrm{d}\mu \leq S\overline{\int} |g_n - h_n|\,\mathrm{d}\mu = S\int (g_n - h_n)\,\mathrm{d}\mu < \frac{1}{n}.$$

Using the coherence (sub-additivity, see Theorem 4.13(v)$_{53}$) of the upper S-integral again, we find that

$$S\int |g_n - g_m|\,\mathrm{d}\mu \leq S\overline{\int} |g_n - f|\,\mathrm{d}\mu + S\overline{\int} |f - g_m|\,\mathrm{d}\mu < \frac{1}{n} + \frac{1}{m},$$

so $\lim_{m,n\to+\infty} S\int |g_n - g_m|\,\mathrm{d}\mu = 0$. Each $|g_n - g_m|$ is $\mathscr{F}$-simple, so $S\int |g_n - g_m|\,\mathrm{d}\mu = E_\mu(|g_n - g_m|) = D\int |g_n - g_m|\,\mathrm{d}\mu$. Hence, $\lim_{m,n\to+\infty} D\int |g_n - g_m|\,\mathrm{d}\mu = 0$ as well. Also observe that $\sup_{n\in\mathbb{N}} \sup|f - g_n| \leq 2\sup|f|$, and hence, Lemma 8.31$_{173}$ applies for the sequence of bounded gambles $f - g_n$ and guarantees that for any $\varepsilon > 0$, the sequence of real numbers $S\overline{\int}\{|f - g_n| > \varepsilon\}\,\mathrm{d}\mu$ converges to zero. But Theorems 8.18$_{163}$ and 8.25$_{169}$ then tell us that the sequence of real numbers

$$S\overline{\int}\{|f - g_n| > \varepsilon\}\,\mathrm{d}\mu = \overline{E}_\mu\left(\{|f - g_n| > \varepsilon\}\right) = \mu^*\left(\{|f - g_n| > \varepsilon\}\right)$$

converges to zero. This demonstrates that f is Dunford integrable with respect to μ and that $D\int f\,\mathrm{d}\mu = \lim_{n\to+\infty} D\int g_n\,\mathrm{d}\mu$.

We now infer from Equation (8.17) that

$$S\int f\,\mathrm{d}\mu = \lim_{n\to+\infty} \sum_{B\in\mathscr{B}_n} \overline{P}_B(f)\mu(B) = \lim_{n\to+\infty} D\int g_n\,\mathrm{d}\mu = D\int f\,\mathrm{d}\mu,$$

so the corrcsponding integrals are equal as well.

Conversely, assume that f is Dunford integrable with respect to μ. Then there is some sequence f_n of $\mathscr{F}$-simple gambles such that for every $\varepsilon > 0$, the sequence of real numbers $\mu^*\left(\{|f - f_n| > \varepsilon\}\right)$ converges to zero. In particular, this implies that for every $\varepsilon > 0$, there is some $\mathscr{F}$-simple gamble f_ε such that

$$\mu^*\left(\{|f - f_\varepsilon| > \varepsilon\}\right) < \varepsilon.$$

Now fix any $\varepsilon > 0$ and let $A_\varepsilon := \{|f - f_\varepsilon| > \varepsilon\}$, then we see that $\mu^*(A_\varepsilon) = \inf\{\mu(B) : A_\varepsilon \subseteq B \in \mathscr{F}\} < \varepsilon$. This means that there must be some $B_\varepsilon \in \mathscr{F}$ such that $A_\varepsilon \subseteq B_\varepsilon$ but still $\mu(B_\varepsilon) < \varepsilon$. As f_ε is $\mathscr{F}$-simple, there is some finite partition $\mathscr{A}_\varepsilon \in \mathfrak{P}_{\mathscr{F}}$ on whose elements f_ε is constant. Define the finite partition[3]

$$\mathscr{B}_\varepsilon := \{B_\varepsilon\} \cup \{A \cap B_\varepsilon^c : A \in \mathscr{A}_\varepsilon\}.$$

Fix any $A \in \mathscr{A}_\varepsilon$. For any $x \in A \cap B_\varepsilon^c$, it holds in particular that $x \notin A_\varepsilon$ and, hence, $|f(x) - f_\varepsilon(x)| \leq \varepsilon$. As f_ε is constant on $A \cap B_\varepsilon^c$, we also find that, because, moreover, the (vacuous) lower and upper previsions $\underline{P}_{A\cap B_\varepsilon^c}$ and $\overline{P}_{A\cap B_\varepsilon^c}$ are coherent,

$$\overline{P}_{A\cap B_\varepsilon^c}(f) - \underline{P}_{A\cap B_\varepsilon^c}(f) = \overline{P}_{A\cap B_\varepsilon^c}(f - f_\varepsilon) - \underline{P}_{A\cap B_\varepsilon^c}(f - f_\varepsilon) \leq 2\overline{P}_{A\cap B_\varepsilon^c}(|f - f_\varepsilon|) \leq 2\varepsilon, \tag{8.18}$$

[3] We are allowing ourselves some leeway here in calling this a partition, as some of its elements might be empty.

where the first inequality follows from Theorem 4.13(ix)$_{53}$. Now define the $\mathscr{F}$-simple gambles $g_\varepsilon := \sum_{B\in\mathscr{B}_\varepsilon} \overline{P}_B(f)I_B$ and $h_\varepsilon := \sum_{B\in\mathscr{B}_\varepsilon} \underline{P}_B(f)I_B$. If we can show that $S\int(g_\varepsilon - h_\varepsilon)\,\mathrm{d}\mu \leq K\varepsilon$ for some $K > 0$ that may depend on f but not on ε, then we will have proved that f is S-integrable. Indeed, in that case, because $g_\varepsilon \geq f \geq h_\varepsilon$, and by the coherence (monotonicity, use Theorem 4.13(iv)$_{53}$) of the lower and upper S-integrals,

$$S\int g_\varepsilon\,\mathrm{d}\mu \geq S\overline{\int} f\,\mathrm{d}\mu \geq S\underline{\int} f\,\mathrm{d}\mu \geq S\int h_\varepsilon\,\mathrm{d}\mu \geq S\int g_\varepsilon\,\mathrm{d}\mu + K\varepsilon, \tag{8.19}$$

where we have also used that the $\mathscr{F}$-simple gambles g_ε and h_ε are S-integrable. This implies that $S\overline{\int} f\,\mathrm{d}\mu - S\underline{\int} f\,\mathrm{d}\mu \leq K\varepsilon$, and because this holds for all $\varepsilon > 0$, we see that $S\overline{\int} f\,\mathrm{d}\mu = S\underline{\int} f\,\mathrm{d}\mu$, so f will be S-integrable. Now

$$\begin{aligned}
S\int(g_\varepsilon - h_\varepsilon)\,\mathrm{d}\mu &= \sum_{B\in\mathscr{B}_\varepsilon}\left[\overline{P}_B(f) - \underline{P}_B(f)\right]\mu(B)\\
&= \left[\overline{P}_{B_\varepsilon}(f) - \underline{P}_{B_\varepsilon}(f)\right]\mu(B_\varepsilon)\\
&\qquad + \sum_{A\in\mathscr{A}_\varepsilon}\left[\overline{P}_{A\cap B_\varepsilon^c}(f) - \underline{P}_{A\cap B_\varepsilon^c}(f)\right]\mu(A\cap B_\varepsilon^c)\\
&\leq 2\varepsilon\sup|f| + 2\varepsilon\sum_{A\in\mathscr{A}_\varepsilon}\mu(A\cap B_\varepsilon^c) \leq 2(\sup|f| + 1)\varepsilon,
\end{aligned}$$

where we have also used the inequalities (8.18)$_\frown$. So the desired inequality is satisfied for $K = 2(\sup|f| + 1)$. This means that f is S-integrable, and we infer from the inequalities (8.19) that its S-integral is given by $S\int f\,\mathrm{d}\mu = \lim_{n\to+\infty} S\int g_{1/n}\,\mathrm{d}\mu$. From this last equality, which can also be written as $\lim_{n\to+\infty} S\int\left|f - g_{1/n}\right|\mathrm{d}\mu = 0$, we can infer two things. First of all,

$$\begin{aligned}
\lim_{n,m\to+\infty} D\int\left|g_{\frac{1}{n}} - g_{\frac{1}{m}}\right|\mathrm{d}\mu &= \lim_{n,m\to+\infty} S\int\left|g_{\frac{1}{n}} - g_{\frac{1}{m}}\right|\mathrm{d}\mu\\
&\leq \lim_{n\to+\infty} S\int\left|g_{\frac{1}{n}} - f\right|\mathrm{d}\mu + \lim_{m\to+\infty} S\int\left|f - g_{\frac{1}{m}}\right|\mathrm{d}\mu = 0,
\end{aligned}$$

where we have used that $\left|g_{1/n} - g_{1/m}\right|$ is $\mathscr{F}$-simple and that the S-integral is a linear prevision (Theorem 8.26$_{170}$) and, therefore, monotone (Corollary 4.14(iv)$_{55}$) and additive (Corollary 4.14(v)$_{55}$). Secondly, because $\sup_{n\in\mathbb{N}}\left|f - g_{1/n}\right| \leq 2\sup|f|$, the sequence $\left|f - g_{1/n}\right|$ is uniformly bounded. Apply Lemma 8.31$_{173}$ once again, now with the lower S-integral as lower prevision, to find that for all $\varepsilon > 0$, the sequence of real numbers $S\overline{\int}\left\{\left|f - g_{1/n}\right| > \varepsilon\right\}\mathrm{d}\mu$ converges to zero. Now apply Theorems 8.18$_{163}$ and 8.25$_{169}$ to see that for any $\varepsilon > 0$, the sequence $\mu^*(\{\left|f - g_{1/n}\right| > \varepsilon\})$ also converges to zero. Hence, $g_{1/n}$ is a determining sequence for f, and therefore

$$D\int f\,\mathrm{d}\mu = \lim_{n\to+\infty} D\int g_{\frac{1}{n}}\,\mathrm{d}\mu = \lim_{n\to+\infty} S\int g_{\frac{1}{n}}\,\mathrm{d}\mu = S\int f\,\mathrm{d}\mu.$$

In conclusion, not only is f S-integrable but its Dunford integral is also equal to its S-integral. □

This concludes our discussion on Dunford integrability for bounded gambles. Interestingly, it turns out that we can also establish equivalence between Dunford integrability and Choquet integrability – it appears that this is not a well-known fact, at least as far as we could tell from our search in the literature. A far more general equivalence – between previsibility and Choquet integrability for 2-monotone lower previsions – will be established in Chapter 15_{327}; see Theorem 15.33_{353}. A proof of the next theorem will be postponed until the discussion following Equation $(15.28)_{359}$.[4] For more about the Choquet integral, we refer to Appendix C_{376} and Chapter 6_{101}.

Theorem 8.33 *Let μ be a probability charge on $\mathscr{P}$. A bounded gamble f on $\mathcal{X}$ is Choquet integrable with respect to μ if and only if f is Dunford integrable with respect to μ, and in that case,*

$$D\int f\,\mathrm{d}\mu = C\int f\,\mathrm{d}\mu.$$

8.8 Consequences of duality

We conclude this chapter with a result (Theorem $8.35_{\curvearrowright}$) that establishes an isomorphism between coherent lower previsions on a linear subspace $\mathscr{K}$ of $\mathbb{B}$ and certain sets of linear previsions on $\mathscr{K}$. The result is a special case of a more general isomorphism established by Troffaes (2005, Theorem 4.85), who considers general domains $\mathscr{K}$ – that is, not assuming that $\mathscr{K}$ a linear space. Because it is far easier to establish the isomorphism for linear spaces $\mathscr{K}$, we restrict ourselves to that simpler case here. The remainder of this section is based on the relevant parts of Troffaes (2005, Section 4.4), although our notation is slightly different.

The result – and its simple proof, which we include for the sake of completeness – is an immediate generalisation of Walley's weak*-compactness theorem (Walley, 1991, Theorem 3.6.1), which deals with the specific case $\mathscr{K} = \mathbb{B}$. Part of the result also constitutes a direct generalisation of what we have discussed in Section 4.7_{74}, and in particular of Proposition 4.43_{75}.

Let us rehash the topological comments made there, taking into account that the relevant domain of our functionals is now no longer $\mathbb{B}$ but some linear subspace $\mathscr{K}$. We provide this linear subspace $\mathscr{K}$ with the topology of uniform convergence. Its **topological dual** $\mathscr{K}^*$ is the set of all continuous real linear functionals on $\mathscr{K}$. We see that $\mathbb{P}^{\mathscr{K}} \subseteq \mathscr{K}^*$, because linear previsions are coherent lower previsions and, therefore, continuous. The linear space $\mathscr{K}^*$ is provided with the topology of pointwise convergence on $\mathscr{K}$ or, equivalently, the weakest topology on $\mathscr{K}^*$ such that all

[4] Actually, the proof there works for arbitrary – not just for bounded – gambles.

so-called evaluation functionals $f^*, f \in \mathscr{K}$ are continuous. These **evaluation functionals** f^* are defined by $f^*(\Lambda) := \Lambda(f)$ for all Λ in $\mathscr{K}^*$. $\mathscr{K}^*$ is a locally convex and Hausdorff topological linear space under this so-called weak* topology (Schechter, 1997, Section 28.15).

All sets of the type $\{f^* \geq \alpha\} = \{\Lambda \in \mathscr{K}^* : \Lambda(f) \geq \alpha\}$, $\alpha \in \mathbb{R}$ are weak*-closed, as inverse images of the closed set of real numbers $[\alpha, +\infty)$ under the continuous map f^*. They are clearly also convex. This implies that the set

$$\mathbb{P}^{\mathscr{K}} = \bigcap_{f \in \mathscr{K}} \left\{\Lambda \in \mathscr{K}^* : \Lambda(f) \geq \inf f\right\}$$

is weak*-closed and convex too (To see why the equality above holds, recall that all linear previsions are continuous, and use Theorem 4.16_{57}). It then follows from the Banach–Alaoglu–Bourbaki theorem (Schechter, 1997, Section 28.29) that $\mathbb{P}^{\mathscr{K}}$ is weak*-compact. Because $\mathscr{K}^*$ is Hausdorff, all weak*-closed subsets of $\mathbb{P}^{\mathscr{K}}$ are weak*-compact and *vice versa*.

We begin the discussion by generalising the duality maps introduced in Definition 4.36_{71}:

Definition 8.34 (Generalised duality maps) *Let $\mathscr{K}$ be a linear subspace of $\mathbb{B}$. We first define a* ***duality map*** *from lower previsions to sets of linear previsions. With any lower prevision $\underline{P}$ on $\mathscr{K}$, we can associate the set of all linear previsions on $\mathscr{K}$ that dominate $\underline{P}$,*

$$\operatorname{lins}^{\mathscr{K}}(\underline{P}) := \left\{Q \in \mathbb{P}^{\mathscr{K}} : (\forall f \in \mathscr{K})\left(Q(f) \geq \underline{P}(f)\right)\right\},$$

called a ***dual model*** *of $\underline{P}$.*

Conversely, we can also define a ***duality map*** *from sets of linear previsions to lower previsions. With any set $\mathscr{M}$ of linear previsions on $\mathscr{K}$, we can associate a lower prevision $\operatorname{lpr}^{\mathscr{K}}(\mathscr{M})$ on $\mathscr{K}$, defined by*

$$\operatorname{lpr}^{\mathscr{K}}(\mathscr{M})(f) := \inf\{Q(f) : Q \in \mathscr{M}\} \text{ for all } f \in \mathscr{K},$$

called a ***dual model*** *of $\mathscr{M}$.*

Theorem 8.35 (Duality) *Let $\mathscr{K}$ be a linear subspace of $\mathbb{B}$. Then the duality maps $\operatorname{lpr}^{\mathscr{K}}$ and $\operatorname{lins}^{\mathscr{K}}$ are bijective – onto and one-to-one – maps between the non-empty weak*-compact convex sets of linear previsions on $\mathscr{K}$ and the coherent lower previsions on $\mathscr{K}$.*

Proof. Let $\mathscr{M}$ be a non-empty weak*-compact convex set of linear previsions on $\mathscr{K}$, and let $\underline{P}$ be a coherent lower prevision on $\mathscr{K}$. It suffices to show that

(a) $\operatorname{lpr}^{\mathscr{K}}(\mathscr{M})$ is a coherent lower prevision on $\mathscr{K}$,

(b) $\operatorname{lins}^{\mathscr{K}}(\underline{P})$ is a weak*-compact convex set of linear previsions on $\mathscr{K}$,

(c) $\operatorname{lins}^{\mathscr{K}}(\operatorname{lpr}^{\mathscr{K}}(\mathscr{M})) = \mathscr{M}$ and $\operatorname{lpr}^{\mathscr{K}}(\operatorname{lins}^{\mathscr{K}}(\underline{P})) = \underline{P}$.

Invoke Proposition 4.20(ii)$_{59}$ to prove (a): $\mathrm{lpr}^{\mathscr{K}}(\mathscr{M})$ is the lower envelope of the set $\mathscr{M}$ of coherent (lower) previsions on $\mathscr{K}$. To prove (b), use a course of reasoning that is completely similar to the proof of Proposition 4.43$_{75}$.

We now turn to the proof of (c). To see that $\mathrm{lins}^{\mathscr{K}}(\mathrm{lpr}^{\mathscr{K}}(\mathscr{M})) \supseteq \mathscr{M}$, observe that any linear prevision in $\mathscr{M}$ dominates $\mathrm{lpr}^{\mathscr{K}}(\mathscr{M})$.

Conversely, to establish that $\mathrm{lins}^{\mathscr{K}}(\mathrm{lpr}^{\mathscr{K}}(\mathscr{M})) \subseteq \mathscr{M}$, we show that if $Q \in \mathbb{P}^{\mathscr{K}}$ but $Q \notin \mathscr{M}$, then Q does not dominate $\mathrm{lpr}^{\mathscr{K}}(\mathscr{M})$. Assume that $Q \notin \mathscr{M}$, so $\{Q\}$ and $\mathscr{M}$ are disjoint non-empty weak*-compact and convex subsets of the set $\mathscr{K}^*$ of continuous real-valued linear maps on $\mathscr{K}$. By a version of the Hahn–Banach theorem (Schechter, 1997, 28.4(HB19)), there is a weak*-continuous linear functional Λ defined on $\mathscr{K}^*$ such that $\Lambda(Q) < \min\{\Lambda(R) \colon R \in \mathscr{M}\}$. As Λ is weak*-continuous, it must be an evaluation map on $\mathscr{K}^*$ (Schechter, 1997, 28.15(c)), so there is some $f \in \mathscr{K}$ such that $\Lambda(R) = R(f)$ for all linear previsions R on $\mathscr{K}$. Hence, $Q(f) < \min\{R(f) \colon R \in \mathscr{M}\}$, so Q does not dominate $\mathrm{lpr}^{\mathscr{K}}(\mathscr{M})$.

The proof of the remaining equality $\mathrm{lpr}^{\mathscr{K}}(\mathrm{lins}^{\mathscr{K}}(\underline{P})) = \underline{P}$ is an immediate generalisation of the proof we have given for the Lower Envelope Theorem 4.38$_{71}$, so we will not include it here. □

In other words, the set of non-empty weak*-compact and convex sets of linear previsions on a linear subspace $\mathscr{K}$ equipped with $\mathrm{lpr}^{\mathscr{K}}$ is isomorphic to the set of coherent lower previsions on $\mathscr{K}$ equipped with the identity map.

In particular, if the domain $\mathscr{K}$ is the set $\mathbb{B}_{\mathscr{F}}$ of measurable bounded gambles with respect to a field $\mathscr{F}$, we can combine the duality isomorphisms with the isomorphism established in Theorem 8.22$_{167}$. We endow the set of probability charges on $\mathscr{F}$ with the topology of point-wise convergence: it is immediate that this space is homeomorphic to the set of linear previsions on $I_{\mathscr{F}} \cup -I_{\mathscr{F}}$ endowed with the topology of point-wise convergence; the homeomorphism is simply $P_{\bullet}$.

In the following corollary, we view $\mathrm{lin}^{\mathbb{B}_{\mathscr{F}}}$ as a map from sets of probability charges on a field $\mathscr{F}$ to sets of linear previsions on $\mathbb{B}_{\mathscr{F}}$,

$$\mathrm{lin}^{\mathbb{B}_{\mathscr{F}}}(m) := \left\{\mathrm{lin}^{\mathbb{B}_{\mathscr{F}}}(\mu) \colon \mu \in m\right\},$$

where m is an arbitrary set of probability charges on $\mathscr{F}$. Similarly, we view $\mathrm{ch}^{\mathscr{F}}$ as a map from sets of linear previsions on $\mathbb{B}_{\mathscr{F}}$ to sets of probability charges on $\mathscr{F}$,

$$\mathrm{ch}^{\mathscr{F}}(\mathscr{M}) := \left\{\mathrm{ch}^{\mathscr{F}}(Q) \colon Q \in \mathscr{M}\right\},$$

where $\mathscr{M}$ is an arbitrary set of linear previsions on $\mathbb{B}_{\mathscr{F}}$.

Corollary 8.36 *Let $\mathscr{F}$ be a field on $\mathscr{X}$. Then $\mathrm{lpr}^{\mathbb{B}_{\mathscr{F}}} \circ \mathrm{lin}^{\mathbb{B}_{\mathscr{F}}}$ and $\mathrm{ch}^{\mathscr{F}} \circ \mathrm{lins}^{\mathbb{B}_{\mathscr{F}}}$ establish one-to-one and onto maps between non-empty compact and convex sets of probability charges on $\mathscr{F}$ and coherent lower previsions on $\mathbb{B}_{\mathscr{F}}$.*

Proof. Immediately from Theorems 8.22$_{167}$ and 8.35. □

Corollary 8.36$_\frown$ leads to another characterisation of natural extension, for instance, through the S-integral.[5] If $\mathscr{F}$ is a field such that $\operatorname{dom}\underline{P}$ consists of $\mathscr{F}$-measurable bounded gambles only, let us denote by $\operatorname{chs}^{\mathscr{F}}(\underline{P})$ the set of charges on $\mathscr{F}$, whose linear extensions to $\mathbb{B}_{\mathscr{F}}$ dominate $\underline{P}$:

$$\operatorname{chs}^{\mathscr{F}}(\underline{P}) := \operatorname{ch}^{\mathscr{F}} \circ \operatorname{lins}^{\mathbb{B}_{\mathscr{F}}}(\underline{P}) = \operatorname{ch}^{\mathscr{F}}(\operatorname{lins}^{\mathbb{B}_{\mathscr{F}}}(\underline{P})), \tag{8.20}$$

The following proposition provides equivalent expressions for $\operatorname{chs}^{\mathscr{F}}(\underline{P})$.

Proposition 8.37 *Let $\underline{P}$ be any lower prevision, let $\mathscr{F}$ be a field on $\mathscr{X}$ and assume that* $\operatorname{dom}\underline{P} \subseteq \mathbb{B}_{\mathscr{F}}$. *Then*

$$\operatorname{chs}^{\mathscr{F}}(\underline{P}) = \left\{ \mu \in \mathbb{P}(\mathscr{F}) \colon (\forall f \in \operatorname{dom}\underline{P})(\underline{P}(f) \leq S\int f \,\mathrm{d}\mu) \right\}.$$

Proof. The equality is a consequence of Equation (8.20) and Theorem 8.26$_{170}$, which states that the natural extension $\underline{E}_\mu$ of μ coincides with the lower S-integral with respect to μ. □

Corollary 8.38 *Let $\underline{P}$ be any lower prevision, let $\mathscr{F}$ be a field on $\mathscr{X}$ and assume that* $\operatorname{dom}\underline{P} \subseteq \mathbb{B}_{\mathscr{F}}$. *Then the following statements hold:*

(i) *$\underline{P}$ avoids sure loss if and only if there is a probability charge μ on $\mathscr{F}$ such that*

$$\underline{P}(f) \leq S\int f \,\mathrm{d}\mu \textit{ for all } f \in \operatorname{dom}\underline{P}.$$

(ii) *If $\underline{P}$ avoids sure loss, then*

$$\underline{E}_{\underline{P}}(f) = \min\left\{ S\underline{\int} f \,\mathrm{d}\mu \colon \mu \in \operatorname{chs}^{\mathscr{F}}(\underline{P}) \right\} \textit{ for any } f \in \mathbb{B} \textit{ hence,}$$

$$= \min\left\{ S\int f \,\mathrm{d}\mu \colon \mu \in \operatorname{chs}^{\mathscr{F}}(\underline{P}) \right\} \textit{ for any } f \in \mathbb{B}_{\mathscr{F}}.$$

(iii) *$\underline{P}$ is coherent if and only if*

$$\underline{P}(f) = \min\left\{ S\int f \,\mathrm{d}\mu \colon \mu \in \operatorname{chs}^{\mathscr{F}}(\underline{P}) \right\} \textit{ for all } f \in \operatorname{dom}\underline{P}.$$

[5] We mention the S-integral here explicitly, but Theorems 8.32$_{174}$ and 8.33$_{177}$ guarantee that Proposition 8.37 and Corollary 8.38 in the following text can also be formulated in terms of Dunford, Lebesgue or Choquet integrals.

9

Examples of linear extension

There are a number of interesting and surprisingly familiar instances of the notion of linear extension of a coherent lower prevision that go beyond integration associated with a probability charge, as studied in Chapter 8. To illustrate this, we look at the form linear extension takes in some of the examples discussed in Chapters 5_{76} and 7_{122}. We also look at the particular case of integration associated with a probability measure rather than just a probability charge.

9.1 Distribution functions

In Section $7.9.1_{147}$, and in particular Theorem 7.15_{149}, we have discussed distribution functions and to what extent these determine precise probability models. Here, we want to use the ideas of Chapter 8 to shed more light on that discussion. We rely on the results and notations introduced in Section 7.9_{147}.

With a distribution function F on the space $\mathcal{X}$ provided with the total preorder $\preceq$, we can associate a set function μ_F on the chain of intervals $\{[0_{\mathcal{X}}, x] \colon x \in \mathcal{X}\}$ defined by

$$\mu([0_{\mathcal{X}}, x]) := F(x) \text{ for all } x \in \mathcal{X}.$$

We can now use Definition 8.12_{159} to associate a prevision P_{μ_F} with this set function, by letting

$$P_{\mu_F}(I_{[0_{\mathcal{X}},x]}) := -P_{\mu_F}(-I_{[0_{\mathcal{X}},x]}) := \mu_F([0_{\mathcal{X}}, x]) = F(x) \text{ for all } x \in \mathcal{X}.$$

As the domain of μ_F is not generally a field of events, we cannot invoke Theorem 8.15_{161} to determine whether or not P_{μ_F} is a linear prevision. But Theorem 7.15_{149} in effect tells us that P_{μ_F} is indeed a linear prevision and that, moreover, $\underline{E}_F := \underline{E}_{\mu_F} = \underline{E}_{\alpha_F}$ and, therefore, also $E_F := E_{\mu_F} = E_{\alpha_F}$.

This allows us to associate a notion of integration with a distribution function F. We call a bounded gamble f **F-integrable** if it is α_F-integrable, and then, using

Lower Previsions, First Edition. Matthias C.M. Troffaes and Gert de Cooman.

Theorems 8.18_{163}, 8.32_{174} and 8.33_{177},

$$E_F(f) = D\int f\,\mathrm{d}\alpha_F = L\int f\,\mathrm{d}\alpha_F = S\int f\,\mathrm{d}\alpha_F = C\int f\,\mathrm{d}(\alpha_F)_*.$$

An in-depth discussion relating integrals such as these to Stieltjes-type integrals can be found in Troffaes (2005, Section 4.3.6, pp. 132–158).

9.2 Limits inferior

We found in Proposition 5.4_{81} that if $\mathscr{A}$ is a directed set, then the limit inferior operator lim inf is a coherent lower prevision defined on the set $\mathbb{B}(\mathscr{A})$ of all bounded real nets on $\mathscr{A}$, whose conjugate upper prevision is the limit superior operator lim sup. Let us now continue the discussion of Section 5.3_{80} in the light of the results in Chapter 8 by considering the lim inf-integral $E_{\liminf}$, defined on the linear space $\mathscr{I}_{\liminf}(\mathscr{A})$ of all lim inf-integrable bounded gambles.

Proposition 9.1 *$\mathscr{I}_{\liminf}(\mathscr{A}) = \mathscr{I}_{\lim}(\mathscr{A})$ is the uniformly closed linear lattice of convergent bounded nets on $\mathscr{A}$ and $E_{\liminf} = \lim$.*

Proof. Consider any bounded real net f on $\mathscr{A}$. We first show that f converges if and only if $\liminf f = \limsup f$. This will then imply immediately that $\mathscr{I}_{\liminf}(\mathscr{A}) = \mathscr{I}_{\lim}(\mathscr{A})$ is the set of all convergent bounded real nets and that $E_{\liminf} = \lim$.

'if'. It follows readily from $\liminf f \geq \limsup f$ and the fact that $\mathscr{A}$ is directed that

$$(\forall \varepsilon > 0)(\exists \gamma \in \mathscr{A})(\forall \alpha, \beta \geq \gamma)|f(\alpha) - f(\beta)| < \varepsilon,$$

so f is a Cauchy net. As the real line provided with the Euclidean norm $|\bullet|$ is a complete space, it follows that f converges (see for instance Schechter (1997, Sections 19.9–11)).

'only if'. Assume that f converges, with limit $\lim f$. If we fix any $\varepsilon > 0$, then we know that there is some $\beta_\varepsilon \in \mathscr{A}$ such that for all $\alpha \geq \beta_\varepsilon$, $\lim f - \varepsilon < f(\alpha) < \lim f + \varepsilon$, whence

$$\lim f - \varepsilon \leq \inf_{\alpha \geq \beta_\varepsilon} f(\alpha) \leq \sup_{\alpha \geq \beta_\varepsilon} f(\alpha) \leq \lim f + \varepsilon$$

and, therefore, also $\lim f - \varepsilon \leq \liminf f \leq \limsup f \leq \lim f + \varepsilon$. As this holds for all $\varepsilon > 0$, we conclude that $\liminf f = \limsup f = \lim f$.

We already know from Proposition $8.8(\mathrm{v})_{155}$ that $\mathscr{I}_{\liminf}(\mathscr{A})$ is a uniformly closed linear space that contains all constant gambles. That it is also a lattice follows from Proposition $1.13(\mathrm{vi})_9$. □

This proposition looks innocent enough but has interesting consequences. As lim is a linear prevision defined on the linear lattice of bounded gambles $\mathscr{I}_{\lim}(\mathscr{A})$ that contains all constant nets, it is completely monotone (and completely alternating) by Theorem 6.6_{105}, and by Corollary 6.23_{118}, its natural extension $\underline{E}_{\lim} = \liminf$ is a *completely monotone* and coherent lower prevision, that is also *comonotone additive* and equal to the Choquet functional associated with its restriction to $\{0,1\}$-valued nets.

What can we say about the lim inf and lim operators on such $\{0,1\}$-valued nets (events)? Obviously, the coherent lower prevision lim inf is $\{0,1\}$-valued on events, so we can use the ideas introduced in Section 5.5_{82}. We see that there is a proper filter $\mathscr{F}_{\mathrm{evt}}$ of events such that

$$\mathscr{F}_{\mathrm{evt}} := \left\{A \subseteq \mathscr{A} : \liminf I_A = 1\right\} = \{A \subseteq \mathscr{A} : (\exists \beta \in \mathscr{A})(\forall \alpha \geq \beta)\alpha \in A\},$$

and lim inf is the lower prevision associated with this proper filter: $\liminf = \underline{P}_{\mathscr{F}_{\mathrm{evt}}}$. $\mathscr{F}_{\mathrm{evt}}$ is called the **eventuality filter** of the directed set $\mathscr{A}$, and its elements are called **eventual events** or events that **occur eventually** (see for instance Schechter (1997, Section 7.7)). $\lim I_A$ exists if and only if either $A \in \mathscr{F}_{\mathrm{evt}}$ or $A^c \in \mathscr{F}_{\mathrm{evt}}$, that is, if either A or A^c eventually occurs, and then $\lim I_A = 1$ or $\lim I_A = 0$, respectively. We also gather from Proposition $8.8(\mathrm{vi})_{155}$ that a net f converges if and only if either $\{f \geq t\}$ or $\{f < t\}$ eventually occurs for all but a countable number of real t in $[\inf f, \sup f]$ and

$$\liminf f = C_{\liminf}(f) = \inf f + R\int_{\inf f}^{\sup f} \liminf I_{\{f \geq t\}}\,\mathrm{d}t = \sup\left\{t : \{f \geq t\} \in \mathscr{F}_{\mathrm{evt}}\right\}.$$

The discussion in Section 11.6_{225} will tell us that, because lim inf is a lower prevision associated with a proper filter, it is an extreme point of the set of all coherent and completely monotone lower previsions on $\mathbb{B}(\mathscr{A})$.

9.3 Lower and upper oscillations

Consider a topological space $\mathscr{X}$ provided with a collection of open sets $\mathscr{T}$. We focus our attention on some single element x of $\mathscr{X}$. As discussed in Section $5.5.5_{93}$ and in particular Definition 5.14_{94}, we can associate with the neighbourhood filter $\mathscr{N}_x$ of x a lower prevision $\underline{\mathrm{osc}}_x = \underline{P}_{\mathscr{N}_x}$, called the lower oscillation in x. It is coherent and completely monotone by Proposition 7.7_{130}.

A bounded gamble f is, by Equation $(5.11)_{94}$, $\underline{\mathrm{osc}}_x$-integrable if and only if its oscillation $\mathrm{osc}_x(f)$ in x is zero or, in other words, if f is continuous in x:

$$\mathscr{I}_{\underline{\mathrm{osc}}_x}(\mathscr{X}) = \{f \in \mathbb{B}(\mathscr{X}) : f \text{ is continuous in } x\},$$

and because, generally, $\underline{\mathrm{osc}}_x(f) \leq f(x) \leq \overline{\mathrm{osc}}_x(f)$, the $\underline{\mathrm{osc}}_x$-integral coincides with the degenerate linear prevision P_x on its domain $\mathscr{I}_{\underline{\mathrm{osc}}_x}(\mathscr{X})$:

$$E_{\underline{\mathrm{osc}}_x}(f) = f(x) \text{ for all } f \text{ continuous in } x.$$

9.4 Linear extension of a probability measure

Consider a σ-field $\mathscr{F}$ of subsets of $\mathscr{X}$ and a probability measure μ on $\mathscr{F}$.

In particular, μ is a positive bounded charge, so the results of Chapter 8_{151} also apply here. For instance, the lower expectation of a bounded gamble f with respect to μ is given by natural extension, the Choquet integral with respect to the inner set function μ_*, the lower S-integral or the lower Lebesgue integral – all are equal (see Equations $(8.7)_{164}$, $(8.12)_{169}$ and $(8.16)_{171}$):

$$\underline{E}_\mu(f) = C\int f\,\mathrm{d}\mu_* = \underline{S\int} f\,\mathrm{d}\mu = \underline{L\int} f\,\mathrm{d}\mu.$$

If a bounded gamble f is S-integrable, Lebesgue integrable, or Dunford integrable with respect to μ, or μ-integrable (it does not matter which, as all these notions of integrability are equivalent for bounded gambles), then its **expectation** is (see Theorems 8.18_{163} and 8.32_{174})

$$E_\mu(f) = C\int f\,\mathrm{d}\mu_* = S\int f\,\mathrm{d}\mu = L\int f\,\mathrm{d}\mu = D\int f\,\mathrm{d}\mu.$$

Interestingly, this expectation operator associated with a probability measure also satisfies monotone convergence. We first prove that monotone convergence holds for $\mathscr{F}$-measurable bounded gambles and then generalise it to all μ-integrable bounded gambles.

The monotone convergence result for $\mathscr{F}$-measurable bounded gambles is well known in measure theory for the $L\int \bullet\,\mathrm{d}\mu$ form of the expectation operator (see, for instance, Kallenberg, 2002, Theorem 1.19). Our proof uses the $C\int \bullet\,\mathrm{d}\mu_*$ form and is based on an idea by Denneberg (1994, Theorem 8.1).

Theorem 9.2 (Monotone convergence for measurable bounded gambles) *Let μ be a probability measure defined on a σ-field $\mathscr{F}$ of subsets of $\mathscr{X}$. Then the associated linear expectation operator E_μ satisfies monotone convergence on the uniformly closed linear lattice $\mathbb{B}_{\mathscr{F}}$ of all $\mathscr{F}$-measurable bounded gambles.*

Proof. By Proposition 1.26_{19}, it suffices to prove that there is upward monotone convergence. So consider a non-decreasing sequence f_n of $\mathscr{F}$-measurable bounded gambles that converges point-wise to some $\mathscr{F}$-measurable bounded gamble $f := \sup_{n\in\mathbb{N}} f_n$. Because the linear prevision E_μ satisfies constant additivity (Corollary $4.14(\mathrm{iii})_{55}$), we may assume without loss of generality that $0 \leq f_n \leq f$. For any real t, we have that $\{f > t\} = \bigcup_{n\in\mathbb{N}}\{f_n > t\}$, and it follows from the assumptions that the sequence of non-negative and non-increasing maps F_n defined on $\mathbb{R}_{\geq 0}$ by $F_n(t) := \mu\left(\{f_n > t\}\right)$ for all $t \in \mathbb{R}_{\geq 0}$ is non-decreasing. Moreover, if we define the non-negative and non-increasing map F on $\mathbb{R}_{\geq 0}$ by $F(t) := \mu(\{f > t\})$ for all $t \in \mathbb{R}_{\geq 0}$, then we get that

$$F(t) = \mu(\{f > t\}) = \mu\left(\bigcup_{n\in\mathbb{N}}\{f_n > t\}\right) = \lim_{n\to+\infty}\mu(\{f_n > t\}) = \lim_{n\to+\infty}F_n(t),$$

where the third equality follows from the fact that the probability measure μ is σ-additive. So the F_n constitute a non-decreasing sequence of real-valued functions that converges point-wise to F. Now use Theorem 8.18$_{163}$, Proposition C.3(i)$_{379}$ and Proposition C.1$_{377}$ to find that

$$\begin{aligned}
E_\mu(f) = C_{\mu_*}(f) &= R\int_0^{+\infty} \mu_*(\{f > t\})\,\mathrm{d}t = R\int_0^{+\infty} F(t)\,\mathrm{d}t \\
&= \lim_{n\to+\infty} R\int_0^{+\infty} F_n(t)\,\mathrm{d}t = \lim_{n\to+\infty} R\int_0^{+\infty} \mu_*(\{f_n > t\})\,\mathrm{d}t = \lim_{n\to+\infty} C_{\mu_*}(f_n) \\
&= \lim_{n\to+\infty} E_\mu(f_n),
\end{aligned}$$

which shows that E_μ satisfies upward monotone convergence on $\mathbb{B}_{\mathscr{F}}$. □

It is now quite easy to prove a version of Fatou's lemma for $\mathscr{F}$-measurable bounded gambles (see, for instance, Kallenberg, 2002). In the following, lim inf and lim sup on a sequence of bounded gambles are taken point-wise. Specifically, for any bounded sequence f_n of bounded gambles, the bounded gambles $\liminf_{n\to+\infty} f_n$ and $\limsup_{n\to+\infty} f_n$ are defined by

$$\Big(\liminf_{n\to+\infty} f_n\Big)(x) := \liminf_{n\to+\infty} f_n(x) \text{ and } \Big(\limsup_{n\to+\infty} f_n\Big)(x) := \limsup_{n\to+\infty} f_n(x) \text{ for all } x \in \mathscr{X}.$$

Corollary 9.3 (Fatou) *Let μ be a probability measure defined on a σ-field $\mathscr{F}$ of subsets of $\mathscr{X}$. Let g_n and h_n be bounded sequences of $\mathscr{F}$-measurable bounded gambles. Then $g := \liminf_{n\to+\infty} g_n$ and $h := \limsup_{n\to+\infty} h_n$ are $\mathscr{F}$-measurable bounded gambles and, moreover,*

$$E_\mu(g) \leq \liminf_{n\to+\infty} E_\mu(g_n) \text{ and } E_\mu(h) \geq \limsup_{n\to+\infty} E_\mu(h_n).$$

In particular, for any bounded sequence f_n of $\mathscr{F}$-measurable bounded gambles whose point-wise limit exists, $f := \lim f_n$ is $\mathscr{F}$-measurable and, moreover,

$$E_\mu(f) = \lim_{n\to+\infty} E_\mu(f_n).$$

Proof. We give the proof for the g_n, the proof for the h_n is similar (or follows by conjugacy). Let $g'_n := \inf_{m\geq n} g_m$, then for all real t, $\{g'_n \geq t\} = \bigcap_{m\geq n}\{g_m \geq t\}$, so it follows from Proposition 1.19$_{15}$ that the g'_n constitute a non-decreasing sequence of $\mathscr{F}$-measurable bounded gambles. Moreover, $g = \liminf_{n\to+\infty} g_n = \sup_{n\in\mathbb{N}} g'_n = \lim_{n\to+\infty} g'_n$, so we infer from the Monotone Convergence Theorem 9.2 that, because also $E_\mu(g'_n) \leq \inf_{m\geq n} E_\mu(g_m)$ (use Corollary 4.14(iv)$_{55}$),

$$E_\mu(g) = \lim_{n\to+\infty} E_\mu(g'_n) = \sup_{n\in\mathbb{N}} E_\mu(g'_n) \leq \sup_{n\in\mathbb{N}} \inf_{m\geq n} E_\mu(g_m) = \liminf_{n\to+\infty} E_\mu(g_n),$$

which completes the proof. □

These results for $\mathscr{F}$-measurable bounded gambles can now be extended to μ-integrable bounded gambles, and we believe this extension – and its proof – to be new. To see how this works, we first look at the relation between $\mathscr{F}$-measurability and μ-integrability. We infer from Proposition 8.17$_{163}$ that all $\mathscr{F}$-measurable bounded gambles are μ-integrable (S-integrable, Lebesgue integrable, Dunford integrable, with respect to μ). But for σ-fields $\mathscr{F}$, there is also an interesting characterisation of μ-integrable bounded gambles in terms of $\mathscr{F}$-measurable ones.

Theorem 9.4 *Let μ be a probability measure defined on a σ-field $\mathscr{F}$ of subsets of $\mathscr{X}$. Then a bounded gamble f is μ-integrable if and only if there are $\mathscr{F}$-measurable bounded gambles g and h such that* $\inf f \leq g \leq f \leq h \leq \sup f$ *and* $E_\mu(g) = E_\mu(h)$.

Proof. 'if'. It follows from the coherence of $\underline{E}_\mu$ and $\overline{E}_\mu$ (Theorem 4.13(iv)$_{53}$) that both $E_\mu(g) \leq \underline{E}_\mu(f) \leq E_\mu(h)$ and $E_\mu(g) \leq \overline{E}_\mu(f) \leq E_\mu(h)$, and therefore $\underline{E}_\mu(f) = \overline{E}_\mu(f)$, so $f \in \mathscr{I}_\mu$.

'only if'. We show that there is an $\mathscr{F}$-measurable bounded gamble g such that $\inf f \leq g \leq f$ and $E_\mu(g) = E_\mu(f)$. The proof for the h is then completely similar (or based on conjugacy). It will suffice to show that there is a non-decreasing sequence of $\mathscr{F}$-measurable bounded gambles g_n such that

$$\inf f \leq g_n \leq f \text{ and } E_\mu(g_n) \leq E_\mu(f) < E_\mu(g_n) + \frac{1}{n} \text{ for all } n \in \mathbb{N}. \tag{9.1}$$

Indeed, in that case, we will know that the g_n converge point-wise to the $\mathscr{F}$-measurable bounded gamble $g := \sup_{n \in \mathbb{N}} g_n$ with $\inf f \leq g \leq f$, so we will be able to infer from the Monotone Convergence Theorem 9.2$_{184}$ that $E_\mu(g) = \lim_{n \to +\infty} E_\mu(g_n)$. Taking the limit for $n \to +\infty$ in Equation (9.1) will then yield that $E_\mu(g) \leq E_\mu(f) \leq E_\mu(g)$ and therefore that indeed $E_\mu(g) = E_\mu(f)$.

We give a proof by induction. Clearly, the statement holds for $n = 0$: simply take $g_0 := \inf f$ and note that a constant gamble is always $\mathscr{F}$-measurable.

Next, suppose that the statement holds for $n = m$. We show that it also holds for $n = m + 1$. We infer from Equation (8.5)$_{163}$ that there is some $g'_{m+1} \in \mathbb{B}_{\mathscr{F}}$ such that

$$g'_{m+1} \leq f \text{ and } E_\mu(g'_{m+1}) \leq E_\mu(f) < E_\mu(g'_{m+1}) + \frac{1}{m+1}.$$

Let $g_{m+1} := g'_{m+1} \vee g_m$, then it follows from the assumptions that g_{m+1} is $\mathscr{F}$-measurable, that $\inf f \leq g_m \leq g_{m+1}$ and that $g'_{m+1} \leq g_{m+1} \leq f$. So it follows from the monotonicity of the linear prevision E_μ (Corollary 4.14(iv)$_{55}$) that $E_\mu(g'_{m+1}) \leq E_\mu(g_{m+1}) \leq E_\mu(f)$ and therefore also that

$$E_\mu(g_{m+1}) \leq E_\mu(f) < E_\mu(g_{m+1}) + \frac{1}{m+1},$$

which completes the proof. □

This characterisation of μ-integrability now allows for an easy proof of monotone convergence for all μ-integrable bounded gambles.

Theorem 9.5 (Monotone convergence for integrable bounded gambles) *Let μ be a probability measure defined on a σ-field $\mathscr{F}$ of subsets of $\mathscr{X}$. Then the associated linear expectation operator E_μ satisfies monotone convergence on the uniformly closed linear lattice $\mathscr{I}_\mu$ of all μ-integrable bounded gambles.*

Proof. By Proposition 1.26_{19}, it suffices to prove that there is upward monotone convergence. So consider a non-decreasing sequence f_n of μ-integrable bounded gambles that converges point-wise to some μ-integrable bounded gamble $f := \sup_{n\in\mathbb{N}} f_n$. Because the linear prevision E_μ satisfies constant additivity (Corollary 4.14(iii)$_{55}$), we may assume without loss of generality that $0 \le f_n \le f$. $f_n \in \mathscr{I}_\mu$ implies by Theorem 9.4 that there are $\mathscr{F}$-measurable bounded gambles g_n and h_n such that $0 \le \inf f_n \le g_n \le f_n \le h_n \le \sup f_n \le \sup f$ such that $E_\mu(g_n) = E_\mu(h_n) = E_\mu(f_n)$. Hence, also $0 \le \limsup_{n\to+\infty} g_n \le f \le \liminf_{n\to+\infty} h_n \le \sup f$, and therefore, invoking the monotonicity of the linear prevision E_μ (Corollary 4.14(iv)$_{55}$) and Corollary 9.3$_{185}$:

$$
\begin{aligned}
\lim_{n\to+\infty} E_\mu(f_n) &= \limsup_{n\to+\infty} E_\mu(f_n) = \limsup_{n\to+\infty} E_\mu(g_n) \\
&\le E_\mu\left(\limsup_{n\to+\infty} g_n\right) \le E_\mu(f) \le E_\mu\left(\liminf_{n\to+\infty} h_n\right) \\
&\le \liminf_{n\to+\infty} E_\mu(h_n) = \liminf_{n\to+\infty} E_\mu(f_n) = \lim_{n\to+\infty} E_\mu(f_n),
\end{aligned}
$$

whence indeed $\lim_{n\to+\infty} E_\mu(f_n) = E_\mu(f)$. □

9.5 Extending a linear prevision from continuous bounded gambles

In Section 5.5.6$_{98}$, we have considered the natural extension $\underline{E}_{\underline{P}}$ of a coherent lower prevision $\underline{P}$ defined on the set $\mathscr{C}$ of all continuous bounded gambles on a compact topological space $\mathscr{X}$. Here, we concentrate on the special case that the coherent lower prevision $\underline{P}$ on the linear lattice $\mathscr{C}$ is a linear prevision P. In that case, the natural extension $\underline{E}_P$ is a completely monotone coherent lower prevision on $\mathbb{B}$, by Theorems 6.6$_{105}$ and 6.19$_{114}$. If we also assume that $\mathscr{X}$ is compact and metrisable, then Proposition 5.21$_{100}$ and Corollary 5.22$_{100}$ specialise to the following result.

Corollary 9.6 *Let $\mathscr{X}$ be a compact and metrisable topological space, and let P be a linear prevision on $\mathscr{C}$. Then P has a unique extension $\hat{P}$ to $\underline{\mathscr{C}}$ that satisfies upward monotone convergence. This unique extension $\hat{P}$ coincides with the natural extension $\underline{E}_P$ on $\underline{\mathscr{C}}$ and is, therefore, also the point-wise smallest coherent lower prevision that extends P to $\underline{\mathscr{C}}$. Finally, for any bounded gamble f on $\mathscr{X}$, $\underline{E}_P(f) = \hat{P}(\underline{\mathrm{osc}}(f))$.*

From Proposition 8.8(vi)$_{155}$, we infer that the set $\mathscr{I}_P$ of P-integrable bounded gambles is a uniformly closed linear lattice. We can make an even stronger claim if we recall the *Riesz representation theorem* (see, for instance, Schechter (1997, Section 29.35)), which guarantees that there is a *unique* probability measure μ_P defined on the σ-field $\mathscr{B}$ of all Borel subsets of $\mathscr{X}$ such that

$$P(f) = L\int f\,\mathrm{d}\mu_P \text{ for all } f \in \mathscr{C}, \tag{9.2}$$

where the integral is the Lebesgue integral associated with the probability measure μ_P, as discussed in Section 9.4_{183}. We have seen there that we can consider the linear expectation E_{μ_P} associated with this probability measure, which can be identified with the Lebesgue integral L_{μ_P} defined on the uniformly closed linear lattice $\mathscr{I}_{\mu_P}$ of all μ_P-integrable bounded gambles by

$$L_{\mu_P}(f) := L\int f\,\mathrm{d}\mu_P \text{ for all } f \in \mathscr{I}_{\mu_P}.$$

As any lower semi-continuous bounded gamble f has open strict level sets $\{f > t\}$, it is Borel measurable and, therefore, also μ_P-integrable, by Proposition 8.17_{163}: $\underline{\mathscr{C}} \subseteq \mathscr{I}_{\mu_P}$. As we can also infer from Theorem 9.5_{186} that the expectation operator L_{μ_P} satisfies monotone convergence on $\mathscr{I}_{\mu_P}$ and extends P by Equation $(9.2)_{\frown}$, it must, by Corollary $9.6_{\frown}$, coincide with $\hat{P}$ on $\underline{\mathscr{C}}$ and, moreover,

$$\underline{E}_P(f) = L_{\mu_P}(\underline{\mathrm{osc}}(f)) = L\int \underline{\mathrm{osc}}(f)\,\mathrm{d}\mu_P \text{ for all bounded gambles } f \text{ on } \mathscr{X}. \quad (9.3)$$

This also leads to a characterisation of the P-integrable bounded gambles in terms of the Riesz measure μ_P (see also De Cooman and Miranda, 2008):

Proposition 9.7 *Let $\mathscr{X}$ be a compact and metrisable topological space, and let P be a linear prevision on $\mathscr{C}$. Then $\mathscr{I}_P$ is the set of all bounded gambles that are almost surely continuous (μ_P): for any bounded gamble f on $\mathscr{X}$,*

$$f \in \mathscr{I}_P \Leftrightarrow L\int \mathrm{osc}(f)\,\mathrm{d}\mu_P = 0 \Leftrightarrow \mu_P(\{\mathrm{osc}(f) > 0\}) = 0.$$

Proof. For the first equivalence, use Equation (9.3), conjugacy and Equation $(5.11)_{94}$ to find that

$$\overline{E}_P(f) - \underline{E}_P(f) = L\int \overline{\mathrm{osc}}(f)\,\mathrm{d}\mu_P - L\int \underline{\mathrm{osc}}(f)\,\mathrm{d}\mu_P = L\int \mathrm{osc}(f)\,\mathrm{d}\mu_P.$$

(Recall that both lower and upper semi-continuous bounded gambles are Borel measurable.)

The second equivalence is a standard result in measure theory (see, for instance, Kallenberg, 2002, Lemma 1.23). It can also be derived directly from the discussion of null sets and gambles in Section 14.1_{305}: start with Definition 14.6_{307} and Proposition $14.14(\mathrm{i})_{310}$ and use the fact that the measure μ_P satisfies monotone convergence. □

9.6 Induced lower previsions and random sets

In this section, we take up the results about induced lower previsions and belief functions in Section 7.6_{138} and briefly discuss their extension to infinite spaces. We refer back to that section for motivation and notation.

Consider a probability charge μ defined on a field $\mathscr{F}$ of subsets of $\mathscr{Y}$ and a multi-valued map $\Gamma : \mathscr{Y} \to \mathscr{P}(\mathscr{X})$. We have seen in Chapter 8 – and also in Section 9.4_{183} for the special case that μ is a probability measure – that there is a way to extend μ, or rather the linear prevision P_μ induced by it, to a coherent lower prevision $\underline{E}_\mu$ defined on all bounded gambles: through natural extension.

This implies that we can look at the *induced lower prevision* $(P_\mu)_\bullet$, given by

$$(P_\mu)_\bullet(f) = \underline{E}_\mu(f_\bullet) = C\int f_\bullet \,\mathrm{d}\mu_* \text{ for all bounded gambles } f \text{ on } \mathscr{X},$$

where we have used Theorem 8.18_{163} and, as before, $f_\bullet(y) := \inf_{x\in\Gamma(y)} f(x)$ for all $y \in \mathscr{Y}$, and μ_* is the inner set function of μ. For events, this reduces to

$$(P_\mu)_\bullet(A) = \mu_*(A_\bullet) \text{ for all } A \subseteq \mathscr{X},$$

where, as before, $A_\bullet := \{y \in \mathscr{Y} : \Gamma(y) \subseteq A\}$.

As $\underline{E}_\mu$ is completely monotone by Theorem 8.16_{162}, so is the induced lower prevision $(P_\mu)_\bullet$, by Proposition 7.10_{136}.

It may be of some interest (but by no means essential or required) to look for those bounded gambles f for which $f_\bullet$ is μ-integrable, and those events A for which $A_\bullet$ is μ_*-integrable, because then

$$(P_\mu)_\bullet(f) = E_\mu(f_\bullet) = L\int f_\bullet \,\mathrm{d}\mu \text{ and } (P_\mu)_\bullet(A) = \mu(A_\bullet), \tag{9.4}$$

so we can use the S-integral, or equivalently the Lebesgue integral, associated with the probability measure μ to calculate the induced lower prevision $(P_\mu)_\bullet(f)$, see Theorems 8.25_{169} and 8.18_{163} and Section 8.6_{171}. Indeed, this is the approach taken originally by Straßen (1964) and Dempster (1967a,b) to define lower probabilities and expectations induced using a multivalued map.

Proposition 9.8 *For any bounded gamble f on $\mathscr{X}$, the bounded gamble $f_\bullet$ is μ-integrable if and only if the sets $\{f \geq t\}_\bullet$ are for all but a countably infinite number of real numbers t.*

Proof. Observe that $\{f_\bullet \geq t\} = \{f \geq t\}_\bullet$,

$$y \in \{f_\bullet \geq t\} \Leftrightarrow f_\bullet(y) \geq t \Leftrightarrow (\forall x \in \Gamma(y)) f(x) \geq t$$
$$\Leftrightarrow \Gamma(y) \subseteq \{f \geq t\} \Leftrightarrow y \in \{f \geq t\}_\bullet,$$

and use Theorem 8.18_{163}. □

Incidentally, we have for any bounded gamble f on $\mathscr{X}$ that

$$\inf f_\bullet = \inf_{y\in\mathscr{Y}} \inf_{x\in\Gamma(y)} f(x) \geq \inf_{x\in\mathscr{X}} f(x) = \inf f$$

and

$$\sup f_\bullet = \sup_{y\in\mathscr{Y}} \inf_{x\in\Gamma(y)} f(x) \leq \sup_{y\in\mathscr{Y}} \sup_{x\in\Gamma(y)} f(x) \leq \sup_{x\in\mathscr{X}} f(x) = \sup f,$$

and therefore

$$
\begin{aligned}
C\int f_\bullet \,\mathrm{d}\mu_* &= \inf f_\bullet + R\int_{\inf f_\bullet}^{\sup f_\bullet} \mu_*\left(\{f_\bullet \geq t\}\right)\mathrm{d}t = \inf f + R\int_{\inf f}^{\sup f} \mu_*\left(\{f_\bullet \geq t\}\right)\mathrm{d}t \\
&= \inf f + R\int_{\inf f}^{\sup f} \mu_*(\{f \geq t\}_\bullet)\,\mathrm{d}t = \inf f + R\int_{\inf f}^{\sup f} (P_\mu)_\bullet(\{f \geq t\})\,\mathrm{d}t \\
&= C\int f\,\mathrm{d}(P_\mu)_\bullet,
\end{aligned}
$$

which provides, for this particular case, an alternative proof for the second part of Proposition 7.10$_{136}$. If $f_\bullet$ is μ-integrable, we can use Equation (9.4)$_\curvearrowleft$ to rewrite the chain of equalities mentioned earlier as follows:

$$
L\int f_\bullet \,\mathrm{d}\mu = \inf f + R\int_{\inf f}^{\sup f} (P_\mu)_\bullet(\{f \geq t\})\,\mathrm{d}t = C\int f\,\mathrm{d}(P_\mu)_\bullet.
$$

In the special case that μ is a probability *measure* μ defined on a σ-field $\mathscr{F}$ of subsets of $\mathscr{Y}$, this is, in essence, a result mentioned several times by Wasserman (1990a,b), see also Miranda et al. (2010, Theorem 14).

10

Lower previsions and symmetry

Throughout the development of probability theory, considerations of symmetry have always been of crucial importance: think about Laplace's historically very important classical definition of probability, which seeks to divide the space of all possible outcomes into 'equiprobable' regions (see, for instance, Laplace, 1951), or the notion of exchangeability that de Finetti (1937, 1974–1975) developed to provide (what is now called) Bayesian inference with a better behavioural and operational foundation, or the notions of stationarity and ergodicity (see, for instance, Kallenberg, 2005) that were at some point deemed crucial for the development of statistical mechanics.

For this reason, it behooves us to pay attention to the notion of symmetry in the context of the generalised probability models that we study in this book. Two questions come to the fore and are addressed in this chapter: (i) if a subject believes that there is some basic symmetry lurking behind a random variable X, how can he represent this type of *structural assessment* using lower previsions, and (ii) how can he combine this structural symmetry assessment with *local assessments*, which assign a lower prevision to certain bounded gambles?

Our developments here are mainly based on the work by De Cooman and Miranda (2007), but we have streamlined the presentation and simplified many of the proofs. Indeed, unless we explicitly state the contrary further on, the reader will find earlier statements of results and proofs for this chapter in that paper, where the authors extended the groundwork laid by Walley (1991, Section 3.5) for the developments in this field.

We begin by providing a systematic and detailed answer to the two questions mentioned earlier in Section 10.1$_{\curvearrowright}$, where we introduce the strong invariance of a lower prevision with respect to a monoid of transformations coding for a specific type of symmetry. This discussion can be taken considerably further when the symmetry monoid satisfies what we call the Moore–Smith property. This is

Lower Previsions, First Edition. Matthias C.M. Troffaes and Gert de Cooman.

explained in some detail in Section 10.2_{200}. The rest of this chapter is devoted to important examples: permutation symmetry – with exchangeability as a special case – in Section $10.3.1_{205}$, shift invariance in Section $10.3.2_{208}$ and stationarity in Section $10.3.3_{210}$.

10.1 Invariance for lower previsions

10.1.1 Definition

We consider a **monoid** $\mathbb{T}$ of transformations T of $\mathcal{X}$, meaning that (i) $\mathbb{T}$ is closed under composition, $ST := S \circ T \in \mathbb{T}$ for all $S, T \in \mathbb{T}$, and (ii) the identity transformation $\mathrm{id}_{\mathcal{X}}$ belongs to $\mathbb{T}$.

We can lift any transformation $T \in \mathbb{T}$ to a *linear transformation* T^t on the linear space $\mathbb{B}$ by letting $T^t f := f \circ T$ for any bounded gamble f on $\mathcal{X}$. In other words, we have $T^t f(x) = f(Tx)$ for all $x \in \mathcal{X}$. Observe that $(ST)^t = T^t S^t$ for all $S, T \in \mathbb{T}$, which explains why we use the 'transposition' superscript $\bullet^t$. We denote by $\mathbb{T}^t$ the set of all such T^t: $\mathbb{T}^t := \{T^t : T \in \mathbb{T}\}$.

Definition 10.1 (Invariant lower previsions) *Let $\mathbb{T}$ be a monoid of transformations of $\mathcal{X}$. We call a lower prevision $\underline{P}$ on $\mathbb{B}$:*

(i) ***weakly** $\mathbb{T}$**-invariant** if $\underline{P}(T^t f) \geq \underline{P}(f)$ for all $f \in \mathbb{B}$ and all $T \in \mathbb{T}$; and*

(ii) ***strongly** $\mathbb{T}$**-invariant** if $\underline{P}(T^t f - f) \geq 0$ for all $f \in \mathbb{B}$ and all $T \in \mathbb{T}$.*

Our definition of invariance is less restrictive than Walley's (1991, Section 3.5.1, p. 140, ll. 1–2), which requires the equality $\underline{P}(T^t f) = \underline{P}(f)$ rather than an inequality. We prefer our version for reasons that are mainly mathematical: (i) it seems more productive and leads to a more elegant formulation of results – for one thing, Walley's invariance would not be preserved by natural extension (see Proposition 10.3_{194}) – and (ii) because it can be more closely connected to the common definition of invariance of a set under linear transformation.[1]

Because the domain of the lower prevision $\underline{P}$ is negation-invariant, and $f - T^t f = T^t(-f) - (-f)$, the condition for strong $\mathbb{T}$-invariance is also equivalent to $\underline{P}(f - T^t f) \geq 0$ for all $f \in \mathbb{B}$ and all $T \in \mathbb{T}$.

Moreover, it is clear from their mixed sub/super-additivity properties (see Theorem 4.13(v)$_{53}$) that for *coherent* lower previsions $\underline{P}$, strong invariance implies invariance: $\underline{P}(T^t f - f) \leq \underline{P}(T^t f) + \overline{P}(-f) = \underline{P}(T^t f) - \underline{P}(f)$.

For coherent lower previsions $\underline{P}$, strong $\mathbb{T}$-invariance is also equivalent with the following condition:

$$\underline{P}(T^t f - f) = \overline{P}(T^t f - f) = 0 \text{ for all } f \in \mathbb{B} \text{ and all } T \in \mathbb{T}.$$

We emphasise that a strongly invariant lower prevision is a *model of symmetry*: if a subject believes that there is symmetry lurking behind the variable X (captured by the transformations in the monoid $\mathbb{T}$), then he should be **indifferent** between the

[1] Refer to De Cooman and Miranda (2007, Section 4.1) for a more detailed discussion on this issue.

bounded gambles f and $T^t f$: he should be willing to exchange f for $T^t f$ in return for any arbitrarily small amount of utility (which is expressed by $\underline{P}(T^t f - f) \geq 0$), and the other way around (as expressed by $\underline{P}(f - T^t f) \geq 0$), leading to the two equalities mentioned earlier.

If, on the other hand, a coherent lower prevision is merely $\mathbb{T}$-invariant and not strongly so, this means that the subject's beliefs are in some sense invariant: there is a *symmetry of beliefs* (or of the model reflecting them) rather than a *belief of symmetry* (captured in a model of symmetry).

It is clear from the definition that if a lower prevision is strongly invariant, then all the dominating lower previsions are as well: an assessment of strong invariance will not be lost if additional assessments are made that tend to make the model more precise – raise the lower prevision.

For linear previsions, there is no difference between invariance and strong invariance. Moreover, we can characterise the strong invariance of a coherent lower prevision in terms of the invariance of its dominating linear previsions.

Proposition 10.2 *Let $\mathbb{T}$ be a monoid of transformations of $\mathcal{X}$, and let P be a linear prevision on $\mathbb{B}$. Then the following statements are equivalent:*

(i) *P is $\mathbb{T}$-invariant;*

(ii) *P is strongly $\mathbb{T}$-invariant;*

(iii) *$P(T^t f) = P(f)$ for all $f \in \mathbb{B}$ and all $T \in \mathbb{T}$;*

(iv) *$P(A) = P(T^{-1}(A))$ for all $A \subseteq \mathcal{X}$ and all $T \in \mathbb{T}$.*[2]

Moreover, any linear prevision that dominates a strongly $\mathbb{T}$-invariant coherent lower prevision is (strongly) $\mathbb{T}$-invariant. Finally, a coherent lower prevision $\underline{P}$ on $\mathbb{B}$ is strongly $\mathbb{T}$-invariant if and only if all its dominating linear previsions are (strongly) $\mathbb{T}$-invariant.

Proof. We begin with a proof of the equivalences. That (i)⇔(ii)⇔(iii) follows at once from the linearity of P.

(iii)⇒(iv). Observe that $T^t I_A = I_{T^{-1}(A)}$ for all events $A \subseteq \mathcal{X}$.

(iv)⇒(iii). We proceed in a number of steps. First of all, for any simple gamble $f = \sum_{k=1}^n \lambda_k I_{A_k}$, we derive from the linearity of T^t that $T^t f = \sum_{k=1}^n \lambda_k I_{T^{-1}(A_k)}$ and, therefore, using the linearity of P, $P(T^t f) = \sum_{k=1}^n \lambda_k P(T^{-1}(A_k)) = \sum_{k=1}^n \lambda_k P(A_k) = P(f)$. So we see that the desired equality holds for simple gambles. Next, because any bounded gamble is a uniform limit of simple gambles (see Proposition 1.20_{15}), and linear previsions are continuous with respect to the topology of uniform convergence (Corollary $4.14(\text{xii})_{55}$), we are done if we can show that the linear transformation T^t is continuous with respect to this topology as well. So assume that we have a sequence of bounded gambles f_n and a bounded gamble f such that $\lim_{n\to+\infty} \sup |f_n - f| = 0$. As

$$0 \leq \sup |T^t f_n - T^t f| = \sup_{x\in\mathcal{X}} |f_n(Tx) - f(Tx)| \leq \sup_{x\in\mathcal{X}} |f_n(x) - f(x)| = \sup |f_n - f|,$$

we see that $\lim_{n\to+\infty} \sup |T^t f_n - T^t f| = 0$ as well.

[2] This minor part of the proposition is new relative to De Cooman and Miranda (2007).

The proof of the remaining statements is immediate, if we take into account the Lower Envelope Theorem 4.38_{71}. □

We conclude from this that linear previsions – precise probability models – cannot distinguish between symmetry of beliefs and beliefs of symmetry. It is this failure that we believe is responsible for the havoc wreaked by the *Principle of Indifference*: complete ignorance about X is symmetrical with respect to any type of transformation of $\mathcal{X}$, so a model representing such ignorance should be invariant with respect to any such transformation. Invariant indeed, but not strongly so! Complete ignorance about a phenomenon cannot be invoked to justify using a strongly invariant model, which embodies a positive belief that there is some symmetry behind the phenomenon that produces the observation X. If we restrict ourselves models that are linear previsions, there is no difference between invariance and strong invariance, so using an invariant model – because it represents symmetry of beliefs – automatically leads to using a strongly invariant model, representing beliefs of symmetry. On our view, the solution should be clear: linear previsions should not be used as models for complete ignorance, because they imply for too many commitments. Rather, we should use the vacuous lower prevision inf, which can be shown to be the only coherent lower prevision that is invariant with respect to the monoid of *all* transformations. For a much more extensive argumentation of this point of view, refer to De Cooman and Miranda (2007).

10.1.2 Existence of invariant lower previsions

For any monoid $\mathbb{T}$, there always are $\mathbb{T}$-invariant coherent lower previsions: the vacuous lower prevision inf is guaranteed to be one of them. This is also related to the following simple results, which guarantees that natural extension preserves invariance. We shall see further on that this does not extend to *strong* invariance.

Proposition 10.3 *Consider a lower prevision $\underline{P}$ that is $\mathbb{T}$-**invariant** in the sense that*

IN1. $T^t f \in \operatorname{dom} \underline{P}$ *for all* $f \in \operatorname{dom} \underline{P}$;

IN2. $\underline{P}(T^t f) \geq \underline{P}(f)$ *for all* $f \in \operatorname{dom} \underline{P}$.

If $\underline{P}$ avoids sure loss, then its natural extension $\underline{E}_{\underline{P}}$ is $\mathbb{T}$-invariant and is, therefore, the point-wise smallest $\mathbb{T}$-invariant coherent lower prevision on $\mathbb{B}$ that dominates $\underline{P}$.

Proof. Consider any bounded gamble f on $\mathcal{X}$ and any T in $\mathbb{T}$. We infer from the expression $(4.10\text{a})_{68}$ for natural extension and IN1 that

$$\underline{E}_{\underline{P}}(T^t f) = \sup_{\substack{\lambda_k \geq 0, f_k \in \mathcal{K}\\ n \geq 0}} \left\{ \alpha : T^t f - \alpha \geq \sum_{k=1}^{n} \lambda_k \left[f_k - \underline{P}(f_k) \right] \right\}$$

$$\geq \sup_{\substack{\lambda_k \geq 0, g_k \in \mathcal{K}\\ n \geq 0}} \left\{ \alpha : T^t f - \alpha \geq \sum_{k=1}^{n} \lambda_k \left[T^t g_k - \underline{P}(T^t g_k) \right] \right\}. \quad (10.1)$$

Now it follows from IN2 that $\underline{P}(T^t g_k) \geq \underline{P}(g_k)$, whence

$$\sum_{k=1}^{n} \lambda_k \left[T^t g_k - \underline{P}(T^t g_k)\right] \leq T^t \sum_{k=1}^{n} \lambda_k \left[g_k - \underline{P}(g_k)\right],$$

and, consequently, $f - \alpha \geq \sum_{k=1}^{n} \lambda_k [g_k - \underline{P}(g_k)]$ implies that

$$T^t f - \alpha \geq T^t \sum_{k=1}^{n} \lambda_k \left[g_k - \underline{P}(g_k)\right] \geq \sum_{k=1}^{n} \lambda_k \left[T^t g_k - \underline{P}(T^t g_k)\right].$$

So we may infer from the inequality (10.1) that indeed

$$\underline{E}_{\underline{P}}(T^t f) \geq \sup_{\substack{\lambda_k \geq 0, g_k \in \mathscr{K} \\ n \geq 0}} \left\{ \alpha : f - \alpha \geq \sum_{k=1}^{n} \lambda_k \left[g_k - \underline{P}(g_k)\right] \right\} = \underline{E}_{\underline{P}}(f).$$

We conclude that $\underline{E}_{\underline{P}}$ is indeed $\mathbb{T}$-invariant. The rest of the proof is now immediate, because the natural extension is the point-wise smallest dominating coherent lower prevision by Theorem 4.26(i)$_{65}$. □

10.1.3 Existence of strongly invariant lower previsions

We now turn to the slightly more involved question of whether there are strongly invariant coherent lower previsions that dominate a given lower prevision. Owing to the dominance and lower envelope results in Proposition 10.2$_{193}$, this question is very closely linked with that of the existence of an invariant linear prevision dominating that lower prevision.

A *convex combination* T^* of elements of $\mathbb{T}^t$ is a transformation of $\mathbb{B}$ of the form

$$T^* := \sum_{k=1}^{n} \lambda_k T_k^t,$$

where $n \geq 1$, $\lambda_1, \ldots, \lambda_n$ are non-negative real numbers that sum to one and, of course, $T^* f := \sum_{k=1}^{n} \lambda_k T_k^t f$ for all bounded gambles f on $\mathcal{X}$. It is not difficult to verify that all such transformations are both linear and continuous with respect to the topology of uniform convergence on $\mathbb{B}$. We denote by $\mathbb{T}^*$ the set of all convex combinations of elements of $\mathbb{T}^t$. For any two elements $T_1^* := \sum_{\ell=1}^{m} \lambda_\ell U_\ell^t$ and $T_2^* := \sum_{k=1}^{n} \mu_k V_k^t$ of $\mathbb{T}^*$, their composition

$$T_2^* T_1^* = \sum_{k=1}^{n} \mu_k V_k^t \left(\sum_{\ell=1}^{m} \lambda_\ell U_\ell^t \right) = \sum_{k=1}^{n} \sum_{\ell=1}^{m} \mu_k \lambda_\ell V_k^t U_\ell^t = \sum_{k=1}^{n} \sum_{\ell=1}^{m} \lambda_\ell \mu_k (U_\ell V_k)^t$$

again belongs to $\mathbb{T}^*$. This implies that $\mathbb{T}^*$ is a monoid of continuous linear transformations of $\mathbb{B}$.

Proposition 10.4 *Let $\mathbb{T}$ be a monoid of transformations of $\mathcal{X}$, and let $\underline{P}$ be a coherent lower prevision on $\mathbb{B}$. Consider any $T^* \in \mathbb{T}^*$, then the lower prevision $T^* \underline{P}$ on $\mathbb{B}$, defined by $T^* \underline{P}(f) := \underline{P}(T^* f)$ for all bounded gambles f on $\mathcal{X}$, is coherent.*

Proof. We use the characterisation in Theorem 4.15_{56} for coherence on linear spaces. We let, with obvious notations, $T^* := \sum_{k=1}^{n} \lambda_k T_k^t$.

$(\mathrm{i})_{56}$. Consider any bounded gamble f on $\mathcal{X}$, then

$$T^*\underline{P}(f) = \underline{P}(T^*f) \geq \inf(T^*f) = \inf \sum_{k=1}^{n} \lambda_k T_k^t f \geq \sum_{k=1}^{n} \lambda_k \inf T_k^t f \geq \sum_{k=1}^{n} \lambda_k \inf f = \inf f,$$

where the first inequality follows from the coherence of $\underline{P}$ (Theorem $4.13(\mathrm{i})_{53}$) and the last one from $\inf(T_k^t f) = \inf(f \circ T_k) \geq \inf f$.

$(\mathrm{ii})_{57}$. Consider any bounded gamble f on $\mathcal{X}$ and any real $\lambda \geq 0$, then

$$T^*\underline{P}(\lambda f) = \underline{P}(T^*(\lambda f)) = \underline{P}(\lambda T^* f) = \lambda \underline{P}(T^* f) = \lambda T^* \underline{P}(f),$$

where the second equality follows from the linearity of the transformation T^* and the third one from the coherence of $\underline{P}$ (Theorem $4.13(\mathrm{vi})_{53}$).

$(\mathrm{iii})_{57}$. Consider any bounded gambles f_1 and f_2 on $\mathcal{X}$, then

$$\begin{aligned} T^*\underline{P}(f_1 + f_2) &= \underline{P}(T^*(f_1 + f_2)) = \underline{P}(T^* f_1 + T^* f_2) \\ &\geq \underline{P}(T^* f_1) + \underline{P}(T^* f_2) = T^*\underline{P}(f_1) + T^*\underline{P}(f_2), \end{aligned}$$

where the second equality follows from the linearity of the transformation T^* and the inequality from the coherence of $\underline{P}$ (Theorem $4.13(\mathrm{v})_{53}$). □

We can characterise invariance and strong invariance also in terms of the elements of $\mathbb{T}^*$ rather than $\mathbb{T}$.

Proposition 10.5 *Let $\mathbb{T}$ be a monoid of transformations of $\mathcal{X}$, and let $\underline{P}$ be a coherent lower prevision on $\mathbb{B}$.*

(i) *$\underline{P}$ is $\mathbb{T}$-invariant if and only if $\underline{P}(T^*f) \geq \underline{P}(f)$ for all $f \in \mathbb{B}$ and all $T^* \in \mathbb{T}^*$.*

(ii) *$\underline{P}$ is strongly $\mathbb{T}$-invariant if and only if $\underline{P}(T^*f - f) \geq 0$ for all $f \in \mathbb{B}$ and all $T^* \in \mathbb{T}^*$.*

Proof. We only prove the second statement, as an example. The proof of the first statement is similar. It clearly suffices to prove the direct implications. Consider an arbitrary $T^* = \sum_k \lambda_k T_k \in \mathbb{T}^*$. For any bounded gamble f on $\mathcal{X}$,

$$\underline{P}(T_k^* f - f) = \underline{P}\left(\sum_k \lambda_k T_k^t f - f\right) = \underline{P}\left(\sum_k \lambda_k (T_k^t f - f)\right) \geq \sum_k \lambda_k \underline{P}(T_k^t f - f) \geq 0,$$

where the first inequality follows from the coherence of $\underline{P}$ (see for instance coherence criterion $(\mathrm{F})_{48}$) and the second one from the strong $\mathbb{T}$-invariance of $\underline{P}$. □

If we consider the following linear (sub)space of bounded gambles

$$\mathcal{D}_{\mathbb{T}} := \operatorname{span}\left(\{T^t f - f : f \in \mathbb{B}, T \in \mathbb{T}\}\right) = \operatorname{span}\left(\{T^* f - f : f \in \mathbb{B}, T^* \in \mathbb{T}^*\}\right),$$

then we can give the following alternative characterisation for the strong invariance of a coherent lower prevision.

Proposition 10.6 *Let $\mathbb{T}$ be a monoid of transformations of $\mathcal{X}$. Then a coherent lower prevision $\underline{P}$ on $\mathbb{B}$ is strongly $\mathbb{T}$-invariant if and only if $\underline{P}(g) = \overline{P}(g) = 0$ for all $g \in \mathcal{D}_{\mathbb{T}}$.*

Proof. It clearly suffices to prove that the condition is necessary. Consider any $g \in \mathcal{D}_{\mathbb{T}}$, then there are $n \geq 0$, λ_1, ..., λ_n in $\mathbb{R}$, f_1, ..., f_n in $\mathbb{B}$ and T_1, ..., T_n in $\mathbb{T}$ such that $g = \sum_{k=1}^n \lambda_k(T_k^t f_k - f_k)$. We may clearly assume, without loss of generality, that all λ_k are non-negative (for negative λ_k, simply observe that we may rewrite $\lambda_k(T_k^t f_k - f_k)$ as $(-\lambda_k)[T_k^t(-f_k) - (-f_k)]$). By the non-negative homogeneity and super-additivity of the coherent $\underline{P}$ (see for instance coherence criterion $(\mathrm{F})_{48}$), $\underline{P}(g) \geq \sum_{k=1}^n \lambda_k \underline{P}(T_k^t f_k - f_k) \geq 0$, where the last inequality follows from the $\mathbb{T}$-invariance of $\underline{P}$. As also $-g \in \mathcal{D}_{\mathbb{T}}$, we infer by a similar argument that $-\overline{P}(g) = \underline{P}(-g) \geq 0$. We conclude that $0 \leq \underline{P}(g) \leq \overline{P}(g) \leq 0$, where the second inequality follows from the coherence of $\underline{P}$ (Theorem $4.13(\mathrm{i})_{53}$). □

We can use the notions and results introduced earlier to answer the existence question: is there a coherent lower prevision that is strongly invariant and dominates a given coherent lower prevision on its domain?

Theorem 10.7 (Invariant natural extension) *Consider any monoid $\mathbb{T}$ of transformations of $\mathcal{X}$ and any lower prevision $\underline{P}$. There is a strongly $\mathbb{T}$-invariant coherent lower prevision on $\mathbb{B}$ that dominates $\underline{P}$ if and only if*

$$\sup\left(g + \sum_{k=1}^{n} \lambda_k \left[f_k - \underline{P}(f_k)\right]\right) \geq 0$$

for all $g \in \mathcal{D}_{\mathbb{T}}$, $n \geq 0$, f_1, ..., f_n in $\operatorname{dom} \underline{P}$ and non-negative λ_1, ..., λ_n in $\mathbb{R}$. (10.2)

In that case, there is a point-wise smallest strongly $\mathbb{T}$-invariant coherent lower prevision $\underline{E}_{\underline{P},\mathbb{T}}$ on $\mathbb{B}$ that dominates $\underline{P}$, and we call this $\underline{E}_{\underline{P},\mathbb{T}}$ the ***strongly $\mathbb{T}$-invariant natural extension*** *of $\underline{P}$. For any bounded gamble f on $\mathcal{X}$, $\underline{E}_{\underline{P},\mathbb{T}}(f)$ is given by*

$$\sup\left\{\alpha \in \mathbb{R} : f - \alpha \geq g + \sum_{k=1}^{n} \lambda_k[f_k - \underline{P}(f_k)], g \in \mathcal{D}_{\mathbb{T}}, n \geq 0, f_k \in \operatorname{dom} \underline{P}, \lambda_k \in \mathbb{R}_{\geq 0}\right\}$$

$$= \sup\left\{\inf\left(f - g - \sum_{k=1}^{n} \lambda_k[f_k - \underline{P}(f_k)]\right) : g \in \mathcal{D}_{\mathbb{T}}, n \geq 0, f_k \in \operatorname{dom} \underline{P}, \lambda_k \in \mathbb{R}_{\geq 0}\right\}.$$

(10.3)

When the condition (10.2) is satisfied, we say that $\underline{P}$ **avoids sure loss under strong $\mathbb{T}$-invariance.**

Proof. We begin with the first statement and first show that the condition (10.2) is necessary. Assume, therefore, that there is some strongly $\mathbb{T}$-invariant coherent lower prevision $\underline{Q}$ on $\mathbb{B}$ that dominates $\underline{P}$, and consider arbitrary $g \in \mathcal{D}_{\mathbb{T}}$, $n \geq 0$, f_1, ..., f_n in $\operatorname{dom} \underline{P}$ and non-negative λ_1, ..., λ_n in $\mathbb{R}$. Then it follows from the coherence (Theorem 4.13(i), (v), (vi) and $(\text{iii})_{53}$) and the strong $\mathbb{T}$-invariance (Proposition $10.6_{\curvearrowleft}$) of the dominating $\underline{Q}$ that indeed

$$\sup\left(g+\sum_{k=1}^{n}\lambda_k\left[f_k-\underline{P}(f_k)\right]\right)\geq\underline{Q}\left(g+\sum_{k=1}^{n}\lambda_k\left[f_k-\underline{P}(f_k)\right]\right)$$
$$\geq\underline{Q}(g)+\sum_{k=1}^{n}\lambda_k\underline{Q}(f_k-\underline{P}(f_k))\geq 0+0.$$

Next, we prove that the condition $(10.2)_{\curvearrowleft}$ is sufficient. Denote, for any bounded gamble f on $\mathcal{X}$, the supremum in the expression $(10.3)_{\curvearrowleft}$ by $\underline{E}(f)$. Then it suffices to show that $\underline{E}$ is a strongly $\mathbb{T}$-invariant coherent lower prevision on $\mathbb{B}$ that dominates $\underline{P}$. To show that $\underline{E}$ is a coherent lower prevision, we use Theorem 4.15_{56}. Consider arbitrary $g\in\mathscr{D}_{\mathbb{T}}$, $n\geq 0$, f_1, …, f_n in dom $\underline{P}$ and non-negative λ_1, …, λ_n in $\mathbb{R}$. Also consider any bounded gamble f on $\mathcal{X}$ and any real α such that

$$f-\alpha\geq g+\sum_{k=1}^{n}\lambda_k[f_k-\underline{P}(f_k)],$$

then it follows from condition $(10.2)_{\curvearrowleft}$ that $\sup(f-\alpha)\geq 0$ and, therefore, $\alpha\leq\sup f$, whence $\underline{E}(f)\leq\sup f$. This already tells us that $\underline{E}(f)$ is a real number. It is now easy to see directly from its definition that $\underline{E}$ satisfies conditions (i)–$(\text{iii})_{57}$ of Theorem 4.15_{56}.

To show that $\underline{E}$ is strongly $\mathbb{T}$-invariant, it suffices to consider any bounded gamble f on $\mathcal{X}$ and any transformation $T\in\mathbb{T}$ and to show that $\underline{E}(T^tf-f)\geq 0$. But if we consider $g:=T^tf-f\in\mathscr{D}_{\mathbb{T}}$ and $n=0$ in the expression $(10.3)_{\curvearrowleft}$, then we see that $T^tf-f-\alpha\geq g$ for $\alpha=0$ and, therefore, indeed $\underline{E}(T^tf-f)\geq 0$.

To complete the proof, consider any strongly $\mathbb{T}$-invariant coherent lower prevision $\underline{Q}$ on $\mathbb{B}$ that dominates $\underline{P}$. Then we have to show that $\underline{Q}$ dominates $\underline{E}$. Consider arbitrary $g\in\mathscr{D}_{\mathbb{T}}$, $n\geq 0$, f_1, …, f_n in dom $\underline{P}$ and non-negative λ_1, …, λ_n in $\mathbb{R}$, and consider any real α such that

$$f-\alpha\geq g+\sum_{k=1}^{n}\lambda_k[f_k-\underline{P}(f_k)].\tag{10.4}$$

Then it follows from the coherence (Theorem 4.13(iii)–$(\text{iv})_{53}$) and the strong $\mathbb{T}$-invariance (Proposition $10.6_{\curvearrowleft}$) of $\underline{Q}$ that

$$\underline{Q}(f)-\alpha\geq\underline{Q}\left(g+\sum_{k=1}^{n}\lambda_k\left[f_k-\underline{P}(f_k)\right]\right)\geq\underline{Q}(g)+\sum_{k=1}^{n}\lambda_k\underline{Q}(f_k-\underline{P}(f_k))\geq 0+0,$$

and, therefore, $\underline{Q}(f)\geq\alpha$. It now follows from the definition of $\underline{E}$, by taking the supremum over all real α that satisfy Equation (10.4), that $\underline{Q}(f)\geq\underline{E}(f)$. □

As an immediate consequence, we can infer a simple criterion for the existence of strongly $\mathbb{T}$-invariant coherent lower previsions on $\mathbb{B}$. A related result was given by Walley (1991, Corollary 3.5.4(a)).

Corollary 10.8 *Consider a monoid $\mathbb{T}$ of transformations of $\mathscr{X}$. There is a strongly $\mathbb{T}$-invariant coherent lower prevision on $\mathbb{B}$ if and only if*

$$\sup g \geq 0 \text{ for all } g \in \mathscr{D}_{\mathbb{T}}. \tag{10.5}$$

In that case, there is a point-wise smallest strongly $\mathbb{T}$-invariant coherent lower prevision $\underline{E}_{\mathbb{T}}$ on $\mathbb{B}$, and for any bounded gamble f on $\mathscr{X}$,

$$\underline{E}_{\mathbb{T}}(f) = \sup\left\{\alpha \in \mathbb{R} : f - \alpha \geq g, g \in \mathscr{D}_{\mathbb{T}}\right\} = \sup_{g \in \mathscr{D}_{\mathbb{T}}} \inf(f - g). \tag{10.6}$$

Let us denote by $\operatorname{lins}(\mathbb{T})$ the set of all (strongly) $\mathbb{T}$-invariant *linear previsions* on $\mathbb{B}$, then it follows from Proposition 10.2_{193} that

$$\operatorname{lins}(\mathbb{T}) = \left\{P \in \mathbb{P} : (\forall g \in \mathscr{D}_{\mathbb{T}})(P(g) \geq 0)\right\} = \left\{P \in \mathbb{P} : (\forall g \in \mathscr{D}_{\mathbb{T}})(P(g) = 0)\right\}.$$

Also, as already observed by Walley (1991, Section 3.5, Note 3), the condition (10.5) is necessary and sufficient for the existence of (strongly) $\mathbb{T}$-invariant linear previsions on $\mathbb{B}$, meaning that $\operatorname{lins}(\mathbb{T}) \neq \emptyset$. Moreover, $\underline{E}_{\mathbb{T}}$ is the lower envelope of all $\mathbb{T}$-invariant linear previsions:

$$\underline{E}_{\mathbb{T}}(f) = \min\left\{P(f) : P \in \operatorname{lins}(\mathbb{T})\right\}.$$

We can use Corollary 10.8 to derive alternative forms for the condition $(10.2)_{197}$ for avoiding sure loss under strong $\mathbb{I}$-invariance and the expression $(10.3)_{197}$ for the strongly $\mathbb{T}$-invariant natural extension. They are very closely related to the corresponding forms for avoiding sure loss and natural extension, where the infimum and supremum operators are replaced by $\underline{E}_{\mathbb{T}}$ and $\overline{E}_{\mathbb{T}}$, respectively. Related but different forms were given by Walley (1991, Theorems 3.5.2 and 3.5.3).

Corollary 10.9 *Consider a monoid $\mathbb{T}$ of transformations of $\mathscr{X}$ such that there are strongly $\mathbb{T}$-invariant coherent lower previsions on $\mathbb{B}$, meaning that the condition* (10.5) *is satisfied. Let $\underline{P}$ be a lower prevision. Then there is a strongly $\mathbb{T}$-invariant coherent lower prevision on $\mathbb{B}$ that dominates $\underline{P}$ if and only if*

$$\overline{E}_{\mathbb{T}}\left(\sum_{k=1}^{n} \lambda_k [f_k - \underline{P}(f_k)]\right) \geq 0$$
$$\text{for all } n \geq 0, f_1, \ldots, f_n \text{ in } \operatorname{dom} \underline{P} \text{ and non-negative } \lambda_1, \ldots, \lambda_n \text{ in } \mathbb{R}. \tag{10.7}$$

In that case, the strongly $\mathbb{T}$-invariant natural extension $\underline{E}_{\underline{P},\mathbb{T}}$ of $\underline{P}$ is given by

$$\underline{E}_{\underline{P},\mathbb{T}}(f) = \sup\left\{\underline{E}_{\mathbb{T}}\left(f - \sum_{k=1}^{n} \lambda_k [f_k - \underline{P}(f_k)]\right) : n \geq 0, f_k \in \operatorname{dom} \underline{P}, \lambda_k \in \mathbb{R}_{\geq 0}\right\} \tag{10.8}$$

for all bounded gambles f on $\mathscr{X}$.

Proof. We give the proof for condition (10.7)$_{\curvearrowleft}$ only; the proof for Equation (10.8)$_{\curvearrowleft}$ is similar. Fix $n \geq 0, f_1, \dots, f_n$ in dom $\underline{P}$ and non-negative $\lambda_1, \dots, \lambda_n$ in $\mathbb{R}$. Then we infer from Equation (10.6)$_{\curvearrowleft}$ that

$$\begin{aligned}\overline{E}_{\mathbb{T}}\left(\sum_{k=1}^{n} \lambda_k[f_k - \underline{P}(f_k)]\right) &= -\underline{E}_{\mathbb{T}}\left(-\sum_{k=1}^{n} \lambda_k[f_k - \underline{P}(f_k)]\right)\\ &= -\sup_{g\in\mathscr{D}_{\mathbb{T}}} \inf\left(-\sum_{k=1}^{n} \lambda_k[f_k - \underline{P}(f_k)] - g\right)\\ &= \inf_{g\in\mathscr{D}_{\mathbb{T}}} \sup\left(\sum_{k=1}^{n} \lambda_k[f_k - \underline{P}(f_k)] + g\right),\end{aligned}$$

which makes the equivalence of the conditions (10.2)$_{197}$ and (10.7)$_{\curvearrowleft}$ obvious. □

It is clear from the Separation Lemma 4.40$_{72}$ that the equivalent conditions (10.2)$_{197}$ and (10.7)$_{\curvearrowleft}$ are in turn equivalent to the existence of a (strongly) $\mathbb{T}$-invariant linear prevision on $\mathbb{B}$ that dominates $\underline{P}$:

$$\text{lins}(\underline{P}) \cap \text{lins}(\mathbb{T}) \neq \emptyset.$$

10.2 An important special case

There is a beautiful and simple argument to show that for some types of monoids $\mathbb{T}$, there always are strongly $\mathbb{T}$-invariant lower previsions that dominate a given lower prevision that is $\mathbb{T}$-invariant and avoids sure loss. It is based on the combination of a number of ideas in the literature: (i) Agnew and Morse (1938, Section 2) constructed some specific type of Minkowski functional and used this together with a Hahn–Banach extension result to prove the existence of linear functionals that are invariant with respect to certain groups of permutations; (ii) Day (1942, Theorem 3) showed, in a discussion of ergodic theorems, that a similar construction always works for Abelian semigroups of transformations; (iii) with crucially important insight, Walley (1991, Theorems 3.5.2 and 3.5.3) recognised that the Minkowski functional in the existence proofs of Agnew and Morse, and Day, is actually what we have called a strongly invariant lower prevision, and he used the ideas behind this construction to introduce what we will call *mixture lower previsions* in Theorem 10.10; (iv) in another seminal discussion of mean ergodic theorems, Alaoglu and Birkhoff (1940) show that (Moore–Smith-like) convergence of convex mixtures of linear transformations is instrumental in characterising ergodicity. In this section, we combine and extend these ideas to prove more general existence results for strongly invariant coherent lower previsions. Walley's (1991, Section 3.5) aforementioned results for Abelian semigroups can then be derived directly – with different proofs – from our more general treatment.

We define the following binary relation $\leq$ on $\mathbb{T}^*$: for T_1^* and T_2^* in $\mathbb{T}^*$, we say that T_2^* *is a successor* of T_1^* and we write $T_1^* \leq T_2^*$, if and only if there is some T^* in $\mathbb{T}^*$ such that $T_2^* = T^*T_1^*$. Clearly, $\leq$ is a reflexive and transitive relation, because $\mathbb{T}^*$

is a monoid. Using the ideas in Section 1.5_7, we say that $\mathbb{T}^*$ has the **Moore–Smith property**, or is **directed by** $\leq$, if any two elements of $\mathbb{T}^*$ have a common successor, that is, for any T_1^* and T_2^* in $\mathbb{T}^*$, there is some T^* in $\mathbb{T}^*$ such that $T_1^* \leq T^*$ and $T_2^* \leq T^*$.

If $\mathbb{T}$ is Abelian, or a finite group, then it is a straightforward exercise to show that $\mathbb{T}^*$ is always directed by $\leq$. This is not necessarily the case if $\mathbb{T}$ is an infinite group or only a monoid, however.

In the rest of this chapter, we assume that $\mathbb{T}^*$ is indeed directed by $\leq$. The reason is that there is then an elegant argument that shows that strongly $\mathbb{T}$-invariant coherent lower previsions are easy to come by.

Theorem 10.10 *Consider a monoid $\mathbb{T}$ of transformations of $\mathcal{X}$ such that $\mathbb{T}^*$ has the Moore–Smith property. Let $\underline{P}$ be a coherent and $\mathbb{T}$-invariant lower prevision on $\mathbb{B}$. Then for any bounded gamble f on $\mathcal{X}$, the real net $\ell_f(T^*) := \underline{P}(T^*f)$ converges to a real number $\lim \ell_f = \lim_{T^*\in\mathbb{T}^*} \underline{P}(T^*f) =: \underline{Q}_{\underline{P},\mathbb{T}}(f)$. Moreover, the* ***mixture lower prevision*** *$\underline{Q}_{\underline{P},\mathbb{T}}$ thus defined is the point-wise smallest strongly $\mathbb{T}$-invariant coherent lower prevision on $\mathbb{B}$ that dominates $\underline{P}$ on $\mathbb{B}$, and for all $f \in \mathbb{B}$,*

$$\underline{Q}_{\underline{P},\mathbb{T}}(f) = \sup\left\{\underline{P}(T^*f) : T^* \in \mathbb{T}^*\right\}$$
$$= \sup\left\{\underline{P}\left(\frac{1}{n}\sum_{k=1}^{n} T_k^t f\right) : n \geq 1, T_1, \dots, T_n \in \mathbb{T}\right\}. \tag{10.9}$$

Proof. First, fix f in $\mathbb{B}$. Consider T_1^* and T_2^* in $\mathbb{T}^*$, and assume that $T_1^* \leq T_2^*$. This means that there is some T^* in $\mathbb{T}^*$ such that $T_2^* = T^*T_1^*$, and, consequently, we find that $\underline{P}(T_2^*f) = \underline{P}(T^*(T_1^*f)) \geq \underline{P}(T_1^*f)$, where the inequality follows from the fact that $\underline{P}$ is $\mathbb{T}$-invariant (apply Proposition 10.5_{196}). This means that the net $\ell_f(T^*) := \underline{P}(T^*f)$, $T^* \in \mathbb{T}^*$ is non-decreasing. As this net is moreover bounded above (by $\sup f$, because $\underline{P}$ is coherent), it converges to some real number $\underline{Q}_{\underline{P},\mathbb{T}}(f)$ (Proposition 1.13(i) and $(\mathrm{ii})_9$) and, clearly,

$$\underline{Q}_{\underline{P},\mathbb{T}}(f) = \lim_{T^*\in\mathbb{T}^*} \underline{P}(T^*f) = \sup\left\{\underline{P}(T^*f) : T^* \in \mathbb{T}^*\right\}. \tag{10.10}$$

If we also recall Proposition 10.4_{195}, we see that the net of coherent lower previsions $T^*\underline{P}$, $T^* \in \mathbb{T}^*$ converges point-wise to the lower prevision $\underline{Q}_{\underline{P},\mathbb{T}}$; so $\underline{Q}_{\underline{P},\mathbb{T}}$ is a coherent lower prevision as well, by Proposition $4.21(\mathrm{ii})_{60}$. As $\mathrm{id}_{\mathcal{X}}^t \in \mathbb{T}^*$, it follows from Equation (10.10) that $\underline{Q}_{\underline{P},\mathbb{T}}(f) \geq \underline{P}(\mathrm{id}_{\mathcal{X}}^t f) = \underline{P}(f)$, so $\underline{Q}_{\underline{P},\mathbb{T}}$ dominates $\underline{P}$ on $\mathbb{B}$.

We now show that $\underline{Q}_{\underline{P},\mathbb{T}}$ is strongly $\mathbb{T}$-invariant.[3] Consider any f in $\mathbb{B}$ and T in $\mathbb{T}$. Then for any $n \geq 1$, $T_n^* := \frac{1}{n}\sum_{k=1}^{n}(T^k)^t$ belongs to $\mathbb{T}^*$, and it follows from the coherence of $\underline{P}$ (Theorem 4.13(vi) and $(\mathrm{i})_{53}$) that

$$\underline{P}(T_n^*(f - T^tf)) = \frac{1}{n}\underline{P}\left(T^tf - (T^{n+1})^tf\right) \geq \frac{1}{n}\inf\left[T^tf - (T^{n+1})^tf\right]$$
$$= -\frac{1}{n}\sup\left[(T^{n+1})^tf - T^tf\right] \geq -\frac{2}{n}\sup|f|,$$

[3] The idea for this part of the proof is due to Walley (1991, Point (iv) of the proof of Theorem 3.5.3).

and, consequently, we infer from Equation (10.10)↶ that

$$\underline{Q}_{\underline{P},\mathbb{T}}(f - T^t f) \geq \sup\left\{-\frac{2}{n}\sup|f| : n \geq 1\right\} = 0.$$

A similar argument can be given for $\underline{Q}_{\underline{P},\mathbb{T}}(T^t f - f) \geq 0$, so $\underline{Q}_{\underline{P},\mathbb{T}}$ is indeed strongly $\mathbb{T}$-invariant.

Next, consider any strongly $\mathbb{T}$-invariant coherent lower prevision $\underline{Q}$ on $\mathbb{B}$ that dominates $\underline{P}$. Then we get for any bounded gamble f on $\mathcal{X}$ and any T^* in $\mathbb{T}^*$

$$\underline{Q}(f) = \underline{Q}(f - T^* f + T^* f) \geq \underline{Q}(f - T^* f) + \underline{Q}(T^* f) \geq \underline{Q}(T^* f) \geq \underline{P}(T^* f),$$

where the first inequality follows from the coherence of $\underline{Q}$ (super-additivity, use Theorem 4.13(v)$_{53}$), the second inequality from its strong $\mathbb{T}$-invariance (use Proposition 10.5$_{196}$) and the last inequality from the fact that $\underline{Q}$ dominates $\underline{P}$. We then deduce from Equation (10.10)↶ that $\underline{Q}$ dominates $\underline{Q}_{\underline{P},\mathbb{T}}$. So $\underline{Q}_{\underline{P},\mathbb{T}}$ is indeed the point-wise smallest strongly $\mathbb{T}$-invariant coherent lower prevision on $\mathbb{B}$ that dominates $\underline{P}$ on $\mathbb{B}$.

We finish by proving the second equality in Equation (10.9)↶. Consider a bounded gamble f and any $\varepsilon > 0$. Then, by Equation (10.10)↶, there is some T^* in $\mathbb{T}^*$ such that $\underline{Q}_{\underline{P},\mathbb{T}}(f) \leq \underline{P}(T^* f) + \varepsilon/2$. For this T^*, there are $n \geq 1$, $T_1, \dots, T_n$ in $\mathbb{T}$ and $\lambda_1, \dots, \lambda_n \geq 0$ that sum to 1, such that $T^* = \sum_{k=1}^n \lambda_k T_k^t$. Let $\rho_1, \dots, \rho_n$ be non-negative rational numbers satisfying $|\rho_i - \lambda_i| \leq \frac{\varepsilon}{2n\sup|f|}$ such that, moreover, $\sum_{i=1}^n \rho_i = 1$.[4] It now follows from the coherence of $\underline{P}$ (use Theorem 4.13(v)$_{53}$) that

$$\underline{P}(T^* f) = \underline{P}\left(\sum_{i=1}^n \lambda_i T_i^t f\right) \leq \underline{P}\left(\sum_{i=1}^n \rho_i T_i^t f\right) - \underline{P}\left(\sum_{i=1}^n (\rho_i - \lambda_i) T_i^t f\right),$$

and also (use Theorem 4.13(v) and (i)$_{53}$) that

$$\underline{P}\left(\sum_{i=1}^n (\rho_i - \lambda_i) T_i^t f\right) \geq \sum_{i=1}^n \underline{P}\left((\rho_i - \lambda_i) T_i^t f\right) \geq \sum_{i=1}^n \inf(\rho_i - \lambda_i) T_i^t f$$

$$\geq \sum_{i=1}^n -\frac{\varepsilon}{2n\sup|f|}\sup|f| = -\frac{\varepsilon}{2},$$

whence

$$\underline{Q}_{\underline{P},\mathbb{T}}(f) \leq \underline{P}(T^* f) + \frac{\varepsilon}{2} \leq \underline{P}\left(\sum_{i=1}^n \rho_i T_i^t f\right) + \varepsilon,$$

[4] To see that there are such rational numbers, it suffices to consider non-negative rational numbers $\rho_1, \dots, \rho_{n-1}$ such that $0 \leq \rho_i \leq \lambda_i \leq 1$ and $|\rho_i - \lambda_i| \leq \frac{\varepsilon}{2n^2\sup|f|}$ for $i = 1, \dots, n-1$ and to let $\rho_n := 1 - \sum_{i=1}^{n-1} \rho_i \geq 1 - \sum_{i=1}^{n-1} \lambda_i = \lambda_n \geq 0$. Then $\rho_n \in [0, 1]$, and for n big enough, and unless we are in the trivial case where $\lambda_i = 1$ for some i, we get $|\rho_n - \lambda_n| \leq \frac{\varepsilon}{2n\sup|f|}$.

and, consequently,

$$\underline{Q}_{\underline{P},\mathbb{T}}(f)$$

$$= \sup\left\{ \underline{P}\left(\sum_{i=1}^{n} \rho_i T_i^t f\right) : n \geq 1, T_1, \ldots, T_n \in \mathbb{T}, \rho_1, \ldots, \rho_n \in \mathbb{Q}^+, \sum_{i=1}^{n} \rho_i = 1 \right\},$$

where $\mathbb{Q}^+$ denotes the set of non-negative rational numbers. It is easy to see (just consider the least common multiple of the denominators of $\rho_1, \ldots, \rho_n$) that this supremum coincides with the right-hand side of Equation $(10.9)_{201}$. □

This result allows us to establish the following corollary. It gives a sufficient condition for the existence of strongly $\mathbb{T}$-invariant lower previsions dominating a given coherent lower prevision $\underline{P}$. The smallest such lower prevision reflects how initial behavioural dispositions, reflected in $\underline{P}$, are modified (strengthened) to $\underline{E}_{\underline{P},\mathbb{T}}$ when we add the extra assessment of strong invariance with respect to a monoid $\mathbb{T}$ of transformations.

We call a bounded gamble f on $\mathcal{X}$ **$\mathbb{T}$-invariant** if $T^t f = f$ for all $T \in \mathbb{T}$ or, equivalently, if $T^* f = f$ for all $T^* \in \mathbb{T}^*$. We call an event A **$\mathbb{T}$-invariant** if its indicator I_A is.

Corollary 10.11 (Invariant natural extension) *Consider a monoid $\mathbb{T}$ of transformations of $\mathcal{X}$ such that $\mathbb{T}^*$ has the Moore–Smith property. Let $\underline{P}$ be a $\mathbb{T}$-invariant lower prevision[5] that avoids sure loss. Then there are strongly $\mathbb{T}$-invariant coherent lower previsions on $\mathbb{B}$ that dominate $\underline{P}$, and the smallest such lower prevision, the* ***strongly $\mathbb{T}$-invariant natural extension*** *of $\underline{P}$, is given by $\underline{E}_{\underline{P},\mathbb{T}} := \underline{Q}_{\underline{E}_{\underline{P}},\mathbb{T}}$. Moreover, for every $\mathbb{T}$-invariant bounded gamble f, we have that $\underline{E}_{\underline{P},\mathbb{T}}(f) = \underline{E}_{\underline{P}}(f)$.*

Proof. The first part of the proof follows at once from Theorem 10.10_{201} and the observation that a coherent lower prevision $\underline{Q}$ on $\mathbb{B}$ dominates $\underline{P}$ on $\mathcal{K}$ if and only if it dominates $\underline{E}_{\underline{P}}$ on all bounded gambles (Theorem $4.26(\text{ii})_{65}$).

For the second part of the proof, simply observe that if f is a $\mathbb{T}$-invariant bounded gamble, then $T^* f = f$ and therefore $\underline{E}_{\underline{P}}(T^* f) = \underline{E}_{\underline{P}}(f)$ for all T^* in $\mathbb{T}^*$. □

Let us show in particular how this result applies when we consider the monoid $\mathbb{T}_T$ generated by a single transformation T.

Corollary 10.12 *Let T be a transformation of $\mathcal{X}$ and consider the Abelian monoid $\mathbb{T}_T := \{T^n : n \geq 0\}$. Then for any $\mathbb{T}_T$-invariant lower prevision $\underline{P}$ on some set of bounded gambles $\mathcal{K}$ that avoids sure loss, there are strongly $\mathbb{T}_T$-invariant coherent lower previsions on $\mathbb{B}$ that dominate $\underline{P}$, and the point-wise smallest such lower prevision $\underline{E}_{\underline{P},T}$ is given by*

$$\underline{E}_{\underline{P},T}(f) = \lim_{n\to+\infty} \underline{E}_{\underline{P}}\left(\frac{1}{n}\sum_{k=0}^{n-1}(T^k)^t f\right) = \sup_{n\geq 1} \underline{E}_{\underline{P}}\left(\frac{1}{n}\sum_{k=0}^{n-1}(T^k)^t f\right) \textit{ for all } f \in \mathbb{B}.$$

[5] This means that conditions IN1_{194} and IN2_{194} of Proposition 10.3_{194} are satisfied.

Proof. The existence of strongly $\mathbb{T}_T$-invariant coherent lower previsions that dominate $\underline{P}$ follows from Corollary $10.11_{\curvearrowleft}$ and the fact that for any Abelian monoid $\mathbb{T}$, $\mathbb{T}^*$ has the Moore–Smith property. It also follows from this corollary that for any bounded gamble f on $\mathcal{X}$,

$$\underline{E}_{\underline{P},T}(f) = \sup\left\{\underline{E}_{\underline{P}}(T^*f) : T^* \in \mathbb{T}_T^*\right\} \geq \sup_{n\geq 1} \underline{E}_{\underline{P}}\left(\frac{1}{n}\sum_{k=0}^{n-1}(T^k)^t f\right).$$

To prove the converse inequality, fix any T^* in $\mathbb{T}_T^*$ and any bounded gamble f on $\mathcal{X}$. Then there are $N \geq 1$ and non-negative $\lambda_0, \ldots, \lambda_{N-1}$ that sum to 1, such that $T^* = \sum_{k=0}^{N-1} \lambda_k (T^k)^t$. Consider the element $S_M^* = \frac{1}{M}\sum_{\ell=0}^{M-1}(T^\ell)^t$ of $\mathbb{T}^*$, where M is any natural number such that $M \geq N$. Observe that

$$S_M^* T^* = \frac{1}{M}\sum_{\ell=0}^{M-1}(T^\ell)^t\left(\sum_{k=0}^{N-1}\lambda_k (T^k)^t\right) = \sum_{\ell=0}^{M-1}\sum_{k=0}^{N-1}\frac{\lambda_k}{M}(T^{k+\ell})^t = \sum_{m=0}^{M+N-2}\mu_m (T^m)^t,$$

where we let, for $0 \leq m \leq M+N-2$,

$$\mu_m := \sum_{k=0}^{N-1}\sum_{\ell=0}^{M-1}\frac{\lambda_k}{M}\delta_{m,k+\ell} = \begin{cases} \sum_{k=0}^{m}\frac{\lambda_k}{M} & \text{if } 0 \leq m \leq N-2 \\ \frac{1}{M} & \text{if } N-1 \leq m \leq M-1 \\ \sum_{k=m-M+1}^{N-1}\frac{\lambda_k}{M} & \text{if } M \leq m \leq M+N-2. \end{cases}$$

This tells us that $\mu_m = 1/M$ for $N-1 \leq m \leq M-1$, and $0 \leq \mu_m \leq 1/M$ for all other m. If we let $\delta_m := \mu_m - [1/(N+M-1)]$, it follows at once that $S_M^* T^* - S_{N+M-1}^* = \sum_{m=0}^{M+N-2}\delta_m (T^m)^t$, with

$$|\delta_m| \leq \begin{cases} \dfrac{N-1}{M(M+N-1)} & \text{if } N-1 \leq m \leq M-1 \\ \dfrac{1}{M+N-1} & \text{if } 0 \leq m \leq N-2 \text{ or } M \leq m \leq M+N-2. \end{cases}$$

Consequently, it follows from the $\mathbb{T}$-invariance (Proposition 10.3_{194}) and the coherence of $\underline{E}_{\underline{P}}$ (use Theorem 4.13(v) and $(\mathrm{i})_{53}$) that

$$\begin{aligned}
&\underline{E}_{\underline{P}}(T^*f) \\
&\leq \underline{E}_{\underline{P}}(S_M^* T^* f) \\
&= \underline{E}_{\underline{P}}\left(S_{M+N-1}^* f + \sum_{m=0}^{M+N-2}\delta_m (T^m)^t f\right) \leq \underline{E}_{\underline{P}}(S_{M+N-1}^* f) + \sum_{m=0}^{M+N-2}|\delta_m| \sup|f| \\
&\leq \underline{E}_{\underline{P}}(S_{M+N-1}^* f) + \sup|f|\left[\frac{N-1}{M(M+N-1)}(M-N+1) + \frac{1}{M+N-1}(2N-2)\right] \\
&= \underline{E}_{\underline{P}}(S_{M+N-1}^* f) + \sup|f|\frac{(N-1)(3M-N+1)}{M(M+N-1)}.
\end{aligned}$$

Recall that f and T^*, and therefore also N are fixed. Consider any $\varepsilon > 0$, then there is some $M_\varepsilon \geq N$ such that $\sup |f|(N-1)(3M-N+1)/M(M+N-1) < \varepsilon$ for all $M \geq M_\varepsilon$, whence

$$\underline{E}_{\underline{P}}(T^*f) \leq \underline{E}_{\underline{P}}(S^*_{M_\varepsilon+N-1}f) + \varepsilon \leq \sup_{n\geq 1} \underline{E}_{\underline{P}}(S^*_n f) + \varepsilon.$$

As this holds for all $\varepsilon > 0$, we get $\underline{E}_{\underline{P}}(T^*f) \leq \sup_{n\geq 1} \underline{E}_{\underline{P}}(S^*_n f)$. Taking the supremum over all T^* in $\mathbb{T}^*$ leads to the desired inequality. □

10.3 Interesting examples

We conclude our discussion of symmetry in this chapter with two simple yet interesting examples. We begin in Section 10.3.1 with an account of permutation invariance on finite spaces, which has de Finetti's (1937; 1974–1975) notion of exchangeability as a special case. It is based on, but streamlines and simplifies, the developments by De Cooman and Miranda (2007, Section 9). The material on time (or shift) invariance of sequences in Section 10.3.2208 summarises the more extensive discussion by De Cooman and Miranda (2007, Section 8), but the related extension to stationary random processes in Section 10.3.3210 is new.

10.3.1 Permutation invariance on finite spaces

Suppose the space $\mathcal{X}$ is finite, and consider a group $\mathbb{T}$ of *permutations* of $\mathcal{X}$. For any x in $\mathcal{X}$, we consider the **$\mathbb{T}$-invariant atom**

$$[x]_{\mathbb{T}} := \{\pi x \colon \pi \in \mathbb{T}\},$$

which is the smallest $\mathbb{T}$-invariant event containing x. We denote the set of all $\mathbb{T}$-invariant atoms by $\mathscr{A}_{\mathbb{T}}$: $\mathscr{A}_{\mathbb{T}} := \left\{[x]_{\mathbb{T}} \,:\, x \in \mathcal{X}\right\}$. $\mathscr{A}_{\mathbb{T}}$ is a partition of $\mathcal{X}$, and a bounded gamble f on $\mathcal{X}$ is $\mathbb{T}$-invariant if and only if it is constant on the elements of this partition.

We now define an interesting linear transformation $\mathrm{inv}_{\mathbb{T}}$ on the linear space $\mathbb{B}$ as follows:

$$\mathrm{inv}_{\mathbb{T}}(f) := \frac{1}{|\mathbb{T}|}\sum_{\pi\in\mathbb{T}} \pi^t f \text{ for all bounded gambles } f \text{ on } \mathcal{X},$$

so $\mathrm{inv}_{\mathbb{T}}$ is a uniform convex mixture of all the available permutations in $\mathbb{T}$. Clearly, for any $\pi \in \mathbb{T}$, because $\mathbb{T}$ is a group of permutations,

$$\mathrm{inv}_{\mathbb{T}} \circ \pi^t = \mathrm{inv}_{\mathbb{T}} \text{ and } \pi^t \circ \mathrm{inv}_{\mathbb{T}} = \mathrm{inv}_{\mathbb{T}},$$

which tells us that $\mathrm{inv}_{\mathbb{T}}$ is insensitive to permuting bounded gambles and that any $\mathrm{inv}_{\mathbb{T}}(f)$ is a permutation invariant bounded gamble and, therefore, constant on the elements of the partition $\mathscr{A}_{\mathbb{T}}$. The constant value that is assumed by $\mathrm{inv}_{\mathbb{T}}(f)$ on the invariant atom $[x]_{\mathbb{T}}$ is given by

$$\frac{1}{|\mathbb{T}|}\sum_{\pi\in\mathbb{T}}\pi^t f(x)=\frac{1}{|\mathbb{T}|}\sum_{\pi\in\mathbb{T}}f(\pi x)=\frac{1}{|\mathbb{T}|}\sum_{y\in[x]_{\mathbb{T}}}\sum_{\substack{\pi\in\mathbb{T}\\ \pi x=y}}f(y)=\frac{1}{|\mathbb{T}|}\sum_{y\in[x]_{\mathbb{T}}}f(y)\frac{|\mathbb{T}|}{|[x]_{\mathbb{T}}|}$$

$$=\frac{1}{|[x]_{\mathbb{T}}|}\sum_{y\in[x]_{\mathbb{T}}}f(y)=U(f|\,[x]_{\mathbb{T}})$$

or, in other words,

$$\mathrm{inv}_{\mathbb{T}}(f)=U(f|\,[\bullet]_{\mathbb{T}}), \tag{10.11}$$

where we generally denote by $U(\bullet|A)$ the linear prevision (or expectation) operator associated with the uniform probability distribution on the finite set A.

The following theorem tells us that if a coherent lower prevision $\underline{P}$ is strongly $\mathbb{T}$-invariant, then

$$\underline{P}(f)=\underline{P}(\mathrm{inv}_{\mathbb{T}}(f)) \text{ for all bounded gambles } f \text{ on } \mathscr{X},$$

so $\underline{P}$ is essentially determined by its behaviour on the linear subspace $\mathrm{inv}_{\mathbb{T}}(\mathbb{B})$ of $\mathbb{B}$ consisting of all $\mathbb{T}$-invariant bounded gambles.

Theorem 10.13 (Representation Theorem) *Let $\mathbb{T}$ be a group of permutations of the finite set $\mathscr{X}$. A coherent lower prevision on $\mathbb{B}$ is strongly $\mathbb{T}$-invariant if and only if there is a coherent lower prevision $\underline{Q}$ on $\mathbb{B}(\mathscr{A}_{\mathbb{T}})$ such that $\underline{P}(f)=\underline{Q}(U(f|\,[\bullet]_{\mathbb{T}}))$ for all f in $\mathbb{B}$. In that case, the* ***representation*** *$\underline{Q}$ is uniquely determined by $\underline{Q}(g)=\underline{P}(g\circ[\bullet]_{\mathbb{T}})$ for all bounded gambles g on $\mathscr{A}_{\mathbb{T}}$.*

Proof. We begin with the 'if' part. Let $\underline{Q}$ be an arbitrary coherent lower prevision on $\mathbb{B}(\mathscr{A}_{\mathbb{T}})$, and assume that $\underline{P}(f)=\underline{Q}(U(f|\,[\bullet]_{\mathbb{T}}))$ for all $f\in\mathbb{B}$. Then it is easy to see that $\underline{P}$ is a coherent lower prevision (use Theorems 4.15_{56} and 4.16_{57}). We show that $\underline{P}$ is strongly $\mathbb{T}$-invariant. Consider any bounded gamble f on $\mathscr{X}$ and any $\pi\in\mathbb{T}$. Then for any A in $\mathscr{A}_{\mathbb{T}}$,

$$U(f-\pi^t f|A)=\frac{1}{|A|}\sum_{x\in A}[f(x)-f(\pi x)]=0,$$

because $x\in A$ is equivalent to $\pi x\in A$. So we see that $\underline{P}(f-\pi^t f)=\underline{Q}(0)=0$, because $\underline{Q}$ is coherent (Theorem $4.13(\mathrm{ii})_{53}$). In a similar way, we can prove that $\underline{P}(\pi^t f-f)=0$, so $\underline{P}$ is indeed strongly $\mathbb{T}$-invariant. Moreover, for any bounded gamble g on $\mathscr{A}_{\mathbb{T}}$ and any $x\in\mathscr{X}$, we see that, with $h\colon\mathscr{X}\to\mathbb{R}: x\to g([x]_{\mathbb{T}})$ or in other words $h:=g\circ[\bullet]_{\mathbb{T}}$,

$$U(h|\,[x]_{\mathbb{T}})=\frac{1}{|[x]_{\mathbb{T}}|}\sum_{y\in[x]_{\mathbb{T}}}h(y)=\frac{1}{|[x]_{\mathbb{T}}|}\sum_{y\in[x]_{\mathbb{T}}}g([y]_{\mathbb{T}})=g([x]_{\mathbb{T}}),$$

where the last equality holds because $y=\pi x$ is equivalent to $[y]_{\mathbb{T}}=[x]_{\mathbb{T}}$. Hence, indeed

$$\underline{Q}(g)=\underline{Q}(U(h|\,[\bullet]_{\mathbb{T}}))=\underline{P}(h)=\underline{P}(g\circ[\bullet]_{\mathbb{T}}).$$

To prove the 'only if' part, consider any bounded gamble f on $\mathscr{X}$, observe that

$$f - \mathrm{inv}_{\mathbb{T}}(f) = f - \frac{1}{|\mathbb{T}|}\sum_{\pi\in\mathbb{T}} \pi^t f = \frac{1}{|\mathbb{T}|}\sum_{\pi\in\mathbb{T}} [f - \pi^t f] \in \mathscr{D}_{\mathbb{T}},$$

and invoke Proposition 10.6_{197} to find that $\underline{P}(f - \mathrm{inv}_{\mathbb{T}}(f)) = \overline{P}(f - \mathrm{inv}_{\mathbb{T}}(f)) = 0$. Use coherence (Theorem 4.13(v)$_{53}$) and Equation (10.11) to deduce from this that

$$\underline{P}(f) = \underline{P}(\mathrm{inv}_{\mathbb{T}}(f)) = \underline{P}(U(f|\,[\bullet]_{\mathbb{T}})).$$

If we now define the lower prevision $\underline{Q}$ on $\mathbb{B}(\mathscr{A}_{\mathbb{T}})$ by $\underline{Q}(g) := \underline{P}(g \circ [\bullet]_{\mathbb{T}})$ for all bounded gambles g on $\mathscr{A}_{\mathbb{T}}$, then it is clear that this lower prevision is coherent (use Theorem 4.15_{56}) and that $\underline{P}(f) = \underline{Q}(U(f|\,[\bullet]_{\mathbb{T}}))$. □

A special case of particular interest is where we consider a number of random variables $X_1, X_2, \dots, X_n$ assuming values in the same finite set $\mathscr{X}$. Suppose our subject makes the assessment that these variables are *exchangeable*: he believes that the order in which they are observed is of no consequence to further inferences. We model such an assessment by considering the permutations π of the index set $\{1, 2, \dots, n\}$, and lifting these to permutations of the value space $\mathscr{X}^n$ of the joint variable $(X_1, X_2, \dots, X_n)$ by letting

$$\pi(x_1, x_2, \dots, x_n) := (x_{\pi(1)}, x_{\pi(2)}, \dots, x_{\pi(n)}) \text{ for all } (x_1, x_2, \dots, x_n) \in \mathscr{X}^n.$$

Then the collection $\mathbb{T}$ of all such permutations of $\mathscr{X}^n$ is a group, and we call a coherent lower prevision on $\mathbb{B}(\mathscr{X}^n)$ **exchangeable** if it is strongly $\mathbb{T}$-invariant.

With any sample $x := (x_1, x_2, \dots, x_n)$, we can associate a **count vector** $m = C(x) \in \mathbb{N}^{\mathscr{X}}$, which is a tuple with as many components as there are elements in $\mathscr{X}$, and such that for each $y \in \mathscr{X}$, the corresponding component

$$m_y = C_y(x) := \left|\left\{k \in \{1, 2, \dots, n\} : x_k = y\right\}\right|$$

represents the number of times the value y is observed in the sample x. Then the $\mathbb{T}$-invariant atoms of $\mathscr{X}^n$ are completely determined by these count vectors:

$$z \in [x]_{\mathbb{T}} \Leftrightarrow C(z) = C(x) \text{ for all } x, z \in \mathscr{X}^n.$$

Moreover, the linear prevision $U(\bullet|\,[x]_{\mathbb{T}})$ on $\mathbb{B}(\mathscr{X}^n)$ is the expectation operator associated with the *multiple hypergeometric distribution* with parameters n and $C(x)$ (Johnson et al., 1997, Chapter 39) and models *sampling without replacement* from an urn with n balls of as many types as there are elements in $\mathscr{X}$ and whose composition is characterised by the count vector $C(x)$: there are $C_y(x)$ balls of each possible type $y \in \mathscr{X}$.

The Representation Theorem 10.13 then tells us that we can interpret an exchangeable lower prevision $\underline{P}$ as a model for sampling without replacement from an urn whose composition is unknown and where the information about the composition is modelled by the representation $\underline{Q}$. It is, therefore, a generalisation to coherent lower previsions of de Finetti's (1937; 1974–1975) Representation

Theorem for finite exchangeable sequences. It is also possible to give a similar representation for infinite sequences in terms of sampling with replacement (see De Cooman et al. (2009) and De Cooman and Quaeghebeur (2012) for detailed expositions), in terms of lower previsions and in terms of sets of acceptable (or desirable) bounded gambles.

10.3.2 Shift invariance and Banach limits

As another example, consider the shift operator ϑ on $\mathbb{N}$ defined by $\vartheta(k) := k + 1$ and the Abelian monoid $\mathbb{T}_\vartheta$ generated by ϑ,

$$\mathbb{T}_\vartheta := \{\vartheta^n : n \in \mathbb{N}\},$$

with of course $\vartheta^n(k) = k + n$. The **shift-invariant**, that is, $\mathbb{T}_\vartheta$-invariant, linear previsions on $\mathbb{B}(\mathbb{N})$ are usually called **Banach limits** in the literature, see, for instance, Bhaskara Rao and Bhaskara Rao (1983, Section 2.1.3) or Walley (1991, Sections 2.9.5 and 3.5.7). As $\mathbb{T}_\vartheta$ is Abelian, we know from Corollary 10.12_{203} that there are always Banach limits that dominate a given shift-invariant lower prevision – so we know that there actually are Banach limits.

Let us denote by $\mathrm{lins}(\mathbb{T}_\vartheta)$ the set of all Banach limits. We also know (from Proposition 10.2_{193}) that a coherent lower prevision on $\mathbb{B}(\mathbb{N})$ is strongly shift-invariant if and only if it is a lower envelope of such Banach limits. The smallest strongly shift-invariant coherent lower prevision $\underline{E}_\vartheta := \underline{E}_{\mathbb{T}_\vartheta}$ on $\mathbb{B}(\mathbb{N})$ is the lower envelope of all Banach limits, and it is given by[6]

$$\underline{E}_\vartheta(f) = \sup_{\substack{m_1,\dots m_n \geq 0 \\ n \geq 0}} \inf_{k \geq 0} \frac{1}{n} \sum_{\ell=1}^{n} f(k + m_\ell) = \lim_{n \to +\infty} \inf_{k \geq 0} \frac{1}{n} \sum_{\ell=k}^{k+n-1} f(\ell)$$

$$\text{for any bounded gamble } f \text{ on } \mathbb{N} \quad (10.12)$$

or, in other words, for any bounded sequence $f(n)_{n\in\mathbb{N}}$ of real numbers. The first equality follows from Theorem 10.10_{201}, and the second from Corollary 10.12_{203}. $\underline{E}_\vartheta(f)$ is obtained by taking the infimum sample mean of f over 'moving windows' of length n and then letting the window length n go to infinity. As this is the lower prevision on $\mathbb{B}(\mathbb{N})$ that can be derived using only considerations of coherence and evidence of shift invariance, it is a natural candidate for a '*uniform distribution*' on $\mathbb{N}$.

We could also sample f over the set $\{1, \dots, n\}$ leading to a 'sampling' linear prevision defined by

$$S_n(f) := \frac{1}{n} \sum_{\ell=0}^{n-1} f(\ell) \text{ for any bounded gamble } f \text{ on } \mathbb{N}$$

but for any given f the sequence of sampling averages, $S_n(f)$ is not guaranteed to converge. Taking the point-wise limit inferior of these linear previsions S_n, however,

[6] See also Walley (1991, Section 3.5.7), who was the first to suggest this as a 'uniform distribution'. The expression on the right-hand side is not a limit inferior!

yields a coherent lower prevision (use Corollary 4.22(iii)$_{61}$) $\underline{S}_\vartheta$ given by

$$\underline{S}_\vartheta(f) = \liminf_{n\to+\infty} S_n(f) = \liminf_{n\to+\infty} \frac{1}{n}\sum_{\ell=0}^{n-1} f(\ell) \text{ for any bounded gamble } f \text{ on } \mathbb{N}.$$

For any event $A \subseteq \mathbb{N}$, or equivalently, any $\{0,1\}$-valued sequence, we have that $S_n(A) = \frac{1}{n}|A \cap \{0,\dots,n-1\}|$ is the 'relative frequency' of ones in the sequence I_A up to position $n-1$ and

$$\underline{S}_\vartheta(A) = \liminf_{n\to+\infty} S_n(A) = \liminf_{n\to+\infty} \frac{1}{n}|A \cap \{0,\dots,n-1\}|.$$

Let $\overline{S}_\vartheta$ denote the conjugate upper prevision of $\underline{S}_\vartheta$, given by $\overline{S}_\vartheta(f) = \limsup_n S_n(f)$. Those events A for which $\underline{S}_\vartheta(A) = \overline{S}_\vartheta(A)$ have a 'limiting relative frequency' equal to this common value. The coherent 'limiting relative frequency' lower prevision $\underline{S}_\vartheta$ is also strongly shift-invariant.[7] This implies that all the linear previsions that dominate $\underline{S}_\vartheta$ are strongly shift-invariant. But $\underline{E}_\vartheta$ is strictly dominated by $\underline{S}_\vartheta$,[8] so there are Banach limits that do not dominate $\underline{S}_\vartheta$.

Proposition 10.14 *Let L be any Banach limit on $\mathbb{B}(\mathbb{N})$, and let f be any bounded gamble on $\mathbb{N}$. Then the following statements hold:*

(i) $\liminf_{n\to+\infty} f(n) \leq \underline{E}_\vartheta(f) \leq \underline{S}_\vartheta(f) \leq \overline{S}_\vartheta(f) \leq \overline{E}_\vartheta(f) \leq \limsup_{n\to+\infty} f(n)$.

(ii) *If $\lim_{n\to+\infty} f(n)$ exists, then*

$$\underline{E}_\vartheta(f) = \underline{S}_\vartheta(f) = \overline{E}_\vartheta(f) = \overline{S}_\vartheta(f) = L(f) = \lim_{n\to+\infty} f(n).$$

(iii) *If f is ϑ^m-invariant – has period $m \geq 1$ – then*

$$\underline{E}_\vartheta(f) = \underline{S}_\vartheta(f) = \overline{E}_\vartheta(f) = \overline{S}_\vartheta(f) = L(f) = \frac{1}{m}\sum_{r=1}^{m-1} f(r).$$

(iv) *If f is zero except in a finite number of elements of $\mathbb{N}$, then $\underline{E}_\vartheta(f) = \underline{S}_\vartheta(f) = \overline{E}_\vartheta(f) = \overline{S}_\vartheta(f) = L(f) = 0$. In particular, this holds for the indicator of any finite subset A of $\mathbb{N}$.*

Proof. We begin with the first statement. By conjugacy, we can concentrate on the lower previsions. We have already argued that $\underline{S}_\vartheta$ is a strongly shift-invariant coherent lower prevision, so $\underline{S}_\vartheta$ will dominate the smallest strongly shift-invariant coherent lower prevision $\underline{E}_\vartheta$. It remains to prove that $\underline{E}_\vartheta$ dominates the limit inferior. Consider the first equality in Equation (10.12). Fix the natural numbers $n \geq 1$, m_1,

[7] The following simple proof is due to Walley (1991, Section 3.5.7). Observe that $S_n(\vartheta^t f - f) = [f(n) - f(0)]/n \to 0$ as $n \to +\infty$, so $\underline{S}_\vartheta(\vartheta^t f - f) = \overline{S}_\vartheta(\vartheta^t f - f) = 0$.

[8] Consider the event $A := \{n^2 + k : n \in \mathbb{N}_{>0} \text{ and } k = 0, 1, \dots, n-1\}$, for which $\underline{S}_\vartheta(A) = \overline{S}_\vartheta(A) = 1/2$, but $\underline{E}_\vartheta(A) = 0$ and $\overline{E}_\vartheta(A) = 1$ (De Cooman and Miranda, 2007).

…, m_n. We can assume without loss of generality that the m_1 is the smallest of all the m_ℓ. Observe that

$$\inf_{k\geq 0}\frac{1}{n}\sum_{\ell=1}^{n} f(k+m_\ell) \geq \inf_{k\geq 0}\min_{\ell=1}^{n} f(k+m_\ell) = \min_{\ell=1}^{n}\inf_{k\geq m_\ell} f(k) = \inf_{k\geq m_1} f(k),$$

and therefore

$$\underline{E}_\vartheta(f) \geq \sup_{m_1\geq 0}\inf_{k\geq m_1} f(k) = \liminf_{n\to+\infty} f(n).$$

The second statement is an immediate consequence of the first, and the third follows easily from the definition of $\underline{E}_\vartheta$ and $\overline{E}_\vartheta$. Finally, the fourth statement follows at once from the second. □

10.3.3 Stationary random processes

Next, consider a countable family of random variables $X_0, X_1, \ldots, X_n, \ldots$ each assuming values in the – not necessarily finite – space $\mathcal{X}$. The joint random variable $X := (X_0, X_1, \ldots, X_n, \ldots)$ then assumes values in the space $\mathcal{X}^{\mathbb{N}}$ and is called a (discrete-time) **random process**, and X_k is the (uncertain) value of the process at 'time' k.

Suppose our subject makes an assessment of *stationarity*: he believes that the behaviour of the process X and the delayed process Y with $Y_k = X_{k+\ell}$ are the same, for any delay $\ell \in \mathbb{N}$. Such an assessment can be modelled as follows: consider the shift operator ϑ on $\mathbb{N}$ defined in the previous section, and use it to define a shift operator θ on $\mathcal{X}^{\mathbb{N}}$ by letting $(\theta x)_k := x_{\vartheta(k)} = x_{k+1}$, so $\theta x := x \circ \vartheta$ for all sequences $x \in \mathcal{X}^{\mathbb{N}}$. Consider the Abelian monoid $\mathbb{T}_\theta$ generated by θ: $\mathbb{T}_\theta := \{\theta^n : n \in \mathbb{N}\}$, then we call a lower prevision $\underline{P}$ on $\mathbb{B}(\mathcal{X}^{\mathbb{N}})$ **stationary** if it is strongly $\mathbb{T}_\theta$-invariant.

We now show that it is possible to characterise the stationary coherent lower previsions on $\mathbb{B}(\mathcal{X}^{\mathbb{N}})$ using the Banach limits on $\mathbb{B}(\mathbb{N})$.

First of all, consider any coherent lower prevision $\underline{P}$ on $\mathbb{B}(\mathcal{X}^{\mathbb{N}})$ and any bounded gamble f on $\mathcal{X}^{\mathbb{N}}$. Define the bounded gamble $f_{\underline{P}}$ on $\mathbb{N}$ by

$$f_{\underline{P}}(n) := \underline{P}((\theta^n)^t f) = \underline{P}(f \circ \theta^n) \text{ for all } n \in \mathbb{N}. \tag{10.13}$$

This is indeed a bounded gamble, as for all $n \in \mathbb{N}$ we deduce from the coherence (bounds, Theorem 4.13(i)$_{53}$) of $\underline{P}$ that $f_{\underline{P}}(n) = \underline{P}(f \circ \theta^n) \leq \sup(f \circ \theta^n) \leq \sup f$ and similarly $f_{\underline{P}}(n) \geq \inf f$. Then $(\theta^t f)_{\underline{P}}(n) = \underline{P}(f \circ \theta^{n+1}) = f_{\underline{P}}(n+1) = f_{\underline{P}}(\vartheta(n))$, so we get to the interesting relation

$$(\theta^t f)_{\underline{P}} = \vartheta^t f_{\underline{P}}. \tag{10.14}$$

This observation allows us to establish a link between the transformation θ on $\mathcal{X}^{\mathbb{N}}$ and the shift transformation ϑ on $\mathbb{N}$ that makes us think of the following trick, inspired by what Bhaskara Rao and Bhaskara Rao (1983, Section 2.1.3(9)) do for probability charges rather than lower previsions. Let L be any shift-invariant linear prevision on $\mathbb{B}(\mathbb{N})$ or, in other words, a Banach limit. Define the real-valued functional $\underline{P}_L$ on $\mathbb{B}(\mathcal{X}^{\mathbb{N}})$ by $\underline{P}_L(f) := L(f_{\underline{P}})$. We show that this functional has very special properties.

Proposition 10.15 *Let L be a shift-invariant linear prevision on $\mathbb{B}(\mathbb{N})$, and let $\underline{P}$ be a coherent lower prevision on $\mathbb{B}(\mathscr{X}^{\mathbb{N}})$. Then the following statements hold:*

(i) *$\underline{P}_L$ is a $\mathbb{T}_\theta$-invariant coherent lower prevision on $\mathbb{B}(\mathscr{X}^{\mathbb{N}})$ (with equality).*

(ii) *If $\underline{P}$ dominates a $\mathbb{T}_\theta$-invariant coherent lower prevision $\underline{Q}$ on $\mathbb{B}(\mathscr{X}^{\mathbb{N}})$, then $\underline{P}_L$ dominates $\underline{Q}$ as well.*

(iii) *If $\underline{P} = P$ is a linear prevision, then P_L is a (strongly) $\mathbb{T}_\theta$-invariant linear prevision on $\mathbb{B}(\mathscr{X}^{\mathbb{N}})$.*

(iv) *If $\underline{Q}$ is a $\mathbb{T}_\theta$-invariant coherent lower prevision on $\mathbb{B}(\mathscr{X}^{\mathbb{N}})$, then the (strongly) $\mathbb{T}_\theta$-invariant linear prevision P_L dominates $\underline{Q}$ for any P in* $\mathrm{lins}(\underline{Q})$.

(v) *If $\underline{P} = P$ is a $\mathbb{T}_\theta$-invariant linear prevision, then $P_L = P$.*

Proof. We begin with the first statement. Consider bounded gambles f and g on $\mathscr{X}^{\mathbb{N}}$. As $\inf f \leq f_{\underline{P}}$, it follows from the coherence (monotonicity, Theorem 4.13(iv)$_{53}$) of L that $\inf f \leq L(f_{\underline{P}}) = \underline{P}_L(f)$. Moreover, we have for any n in $\mathbb{N}$ that

$$\begin{aligned}(f+g)_{\underline{P}}(n) &= \underline{P}((f+g)\circ\theta^n) = \underline{P}(f\circ\theta^n + g\circ\theta^n)\\ &\geq \underline{P}(f\circ\theta^n) + \underline{P}(g\circ\theta^n) = f_{\underline{P}}(n) + g_{\underline{P}}(n),\end{aligned}$$

where the inequality follows from the coherence (super-additivity, Theorem 4.13(v)$_{53}$) of $\underline{P}$. As L is a linear prevision and therefore monotone (Corollary 4.14(iv)$_{55}$), we infer that $\underline{P}_L(f+g) \geq L(f_{\underline{P}}) + L(g_{\underline{P}}) = \underline{P}_L(f) + \underline{P}_L(g)$. Finally, for any $\lambda \geq 0$, we have that $(\lambda f)_{\underline{P}}(n) = \underline{P}((\lambda f)\circ\theta^n) = \underline{P}(\lambda(f\circ\theta^n)) = \lambda\underline{P}(f\circ\theta^n) = \lambda f_{\underline{P}}(n)$, because $\underline{P}$ is coherent and therefore non-negatively homogeneous (Theorem 4.13(vi)$_{53}$). Consequently, $\underline{P}_L(\lambda f) = L(\lambda f_{\underline{P}}) = \lambda L(f_{\underline{P}}) = \lambda\underline{P}_L(f)$, because L is a linear prevision and therefore non-negatively homogeneous too (Corollary 4.14(vi)$_{55}$). This proves that $\underline{P}_L$ is a coherent lower prevision on $\mathbb{B}(\mathscr{X}^{\mathbb{N}})$ (because the three requirements of Theorem 4.15$_{56}$ are satisfied). To show that it is $\mathbb{T}_\theta$-invariant, recall that $(\theta^t f)_{\underline{P}} = \vartheta^t f_{\underline{P}}$ by Equation (10.14), whence

$$\underline{P}_L(\theta^t f) = L((\theta^t f)_{\underline{P}}) = L(\vartheta^t f_{\underline{P}}) = L(f_{\underline{P}}) = \underline{P}_L(f),$$

because L is shift-invariant.

To prove the second statement, assume that $\underline{P}$ dominates the $\mathbb{T}_\theta$-invariant coherent lower prevision $\underline{Q}$ on $\mathbb{B}(\mathscr{X}^{\mathbb{N}})$. Then for any bounded gamble f on $\mathscr{X}^{\mathbb{N}}$, we see that

$$f_{\underline{P}}(n) = \underline{P}(f\circ\theta^n) \geq \underline{Q}(f\circ\theta^n) \geq \underline{Q}(f),$$

where the last inequality follows from the $\mathbb{T}_\theta$-invariance of $\underline{Q}$. Consequently, because L is a linear prevision (use Corollary 4.14(iv) and (ii)$_{55}$), we get $\underline{P}_L(f) = L(f_{\underline{P}}) \geq \underline{Q}(f)$.

The third statement follows immediately from the first and the fact that P_L is a self-conjugate coherent lower prevision (and therefore a linear prevision) because P and L are.

The fourth statement follows at once from the second and the third. The fifth is an immediate consequence of the definition of P_L. □

We can use the results in this proposition to characterise all strongly $\mathbb{T}_\theta$-invariant coherent lower previsions using Banach limits.

Theorem 10.16 *Let $\underline{P}$ be a $\mathbb{T}_\theta$-invariant coherent lower prevision that avoids sure loss. Then the set of all $\mathbb{T}_\theta$-invariant linear previsions on $\mathbb{B}(\mathscr{X}^{\mathbb{N}})$ that dominate $\underline{P}$ on* dom $\underline{P}$ *is given by $\{P_L \colon P \in \mathrm{lins}(\underline{P}) \text{ and } L \in \mathrm{lins}(\mathbb{T}_\vartheta)\}$, so the smallest stationary coherent lower prevision $\underline{E}_{\underline{P},\theta}$ on $\mathbb{B}(\mathscr{X}^{\mathbb{N}})$ that dominates $\underline{P}$ – the strongly $\mathbb{T}_\theta$-invariant (or **stationary**) natural extension of $\underline{P}$ – is the lower envelope of this set and also given by*

$$\underline{E}_{\underline{P},\theta}(f) = \inf_{P\in\mathrm{lins}(\underline{P})} \underline{E}_\vartheta(f_P) = \inf_{P\in\mathrm{lins}(\underline{P})} \sup_{n\geq 1} \inf_{k\geq 0} \left(\frac{1}{n} \sum_{\ell=k}^{k+n-1} P(f\circ\theta^\ell) \right)$$

for any bounded gamble f on $\mathscr{X}^{\mathbb{N}}$.

As a consequence, the set $\mathrm{lins}(\mathbb{T}_\theta)$ *of all stationary linear previsions on $\mathbb{B}(\mathscr{X}^{\mathbb{N}})$ is given by*

$$\mathrm{lins}(\mathbb{T}_\theta) = \{P_L \colon P \in \mathbb{P}(\mathscr{X}^{\mathbb{N}}) \text{ and } L \in \mathrm{lins}(\mathbb{T}_\vartheta)\}.$$

This tells us that all $\mathbb{T}_\theta$-invariant linear previsions can be constructed using Banach limits. The smallest stationary coherent lower prevision $\underline{E}_\theta$ on $\mathbb{B}(\mathscr{X}^{\mathbb{N}})$ is the lower envelope of this set and also given by

$$\underline{E}_\theta(f) = \inf_{P\in\mathbb{P}(\mathscr{X}^{\mathbb{N}})} \underline{E}_\vartheta(f_P) = \inf_{P\in\mathbb{P}(\mathscr{X}^{\mathbb{N}})} \sup_{n\geq 1} \inf_{k\geq 0} \left(\frac{1}{n} \sum_{\ell=k}^{k+n-1} P(f\circ\theta^\ell) \right)$$

for any bounded gambles f on $\mathscr{X}^{\mathbb{N}}$.

It should be mentioned here that these expressions are not constructive, as they make use of the inconstructible set of linear previsions $\mathbb{P}(\mathscr{X}^{\mathbb{N}})$. Constructive expressions can be found using Corollary 10.12_{203}.

Proof. First of all, a linear prevision P on $\mathbb{B}(\mathscr{X}^{\mathbb{N}})$ belongs to $\mathrm{lins}(\underline{P})$, that is, dominates $\underline{P}$ on its domain dom $\underline{P}$, if and only if P dominates the natural extension $\underline{E}_{\underline{P}}$ on all bounded gambles (use Theorem 4.26_{65}). Moreover, $\underline{E}_{\underline{P}}$ is $\mathbb{T}_\theta$-invariant by Proposition 10.3_{194}. Now consider any $P \in \mathrm{lins}(\underline{P})$. Use the earlier observations together with Proposition 10.15(iii) and (iv)$_\curvearrowleft$ to show that for any Banach limit L, P_L is a $\mathbb{T}_\theta$-invariant linear prevision that dominates $\underline{P}$. Conversely, if P is a $\mathbb{T}_\theta$-invariant linear prevision on $\mathbb{B}(\mathscr{X}^{\mathbb{N}})$ that dominates $\underline{P}$ on its domain, then by Proposition 10.15(v)$_\curvearrowleft$, $P = P_L$ for any Banach limit L. This shows that $\{P_L \colon P \in \mathrm{lins}(\underline{P}), L \in \mathrm{lins}(\mathbb{T}_\vartheta)\}$ is indeed the set of $\mathbb{T}_\theta$-invariant linear previsions on $\mathbb{B}(\mathscr{X}^{\mathbb{N}})$ that dominate $\underline{P}$ on its domain. Consequently, $\underline{E}_{\underline{P},\theta}$ is the lower envelope of this set, and for any bounded gamble f on $\mathscr{X}^{\mathbb{N}}$,

$$\begin{aligned}\underline{E}_{\underline{P},\theta}(f) &= \inf_{P\in\mathrm{lins}(\underline{P})}\ \inf_{L\in\mathrm{lins}(\mathbb{T}_\vartheta)} P_L(f) = \inf_{P\in\mathrm{lins}(\underline{P})}\ \inf_{L\in\mathrm{lins}(\mathbb{T}_\vartheta)} L(f_P)\\ &= \inf_{P\in\mathrm{lins}(\underline{P})} \underline{E}_\theta(f_P) = \inf_{P\in\mathrm{lins}(\underline{P})}\ \sup_{n\geq 1}\ \inf_{k\geq 0}\left(\frac{1}{n}\sum_{\ell=k}^{k+n-1} P(f\circ\theta^\ell)\right),\end{aligned}$$

where the third equality follows because $\underline{E}_\theta$ is the lower envelope of $\mathrm{lins}(\mathbb{T}_\vartheta)$, and the last equality follows from Equations $(10.12)_{208}$ and $(10.13)_{210}$. The rest of the proof is now immediate. □

11

Extreme lower previsions

In a famous paper, Choquet (1953–1954) proved, besides many other things, that any – what we have called – completely monotone coherent lower probability can be written as a 'σ-additive convex combination' of the extreme ones, which he identified as the lower probabilities associated with proper filters; see also our discussion in Sections $5.5.1_{82}$ and 7.4_{129}. This so-called Choquet representation theorem was later significantly extended by Bishop and De Leeuw (1959).

In his PhD thesis, Maaß (2002) showed that the Bishop–De Leeuw theorem can be used to prove generic Choquet-like representation results for what he called classes of inequality preserving functionals. All interesting types of coherent lower previsions that we have considered in previous chapters can be made to fit nicely within the framework created by Maaß, and in this chapter, we explain exactly how to do that.

The starting point for much of the material in the earlier Sections 11.1–11.5_{224} is the work by Maaß (2002, 2003): we have reproduced many of his results there, but we have streamlined, added and simplified proofs, stressed and extensively elucidated the relationships with coherent lower previsions, provided much more detailed motivation, and added comments to establish links with results in previous chapters.

Section 11.1 lays out the basic topological groundwork to set the stage for applying the Bishop–De Leeuw theorem, which is done in Section 11.5_{224}, after introducing inequality preserving functionals in Section 11.2_{217}, where we also show how these are related to various types of coherent lower previsions discussed in previous chapters. We comment on the connection with Choquet's work in Section 11.6_{225} and draw attention to the special case of strongly invariant coherent lower previsions in Section 11.7_{228}.

Lower Previsions, First Edition. Matthias C.M. Troffaes and Gert de Cooman.

11.1 Preliminary results concerning real functionals

Throughout this chapter, we consider an arbitrary non-empty subset $\mathscr{K}$ of the set $\mathbb{B}$ of all bounded gambles on $\mathscr{X}$. As announced in Section 1.9_{17}, we call any map $\Gamma\colon \mathscr{K} \to \mathbb{R}$ a *real functional* on $\mathscr{K}$. For any such functional Γ, we define its **operator norm** $\|\Gamma\|_{\text{op}}$ as the extended real number

$$\|\Gamma\|_{\text{op}} := \sup_{f\in\mathscr{K}\setminus\{0\}} \frac{|\Gamma(f)|}{\|f\|_{\inf}}, \tag{11.1}$$

where $\|\bullet\|_{\inf} = \sup|\bullet|$ is the supremum norm introduced in Section 4.7_{74}. The set of all **bounded** real functionals on $\mathscr{K}$ is then given by

$$B(\mathscr{K}) := \left\{\Gamma \in \mathbb{R}^{\mathscr{K}} : \|\Gamma\|_{\text{op}} < +\infty\right\}.$$

Under point-wise addition and point-wise scalar multiplication, $B(\mathscr{K})$ is a linear space. We could use the operator norm to turn $B(\mathscr{K})$ into a topological linear space, but for the purposes of this chapter, we actually need a very different topology, which we introduce next.

First, for any $f \in \mathscr{K}$, we can define an **evaluation functional** $\tilde{f}\colon B(\mathscr{K}) \to \mathbb{R}$ by

$$\tilde{f}(\Gamma) := \Gamma(f) \text{ for all } \Gamma \in B(\mathscr{K}). \tag{11.2}$$

Evaluation functionals are real linear functionals on the linear space $B(\mathscr{K})$. Indeed, for all $\Gamma, \Psi \in B(\mathscr{K})$ and all $\lambda \in \mathbb{R}$,

$$\tilde{f}(\Gamma+\Psi) = (\Gamma+\Psi)(f) = \Gamma(f) + \Psi(f) = \tilde{f}(\Gamma) + \tilde{f}(\Psi) \tag{11.3}$$

$$\tilde{f}(\lambda\Gamma) = (\lambda\Gamma)(f) = \lambda\Gamma(f) = \lambda\tilde{f}(\Gamma). \tag{11.4}$$

We denote by $\tilde{\mathscr{K}}$ the set of all evaluation functionals:

$$\tilde{\mathscr{K}} := \left\{\tilde{f} : f \in \mathscr{K}\right\}.$$

Under point-wise addition and point-wise scalar multiplication, the set

$$\operatorname{span}(\tilde{\mathscr{K}}) := \left\{\sum_{k=1}^{n} \lambda_k \tilde{f}_k : n \geq 0, \lambda_k \in \mathbb{R}, f_k \in \mathscr{K}\right\}$$

of linear combinations of its elements is then clearly a linear space of real linear functionals on $B(\mathscr{K})$.

We endow the linear space $B(\mathscr{K})$ with the **topology of point-wise convergence** $\mathscr{T}^*$. Under this topology, a net Γ_α converges to Γ whenever $\lim_\alpha \Gamma_\alpha(f) = \Gamma(f)$ for all $f \in \mathscr{K}$. Consequently, it is also the smallest topology making all evaluation functionals continuous. A sub-base for this topology is made up of the sets:

$$B(\Gamma, f, \varepsilon) := \left\{\Gamma' \in B(\mathscr{K}) : \left|\Gamma(f) - \Gamma'(f)\right| < \varepsilon\right\} \text{ for } \Gamma \in B(\mathscr{K}), f \in \mathscr{K} \text{ and } \varepsilon > 0; \tag{11.5}$$

see, for instance, Bollobás (1999, Chapter 8), or Schechter (1997, 28.11(c)). This implies that for a fixed Γ_0, the sets $B(\Gamma_0, f, \varepsilon)$, $f \in \mathscr{K}$ and $\varepsilon > 0$, constitute a filter sub-base for the filter of local (open) neighbourhoods of Γ_0.

The following propositions lay the foundations for the work in Section 11.5_{224}. They go back to Maaß (2003, Section 2.2).

Proposition 11.1 *The linear space $B(\mathscr{K})$ provided with the topology $\mathscr{T}^*$ of point-wise convergence is a topological linear space that is locally convex and Hausdorff.*

Proof. See also Schechter (1997, Section 28.12(a) and (f)); we provide a sketch of the proof for the sake of completeness.

To show that we have a topological linear space that is locally convex, it suffices (see, for instance, Holmes (1975, Section 10.A)) to show that $\mathscr{T}^*$ has a local neighbourhood sub-base consisting of convex sets or, in other words, that for any $\Gamma_0 \in B(\mathscr{K})$, $f \in \mathscr{K}$ and $\varepsilon > 0$, the set $B(\Gamma_0, f, \varepsilon)$ is a convex subset of $B(\mathscr{K})$. But it follows from Equations $(11.2)_{\curvearrowleft}$ and $(11.5)_{\curvearrowleft}$ that

$$B(\Gamma_0, f, \varepsilon) = \tilde{f}^{-1}\left((\Gamma_0(f) - \varepsilon, \Gamma_0(f) + \varepsilon)\right),$$

and so we see that $B(\Gamma_0, f, \varepsilon)$ is indeed convex, because the inverse image of a convex set under a linear map always is.

To show that $\mathscr{T}^*$ is Hausdorff, it suffices to show that (see the lemma in Holmes (1975, Section 9.A)) $\{\Gamma_0\} = \bigcap_{f \in \mathscr{K}, \varepsilon > 0} B(\Gamma_0, f, \varepsilon)$ for all $\Gamma_0 \in B(\mathscr{K})$, which is trivial. □

Of course, $B(\mathscr{K})$ also constitutes a metric space when provided with the operator norm $\|\bullet\|_{\mathrm{op}}$. For the unit ball

$$B_1(\mathscr{K}) := \left\{\Gamma \in B(\mathscr{K}) \colon \|\Gamma\|_{\mathrm{op}} \leq 1\right\}$$

in this metric space, we can prove the following version of the Banach–Alaoglu–Bourbaki theorem (see, for instance, Schechter, 1997, Section 28.29).

Proposition 11.2 *The unit ball $B_1(\mathscr{K})$ is compact in the topology $\mathscr{T}^*$ of point-wise convergence on $B(\mathscr{K})$.*

Proof. We give a sketch of the proof, which consists of two parts and is very closely related to a common proof for the original version of the Banach–Alaoglu–Bourbaki theorem. It is not difficult to show that the map $\phi \colon B_1(\mathscr{K}) \to [-1, 1]^{\mathscr{K}}$, defined by $\phi(\Gamma)(f) := \Gamma(f)/\|f\|_{\inf}$ for all $f \in \mathscr{K}$, homeomorphically embeds $B_1(\mathscr{K})$ provided with the relativisation of $\mathscr{T}^*$ into the space $[-1, 1]^{\mathscr{K}}$ provided with the product topology, which is compact (use Tychonov's theorem (Willard, 1970, Section 17.8)). It, therefore, suffices to show that $B_1(\mathscr{K})$ is closed in $\mathscr{T}^*$, if we want to show that $B_1(\mathscr{K})$ is compact in $\mathscr{T}^*$. We now show that $B(\mathscr{K}) \setminus B_1(\mathscr{K})$ is open. Consider any Γ in $B(\mathscr{K})$ such that $\|\Gamma\|_{\mathrm{op}} > 1$, then it follows from Equation $(11.1)_{\curvearrowleft}$ that there are $\varepsilon > 0$ and $f \in \mathscr{K}$ such that $|\Gamma(f)| > \|f\|_{\inf} + \varepsilon$. Now take any Γ' in the open neighbourhood $B(\Gamma, f, \varepsilon/2)$, then $|\Gamma'(f)| > |\Gamma(f)| - \varepsilon/2$ and therefore $|\Gamma'(f)| > \|f\|_{\inf} + \varepsilon/2$, which implies that $\|\Gamma'\|_{\mathrm{op}} > 1$ as well. This shows that there is an open neighbourhood in $B(\mathscr{K}) \setminus B_1(\mathscr{K})$ around any of its elements, so $B(\mathscr{K}) \setminus B_1(\mathscr{K})$ is indeed open and $B_1(\mathscr{K})$ closed. □

11.2 Inequality preserving functionals

11.2.1 Definition

We denote by $\mathbb{I}(\mathscr{K})$ the collection of all tuples $(\lambda_k, f_k)_{k=1,\dots,n}$, $n \in \mathbb{N}_{>0}$ of elements of $\mathbb{R} \times \mathscr{K}$ satisfying

$$\sum_{k=1}^{n} \lambda_k f_k \geq 0, \tag{11.6}$$

and we call any subset $\mathscr{S}$ of $\mathbb{I}(\mathscr{K})$ a **system of linear inequalities** on $\mathscr{K}$.

With a system of linear inequalities $\mathscr{S}$, we associate the following class of bounded real functionals on $\mathscr{K}$:

$$\mathscr{C}\left(\mathscr{S}\right) := \left\{\Gamma \in B(\mathscr{K}) \colon (\forall (\lambda_k, f_k)_{k=1,\dots,n} \in \mathscr{S}) \sum_{k=1}^{n} \lambda_k \Gamma(f_k) \geq 0\right\}.$$

We call $\mathscr{C}\left(\mathscr{S}\right)$ the set of **linear inequality preserving functionals** associated with the system $\mathscr{S}$. Obviously, if $\mathscr{S}_i$, $i \in I$ is any family of systems of linear inequalities, then so is its union $\bigcup_{i \in I} \mathscr{S}_i$ and, moreover,

$$\mathscr{C}\left(\bigcup_{i \in I} \mathscr{S}_i\right) = \bigcap_{i \in I} \mathscr{C}\left(\mathscr{S}_i\right). \tag{11.7}$$

In particular, for $I = \emptyset$, we get $\mathscr{C}\left(\emptyset\right) = B(\mathscr{K})$.

For certain specific choices of $\mathscr{S}$, the associated classes $\mathscr{C}\left(\mathscr{S}\right)$ turn out to be quite interesting and to correspond to specific types of functionals we have been studying in the previous chapters. We give a few examples, most of which go back to Maaß (2003, Example 2.3.1).

11.2.2 Linear functionals

Let us call a real functional on $\mathscr{K}$ **linear** if it can be extended to a linear functional on span($\mathscr{K}$) or, what is equivalent, on some linear space including $\mathscr{K}$. Such an extension is then necessarily unique on span($\mathscr{K}$). The set of all linear functionals on $\mathscr{K}$ is then the class $\mathscr{C}\left(\mathscr{S}_{\text{lin}}\right)$ of linear inequality preserving functionals associated with the system

$$\mathscr{S}_{\text{lin}} := \left\{(\lambda_k, f_k)_{k=1,\dots,n} \colon n \in \mathbb{N}_{>0} \text{ and } \sum_{k=1}^{n} \lambda_k f_k = 0\right\}.$$

This is an immediate consequence of a well-known linear extension result, which we mentioned in Proposition 1.24_{18} (see also Schechter, 1997, Proposition 11.10).[1]

[1] Compare this with Equation $(8.2)_{160}$ in the proof of unicity associated with Definition 8.13_{160}.

11.2.3 Monotone functionals

The set of all monotone functionals on $\mathscr{K}$ is the class $\mathscr{C}\left(\mathscr{S}_{\text{mon}}\right)$ of linear inequality preserving functionals associated with the system

$$\mathscr{S}_{\text{mon}} := \{\{(1,f),(-1,g)\} : f \geq g\}.$$

11.2.4 *n*-Monotone functionals

Maaß (2003, Example 2.3.1) discussed n-monotone lower probabilities as linear inequality preserving functionals; here we consider the more general case of n-monotone functionals.

Suppose $\mathscr{K}$ is a $\wedge$-semilattice of bounded gambles on $\mathscr{X}$ and let $n \in \mathbb{N}^*_{>0}$. The set of all n-monotone functionals on $\mathscr{K}$ is the class $\mathscr{C}\left(\mathscr{S}_{n\text{-mon}}\right)$ of linear inequality preserving functionals associated with the system

$$\mathscr{S}_{n\text{-mon}} := \left\{ \left((-1)^{|I|}, f \wedge \bigwedge_{i\in I} f_i \right)_{I\subseteq\{1,\ldots,p\}} : p \in \mathbb{N}_{>0}, p \leq n \text{ and } f, f_1, \ldots f_p \in \mathscr{K} \right\},$$

where it is understood, as in Definition 6.1_{102}, that $\bigwedge_{i\in\emptyset} f_i = +\infty$. That $\mathscr{S}_{n\text{-mon}}$ is a system of linear inequalities – satisfies Equation $(11.6)_{\curvearrowleft}$ – follows from the following alternative formulation of the *inclusion–exclusion principle* (see also Equation $(6.2)_{106}$):

$$\sum_{I\subseteq\{1,\ldots,p\}} (-1)^{|I|} \left(f \wedge \bigwedge_{i\in I} f_i \right) \geq 0 \text{ for all } p \in \mathbb{N}_{>0} \text{ and all } f, f_1, \ldots, f_p \in \mathbb{B}. \quad (11.8)$$

Proof. Fix any x in $\mathscr{X}$ and assume, without loss of generality, that $f_1(x) \leq f_2(x) \leq \cdots \leq f_p(x)$. Then indeed, letting $a := f(x)$ and $a_i := f_i(x)$ for notational simplicity:

$$\sum_{I\subseteq\{1,\ldots,p\}} (-1)^{|I|} \min\left\{ a, \min_{i\in I} a_i \right\}$$

$$= a + \sum_{\emptyset\neq I\subseteq\{1,\ldots,p\}} (-1)^{|I|} \min\{a, a_{\min I}\} = a + \sum_{i=1}^{p} \sum_{\substack{\emptyset\neq I\subseteq\{1,\ldots,p\} \\ \min I = i}} (-1)^{|I|} \min\{a, a_i\}$$

$$= a - \min\{a, a_p\} + \sum_{i=1}^{p-1} \sum_{\substack{\emptyset\neq I\subseteq\{1,\ldots,p\} \\ \min I = i}} (-1)^{|I|} \min\{a, a_i\}$$

$$= a - \min\{a, a_p\} + \sum_{i=1}^{p-1} \min\{a, a_i\} \sum_{\substack{I=\{i\}\cup J \\ J\subseteq\{i+1,\ldots,p\}}} (-1)^{|I|}$$

$$= a - \min\{a, a_p\} + \sum_{i=1}^{p-1} \min\{a, a_i\} \sum_{J\subseteq\{i+1,\ldots,p\}} (-1)^{|J|+1} = a - \min\{a, a_p\} \geq 0,$$

where the last equality follows from

$$\sum_{J\subseteq\{i+1,\dots,p\}}(-1)^{|J|+1} = -\sum_{J\subseteq\{i+1,\dots,p\}}(-1)^{|J|} = -\sum_{k=0}^{p-i}\binom{p-i}{k}(-1)^k 1^{p-i-k} = 0,$$

where we have used the binomial theorem. □

The set of all completely monotone functionals on $\mathscr{K}$ is the class $\mathscr{C}\left(\mathscr{S}_{\infty\text{-mon}}\right)$ of linear inequality preserving functionals associated with the system

$$\mathscr{S}_{\infty\text{-mon}} := \left\{\left((-1)^{|I|}, f \wedge \bigwedge_{i\in I} f_i\right)_{I\subseteq\{1,\dots,p\}} : p \in \mathbb{N}_{>0} \text{ and } f, f_1, \dots f_p \in \mathscr{K}\right\}.$$

11.2.5 Coherent lower previsions

There is an interesting way of associating the set $\underline{\mathbb{P}}^{\mathscr{K}}$ of all coherent lower previsions on $\mathscr{K}$ with a particular class of linear inequality preserving functionals. It is easiest to assume that the domain $\mathscr{K}$ contains the constant gamble 1. This is, of course, no real restriction, because the natural extension of any coherent lower prevision to this constant gamble assumes the value 1. Consider the system of linear inequalities

$$\mathscr{S}_{\text{ex}} := \left\{ \{(1, f), (-\lambda_0, 1), (-\lambda_1, f_1), \dots, (-\lambda_n, f_n)\} : n \in \mathbb{N}, \lambda_1, \dots, \lambda_n \geq 0 \text{ and } f \geq \lambda_0 + \sum_{k=1}^{n} \lambda_k f_k \right\},$$

then the functionals in the corresponding $\mathscr{C}\left(\mathscr{S}_{\text{ex}}\right)$ are the so-called **exact** functionals (Maaß, 2002, 2003), that is, those real functionals Γ on $\mathscr{K}$ that satisfy

$$f \geq \lambda_0 + \sum_{k=1}^{n} \lambda_k f_k \Rightarrow \Gamma(f) \geq \lambda_0 \Gamma(1) + \sum_{k=1}^{n} \lambda_k \Gamma(f_k) \tag{11.9}$$

for all n in $\mathbb{N}$, λ_0 in $\mathbb{R}$, non-negative $\lambda_1, \dots, \lambda_n$ in $\mathbb{R}$ and bounded gambles $f, f_1, \dots, f_n$ in $\mathscr{K}$.

It turns out that the operator norm of an exact functional is easy to find (Maaß, 2003, Proposition 1.2.4).

Proposition 11.3 *Assume that the set of bounded gambles $\mathscr{K}$ contains the constant gamble* 1, *and let Γ be an exact functional on $\mathscr{K}$. Then $\|\Gamma\|_{\text{op}} = \Gamma(1)$.*

Proof. We infer from $1 \in \mathscr{K}$ and $\|1\|_{\inf} = 1$ that $\|\Gamma\|_{\text{op}} \geq |\Gamma(1)| \geq \Gamma(1)$. For the converse inequality, consider any $f \in \mathscr{K}$ such that $f \neq 0$. Then, on the one hand, it follows from $f \geq -\sup|f|$ and the exactness of Γ that $\Gamma(f) \geq -\Gamma(1)\sup|f|$. On the other hand, we infer from $f \leq \sup|f|$ that $f \geq -\sup|f| + 2f$, and therefore, the exactness of Γ guarantees that $\Gamma(f) \geq -\Gamma(1)\sup|f| + 2\Gamma(f)$, whence $\Gamma(f) \leq \Gamma(1)\sup|f|$. In summary, $|\Gamma(f)| \leq \Gamma(1)\sup|f|$, whence indeed $\|\Gamma\|_{\text{op}} \leq \Gamma(1)$. □

If we compare the exactness condition $(11.9)_{\curvearrowleft}$ with the coherence criterion $(F)_{48}$, we see that the coherent lower previsions on $\mathscr{K}$ are precisely those exact functionals Γ that satisfy the extra normalisation condition $\Gamma(1) = 1$:

$$\underline{\mathbb{P}}^{\mathscr{K}} = \mathscr{C}\left(\mathscr{S}_{\text{ex}}\right) \cap \{\Gamma \in B(\mathscr{K}) : \Gamma(1) = 1\} \subseteq B_1(\mathscr{K}). \tag{11.10}$$

11.2.6 Combinations

We infer from Equation $(11.7)_{217}$ that intersections of the above-mentioned sets can still be seen as sets of inequality preserving functionals. As an example, the set of all n-monotone exact functionals on the $\wedge$-semilattice of bounded gambles $\mathscr{K}$, where $1 \in \mathscr{K}$ and $n \in \mathbb{N}^*_{>0}$, is given by

$$\mathscr{C}\left(\mathscr{S}_{\text{ex}}\right) \cap \mathscr{C}\left(\mathscr{S}_{n\text{-mon}}\right) = \mathscr{C}\left(\mathscr{S}_{\text{ex}} \cup \mathscr{S}_{n\text{-mon}}\right),$$

and the set $\underline{\mathbb{P}}_n^{\mathscr{K}}$ of all n-monotone coherent lower previsions on $\mathscr{K}$ is given by

$$\underline{\mathbb{P}}_n^{\mathscr{K}} := \mathscr{C}\left(\mathscr{S}_{\text{ex}} \cup \mathscr{S}_{n\text{-mon}}\right) \cap \{\Gamma \in B(\mathscr{K}) : \Gamma(1) = 1\} \subseteq B_1(\mathscr{K}). \tag{11.11}$$

11.3 Properties of inequality preserving functionals

We now turn to a number of useful general properties for sets of inequality preserving functionals. The material here is based on, but further elaborates and extends, the discussion by Maaß (2003, Section 2.3).

Proposition 11.4 *Let $\mathscr{S} \subseteq \mathbb{I}(\mathscr{K})$ be any system of linear inequalities. Then the corresponding set of linear inequality preserving functionals $\mathscr{C}\left(\mathscr{S}\right)$ is a convex cone that is closed in the topology $\mathscr{T}^*$ of point-wise convergence.*

Proof. It is obvious that $\mathscr{C}\left(\mathscr{S}\right)$ is closed under non-negative linear combinations and therefore a convex cone. To prove that $\mathscr{C}\left(\mathscr{S}\right)$ is $\mathscr{T}^*$-closed, we only need to show that every converging net in $\mathscr{C}\left(\mathscr{S}\right)$ converges to an element of $\mathscr{C}\left(\mathscr{S}\right)$. But this is obvious as well, because point-wise linear inequalities are preserved when taking point-wise limits. □

Proposition 11.5 *Let $\mathscr{S} \subseteq \mathbb{I}(\mathscr{K})$ be any system of linear inequalities. Then $\mathscr{C}\left(\mathscr{S}\right) \cap B_1(\mathscr{K})$ and any of its $\mathscr{T}^*$-closed subsets is compact in the topology $\mathscr{T}^*$ of point-wise convergence.*

Proof. Any closed subset of a compact set is compact, and in a Hausdorff space, all compact sets are closed. By Proposition 11.1_{216}, $\mathscr{T}^*$ is Hausdorff, and by Proposition 11.2_{216}, $B_1(\mathscr{K})$ is $\mathscr{T}^*$-compact and therefore also $\mathscr{T}^*$-closed. By Proposition 11.4, $\mathscr{C}\left(\mathscr{S}\right)$ and therefore also $\mathscr{C}\left(\mathscr{S}\right) \cap B_1(\mathscr{K})$ is $\mathscr{T}^*$-closed. We conclude that $\mathscr{C}\left(\mathscr{S}\right) \cap B_1(\mathscr{K})$ and all its $\mathscr{T}^*$-closed subsets are $\mathscr{T}^*$-compact. □

We infer from Proposition 11.4 that $\mathscr{C}\left(\mathscr{S}\right)$ is closed under (finite) non-negative linear combinations. We now work towards a result (Proposition 11.6_{223} further on) that takes this a considerable step further: $\mathscr{C}\left(\mathscr{S}\right)$ is even closed under 'infinite' non-negative linear combinations.

11.4 Infinite non-negative linear combinations of inequality preserving functionals

11.4.1 Definition

First, we must define precisely what we mean by 'infinite non-negative linear combinations'. We can draw an analogy with a similar concept that we employed in the theory of coherent lower previsions: namely, linear previsions, which can be interpreted as a way of taking infinite *convex* combinations. Indeed, a convex combination is a very simple type of linear prevision, namely, that type which corresponds to a probability mass function. So a linear prevision can be seen as a natural generalisation of specifying convex combinations over $\mathscr{C}\left(\mathscr{S}\right)$, where $\mathscr{C}\left(\mathscr{S}\right)$ is infinite. One way of specifying infinite *convex* combinations over $\mathscr{C}\left(\mathscr{S}\right)$ is, therefore, to define a linear prevision on bounded gambles on $\mathscr{C}\left(\mathscr{S}\right)$.

Consequently, in order to define infinite non-negative *linear* combinations over $\mathscr{C}\left(\mathscr{S}\right)$, we could simply specify a *non-negative linear functional* on bounded gambles on $\mathscr{C}\left(\mathscr{S}\right)$. It turns out that this procedure is slightly too general for what we will need. We simplify it in two ways.

Firstly, we focus only on a non-empty subset $\mathscr{Z}$ of $\mathscr{C}\left(\mathscr{S}\right)$. In particular, we will not attempt to take non-negative linear combinations over all of $\mathscr{C}\left(\mathscr{S}\right)$ at once and restrict to those subsets $\mathscr{Z}$ of $\mathscr{C}\left(\mathscr{S}\right)$ that can be bounded, as explained in the next step.

Secondly, we specify our non-negative linear functional only over a particular subset of bounded gambles on the set of real functionals $\mathscr{Z}$, namely, the set

$$\tilde{\mathscr{K}}|_{\mathscr{Z}} := \left\{\tilde{f}|_{\mathscr{Z}} : f \in \mathscr{K}\right\} \tag{11.12}$$

of restrictions $\tilde{f}|_{\mathscr{Z}}$ to $\mathscr{Z}$ of the evaluation functionals $\tilde{f}$ defined on $B(\mathscr{K})$. If we observe that, using Equation $(11.1)_{215}$,

$$\sup\left|\tilde{f}|_{\mathscr{Z}}\right| = \sup_{\Gamma\in\mathscr{Z}}\left|\tilde{f}(\Gamma)\right| = \sup_{\Gamma\in\mathscr{Z}}|\Gamma(f)| \leq \sup_{\Gamma\in\mathscr{Z}}\|\Gamma\|_{\text{op}}\|f\|_{\text{inf}} = \|f\|_{\text{inf}}\sup_{\Gamma\in\mathscr{Z}}\|\Gamma\|_{\text{op}},$$

then we can guarantee that $\tilde{f}|_{\mathscr{Z}}$ is a bounded gamble on $\mathscr{Z}$ by requiring that $\mathscr{Z}$ should be **uniformly bounded**, meaning that there should be some real L such that $\|\Gamma\|_{\text{op}} \leq L$ for all $\Gamma \in \mathscr{Z}$. This is what we do in the rest of this section.

So, in summary, we are about to consider non-negative linear functionals defined on the set $\tilde{\mathscr{K}}|_{\mathscr{Z}}$ of bounded gambles on $\mathscr{Z}$. As is usually done, we call a real functional Λ on $\tilde{\mathscr{K}}|_{\mathscr{Z}}$ **linear** if it can be extended to some linear real functional Λ' on the linear space $\operatorname{span}(\tilde{\mathscr{K}})|_{\mathscr{Z}}$ of restrictions to $\mathscr{Z}$ of the elements of $\operatorname{span}(\tilde{\mathscr{K}})$. We call Λ a **positive linear functional** if it has some linear extension Λ' to $\operatorname{span}(\tilde{\mathscr{K}})|_{\mathscr{Z}}$ that is **positive**, meaning that $\Lambda'(g) \geq 0$ for all $g \geq 0$ in $\operatorname{span}(\tilde{\mathscr{K}})|_{\mathscr{Z}}$.

Suppose, then, that Λ is a positive linear real functional on $\tilde{\mathscr{K}}|_{\mathscr{Z}}$, and define the corresponding positive linear real functional Γ_Λ on $\mathscr{K}$ by

$$\Gamma_\Lambda(f) := \Lambda(\tilde{f}|_{\mathscr{Z}}) \text{ for all } f \in \mathscr{K}. \tag{11.13}$$

This functional Γ_Λ is the *generalisation of the notion of a non-negative linear combination of elements of $\mathcal{Z}$* we have been looking for.

11.4.2 Examples

Let us give a few examples to illustrate what this generalisation encompasses. For a start, assume that $\mathcal{Z} = \{\Gamma_1, \dots, \Gamma_n\}$ is a finite subset of $\mathscr{C}\left(\mathscr{S}\right)$, and let $\lambda_1, \dots, \lambda_n$ be non-negative real numbers. Consider the degenerate linear previsions P_{Γ_k} on $\mathbb{B}(\mathcal{Z})$ defined by $P_{\Gamma_k}(z) := z(\Gamma_k)$ for all $z \in \mathbb{B}(\mathcal{Z})$ (see also Section 5.4_{81} for the definition of such degenerate linear previsions). Then

$$\Lambda_0 := \sum_{k=1}^{n} \lambda_k P_{\Gamma_k}$$

is a positive linear functional on $\mathbb{B}(\mathcal{Z})$. But $\tilde{\mathscr{K}}|_{\mathcal{Z}}$ is a subset of $\mathbb{B}(\mathcal{Z})$, hence, Λ_0 can also be considered as a positive linear functional on $\tilde{\mathscr{K}}|_{\mathcal{Z}}$. Moreover, we see that for any $f \in \mathscr{K}$,

$$\Gamma_{\Lambda_0}(f) = \Lambda_0(\tilde{f}|_{\mathcal{Z}}) = \sum_{k=1}^{n} \lambda_k P_{\Gamma_k}(\tilde{f}|_{\mathcal{Z}}) = \sum_{k=1}^{n} \lambda_k \tilde{f}|_{\mathcal{Z}}(\Gamma_k) = \sum_{k=1}^{n} \lambda_k \Gamma_k(f),$$

so Γ_{Λ_0} is a non-negative linear combination of the elements of $\mathcal{Z}$.

Besides the special case of finite non-negative linear combinations – and in particular that of finite convex combinations – there is another one that deserves special mention in view of the representation theorem that we are about to prove in Section 11.5_{224}. Let $\mathscr{F}$ be a field of subsets of $\mathcal{Z}$, and let μ be a probability charge on $\mathscr{F}$. Suppose that all elements $\tilde{f}|_{\mathcal{Z}}$ of $\tilde{\mathscr{K}}|_{\mathcal{Z}}$ are μ-integrable, so $\tilde{\mathscr{K}}|_{\mathcal{Z}} \subseteq \mathscr{I}_\mu(\mathcal{Z})$. We define the linear prevision (and therefore positive linear functional) Λ_μ on $\tilde{\mathscr{K}}|_{\mathcal{Z}}$ and the corresponding real functional Γ_{Λ_μ} on $\mathscr{K}$ by (see also Theorem 8.32_{174})

$$\Gamma_{\Lambda_\mu}(f) := \Lambda_\mu(\tilde{f}|_{\mathcal{Z}}) = E_\mu(\tilde{f}|_{\mathcal{Z}}) = L\int \tilde{f}|_{\mathcal{Z}}\, \mathrm{d}\mu \text{ for all } f \in \mathscr{K}. \tag{11.14}$$

In Section 11.5_{224}, we shall come across the special case where $\mathscr{F}$ is a σ-field of subsets of $\mathcal{Z}$ that makes all $\tilde{f}|_{\mathcal{Z}}$ $\mathscr{F}$-measurable, and where μ is a probability measure. The integral in Equation (11.14) then coincides with the usual (Lebesgue) integral associated with the probability measure μ (see also Section 9.4_{183}).

One more special case deserves to be elucidated. Suppose that the set $\mathcal{Z}$ is a net of real functionals Γ_α in $\mathscr{C}\left(\mathscr{S}\right)$ such that the real net $\Gamma_\alpha(f)$ converges to some real number $\Gamma(f)$ for every $f \in \mathscr{K}$. Then, drawing from the discussion in Sections 5.3_{80} and 9.2_{182}, we can interpret each $\tilde{f}|_{\mathcal{Z}}$ in $\tilde{\mathscr{K}}|_{\mathcal{Z}}$ as a convergent bounded net on $\mathcal{Z}$, and this allows us to consider the limit operator lim as a linear prevision (and therefore positive linear functional) on $\tilde{\mathscr{K}}|_{\mathcal{Z}}$, with

$$\Lambda_{\lim}(\tilde{f}|_{\mathcal{Z}}) := \lim \tilde{f}|_{\mathcal{Z}} = \lim_\alpha \tilde{f}|_{\mathcal{Z}}(\Gamma_\alpha) = \lim_\alpha \Gamma_\alpha(f) = \Gamma(f),$$

and then the corresponding real functional $\Gamma_{\Lambda_{\lim}}$ on $\mathscr{K}$ is the limit functional Γ: $\Gamma_{\Lambda_{\lim}}(f) = \Lambda_{\lim}(\tilde{f}|_{\mathscr{Z}}) = \lim \tilde{f}|_{\mathscr{Z}} = \Gamma(f)$ for all $f \in \mathscr{K}$.

To summarise, non-negative linear combinations, finitely additive integrals – including the σ-additive ones – and limits of convergent nets of elements of $\mathscr{C}\left(\mathscr{S}\right)$ are all special cases of the Γ_Λ defined by Equation $(11.13)_{221}$.

11.4.3 Main result

We now show that all such Γ_Λ still belong to $\mathscr{C}\left(\mathscr{S}\right)$.

Proposition 11.6 *Let $\mathscr{S} \subseteq \mathbb{I}(\mathscr{K})$ be any system of linear inequalities, and consider any uniformly bounded non-empty subset $\mathscr{Z} \subseteq \mathscr{C}\left(\mathscr{S}\right)$. Then for any positive linear functional Λ on the set $\tilde{\mathscr{K}}|_{\mathscr{Z}}$ defined by Equation* $(11.12)_{221}$*, the real functional Γ_Λ on $\mathscr{K}$ defined by Equation* $(11.13)_{221}$ *belongs to $\mathscr{C}\left(\mathscr{S}\right)$.*

Proof. Consider any $(\lambda_k, f_k)_{k=1,\dots,n} \in \mathscr{S}$, then we know that $\sum_{k=1}^n \lambda_k \Gamma(f_k) \geq 0$ for all $\Gamma \in \mathscr{C}\left(\mathscr{S}\right)$ and, therefore, $\sum_{k=1}^n \lambda_k \tilde{f}_k \geq 0$. Hence, with notation Λ' for the positive and unique linear extension to $\operatorname{span}(\tilde{\mathscr{K}})|_{\mathscr{Z}}$ of the linear functional Λ,

$$\sum_{k=1}^n \lambda_k \Gamma_\Lambda(f_k) = \sum_{k=1}^n \lambda_k \Lambda(\tilde{f}_k|_{\mathscr{Z}}) = \Lambda'\left(\sum_{k=1}^n \lambda_k \tilde{f}_k|_{\mathscr{Z}}\right) \geq 0,$$

where the second equality follows from the linearity of Λ' and the inequality from its positivity. □

As we took some trouble to explain above, the following corollaries generalise Propositions 4.19_{59} and 4.21_{60}, and Propositions 6.7_{106} and 6.8_{107}, respectively: a linear prevision 'of' coherent lower previsions is a coherent lower prevision and a linear prevision 'of' n-monotone coherent lower previsions is n-monotone.

Corollary 11.7 *Consider any set of bounded gambles $\mathscr{K}$ on $\mathscr{X}$ that contains the constant gamble 1, and consider any linear prevision P on the set $\tilde{\mathscr{K}}|_{\underline{\mathbb{P}}^{\mathscr{K}}}$. Then the real functional $\underline{P}_P$ on $\mathscr{K}$ defined by*

$$\underline{P}_P(f) := P\left(\tilde{f}|_{\underline{\mathbb{P}}^{\mathscr{K}}}\right) \text{ for all } f \in \mathscr{K}$$

is a coherent lower prevision on $\mathscr{K}$.

Proof. Consider the system of inequalities $\mathscr{S}_{\text{ex}}$, then, with $\mathscr{Z} = \underline{\mathbb{P}}^{\mathscr{K}}$, we infer from Equation $(11.10)_{220}$ that $\mathscr{Z} \subseteq \mathscr{C}\left(\mathscr{S}_{\text{ex}}\right)$. Moreover, $\mathscr{Z}$ is uniformly bounded because $\|\underline{P}\|_{\text{op}} = \underline{P}(1) = 1$ for all $\underline{P} \in \mathscr{Z}$. To complete the proof, observe that any linear prevision is a positive real linear functional and apply Proposition 11.6. □

Corollary 11.8 *Let $n \in \mathbb{N}^*_{>0}$. Consider any $\wedge$-semilattice of bounded gambles $\mathscr{K}$ on $\mathscr{X}$ that includes the constant gamble 1 and any linear prevision P on the set $\tilde{\mathscr{K}}|_{\underline{\mathbb{P}}_n^{\mathscr{K}}}$. Then the real functional $\underline{P}_P$ on $\mathscr{K}$ defined by*

$$\underline{P}_P(f) := P\left(\tilde{f}|_{\underline{\mathbb{P}}_n^{\mathscr{K}}}\right) \text{ for all } f \in \mathscr{K}$$

is an n-monotone coherent lower prevision on $\mathscr{K}$.

Proof. The proof is completely analogous to the proof of Corollary 11.7 but now with the system of linear inequalities $\mathscr{S}_{\mathrm{ex}} \cup \mathscr{S}_{n\text{-mon}}$ and $\mathcal{Z} = \underline{\mathbb{P}}_n^{\mathscr{K}}$. □

11.5 Representation results

Let us now consider an arbitrary convex subset $\mathcal{Z}$ of $\mathscr{C}\left(\mathscr{S}\right) \cap B_1(\mathscr{K})$ that is closed, and therefore compact, in the topology $\mathscr{T}^*$ of point-wise convergence. It will then, of course, also be uniformly bounded (with bound 1).

We consider the relativisation of the topology $\mathscr{T}^*$ to $\mathcal{Z}$ – which is simply the topology of point-wise convergence on $\mathcal{Z}$ – and consider the Baire σ-field $\mathscr{B}_0(\mathcal{Z})$ on $\mathcal{Z}$, introduced in Definition 1.8_6.

We denote by $\operatorname{ext}(\mathcal{Z})$ the set of extreme points of $\mathcal{Z}$. As $B(\mathscr{K})$ is a locally convex topological linear space (Proposition 11.1_{216}), we infer from the Krein–Milman Theorem $B.8_{374}$ that $\operatorname{ext}(\mathcal{Z})$ is non-empty. The set $\left\{\operatorname{ext}(\mathcal{Z}) \cap B : B \in \mathscr{B}_0(\mathcal{Z})\right\}$ is then a σ-field on $\operatorname{ext}(\mathcal{Z})$, called the **relative Baire σ-field**. Applying the Bishop–De Leeuw Theorem $B.9_{375}$ brings us to Maaß's (2003, Lemma 2.4.1) main result.

Theorem 11.9 (Representation Theorem) *Consider an arbitrary non-empty convex and $\mathscr{T}^*$-closed subset $\mathcal{Z}$ of $\mathscr{C}\left(\mathscr{S}\right) \cap B_1(\mathscr{K})$. Then for every Γ in $\mathcal{Z}$, there is a probability measure μ_Γ, defined on the σ-field $\left\{\operatorname{ext}(\mathcal{Z}) \cap B : B \in \mathscr{B}_0(\mathcal{Z})\right\}$, that* ***represents*** *Γ, meaning that*

$$\Gamma(f) = L\int \tilde{f}|_{\operatorname{ext}(\mathcal{Z})} \,\mathrm{d}\mu_\Gamma \text{ for all } f \in \mathscr{K}.$$

Proof. We apply the Bishop–De Leeuw Theorem $B.9_{375}$ with $B(\mathscr{K})$ as the locally convex and Hausdorff topological linear space V (Proposition 11.1_{216}), and $\mathcal{Z}$ as the non-empty convex and compact subset W (Proposition 11.5_{220}). Fix any $\Gamma \in \mathcal{Z}$. All evaluation functionals $\tilde{f}$, $f \in \mathscr{K}$ are continuous linear, and therefore affine, real functionals on $\mathcal{Z}$, so we infer from the theorem that there is some probability measure μ_Γ on $\left\{\operatorname{ext}(\mathcal{Z}) \cap B : B \in \mathscr{B}_0(\mathcal{Z})\right\}$ such that $\tilde{f}(\Gamma) = L\int \tilde{f}|_{\operatorname{ext}(\mathcal{Z})} \,\mathrm{d}\mu_\Gamma$ for all $f \in \mathscr{K}$. Now use $\tilde{f}(\Gamma) = \Gamma(f)$. □

It is clear from Equation $(11.10)_{220}$ that the convex set $\underline{\mathbb{P}}^{\mathscr{K}}$ of all coherent lower previsions on $\mathscr{K}$ is a $\mathscr{T}^*$-closed subset of $\mathscr{C}\left(\mathscr{S}_{\mathrm{ex}}\right) \cap B_1(\mathscr{K})$. This leads to Corollary 11.10.

Corollary 11.10 *Let $\mathscr{K}$ be any set of bounded gambles on $\mathcal{X}$ that contains the constant gamble* 1. *For every coherent lower prevision $\underline{P}$ on $\mathscr{K}$, there is some probability measure $\mu_{\underline{P}}$, defined on the σ-field $\left\{\operatorname{ext}(\underline{\mathbb{P}}^{\mathscr{K}}) \cap B : B \in \mathscr{B}_0(\underline{\mathbb{P}}^{\mathscr{K}})\right\}$, that represents $\underline{P}$, meaning that*

$$\underline{P}(f) = L\int \tilde{f}|_{\operatorname{ext}(\underline{\mathbb{P}}^{\mathscr{K}})} \,\mathrm{d}\mu_{\underline{P}} \text{ for all } f \in \mathscr{K}.$$

Similarly, we infer from Equation $(11.11)_{220}$ that the convex set $\underline{\mathbb{P}}_n^{\mathscr{K}}$ of all n-monotone coherent lower previsions on $\mathscr{K}$ is a $\mathscr{T}^*$-closed subset of $\mathscr{C}\left(\mathscr{S}_{\mathrm{ex}} \cup \mathscr{S}_{n\text{-mon}}\right) \cap B_1(\mathscr{K})$. This leads to Corollary 11.11.

Corollary 11.11 *Let $\mathscr{K}$ be any $\wedge$-semilattice of bounded gambles on $\mathscr{X}$ that contains the constant gamble* 1, *and let $n \in \mathbb{N}^*_{>0}$. For every n-monotone coherent lower prevision $\underline{P}$ on $\mathscr{K}$, there is some probability measure $\mu_{\underline{P}}$, defined on the σ-field $\{\mathrm{ext}(\underline{\mathbb{P}}_n^{\mathscr{K}}) \cap B : B \in \mathscr{B}_0(\underline{\mathbb{P}}_n^{\mathscr{K}})\}$, that represents $\underline{P}$, meaning that*

$$\underline{P}(f) = L\int \tilde{f}|_{\mathrm{ext}(\underline{\mathbb{P}}_n^{\mathscr{K}})}\,\mathrm{d}\mu_{\underline{P}} \text{ for all } f \in \mathscr{K}.$$

For general $\mathscr{X}$, $\mathscr{K}$ and n, we are still lacking elegant characterisations or descriptions of what the extreme coherent lower previsions in $\mathrm{ext}(\underline{\mathbb{P}}^{\mathscr{K}})$ and $\mathrm{ext}(\underline{\mathbb{P}}_n^{\mathscr{K}})$ look like, although some interesting work has been done for the special case that $\mathscr{X}$ is finite (De Bock and De Cooman, 2013; Quaeghebeur and De Cooman, 2008) and for the case that $\mathscr{K}$ is finite (Quaeghebeur, 2010). There is one interesting special case, however, where we know all extreme points, thanks to Choquet's (1953–1954) work. This is what we turn to in the next section.

11.6 Lower previsions associated with proper filters

As an important illustration, we look at the set $\underline{\mathbb{P}}_\infty = \underline{\mathbb{P}}_\infty^{\mathbb{B}}$ of all coherent and completely monotone lower previsions on the lattice of bounded gambles $\mathbb{B}$. We infer from Corollary 11.11 that for every coherent and completely monotone lower prevision $\underline{P}$ on $\mathbb{B}$, there is some probability measure $\mu_{\underline{P}}$, defined on the relative Baire σ-field $\{\mathrm{ext}(\underline{\mathbb{P}}_\infty) \cap B : B \in \mathscr{B}_0(\underline{\mathbb{P}}_\infty)\}$ on the set $\mathrm{ext}(\underline{\mathbb{P}}_\infty)$ of all extreme coherent and completely monotone lower previsions on $\mathbb{B}$, such that

$$\underline{P}(f) = L\int \tilde{f}|_{\mathrm{ext}(\underline{\mathbb{P}}_\infty)}\,\mathrm{d}\mu_{\underline{P}} \text{ for all bounded gambles } f \text{ on } \mathscr{X}. \tag{11.15}$$

But we have already identified the extreme points in Proposition 7.8_{130}; they are the lower previsions associated with proper filters that we have studied in Sections $5.5.1_{82}$ and 7.4_{129}: $\mathrm{ext}(\underline{\mathbb{P}}_\infty) = \{\underline{P}_{\mathscr{F}} : \mathscr{F} \in \mathbb{F}(\mathscr{X})\}$. If we restrict this equality to (indicators of) events, this is essentially Choquet's (1953–1954, Sections 43–45) representation theorem; see also Shafer (1979) for additional discussion.

We now discuss three special cases of this representation result, where we are actually able to identify the representing probability measures as well. In addition, we show that any completely monotone coherent lower prevision can be seen as an induced lower prevision in the sense of Section 7.5_{131}.

11.6.1 Belief functions

First, look at the case of finite $\mathscr{X}$. Any proper filter $\mathscr{F}$ is closed under finite intersections, and because it may only contain a finite number of elements, we see that $\bigcap \mathscr{F} \in \mathscr{F}$. This tells us that $\mathscr{F}$ is **fixed**, meaning that $\bigcap \mathscr{F} \neq \emptyset$ and therefore $\mathscr{F} = \{A \subseteq \mathscr{X} : \bigcap \mathscr{F} \subseteq A\}$. We conclude that the extreme coherent and completely monotone lower previsions on $\mathbb{B}$ are precisely the vacuous ones:

$$\mathrm{ext}(\underline{\mathbb{P}}_\infty) = \{\underline{P}_A : \emptyset \neq A \subseteq \mathscr{X}\}.$$

Any probability measure on the finite set of extreme points $\operatorname{ext}(\underline{\mathbb{P}}_\infty)$ is therefore completely determined by its probability mass function m, which is an element of the simplex $\Sigma_{\mathscr{P}(\mathscr{X})}$ with $m(\emptyset) = 0$.[2] We then infer from the earlier representation results that for every coherent and completely monotone lower prevision $\underline{P}$ on $\mathbb{B}$, there is some such mass function m for which we have the following representation:

$$\underline{P}(f) = \sum_{F \subseteq \mathscr{X}} m(F)\underline{P}_F(f) \text{ for all bounded gambles } f \text{ on } \mathscr{X}.$$

In other words, *any coherent and completely monotone lower prevision on* $\mathbb{B}$ *is a convex combination of vacuous lower previsions and therefore the natural extension of a belief function*, as introduced in Section 7.3_{128}. The basic probability assignment for this belief function is essentially the mass function of the representing probability measure on extreme points.

11.6.2 Possibility measures

For the next example, we turn to a possibility measure Π on a non-empty set $\mathscr{X}$, with possibility distribution π, as defined in Section 7.8_{143}. If we assume that $\sup \pi = 1$, then we infer from Proposition 7.14_{144} that Π is a coherent upper probability on $\mathscr{P}(\mathscr{X})$, whose natural extension to $\mathbb{B}$ is given by

$$\overline{E}_\Pi(f) = R\int_0^1 \sup\left\{f(x) : x \in \pi_\alpha\right\} \mathrm{d}\alpha = R\int_0^1 \overline{P}_{\pi_\alpha}(f)\,\mathrm{d}\alpha = R\int_0^1 \tilde{f}(\overline{P}_{\pi_\alpha})\,\mathrm{d}\alpha, \tag{11.16}$$

where we let $\pi_\alpha := \{x \in \mathscr{X} : \pi(x) \geq \alpha\}$. Consider the probability measure on the extreme points defined by

$$\mu_{\overline{E}_\Pi}(A) = \lambda\left(\left\{\alpha \in [0,1] : \overline{P}_{\pi_\alpha} \in A\right\}\right)$$

for all $A \subseteq \operatorname{ext}(\underline{\mathbb{P}}_\infty)$ such that $\{\alpha \in [0,1] : \overline{P}_{\pi_\alpha} \in A\} \in \mathscr{B}([0,1])$, where λ is the Lebesgue measure on the real unit interval $[0,1]$. This probability measure is a representation of the upper prevision $\overline{E}_\Pi$, because Equation (11.16) can be rewritten as $\overline{E}_\Pi(f) = L\int \tilde{f}|_{\operatorname{ext}(\underline{\mathbb{P}}_\infty)}\,\mathrm{d}\mu_{\overline{E}_\Pi}$ (just observe that the function $\overline{P}_{\pi_\alpha}$ in α is non-increasing and, therefore, Riemann integrable; the Lebesgue and Riemann integrals coincide on Riemann integrable functions). The probability measure $\mu_{\overline{E}_\Pi}$ is **supported by** the set of extreme points $\{\overline{P}_{\pi_\alpha} : \alpha \in [0,1]\}$, because

$$\mu_{\overline{E}_\Pi}\left(\{\overline{P}_{\pi_\alpha} : \alpha \in [0,1]\}\right) = \lambda([0,1]) = 1.$$

11.6.3 Extending a linear prevision defined on all continuous bounded gambles

To finish this short list of examples, we return to the discussion started in Section $5.5.6_{98}$ and continued in Section 9.5_{187}, and consider a linear prevision P

[2] It is easy to see that all subsets of the finite set $\operatorname{ext}(\underline{\mathbb{P}}_\infty)$ are relative Baire, because all singletons are.

defined on the set $\mathscr{C}(\mathcal{X})$ of all continuous real functions on $\mathcal{X}$. We assume that there is a topology $\mathscr{T}$ of open sets such that $\mathcal{X}$ is compact and metrisable.

We are interested in the natural extension $\underline{E}_P$ of the linear prevision P to all bounded gambles. As P is completely monotone by Theorem 6.6_{105}, we infer from Corollary 6.23_{118} that its natural extension $\underline{E}_P$ is completely monotone as well.

On the other hand, the Riesz representation theorem (see, for instance, Schechter, 1997, Section 29.35) guarantees that there is a *unique* probability measure μ_P defined on the σ-field $\mathscr{B}(\mathcal{X})$ of all Borel subsets of $\mathcal{X}$ such that $P(f) = L\int f\,\mathrm{d}\mu_P$ for all $f \in \mathscr{C}(\mathcal{X})$. We then infer from the discussion in Section 9.5_{187} that

$$\underline{E}_P(f) = L\int \underline{\mathrm{osc}}(f)\,\mathrm{d}\mu_P \text{ for all bounded gambles } f \text{ on } \mathcal{X}. \tag{11.17}$$

If we recall from Definition 5.14_{94} that $\underline{\mathrm{osc}}_x(f) = \underline{P}_{\mathcal{N}_x}(f) = \tilde{f}(\underline{P}_{\mathcal{N}_x})$, where $\underline{P}_{\mathcal{N}_x}$ is the lower prevision associated with the neighbourhood filter $\mathcal{N}_x$ of the element x of $\mathcal{X}$, then we see that *Equation* (11.17) *is essentially the representation formula* $\underline{E}_P(f) = L\int \tilde{f}|_{\mathrm{ext}(\underline{\mathbb{P}}_\infty)}\,\mathrm{d}\mu_{\underline{E}_P}$, where the representing probability measure on the extreme points is essentially the Riesz measure μ_P:

$$\mu_{\underline{E}_P}(A) := \mu_P\left(\left\{x \in \mathcal{X} : \underline{P}_{\mathcal{N}_x} \in A\right\}\right)$$

for all $A \subseteq \mathrm{ext}(\underline{\mathbb{P}}_\infty)$ such that $\{x \in \mathcal{X} : \underline{P}_{\mathcal{N}_x} \in A\} \in \mathscr{B}(\mathcal{X})$. Clearly, $\mu_{\underline{E}_P}$ is *supported by* the set of extreme points $\{\underline{P}_{\mathcal{N}_x} : x \in \mathcal{X}\}$, because

$$\mu_{\underline{E}_P}\left(\left\{\underline{P}_{\mathcal{N}_x} : x \in \mathcal{X}\right\}\right) = \mu_P(\mathcal{X}) = 1.$$

11.6.4 The connection with induced lower previsions

There is an interesting connection between the results in this chapter and the discussion of induced lower previsions in Section 7.5_{131}; we refer to that section for an introduction to the notions and notations used in the following discussion. Consider, besides the space $\mathcal{X}$, an arbitrary space $\mathcal{Y}$, any filter map $\Phi : \mathcal{Y} \to \mathbb{F}(\mathcal{X})$ and any linear prevision P on $\mathbb{B}(\mathcal{Y})$.

The filter map Φ can be used to turn the linear prevision P into the (coherent) induced lower prevision $P_\bullet$ on $\mathbb{B}(\mathcal{X})$, where

$$P_\bullet(f) = P(f_\bullet) \text{ and } f_\bullet(y) = \underline{P}_{\Phi(y)}(f) \text{ for all } f \in \mathbb{B}(\mathcal{X}) \text{ and all } y \in \mathcal{Y}.$$

$\underline{P}_{\Phi(y)}$ is, of course, the coherent lower prevision associated with the proper filter $\Phi(y)$ of subsets of $\mathcal{Y}$. As the linear prevision P is completely monotone (by Theorem 6.6_{105}) and inducing preserves complete monotonicity (by Proposition $7.9(\mathrm{iii})_{135}$), the induced lower prevision $P_\bullet$ is not only coherent but also completely monotone.

Conversely, any completely monotone coherent lower prevision $\underline{P}$ on $\mathbb{B}(\mathcal{X})$ can be seen as an induced lower prevision, and trivially so, by taking $X = Y$, $\Phi(y) =$

$\{B \subseteq \mathcal{X} : y \in B\}$, because then $\underline{P}_\bullet = \underline{P}$. But the results in this section bear out that there is a less trivial way: Let $\mathcal{Y}$ be the set

$$\underline{\mathbb{P}}_\infty = \underline{\mathbb{P}}_\infty^{\mathbb{B}(\mathcal{X})} = \left\{\underline{P}_{\mathcal{F}} : \mathcal{F} \in \mathbb{F}(\mathcal{X})\right\}$$

of all extreme completely monotone coherent lower prevision on $\mathbb{B}(\mathcal{X})$ and define the corresponding filter map $\Phi : \mathcal{Y} \to \mathbb{F}(\mathcal{X})$ by

$$\Phi\left(\underline{P}_{\mathcal{F}}\right) := \mathcal{F} = \left\{F \subseteq \mathcal{X} : \underline{P}_{\mathcal{F}}(F) = 1\right\},$$

then the corresponding lower inverse $f_\bullet$ of any bounded gamble f on $\mathcal{X}$ is given by

$$f_\bullet\left(\underline{P}_{\mathcal{F}}\right) = \underline{P}_{\Phi(\underline{P}_{\mathcal{F}})}(f) = \underline{P}_{\mathcal{F}}(f) = \tilde{f}\left(\underline{P}_{\mathcal{F}}\right),$$

showing that $f_\bullet = \tilde{f}|_{\mathrm{ext}(\underline{\mathbb{P}}_\infty)}$. Consequently, we can rewrite Equation $(11.15)_{225}$ as

$$\underline{P}(f) = L\int \tilde{f}|_{\mathrm{ext}(\underline{\mathbb{P}}_\infty)}\,\mathrm{d}\mu_{\underline{P}} = L\int f_\bullet\,\mathrm{d}\mu_{\underline{P}} = E_{\mu_{\underline{P}}}(f_\bullet)$$

or, using the developments of Section 9.6_{188}, as

$$\underline{P} = (P_{\mu_{\underline{P}}})_\bullet,$$

where $P_{\mu_{\underline{P}}}$ is the linear prevision associated with the probability measure $\mu_{\underline{P}}$. Such linear previsions, and their relation with integration, were introduced and discussed in detail in Chapter 8_{151} and Section 9.4_{183}.

11.7 Strongly invariant coherent lower previsions

To finish this chapter, we briefly discuss the notion of *ergodicity* in relation to coherent lower previsions. Consider a monoid $\mathbb{T}$ of transformations of the non-empty set $\mathcal{X}$ and the set $\underline{\mathbb{P}}_{\mathbb{T}}(\mathcal{X})$ of all strongly $\mathbb{T}$-invariant coherent lower previsions on $\mathbb{B}$. We know from Corollary 10.8_{199} that this set will be non-empty provided that $\sup g \geq 0$ for all $g \in \mathcal{D}_{\mathbb{T}}$. This will, for instance, be the case if $\mathbb{T}^*$ has the Moore–Smith property, as we made clear in Theorem 10.10_{201}.

It is clear from Proposition 10.6_{197} that a coherent lower prevision $\underline{P} \in \underline{\mathbb{P}}(\mathcal{X})$ is strongly $\mathbb{T}$-invariant if and only

$$\underline{P}(g) = \tilde{g}(\underline{P}) = 0 \text{ for all } g \in \mathcal{D}_{\mathbb{T}},$$

which implies that $\underline{\mathbb{P}}_{\mathbb{T}}(\mathcal{X})$ is a $\mathcal{T}^*$-closed subset of $\mathcal{C}\left(\mathcal{S}_{\mathrm{ex}}\right) \cap B_1(\mathcal{K})$: the intersection of this set with the $\mathcal{T}^*$-closed set $\bigcap_{g\in\mathcal{D}_{\mathbb{T}}} \tilde{g}^{-1}(\{0\})$. This leads to the following corollary to Theorem 11.9_{224}.

Corollary 11.12 *Consider a monoid $\mathbb{T}$ of transformations of the non-empty set $\mathcal{X}$, such that the set $\underline{\mathbb{P}}_{\mathbb{T}}(\mathcal{X})$ of all strongly $\mathbb{T}$-invariant coherent lower previsions on $\mathbb{B}$ is non-empty. For every strongly $\mathbb{T}$-invariant coherent lower prevision $\underline{P}$ on $\mathbb{B}$, there is some probability measure $\mu_{\underline{P}}$, defined on the σ-field $\{\text{ext}(\underline{\mathbb{P}}_{\mathbb{T}}(\mathcal{X})) \cap B : B \in \mathscr{B}_0(\underline{\mathbb{P}}_{\mathbb{T}}(\mathcal{X}))\}$, that represents $\underline{P}$, meaning that*

$$\underline{P}(f) = L \int \tilde{f}|_{\text{ext}(\underline{\mathbb{P}}_{\mathbb{T}}(\mathcal{X}))} \, \mathrm{d}\mu_{\underline{P}} \text{ for all bounded gambles } f \text{ on } \mathcal{X}.$$

In other words, every strongly $\mathbb{T}$-invariant coherent lower prevision can be written as a 'σ-additive convex combination' of the extreme strongly $\mathbb{T}$-invariant coherent lower previsions.

There are several special instances of this theorem that definitely merit further attention. We restrict ourselves here to a succinct discussion of two of them: the cases of exchangeability and stationarity.

Let us first consider the case of *exchangeability* for a countable number of finite-valued random variables, which was mentioned very briefly near the end of Section 10.3.1$_{205}$ and discussed in depth by De Cooman et al. (2009) and De Cooman and Quaeghebeur (2012). Here, Corollary 11.12 extends a well-known result by de Finetti (1937) (see also Hewitt and Savage, 1955) stating roughly that any precise probability model describing an exchangeable sequence of finite-valued random variables can be seen as a 'sigma-additive convex mixture' of precise models for independent identically distributed (iid) sequences. These iid models are the extreme points of the convex set of exchangeable models.

The necessary background for the case of *stationarity* was presented in Section 10.3.3$_{210}$. Here, Corollary 11.12 extends the ergodic decomposition theorem by Krylov and Bogolioubov (1937) (see also Kallenberg, 2002, Theorem 10.26), which states roughly that any stationary probability measure is a 'sigma-additive convex mixture' of extreme stationary measures, called *ergodic measures*. We could also call **ergodic** the extreme points of the set of all stationary coherent lower previsions, but as is the case with many other types of extreme lower previsions, we are still lacking a good characterisation of ergodicity for coherent lower previsions.

Part II

EXTENDING THE THEORY TO UNBOUNDED GAMBLES

12

Introduction

As we have seen in the first part, the existing theory of coherent lower previsions deals exclusively with bounded gambles, which makes it quite difficult to apply in many applications, such as statistical decision-making and optimal control problems, as these often involve unbounded loss or gain functions.

Our motivation for extending the existing theory to unbounded gambles is, therefore, mainly practical. However, in order to keep this book from becoming unmanageably large, we have decided to concentrate most on discussing and describing the mathematical details and underpinnings of our extension method.

In order to generalise Walley's (1991) theory of lower previsions, which are real-valued maps on bounded gambles, to arbitrary gambles, we introduce *(extended) lower previsions* as $\mathbb{R} \cup \{-\infty, +\infty\}$-valued maps on arbitrary, not necessarily bounded, gambles. Similar to its counterpart for bounded gambles, a lower prevision for a gamble represents a supremum buying price; also, when its lower prevision is $-\infty$, it will never be bought at any price, and when its lower prevision is $+\infty$, it will be bought at any price.

In Chapter 13_{235}, we suggest and motivate conditions for *avoiding sure loss* and *coherence* of such lower previsions for gambles and we construct a *natural extension* for them, which turns out to be the smallest dominating coherent lower prevision on all gambles.

In the remaining two chapters, we investigate how we can extend given lower and upper previsions from bounded gambles to a larger set of gambles. Our approach is based on the theory of integration (see Halmos, 1974; Bhaskara Rao and Bhaskara Rao, 1983; and references therein). We introduce *null events* as sets that a subject is practically certain do not occur, and *null gambles* – essentially – as gambles that the subject is practically will only assume the zero value. An immediate first step, taken in Chapter 14_{304}, is to extend a lower prevision to gambles that are bounded on the complement of a null set – *essentially bounded gambles*. Indeed, we show that buying and selling prices are equal when bounded gambles differ only on

Lower Previsions, First Edition. Matthias C.M. Troffaes and Gert de Cooman.

a null set. Consequently, we can extend the domain of a lower prevision to the set of its essentially bounded gambles. This extension is achieved by a natural extension procedure that can be motivated by an axiom stating that 'adding null gambles does not affect acceptability'.

In Chapter 15_{327} then, in a further step towards more generality, we consider approximating an unbounded gamble by a sequence of bounded ones. Its lower prevision is then defined as the limit of the sequence of the lower previsions of its approximating bounded gambles. We identify gambles for which this limit does not depend on the details of the approximation. An approximating sequence of bounded gambles is called a *determining sequence* and gambles for which a determining sequence exists are called *previsible*. We show that a gamble is previsible if and only if a sequence of *cuts* is a determining sequence. This gives rise to a simple sufficient condition for a gamble to be previsible. If a lower prevision is 2-monotone, we find an integral representation for the lower prevision of previsible gambles, and we prove that previsibility coincides with Choquet integrability.

Both types of extension just described are coherent in the sense of Chapter 13.

Much of the material in this part has not been published elsewhere in much detail. The foundations for Chapter 13 were laid in Troffaes and De Cooman (2002b), which was developed more extensively in Troffaes (2005, Section 5.2). Those works only deal with the unconditional case. The more general conditional case we treat here has already been outlined succinctly in Troffaes (2006), and we mostly build on and extend that work. A very brief version of Chapters 14_{304} and 15_{327}, with an outline of only the essential ideas, can be found in Troffaes and De Cooman (2002a, 2003). A more detailed account is given in Troffaes (2005, Sections 5.3–5.6), which we will follow closely, adding various improvements and extensions as we go along.

13

Conditional lower previsions

This chapter has two purposes: (i) to extend the theory of lower previsions in Part I_{21} to account also for unbounded gambles and (ii) to extend the theory to deal with conditioning.

Walley's (1991) theory of conditional lower previsions unifies many of the imprecise probability models in the literature, and from a foundational point of view, it seems to be quite satisfactory, as we have argued for the unconditional case in Chapter 4_{37}. However, one technical problem with Walley's theory is that his conditional lower previsions are only defined on bounded gambles, whereas in many applications, unbounded gambles are legion. A few examples are optimisation using an imprecise cost criterion, with an unbounded – for example, quadratic – cost (Chevé and Congar, 2000; De Cooman and Troffaes, 2003), or the estimation of unbounded quantities that depend on parameters that are not well known, such as time to failure in reliability theory (Utkin and Gurov, 1999).

So the question arises whether conditional lower previsions can be extended from bounded to arbitrary gambles. The work of Crisma, Gigante and Millossovich (1997b) is particularly relevant and interesting in this context: they introduce unconditional *linear* previsions for arbitrary real-valued gambles, and these linear previsions may also assume the values $+\infty$ and $-\infty$. As this work indicates that the domain of *linear previsions* can be extended to all gambles by including $\pm\infty$ in the range of the prevision, it is now a natural question whether something similar can also be achieved for *coherent conditional lower previsions*.

In this chapter, we show that this is indeed possible. Unconditional lower previsions for arbitrary gambles were already discussed in Troffaes and De Cooman (2002b) and treated in detail in Troffaes (2005, Section 5.2). That work was extended to the conditional case – without proofs – in Troffaes (2006), on which this chapter builds. However, in contrast with this earlier work, here we will start out from sets of acceptable gambles and use an approach that is similar to that in Chapters 3_{25} and 4_{37}.

Lower Previsions, First Edition. Matthias C.M. Troffaes and Gert de Cooman.

13.1 Gambles

We first relax the assumption that rewards have bounded utility; therefore, we need to start out with a fixed *unbounded* utility scale. Necessary and sufficient conditions under which an unbounded utility scale can be constructed from an order on lotteries over some set of rewards have been given by Herstein and Milnor (1953).

Let X be a random variable taking values in a set $\mathscr{X}$.[1] A **gamble** f on $\mathscr{X}$ is a real-valued gain, expressed in our fixed utility scale, that is a function of X. Mathematically, it is an $\mathscr{X} - \mathbb{R}$-map, interpreted as an uncertain gain: if x turns out to be the realisation of X, then we receive an amount $f(x)$ of utility. When working with a number of different random variables, we may write $f(X)$ in order to emphasise that f is a function of X.

Recall from Chapter 1_1 that the set of all gambles on $\mathscr{X}$ is denoted by $\mathbb{G}(\mathscr{X})$ or, more simply, by $\mathbb{G}$. The set $\mathbb{B}(\mathscr{X})$ of bounded gambles on $\mathscr{X}$ is a subset of $\mathbb{G}(\mathscr{X})$, and, similar to $\mathbb{B}(\mathscr{X})$, $\mathbb{G}(\mathscr{X})$ is a linear lattice with respect to the point-wise addition, the point-wise scalar multiplication and the point-wise ordering of gambles. We emphasise that gambles assume only values in $\mathbb{R}$: they are never infinite.

13.2 Sets of acceptable gambles

As we argued earlier in Section 3.4_{29}, beliefs a subject has about X will lead him to accept or reject transactions whose reward depends on X. Therefore, we can take the subject's **set of acceptable gambles** $\mathscr{D}$ – the set of gambles he is willing to accept – to be a model for a subject's beliefs about X.

If we consider two subjects, one with a set of acceptable gambles $\mathscr{D}_1$ and another with $\mathscr{D}_2$, then the inclusion relation $\mathscr{D}_1 \subseteq \mathscr{D}_2$ means that the second subject will at least accept all gambles that the first one does. Here too, the inclusion relation '$\subseteq$' between sets of acceptable gambles can therefore be interpreted as '*is at most as informative as*', '*is at most as committal as*' or '*is at least as conservative as*'.

13.2.1 Rationality criteria

Similarly to what we did for acceptable bounded gambles in Section $3.4.1_{29}$, we introduce the following rationality criteria for the acceptability of gambles.

Axiom 13.1 (Rationality for sets of acceptable gambles) *For all gambles f and g, and all non-negative real numbers λ,*

A1. *If $f < 0$, then $f \notin \mathscr{D}$: a subject should not be disposed to accept any gamble he cannot win from, and that may actually make him lose.* **(avoiding partial loss).**

A2. *If $f \geq 0$, then $f \in \mathscr{D}$: a subject should be disposed to accept any gamble in which he cannot lose.* **(accepting partial gain).**

[1] Random variables were introduced in Section 3.1_{26}.

A3. *If $f \in \mathcal{D}$, then $\lambda f \in \mathcal{D}$: a subject disposed to accept f, should also be disposed to accept λf.* **(scale invariance)**.

A4. *If $f \in \mathcal{D}$ and $g \in \mathcal{D}$, then $f + g \in \mathcal{D}$: a subject disposed to accept f and g, should also be disposed to accept their combination $f + g$.* **(combination)**.

The justification for these axioms is exactly as for bounded gambles (see Section 3.4.1$_{29}$). Again, we naturally arrive at the following definition.

Definition 13.2 (Coherence for sets of acceptable gambles) *Any set of acceptable gambles $\mathcal{D}$ that satisfies the rationality criteria A1–A4 of Axiom 13.1 is called* ***coherent****. We denote the set of all coherent sets of acceptable gambles on $\mathcal{X}$ by $\mathbb{D}(\mathcal{X})$, or simply $\mathbb{D}$ when no confusion can arise.*

Here too, coherent sets of acceptable gambles are convex cones that include the non-negative orthant and that have no point in common with the negative orthant.

Example 13.3 (The vacuous belief model) A very special case of a coherent set of acceptable gambles is the non-negative orthant of $\mathbb{G}$:

$$\mathbb{G}_{\geq 0} := \{f \in \mathbb{G} : f \geq 0\}.$$

This set is included in any other element of $\mathbb{D}$ and therefore the smallest, most conservative or least informative element of $\mathbb{D}$. The set $\mathbb{G}_{\geq 0}$ can be taken as a model of a subject's belief if he has no information whatsoever regarding X: adopting $\mathbb{G}_{\geq 0}$ means accepting only those bounded gambles that will not lead to a partial loss.

More generally, for any non-empty subset B of $\mathcal{X}$, consider

$$\mathcal{D}_B := \left\{f \in \mathbb{G} : fI_B > 0\right\} \cup \mathbb{G}_{\geq 0}.$$

This model reflects a subject's belief that no outcome outside B will obtain, and nothing more. Indeed, adopting $\mathcal{D}_B$ means accepting any gamble that yields either a positive utility for some x in B or a non-negative utility regardless of the outcome of x. It is easy to check that $\mathcal{D}_B$ is a coherent set of acceptable gambles.

The set $\left\{f \in \mathbb{G} : fI_B \geq 0\right\}$ is *not coherent* (unless $B = \mathcal{X}$): this set contains, for example, the bounded gamble $-I_{B^c}$, which incurs partial loss and therefore violates axiom A1. ◆

These axioms of rationality, and the corresponding definition of coherence, which lay the foundations for the remainder of this chapter, are not immediately compatible with their counterparts from Chapter 4$_{37}$: for one thing, a coherent set of acceptable bounded gambles is not coherent when considered as a set of gambles, as it does not contain the unbounded non-negative gambles. This incompatibility is not a problem, provided we know which definitions we are referring to: when we know whether we are working only with bounded gambles or with all gambles. Fortunately, both definitions can be unified and generalised by using the notion of coherence *relative to a cone*. As a detailed discussion of these issues is not of crucial importance to the developments in this book, we simply refer to Walley (1991, Section 3.7) and, for a more general treatment, to De Cooman and Quaeghebeur (2012, Section 2).

13.2.2 Inference

In general, we cannot expect a subject's specification of a set $\mathscr{A}$ of acceptable gambles to be coherent. For instance, when we are performing an elicitation of a subject's beliefs by looking at which gambles are acceptable to him, his assessment $\mathscr{A}$ will often only be finite, for practical reasons.

However, we can always try to extend $\mathscr{A}$ to a coherent set of acceptable gambles, in a way that is as conservative as possible:

$$\mathscr{E}_{\mathscr{A}} := \left\{ g + \sum_{k=1}^{n} \lambda_k f_k : g \geq 0, n \geq 0, f_k \in \mathscr{A}, \lambda_k \in \mathbb{R}_{\geq 0}, k = 1, \ldots, n \right\}, \quad (13.1)$$

$$= \text{nonneg}(\mathbb{G}_{\geq 0} \cup \mathscr{A}) = \mathbb{G}_{\geq 0} + \text{nonneg}(\mathscr{A}).$$

Clearly, $\mathscr{E}_{\mathscr{A}}$ includes $\mathscr{A}$ and satisfies the coherence conditions A2–A4$_{\curvearrowleft}$. However, $\mathscr{E}_{\mathscr{A}}$ will not necessarily satisfy condition A1$_{236}$. This leads us to the following definition.

Definition 13.4 (Consistency) *A set $\mathscr{A}$ of acceptable gambles is called* **consistent**, *or* **avoids partial loss**, *if one (and hence all) of the following equivalent conditions is satisfied:*

(A) *$\mathscr{A}$ is included in some coherent set of acceptable gambles:*

$$\{\mathscr{D} \in \mathbb{D} : \mathscr{A} \subseteq \mathscr{D}\} \neq \emptyset.$$

(B) *For all n in $\mathbb{N}$, non-negative λ_1, ..., λ_n in $\mathbb{R}$ and gambles f_1, ..., f_n in $\mathscr{A}$,*

$$\sum_{k=1}^{n} \lambda_k f_k \nless 0.$$

Recall that '$f < g$' stands for '$f \leq g$ and not $f = g$'. Condition (B) says that no non-negative linear combination of gambles in $\mathscr{A}$ incurs partial loss in the sense of condition A1$_{236}$, and this is why the consistency condition is also called '*avoiding partial loss*'.

Let us first prove that these conditions are indeed equivalent.

Proof. (A)⇒(B). Assume that there is some coherent $\mathscr{D}$ that includes $\mathscr{A}$. Consider arbitrary $n \geq 0$, real $\lambda_k \geq 0$ and f_k in $\mathscr{A}$. Then $g := \sum_{k=1}^{n} \lambda_k f_k$ belongs to $\mathscr{D}$, by coherence conditions A2–A4$_{\curvearrowleft}$. But then $g \nless 0$ by coherence condition A1$_{236}$.

(B)⇒(A). Recall that $\mathscr{E}_{\mathscr{A}}$ is a set of gambles that satisfies the coherence conditions A2–A4$_{\curvearrowleft}$. We infer from (B) that $h \nless 0$ for all h in $\mathscr{E}_{\mathscr{A}}$, so $\mathscr{E}_{\mathscr{A}}$ also satisfies coherence condition A1$_{236}$ and is therefore a coherent set of desirable gambles that includes $\mathscr{A}$. □

If a subject can provide us with a consistent set of acceptable gambles $\mathscr{A}$, then, by definition, $\mathscr{A}$ can be extended to a coherent set of acceptable gambles. Of all those extensions, can we find one that is as conservative as possible? This turns out to be very straightforward. First, we observe that, trivially, any intersection of coherent sets of acceptable gambles is coherent as well. The most conservative coherent extension is, therefore, given by the intersection of all coherent extensions.

Proposition 13.5 *Consider a non-empty family $\mathscr{D}_i$, $i \in I$ of sets of acceptable gambles. If all $\mathscr{D}_i$ are coherent, then so is their intersection $\bigcap_{i\in I} \mathscr{D}_i$.*

Again, similarly to what we have done with bounded gambles, for any consistent assessment $\mathscr{A}$, we can now consider the non-empty collection $\{\mathscr{D} \in \mathbb{D} : \mathscr{A} \subseteq \mathscr{D}\}$ of all coherent sets of acceptable gambles that include $\mathscr{A}$ and we define the intersection of this collection to be the **closure** $\mathrm{Cl}_{\mathbb{D}}(\mathscr{A})$ of $\mathscr{A}$:

$$\mathrm{Cl}_{\mathbb{D}}(\mathscr{A}) := \bigcap \{\mathscr{D} \in \mathbb{D} : \mathscr{A} \subseteq \mathscr{D}\}.$$

In the expression above, we take the intersection of the empty collection to be equal to the set $\mathbb{G}$ of all gambles: $\bigcap \emptyset := \mathbb{G}$. The closure operator $\mathrm{Cl}_{\mathbb{D}}$ has the following interesting properties.

Proposition 13.6 *Let $\mathscr{A}$, $\mathscr{A}_1$ and $\mathscr{A}_2$ be consistent sets of acceptable gambles. Then the following statements hold:*

(i) $\mathscr{A} \subseteq \mathrm{Cl}_{\mathbb{D}}(\mathscr{A})$.

(ii) *If* $\mathscr{A}_1 \subseteq \mathscr{A}_2$, *then* $\mathrm{Cl}_{\mathbb{D}}(\mathscr{A}_1) \subseteq \mathrm{Cl}_{\mathbb{D}}(\mathscr{A}_2)$.

(iii) $\mathrm{Cl}_{\mathbb{D}}(\mathrm{Cl}_{\mathbb{D}}(\mathscr{A})) = \mathrm{Cl}_{\mathbb{D}}(\mathscr{A})$.

(iv) *If* $\mathscr{A} \subseteq \mathbb{G}_{\geq 0}$, *then* $\mathrm{Cl}_{\mathbb{D}}(\mathscr{A}) = \mathbb{G}_{\geq 0}$.

(v) $\mathscr{A}$ *is consistent if and only if* $\mathrm{Cl}_{\mathbb{D}}(\mathscr{A}) \neq \mathbb{G}$.

(vi) $\mathscr{A}$ *is a coherent set of acceptable gambles if and only if* $\mathscr{A} = \mathrm{Cl}_{\mathbb{D}}(\mathscr{A})$.

Proof. The proof is completely analogous to that of Proposition 3.6_{33}. □

Theorem 13.7 (Natural extension) *If a set $\mathscr{A}$ of acceptable gambles is consistent, then there is a smallest coherent set of acceptable gambles that includes $\mathscr{A}$. It is given by*

$$\mathrm{Cl}_{\mathbb{D}}(\mathscr{A}) = \mathscr{E}_{\mathscr{A}} = \left\{ g + \sum_{k=1}^{n} \lambda_k f_k : g \geq 0, n \geq 0, f_k \in \mathscr{A}, \lambda_k \in \mathbb{R}_{\geq 0} \right\}$$

$$= \left\{ h \in \mathbb{G} : h \geq \sum_{k=1}^{n} \lambda_k f_k \text{ for some } n \geq 0, f_k \in \mathscr{A}, \lambda_k \in \mathbb{R}_{\geq 0} \right\},$$

and it is called the ***natural extension*** *of* $\mathscr{A}$.

Proof. The proof is analogous to that of Theorem 3.7_{34}. □

This theorem shows that, as with bounded gambles, the closure operator $\mathrm{Cl}_{\mathbb{D}}$ simply adds those gambles to $\mathscr{A}$ that can be obtained from gambles in $\mathscr{A}$ using the 'production rules' A2–A4$_{237}$ in Axiom 13.1$_{236}$ and no other gambles.

It is clear from the discussion above that, when working with acceptability as the primitive notion for modelling uncertainty, generalising from bounded gambles to all gambles is trivial.

Let us now turn our attention to generalising from lower previsions for bounded gambles to conditional lower previsions for arbitrary gambles. This generalisation is not nearly as trivial and involves more than a few technicalities. The results we have derived so far for acceptability will enable us to hide, or rather circumvent, many of them, particularly in proofs.

13.3 Conditional lower previsions

13.3.1 Going from sets of acceptable gambles to conditional lower previsions

As in Chapter 4$_{37}$, acceptability is the fundamental notion we start out from. Let us assume that our subject has specified a consistent set of acceptable gambles, from which we can infer a coherent set $\mathscr{D}$ of acceptable gambles through natural extension or conservative inference. Recall that $\mathbb{R}^* := \mathbb{R} \cup \{-\infty, +\infty\}$ denotes the set of extended real numbers, and let $\mathscr{P}^{\circ}(\mathscr{X})$, or simply $\mathscr{P}^{\circ}$, denote the power set of $\mathscr{X}$ excluding the empty set. We discuss how to work with extended real numbers in Appendix D$_{391}$.

We can associate with $\mathscr{D}$ two special functionals defined on $\mathbb{G} \times \mathscr{P}^{\circ}$: the **conditional lower prevision** $\mathrm{lpr}(\mathscr{D})(\bullet|\bullet) : \mathbb{G} \times \mathscr{P}^{\circ} \to \mathbb{R}^*$ defined by

$$\mathrm{lpr}(\mathscr{D})(f|A) := \sup\left\{\mu \in \mathbb{R} : (f - \mu)I_A \in \mathscr{D}\right\} \tag{13.2}$$

and the **conditional upper prevision** $\mathrm{upr}(\mathscr{D})(\bullet|\bullet) : \mathbb{G} \times \mathscr{P}^{\circ} \to \mathbb{R}^*$ defined by

$$\mathrm{upr}(\mathscr{D})(f|A) := \inf\left\{\mu \in \mathbb{R} : (\mu - f)I_A \in \mathscr{D}\right\}$$

for any gamble f on $\mathscr{X}$ and any event A in $\mathscr{P}^{\circ}$.

So lpr can be seen as an operator that turns any coherent set of acceptable gambles into an $\mathbb{G} \times \mathscr{P}^{\circ} - \mathbb{R}^*$ map, which we call a conditional lower prevision. Similarly, upr turns any coherent set of acceptable gambles into an $\mathbb{G} \times \mathscr{P}^{\circ} - \mathbb{R}^*$ map, which we call a conditional upper prevision. What interpretation can we give to these two $\mathbb{G} \times \mathscr{P}^{\circ} - \mathbb{R}^*$ maps?

Clearly, $\mathrm{lpr}(\mathscr{D})(f|A)$ is the supremum price μ such that the subject accepts to **buy f for μ if A obtains**, that is, to exchange the called-off reward μI_A for the called-off random reward fI_A. In other words, the conditional lower prevision $\mathrm{lpr}(\mathscr{D})(f|A)$ of f given A is the *supremum acceptable called-off buying price* for f if A obtains, associated with the coherent set of acceptable gambles $\mathscr{D}$.

Similarly, $\mathrm{upr}(\mathscr{D})(f|A)$ is the infimum price μ such that the subject accepts to **sell f for μ if A obtains**, that is, to exchange the called-off random reward fI_A for the called-off reward μI_A. In other words, the conditional upper prevision $\mathrm{upr}(\mathscr{D})(f|A)$ of f given A is the *infimum acceptable called-off selling price* for f if A obtains, associated with the coherent set of acceptable gambles $\mathscr{D}$.

As the $\mathbb{R}^*$-valued functionals $\mathrm{lpr}(\mathscr{D})(\bullet|\bullet)$ and $\mathrm{upr}(\mathscr{D})(\bullet|\bullet)$ satisfy the following **conjugacy relationship**

$$\mathrm{upr}(\mathscr{D})(f|A) = -\,\mathrm{lpr}(\mathscr{D})(-f|A) \text{ for all } (f, A) \text{ in } \mathbb{G} \times \mathscr{P}^{\circ},$$

we can always express one type of functional in terms of the other. For this reason, we will concentrate on conditional lower previsions.

For lower previsions on bounded gambles, we have got by quite well without infinite values, so a natural question at this point is whether we really need to allow $\mathrm{lpr}(\mathscr{D})(\bullet|\bullet)$ to assume infinite values. It turns out that we do. The next two examples are adapted from Troffaes (2005, pp. 214–215); we will come back to similar examples in Section 13.11_{301}.

First observe that, even when $\mathscr{D}$ is coherent, the set $\{\mu \in \mathbb{R} : (f - \mu)I_A \in \mathscr{D}\}$ may be empty for certain unbounded gambles f and events A, in which case $\mathrm{lpr}(\mathscr{D})(f|A) = -\infty$.

Example 13.8 Consider the coherent set $\mathbb{G}_{\geq 0}$ of all non-negative gambles. If f is unbounded below on A, then $(f - \mu)I_A$ can never be non-negative, whatever the value of $\mu \in \mathbb{R}$. Hence, $\mathrm{lpr}(\mathbb{G}_{\geq 0})(f|A) = -\infty$. ♦

Secondly, it may also happen that, for certain unbounded gambles f, all buying prices are acceptable, meaning that $\{\mu \in \mathbb{R} : f - \mu \in \mathscr{D}\} = \mathbb{R}$, in which case $\mathrm{lpr}(\mathscr{D})(f|\mathscr{X}) = +\infty$.

Example 13.9 Assume that X takes values in $\mathbb{N}$, and consider $\mathscr{A} := \{f_n : n \in \mathbb{N}\}$ with[2]

$$f_n(X) := \min\{X - n + 1, 1\}.$$

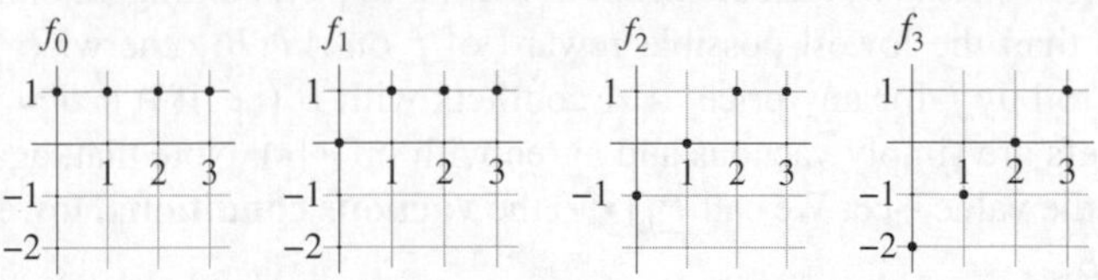

This set $\mathscr{A}$ of acceptable gambles is consistent, because any non-trivial non-negative linear combination of them is strictly positive for large enough x. Consider $\mathscr{D} := \mathrm{Cl}_{\mathbb{D}}(\mathscr{A})$. By Theorem 13.7_{239}, $\mathscr{D}$ is coherent. Then

$$\begin{aligned}\mathrm{lpr}(\mathscr{D})(X|\mathscr{X}) &= \sup\{\mu \in \mathbb{R} : X - \mu \in \mathscr{D}\}\\ &\geq \sup\{\mu \in \mathbb{R} : (\exists n \in \mathbb{N})X - \mu \geq f_n(X)\}\end{aligned}$$

but $X - n + 1 \geq f_n(X)$, so the supremum is $+\infty$. ♦

[2] Remember that we write X also for the identity gamble.

If fI_A is a bounded gamble, then it can be shown quite easily that $\operatorname{lpr}(\mathscr{D})(f|A)$ and $\operatorname{upr}(\mathscr{D})(f|A)$ cannot be infinite. However, as soon as we admit unbounded gambles, we now see that coherent sets of acceptable gambles may induce infinite supremum buying prices and infinite infimum selling prices. This will lead to some technical differences between lower previsions on bounded gambles and their counterparts on gambles.

Example 13.10 We consider the conditional lower prevision induced by the non-negative orthant of $\mathbb{G}$, namely, $\mathbb{G}_{\geq 0}$. As we discussed earlier in Example 13.3_{237}, $\mathbb{G}_{\geq 0}$ is the smallest element of $\mathbb{D}$ and models vacuous beliefs: a subject can adopt it as a belief model if he has no information whatsoever regarding X. We easily find that

$$\inf(f|A) := \operatorname{lpr}(\mathbb{G}_{\geq 0})(f|A) = \inf_{x \in A} f(x) \text{ for all } (f, A) \in \mathbb{G} \times \mathscr{P}^{\circ}.$$

So, with the conditional lower prevision $\inf(\bullet|\bullet)$, if f is bounded below on A, then the agent is willing to pay, contingent on A, any price strictly less than the lowest possible value of the reward f on A; otherwise the agent is not willing to buy f for any price. Note that $\inf(\bullet|\bullet)$ never takes the value $+\infty$. We call $\inf(\bullet|\bullet)$ the **vacuous conditional lower prevision**. The conjugate **vacuous conditional upper prevision** $\sup(\bullet|\bullet)$ is given by

$$\sup(f|A) := \operatorname{upr}(\mathbb{G}_{\geq 0})(f|A) = \sup_{x \in A} f(x) \text{ for all } (f, A) \in \mathbb{G} \times \mathscr{P}^{\circ}.$$

Next, consider a non-empty set B, and let us calculate the conditional lower prevision induced by the set $\mathscr{D}_B$, also introduced in Example 13.3_{237}:

$$\underline{P}_B(f|A) := \operatorname{lpr}(\mathscr{D}_B)(f|A) = \begin{cases} \inf(f|A \cap B) & \text{if } A \cap B \neq \emptyset \\ \inf(f|A) & \text{otherwise} \end{cases}$$

for all $(f, A) \in \mathbb{G} \times \mathscr{P}^{\circ}$. So if A does not conflict with B (i.e. if $A \cap B \neq \emptyset$) and f is bounded below on $A \cap B$, then the agent is willing to pay, contingent on B, any price strictly less than the lowest possible reward of f on $A \cap B$; otherwise the agent is not willing to buy f for any price. If A conflicts with B (i.e. if $A \cap B = \emptyset$), then the agent's beliefs are simply vacuous and agree with $\inf(\bullet|\bullet)$. Note that, again, $\underline{P}_B(\bullet|\bullet)$ never takes the value $+\infty$. We call $\underline{P}_B(\bullet|\bullet)$ the **vacuous conditional lower prevision relative to B.** ♦

The next theorem shows that, similarly to the lower previsions on bounded gambles discussed in Chapter 4_{37}, conditional lower previsions derived from coherent sets of acceptable gambles have a number of fundamental and important properties. Why they are so important will become clear in the subsequent Theorem 13.12_{244}. The proof of both theorems, in the bounded gamble case, goes back to Williams (1975b).

Theorem 13.11 *Let $\operatorname{lpr}(\mathscr{D})(\bullet|\bullet)$ be the conditional lower prevision associated with a coherent set of acceptable gambles $\mathscr{D}$. Then the following statements hold for all gambles f and g, all non-negative real numbers λ, all real numbers a and all non-empty events A and B such that $A \subseteq B$:*

CLP1. $\inf(f|A) \leq \mathrm{lpr}(\mathscr{D})(f|A)$. **(bounds)**.

CLP2. $\mathrm{lpr}(\mathscr{D})(f+g|A) \geq \mathrm{lpr}(\mathscr{D})(f|A) + \mathrm{lpr}(\mathscr{D})(g|A)$ *whenever the right-hand side is well defined.* **(super-additivity)**.

CLP3. $\mathrm{lpr}(\mathscr{D})(\lambda f|A) = \lambda\,\mathrm{lpr}(\mathscr{D})(f|A)$. **(non-negative homogeneity)**.

CLP4. $\mathrm{lpr}(\mathscr{D})((f-a)I_A|B) \begin{cases} \geq 0 \text{ if } a < \mathrm{lpr}(\mathscr{D})(f|A) \\ \leq 0 \text{ if } a > \mathrm{lpr}(\mathscr{D})(f|A). \end{cases}$ **(Bayes's rule)**.

Proof. The first statement is trivial if $\inf(f|A) = -\infty$. When $\inf(f|A) \neq -\infty$, observe that $\inf(f|A)$ must be in $\mathbb{R}$ because the infimum can never be $+\infty$. As then $(f - \inf(f|A))I_A \geq 0$, it follows that $(f - \inf(f|A))I_A \in \mathscr{D}$, by $\mathrm{A2}_{236}$. The desired inequality now follows from Equation $(13.2)_{240}$.

The second statement is immediate when $\mathrm{lpr}(\mathscr{D})(f|A)$ or $\mathrm{lpr}(\mathscr{D})(g|A)$ are equal to $-\infty$. So suppose both are not equal to $-\infty$. Consider any real $\alpha < \mathrm{lpr}(\mathscr{D})(f|A)$ and any real $\beta < \mathrm{lpr}(\mathscr{D})(g|A)$. By Equation $(13.2)_{240}$, both $(f-\alpha)I_A \in \mathscr{D}$ and $(g-\beta)I_A \in \mathscr{D}$. Use $\mathrm{A4}_{237}$ to conclude that $[(f+g)-(\alpha+\beta)]I_A \in \mathscr{D}$, whence $\mathrm{lpr}(\mathscr{D})(f+g|A) \geq \alpha+\beta$. Taking the supremum over all real $\alpha < \mathrm{lpr}(\mathscr{D})(f|A)$ and all real $\beta < \mathrm{lpr}(\mathscr{D})(g|A)$ yields the desired inequality.

For the third statement, assume first that $\lambda > 0$. It follows from $\mathrm{A3}_{237}$ that $(\lambda f - \mu)I_A \in \mathscr{D}$ if and only if $(f - \mu/\lambda)I_A \in \mathscr{D}$. Hence indeed

$$\begin{aligned} \mathrm{lpr}(\mathscr{D})(\lambda f|A) &= \sup\left\{\mu \in \mathbb{R} : (\lambda f - \mu)I_A \in \mathscr{D}\right\} \\ &= \sup\left\{\mu \in \mathbb{R} : \left(\frac{f-\mu}{\lambda}\right)I_A \in \mathscr{D}\right\} \\ &= \sup\left\{\lambda\mu \in \mathbb{R} : (f-\mu)I_A \in \mathscr{D}\right\} \\ &= \lambda \sup\left\{\mu \in \mathbb{R} : (f-\mu)I_A \in \mathscr{D}\right\} \\ &= \lambda\,\mathrm{lpr}(\mathscr{D})(f|A). \end{aligned}$$

For $\lambda = 0$, consider that $\mathrm{lpr}(\mathscr{D})(0|A) = \sup\left\{\mu \in \mathbb{R} : -\mu I_A \in \mathscr{D}\right\} = 0$, where the last equality follows from $\mathrm{A1}_{236}$ and $\mathrm{A2}_{236}$ because $A \neq \emptyset$.

We are left with the proof of the last statement. First, we let $a < \mathrm{lpr}(\mathscr{D})(f|A)$ and show that $\mathrm{lpr}(\mathscr{D})((f-a)I_A|B) \geq 0$. By Equation $(13.2)_{240}$, it clearly suffices to show that $((f-a)I_A - 0)I_B \in \mathscr{D}$. As $A \subseteq B$, it follows that $((f-a)I_A - 0)I_B = (f-a)I_A$, and again by Equation $(13.2)_{240}$, we know that $(f-a)I_A \in \mathscr{D}$ because $a < \mathrm{lpr}(\mathscr{D})(f|A)$ by assumption.

Next, we let $a > \mathrm{lpr}(\mathscr{D})(f|A)$. Then $(f-a)I_A \notin \mathscr{D}$ by Equation $(13.2)_{240}$. If we can show that also $((f-a)I_A - b)I_B \notin \mathscr{D}$ for all $b > 0$, Equation $(13.2)_{240}$ will tell us that then indeed $\mathrm{lpr}(\mathscr{D})((f-a)I_A|B) \leq 0$. This is what we now set out to do. As $bI_B \geq 0$, it follows that $bI_B \in \mathscr{D}$ (use $\mathrm{A2}_{236}$), and so if $((f-a)I_A - b)I_B$ were to belong to $\mathscr{D}$, then so would $((f-a)I_A - b)I_B + bI_B = (f-a)I_A$, by $\mathrm{A4}_{237}$. This is a contradiction. □

If, conversely, we have an $\mathbb{R}^*$-valued functional $\underline{P}(\bullet|\bullet)$ defined on $\mathbb{G} \times \mathscr{P}^\circ$, we can ask ourselves under what conditions it can be seen as a conditional lower prevision associated with some coherent set of acceptable gambles.

Theorem 13.12 *Consider any $\mathbb{R}^*$-valued functional $\underline{P}(\bullet|\bullet)$ on $\mathbb{G}\times\mathscr{P}^\circ$. Then there is some coherent set of acceptable gambles $\mathscr{D}$ such that $\underline{P}(\bullet|\bullet) = \operatorname{lpr}(\mathscr{D})(\bullet|\bullet)$ if and only if $\underline{P}(\bullet|\bullet)$ satisfies the properties CLP1–CLP4$_\frown$.*

Proof. The 'only if' part was proved in Theorem 13.11$_{242}$. Let us prove the 'if' part: assuming that $\underline{P}(\bullet|\bullet)$ satisfies the properties CLP1–CLP4$_\frown$, we first show that

$$\mathscr{A} := \left\{(f-\mu)I_A : (f,A)\in\mathbb{G}\times\mathscr{P}^\circ,\ \mu\in\mathbb{R}\text{ and }\mu<\underline{P}(f|A)\right\}$$

is consistent, and then we show that $\underline{P}(\bullet|\bullet) = \operatorname{lpr}(\mathscr{D})(\bullet|\bullet)$ for $\mathscr{D} := \operatorname{Cl}_{\mathbb{D}}(\mathscr{A})$.

Consider n in $\mathbb{N}$, non-negative $\lambda_1, \dots, \lambda_n$ in $\mathbb{R}$, $(f_1,A_1), \dots, (f_n,A_n)$ in $\mathbb{G}\times\mathscr{P}^\circ$ and $\mu_1, \dots, \mu_n \in \mathbb{R}$ such that $\mu_k < \underline{P}(f_k|A_k)$, and assume *ex absurdo* that

$$\sum_{k=1}^{n} \lambda_k(f_k-\mu_k)I_{A_k} < 0. \tag{13.3}$$

Then, clearly, $n>0$, and there is at least one k for which $\lambda_k>0$. Because terms for which the $\lambda_k=0$ do not contribute to the sum above, we may assume without loss of generality that $\lambda := \min_{k=1}^n \lambda_k > 0$. Also, because $\mu_k < \underline{P}(f_k|A_k)$, we can find some $\varepsilon>0$ small enough such that still $a_k := \mu_k+\varepsilon < \underline{P}(f_k|A_k)$ for all k. We infer from Equation (13.3) that $\sum_{k=1}^n \lambda_k(f_k-a_k)I_{A_k} + \lambda\varepsilon I_A < 0$, where we let $A := \bigcup_{k=1}^n A_k$. If we now apply Lemma 13.13, (iv), (vi), (v) and (i), we are led to

$$\sum_{k=1}^{n} \lambda_k\underline{P}((f_k-a_k)I_{A_k}|A) \le \underline{P}\left(\sum_{k=1}^{n}\lambda_k(f_k-a_k)I_{A_k}\middle|A\right) \le \underline{P}(-\lambda\varepsilon I_A|A) = -\lambda\varepsilon < 0.$$

But because all $a_k < \underline{P}(f_k|A_k)$, we infer from Lemma 13.13(xiv) that the sum on the left-hand side must be non-negative, a contradiction.

Next, because $\mathscr{A}$ is indeed consistent, we can consider the coherent set of acceptable gambles $\mathscr{D} := \operatorname{Cl}_{\mathbb{D}}(\mathscr{A})$. To complete the proof, we show that $\underline{P}(\bullet|\bullet) = \operatorname{lpr}(\mathscr{D})(\bullet|\bullet)$. Consider any $(f,A)\in\mathbb{G}\times\mathscr{P}^\circ$.

We first prove that $\underline{P}(f|A) \le \operatorname{lpr}(\mathscr{D})(f|A)$. We may clearly assume without loss of generality that $\underline{P}(f|A) > -\infty$. Consider any real $\mu < \underline{P}(f|A)$, then obviously $(f-\mu)I_A \in \mathscr{A} \subseteq \mathscr{D}$ and, therefore, $\mu \le \operatorname{lpr}(\mathscr{D})(f|A)$, by Equation (13.2)$_{240}$. Hence indeed $\underline{P}(f|A) \le \operatorname{lpr}(\mathscr{D})(f|A)$.

Finally, we prove that $\underline{P}(f|A) \ge \operatorname{lpr}(\mathscr{D})(f|A)$. We may clearly assume without loss of generality that $\operatorname{lpr}(\mathscr{D})(f|A) > -\infty$ and $\underline{P}(f|A) < +\infty$, so we infer from Equation (13.2)$_{240}$ that there are real μ and a such that $(f-\mu)I_A\in\mathscr{D}$ and $a > \underline{P}(f|A)$. Consider any such μ and a. Then it follows from the definition of $\mathscr{D}$ that there are n in $\mathbb{N}$, non-negative $\lambda_1, \dots, \lambda_n$ in $\mathbb{R}$, $(f_1,A_1), \dots, (f_n,A_n)$ in $\mathbb{G}\times\mathscr{P}^\circ$ and $\mu_1, \dots, \mu_n\in\mathbb{R}$ such that $\mu_k < \underline{P}(f_k|A_k)$ and

$$(f-\mu)I_A \ge \sum_{k=1}^{n}\lambda_k(f_k-\mu_k)I_{A_k}. \tag{13.4}$$

There are two possibilities.

Either $\sum_{k=1}^{n} \lambda_k = 0$ (this includes the case $n = 0$), so we have that $(f - \mu)I_A \geq 0$ and, therefore, $\mu \leq \inf(f|A) \leq \underline{P}(f|A) < a$, also using CLP1$_{243}$.

Or $\sum_{k=1}^{n} \lambda_k > 0$. Because all $\mu_k < \underline{P}(f_k|A_k)$, we can find an $\varepsilon > 0$ small enough such that still all $a_k := \mu_k + \varepsilon < \underline{P}(f_k|A_k)$. Equation (13.4) clearly remains valid if we drop the terms for which $\lambda_k = 0$ (we will want to do so for reasons that will become apparent later in the proof); for simplicity of notation, we simply assume that $\lambda_k > 0$ for all k. We then infer from Equation (13.4) that

$$-\varepsilon \sum_{k=1}^{n} \lambda_k I_{A_k} + (a - \mu)I_A \geq \sum_{k=1}^{n} \lambda_k (f_k - a_k) I_{A_k} - (f - a)I_A$$

and therefore, invoking Lemma 13.14$_{251}$, that

$$\sup\left(-\varepsilon \sum_{k=1}^{n} \lambda_k I_{A_k} + (a - \mu)I_A \middle| A \cup A_1 \cup \cdots \cup A_n\right) \geq 0,$$

whence $a \geq \mu$, because if not, the function inside the supremum would be uniformly negative on $A \cup A_1 \cup \cdots \cup A_n$ (remember that $\lambda_k > 0$ for all k), and hence, the inequality could not hold.

So, in both cases, we have shown that $a \geq \mu$ for all real μ and a such that $(f - \mu)I_A \in \mathscr{D}$ and $a > \underline{P}(f|A)$. This shows that indeed $\underline{P}(f|A) \geq \mathrm{lpr}(\mathscr{D})(f|A)$. □

Lemma 13.13 *Let $\underline{P}(\bullet|\bullet)$ be any restriction of an $\mathbb{R}^*$-valued functional $\underline{Q}(\bullet|\bullet)$ defined on $\mathbb{G} \times \mathscr{P}^\circ$, where $\underline{Q}(\bullet|\bullet)$ satisfies the properties* CLP1–CLP4$_{243}$.[3] *Then for all gambles f and g, nets of gambles f_α, real numbers a, non-negative real numbers λ and events A, $B \in \mathscr{P}^\circ$ such that $A \subseteq B$, each of the following statements holds whenever every argument in the statement belongs to the domain of its corresponding function:*

(i) $\inf(f|A) \leq \underline{P}(f|A) \leq \overline{P}(f|A) \leq \sup(f|A)$. **(bounds)**

(ii) $\underline{P}(a|A) = \overline{P}(a|A) = a$. **(normality)**

(iii) $\underline{P}(f + a|A) = \underline{P}(f|A) + a$ *and* $\overline{P}(f + a|A) = \overline{P}(f|A) + a$. **(constant additivity)**

(iv) $\underline{\underline{P}}(f|A) + \underline{\underline{P}}(g|A) \leq \underline{P}(f + g|A) \leq \underline{P}(f|A) + \overline{P}(g|A) \leq \overline{P}(f + g|A) \leq \overline{\overline{P}}(f|A) + \overline{\overline{P}}(g|A)$, *where each inequality holds if both sides are well defined.* **(mixed super-/sub-additivity)**

(v) $\underline{P}(\lambda f|A) = \lambda \underline{P}(f|A)$ *and* $\overline{P}(\lambda f|A) = \lambda \overline{P}(f|A)$. **(non-negative homogeneity)**

(vi) $f \leq g + a \Rightarrow \underline{P}(f|A) \leq \underline{P}(g|A) + a$ *and* $\overline{P}(f|A) \leq \overline{P}(g|A) + a$. **(monotonicity)**

[3] In other words, by condition (B)$_{260}$ of Definition 13.25$_{260}$, let $\underline{P}(\bullet|\bullet)$ be any coherent conditional lower prevision.

(vii) $\underline{P}\big(|f|\big|A\big) \geq \underline{P}(f|A)$ *and* $\overline{P}\big(|f|\big|A\big) \geq \overline{P}(f|A)$.

(viii) $\big|\underline{P}(f|A) - \underline{P}(g|A)\big| \leq \overline{P}\big(|f-g|\big|A\big)$ *if* $\underline{P}(f|A) - \underline{P}(g|A)$ *is well defined, and* $\Big|\overline{P}(f|A) - \overline{P}(g|A)\Big| \leq \overline{P}\big(|f-g|\big|A\big)$ *if* $\overline{P}(f|A) - \overline{P}(g|A)$ *is well defined.*

(ix) $\underline{P}\big(|f+g|\big|A\big) \leq \underline{P}\big(|f|\big|A\big) + \overline{P}\big(|g|\big|A\big)$ *and* $\overline{P}\big(|f+g|\big|A\big) \leq \overline{P}\big(|f|\big|A\big) + \overline{P}\big(|g|\big|A\big)$. **(mixed Cauchy–Schwartz inequalities)**

(x) $\underline{P}(f \vee g|A) + \underline{P}(f \wedge g|A) \leq \underline{P}(f|A) + \overline{P}(g|A) \leq \overline{P}(f \vee g|A) + \overline{P}(f \wedge g|A)$, $\underline{P}(f|A) + \underline{P}(g|A) \leq \underline{P}(f \vee g|A) + \overline{P}(f \wedge g|A) \leq \overline{P}(f|A) + \overline{P}(g|A)$ *and* $\underline{P}(f|A) + \underline{P}(g|A) \leq \overline{P}(f \vee g|A) + \underline{P}(f \wedge g|A) \leq \overline{P}(f|A) + \overline{P}(g|A)$, *where each inequality holds if both sides are well defined.*

(xi) *Assume that* $R = \overline{P}\big(|f-g|\big|A\big)$ *is a real number. Then* $\underline{P}(g|A) - R \leq \underline{P}(f|A) \leq \underline{P}(g|A) + R$ *and* $\overline{P}(g|A) - R \leq \overline{P}(f|A) \leq \overline{P}(g|A) + R$.

(xii) $\overline{P}\big(|f_\alpha - f|\big|A\big) \to 0 \Rightarrow \underline{P}(f_\alpha|A) \to \underline{P}(f|A)$ *and* $\overline{P}(f_\alpha|A) \to \overline{P}(f|A)$

(xiii) $\underline{P}(\bullet|A)$ *is uniformly continuous with respect to the topology generated by the seminorm*[4] $\sup\big(|\bullet|\big|A\big)$*: for any* $\varepsilon > 0$, *if* $\sup\big(|f-g|\big|A\big) \leq \varepsilon$, *then* $\underline{P}(f|A) - \varepsilon \leq \underline{P}(g|A) \leq \underline{P}(f|A) + \varepsilon$. **(uniform continuity)**

(xiv) $\phi : \mathbb{R} \to \mathbb{R}^* : a \mapsto \underline{P}((f-a)I_A|B)$ *is a non-increasing and uniformly continuous function and satisfies for all* $a \in \mathbb{R}$*:*

$$\phi(a) \geq 0 \text{ if } a < \underline{P}(f|A),$$
$$\phi(a) = 0 \text{ if } a = \underline{P}(f|A),$$
$$\phi(a) \leq 0 \text{ if } a > \underline{P}(f|A).$$

(xv) $\psi : \mathbb{R} \to \mathbb{R}^* : a \mapsto \overline{P}((f-a)I_A|B)$ *is a non-increasing and uniformly continuous function and satisfies for all* $a \in \mathbb{R}$*:*

$$\psi(a) \geq 0 \text{ if } a < \overline{P}(f|A),$$
$$\psi(a) = 0 \text{ if } a = \overline{P}(f|A),$$
$$\psi(a) \leq 0 \text{ if } a > \overline{P}(f|A).$$

(xvi) *If* $0 \leq \underline{P}(f|A) < +\infty$, *then*

$$\underline{P}(f|A)\underline{P}(A|B) \leq \underline{P}(fI_A|B) \leq \underline{P}(f|A)\overline{P}(A|B) \leq \overline{P}(fI_A|B), \tag{13.5}$$

and if $0 \leq \overline{P}(f|A) < +\infty$, *then*

$$\underline{P}(fI_A|B) \leq \overline{P}(f|A)\underline{P}(A|B) \leq \overline{P}(fI_A|B) \leq \overline{P}(f|A)\overline{P}(A|B). \tag{13.6}$$

---

[4] See Appendix B_{371} and in particular Definition $B.6_{374}$ for a definition and brief discussion of the notion of a seminorm.

If $0 \leq \underline{P}(f|A) \leq +\infty$, *then*

$$\underline{P}(fI_A|B) \geq \underline{P}(f|A)\underline{P}(A|B) \quad (13.7)$$

$$\overline{P}(fI_A|B) \geq \underline{P}(f|A)\overline{P}(A|B) \quad (13.8)$$

so, in particular, when $\underline{P}(f|A) = +\infty$,

if $\underline{P}(A|B)$	*then* $\underline{P}(fI_A|B)$
$+$	$+\infty$
0	≥ 0

and

if $\overline{P}(A|B)$	*then* $\overline{P}(fI_A|B)$
$+$	$+\infty$
0	≥ 0

If $0 \leq \overline{P}(f|A) \leq +\infty$, *then*

$$\overline{P}(fI_A|B) \geq \overline{P}(f|A)\underline{P}(A|B) \quad (13.9)$$

so, in particular, when $\overline{P}(f|A) = +\infty$,

if $\underline{P}(A|B)$	*then* $\overline{P}(fI_A|B)$
$+$	$+\infty$
0	≥ 0

where 0 *means 'is zero',* + *means 'is strictly positive',* $+\infty$ *means 'is plus infinity' and* ≥ 0 *means 'is non-negative'.*

(xvii) $\underline{P}(f|A) = \underline{P}(fI_A|A)$ *and* $\overline{P}(f|A) = \overline{P}(fI_A|A)$.

(xviii) *If* $\overline{P}(A|B) = 0$, *then*

if $\underline{P}(f|A)$	*then* $\underline{P}(fI_A|B)$
$-\infty$	≤ 0
$\mathbb{R}$	0
$+\infty$	≥ 0

and

if $\overline{P}(f|A)$	*then* $\overline{P}(fI_A|B)$
$-\infty$	≤ 0
$\mathbb{R}$	0
$+\infty$	≥ 0

and moreover, the following functions are constant:

$$\left.\begin{aligned} \phi(a) &:= \underline{P}((f-a)I_A|B) = \underline{P}(fI_A|B) \\ \psi(a) &:= \overline{P}((f-a)I_A|B) = \overline{P}(fI_A|B) \end{aligned}\right\} \text{ for all } a \in \mathbb{R}.$$

Proof. Without loss of generality, we can assume in the proof that dom $\underline{P}(\bullet|\bullet) = \mathbb{G} \times \mathscr{P}^{\circ}$.

(i)$_{245}$. The first and third inequalities follow from CLP1$_{243}$ and conjugacy.

For the second, it suffices to consider the case $\underline{P}(f|A) \neq -\infty$. CLP2$_{243}$ implies that $\underline{P}(f|A) + \underline{P}(-f|A) \leq \underline{P}(0|A)$ whenever $\underline{P}(f|A) + \underline{P}(-f|A)$ is well defined. By CLP3$_{243}$, $\underline{P}(0|A) = 0$. Combining all this with conjugacy, we find that $\overline{P}(f|A) = -\underline{P}(-f|A) \geq \underline{P}(f|A)$ whenever $\underline{P}(f|A) + \underline{P}(-f|A)$ is well defined. Consider the case that $\underline{P}(f|A) + \underline{P}(-f|A)$ is not well defined. Then it can only be that $\underline{P}(f|A) = +\infty$ and

$\underline{P}(-f|A) = -\infty$ because we assumed that $\underline{P}(f|A) \neq -\infty$; this completes the proof of the second inequality.

(ii)$_{245}$. Immediately from (i)$_{245}$.

(iii)$_{245}$. CLP2$_{243}$ implies that both $\underline{P}(f + \mu|A) \geq \underline{P}(f|A) + \underline{P}(\mu|A)$ and $\underline{P}(f|A) \geq \underline{P}(f + \mu|A) + \underline{P}(-\mu|A)$ whenever the sums are well defined. Now apply (ii)$_{245}$.

(iv)$_{245}$. The first inequality is exactly CLP2$_{243}$.

We prove the second inequality. By CLP2$_{243}$, $\underline{P}(f + g|A) + \underline{P}(-g|A) \leq \underline{P}(f|A)$ whenever the left-hand side is well defined. Apply Lemma D.9(iv)$_{393}$ and use conjugacy, in particular, $-\underline{P}(-g|A) = \overline{P}(g|A)$.

The third inequality follows from the second one and conjugacy, and the fourth inequality from the first one and conjugacy.

(v)$_{245}$. The first equality is exactly CLP3$_{243}$. The second one follows from the first and conjugacy.

(vi)$_{245}$. It suffices to prove the first inequality, because the second one follows readily from the first and conjugacy. If $f \leq g + a$, then $f - g - a \leq 0$ and, therefore, $\overline{P}(f - g|A) - a = \overline{P}(f - g - a|A) \leq 0$ by (iii)$_{245}$ and (i)$_{245}$. Now use (iv)$_{245}$ and conjugacy to infer that

$$a \geq \overline{P}(f - g|A) \geq \underline{P}(f|A) + \overline{P}(-g|A) \geq \underline{P}(f|A) - \underline{P}(g|A),$$

whenever the right-hand sides are well defined. This proves the first inequality when $\underline{P}(f|A) - \underline{P}(g|A)$ is well defined. The only remaining possibilities are that $\underline{P}(f|A) = \underline{P}(g|A) = -\infty$ and $\underline{P}(f|A) = \underline{P}(g|A) = +\infty$, and then the first inequality is trivially valid.

(vii)$_{246}$. This follows from $f \leq |f|$ and (vi)$_{245}$.

(viii)$_{246}$. It suffices to prove the first inequality, because the second one follows readily from the first and conjugacy. Whenever $\underline{P}(f|A) - \underline{P}(g|A)$ is well defined, it follows from (iv)$_{245}$ and (vi)$_{245}$ that

$$\underline{P}(f|A) - \underline{P}(g|A) = \underline{P}(f|A) + \overline{P}(-g|A) \leq \overline{P}(f - g|A) \leq \overline{P}\big(|f - g|\big|A\big)$$

and, similarly,

$$\underline{P}(g|A) - \underline{P}(f|A) = \underline{P}(g|A) + \overline{P}(-f|A) \leq \overline{P}(g - f|A) \leq \overline{P}\big(|f - g|\big|A\big).$$

(ix)$_{246}$. This follows from $|f + g| \leq |f| + |g|$, (vi)$_{245}$ and (iv)$_{245}$.

(x)$_{246}$. This follows from $f \vee g + f \wedge g = f + g$ and (iv)$_{245}$. For example (the other inequalities follow similarly), provided all occurring sums are well defined,

$$\underline{P}(f \vee g|A) + \underline{P}(f \wedge g|A) \leq \underline{P}(f \vee g + f \wedge g|A) = \underline{P}(f + g|A) \leq \underline{P}(f|A) + \overline{P}(g|A)$$

and

$$\underline{P}(f|A) + \underline{P}(g|A) \leq \underline{P}(f + g|A) = \underline{P}(f \vee g + f \wedge g|A) \leq \underline{P}(f \vee g|A) + \overline{P}(f \wedge g|A).$$

(xi)$_{246}$. If $\overline{P}\big(|f - g|\big|A\big) = R$ is a real number, then $\underline{P}(f - g|A)$ and $\overline{P}(f - g|A)$ must be real numbers as well, because it follows from (i)$_{245}$, (vii)$_{246}$ and conjugacy that

$$-R = -\overline{P}(|f-g||A) \le -\overline{P}(f-g|A) \le -\underline{P}(f-g|A)$$
$$= \overline{P}(g-f|A) \le \overline{P}(|g-f||A) = R.$$

This allows us to infer from (iv)$_{245}$ and (vii)$_{246}$ that

$$\begin{aligned}\underline{P}(g|A) - R = \underline{P}(g|A) - \overline{P}(|g-f||A) &\le \underline{P}(g|A) - \overline{P}(g-f|A) \\ &\le \underline{P}(g-g+f|A) = \underline{P}(f|A) = \underline{P}(g+f-g|A) \\ &\le \underline{P}(g|A) + \overline{P}(f-g|A) \le \underline{P}(g|A) + \overline{P}(|f-g||A) = \underline{P}(g|A) + R.\end{aligned}$$

The remaining inequalities now follow from conjugacy.

(xii)$_{246}$. If $\overline{P}(|f_\alpha - f||A)$ converges to zero, then there must be some α^* such that $\overline{P}(|f_\alpha - f||A)$ is a real number for all $\alpha \ge \alpha^*$. Now apply (xi)$_{246}$ to the subnet restricted to $\alpha \ge \alpha^*$, with $g = f_\alpha$.

(xiii)$_{246}$. Immediately from (xi)$_{246}$ and (i)$_{245}$.

(xiv)$_{246}$. That ϕ is non-increasing follows from (vi)$_{245}$.

To prove that ϕ is uniformly continuous, we show that for all $\varepsilon > 0$, there is some $\delta > 0$ such that for all $a, a' \in \mathbb{R}$, $|a - a'| < \delta$ implies $|\phi(a) - \phi(a')| < \varepsilon$. Fix $\varepsilon > 0$, and let $\delta := \varepsilon/(1 + \overline{P}(A|B))$. Note that $0 \le \overline{P}(A|B) \le 1$ by (i)$_{245}$ and that this choice of δ guarantees that $\overline{P}(A|B)\delta < \varepsilon$. Assume that $|a - a'| < \delta$. Then by (vi)$_{245}$ and (iv)$_{245}$,

$$\underline{P}((f-a)I_A|B) \le \underline{P}\big((f-a'+|a-a'|)I_A|B\big) \le \underline{P}((f-a')I_A|B) + \overline{P}\big(|a-a'|I_A|B\big)$$

so we find, using (v)$_{245}$, that $\phi(a) \le \phi(a') + |a-a'|\overline{P}(A|B)$ and, similarly, $\phi(a') \le \phi(a) + |a-a'|\overline{P}(A|B)$. Hence indeed

$$|\phi(a) - \phi(a')| \le |a-a'|\overline{P}(A|B) < \delta\overline{P}(A|B) < \varepsilon.$$

This establishes the uniform continuity of ϕ.[5]

It follows from CLP4$_{243}$ that we are only left to prove that $\underline{P}((f-a)I_A|B) = 0$ for $a = \underline{P}(f|A)$ whenever $\underline{P}(f|A) \in \mathbb{R}$. So, let $a = \underline{P}(f|A)$, and assume *ex absurdo* that $\phi(a) > 0$. From the continuity of ϕ, it follows that $\phi(a+\delta) > 0$ for sufficiently small $\delta > 0$. But this contradicts CLP4$_{243}$, which states that $\phi(b) \le 0$ for all $b > a$. Hence, $\phi(a)$ cannot be strictly positive. In the same way, we can prove that $\phi(a)$ cannot be strictly negative. We conclude that $\phi(a)$ must be zero.

(xv)$_{246}$. Immediately from (xiv)$_{246}$, once we observe that, by conjugacy,

$$\overline{P}((f-a)I_A|B) = -\underline{P}((-f+a)I_A|B),$$
$$\overline{P}(f|A) = -\underline{P}(-f|A).$$

[5] Actually, this argument establishes that ϕ is even *Lipschitz continuous*, with Lipschitz constant at most $\overline{P}(A|B)$ and, therefore, always non-expansive. The map ϕ is a contraction if $\overline{P}(A|B) < 1$. See Schechter (1997, Section 18.2) for more details.

For example, if $a < \overline{P}(f|A)$, then it follows that $-a > \underline{P}(-f|A)$, and hence, by (xiv)$_{246}$, $\underline{P}((-f + a)I_A|B) \leq 0$, which is equivalent to stating that $\psi(a) \geq 0$.

(xvi)$_{246}$. Let us prove each inequality. Some inequalities are also proved for the $+\infty$ case, where relevant.

(a) $\underline{P}(f|A)\underline{P}(A|B) \leq \underline{P}(fI_A|B)$ for $0 \leq \underline{P}(f|A) \leq +\infty$. As $0 \leq \underline{P}(A|B) \leq 1$ by (i)$_{245}$, it follows from conjugacy, (v)$_{245}$, (iv)$_{245}$ and (xiv)$_{246}$ that

$$\underline{P}(fI_A|B) - a\underline{P}(A|B) = \underline{P}(fI_A|B) + \overline{P}(-aI_A|B) \geq \underline{P}((f - a)I_A|B) \geq 0$$

for every real number a such that $0 \leq a \leq \underline{P}(f|A)$. Hence, $\underline{P}(fI_A|B) \geq \underline{P}(f|A)\underline{P}(A|B)$.

(b) $\underline{P}(fI_A|B) \leq \underline{P}(f|A)\overline{P}(A|B)$ for $0 \leq \underline{P}(f|A) < +\infty$. As $0 \leq \overline{P}(A|B) \leq 1$ by (i)$_{245}$, it also follows from conjugacy, (v)$_{245}$, (iv)$_{245}$ and (xiv)$_{246}$ that

$$\underline{P}(fI_A|B) - a\overline{P}(A|B) = \underline{P}(fI_A|B) + \underline{P}(-aI_A|B) \leq \underline{P}((f - a)I_A|B) \leq 0$$

for every real number a such that $a \geq \underline{P}(f|A)$. Hence, $\underline{P}(fI_A|B) \leq \underline{P}(f|A)\overline{P}(A|B)$.

(c) $\underline{P}(f|A)\overline{P}(A|B) \leq \overline{P}(fI_A|B)$ for $0 \leq \underline{P}(f|A) \leq +\infty$. As $0 \leq \overline{P}(A|B) \leq 1$ by (i)$_{245}$, it also follows from conjugacy, (v)$_{245}$, (iv)$_{245}$ and (xiv)$_{246}$ that

$$\overline{P}(fI_A|B) - a\overline{P}(A|B) = \overline{P}(fI_A|B) + \underline{P}(-aI_A|B) \geq \underline{P}((f - a)I_A|B) \geq 0$$

for every real number a such that $0 \leq a \leq \underline{P}(f|A)$. Hence, $\overline{P}(fI_A|B) \geq \underline{P}(f|A)\overline{P}(A|B)$.

This already tells us that Equations (13.5)$_{246}$, (13.7)$_{247}$ and (13.8)$_{247}$ hold.

(d) $\underline{P}(fI_A|B) \leq \overline{P}(f|A)\underline{P}(A|B)$ for $0 \leq \overline{P}(f|A) < +\infty$. As $0 \leq \underline{P}(A|B) \leq 1$ by (i)$_{245}$, it follows from conjugacy, (v)$_{245}$, (iv)$_{245}$ and (xv)$_{246}$ that

$$\underline{P}(fI_A|B) - a\underline{P}(A|B) = \underline{P}(fI_A|B) + \overline{P}(-aI_A|B) \leq \overline{P}((f - a)I_A|B) \leq 0$$

for every real number a such that $a \geq \overline{P}(f|A)$. Hence, $\underline{P}(fI_A|B) \leq \overline{P}(f|A)\underline{P}(A|B)$.

(e) $\overline{P}(f|A)\underline{P}(A|B) \leq \overline{P}(fI_A|B)$ for $0 \leq \overline{P}(f|A) \leq +\infty$. As $0 \leq \underline{P}(A|B) \leq 1$ by (i)$_{245}$, it also follows from conjugacy, (v)$_{245}$, (iv)$_{245}$ and (xv)$_{246}$ that

$$\overline{P}(fI_A|B) - a\underline{P}(A|B) = \overline{P}(fI_A|B) + \overline{P}(-aI_A|B) \geq \overline{P}((f - a)I_A|B) \geq 0$$

for every real number a such that $0 \leq a \leq \overline{P}(f|A)$. Hence, $\overline{P}(fI_A|B) \geq \overline{P}(f|A)\underline{P}(A|B)$.

(f) $\overline{P}(fI_A|B) \leq \overline{P}(f|A)\overline{P}(A|B)$ for $0 \leq \overline{P}(f|A) < +\infty$. As $0 \leq \overline{P}(A|B) \leq 1$ by (i)$_{245}$, it also follows from conjugacy, (v)$_{245}$, (iv)$_{245}$ and (xv)$_{246}$ that

$$\overline{P}(fI_A|B) - a\overline{P}(A|B) = \overline{P}(fI_A|B) + \underline{P}(-aI_A|B) \leq \overline{P}((f - a)I_A|B) \leq 0$$

for every real number a such that $a \geq \overline{P}(f|A)$. Hence, $\overline{P}(fI_A|B) \leq \overline{P}(f|A)\overline{P}(A|B)$.

This in turn tells us that Equations (13.6)$_{246}$ and (13.9)$_{247}$ hold. The rest is now immediate.

(xvii)$_{247}$. It suffices to prove the equality for $\underline{P}(\bullet|\bullet)$. The equality for $\overline{P}(\bullet|\bullet)$ follows from conjugacy. To show that $\underline{P}(f|A) = \underline{P}(fI_A|A)$ for all $(f, A) \in \mathbb{G} \times \mathscr{P}^\circ$, we use the inequalities in (xvi)$_{246}$, also taking into account that $\underline{P}(A|A) = \overline{P}(A|A) = 1$, by (i)$_{245}$. There are a number of possibilities.

If $\underline{P}(f|A) = +\infty$, use Equation $(13.7)_{247}$ to find that also $\underline{P}(fI_A|A) = +\infty$.

If $0 \le \underline{P}(f|A) < +\infty$, use Equation $(13.5)_{246}$ to find that $\underline{P}(f|A) = \underline{P}(fI_A|A)$.

If $-\infty < \underline{P}(f|A) \le 0$, then by conjugacy $+\infty > \overline{P}(-f|A) \ge 0$. Now use Equation $(13.6)_{246}$ for $-f$ to find that $\overline{P}(-f|A) = \overline{P}(-fI_A|A)$ or, equivalently, $\underline{P}(f|A) = \underline{P}(fI_A|A)$.

If $\underline{P}(f|A) = -\infty$, then by conjugacy $\overline{P}(-f|A) = +\infty$. Now use Equation $(13.9)_{247}$ for $-f$ to find that also $\overline{P}(-fI_A|A) = +\infty$ and therefore $\underline{P}(fI_A|A) = -\infty$.

$(\text{xviii})_{247}$. If $\overline{P}(A|B) = 0$, then also $\underline{P}(A|B) = 0$ because $0 \le \underline{P}(A|B) \le \overline{P}(A|B)$ by $(\text{i})_{245}$. We now apply the inequalities in $(\text{xvi})_{246}$ to the relevant possible cases. We only consider $\underline{P}(f|A)$: the corresponding statements for $\overline{P}(f|A)$ follow by conjugacy.

If $\underline{P}(f|A) = -\infty$, then also $\overline{P}(-f|A) = +\infty$ and the best we can then conclude is that $\overline{P}(-fI_A|A) \ge 0$ and therefore $\underline{P}(fI_A|A) \le 0$.

If $\underline{P}(f|A) \in \mathbb{R}$, then it follows from Equation $(13.5)_{246}$ that $\underline{P}(fI_A|A) = 0$.

If $\underline{P}(f|A) = +\infty$, then the best we can conclude is that $\underline{P}(fI_A|A) \ge 0$.

To complete this proof, we show that ϕ is constant. That ψ is constant then follows from conjugacy. As $0 \le \underline{P}(A|B) \le \overline{P}(A|B) \le 1$ by $(\text{i})_{245}$, we infer from $(\text{iv})_{245}$ that for any $a \in \mathbb{R}$,

$$\underline{P}(fI_A|B) + \underline{P}(-aI_A|B) \le \underline{P}((f-a)I_A|B) \le \underline{P}(fI_A|B) + \overline{P}(-aI_A|B)$$

because the sums are always well defined. But because we can deduce from $(\text{v})_{245}$ that

$$\begin{aligned}\underline{P}(-aI_A|B) &= \max\{-a,0\}\underline{P}(A|B) + \min\{-a,0\}\overline{P}(A|B)\\ \overline{P}(-aI_A|B) &= \max\{-a,0\}\overline{P}(A|B) + \min\{-a,0\}\underline{P}(A|B),\end{aligned}$$

we infer from the assumption that $\underline{P}(-aI_A|B) = \overline{P}(-aI_A|B) = 0$ and, therefore, $\phi(a) = \underline{P}((f-a)I_A|B) = \underline{P}(fI_A|B)$ for all $a \in \mathbb{R}$. □

Lemma 13.14 *Let $\underline{P}(\bullet|\bullet)$ be an $\mathbb{R}^*$-valued functional defined on $\mathbb{G} \times \mathscr{P}^\circ$ that satisfies the properties* CLP1–CLP4$_{243}$. *Then for all n in $\mathbb{N}$, all non-negative real λ_0, λ_1, ..., λ_n, all (f_0, A_0), (f_1, A_1), ..., (f_n, A_n) in $\mathbb{G} \times \mathscr{P}^\circ$ and all real a_0, ..., a_n such that $a_0 > \underline{P}(f_0|A)$ and $a_k < \underline{P}(f_k|A_k)$ for all $k \in \{1, \dots, n\}$:*

$$\sup\left(\sum_{k=1}^{n} \lambda_k(f_k - a_k)I_{A_k} - \lambda_0(f_0 - a_0)I_{A_0}\middle| A_0 \cup A_1 \cup \cdots \cup A_n\right) \ge 0.$$

Proof. Suppose *ex absurdo* that

$$\sup\left(\sum_{k=1}^{n} \lambda_k(f_k - a_k)I_{A_k} - \lambda_0(f_0 - a_0)I_{A_0}\middle| A_0 \cup A_1 \cup \cdots \cup A_n\right) < 0.$$

Then there is some $\varepsilon > 0$ such that $\sum_{k=1}^n \lambda_k(f_k - a_k)I_{A_k} + \varepsilon I_A \le \lambda_0(f_0 - a_0)I_{A_0}$, where we let $A := \bigcup_{k=0}^n A_k$. We infer from this inequality that

$$\lambda_0\underline{P}((f_0 - a_0)I_{A_0}|A) \ge \sum_{k=1}^{n} \lambda_k\underline{P}((f_k - a_k)I_{A_k}|A) + \varepsilon > 0,$$

where the first inequality follows from Lemma 13.13, (vi), (iv), (v) and $(\text{i})_{245}$, and the second one from Lemma $13.13(\text{xiv})_{245}$. But this is a contradiction, because the leftmost side is non-positive because $\lambda_0 \geq 0$ and $a_0 > \underline{P}(f_0|A)$ (use Lemma $13.13(\text{xiv})_{245}$). □

The following lemma will also be useful.

Lemma 13.15 *Let $\underline{P}(\bullet|\bullet)$ be an $\mathbb{R}^*$-valued functional defined on some subset of $\mathbb{G} \times \mathscr{P}^\circ$. Suppose that for all n in $\mathbb{N}$, all non-negative real λ_0, λ_1, ..., λ_n, all (f_0, A_0), (f_1, A_1), ..., (f_n, A_n) in $\operatorname{dom}\underline{P}(\bullet|\bullet)$ and all real a_0, ..., a_n such that $a_0 > \underline{P}(f_0|A)$ and $a_k < \underline{P}(f_k|A_k)$ for all $k \in \{1, \dots, n\}$:*

$$\sup\left(\sum_{k=1}^{n} \lambda_k(f_k - a_k)I_{A_k} - \lambda_0(f_0 - a_0)I_{A_0}\middle| A_0 \cup A_1 \cup \cdots \cup A_n\right) \geq 0.$$

Then for all n in $\mathbb{N}$, all non-negative real λ_1, ..., λ_n, all (f_1, A_1), ..., (f_n, A_n) in $\operatorname{dom}\underline{P}(\bullet|\bullet)$ and all real a_1, ..., a_n such that $a_k < \underline{P}(f_k|A_k)$ for all $k \in \{1, \dots, n\}$:

$$\sup\left(\sum_{k=1}^{n} \lambda_k(f_k - a_k)I_{A_k}\middle| A_1 \cup \cdots \cup A_n\right) \geq 0.$$

Proof. Simply let $\lambda_0 := 0$ and A_0 be any subset of $A_1 \cup \cdots \cup A_n$. □

13.3.2 Conditional lower previsions directly

Up to now, we have introduced sets of acceptable gambles and the rationality constraints such sets ought to satisfy, and we have derived conditional lower previsions purely from acceptability arguments. It is of course possible to introduce conditional lower previsions directly as well, as we did with unconditional lower previsions for bounded gambles in Section $4.1.2_{40}$.

A subject's **conditional lower prevision** $\underline{P}(f|A)$ for a gamble f conditional on a non-empty event A can be defined directly as his called-off supremum buying price for f: $\underline{P}(f|A)$ is the highest extended real number $s \in \mathbb{R}^*$ such that for any real price $t \in \mathbb{R}$ that is strictly lower than s, he (states he) is willing to pay t if he is guaranteed to receive $f(x)$ when observing $X = x$, under the condition that $x \in A$ (otherwise, the subject pays nothing and also receives nothing). So the subject is only required to consider whether he accepts gambles of the type $(f - \mu)I_A$. *By specifying a conditional lower prevision $\underline{P}(f|A)$ for a gamble f and event A, our subject is in effect stating that he accepts the gambles $(f - \mu)I_A$ for all real numbers μ such that $\mu < \underline{P}(f|A)$. He is not making any such statement for higher μ.*

As we have also done for bounded gambles, we stress at this point that we do not give an *exhaustive interpretation* to a lower prevision (Walley, 1991, Section 2.3.1): the subject neither states that he does not accept such gambles for higher μ, nor states that he accepts them. He remains uncommitted and can, therefore, be corrected (without internal contradiction) towards higher values, which is what the notion of natural extension tends to do, as we shall see further on in Section 13.7_{279}.

A subject can of course do this for any number of gambles. Mathematically speaking, we call **conditional lower prevision** an $\mathbb{R}^*$-valued map defined on some subset $\operatorname{dom} \underline{P}(\bullet|\bullet)$ of the set $\mathbb{G} \times \mathscr{P}^{\circ}$ of all gambles and non-empty events. The set $\operatorname{dom} \underline{P}(\bullet|\bullet)$ is called the **domain** of $\underline{P}(\bullet|\bullet)$. We do not require a conditional lower prevision to be defined for all gambles and all non-empty events. We do not even impose a structure on $\operatorname{dom} \underline{P}(\bullet|\bullet)$: it can be any subset of $\mathbb{G} \times \mathscr{P}^{\circ}$. By generalising the notions of avoiding sure loss and coherence in a straightforward way, we will establish that for every conditional lower prevision that avoids sure loss, there is a least-committal coherent lower prevision (i.e. natural extension) on $\mathbb{G} \times \mathscr{P}^{\circ}$ that dominates it.

Conditional lower previsions differ from lower previsions on bounded gambles, as defined in Section $4.1.2_{40}$, in three ways: they include conditioning, they are defined not just for bounded gambles but also for unbounded ones and they take values in a larger set – the set of extended real numbers. If $\underline{P}(f|A) = -\infty$, this means that our subject is not stating a willingness willing to buy f at any price $t \in \mathbb{R}$, contingent on A; this can be reasonable if f is unbounded below on A. If $\underline{P}(f|A) = +\infty$, this means that our subject is stating a willingness to buy f at any price, contingent on A; whether or not this is reasonable is of course debatable. We will pay more attention to this issue in Section 13.11_{301}. For now, let us simply see where these conditional lower previsions lead us.

Similarly to what we have done for bounded gambles, we can also interpret a gamble f as an uncertain loss: if x turns out to be the true value of X, we lose an amount of utility $f(x)$. The subject's **conditional upper prevision** $\overline{P}(f|A)$ of a gamble f conditional on a non-empty event A is then the called-off infimum acceptable selling price for f: it is the lowest extended real number s, such that for any real price $t \in \mathbb{R}$ strictly larger than s, he (states he) is willing to receive t if he is guaranteed to lose $f(x)$ when observing $X = x$, under the condition that $x \in A$ (otherwise, the subject pays nothing and also receives nothing). As a gain r is equivalent to a loss $-r$, we see that $\overline{P}(f|A) = -\underline{P}(-f|A)$: from any conditional lower prevision $\underline{P}(\bullet|\bullet)$, we can infer a so-called **conjugate** conditional upper prevision $\overline{P}(\bullet|\bullet)$ on $\operatorname{dom} \overline{P}(\bullet|\bullet) := \{(-f, A) : (f, A) \in \operatorname{dom} \underline{P}(\bullet|\bullet)\}$ which represents the same behavioural dispositions. We can, therefore, restrict our attention to the study of conditional lower previsions only, without any loss of generality. Also, if we use the notation $\underline{P}(\bullet|\bullet)$ for a conditional lower prevision, $\overline{P}(\bullet|\bullet)$ will always denote its conjugate.

When the domain of $\underline{P}(\bullet|\bullet)$ is a subset $\mathscr{K} \times \{\mathscr{X}\}$ of $\mathbb{G} \times \{\mathscr{X}\}$, then we also use the simpler notation:

$$\underline{P}(f) := \underline{P}(f|\mathscr{X}) \text{ for all } f \text{ in } \mathscr{K},$$

and we call $\underline{P}(\bullet|\bullet)$ or $\underline{P}$ an (unconditional) **lower prevision**, whose conjugate **upper prevision** is also denoted by $\overline{P}$. We also call $\mathscr{K}$ the domain of $\underline{P}$ and write $\mathscr{K} = \operatorname{dom} \underline{P}$. *This ensures that the lower and upper previsions for bounded gambles studied in Part* I_{21} *are special cases of conditional lower and upper previsions.*

It may happen that $\underline{P}(\bullet|\bullet)$ is **self-conjugate**, meaning that $\operatorname{dom} \underline{P}(\bullet|\bullet) = \operatorname{dom} \overline{P}(\bullet|\bullet)$ and $\underline{P}(f|A) = \overline{P}(f|A)$ for all $(f, A) \in \operatorname{dom} \underline{P}(\bullet|\bullet)$. We then simply write $P(\bullet|\bullet)$ instead of $\underline{P}(\bullet|\bullet)$ or $\overline{P}(\bullet|\bullet)$, provided that it is clear from the context whether we are considering either buying or selling prices (or both). We call a self-conjugate conditional lower prevision $P(\bullet|\bullet)$ simply a **conditional prevision**, and $P(f|A)$ represents a so-called **fair price** for the gamble f conditional on A: our subject is willing to buy f for any price $t < P(f|A)$ and willing to sell f for any price $t > P(f|A)$, contingent on A.

Unconditional previsions for arbitrary gambles, interpreted as fair prices, were considered by Crisma, Gigante and Millossovich (1997a, 1997b), as an extension of the work of de Finetti (1937). Unconditional lower previsions for arbitrary gambles were discussed in Troffaes and De Cooman (2002b) and treated in detail in Troffaes (2005, Section 5.2). Conditional lower previsions (defined on bounded gambles only) were considered by both Williams (1975b, 1976) and Walley (1991). In the rest of this chapter, we extend Williams' approach from bounded gambles to gambles and our earlier approach from the unconditional case to the conditional case.

Walley's (1991) approach differs from what we will do here, because he insists on imposing, besides the coherence Axioms A1–A4$_{237}$, an additional requirement of *conglomerability*. We do not find the arguments for conglomerability as a rationality requirement all that compelling. Moreover, imposing conglomerability tends to complicate matters significantly. Our approach resembles to some extent the one we take with countable additivity: we try to build a theory without imposing it *ab initio*, and then, if necessary or useful, try and find out what are, in specific cases, the mathematical consequences of adding the extra requirement. This is, for instance, what we will do in Section 13.9$_{288}$ when dealing with marginal extension.

13.4 Consistency for conditional lower previsions

13.4.1 Definition and justification

Given a conditional lower prevision, which specifies supremum buying prices for some gambles conditional on some events, can we also infer supremum buying prices for any other gamble conditional on any other event? As we did previously in Section 4.2$_{41}$, a first step towards extending conditional lower previsions is to translate the inclusion relation for sets of acceptable gambles to conditional lower previsions. Inclusion for sets of acceptable gambles amounts to implying at least as many behavioural dispositions, that is, accepting at least as many gambles as uncertain rewards. All behavioural dispositions implied by a conditional lower prevision $\underline{P}(\bullet|\bullet)$ are also implied by a conditional lower prevision $\underline{Q}(\bullet|\bullet)$ exactly when every buying price implied by $\underline{P}(\bullet|\bullet)$ is also a buying price implied by $\underline{Q}(\bullet|\bullet)$:

$$\underline{P}(f|A) > s \Rightarrow \underline{Q}(f|A) > s \text{ for all } s \in \mathbb{R} \text{ and } (f, A) \text{ in } \operatorname{dom} \underline{P}(\bullet|\bullet).$$

This is equivalent to the following definition. The proof of the equivalence is easy and is left to the reader.

Definition 13.16 *We say that a conditional lower prevision $\underline{Q}(\bullet|\bullet)$ **dominates** a conditional lower prevision $\underline{P}(\bullet|\bullet)$ if $\operatorname{dom} \underline{P}(\bullet|\bullet) \subseteq \operatorname{dom} \underline{Q}(\bullet|\bullet)$ and $\underline{P}(f|A) \leq \underline{Q}(f|A)$ for all $(f, A) \in \operatorname{dom} \underline{P}(\bullet|\bullet)$.*

The relation between the inclusion relation for sets of acceptable gambles on the one hand and the notion of dominance on the other hand, as explained in the following proposition, is again immediate.

Proposition 13.17 *Let $\mathscr{A}_1$ and $\mathscr{A}_2$ be consistent sets of acceptable gambles. If $\mathscr{A}_1 \subseteq \mathscr{A}_2$, then $\mathrm{Cl}_{\mathbb{D}}(\mathscr{A}_1) \subseteq \mathrm{Cl}_{\mathbb{D}}(\mathscr{A}_2)$ and $\mathrm{lpr}(\mathrm{Cl}_{\mathbb{D}}(\mathscr{A}_2))(\bullet|\bullet)$ dominates $\mathrm{lpr}(\mathrm{Cl}_{\mathbb{D}}(\mathscr{A}_1))(\bullet|\bullet)$.*

As we have seen in Section 13.3, the subject's conditional lower prevision $\underline{P}(\bullet|\bullet)$ corresponds to his specifying a set of acceptable gambles:

$$\mathscr{A}_{\underline{P}(\bullet|\bullet)} := \left\{ (f-\mu)I_A : (f,A) \in \mathrm{dom}\,\underline{P}(\bullet|\bullet),\ \mu \in \mathbb{R} \text{ and } \mu < \underline{P}(f|A) \right\}. \quad (13.10)$$

As with lower previsions for bounded gambles, when this set of acceptable gambles is *consistent*, we say that the conditional lower prevision $\underline{P}(\bullet|\bullet)$ avoids sure loss. This leads to the following definition.

Definition 13.18 (Avoiding sure loss) *Let $\underline{P}(\bullet|\bullet)$ be a conditional lower prevision. The following conditions are equivalent; if any (and hence all) of them are satisfied, we say that $\underline{P}(\bullet|\bullet)$ **avoids sure loss**.*

(A) *The set of acceptable gambles $\mathscr{A}_{\underline{P}(\bullet|\bullet)}$ is consistent.*

(B) *There is some coherent set of acceptable gambles $\mathscr{D}$ such that $\mathrm{lpr}(\mathscr{D})(\bullet|\bullet)$ dominates $\underline{P}(\bullet|\bullet)$.*

(C) *There is some conditional lower prevision $\underline{Q}(\bullet|\bullet)$ defined on $\mathbb{G} \times \mathscr{P}^\circ$ that satisfies the properties CLP1–CLP4$_{243}$ and that dominates $\underline{P}(\bullet|\bullet)$.*

(D) *For all n in $\mathbb{N}$, non-negative $\lambda_1, \dots, \lambda_n$ in $\mathbb{R}$, $(f_1, A_1), \dots, (f_n, A_n)$ in $\mathrm{dom}\,\underline{P}(\bullet|\bullet)$ and $a_1, \dots, a_n \in \mathbb{R}$ such that $a_i < \underline{P}(f_i|A_i)$, we have that*

$$\sup\left(\sum_{k=1}^{n} \lambda_k (f_k - a_k) I_{A_k} \middle| A_1 \cup \cdots \cup A_n \right) \geq 0. \quad (13.11)$$

Condition (D) was already given by Troffaes (2006, Definition 2). Let us first prove that these conditions are indeed equivalent.

Proof. We prove that (A)⇒(B)⇒(C)⇒(D)⇒(A).

(A)⇒(B). We infer from Definition 13.4(A)$_{238}$ that the consistent $\mathscr{A}_{\underline{P}(\bullet|\bullet)}$ is included in some coherent set $\mathscr{D}$ of acceptable gambles. We show that $\mathrm{lpr}(\mathscr{D})(\bullet|\bullet)$ dominates $\underline{P}(\bullet|\bullet)$. Indeed, for any $(f,A) \in \mathrm{dom}\,\underline{P}(\bullet|\bullet)$,

$$\begin{aligned} \mathrm{lpr}(\mathscr{D})(f|A) &= \sup\left\{ \mu \in \mathbb{R} : (f-\mu)I_A \in \mathscr{D} \right\} \\ &\geq \sup\left\{ \mu \in \mathbb{R} : (f-\mu)I_A \in \mathscr{A}_{\underline{P}(\bullet|\bullet)} \right\} \\ &\geq \sup\left\{ \mu \in \mathbb{R} : \mu < \underline{P}(f|A) \right\} = \underline{P}(f|A). \end{aligned}$$

(B)⇒(C). Take $\underline{Q}(\bullet|\bullet) := \mathrm{lpr}(\mathscr{D})(\bullet|\bullet)$ and use Theorem 13.11$_{242}$.

(C)$_{\curvearrowleft}\Rightarrow$(D)$_{\curvearrowleft}$. Let $\underline{Q}(\bullet|\bullet)$ be any conditional lower prevision defined on $\mathbb{G}\times\mathscr{P}^\circ$ that satisfies the properties CLP1–CLP4$_{243}$ and that dominates $\underline{P}(\bullet|\bullet)$. Consider n in $\mathbb{N}$, non-negative λ_1, …, λ_n in $\mathbb{R}$, (f_1,A_1), …, (f_n,A_n) in dom $\underline{P}(\bullet|\bullet)$ and a_1, …, $a_n \in \mathbb{R}$ such that $a_k < \underline{P}(f_k|A_k)$ and therefore $a_k < \underline{Q}(f|A)$. Now use Lemmas 13.14$_{251}$ and 13.15$_{252}$.

(D)$_{\curvearrowleft}\Rightarrow$(A)$_{\curvearrowleft}$. We prove that Condition (B)$_{238}$ of Definition 13.4$_{238}$ is satisfied. Every element of $\mathscr{A}_{\underline{P}(\bullet|\bullet)}$ is of the form $(f-a)I_A$ with $(f,A)\in$ dom $\underline{P}(\bullet|\bullet)$, $a\in\mathbb{R}$ and $a<\underline{P}(f|A)$. So we must show that $\sum_{k=1}^n \lambda_k(f_k-a_k)I_{A_k} \nless 0$ for all n in $\mathbb{N}$, non-negative λ_1, …, λ_n in $\mathbb{R}$, (f_1,A_1), …, (f_n,A_n) in dom $\underline{P}(\bullet|\bullet)$ and a_1, …, $a_n\in\mathbb{R}$ such that $a_i<\underline{P}(f_i|A_i)$. Assume *ex absurdo* that $\sum_{k=1}^n \lambda_k(f_k-a_k)I_{A_k}<0$. This already implies that $n>0$ and that there is some $\lambda_k>0$. Because terms for which the $\lambda_k=0$ do not contribute to the sum above, we may assume without loss of generality that $\lambda := \min_{k=1}^n \lambda_k > 0$. Let $A := A_1\cup\cdots\cup A_n$. Observe that there is some $\varepsilon>0$ small enough such that still $a_k+\varepsilon<\underline{P}(f_k|A_k)$ for all k, and therefore, applying Theorem 13.12$_{244}$ and Lemma 13.13(iv) and (v)$_{245}$ for the conditional lower prevision $\inf(\bullet|\bullet) := \mathrm{lpr}(\mathbb{G}_{\geq 0})(\bullet|\bullet)$ (see the discussion in Example 13.10$_{242}$)

$$\begin{aligned}\sup\left(\sum_{k=1}^n \lambda_k(f_k-a_k)I_{A_k}\middle|A\right) &\geq \sup\left(\sum_{k=1}^n \lambda_k(f_k-a_k-\varepsilon)I_{A_k}\middle|A\right)\\ &\quad+\varepsilon\inf\left(\sum_{k=1}^n \lambda_k I_{A_k}\middle|A\right)\\ &\geq \varepsilon\inf\left(\sum_{k=1}^n \lambda_k I_{A_k}\middle|A\right)\geq\varepsilon\lambda\inf\left(\sum_{k=1}^n I_{A_k}\middle|A\right)\\ &\geq\varepsilon\lambda>0,\end{aligned}$$

where the second inequality follows from assumption (D)$_{\curvearrowleft}$. We have arrived at a contradiction. □

In the case of unconditional lower previsions, there are simpler conditions for avoiding sure loss. Condition (F) was already given by Troffaes (2005, Definition 5.2).

Proposition 13.19 *Let $\underline{P}(\bullet|\bullet)$ be a conditional lower prevision defined on a subset of $\mathbb{G}\times\{\mathscr{X}\}$. Then $\underline{P}(\bullet|\bullet)$ avoids sure loss if and only if any (and hence all) of the following equivalent conditions are satisfied:*

(E) *There is some conditional lower prevision $\underline{Q}(\bullet|\bullet)$ defined on $\mathbb{G}\times\{\mathscr{X}\}$ that satisfies the properties* CLP1–CLP3$_{243}$ *and that dominates $\underline{P}(\bullet|\bullet)$.*

(F) *For all n in $\mathbb{N}$, non-negative λ_1, …, λ_n in $\mathbb{R}$ and gambles f_1, …, f_n in* dom $\underline{P}(\bullet|\mathscr{X})$ *such that $\sum_{k=1}^n \lambda_k\underline{P}(f_k|\mathscr{X})$ is well defined, we have that*

$$\sup\left(\sum_{k=1}^n \lambda_k f_k\middle|\mathscr{X}\right)\geq\sum_{k=1}^n \lambda_k\underline{P}(f_k|\mathscr{X}). \tag{13.12}$$

Proof. (C)$_{255}$⇒(E). Simply restrict the $\underline{Q}(\bullet|\bullet)$ that appears in the avoiding sure loss criterion (C)$_{255}$ to $\mathbb{G} \times \{\mathcal{X}\}$.

(E)⇒(F). Let $\underline{Q}(\bullet|\bullet)$ be any conditional lower prevision defined on $\mathbb{G} \times \{\mathcal{X}\}$ that satisfies the properties CLP1–CLP3$_{243}$ and that dominates $\underline{P}(\bullet|\bullet)$, guaranteed to exist by condition (E). Consider any n in $\mathbb{N}$, non-negative $\lambda_1, \dots, \lambda_n$ in $\mathbb{R}$ and gambles f_1, …, f_n in dom $\underline{P}(\bullet|\mathcal{X})$ such that $\sum_{k=1}^n \lambda_k \underline{P}(f_k|\mathcal{X})$ is well defined. If $\sum_{k=1}^n \lambda_k \underline{P}(f_k|\mathcal{X}) = -\infty$, then the condition (13.12) is trivially satisfied. If not, then all $\lambda_k \underline{P}(f_k|\mathcal{X}) > -\infty$ and, hence, also all $\lambda_k \underline{Q}(f_k|\mathcal{X}) > -\infty$ because $\underline{Q}(\bullet|\bullet)$ dominates $\underline{P}(\bullet|\bullet)$. This means that $\sum_{k=1}^n \lambda_k \underline{Q}(f_k|\mathcal{X})$ is also well defined. Now observe that

$$\sum_{k=1}^{n} \lambda_k \underline{P}(f_k|\mathcal{X}) \leq \sum_{k=1}^{n} \lambda_k \underline{Q}(f_k|\mathcal{X}) \leq \underline{Q}\left(\sum_{k=1}^{n} \lambda_k f_k \middle| \mathcal{X}\right) \leq \sup\left(\sum_{k=1}^{n} \lambda_k f_k \middle| \mathcal{X}\right),$$

where the second inequality follows from CLP2$_{243}$ and CLP3$_{243}$ and the third from Lemma 13.13(i)$_{245}$.

(F)⇒(D)$_{255}$. Consider any n in $\mathbb{N}$, non-negative $\lambda_1, \dots, \lambda_n$ in $\mathbb{R}$, $f_1, \dots, f_n$ in dom $\underline{P}(\bullet|\mathcal{X})$ and $a_1, \dots, a_n \in \mathbb{R}$ such that $a_k < \underline{P}(f_k|\mathcal{X})$. Then $\underline{P}(f_k|\mathcal{X}) > -\infty$ for all k, and hence, $\sum_{k=1}^n \lambda_k \underline{P}(f_k|\mathcal{X})$ is well defined, so Equation (13.12) holds, and therefore,

$$\sup\left(\sum_{k=1}^{n} \lambda_k f_k \middle| \mathcal{X}\right) \geq \sum_{k=1}^{n} \lambda_k \underline{P}(f_k|\mathcal{X}) \geq \sum_{k=1}^{n} \lambda_k a_k.$$

This implies that Equation (13.11)$_{255}$ holds. □

Avoiding sure loss has many interesting consequences. One we wish to draw particular attention to, and which is similar to Proposition 4.7$_{46}$, is the following.

Proposition 13.20 *If the conditional lower prevision $\underline{P}(\bullet|\bullet)$ avoids sure loss, then $\underline{P}(f|A) \leq \overline{P}(f|A)$ for all $(f, A) \in$ dom $\underline{P}(\bullet|\bullet)$ such that also $(-f, A) \in$ dom $\underline{P}(\bullet|\bullet)$.*

Proof. By the avoiding sure loss criterion (C)$_{255}$, there is some conditional lower prevision $\underline{Q}(\bullet|\bullet)$ defined on $\mathbb{G} \times \mathcal{P}^\circ$ that satisfies the properties CLP1–CLP4$_{243}$ and that dominates $\underline{P}(\bullet|\bullet)$. Hence, for any $(f, A) \in$ dom $\underline{P}(\bullet|\bullet)$ such that also $(-f, A) \in$ dom $\underline{P}(\bullet|\bullet)$,

$$\underline{P}(f|A) \leq \underline{Q}(f|A) \leq \overline{Q}(f|A) \leq \overline{P}(f|A),$$

where we used Lemma 13.13(i)$_{245}$ for the second inequality. □

Again, this justifies our calling $\underline{P}(\bullet|\bullet)$ a conditional 'lower' prevision and $\overline{P}(\bullet|\bullet)$ a conditional 'upper' prevision.

In the rest of this book, we wish to focus mostly on the stronger condition of *coherence*, to be discussed in Section 13.5$_{259}$.

13.4.2 Avoiding sure loss and avoiding partial loss

For conditional lower previsions, some authors, such as Walley (1991, Sections 7.1.2 and 7.1.3) and Troffaes (2006, Section 3.1), use the term *avoiding partial loss* for what we have been calling 'avoiding sure loss'. Here, we prefer to stick to the latter terminology, with the understanding that the loss is sure[6] *contingent on all participating conditioning events*.

13.4.3 Compatibility with the definition for lower previsions on bounded gambles

Before we conclude this discussion on avoiding sure loss, it behoves us to check whether our new notion of avoiding sure loss for conditional lower previsions reduces to the one for lower previsions on bounded gambles, which we studied in Section 4.2_{41}.

Theorem 13.21 *Consider a lower prevision* $\underline{Q}$ *on bounded gambles, so* $\operatorname{dom}\underline{Q} \subseteq \mathbb{B}$, *and define the corresponding (unconditional) lower prevision* $\underline{P}(\bullet|\bullet)$ *on* $\operatorname{dom}\underline{Q} \times \{\mathcal{X}\}$ *by letting* $\underline{P}(f|\mathcal{X}) := \underline{Q}(f)$ *for all* f *in* $\operatorname{dom}\underline{Q}$. *Then the lower prevision* $\underline{P}(\bullet|\bullet)$ *avoids sure loss in the sense of Definition* 13.18_{255} *if and only the lower prevision* $\underline{Q}$ *avoids sure loss in the sense of Definition* 4.6_{42}.

Proof. Compare criteria $(\mathrm{E})_{43}$ and $(\mathrm{F})_{256}$ for avoiding sure loss, and observe that $\sum_{k=1}^{n} \lambda_k \underline{P}(f_k|\mathcal{X}) = \sum_{k=1}^{n} \lambda_k \underline{Q}(f_k)$ is well defined for all $n \in \mathbb{N}$, all real non-negative λ_k and all bounded gambles $f_k \in \operatorname{dom}\underline{Q}$. □

13.4.4 Comparison with avoiding sure loss for lower previsions on bounded gambles

When comparing Proposition 13.19_{256} with Definition 4.6_{42}, we observe that, contrary to the case of lower previsions on bounded gambles, we can no longer restrict the coefficients $\lambda_1, \dots, \lambda_n$ to integer values in order to characterise avoiding sure loss for (unconditional) lower previsions on arbitrary gambles: roughly speaking, Definition 4.6(D) and $(\mathrm{E})_{42}$ are no longer equivalent when generalised to conditional lower previsions on arbitrary gambles, so we need to resort to using the stronger condition. Of course, as we have just seen in Theorem 13.21, Definition 13.18_{255} generalises Definition 4.6_{42}: a lower prevision on bounded gambles avoids sure loss according to Definition 4.6_{42} if and only if it avoids sure loss according to Definition 13.18_{255}.

Why does it not suffice to consider only integer combinations in Definition 13.18_{255}? Consider the following counterexample, taken from Troffaes (2005, p. 201).

Example 13.22 Let $\mathcal{X} := \mathbb{R}$, and consider the (unconditional) lower prevision $\underline{P}(\bullet|\bullet)$ with domain $\{X, -\sqrt{2}X\} \times \{\mathbb{R}\}$, defined by $\underline{P}(X|\mathbb{R}) := 1$ and $\underline{P}(-\sqrt{2}X|\mathbb{R}) := 2$.[7] As $n - m\sqrt{2} \neq 0$ for every n and m in $\mathbb{N}$ not both zero, we find that $\sup[nX - m\sqrt{2}X] = +\infty$ for every n and m in $\mathbb{N}$ not both zero. Consequently, the inequality

$$\sup\left[nX - m\sqrt{2}X\right] \geq n + 2m$$

[6] See the discussion in Section $4.2.3_{45}$.

[7] Remember that we write X also for the identity gamble.

holds for every n and m in $\mathbb{N}$ not both zero, and if n and m are both zero, then the inequality also holds: we have constructed a lower prevision $\underline{P}(\bullet|\bullet)$ such that

$$\sup\left[\sum_{k=1}^{n} f_k\right] \geq \sum_{k=1}^{n} \underline{P}(f_k|\mathcal{X})$$

for all n in $\mathbb{N}$ and gambles $f_1, \ldots, f_n$ in $\operatorname{dom}\underline{P}(\bullet|\mathcal{X})$ such that $\sum_{k=1}^{n} \underline{P}(f_k|\mathcal{X})$ is well defined. But $\underline{P}(\bullet|\bullet)$ does not avoid sure loss: for $\lambda_1 := \sqrt{2}$, $f_1(X) := X$, $\lambda_2 := 1$, and $f_2(X) := -\sqrt{2}X$, it holds that

$$\sup\left[\lambda_1 f_1 + \lambda_2 f_2\right] = 0 < \sqrt{2} + 2 = \lambda_1 + 2\lambda_2 = \lambda_1 \underline{P}(f_1|\mathcal{X}) + \lambda_2 \underline{P}(f_2|\mathcal{X}),$$

so avoiding sure loss criterion $(\mathrm{F})_{256}$ is not satisfied. ♦

13.5 Coherence for conditional lower previsions

13.5.1 Definition and justification

How can we infer supremum buying prices for arbitrary gambles, given a subject's conditional lower prevision assessments? As we discussed earlier when dealing with bounded gambles, a simple way to arrive at an extension is to use the natural extension of the set of acceptable gambles induced by the subject's conditional lower prevision.

More explicitly, given a conditional lower prevision $\underline{P}(\bullet|\bullet)$, the smallest set of acceptable gambles that takes into account coherence and the assessments embodied by $\underline{P}(\bullet|\bullet)$ is the natural extension of the set $\mathscr{A}_{\underline{P}(\bullet|\bullet)}$ defined by Equation $(13.10)_{255}$:

$$\begin{aligned}\mathrm{Cl}_{\mathbb{D}}(\mathscr{A}_{\underline{P}(\bullet|\bullet)}) = \mathscr{E}_{\mathscr{A}_{\underline{P}(\bullet|\bullet)}} &= \operatorname{nonneg}(\mathbb{G}_{\geq 0} \cup \mathscr{A}_{\underline{P}(\bullet|\bullet)}) \\ &= \Bigg\{ h \in \mathbb{G} : h \geq \sum_{k=1}^{n} \lambda_k (f_k - a_k) I_{A_k},\ n \in \mathbb{N},\ (f_k, A_k) \in \operatorname{dom}\underline{P}(\bullet|\bullet), \\ &\qquad \lambda_k \in \mathbb{R}_{\geq 0},\ a_k \in \mathbb{R},\ a_k < \underline{P}(f_k|A_k) \Bigg\} \qquad (13.13)\end{aligned}$$

We can now use this set $\mathscr{E}_{\mathscr{A}_{\underline{P}(\bullet|\bullet)}}$ of acceptable gambles to construct a conditional lower prevision $\operatorname{lpr}(\mathscr{E}_{\mathscr{A}_{\underline{P}(\bullet|\bullet)}})(\bullet|\bullet)$ on all of $\mathbb{G} \times \mathscr{P}^{\circ}$.

Definition 13.23 (Natural extension) *Consider a conditional lower prevision $\underline{P}(\bullet|\bullet)$ that avoids sure loss. Then the conditional lower prevision $\underline{E}_{\underline{P}(\bullet|\bullet)}(\bullet|\bullet) := \operatorname{lpr}(\mathrm{Cl}_{\mathbb{D}}(\mathscr{A}_{\underline{P}(\bullet|\bullet)}))(\bullet|\bullet) = \operatorname{lpr}(\mathscr{E}_{\mathscr{A}_{\underline{P}(\bullet|\bullet)}})(\bullet|\bullet)$ is called the* ***natural extension*** *of $\underline{P}(\bullet|\bullet)$. It is given for all gambles $f \in \mathbb{G}$ and events $A \in \mathscr{P}^{\circ}$ by*

$$\underline{E}_{\underline{P}(\bullet|\bullet)}(f|A) := \sup\Bigg\{ \alpha \in \mathbb{R} : n \in \mathbb{N},\ a_k \in \mathbb{R},\ a_k < \underline{P}(f_k|A_k),$$

$$\left.\begin{array}{c}\lambda_k \geq 0, (f_1, A_1), \ldots, (f_n, A_n) \in \operatorname{dom} \underline{P}(\bullet|\bullet), \\ (f-\alpha)I_A \geq \sum_{k=1}^{n} \lambda_k (f_k - a_k) I_{A_k}\end{array}\right\} \tag{13.14}$$

Remember that if $\underline{P}(\bullet|\bullet)$ avoids sure loss, then $\mathscr{A}_{\underline{P}(\bullet|\bullet)}$ is a consistent and $\mathscr{E}_{\mathscr{A}_{\underline{P}(\bullet|\bullet)}}$ a coherent set of acceptable gambles, by Theorem 13.7$_{239}$. Theorem 13.11$_{242}$ then guarantees that $\operatorname{lpr}(\mathscr{E}_{\mathscr{A}_{\underline{P}(\bullet|\bullet)}})(\bullet|\bullet)$ satisfies CLP1–CLP4$_{243}$. Equation (13.14) for the natural extension follows immediately by combining Equations (13.2)$_{240}$ and (13.13)$_{\curvearrowleft}$.

It may happen that our subject's $\underline{P}(f|A)$ does not agree with $\underline{E}_{\underline{P}(\bullet|\bullet)}(f|A)$ for certain gambles f and events A. As $\underline{E}_{\underline{P}(\bullet|\bullet)}(f|A)$ is the most conservative supremum called-off buying price for f that can be derived from his assessments $\underline{P}(g|B)$, $(g, B) \in \operatorname{dom} \underline{P}(\bullet|\bullet)$, this means that the subject has not fully taken into account all behavioural implications of his assessments and coherence.

If a subject's conditional lower prevision assessment does not show this kind of imperfection, we call it coherent. This is formalised in Definition 13.25, which generalises Definition 4.10$_{47}$ to conditional lower previsions.

Definition 13.24 *Let $\underline{P}(\bullet|\bullet)$ and $\underline{Q}(\bullet|\bullet)$ be a conditional lower previsions. Then $\underline{P}(\bullet|\bullet)$ is called a* ***restriction*** *of $\underline{Q}(\bullet|\bullet)$ and $\underline{Q}(\bullet|\bullet)$ is called an* ***extension*** *of $\underline{P}(\bullet|\bullet)$, whenever $\operatorname{dom} \underline{P}(\bullet|\bullet) \subseteq \operatorname{dom} \underline{Q}(\bullet|\bullet)$ and $\underline{P}(f|A) = \underline{Q}(f|A)$ for all $(f, A) \in \operatorname{dom} \underline{P}(\bullet|\bullet)$.*

Definition 13.25 (Coherence) *Let $\underline{P}(\bullet|\bullet)$ be a conditional lower prevision. The following conditions are equivalent; if any (and hence all) of them are satisfied, then we say that $\underline{P}(\bullet|\bullet)$ is* ***coherent****.*

(A) *There is some coherent set of acceptable gambles $\mathscr{D}$ such that $\operatorname{lpr}(\mathscr{D})(\bullet|\bullet)$ is an extension of $\underline{P}(\bullet|\bullet)$.*

(B) *There is some conditional lower prevision $\underline{Q}(\bullet|\bullet)$ defined on $\mathbb{G} \times \mathscr{P}^\circ$ that satisfies the properties CLP1–CLP4$_{243}$ and that is an extension of $\underline{P}(\bullet|\bullet)$.*

(C) *$\underline{P}(\bullet|\bullet)$ avoids sure loss and is a restriction of its natural extension $\underline{E}_{\underline{P}(\bullet|\bullet)}(\bullet|\bullet)$.*

(D) *For all n in $\mathbb{N}$, non-negative $\lambda_0, \lambda_1, \ldots, \lambda_n$ in $\mathbb{R}$, $(f_0, A_0), (f_1, A_1), \ldots, (f_n, A_n)$ in $\operatorname{dom} \underline{P}(\bullet|\bullet)$ and $a_0, \ldots, a_n \in \mathbb{R}$ such that $a_0 > \underline{P}(f_0|A_0)$ and $a_k < \underline{P}(f_k|A_k)$ for all $k \in \{1, \ldots, n\}$, we have that*

$$\sup\left(\sum_{k=1}^{n} \lambda_k (f_k - a_k) I_{A_k} - \lambda_0 (f_0 - a_0) I_{A_0} \middle| A_0 \cup A_1 \cup \cdots \cup A_n\right) \geq 0. \tag{13.15}$$

Condition (D) was already given by Troffaes (2006, Definition 3). Let us show that these conditions are indeed equivalent.

Proof. We prove that (A)⇒(B)⇒(C)⇒(D)⇒(A).

(A)⇒(B). The conditional lower prevision $\underline{Q}(\bullet|\bullet) := \mathrm{lpr}(\mathscr{D})(\bullet|\bullet)$ on $\mathbb{G} \times \mathscr{P}^{\circ}$ is an extension of $\underline{P}(\bullet|\bullet)$, and it satisfies CLP1–CLP4$_{243}$ by Theorem 13.11$_{242}$.

(B)⇒(C). As $\underline{Q}(\bullet|\bullet)$ in particular also dominates $\underline{P}(\bullet|\bullet)$, we infer from condition (C)$_{255}$ in Definition 13.18$_{255}$ that $\underline{P}(\bullet|\bullet)$ avoids sure loss. We can therefore consider its natural extension $\underline{E}_{\underline{P}(\bullet|\bullet)}(\bullet|\bullet) = \mathrm{lpr}(\mathrm{Cl}_{\mathbb{D}}(\mathscr{A}_{\underline{P}(\bullet|\bullet)}))(\bullet|\bullet)$. It remains to show that $\underline{E}_{\underline{P}(\bullet|\bullet)}(\bullet|\bullet)$ coincides with $\underline{P}(\bullet|\bullet)$ on dom $\underline{P}(\bullet|\bullet)$.

We first show that $\underline{E}_{\underline{P}(\bullet|\bullet)}(\bullet|\bullet)$ dominates $\underline{P}(\bullet|\bullet)$. Consider any $(f, A) \in$ dom $\underline{P}(\bullet|\bullet)$ and any real $\alpha < \underline{P}(f|A)$, then it follows from Equation (13.10)$_{255}$ that $(f - \alpha)I_A \in \mathscr{A}_{\underline{P}(\bullet|\bullet)}$ and, therefore, also $(f - \alpha)I_A \in \mathrm{Cl}_{\mathbb{D}}(\mathscr{A}_{\underline{P}(\bullet|\bullet)})$, so $\alpha \leq \underline{E}_{\underline{P}(\bullet|\bullet)}(f|A)$. Hence indeed $\underline{P}(f|A) \leq \underline{E}_{\underline{P}(\bullet|\bullet)}(f|A)$.

It now suffices to prove that $\underline{Q}(\bullet|\bullet)$ dominates $\underline{E}_{\underline{P}(\bullet|\bullet)}(\bullet|\bullet)$, because we know that $\underline{Q}(\bullet|\bullet)$ coincides with $\underline{P}(\bullet|\bullet)$ on dom $\underline{P}(\bullet|\bullet)$. Because $\underline{Q}(\bullet|\bullet)$ coincides with $\underline{P}(\bullet|\bullet)$ on dom $\underline{P}(\bullet|\bullet)$, it follows that $\mathscr{A}_{\underline{P}(\bullet|\bullet)} \subseteq \mathscr{A}_{\underline{Q}(\bullet|\bullet)}$ and, therefore, also $\mathrm{Cl}_{\mathbb{D}}(\mathscr{A}_{\underline{P}(\bullet|\bullet)}) \subseteq \mathrm{Cl}_{\mathbb{D}}(\mathscr{A}_{\underline{Q}(\bullet|\bullet)})$. By avoiding sure loss criterion (C)$_{255}$, $\underline{Q}(\bullet|\bullet)$ avoids sure loss, and therefore, $\mathscr{A}_{\underline{Q}(\bullet|\bullet)}$ is consistent by avoiding sure loss criterion (A)$_{255}$. We then infer from Proposition 13.17$_{255}$ that

$$\underline{E}_{\underline{P}(\bullet|\bullet)}(\bullet|\bullet) = \mathrm{lpr}(\mathrm{Cl}_{\mathbb{D}}(\mathscr{A}_{\underline{P}(\bullet|\bullet)}))(\bullet|\bullet) \leq \mathrm{lpr}(\mathrm{Cl}_{\mathbb{D}}(\mathscr{A}_{\underline{Q}(\bullet|\bullet)}))(\bullet|\bullet) = \underline{Q}(\bullet|\bullet),$$

where the last equality follows from the argument in the proof of Theorem 13.12$_{244}$. This establishes the desired result.

(C)⇒(D). As $\underline{P}(\bullet|\bullet)$ avoids sure loss, $\mathscr{A}_{\underline{P}(\bullet|\bullet)}$ is consistent and, therefore, $\mathscr{E}_{\mathscr{A}_{\underline{P}(\bullet|\bullet)}}$ is a coherent set of acceptable gambles. This implies that the natural extension $\underline{E}_{\underline{P}(\bullet|\bullet)}(\bullet|\bullet) = \mathrm{lpr}(\mathscr{E}_{\mathscr{A}_{\underline{P}(\bullet|\bullet)}})(\bullet|\bullet)$ satisfies CLP1–CLP4$_{243}$ by Theorem 13.11$_{242}$. Now use Lemma 13.14$_{251}$ and the fact that $\underline{E}_{\underline{P}(\bullet|\bullet)}(\bullet|\bullet)$ and $\underline{P}(\bullet|\bullet)$ coincide on dom $\underline{P}(\bullet|\bullet)$.

(D)⇒(A). First, we have seen in Lemma 13.15$_{252}$ that the present condition (D) implies condition (D)$_{255}$ for avoiding sure loss in Definition 13.18$_{255}$, which we know is equivalent to condition (A)$_{255}$ there. Hence, $\mathscr{A}_{\underline{P}(\bullet|\bullet)}$ is a consistent set of acceptable gambles. If we invoke Theorem 13.7$_{239}$, we find that the natural extension $\mathrm{Cl}_{\mathbb{D}}(\mathscr{A}_{\underline{P}(\bullet|\bullet)})$ of $\mathscr{A}_{\underline{P}(\bullet|\bullet)}$ is the smallest coherent set of acceptable gambles that includes $\mathscr{A}_{\underline{P}(\bullet|\bullet)}$. We let $\mathscr{D} := \mathrm{Cl}_{\mathbb{D}}(\mathscr{A}_{\underline{P}(\bullet|\bullet)})$ and show that $\mathrm{lpr}(\mathscr{D})(\bullet|\bullet)$ is an extension of $\underline{P}(\bullet|\bullet)$.

Consider any $(f_0, A_0) \in$ dom $\underline{P}(\bullet|\bullet)$ and any $\alpha < \underline{P}(f_0|A_0)$. As $(f_0 - \alpha)I_{A_0} \in \mathscr{A}_{\underline{P}(\bullet|\bullet)} \subseteq \mathscr{D}$ and therefore, by Equation (13.2)$_{240}$, $\alpha \leq \mathrm{lpr}(\mathscr{D})(f_0|A_0)$, we infer that $\mathrm{lpr}(\mathscr{D})(f_0|A_0) \geq \underline{P}(f_0|A_0)$. Assume *ex absurdo* that $\mathrm{lpr}(\mathscr{D})(f_0|A_0) > \underline{P}(f_0|A_0)$. Then there is some $\varepsilon > 0$ sufficiently small such that still $\mathrm{lpr}(\mathscr{D})(f_0|A_0) > \underline{P}(f_0|A_0) + \varepsilon$. Recall from Definition 13.23$_{259}$ that $\mathrm{lpr}(\mathscr{D})(\bullet|\bullet) = \underline{E}_{\underline{P}(\bullet|\bullet)}(\bullet|\bullet)$, so it follows from Equation (13.14) that there are $\alpha \in \mathbb{R}$, n in $\mathbb{N}$, non-negative λ_1, …, λ_n in $\mathbb{R}$, (f_1, A_1), …, (f_n, A_n) in dom $\underline{P}(\bullet|\bullet)$ and a_1, …, $a_n \in \mathbb{R}$ such that $a_k < \underline{P}(f_k|A_k)$, $\alpha > \underline{P}(f_0|A_0) + \varepsilon$ and $(f_0 - \alpha)I_{A_0} \geq \sum_{k=1}^{n} \lambda_k (f_k - a_k) I_{A_k}$. We may assume without loss of generality that $\lambda_k > 0$ for all $k \in \{1, \dots, n\}$.

If $n = 0$, this implies that $[f_0 - (\alpha - \varepsilon)]I_{A_0} \geq \varepsilon I_{A_0}$ and therefore also that

$$\sup(-[f_0 - (\alpha - \varepsilon)]I_{A_0}|A_0) \leq \sup(-\varepsilon I_{A_0}|A_0) = -\varepsilon < 0,$$

which contradicts the assumption (D)$_{260}$ (where we let $n = 0$, $a_0 = \alpha - \varepsilon > \underline{P}(f_0|A_0)$ and $\lambda_0 = 1$).

If $n \geq 1$, this implies similarly that for all $\delta > 0$,

$$\sup\left(\sum_{k=1}^{n} \lambda_k[f_k - (a_k + \delta)]I_{A_k} - [f_0 - (\alpha - \varepsilon)]I_{A_0}\middle|A_0 \cup A_1 \cup \cdots \cup A_n\right)$$
$$= \sup\left(\sum_{k=1}^{n} \lambda_k(f_k - a_k)I_{A_k} - (f_0 - \alpha)I_{A_0} - \delta\sum_{k=1}^{n} \lambda_k I_{A_k} - \varepsilon I_{A_0}\middle|A_0 \cup A_1 \cup \cdots \cup A_n\right)$$
$$\leq \sup\left(-\delta\sum_{k=1}^{n} \lambda_k I_{A_k} - \varepsilon I_{A_0}\middle|A_0 \cup A_1 \cup \cdots \cup A_n\right) < 0,$$

which contradicts (D)$_{260}$, provided that we choose $\delta > 0$ small enough to ensure that $a_k + \delta < \underline{P}(f_k|A_k)$ for all k. □

In the case of unconditional lower previsions, there are simpler conditions for coherence. Condition (F) was already given by Troffaes (2005, Definition 5.3).

Proposition 13.26 *Let $\underline{P}(\bullet|\bullet)$ be a conditional lower prevision defined on a subset of $\mathbb{G} \times \{\mathcal{X}\}$. Then $\underline{P}(\bullet|\bullet)$ is coherent if and only if any (and hence all) of the following equivalent conditions are satisfied:*

(E) *There is some conditional lower prevision $\underline{Q}(\bullet|\bullet)$ defined on $\mathbb{G} \times \{\mathcal{X}\}$ that satisfies the properties* CLP1–CLP3$_{243}$ *on its domain and that is an extension of $\underline{P}(\bullet|\bullet)$.*

(F) *For every n in $\mathbb{N}$, non-negative $\lambda_0, \lambda_1, \ldots, \lambda_n$ in $\mathbb{R}$ and gambles $f_0, f_1, \ldots, f_n$ in* dom $\underline{P}(\bullet|\mathcal{X})$ *such that $\sum_{k=1}^{n} \lambda_k \underline{P}(f_k|\mathcal{X}) - \lambda_0 \underline{P}(f_0|\mathcal{X})$ is well defined, we have that*

$$\sup\left(\sum_{k=1}^{n} \lambda_k f_k - \lambda_0 f_0\middle|\mathcal{X}\right) \geq \sum_{k=1}^{n} \lambda_k \underline{P}(f_k|\mathcal{X}) - \lambda_0 \underline{P}(f_0|\mathcal{X}). \qquad (13.16)$$

Proof. (B)$_{260}$⇒(E). Simply restrict the $\underline{Q}(\bullet|\bullet)$ that appears in the coherence condition (B)$_{260}$ to $\mathbb{G} \times \{\mathcal{X}\}$.

(E)⇒(F). Let $\underline{Q}(\bullet|\bullet)$ be any conditional lower prevision defined on $\mathbb{G} \times \{\mathcal{X}\}$ that satisfies the properties CLP1–CLP3$_{243}$ and that is an extension of $\underline{P}(\bullet|\bullet)$, guaranteed to exist by condition (E). Consider any n in $\mathbb{N}$, non-negative $\lambda_0, \lambda_1, \ldots, \lambda_n$ in $\mathbb{R}$ and gambles $f_0, f_1, \ldots, f_n$ in dom $\underline{P}(\bullet|\mathcal{X})$ such that $\sum_{k=1}^{n} \lambda_k \underline{P}(f_k|\mathcal{X}) - \lambda_0 \underline{P}(f_0|\mathcal{X})$ is well defined. Because $\underline{Q}(\bullet|\bullet)$ coincides with $\underline{P}(\bullet|\bullet)$ on its domain, this implies that $\sum_{k=1}^{n} \lambda_k \underline{Q}(f_k|\mathcal{X}) - \lambda_0 \underline{Q}(f_0|\mathcal{X})$ is well defined too.

If we can show that

$$\sum_{k=1}^{n} \lambda_k \underline{Q}(f_k|\mathcal{X}) - \lambda_0 \underline{Q}(f_0|\mathcal{X}) \leq \sup\left(\sum_{k=1}^{n} \lambda_k f_k - \lambda_0 f_0 \middle| \mathcal{X}\right), \tag{13.17}$$

then the desired result follows, again using the fact that $\underline{Q}(\bullet|\bullet)$ coincides with $\underline{P}(\bullet|\bullet)$ on $\operatorname{dom}\underline{P}(\bullet|\bullet)$. There are two possible cases.

In the first case, the sum $\underline{Q}\big(\sum_{k=1}^{n} \lambda_k f_k|\mathcal{X}\big) + \overline{Q}(-\lambda_0 f_0|\mathcal{X})$ is well defined, and then indeed

$$\begin{aligned}
\sum_{k=1}^{n} \lambda_k \underline{Q}(f_k|\mathcal{X}) - \lambda_0 \underline{Q}(f_0|\mathcal{X}) &= \sum_{k=1}^{n} \underline{Q}(\lambda_k f_k|\mathcal{X}) + \overline{Q}(-\lambda_0 f_0|\mathcal{X}) \\
&\leq \underline{Q}\left(\sum_{k=1}^{n} \lambda_k f_k \middle| \mathcal{X}\right) + \overline{Q}(-\lambda_0 f_0|\mathcal{X}) \\
&\leq \overline{Q}\left(\sum_{k=1}^{n} \lambda_k f_k - \lambda_0 f_0 \middle| \mathcal{X}\right) \\
&\leq \sup\left(\sum_{k=1}^{n} \lambda_k f_k - \lambda_0 f_0 \middle| \mathcal{X}\right),
\end{aligned}$$

where the equality follows from conjugacy and CLP3_{243}, the first inequality from CLP2_{243}, the second one from conjugacy and CLP2_{243} and the last one from conjugacy and CLP1_{243}. So Equation (13.17) holds. (Observe that if $\underline{Q}\big(\sum_{k=1}^{n} \lambda_k f_k|\mathcal{X}\big) + \overline{Q}(-\lambda_0 f_0|\mathcal{X})$ is not well defined, the first and second inequalities do not follow from CLP2_{243}.)

Now consider the second possible case, where the sum $\underline{Q}\big(\sum_{k=1}^{n} \lambda_k f_k|\mathcal{X}\big) + \overline{Q}(-\lambda_0 f_0|\mathcal{X})$ is not well defined. On the one hand, if $\underline{Q}\big(\sum_{k=1}^{n} \lambda_k f_k|\mathcal{X}\big) = -\infty$, then by CLP2_{243} clearly also $\sum_{k=1}^{n} \lambda_k \underline{Q}(f_k|\mathcal{X}) = -\infty$ (the latter sum is well defined by assumption), and therefore, the (again, by assumption well defined) left-hand side of Equation (13.17) is $-\infty$ as well, so Equation (13.17) holds. On the other hand, if $\overline{Q}(-\lambda_0 f_0|\mathcal{X}) = -\infty$, then by conjugacy and CLP3_{243} also $-\lambda_0 \underline{Q}(f_0|\mathcal{X}) = -\infty$, and therefore, the (once more, by assumption well defined) left-hand side of Equation (13.17) is $-\infty$ as well, so Equation (13.17) holds here too.

$(\text{F}) \Rightarrow (\text{D})_{260}$. Consider arbitrary n in $\mathbb{N}$, non-negative $\lambda_0, \lambda_1, \dots, \lambda_n$ in $\mathbb{R}$, gambles $f_0, f_1, \dots, f_n$ in $\operatorname{dom}\underline{P}(\bullet|\mathcal{X})$ and $a_0, \dots, a_n \in \mathbb{R}$ such that $a_0 > \underline{P}(f_0|\mathcal{X})$ and $a_k < \underline{P}(f_k|\mathcal{X})$ for all $k \in \{1, \dots, n\}$. This already implies that $\underline{P}(f_0|\mathcal{X}) < +\infty$ and all $\underline{P}(f_k|\mathcal{X}) > -\infty$ and, therefore, $\sum_{k=1}^{n} \lambda_k \underline{P}(f_k|\mathcal{X}) - \lambda_0 \underline{P}(f_0|\mathcal{X})$ is well defined. It now follows from Equation (13.16) that

$$\sup\left(\sum_{k=1}^{n} \lambda_k f_k - \lambda_0 f_0 \middle| \mathcal{X}\right) \geq \sum_{k=1}^{n} \lambda_k \underline{P}(f_k|\mathcal{X}) - \lambda_0 \underline{P}(f_0|\mathcal{X}) \geq \sum_{k=1}^{n} \lambda_k a_k - \lambda_0 a_0,$$

so Equation $(13.15)_{260}$ follows. □

13.5.2 Compatibility with the definition for lower previsions on bounded gambles

As for avoiding sure loss, we need to investigate whether our new notion of coherence for conditional lower previsions reduces to the one for lower previsions on bounded gambles we studied in Section 4.3_{46}.

Theorem 13.27 *Consider a lower prevision $\underline{Q}$ on bounded gambles, so* $\operatorname{dom}\underline{Q} \subseteq \mathbb{B}$, *and define the corresponding (unconditional) lower prevision* $\underline{P}(\bullet|\bullet)$ *on* $\operatorname{dom}\underline{Q} \times \{\mathcal{X}\}$ *by letting* $\underline{P}(f|\mathcal{X}) := \underline{Q}(f)$ *for all* f *in* $\operatorname{dom}\underline{Q}$. *Then the lower prevision* $\underline{P}(\bullet|\bullet)$ *is coherent in the sense of Definition* 13.25_{260} *if and only the lower prevision* $\underline{Q}$ *is coherent in the sense of Definition* 4.10_{47}.

Proof. Compare criteria $(E)_{47}$ and $(F)_{262}$ for coherence, and take into account that $\sum_{k=1}^{n} \lambda_k \underline{P}(f_k|\mathcal{X}) - \lambda_0 \underline{P}(f_0|\mathcal{X}) = \sum_{k=1}^{n} \lambda_k \underline{Q}(f_k) - \lambda_0 \underline{Q}(f_0)$ is well defined for all $n \in \mathbb{N}$, all real non-negative $\lambda_0, \lambda_1, \ldots, \lambda_n$ and all bounded gambles $f_0, f_1, \ldots f_n$ in $\operatorname{dom}\underline{Q}$. □

13.5.3 Comparison with coherence for lower previsions on bounded gambles

Compare Proposition 13.26_{262} with Definition 4.10_{47}: as in the case of avoiding sure loss, and contrary to lower previsions on bounded gambles, we cannot restrict the coefficients $\lambda_0, \lambda_1, \ldots, \lambda_n$ to integer values in order to characterise coherence for (unconditional) lower previsions on arbitrary gambles. Definition 4.10(D) and $(E)_{47}$ are no longer equivalent when generalised to conditional lower previsions, so we take the strongest condition. Of course, as we have just seen earlier in Theorem 13.27, Definition 13.25_{260} generalises Definition 4.10_{47}.

13.5.4 Linear previsions

Similarly to what we did in Section $4.3.4_{51}$, we now study coherence for conditional previsions, that is, self-conjugate conditional lower previsions. Recall from the discussion near the end of Section $13.3.2_{252}$ that a conditional lower prevision $\underline{P}(\bullet|\bullet)$ is *self-conjugate*, and then called a *conditional prevision*, if $\operatorname{dom}\underline{P}(\bullet|\bullet) = \operatorname{dom}\overline{P}(\bullet|\bullet)$ and $\underline{P}(f|A) = \overline{P}(f|A)$ for all $(f, A) \in \operatorname{dom}\underline{P}(\bullet|\bullet)$, and that we then simply write $P(\bullet|\bullet)$ instead of $\underline{P}(\bullet|\bullet)$ or $\overline{P}(\bullet|\bullet)$.

Definition 13.28 (Linear conditional previsions) *We say that a conditional lower prevision is* ***linear*** *or that it is a* ***linear conditional prevision****, or a* ***coherent conditional prevision****, if it is self-conjugate (and therefore a conditional prevision) and coherent. We denote the set of all linear conditional previsions with domain* $\mathcal{K} \subseteq \mathbb{G} \times \mathcal{P}^{\circ}$ *by* $\mathbb{CP}^{\mathcal{K}}(\mathcal{X})$, *or simply by* $\mathbb{CP}^{\mathcal{K}}$. *The set* $\mathbb{CP}^{\mathbb{G}\times\mathcal{P}^{\circ}}(\mathcal{X})$ *is also denoted by* $\mathbb{CP}(\mathcal{X})$, *or simply by* $\mathbb{CP}$.

Theorem 13.29 *A conditional prevision* $P(\bullet|\bullet)$ *is linear if and only if any (and hence all) of the following equivalent conditions is satisfied:*

(A) *$P(\bullet|\bullet)$ is coherent (as a conditional lower prevision).*

(B) *$P(\bullet|\bullet)$ avoids sure loss (as a conditional lower prevision).*

(C) *for all n in $\mathbb{N}$, λ_1, …, λ_n in $\mathbb{R}$, (f_1,A_1), …, (f_n,A_n) in* dom *$P(\bullet|\bullet)$ and a_1, …, $a_n \in \mathbb{R}$ such that $\lambda_k a_k < \lambda_k P(f_k|A_k)$ for all $k \in \{1,\dots,n\}$, we have that*

$$\sup\left(\sum_{k=1}^{n} \lambda_k(f_k - a_k)I_{A_k}\middle| A_1 \cup \cdots \cup A_n\right) \geq 0. \tag{13.18}$$

Proof. (A) is simply a reformulation of the definition of a linear conditional prevision. We show that (A)⇒(B)⇒(C)⇒(A).

(A)⇒(B). Immediate: coherence implies avoiding sure loss for any conditional lower prevision; compare Definition 13.25_{260} with Definition 13.18_{255}.

(B)⇒(C). Consider arbitrary n in $\mathbb{N}$, λ_1, …, λ_n in $\mathbb{R}$, (f_1,A_1), …, (f_n,A_n) in dom $P(\bullet|\bullet)$ and a_1, …, $a_n \in \mathbb{R}$ such that $\lambda_k a_k < \lambda_k P(f_k|A_k)$ for all $k \in \{1,\dots,n\}$. As $P(\bullet|\bullet)$ avoids sure loss, we infer from the avoiding sure loss criterion $(\mathrm{D})_{255}$ that

$$\sup\left(\sum_{k=1,\lambda_k>0}^{n} \lambda_k(f_k - a_k)I_{A_k} + \sum_{k=1,\lambda_k<0}^{n} (-\lambda_k)(-f_k - (-a_k))I_{A_k}\middle| A_1 \cup \cdots \cup A_n\right) \geq 0,$$

once we observe that it follows from the self-conjugacy of $P(\bullet|\bullet)$ that $-a_k < P(-f_k|A_k)$ whenever $\lambda_k < 0$. This brings us to Equation (13.18).

(C)⇒(A). Immediate: coherence criterion $(\mathrm{D})_{260}$ is weaker than (implied by) the present condition (C). □

Again, conditions for coherence (linearity) simplify in the unconditional case. Condition (D) was already given by Crisma et al. (1997b, Definition 3.1). We also call a linear unconditional prevision simply a **linear prevision**, as there is no possible conflict with the definition of a linear prevision on bounded gambles (see Theorem $4.12(\mathrm{D})_{51}$).

Proposition 13.30 *Let $P(\bullet|\bullet)$ be a conditional prevision defined on a subset of $\mathbb{G} \times \{\mathcal{X}\}$. Then $P(\bullet|\bullet)$ is linear if and only if*

(D) *for all n in $\mathbb{N}$, λ_1, …, λ_n in $\mathbb{R}$ and gambles f_1, …, f_n in* dom *$P(\bullet|\mathcal{X})$ such that $\sum_{k=1}^{n} \lambda_k P(f_k|\mathcal{X})$ is well defined, we have that*

$$\sup\left(\sum_{k=1}^{n} \lambda_k f_k\middle|\mathcal{X}\right) \geq \sum_{k=1}^{n} \lambda_k P(f_k|\mathcal{X}).$$

Proof. (B)⇒(D). Consider any n in $\mathbb{N}$, λ_1, …, λ_n in $\mathbb{R}$ and gambles f_1, …, f_n in dom $P(\bullet|\mathcal{X})$ such that $\sum_{k=1}^{n} \lambda_k P(f_k|\mathcal{X})$ is well defined. Then, because we know from the assumption (B) that $P(\bullet|\bullet)$ avoids sure loss, we infer from the avoiding sure loss criterion $(\mathrm{F})_{256}$ and the self-conjugacy of $P(\bullet|\bullet)$ that indeed

$$\sum_{k=1}^{n} \lambda_k P(f_k|\mathcal{X}) = \sum_{\lambda_k \geq 0} \lambda_k P(f_k|\mathcal{X}) + \sum_{\lambda_k<0} (-\lambda_k) P(-f_k|\mathcal{X})$$

$$\leq \sup\left(\sum_{\lambda_k\geq 0}\lambda_k f_k + \sum_{\lambda_k<0}(-\lambda_k)(-f_k)\middle|\mathcal{X}\right) = \sup\left(\sum_{k=1}^{n}\lambda_k f_k\middle|\mathcal{X}\right).$$

(D)$_\curvearrowleft$⇒(C)$_\curvearrowleft$. Consider any n in $\mathbb{N}$, λ_1, …, λ_n in $\mathbb{R}$, gambles f_1, …, f_n in dom $\underline{P}(\bullet|\mathcal{X})$ and a_1, …, $a_n \in \mathbb{R}$ such that $\lambda_k a_k < \lambda_k \underline{P}(f_k|\mathcal{X})$ for all $k \in \{1, \dots, n\}$. This already implies that all $\lambda_k \underline{P}(f_k|\mathcal{X}) > -\infty$ and, therefore, $\sum_{k=1}^{n} \lambda_k \underline{P}(f_k|\mathcal{X})$ is well defined. So it follows from (D)$_\curvearrowleft$ that

$$\sup\left(\sum_{k=1}^{n}\lambda_k f_k\middle|\mathcal{X}\right) \geq \sum_{k=1}^{n}\lambda_k \underline{P}(f_k|\mathcal{X}) \geq \sum_{k=1}^{n}\lambda_k a_k,$$

so (C)$_\curvearrowleft$ follows. □

13.6 Properties of coherent conditional lower previsions

13.6.1 Interesting consequences of coherence

The following theorem summarises the most important consequences of coherence; it is a straightforward generalisation of Theorem 4.13$_{53}$.

Theorem 13.31 *Let $\underline{P}(\bullet|\bullet)$ be a coherent conditional lower prevision. Then for all gambles f and g, nets of gambles f_α, real numbers a, non-negative real numbers λ and events A, $B \in \mathscr{P}^\circ$ such that $A \subseteq B$, each of the following statements holds whenever every argument in the statement belongs to the domain of its corresponding function:*

(i) $\inf(f|A) \leq \underline{P}(f|A) \leq \overline{P}(f|A) \leq \sup(f|A)$. **(bounds)**

(ii) $\underline{P}(a|A) = \overline{P}(a|A) = a$. **(normality)**

(iii) $\underline{P}(f + a|A) = \underline{P}(f|A) + a$ *and* $\overline{P}(f + a|A) = \overline{P}(f|A) + a$. **(constant additivity)**

(iv) $\underline{P}(f|A) + \underline{P}(g|A) \leq \underline{P}(f + g|A) \leq \underline{P}(f|A) + \overline{P}(g|A) \leq \overline{P}(f + g|A) \leq \overline{P}(f|A) + \overline{P}(g|A)$, *where each inequality holds if both of its sides are well defined.* **(mixed super-/sub-additivity)**

(v) $\underline{P}(\lambda f|A) = \lambda\underline{P}(f|A)$ *and* $\overline{P}(\lambda f|A) = \lambda\overline{P}(f|A)$. **(non-negative homogeneity)**

(vi) $f \leq g + a \Rightarrow \underline{P}(f|A) \leq \underline{P}(g|A) + a$ *and* $\overline{P}(f|A) \leq \overline{P}(g|A) + a$. **(monotonicity)**

(vii) $\underline{P}\big(|f|\big|A\big) \geq \underline{P}(f|A)$ *and* $\overline{P}\big(|f|\big|A\big) \geq \overline{P}(f|A)$.

(viii) $\big|\underline{P}(f|A) - \underline{P}(g|A)\big| \leq \overline{P}\big(|f - g|\big|A\big)$ *if* $\underline{P}(f|A) - \underline{P}(g|A)$ *is well defined, and* $\left|\overline{P}(f|A) - \overline{P}(g|A)\right| \leq \overline{P}\big(|f - g|\big|A\big)$ *if* $\overline{P}(f|A) - \overline{P}(g|A)$ *is well defined.*

(ix) $\underline{P}\big(|f+g|\big|A\big) \leq \underline{P}\big(|f|\big|A\big) + \overline{P}\big(|g|\big|A\big)$ *and* $\overline{P}\big(|f+g|\big|A\big) \leq \overline{P}\big(|f|\big|A\big) + \overline{P}\big(|g|\big|A\big)$. **(mixed Cauchy–Schwartz inequalities)**

(x) $\underline{P}(f \vee g|A) + \underline{P}(f \wedge g|A) \leq \underline{P}(f|A) + \overline{P}(g|A) \leq \overline{P}(f \vee g|A) \quad +\overline{P}(f \wedge g|A)$, $\underline{P}(f|A) + \underline{P}(g|A) \leq \underline{P}(f \vee g|A) + \overline{P}(f \wedge g|A) \leq \overline{P}(f|A) + \overline{P}(g|A)$ *and* $\underline{P}(f|A) + \underline{P}(g|A) \leq \overline{P}(f \vee g|A) + \underline{P}(f \wedge g|A) \leq \overline{P}(f|A) + \overline{P}(g|A)$, *where each inequality holds if both of its sides are well defined.*

(xi) *Assume that* $R = \overline{P}\big(|f-g|\big|A\big)$ *is a real number. Then* $\underline{P}(g|A) - R \leq \underline{P}(f|A) \leq \underline{P}(g|A) + R$ *and* $\overline{P}(g|A) - R \leq \overline{P}(f|A) \leq \overline{P}(g|A) + R$.

(xii) $\overline{P}\big(|f_\alpha - f|\big|A\big) \to 0 \Rightarrow \underline{P}(f_\alpha|A) \to \underline{P}(f|A)$ *and* $\overline{P}(f_\alpha|A) \to \overline{P}(f|A)$.

(xiii) $\underline{P}(\bullet|A)$ *is uniformly continuous with respect to the norm* $\sup\big(|\bullet|\big|A\big)$*: for any* $\varepsilon > 0$, *if* $\sup\big(|f-g|\big|A\big) \leq \varepsilon$, *then* $\underline{P}(f|A) - \varepsilon \leq \underline{P}(g|A) \leq \underline{P}(f|A) + \varepsilon$. **(uniform continuity)**

(xiv) $\phi : \mathbb{R} \to \mathbb{R}^* : a \mapsto \underline{P}((f-a)I_A|B)$ *is a non-increasing and uniformly continuous function and satisfies for all* $a \in \mathbb{R}$*:*

$$\phi(a) \geq 0 \text{ if } a < \underline{P}(f|A),$$
$$\phi(a) = 0 \text{ if } a = \underline{P}(f|A),$$
$$\phi(a) \leq 0 \text{ if } a > \underline{P}(f|A).$$

(xv) $\psi : \mathbb{R} \to \mathbb{R}^* : a \mapsto \overline{P}((f-a)I_A|B)$ *is a non-increasing and uniformly continuous function and satisfies for all* $a \in \mathbb{R}$*:*

$$\psi(a) \geq 0 \text{ if } a < \overline{P}(f|A),$$
$$\psi(a) = 0 \text{ if } a = \overline{P}(f|A),$$
$$\psi(a) \leq 0 \text{ if } a > \overline{P}(f|A).$$

(xvi) *If* $0 \leq \underline{P}(f|A) < +\infty$, *then*

$$\underline{P}(f|A)\underline{P}(A|B) \leq \underline{P}(fI_A|B) \leq \underline{P}(f|A)\overline{P}(A|B) \leq \overline{P}(fI_A|B), \tag{13.19}$$

and if $0 \leq \overline{P}(f|A) < +\infty$, *then*

$$\underline{P}(fI_A|B) \leq \overline{P}(f|A)\underline{P}(A|B) \leq \overline{P}(fI_A|B) \leq \overline{P}(f|A)\overline{P}(A|B). \tag{13.20}$$

If $0 \leq \underline{P}(f|A) \leq +\infty$, *then*

$$\underline{P}(fI_A|B) \geq \underline{P}(f|A)\underline{P}(A|B) \tag{13.21}$$
$$\overline{P}(fI_A|B) \geq \underline{P}(f|A)\overline{P}(A|B) \tag{13.22}$$

so, in particular, when $\underline{P}(f|A) = +\infty$,

if $\underline{P}(A\|B)$	*then* $\underline{P}(fI_A\|B)$
+	$+\infty$
0	≥ 0

and

if $\overline{P}(A\|B)$	*then* $\overline{P}(fI_A\|B)$
+	$+\infty$
0	≥ 0

If $0 \leq \overline{P}(f|A) \leq +\infty$, *then*

$$\overline{P}(fI_A|B) \geq \overline{P}(f|A)\underline{P}(A|B) \tag{13.23}$$

so, in particular, when $\overline{P}(f|A) = +\infty$,

if $\underline{P}(A\|B)$	*then* $\overline{P}(fI_A\|B)$
+	$+\infty$
0	≥ 0

where 0 *means 'is zero',* + *means 'is strictly positive',* $+\infty$ *means 'is plus infinity' and* ≥ 0 *means 'is non-negative'.*

(xvii) $\underline{P}(f|A) = \underline{P}(fI_A|A)$ *and* $\overline{P}(f|A) = \overline{P}(fI_A|A)$.

(xviii) *If* $\overline{P}(A|B) = 0$, *then*

if $\underline{P}(f\|A)$	*then* $\underline{P}(fI_A\|B)$
$-\infty$	≤ 0
$\mathbb{R}$	0
$+\infty$	≥ 0

and

if $\overline{P}(f\|A)$	*then* $\overline{P}(fI_A\|B)$
$-\infty$	≤ 0
$\mathbb{R}$	0
$+\infty$	≥ 0

and, moreover, the following functions are constant:

$$\phi(a) := \underline{P}((f-a)I_A|B) = \underline{P}(fI_A|B)$$
$$\psi(a) := \overline{P}((f-a)I_A|B) = \overline{P}(fI_A|B)$$

for all $a \in \mathbb{R}$.

Proof. See Lemma 13.13_{245} and rely on coherence criterion $(B)_{260}$. □

The self-conjugacy property leads at once to the following properties of linear conditional previsions.

Corollary 13.32 *Let* $P(\bullet|\bullet)$ *be a coherent (or linear) conditional prevision. Let* f *and* g *be gambles. Let* a *and* λ *be real numbers. Let* f_α *be a net of gambles. Then the following statements hold whenever every expression is well defined:*

(i) $\inf(f|A) \leq P(f|A) \leq \sup(f|A)$. **(bounds)**

(ii) $P(a|A) = a$. **(normality)**

(iii) $P(f + a|A) = P(f|A) + a$. **(constant additivity)**

(iv) $P(f|A) + P(g|A) = P(f + g|A)$. **(additivity)**

(v) $P(\lambda f|A) = \lambda P(f|A)$. **(homogeneity)**

(vi) $f \leq g + a \Rightarrow P(f|A) \leq P(g|A) + a$. **(monotonicity)**

(vii) $P\big(|f|\big|A\big) \geq |P(f|A)| \geq P(f|A)$.

(viii) $|P(f|A) - P(g|A)| \leq P\big(|f-g|\big|A\big)$.

(ix) $P\big(|f+g|\big|A\big) \leq P\big(|f|\big|A\big) + P\big(|g|\big|A\big)$. **(Cauchy–Schwartz inequality)**

(x) $P(f \vee g|A) + P(f \wedge g|A) = P(f|A) + P(g|A)$. **(modularity)**

(xi) *Assume that* $R = P\big(|f-g|\big|A\big)$ *is a real number. Then* $P(g|A) - R \leq P(f|A) \leq P(g|A) + R$.

(xii) *If* $\lim_\alpha P\big(|f_\alpha - f|\big|A\big) = 0$, *then* $\lim_\alpha P(f_\alpha|A) = P(f|A)$.

(xiii) $P(\bullet|A)$ *is uniformly continuous with respect to the norm* $\sup\big(|\bullet|\big|A\big)$*: for all* $\varepsilon > 0$, *if* $\sup\big(|f-g|\big|A\big) \leq \varepsilon$, *then* $P(f|A) - \varepsilon \leq P(g|A) \leq P(f|A) + \varepsilon$. **(uniform continuity)**

(xiv) $$P(fI_A|B) \begin{cases} = P(f|A)P(A|B) & \text{if } P(f|A) \in \mathbb{R} \text{ or } P(A|B) > 0 \\ \geq 0 & \text{if } P(f|A) = +\infty \text{ and } P(A|B) = 0 \\ \leq 0 & \text{if } P(f|A) = -\infty \text{ and } P(A|B) = 0. \end{cases}$$

(xv) $P(f|A) = P(fI_A|A)$.

13.6.2 Trivial extension

Consider a conditional lower prevision $\underline{P}(\bullet|\bullet)$. Define, for any non-empty subset A of $\mathcal{X}$, the following set of gambles:

$$\operatorname{dom} \underline{P}(\bullet|A) := \left\{ f \in \mathbb{G} : (f, A) \in \operatorname{dom} \underline{P}(\bullet|\bullet) \right\}.$$

Now suppose that $\underline{P}(\bullet|\bullet)$ is coherent. If $(f,A) \in \operatorname{dom} \underline{P}(\bullet|\bullet)$ and g is any gamble such that $fI_A = gI_A$, then we infer from Theorem 13.31(xvii)$_{266}$ that coherence enforces the equality

$$\underline{P}(f|A) = \underline{P}(g|A), \tag{13.24}$$

in the sense that if (g,A) is not already in the domain of $\underline{P}(\bullet|\bullet)$, then we can add it to the domain and coherence leaves us no choice as to its value. This means that Equation (13.24) allows us to uniquely extend the coherent conditional lower prevision $\underline{P}(\bullet|\bullet)$ to a conditional lower prevision $\underline{P}_\star(\bullet|\bullet)$ on the set

$$\begin{aligned} \operatorname{dom} \underline{P}_\star(\bullet|\bullet) &:= \left\{ (g,A) : (\exists (f,A) \in \operatorname{dom} \underline{P}(\bullet|\bullet)) fI_A = gI_A \right\} \\ &= \left\{ (fI_A + hI_{A^c}, A) : (f,A) \in \operatorname{dom} \underline{P}(\bullet|\bullet) \text{ and } h \in \mathbb{G} \right\} \end{aligned}$$

and, therefore, also

$$\begin{aligned} \operatorname{dom} \underline{P}_\star(\bullet|A) &= \left\{ fI_A + hI_{A^c} : f \in \operatorname{dom} \underline{P}(\bullet|A) \text{ and } h \in \mathbb{G} \right\} \\ &= \operatorname{dom} \underline{P}(\bullet|A) I_A + \mathbb{G} I_{A^c}. \end{aligned}$$

It is immediately clear from coherence criterion $(D)_{260}$ that $\underline{P}_{\star}(\bullet|\bullet)$ is coherent as well – and we have already seen that it is uniquely so.

13.6.3 Easier ways to prove coherence

If the structure of the domain $\operatorname{dom}\underline{P}(\bullet|\bullet)$ of a conditional lower prevision $\underline{P}(\bullet|\bullet)$ is sufficiently rich, then necessary and sufficient conditions for coherence simplify significantly. Compare the following theorem, which can already be found – without proof – in Troffaes (2006, Theorem 3), with Theorem 4.15_{56}.

Theorem 13.33 *Let $\underline{P}(\bullet|\bullet)$ be a conditional lower prevision and assume that $\operatorname{dom}\underline{P}(\bullet|\bullet) = \mathscr{K} \times \mathscr{A} \subseteq \mathbb{G} \times \mathscr{P}^{\circ}$ where $\mathscr{K}$ is a linear space, $\mathscr{A}$ is closed under finite unions, $I_A \in \mathscr{K}$ whenever $A \in \mathscr{A}$ and $fI_A \in \mathscr{K}$ whenever $f \in \mathscr{K}$ and $A \in \mathscr{A}$. Then $\underline{P}(\bullet|\bullet)$ is coherent if and only if the following conditions are met for all gambles f and g in $\mathscr{K}$, all non-negative real numbers λ, all real numbers a, and all events A and B in $\mathscr{A}$ such that $A \subseteq B$:*

(i) $\inf(f|A) \leq \underline{P}(f|A)$. **(bounds)**

(ii) $\underline{P}(f+g|A) \geq \underline{P}(f|A) + \underline{P}(g|A)$ *whenever the sum on the right-hand side is well defined.* **(super-additivity)**

(iii) $\underline{P}(\lambda f|A) = \lambda\underline{P}(f|A)$. **(non-negative homogeneity)**

(iv) $\underline{P}((f-a)I_A|B) \begin{cases} \geq 0 \text{ if } a < \underline{P}(f|A) \\ \leq 0 \text{ if } a > \underline{P}(f|A). \end{cases}$ **(Bayes's rule)**

Proof. Necessity follows at once from coherence criterion $(B)_{260}$.

To prove sufficiency, assume that $\underline{P}(\bullet|\bullet)$ satisfies properties (i)–(iv) on its domain. We prove that coherence criterion $(D)_{260}$ is satisfied. We rely implicitly on the assumptions on the domain of $\underline{P}(\bullet|\bullet)$. Consider any n in $\mathbb{N}$, non-negative λ_0, λ_1, …, λ_n in $\mathbb{R}$, (f_0, A_0), (f_1, A_1), …, (f_n, A_n) in $\operatorname{dom}\underline{P}(\bullet|\bullet)$ and a_0, …, $a_n \in \mathbb{R}$ such that $a_0 > \underline{P}(f_0|A_0)$ and $a_k < \underline{P}(f_k|A_k)$ for all $k \in \{1, \dots, n\}$, and use the shorthand notations $f := \sum_{k=1}^{n} \lambda_k(f_k - a_k)I_{A_k}$, $g := \lambda_0(f_0 - a_0)I_{A_0}$ and $A := A_0 \cup A_1 \cup \cdots \cup A_n$. It follows from the assumptions that $A \in \mathscr{A}$ and that all $(\lambda_k(f_k - a_k)I_{A_k}, A)$, $(\lambda_0(f_0 - a_0)I_{A_o}, A)$, (f, A), (g, A) and $(f - g, A)$ belong to $\operatorname{dom}\underline{P}(\bullet|\bullet)$. We must show that $\sup(f - g|A) \geq 0$. By (i) and conjugacy, $\sup(f - g|A) \geq \overline{P}(f - g|A)$, so it suffices to show that $\overline{P}(f - g|A) \geq 0$. If $\overline{P}(f - g|A) = +\infty$, then the result follows trivially. So, we may assume without loss of generality that $\overline{P}(f - g|A) < +\infty$ or, equivalently, $\underline{P}(g - f|A) > -\infty$.

We infer from (iv) that $\underline{P}((f_k - a_k)I_{A_k}|A) \geq 0$ and $\underline{P}((f_0 - a_0)I_{A_0}|A) \leq 0$. If we now invoke (ii) and (iii), we find that $\underline{P}(f|A) \geq 0$ and $\underline{P}(g|A) \leq 0$. We infer from (ii) that

$$\underline{P}(g|A) = \underline{P}(g - f + f|A) \geq \underline{P}(g - f|A) + \underline{P}(f|A)$$

whenever the right-hand side is well defined, which is clearly the case because we have just found out that $\underline{P}(f|A) \geq 0$ and we have assumed that $\underline{P}(g - f|A) > -\infty$.

Using Lemma D.9$_{393}$, this leads to

$$\underline{P}(g - f|A) \leq \underline{P}(g|A) - \underline{P}(f|A) \leq 0.$$

Now use conjugacy to find that indeed $\overline{P}(f - g|A) \geq 0$. □

For the unconditional case, the coherence conditions are again simpler than for the conditional one (Troffaes, 2005, Theorem 5.6).

Theorem 13.34 *Let $\underline{P}(\bullet|\bullet)$ be a conditional lower prevision and assume that* $\operatorname{dom} \underline{P}(\bullet|\bullet) = \mathscr{K} \times \{\mathcal{X}\} \subseteq \mathbb{G} \times \{\mathcal{X}\}$ *where $\mathscr{K}$ is a linear space. Then $\underline{P}(\bullet|\bullet)$ is coherent if and only if the following conditions are met for all gambles f and g in $\mathscr{K}$ and all non-negative real numbers λ:*

(i) $\inf(f|\mathcal{X}) \leq \underline{P}(f|\mathcal{X})$. **(bounds)**

(ii) $\underline{P}(f + g|\mathcal{X}) \geq \underline{P}(f|\mathcal{X}) + \underline{P}(g|\mathcal{X})$ *whenever the sum on the right-hand side is well defined.* **(super-additivity)**

(iii) $\underline{P}(\lambda f|\mathcal{X}) = \lambda \underline{P}(f|\mathcal{X})$. **(non-negative homogeneity)**

Proof. Repeat the proof of Theorem 13.33, and observe that only properties (i)–(iii) are required. □

There is a subtle difference between Theorem 13.34 and its counterpart for lower previsions on bounded gambles, Theorem 4.15$_{56}$: for the former, we must require explicitly that condition (iii) should also be satisfied for $\lambda = 0$. To see why, consider the following example taken from Troffaes (2005, p. 207).

Example 13.35 Consider the conditional lower prevision defined by $\underline{P}(f|\mathcal{X}) := +\infty$ for all gambles f: $\underline{P}(\bullet|\bullet)$ satisfies the conditions of Theorem 13.34 for all gambles f and g and all non-negative real λ apart from $\lambda = 0$. But this $\underline{P}(\bullet|\bullet)$ is clearly not coherent: it does not even avoid sure loss. ♦

We now turn to conditional lower previsions that are self-conjugate. For these conditional previsions, the following simplified coherence conditions, which are due to Troffaes (2006, Theorem 4), can be inferred readily from Theorem 13.33.

Theorem 13.36 *Let $P(\bullet|\bullet)$ be a conditional prevision, and assume that* $\operatorname{dom} P(\bullet|\bullet) = \mathscr{K} \times \mathscr{A} \subseteq \mathbb{G} \times \mathscr{P}^{\circ}$ *where $\mathscr{K}$ is a linear space, $\mathscr{A}$ is closed under finite unions, $I_A \in \mathscr{K}$ whenever $A \in \mathscr{A}$ and $fI_A \in \mathscr{K}$ whenever $f \in \mathscr{K}$ and $A \in \mathscr{A}$. Then $P(\bullet|\bullet)$ is coherent if and only if the following conditions are met for all gambles f and g in $\mathscr{K}$, all real numbers λ and all events A and B in $\mathscr{A}$ such that $A \subseteq B$:*

(i) $\inf(f|A) \leq P(f|A)$. **(bounds)**

(ii) $P(f + g|A) = P(f|A) + P(g|A)$ *whenever the sum on the right-hand side is well defined.* **(additivity)**

(iii) $P(\lambda f|A) = \lambda P(f|A)$. **(homogeneity)**

$$\text{(iv) } P(fI_A|B)\begin{cases} = P(f|A)P(A|B) & \textit{if } P(f|A) \in \mathbb{R} \textit{ or } P(A|B) > 0 \\ \geq 0 & \textit{if } P(f|A) = +\infty \textit{ and } P(A|B) = 0 \\ \leq 0 & \textit{if } P(f|A) = -\infty \textit{ and } P(A|B) = 0. \end{cases} \quad \textbf{(Bayes's rule)}$$

Proof. 'if'. Assume that (i)–(iv) are satisfied. We use Theorem 13.33_{270} to prove that $P(\bullet|\bullet)$ is coherent. All the conditions of Theorem 13.33_{270} are trivially satisfied (use the fact that $P(\bullet|\bullet)$ is self-conjugate), except perhaps for condition $(\text{iv})_{270}$. We concentrate on that condition. Consider any $a \in \mathbb{R}$.

First, observe that $P(f - a|A) = P(f|A) - a$ by $(\text{ii})_{\curvearrowleft}$, $(\text{i})_{\curvearrowleft}$ and the self-conjugacy of $P(\bullet|\bullet)$. Hence, by (iv), whenever $P(f|A) \in \mathbb{R}$ or $P(A|B) > 0$,

$$P((f - a)I_A|B) = P(f - a|A)P(A|B) = (P(f|A) - a)P(A|B).$$

It follows that the right-hand side of the above expression is non-negative whenever $a < P(f|A)$ and non-positive whenever $a > P(f|A)$.

If $P(f|A) = +\infty$ and $P(A|B) = 0$, then $(\text{ii})_{\curvearrowleft}$, $(\text{i})_{\curvearrowleft}$ and (iv) tell us that

$$P((f - a)I_A|B) = P(fI_A|B) - aP(A|B) = P(fI_A|B) \geq 0,$$

so $P((f - a)I_A|B)$ is certainly non-negative for all $a < P(f|A)$.

If $P(f|A) = -\infty$ and $P(A|B) = 0$, then $(\text{ii})_{\curvearrowleft}$, $(\text{i})_{\curvearrowleft}$ and (iv) tell us that

$$P((f - a)I_A|B) = P(fI_A|B) - aP(A|B) = P(fI_A|B) \leq 0,$$

so $P((f - a)I_A|B)$ is certainly non-positive for all $a > P(f|A)$.

'only if'. The conditions (i)–(iv) follow immediately from the coherence properties (see Corollary 13.32_{268}). □

For unconditional previsions, the following characterisation of coherence (linearity) was first proved by Crisma, Gigante and Millossovich (1997a, Theorem 3.6). It follows without further ado from the previous theorem, or, alternatively, from Theorem $13.34_{\curvearrowleft}$ and the self-conjugacy of $P(\bullet|\bullet)$.

Theorem 13.37 *Let $P(\bullet|\bullet)$ be a conditional prevision, and assume that* $\operatorname{dom} P(\bullet|\bullet) = \mathscr{K} \times \{\mathscr{X}\} \subseteq \mathbb{G} \times \{\mathscr{X}\}$ *where $\mathscr{K}$ is a linear space. Then $P(\bullet|\bullet)$ is coherent if and only if the following conditions are met for all gambles f and g in $\mathscr{K}$ and all real numbers λ:*

(i) $\inf(f|\mathscr{X}) \leq P(f|\mathscr{X})$. **(bounds)**

(ii) $P(f + g|\mathscr{X}) = P(f|\mathscr{X}) + P(g|\mathscr{X})$ *whenever the right-hand side is well defined.* **(additivity)**

(iii) $P(\lambda f|\mathscr{X}) = \lambda P(f|\mathscr{X})$. **(homogeneity)**

In contradistinction with the related Theorem 4.16_{57} for bounded gambles, the homogeneity condition (iii) must be imposed explicitly in the present Theorem 13.37: it is no longer implied by the remaining conditions (i) and (ii). To show this, we present the following counterexample – for once, not from Troffaes (2005).

Example 13.38 Consider any gamble f that is unbounded above and below. It is well known that the set of reals $\mathbb{R}$ is a linear space over the field $\mathbb{Q}$ of rationals. By the Axiom of Choice, there is an uncountable Hamel basis B for $\mathbb{R}$ with respect to this field $\mathbb{Q}$ (see, for instance, Schechter, 1997, Corollary 11.30c). This means that every element λ of $\mathbb{R}$ can be written as a unique rational combination of a finite number of elements of B:

$$\lambda = \sum_{k=1}^{n} q_k b_k \text{ with } q_k \in \mathbb{Q} \text{ and } b_k \in B.$$

We now define the conditional prevision $P(\bullet|\bullet)$ on $\{\lambda f : \lambda \in \mathbb{R}\} \times \{\mathcal{X}\}$ by $P(\lambda f|\mathcal{X}) := \sum_{k=1}^{n} q_k$. Clearly, $\{\lambda f : \lambda \in \mathbb{R}\}$ is a linear space. This conditional prevision satisfies condition (i) of Theorem 13.37 because $\inf(\lambda f|\mathcal{X}) = -\infty$ for every $\lambda \neq 0$ and $P(0|\mathcal{X}) = 0$ by our definition of $P(\bullet|\bullet)$. Condition (ii) of Theorem 13.37 is satisfied as well: for any two real numbers, we may always write

$$\lambda = \sum_{k=1}^{n} q_k b_k \text{ and } \lambda' = \sum_{k=1}^{n} q_k' b_k,$$

so it follows from the basis property of B that

$$P((\lambda + \lambda')f|\mathcal{X}) = \sum_{k=1}^{n} (q_k + q_k') = P(\lambda f|\mathcal{X}) + P(\lambda' f|\mathcal{X}).$$

However, $P(\bullet|\bullet)$ violates condition (iii) of Theorem 13.37: $P(\sqrt{2}f|\mathcal{X})$ can never be equal to $\sqrt{2}P(f|\mathcal{X})$ because $P(\bullet|\bullet)$ is rational valued. ♦

We conclude from the earlier discussion that the conditions for avoiding sure loss, coherence and linearity involving linear subspaces are much easier to check than those for general domains. Therefore, a useful method for proving that a given conditional lower prevision avoid sure loss, is coherent or linear, consists in showing that it is the restriction of some conditional lower prevision whose domain is conveniently large – such as a linear space – and for which it is much easier to check conditions for avoiding sure loss, coherence or linearity.

Proposition 13.39 *The following statements hold:*

(i) *The restriction of a conditional lower prevision that avoids sure loss, also avoids sure loss.*

(ii) *The restriction of a coherent conditional lower prevision is also coherent.*

(iii) *The restriction of a linear conditional prevision to a conditional prevision is also linear.*

Proof. Immediately from Definitions 13.18_{255}, 13.25_{260} and 13.28_{264}. □

As is the case for lower previsions on bounded gambles (see Proposition 4.19_{59}), taking convex combinations of *unconditional* lower previsions preserves avoiding sure loss, coherence and linearity.

Proposition 13.40 *Suppose* $\Gamma = \{\underline{P}_1(\bullet|\bullet), \underline{P}_2(\bullet|\bullet), \dots, \underline{P}_p(\bullet|\bullet)\}$ *is a finite collection of* p *(unconditional) lower previsions defined on a common domain* $\mathscr{O} \subseteq \mathbb{G} \times \{\mathscr{X}\}$. *Let* $\underline{Q}(\bullet|\bullet)$ *be a* **convex combination** *of elements of* Γ, *meaning that*

$$\underline{Q}(f|\mathscr{X}) := \sum_{j=1}^{p} \alpha_j \underline{P}_j(f|\mathscr{X}) \text{ for all } (f, \mathscr{X}) \in \mathscr{O},$$

for some real $\alpha_1, \dots, \alpha_p \geq 0$ *such that* $\sum_{j=1}^{p} \alpha_j = 1$, *where* we require that sum on the right-hand side in the equality above should be well defined on the common domain $\mathscr{O}$. *Then the following statements hold:*

(i) *If all (unconditional) lower previsions in* Γ *avoid sure loss, then* $\underline{Q}(\bullet|\bullet)$ *avoids sure loss as well.*

(ii) *If all (unconditional) lower previsions in* Γ *are coherent, then* $\underline{Q}(\bullet|\bullet)$ *is coherent as well.*

(iii) *If all (unconditional) lower previsions in* Γ *are linear (unconditional) previsions, then* $\underline{Q}(\bullet|\bullet)$ *is a linear (unconditional) prevision as well.*

Proof. We begin with (ii). The proof for (i) is very similar, if slightly less involved. We show that $\underline{Q}(\bullet|\bullet)$ satisfies the coherence criterion $(\mathrm{F})_{262}$. Invoking this coherence criterion and the coherence of every $\underline{P}_j(\bullet|\bullet)$, we find that for all n in $\mathbb{N}$, non-negative $\lambda_0, \lambda_1, \dots, \lambda_n$ in $\mathbb{R}$ and gambles $f_0, f_1, \dots, f_n$ in dom $\underline{P}_j(\bullet|\mathscr{X})$ such that $\sum_{k=1}^{n} \lambda_k \underline{P}_j(f_k|\mathscr{X}) - \lambda_0 \underline{P}_j(f_0|\mathscr{X})$ is well defined:

$$\alpha_j \sup\left(\sum_{k=1}^{n} \lambda_k f_k - \lambda_0 f_0 \middle| \mathscr{X} \right) \geq \sum_{k=1}^{n} \lambda_k \alpha_j \underline{P}_j(f_k|\mathscr{X}) - \lambda_0 \alpha_j \underline{P}_j(f_0|\mathscr{X}).$$

Summing the above expression over all j and using Lemma $\mathrm{D}.7_{392}$, we find that

$$\sup\left(\sum_{k=1}^{n} \lambda_k f_k - \lambda_0 f_0 \middle| \mathscr{X} \right) \geq \sum_{k=1}^{n} \lambda_k \sum_{j=1}^{p} \alpha_j \underline{P}_j(f_k|\mathscr{X}) - \lambda_0 \sum_{j=1}^{p} \alpha_j \underline{P}_j(f_0|\mathscr{X}).$$

whenever the right-hand side is well defined, which establishes the coherence of $\sum_{j=1}^{p} \alpha_j \underline{P}_j(\bullet|\bullet)$ and, hence, of $\underline{Q}(\bullet|\bullet)$.

(iii). It already follows from (ii) that $\underline{Q}(\bullet|\bullet)$ is coherent. To show that $\underline{Q}(\bullet|\bullet)$ is linear, it suffices to establish that $\underline{Q}(\bullet|\bullet)$ is self-conjugate. Now for all $(f, \mathscr{X})$ in $\mathscr{O}$,

$$\underline{Q}(-f|\mathscr{X}) = \sum_{j=1}^{p} \alpha_j \underline{P}_j(-f|\mathscr{X}) = -\sum_{j=1}^{p} \alpha_j \underline{P}_j(f|\mathscr{X}) = -\underline{Q}(f|\mathscr{X}),$$

by the self-conjugacy of each $\underline{P}_j(\bullet|\bullet)$. □

Taking convex combinations does not generally preserve coherence *in the conditional case*, as the following example shows.

Example 13.41 Let $\mathscr{X} := \{a, b, c, d\}$, and consider the (logically independent) events $A := \{a, b\}$ and $B := \{a, c\}$. Let p_1 and p_2 be probability mass functions on $\mathscr{X}$ defined as follows:

$$p_1(a) := \frac{2}{6} \qquad p_1(b) := \frac{2}{6} \qquad p_1(c) := \frac{1}{6} \qquad p_1(d) := \frac{1}{6}$$
$$p_2(a) := \frac{1}{6} \qquad p_2(b) := \frac{2}{6} \qquad p_2(c) := \frac{1}{6} \qquad p_2(d) := \frac{2}{6}.$$

These unconditional probability assignments completely determine linear conditional previsions $P_1(\bullet|\bullet)$ and $P_2(\bullet|\bullet)$ as follows (through Bayes's rule and taking expectations):

$$P_k(f|C) := \frac{\sum_{x\in\mathscr{X}} p_k(x)f(x)}{\sum_{x\in C} p_k(x)} \text{ for all } f \in \mathbb{G} \text{ and all non-empty } C \subseteq \mathscr{X}.$$

However, $P(\bullet|\bullet) := \frac{1}{2}P_1(\bullet|\bullet) + \frac{1}{2}P_2(\bullet|\bullet)$ is not coherent, because

$$P(A|B) = \frac{1}{2}P_1(A|B) + \frac{1}{2}P_2(A|B) = \frac{1}{2}(\frac{2}{3} + \frac{1}{2}) = \frac{7}{12},$$

whereas

$$\frac{P(A \cap B|\mathscr{X})}{P(B|\mathscr{X})} = \frac{\frac{1}{2}P_1(A \cap B|\mathscr{X}) + \frac{1}{2}P_2(A \cap B|\mathscr{X})}{\frac{1}{2}P_1(B|\mathscr{X}) + \frac{1}{2}P_2(B|\mathscr{X})} = \frac{\frac{2}{6} + \frac{1}{6}}{\frac{3}{6} + \frac{2}{6}} = \frac{3}{5},$$

so $P(A|B)P(B|\mathscr{X}) \neq P(A \cap B|\mathscr{X})$, violating the coherence requirement (iv)$_{272}$ in Theorem 13.36$_{271}$. ◆

The following result generalises Proposition 4.20$_{59}$.

Proposition 13.42 *Suppose* Γ *is a non-empty collection of conditional lower previsions defined on a common domain* $\mathscr{O}$. *Let* $\underline{Q}(\bullet|\bullet)$ *be the* ***lower envelope*** *of* Γ, *meaning that*

$$\underline{Q}(f|A) := \inf_{\underline{P}(\bullet|\bullet)\in\Gamma} \underline{P}(f|A) \text{ for all } (f, A) \in \mathscr{O}.$$

Then the following statements hold:

(i) *If some conditional lower prevision in* Γ *avoids sure loss, then so does* $\underline{Q}(\bullet|\bullet)$. *Equivalently, if a conditional lower prevision is dominated by a conditional lower prevision that avoids sure loss, then it avoids sure loss as well.*

(ii) *If all conditional lower previsions in* Γ *are coherent, then so is* $\underline{Q}(\bullet|\bullet)$.

(iii) *If all conditional lower previsions in* Γ *are linear, then* $\underline{Q}(\bullet|\bullet)$ *is coherent but not linear unless* Γ *is a singleton.*

Proof. (i)$_\frown$. This is a direct consequence of avoiding sure loss criterion $(\mathrm{C})_{255}$. Indeed, suppose $\underline{P}(\bullet|\bullet) \in \Gamma$ avoids sure loss. Then there is some coherent conditional lower prevision $\underline{R}(\bullet|\bullet)$ on $\mathbb{G} \times \mathscr{P}^\circ$ that dominates $\underline{P}(\bullet|\bullet)$. Obviously, this $\underline{R}(\bullet|\bullet)$ also dominates $\underline{Q}(\bullet|\bullet)$ because $\underline{P}(\bullet|\bullet)$ dominates $\underline{Q}(\bullet|\bullet)$. Therefore, $\underline{Q}(\bullet|\bullet)$ avoids sure loss as well.

(ii)$_\frown$. We use coherence criterion $(\mathrm{D})_{260}$. Consider arbitrary n in $\mathbb{N}$, non-negative $\lambda_0, \lambda_1, \dots, \lambda_n$ in $\mathbb{R}$, (f_0,A_0), (f_1,A_1), …, (f_n,A_n) in $\operatorname{dom}\underline{Q}(\bullet|\bullet)$ and $a_0, \dots, a_n \in \mathbb{R}$ such that $a_0 > \underline{Q}(f_0|A_0)$ and $a_i < \underline{Q}(f_i|A_i)$ for all $i \in \{1,\dots,n\}$. Then there is some $\underline{P}(\bullet|\bullet)$ in Γ such that also $a_0 > \underline{P}(f_0|A_0)$: if $a_0 \le \underline{P}(f_0|A_0)$ for all $\underline{P}(\bullet|\bullet)$ in Γ, then it would follow that also $a_0 \le \underline{Q}(f_0|A_0)$, a contradiction. For this $\underline{P}(\bullet|\bullet)$ (and for any other conditional lower prevision in Γ) it also holds that $a_k < \underline{P}(f_k|A_k)$ simply because $\underline{Q}(\bullet|\bullet) \le \underline{P}(\bullet|\bullet)$. The coherence of $\underline{P}(\bullet|\bullet)$, therefore, implies that

$$\sup\left(\sum_{k=1}^{n} \lambda_k(f_k - a_k)I_{A_k} - \lambda_0(f_0 - a_0)I_{A_0}\middle| A_0 \cup A_1 \cup \cdots \cup A_n\right) \ge 0,$$

and this leads to the conclusion that $\underline{Q}(\bullet|\bullet)$ satisfies coherence criterion $(\mathrm{D})_{260}$.

(iii)$_\frown$. It suffices to check that $\underline{Q}(\bullet|\bullet)$ is not self-conjugate whenever Γ contains more than one element. Choose two distinct linear conditional previsions $P_1(\bullet|\bullet)$ and $P_2(\bullet|\bullet)$ in Γ. As they are distinct, there is some (f,A) in $\mathscr{O}$ such that, for instance, $P_1(f|A) < P_2(f|A)$. This means that $\underline{Q}(f|A) \le P_1(f|A) < P_2(f|A) \le \overline{Q}(f|A)$, so $\underline{Q}(\bullet|\bullet)$ is not self-conjugate. □

As is the case for lower previsions on bounded gambles (see Proposition 4.21_{60}), limits of nets of conditional lower previsions satisfy similar properties: avoiding sure loss, coherence and linearity are preserved under taking such limits.

Proposition 13.43 *Let $\mathscr{A}$ be a directed set, and let $\underline{P}_\alpha(\bullet|\bullet)$ be a net of conditional lower previsions defined on a common domain $\mathscr{O}$. Suppose that $\underline{P}_\alpha(\bullet|\bullet)$* ***converges point-wise*** *to a conditional lower prevision $\underline{Q}(\bullet|\bullet)$ defined on $\mathscr{O}$, meaning that for all (f,A) in $\mathscr{O}$, the net $\underline{P}_\alpha(f|A)$ converges to an extended real number $\underline{Q}(f|A)$. Then the following statements hold:*

(i) *If $\underline{P}_\alpha(\bullet|\bullet)$ avoids sure loss eventually, then so does $\underline{Q}(\bullet|\bullet)$.*

(ii) *If $\underline{P}_\alpha(\bullet|\bullet)$ is coherent eventually, then so is $\underline{Q}(\bullet|\bullet)$.*

(iii) *If $\underline{P}_\alpha(\bullet|\bullet)$ is linear eventually, then so is $\underline{Q}(\bullet|\bullet)$.*

Proof. We begin with a proof for (ii). The proof for (i) is analogous but simpler. We infer from the assumptions that there is some β such that $\underline{P}_\alpha$ is coherent for all $\alpha \ge \beta$. We now use coherence criterion $(\mathrm{D})_{260}$. Consider arbitrary n in $\mathbb{N}$, non-negative $\lambda_0, \lambda_1, \dots, \lambda_n$ in $\mathbb{R}$, (f_0,A_0), (f_1,A_1), …, (f_n,A_n) in $\operatorname{dom}\underline{Q}(\bullet|\bullet)$ and $a_0, \dots, a_n \in \mathbb{R}$ such that $a_0 > \underline{Q}(f_0|A_0)$ and $a_i < \underline{Q}(f_i|A_i)$ for all $i \in \{1,\dots,n\}$. By Lemma $\mathrm{D}.8_{392}$, there are $\alpha_0, \alpha_1, \dots, \alpha_n$ in $\mathscr{A}$ such that $a_0 > \underline{P}_\alpha(f_0|A_0)$ for all $\alpha \ge \alpha_0$ and $a_k < \underline{P}_\alpha(f_k|A_k)$ for all $\alpha \ge \alpha_k$, for $k \in \{1,\dots,n\}$. Let $\alpha^* \in A$ be such that $\alpha^* \ge \beta$ and $\alpha^* \ge \alpha_k$ for all $k \in \{0,1,\dots,n\}$ (there is such an α^* because $\mathscr{A}$ is a directed set). By the coherence of

$\underline{P}_{\alpha^*}(\bullet|\bullet)$, and the fact that $a_0 > \underline{P}_{\alpha^*}(f_0|A_0)$ and $a_k < \underline{P}_{\alpha^*}(f_k|A_k)$ for all $k \in \{1, \dots, n\}$, it follows that

$$\sup\left(\sum_{k=1}^{n} \lambda_k(f_k - a_k)I_{A_k} - \lambda_0(f_0 - a_0)I_{A_0} \middle| A_0 \cup A_1 \cup \dots \cup A_n\right) \geq 0,$$

and we conclude that $\underline{Q}(\bullet|\bullet)$ is indeed coherent.

(iii). It suffices to check the self-conjugacy of $\underline{Q}(\bullet|\bullet)$. For any $(f, A) \in \mathscr{O}$,

$$\begin{aligned}\underline{Q}(f|A) = \lim_{\alpha} \underline{P}_{\alpha}(f|A) &= \lim_{\alpha} -\underline{P}_{\alpha}(-f|A)\\ &= -\lim_{\alpha} \underline{P}_{\alpha}(-f|A) = -\underline{Q}(-f|A) = \overline{Q}(f|A),\end{aligned}$$

which completes the proof. □

If we combine Propositions 13.42$_{275}$ and 13.43, we find a similar result for the point-wise limit inferior of a net of lower previsions, generalising Corollary 4.22$_{61}$. For a discussion of the limit inferior of a net, see Section 1.5$_7$.

Corollary 13.44 *Suppose $\underline{P}_{\alpha}(\bullet|\bullet)$ is a net of conditional lower previsions defined on a common domain $\mathscr{O}$. Let $\underline{Q}(\bullet|\bullet)$ be the* ***point-wise limit inferior*** *of this net, meaning that for every (f, A) in $\mathscr{O}$,*

$$\underline{Q}(f|A) := \liminf_{\alpha} \underline{P}_{\alpha}(f|A) = \sup_{\alpha} \inf_{\beta \geq \alpha} \underline{P}_{\beta}(f|A).$$

Then the following statements hold:

(i) *If the net $\underline{P}_{\alpha}(\bullet|\bullet)$ avoids sure loss eventually, then $\underline{Q}(\bullet|\bullet)$ avoids sure loss as well.*

(ii) *If the net $\underline{P}_{\alpha}(\bullet|\bullet)$ is coherent eventually, then $\underline{Q}(\bullet|\bullet)$ is coherent as well.*

(iii) *If the net $\underline{P}_{\alpha}(\bullet|\bullet)$ is linear eventually, then $\underline{Q}(\bullet|\bullet)$ is coherent. Moreover, $\underline{Q}(\bullet|\bullet)$ is linear if and only if $\underline{P}_{\alpha}(\bullet|\bullet)$ converges point-wise.*

Proof. (i). By assumption, there is some α^* such that $\underline{P}_{\alpha}(\bullet|\bullet)$ avoids sure loss for all $\alpha \geq \alpha^*$. By Proposition 13.42(i)$_{275}$, the conditional lower prevision $\underline{Q}_{\alpha}(\bullet|\bullet) := \inf_{\beta \geq \alpha} \underline{P}_{\beta}(\bullet|\bullet)$ avoids sure loss for every $\alpha \geq \alpha^*$. By definition of the limit inferior, the net $\underline{Q}_{\alpha}(\bullet|\bullet)$ converges point-wise to $\underline{Q}(\bullet|\bullet)$. Hence, Proposition 13.43(i) tells us that $\underline{Q}(\bullet|\bullet)$ avoid sure loss as well.

(ii). Similar to the proof of (i).

(iii). The proof of the first part (linearity) is similar to that of (i). For the second part, note that if $\underline{P}_{\alpha}(\bullet|\bullet)$ converges, then $\liminf_{\alpha} \underline{P}_{\alpha}(\bullet|\bullet)$ coincides with $\lim_{\alpha} \underline{P}_{\alpha}(\bullet|\bullet)$ and therefore $\underline{Q}(\bullet|\bullet)$ must be linear, by Proposition 13.43(iii). Conversely, if $\underline{Q}(\bullet|\bullet)$ is linear, then for every $(f, A) \in \mathscr{O}$,

$$\begin{aligned}\liminf_{\alpha} \underline{P}_{\alpha}(f|A) = \underline{Q}(f|A) = -\underline{Q}(-f|A) &= -\liminf_{\alpha} \underline{P}_{\alpha}(-f|A)\\ &= \limsup_{\alpha} -\underline{P}_{\alpha}(-f|A) = \limsup_{\alpha} \underline{P}_{\alpha}(f|A),\end{aligned}$$

which means that $\underline{P}_{\alpha}(\bullet|\bullet)$ converges point-wise. □

13.6.4 Separate coherence

Recall from Section 13.6.2$_{269}$ that coherence allows us to uniquely extend the coherent conditional lower prevision $\underline{P}(\bullet|\bullet)$ to a coherent conditional lower prevision $\underline{P}_\star(\bullet|\bullet)$ on the set $\operatorname{dom}\underline{P}_\star(\bullet|\bullet) = \{(fI_A + hI_{A^c}, A) : (f, A) \in \operatorname{dom}\underline{P}(\bullet|\bullet) \text{ and } h \in \mathbb{G}\}$. This implies that we can associate with a coherent conditional lower prevision $\underline{P}(\bullet|\bullet)$ a family of coherent (unconditional) lower previsions $\underline{P}_s(\bullet|A)$, each for a different event A. Indeed, for any non-empty event A such that $\operatorname{dom}\underline{P}(\bullet|A) \neq \emptyset$, define the (unconditional) lower prevision $\underline{P}_s(\bullet|A)$ on $\operatorname{dom}\underline{P}_\star(\bullet|A)|_A \times \{A\}$ by

$$\underline{P}_s(f|_A|A) := \underline{P}_\star(f|A) \text{ for all } f \in \operatorname{dom}\underline{P}_\star(\bullet|A). \tag{13.25}$$

Because $\underline{P}_\star(\bullet|\bullet)$ extends $\underline{P}(\bullet|\bullet)$ and $\operatorname{dom}\underline{P}_\star(\bullet|A)|_A = \operatorname{dom}\underline{P}(\bullet|A)|_A$, we can also define, equivalently,

$$\underline{P}_s(f|_A|A) := \underline{P}(f|A) \text{ for all } f \in \operatorname{dom}\underline{P}(\bullet|A).$$

These lower previsions turn out to be coherent *separately*: for each fixed value of A.

Theorem 13.45 (Separate coherence) *Consider for a coherent conditional lower prevision $\underline{P}(\bullet|\bullet)$ and any non-empty event A such that* $\operatorname{dom}\underline{P}(\bullet|A) \neq \emptyset$, *the corresponding (unconditional) lower prevision $\underline{P}_s(\bullet|A)$ defined on* $\operatorname{dom}\underline{P}_\star(\bullet|A)|_A \times \{A\}$ *by Equation* (13.25). *Then the following statements hold:*

(i) *If $\underline{P}(\bullet|\bullet)$ avoids sure loss, then $\underline{P}_s(\bullet|A)$ avoids sure loss.*

(ii) *If $\underline{P}(\bullet|\bullet)$ is coherent, then $\underline{P}_s(\bullet|A)$ is coherent.*

Proof. We only give a proof for the second statement, as the proof for the first is similar but less involved. Fix any non-empty event A such that $\operatorname{dom}\underline{P}(\bullet|A) \neq \emptyset$. We use coherence criterion (D)$_{260}$. Consider any $n \in \mathbb{N}$, non-negative $\lambda_0, \lambda_1, \ldots, \lambda_n$ in $\mathbb{R}$, $(h_0, A), (h_1, A), \ldots, (h_n, A)$ in $\operatorname{dom}\underline{P}_s(\bullet|A)$ and $a_0, a_1, \ldots, a_n$ such that $a_0 > \underline{P}_s(h_0|A)$ and $a_k < \underline{P}_s(h_k|A)$. Recall from the definition that $\underline{P}_s(h_0|A) = \underline{P}_\star(h_0 I_A|A)$ and $\underline{P}_s(h_k|A) = \underline{P}_\star(h_k I_A|A)$. Hence, also $a_0 > \underline{P}_\star(h_0 I_A|A)$ and $a_k < \underline{P}_\star(h_k I_A|A)$, so it follows from the coherence of $\underline{P}_\star(\bullet|\bullet)$ that in particular

$$\sup\left(\sum_{k=1}^{n} \lambda_k(f_k - a_k)I_A - \lambda_0(f_0 - a_0)I_A \middle| A\right) \geq 0,$$

which proves the coherence of $\underline{P}_s(\bullet|A)$. □

The converse result does not hold in general: for $\underline{P}(\bullet|\bullet)$ to be coherent, it is usually not sufficient that all $\underline{P}_s(\bullet|A)$ are coherent separately, as we also have to impose conditions that relate values of $\underline{P}(\bullet|A)$ for *different* conditioning events A. The following theorem identifies a simple but useful situation where this becomes particularly obvious.

Theorem 13.46 *Let $\mathscr{B}$ be any non-trivial partition of $\mathscr{X}$. Consider any conditional lower prevision $\underline{P}(\bullet|\bullet)$ with domain* $\operatorname{dom}\underline{P}(\bullet|\bullet) = \mathbb{G} \times (\mathscr{B} \cup \{\mathscr{X}\})$. *Then $\underline{P}(\bullet|\bullet)$ is coherent if and only if*

(i) *for all* $A \in \mathscr{B} \cup \{\mathscr{X}\}$, *the (unconditional) lower prevision* $\underline{P}_{\mathrm{s}}(\bullet|A)$ *is coherent;*

(ii) $\underline{P}((f-a)I_A|\mathscr{X}) \begin{cases} \geq 0 \text{ if } a < \underline{P}(f|A) \\ \leq 0 \text{ if } a > \underline{P}(f|A) \end{cases}$ *for all* $a \in \mathbb{R}$, *all* $f \in \mathbb{G}$ *and all* $A \in \mathscr{B}$.

Proof. That conditions (i) and (ii) are necessary, follows from Theorems 13.45 and 13.31(xiv)$_{267}$ respectively.

To prove sufficiency, we use coherence criterion (D)$_{260}$. Consider any n in $\mathbb{N}$, non-negative $\lambda_0, \lambda_1, \ldots, \lambda_n$ in $\mathbb{R}$, $(f_0, A_0), (f_1, A_1), \ldots, (f_n, A_n)$ in $\operatorname{dom} \underline{P}(\bullet|\bullet)$ and $a_0, \ldots, a_n \in \mathbb{R}$ such that $a_0 > \underline{P}(f_0|A_0)$ and $a_k < \underline{P}(f_k|A_k)$ for all $k \in \{1, \ldots, n\}$. If we use the notations $g_\ell := f_\ell|_{A_\ell}$ for $\ell \in \{0, 1, \ldots, n\}$, then this implies that $a_0 > \underline{P}_{\mathrm{s}}(g_0|A_0)$ and $a_k < \underline{P}_{\mathrm{s}}(g_k|A_k)$ for $k \in \{1, \ldots, n\}$. There are then two possibilities. Either $\mathscr{X} \notin \{A_0, A_1, \ldots, A_n\}$, and then we know that all $A_\ell \in \mathscr{B}$, whence

$$\sup\left(\sum_{k=1}^{n} \lambda_k (f_k - a_k) I_{A_k} - \lambda_0 (f_0 - a_0) I_{A_0} \middle| A_0 \cup A_1 \cup \cdots \cup A_n\right)$$
$$= \max_{\ell=0}^{n} \sup\left(\sum_{k=1}^{n} \lambda_k (f_k - a_k) I_{A_k} - \lambda_0 (f_0 - a_0) I_{A_0} \middle| A_\ell\right)$$
$$= \max_{\ell=0}^{n} \sup\left(\sum_{k=1\,:\,A_k = A_\ell}^{n} \lambda_k (g_k - a_k) - \mu_{0,\ell}(g_0 - a_0) \middle| A_\ell\right) \geq 0,$$

where $\mu_{0,\ell}$ is equal to either λ_0 or zero, depending on whether $A_0 = A_\ell$ or not, and the inequality follows from (i) (essentially, the separate coherence of the $\underline{P}_{\mathrm{s}}(\bullet|A)$ for all $A \in \mathscr{B}$ guarantees that the suprema are non-negative). Or $\mathscr{X} \in \{A_0, A_1, \ldots, A_n\}$, so we infer from (i) and Theorem 13.34$_{271}$ that (essentially, we now use the separate coherence of $\underline{P}_{\mathrm{s}}(\bullet|\mathscr{X})$)

$$\sup\left(\sum_{k=1}^{n} \lambda_k (f_k - a_k) I_{A_k} - \lambda_0 (f_0 - a_0) I_{A_0} \middle| \mathscr{X}\right)$$
$$\geq \sum_{k=1}^{n} \lambda_k \underline{P}((f_k - a_k) I_{A_k} | \mathscr{X}) - \lambda_0 \underline{P}((f_0 - a_0) I_{A_0} | \mathscr{X}),$$

where the right-hand side is well defined and non-negative, because (ii) guarantees that $\underline{P}((f_0 - a_0) I_{A_0} | \mathscr{X}) \leq 0$ and all $\underline{P}((f_k - a_k) I_{A_k} | \mathscr{X}) \geq 0$. □

13.7 The natural extension of a conditional lower prevision

In our discussion of the axioms of acceptability and their consequences for conditional lower previsions, we have argued that a sensible minimal requirement for conditional lower previsions, as rational models for uncertainty, is that they should be coherent. In fact, we need only to impose the weaker condition of avoiding sure loss: as described in Section 13.5.1$_{259}$, any conditional lower prevision that avoids sure loss can be converted into a coherent one in a canonical manner via natural extension. Natural extension also embodies inference, as it tells us how to extend assessments on any subset of gambles to the set of all gambles. For this reason, we study natural extension in more detail here.

13.7.1 Natural extension as least-committal extension

First of all, we establish that the natural extension is as conservative as possible. Recall: the avoiding sure loss criterion (C)$_{255}$ essentially states that $\underline{P}(\bullet|\bullet)$ avoids sure loss whenever there is a coherent conditional lower prevision on $\mathbb{G}\times\mathscr{P}^\circ$ that dominates it. The following theorem states amongst other things that the natural extension $\underline{E}_{\underline{P}(\bullet|\bullet)}(\bullet|\bullet)$ of $\underline{P}(\bullet|\bullet)$ is the most conservative such coherent conditional lower prevision: all coherent conditional lower previsions on $\mathbb{G}\times\mathscr{P}^\circ$ that dominate $\underline{P}(\bullet|\bullet)$ also dominate $\underline{E}_{\underline{P}(\bullet|\bullet)}(\bullet|\bullet)$.

Theorem 13.47 (Natural extension) *Let $\underline{P}(\bullet|\bullet)$ be a conditional lower prevision. Then the following statements hold:*

(i) *If $\underline{P}(\bullet|\bullet)$ avoids sure loss, then $\underline{E}_{\underline{P}(\bullet|\bullet)}(\bullet|\bullet)$ is the point-wise smallest coherent conditional lower prevision on $\mathbb{G}\times\mathscr{P}^\circ$ that dominates $\underline{P}(\bullet|\bullet)$.*

(ii) *If $\underline{P}(\bullet|\bullet)$ is coherent, then $\underline{E}_{\underline{P}(\bullet|\bullet)}(\bullet|\bullet)$ is the point-wise smallest coherent conditional lower prevision on $\mathbb{G}\times\mathscr{P}^\circ$ that coincides with $\underline{P}(\bullet|\bullet)$ on* dom $\underline{P}(\bullet|\bullet)$.

Because of this property, $\underline{E}_{\underline{P}(\bullet|\bullet)}(\bullet|\bullet)$ is sometimes also called the **least-committal extension** of $\underline{P}(\bullet|\bullet)$, see, for instance, Walley (1981, p. 28).

Proof. (i). Assume that $\underline{P}(\bullet|\bullet)$ avoids sure loss. Clearly, $\underline{E}_{\underline{P}(\bullet|\bullet)}(\bullet|\bullet)$ dominates $\underline{P}(\bullet|\bullet)$, by Definition 13.23$_{259}$. Let $\underline{Q}(\bullet|\bullet)$ be any coherent conditional lower prevision on $\mathbb{G}\times\mathscr{P}^\circ$ that dominates $\underline{P}(\bullet|\bullet)$. We show that $\underline{Q}(\bullet|\bullet)$ also dominates $\underline{E}_{\underline{P}(\bullet|\bullet)}(\bullet|\bullet)$.

Indeed, because $\underline{Q}(\bullet|\bullet)$ dominates $\underline{P}(\bullet|\bullet)$, it follows that $\mathscr{A}_{\underline{P}(\bullet|\bullet)}\subseteq\mathscr{A}_{\underline{Q}(\bullet|\bullet)}$ (see Equation (13.10)$_{255}$). Hence, by Proposition 13.17$_{255}$, $\operatorname{lpr}(\operatorname{Cl}_{\mathbb{D}}(\mathscr{A}_{\underline{Q}(\bullet|\bullet)}))$ dominates $\operatorname{lpr}(\operatorname{Cl}_{\mathbb{D}}(\mathscr{A}_{\underline{P}(\bullet|\bullet)}))$. Now observe that $\operatorname{lpr}(\operatorname{Cl}_{\mathbb{D}}(\mathscr{A}_{\underline{Q}(\bullet|\bullet)}))=\underline{E}_{\underline{Q}(\bullet|\bullet)}(\bullet|\bullet)=\underline{Q}(\bullet|\bullet)$ by the definition of $\underline{E}_{\underline{Q}(\bullet|\bullet)}(\bullet|\bullet)$ and coherence criterion (C)$_{260}$ and that similarly $\underline{E}_{\underline{P}(\bullet|\bullet)}(\bullet|\bullet)=\operatorname{lpr}(\operatorname{Cl}_{\mathbb{D}}(\mathscr{A}_{\underline{P}(\bullet|\bullet)}))$ by the definition of $\underline{E}_{\underline{P}(\bullet|\bullet)}(\bullet|\bullet)$.

(ii). By coherence criterion (C)$_{260}$, $\underline{E}_{\underline{P}(\bullet|\bullet)}(\bullet|\bullet)$ is indeed a coherent extension of $\underline{P}(\bullet|\bullet)$. By (i), it is also the point-wise smallest one. □

Proposition 4.27$_{65}$ also generalises to conditional lower previsions.

Proposition 13.48 *Let $\underline{P}(\bullet|\bullet)$ and $\underline{Q}(\bullet|\bullet)$ be conditional lower previsions that avoid sure loss. If $\underline{Q}(\bullet|\bullet)$ dominates $\underline{P}(\bullet|\bullet)$, then $\underline{E}_{\underline{Q}(\bullet|\bullet)}(\bullet|\bullet)$ dominates $\underline{E}_{\underline{P}(\bullet|\bullet)}(\bullet|\bullet)$ too: $\underline{E}_{\underline{Q}(\bullet|\bullet)}(f|A)\geq\underline{E}_{\underline{P}(\bullet|\bullet)}(f|A)$ for all (f,A) in $\mathbb{G}\times\mathscr{P}^\circ$.*

Proof. This result is a reformulation, in terms of lower previsions and natural extension, of Proposition 13.17_{255}. If $\underline{Q}(\bullet|\bullet)$ dominates $\underline{P}(\bullet|\bullet)$, then $\mathscr{A}_{\underline{P}(\bullet|\bullet)} \subseteq \mathscr{A}_{\underline{Q}(\bullet|\bullet)}$. Indeed, consider any g in $\mathscr{A}_{\underline{P}(\bullet|\bullet)}$, so there are (f, A) in $\operatorname{dom} \underline{P}(\bullet|\bullet)$ and $\mu < \underline{P}(f|A)$ such that $g = (f - \mu)I_A$. As by assumption then also $(f, A) \in \operatorname{dom} \underline{Q}(\bullet|\bullet)$ and $\mu < \underline{Q}(f|A)$, we see that $g \in \mathscr{A}_{\underline{Q}(\bullet|\bullet)}$ too. Now use Proposition 13.17_{255} to conclude that $\underline{E}_{\underline{Q}(\bullet|\bullet)}(\bullet|\bullet) = \operatorname{lpr}(\operatorname{Cl}_{\mathbb{D}}(\mathscr{A}_{\underline{Q}(\bullet|\bullet)}))(\bullet|\bullet)$ dominates $\underline{E}_{\underline{P}(\bullet|\bullet)}(\bullet|\bullet) = \operatorname{lpr}(\operatorname{Cl}_{\mathbb{D}}(\mathscr{A}_{\underline{P}(\bullet|\bullet)}))(\bullet|\bullet)$.

Alternatively, if $\underline{Q}(\bullet|\bullet)$ dominates $\underline{P}(\bullet|\bullet)$, then any coherent conditional lower prevision that dominates $\underline{Q}(\bullet|\bullet)$ also dominates $\underline{P}(\bullet|\bullet)$. So $\underline{E}_{\underline{Q}(\bullet|\bullet)}(\bullet|\bullet)$ dominates $\underline{Q}(\bullet|\bullet)$ and, therefore, $\underline{P}(\bullet|\bullet)$, by Theorem 13.47(i). And by the same theorem, $\underline{E}_{\underline{P}(\bullet|\bullet)}(\bullet|\bullet)$ is the point-wise smallest coherent conditional lower prevision that dominates $\underline{P}(\bullet|\bullet)$. Hence indeed $\underline{E}_{\underline{Q}(\bullet|\bullet)}(\bullet|\bullet) \geq \underline{E}_{\underline{P}(\bullet|\bullet)}(\bullet|\bullet)$. □

13.7.2 Natural extension and equivalence

We also extend the notion of *equivalence*, introduced in Definition 4.28_{66}, to apply to conditional lower previsions.

Definition 13.49 (Equivalence) *Two conditional lower previsions $\underline{P}(\bullet|\bullet)$ and $\underline{Q}(\bullet|\bullet)$ that avoid sure loss are called* **equivalent** *if their natural extensions are equal: $\underline{E}_{\underline{P}(\bullet|\bullet)}(\bullet|\bullet) = \underline{E}_{\underline{Q}(\bullet|\bullet)}(\bullet|\bullet)$.*

Similarly, two consistent sets of acceptable gambles $\mathscr{A}_1$ and $\mathscr{A}_2$ are called **equivalent** *if their natural extensions are the same:* $\operatorname{Cl}_{\mathbb{D}}(\mathscr{A}_1) = \operatorname{Cl}_{\mathbb{D}}(\mathscr{A}_2)$.

Equivalence on conditional lower previsions is an equivalence relation, and so is equivalence on sets of acceptable gambles. The following proposition gives an alternative characterisation of equivalence.

Proposition 13.50 *Let $\underline{P}(\bullet|\bullet)$ be a conditional lower prevision that avoids sure loss. Let $\underline{Q}(\bullet|\bullet)$ be any coherent conditional lower prevision that dominates $\underline{P}(\bullet|\bullet)$. Then $\underline{P}(\bullet|\bullet)$ is equivalent to $\underline{Q}(\bullet|\bullet)$ if and only if $\underline{E}_{\underline{P}(\bullet|\bullet)}(\bullet|\bullet)$ is an extension of $\underline{Q}(\bullet|\bullet)$ or, in other words, if and only if $\underline{Q}(\bullet|\bullet)$ and $\underline{E}_{\underline{P}(\bullet|\bullet)}(\bullet|\bullet)$ coincide on* $\operatorname{dom} \underline{Q}(\bullet|\bullet)$.

Proof. As $\underline{P}(\bullet|\bullet)$ avoids sure loss and $\underline{Q}(\bullet|\bullet)$ is coherent, both $\underline{E}_{\underline{P}(\bullet|\bullet)}(\bullet|\bullet)$ and $\underline{E}_{\underline{Q}(\bullet|\bullet)}(\bullet|\bullet)$ exist.

'if'. As $\underline{Q}(\bullet|\bullet)$ dominates $\underline{P}(\bullet|\bullet)$, Theorem 13.47 guarantees that $\underline{E}_{\underline{Q}(\bullet|\bullet)}(\bullet|\bullet)$ dominates $\underline{P}(\bullet|\bullet)$ and, therefore, also dominates $\underline{E}_{\underline{P}(\bullet|\bullet)}(\bullet|\bullet)$: $\underline{E}_{\underline{Q}(\bullet|\bullet)}(\bullet|\bullet) \geq \underline{E}_{\underline{P}(\bullet|\bullet)}(\bullet|\bullet)$. To prove the converse inequality, we use the assumption that the coherent $\underline{E}_{\underline{P}(\bullet|\bullet)}(\bullet|\bullet)$ is an extension of $\underline{Q}(\bullet|\bullet)$ and, therefore, clearly, in particular also dominates $\underline{Q}(\bullet|\bullet)$. By Theorem 13.47, it therefore dominates $\underline{E}_{\underline{Q}(\bullet|\bullet)}(\bullet|\bullet)$: $\underline{E}_{\underline{P}(\bullet|\bullet)}(\bullet|\bullet) \geq \underline{E}_{\underline{Q}(\bullet|\bullet)}(\bullet|\bullet)$.

'only if'. Assume that $\underline{E}_{\underline{P}(\bullet|\bullet)}(\bullet|\bullet) = \underline{E}_{\underline{Q}(\bullet|\bullet)}(\bullet|\bullet)$. As $\underline{Q}(\bullet|\bullet)$ is coherent, it follows from coherence criterion $(\mathrm{C})_{260}$ that $\underline{Q}(\bullet|\bullet)$ and $\underline{E}_{\underline{Q}(\bullet|\bullet)}(\bullet|\bullet)$ coincide on $\operatorname{dom} \underline{Q}(\bullet|\bullet)$ and, therefore, $\underline{E}_{\underline{P}(\bullet|\bullet)}(\bullet|\bullet)$ and $\underline{Q}(\bullet|\bullet)$ coincide on $\operatorname{dom} \underline{Q}(\bullet|\bullet)$. □

13.7.3 Natural extension to a specific domain and the transitivity of natural extension

It may be useful to consider the restriction of the coherent conditional lower prevision $\underline{E}_{\underline{P}(\bullet|\bullet)}(\bullet|\bullet)$ to some set $\mathscr{O}$ that includes dom $\underline{P}(\bullet|\bullet)$. This leads to a coherent conditional lower prevision whose domain is $\mathscr{O}$, which we denote by $\underline{E}^{\mathscr{O}}_{\underline{P}(\bullet|\bullet)}(\bullet|\bullet)$ and call the **natural extension of $\underline{P}(\bullet|\bullet)$ to $\mathscr{O}$.** We first generalise Theorem 13.47$_{280}$ to this restricted form of the natural extension.

Theorem 13.51 *Let $\underline{P}(\bullet|\bullet)$ be a conditional lower prevision. Then the following statements hold:*

(i) *If $\underline{P}(\bullet|\bullet)$ avoids sure loss, then $\underline{E}^{\mathscr{O}}_{\underline{P}(\bullet|\bullet)}(\bullet|\bullet)$ is the point-wise smallest coherent conditional lower prevision on $\mathscr{O}$ that dominates $\underline{P}(\bullet|\bullet)$.*

(ii) *If $\underline{P}(\bullet|\bullet)$ is coherent, then $\underline{E}^{\mathscr{O}}_{\underline{P}(\bullet|\bullet)}(\bullet|\bullet)$ is the point-wise smallest coherent conditional lower prevision on $\mathscr{O}$ that coincides with $\underline{P}(\bullet|\bullet)$ on* dom $\underline{P}(\bullet|\bullet)$.

Proof. (i). Assume that $\underline{P}(\bullet|\bullet)$ avoids sure loss. Clearly, $\underline{E}^{\mathscr{O}}_{\underline{P}(\bullet|\bullet)}(\bullet|\bullet)$ dominates $\underline{P}(\bullet|\bullet)$, by Definition 13.23$_{259}$. Let $\underline{Q}(\bullet|\bullet)$ be any coherent conditional lower prevision on $\mathscr{O}$ that dominates $\underline{P}(\bullet|\bullet)$. Then $\underline{E}_{\underline{Q}(\bullet|\bullet)}(\bullet|\bullet)$ dominates $\underline{E}_{\underline{P}(\bullet|\bullet)}(\bullet|\bullet)$ by Proposition 13.48$_{280}$, and hence, the restriction $\underline{Q}(\bullet|\bullet)$ of $\underline{E}_{\underline{Q}(\bullet|\bullet)}(\bullet|\bullet)$ to $\mathscr{O}$ (see Theorem 13.47(ii)$_{280}$) dominates the restriction $\underline{E}^{\mathscr{O}}_{\underline{P}(\bullet|\bullet)}(\bullet|\bullet)$ of $\underline{E}_{\underline{P}(\bullet|\bullet)}(\bullet|\bullet)$ to $\mathscr{O}$.

(ii). By Theorem 13.47(ii)$_{280}$, $\underline{E}^{\mathscr{O}}_{\underline{P}(\bullet|\bullet)}(\bullet|\bullet)$ is a coherent extension of $\underline{P}(\bullet|\bullet)$ to $\mathscr{O}$. By (i), it is the point-wise smallest one. □

We can use this idea of a restricted natural extension to prove that taking natural extensions in several steps to successively larger domains is the same thing as taking a natural extension in one single step.

Corollary 13.52 *Let $\underline{P}(\bullet|\bullet)$ be a conditional lower prevision that avoids sure loss, and let* dom $\underline{P}(\bullet|\bullet) \subseteq \mathscr{O} \subseteq \mathbb{G} \times \mathscr{P}^{\circ}$. *Then*

$$\underline{E}_{\underline{E}^{\mathscr{O}}_{\underline{P}(\bullet|\bullet)}(\bullet|\bullet)}(f|A) = \underline{E}_{\underline{P}(\bullet|\bullet)}(f|A) \textit{ for all } (f,A) \textit{ in } \mathbb{G} \times \mathscr{P}^{\circ}.$$

In other words, $\underline{P}(\bullet|\bullet)$ and $\underline{E}^{\mathscr{O}}_{\underline{P}(\bullet|\bullet)}(\bullet|\bullet)$ are equivalent. In particular, let dom $\underline{P}(\bullet|\bullet) \subseteq \mathscr{Q} \subseteq \mathscr{O} \subseteq \mathbb{G} \times \mathscr{P}^{\circ}$, *then $\underline{E}^{\mathscr{O}}_{\underline{P}(\bullet|\bullet)}(\bullet|\bullet)$ is an extension of $\underline{E}^{\mathscr{Q}}_{\underline{P}(\bullet|\bullet)}(\bullet|\bullet)$ and*

$$\underline{E}^{\mathscr{O}}_{\underline{E}^{\mathscr{Q}}_{\underline{P}(\bullet|\bullet)}(\bullet|\bullet)}(f|A) = \underline{E}^{\mathscr{O}}_{\underline{P}(\bullet|\bullet)}(f|A) \textit{ for all } (f,A) \textit{ in } \mathscr{O}.$$

Proof. As $\underline{E}_{\underline{P}(\bullet|\bullet)}(\bullet|\bullet)$ is an extension of $\underline{E}^{\mathscr{O}}_{\underline{P}(\bullet|\bullet)}(\bullet|\bullet)$ and $\underline{E}^{\mathscr{O}}_{\underline{P}(\bullet|\bullet)}(\bullet|\bullet)$ dominates $\underline{P}(\bullet|\bullet)$, we can invoke Proposition 13.50$_{\curvearrowleft}$ to find that $\underline{P}(\bullet|\bullet)$ and $\underline{E}^{\mathscr{O}}_{\underline{P}(\bullet|\bullet)}(\bullet|\bullet)$ are indeed equivalent. By definition, $\underline{E}^{\mathscr{O}}_{\underline{P}(\bullet|\bullet)}(\bullet|\bullet)$ is an extension of $\underline{E}^{\mathscr{Q}}_{\underline{P}(\bullet|\bullet)}(\bullet|\bullet)$. The last equality follows from the first by taking $\mathscr{Q}$ for $\mathscr{O}$ and restricting to $\mathscr{O}$. □

In summary, let $\underline{P}(\bullet|\bullet)$, $\underline{Q}(\bullet|\bullet)$ and $\underline{R}(\bullet|\bullet)$ be three coherent conditional lower previsions, and assume that $\operatorname{dom}\underline{P}(\bullet|\bullet) \subseteq \operatorname{dom}\underline{Q}(\bullet|\bullet) \subseteq \operatorname{dom}\underline{R}(\bullet|\bullet)$; if $\underline{Q}(\bullet|\bullet)$ is the natural extension of $\underline{P}(\bullet|\bullet)$ to $\operatorname{dom}\underline{Q}(\bullet|\bullet)$ and $\underline{R}(\bullet|\bullet)$ is the natural extension of $\underline{Q}(\bullet|\bullet)$ to $\operatorname{dom}\underline{R}(\bullet|\bullet)$, then we gather from the corollary above that $\underline{R}(\bullet|\bullet)$ is the natural extension of $\underline{P}(\bullet|\bullet)$ to $\operatorname{dom}\underline{R}(\bullet|\bullet)$. In other words, the relation 'is the natural extension of' is transitive.

13.7.4 Natural extension and avoiding sure loss

As a next step in our exploration of natural extension, we delve into its relationship with the criterion of avoiding sure loss. The following theorem, which is roughly based on Troffaes (2006, Theorem 5), provides us with a constructive expression for calculating the natural extension, as well as a number of alternative criteria, in addition to the ones in Definition 13.18$_{255}$, for checking whether a conditional lower prevision avoids sure loss. Note that there are subtle but important differences with its counterpart Theorem 4.33$_{68}$ for lower previsions on bounded gambles.

Theorem 13.53 *Let $\underline{P}(\bullet|\bullet)$ be a conditional lower prevision, and consider any $\operatorname{dom}\underline{P}(\bullet|\bullet) \subseteq \mathcal{O} \subseteq \mathbb{G}\times\mathscr{P}^\circ$. Define the conditional lower prevision $\underline{E}(\bullet|\bullet)$ on $\mathbb{G}\times\mathscr{P}^\circ$ by letting $\underline{E}(f|A)$ be equal to the extended real number:*

$$\sup\Bigg\{\alpha\in\mathbb{R} : n\in\mathbb{N}, a_1,\dots,a_n\in\mathbb{R}, a_k<\underline{P}(f_k|A_k), \\ \lambda_1,\dots,\lambda_n\geq 0, (f_1,A_1),\dots,(f_n,A_n)\in\operatorname{dom}\underline{P}(\bullet|\bullet), \\ (f-\alpha)I_A \geq \sum_{k=1}^{n}\lambda_k(f_k-a_k)I_{A_k}\Bigg\}$$

for all $(f,A)\in\mathbb{G}\times\mathscr{P}^\circ$. Then the following conditions are equivalent.

(i) *$\underline{P}(\bullet|\bullet)$ avoids sure loss.*

(ii) *$\underline{E}(\bullet|\bullet)$ is a coherent conditional lower prevision on $\mathbb{G}\times\mathscr{P}^\circ$.*

(iii) *$-\infty<\underline{E}(f|A)<+\infty$ for all bounded gambles $f\in\mathbb{B}$ and all events $A\in\mathscr{P}^\circ$.*

(iv) *$-\infty<\underline{E}(f_0|\mathcal{X})<+\infty$ for some gamble $f_0\in\mathbb{G}$.*

(v) *There is a point-wise smallest coherent conditional lower prevision on $\mathcal{O}$ that dominates $\underline{P}(\bullet|\bullet)$: its natural extension $\underline{E}^{\mathcal{O}}_{\underline{P}(\bullet|\bullet)}(\bullet|\bullet)$. This conditional lower prevision coincides with $\underline{E}(\bullet|\bullet)$ on $\mathcal{O}$.*

(vi) *There is at least one conditional lower prevision on $\mathcal{O}$ that is coherent and dominates $\underline{P}(\bullet|\bullet)$.*

(vii) *There is at least one conditional lower prevision on $\mathcal{O}$ that avoids sure loss and dominates $\underline{P}(\bullet|\bullet)$.*

Proof. It clearly suffices to establish the two loops (i)⇒(v)⇒(vi)⇒(vii)⇒(i) and (i)⇒(ii)⇒(iii)⇒(iv)⇒(i).

$(\mathrm{i})_{\curvearrowleft}\Rightarrow(\mathrm{v})_{\curvearrowleft}$. The first part is an immediate consequence of Theorem 13.51(i)$_{282}$. For the second part, using Equation (13.2)$_{240}$ and the definition of natural extension, we see that $\underline{E}_{\underline{P}(\bullet|\bullet)}(f|A) = \sup\left\{\alpha \in \mathbb{R} : (f-\alpha)I_A \in \mathrm{Cl}_{\mathbb{D}}(\mathscr{A}_{\underline{P}(\bullet|\bullet)})\right\}$. Now use the expression for $\mathrm{Cl}_{\mathbb{D}}(\mathscr{A}_{\underline{P}(\bullet|\bullet)})$ in Equation (13.13)$_{259}$ to see that $\underline{E}_{\underline{P}(\bullet|\bullet)}(f|A)$ and $\underline{E}(f|A)$ coincide.

$(\mathrm{v})_{\curvearrowleft}\Rightarrow(\mathrm{vi})_{\curvearrowleft}$. Immediate.

$(\mathrm{vi})_{\curvearrowleft}\Rightarrow(\mathrm{vii})_{\curvearrowleft}$. Immediate, because a coherent conditional lower prevision in particular avoids sure loss; see, for instance, coherence criterion (C)$_{260}$.

$(\mathrm{vii})_{\curvearrowleft}\Rightarrow(\mathrm{i})_{\curvearrowleft}$. Let $\underline{Q}(\bullet|\bullet)$ be any conditional lower prevision on $\mathscr{O}$ that avoids sure loss and dominates $\underline{P}(\bullet|\bullet)$. We must show that $\underline{P}(\bullet|\bullet)$ avoids sure loss. Consider the natural extension $\underline{E}_{\underline{Q}(\bullet|\bullet)}(\bullet|\bullet)$ of $\underline{Q}(\bullet|\bullet)$. By Proposition 13.48$_{280}$, $\underline{E}_{\underline{Q}(\bullet|\bullet)}(\bullet|\bullet)$ dominates $\underline{P}(\bullet|\bullet)$, and by Theorem 13.47(i)$_{280}$, $\underline{E}_{\underline{Q}(\bullet|\bullet)}(\bullet|\bullet)$ is coherent. By coherence criterion (B)$_{260}$, $\underline{E}_{\underline{Q}(\bullet|\bullet)}(\bullet|\bullet)$ satisfies the properties CLP1–CLP4$_{243}$. Now apply avoiding sure loss criterion (C)$_{255}$.

$(\mathrm{i})_{\curvearrowleft}\Rightarrow(\mathrm{ii})_{\curvearrowleft}$. Suppose $\underline{P}(\bullet|\bullet)$ avoids sure loss. We have already proved that $(\mathrm{i})_{\curvearrowleft}\Rightarrow(\mathrm{v})_{\curvearrowleft}$, and therefore, $(\mathrm{v})_{\curvearrowleft}$ holds for $\mathscr{O} = \mathbb{G}\times\mathscr{P}^{\circ}$. Hence, $\underline{E}(\bullet|\bullet) = \underline{E}_{\underline{P}(\bullet|\bullet)}(\bullet|\bullet)$ is a coherent conditional lower prevision on $\mathbb{G}\times\mathscr{P}^{\circ}$.

$(\mathrm{ii})_{\curvearrowleft}\Rightarrow(\mathrm{iii})_{\curvearrowleft}$. This is an immediate consequence of Theorem 13.31(i)$_{266}$.

$(\mathrm{iii})_{\curvearrowleft}\Rightarrow(\mathrm{iv})_{\curvearrowleft}$. Immediate.

$(\mathrm{iv})_{\curvearrowleft}\Rightarrow(\mathrm{i})_{\curvearrowleft}$. Assume that $\underline{P}(\bullet|\bullet)$ does not avoid sure loss. We show that in that case $\underline{E}(f|\mathscr{X}) > -\infty$ implies that $\underline{E}(f|\mathscr{X}) = +\infty$ for all gambles f or, equivalently, that $\underline{E}(f|\mathscr{X}) = \pm\infty$ for all gambles f. By contraposition, this will clearly imply that if $-\infty < \underline{E}(f_0|\mathscr{X}) < +\infty$ for at least one gamble f_0, then $\underline{P}(\bullet|\bullet)$ must avoid sure loss.

To simplify our argument, we rewrite the expression for $\underline{E}(f|A)$ as

$$\underline{E}(f|A) = \sup\Bigg\{\alpha \in \mathbb{R} : n \in \mathbb{N},\, a_1,\ldots,a_n \in \mathbb{R},\, a_k < \underline{P}(f_k|A_k),\\ \lambda_1,\ldots,\lambda_n \geq 0,\, (f_1,A_1),\ldots,(f_n,A_n) \in \operatorname{dom}\underline{P}(\bullet|\bullet),\\ \alpha I_A \leq fI_A - \sum_{k=1}^{n}\lambda_k(f_k - a_k)I_{A_k}\Bigg\}. \tag{13.26}$$

Assume that $\underline{E}(f|\mathscr{X}) > -\infty$, then it follows from Equation (13.26) that there are $\beta \in \mathbb{R}$, m in $\mathbb{N}$, non-negative μ_1, …, μ_m in $\mathbb{R}$, (g_1, B_1), …, (g_m, B_m) in dom $\underline{P}(\bullet|\bullet)$ and b_1, …, $b_m \in \mathbb{R}$ such that $b_\ell < \underline{P}(g_\ell|B_\ell)$ and

$$\beta \leq f - \sum_{\ell=1}^{m}\mu_\ell(g_\ell - b_\ell)I_{B_\ell}. \tag{13.27}$$

As $\underline{P}(\bullet|\bullet)$ does not avoid sure loss, we infer from the avoiding sure loss criterion (D)$_{255}$ that there are p in $\mathbb{N}$, non-negative ν_1, …, ν_p in $\mathbb{R}$, (h_1, C_1), …, (h_p, C_p) in dom $\underline{P}(\bullet|\bullet)$ and c_1, …, $c_p \in \mathbb{R}$ such that $c_k < \underline{P}(h_k|C_k)$ and

$$\varepsilon := -\sup\left(\sum_{k=1}^{p}\nu_k(h_k - c_k)I_{C_k}\,\middle|\, C_1\cup\cdots\cup C_p\right) > 0. \tag{13.28}$$

Combining the inequalities (13.27) and (13.28), we find that for all real $\kappa \geq 0$,

$$\beta + \kappa\varepsilon \leq f - \sum_{j=1}^{m} \mu_j (g_j - b_j) I_{B_j} - \sum_{k=1}^{p} \kappa \nu_k (h_k - c_k) I_{C_k},$$

and if we combine this with Equation (13.26), we find that $\underline{E}(f|\mathcal{X}) \geq \beta + \kappa\varepsilon$ for all real $\kappa \geq 0$, whence indeed $\underline{E}(f|\mathcal{X}) = +\infty$. □

As stated earlier, Theorems 4.33_{68} and 13.53_{283} differ in subtle ways. Even if there is a gamble f such that $\underline{E}(f|\mathcal{X}) < +\infty$, $\underline{P}(\bullet|\bullet)$ may still incur sure loss, as the following counterexample shows.

Example 13.54 Let $\mathcal{X} := \mathbb{R}$, and define the conditional lower prevision $\underline{P}(\bullet|\bullet)$ on $\{X, 0\} \times \{\mathcal{X}\}$ by $\underline{P}(X|\mathcal{X}) := -\infty$ and $\underline{P}(0|\mathcal{X}) := 1$.[8] Clearly, $\underline{P}(\bullet|\bullet)$ does not avoid sure loss because $\sup 0 = 0 < 1 = \underline{P}(0|\mathcal{X})$. Nevertheless, $\underline{E}(X|\mathcal{X}) = -\infty < +\infty$. Note, however, that $\underline{E}(g|\mathcal{X}) = \pm\infty$ for every gamble $g \in \mathbb{G}$: $\underline{E}(g|\mathcal{X}) = -\infty$ if g is unbounded below and $\underline{E}(g|\mathcal{X}) = +\infty$ otherwise. ◆

13.7.5 Simpler ways of calculating the natural extension

It is possible to give an alternative and simpler expression for natural extension when $\underline{P}(\bullet|\bullet)$ is a coherent unconditional lower previsions defined on a linear space. Compare the following result with Theorem 4.34_{69}.

Theorem 13.55 *Let $\underline{P}(\bullet|\bullet)$ be a coherent conditional lower prevision defined on $\mathcal{K} \times \{\mathcal{X}\}$ where $\mathcal{K}$ is some linear subspace of $\mathbb{G}$. Then its natural extension $\underline{E}_{\underline{P}(\bullet|\bullet)}(\bullet|\mathcal{X})$ is given by*

$$\underline{E}_{\underline{P}(\bullet|\bullet)}(f|\mathcal{X}) = \sup\left\{a + \underline{P}(g|\mathcal{X}) : a \in \mathbb{R},\ g \in \mathcal{K} \text{ and } a + g \leq f\right\} \text{ for all } f \text{ in } \mathbb{G}.$$

When $\mathcal{K}$ also contains all constant gambles $a \in \mathbb{R}$, this simplifies to

$$\underline{E}_{\underline{P}(\bullet|\bullet)}(f|\mathcal{X}) = \sup\left\{\underline{P}(g|\mathcal{X}) : g \in \mathcal{K} \text{ and } g \leq f\right\} \text{ for all } f \text{ in } \mathbb{G}.$$

Proof. Given the particular form of the domain of $\underline{P}(\bullet|\bullet)$, the definition of the natural extension in Equation $(13.14)_{260}$ yields

$$\underline{E}_{\underline{P}(\bullet|\bullet)}(f|\mathcal{X}) = \sup\left\{\alpha + \sum_{k=1}^{n} \lambda_k a_k : \alpha \in \mathbb{R},\ n \in \mathbb{N},\ a_1, \ldots, a_n \in \mathbb{R},\right.$$
$$a_k < \underline{P}(f_k|\mathcal{X}),$$
$$\lambda_1, \ldots, \lambda_n \geq 0,\ f_1, \ldots, f_n \in \mathcal{K},$$
$$\left. f - \alpha \geq \sum_{k=1}^{n} \lambda_k f_k \right\}. \tag{13.29}$$

[8] Remember that we write X also for the identity gamble.

This expression allows us to infer that (restrict the supremum to $n = 1$ and $\lambda_1 = 1$)

$$\underline{E}_{\underline{P}(\bullet|\bullet)}(f|\mathscr{X}) \geq \sup\left\{a + \underline{P}(g|\mathscr{X}) : a \in \mathbb{R}, g \in \mathscr{K} \text{ and } a + g \leq f\right\}.$$

To prove the converse inequality, observe that when $a_k < \underline{P}(f_k|\mathscr{X})$ for all k, then $\sum_{k=1}^{n} \lambda_k \underline{P}(f_k|\mathscr{X})$ must be well defined and, therefore, by the coherence (use Theorem 13.31(iv) and (v)$_{266}$) of $\underline{P}(\bullet|\bullet)$ on $\mathscr{K} \times \{\mathscr{X}\}$,

$$\sum_{k=1}^{n} \lambda_k a_k \leq \sum_{k=1}^{n} \lambda_k \underline{P}(f_k|\mathscr{X}) \leq \underline{P}\left(\sum_{k=1}^{n} \lambda_k f_k \middle| \mathscr{X}\right).$$

So we find that, indeed, again by Equation (13.29)$_\frown$, taking into account that the linear space $\mathscr{K}$ is in particular closed under non-negative linear combinations:

$$\underline{E}_{\underline{P}(\bullet|\bullet)}(f|\mathscr{X}) \leq \sup\left\{a + \underline{P}(g|\mathscr{X}) : a \in \mathbb{R}, g \in \mathscr{K} \text{ and } f - a \geq g\right\}.$$

The proof of the last statement is now immediate, if we also take into account the constant additivity property (Theorem 13.31(iii)$_{266}$) of the coherent $\underline{P}(\bullet|\bullet)$. □

13.7.6 Compatibility with the definition for lower previsions on bounded gambles

We still need to find out whether our new notion of natural extension for conditional lower previsions reduces to the one for lower previsions on bounded gambles we have already studied in Section 4.5$_{65}$.

Theorem 13.56 *Consider a lower prevision $\underline{Q}$ on bounded gambles that avoids sure loss, and define the corresponding (unconditional) lower prevision $\underline{P}(\bullet|\bullet)$ on* dom $\underline{Q} \times \{\mathscr{X}\}$ *by letting $\underline{P}(f|\mathscr{X}) := \underline{Q}(f)$ for all bounded gambles f in* dom $\underline{Q}$. *Then the restriction to $\mathbb{B} \times \{\mathscr{X}\}$ of the natural extension $\underline{E}_{\underline{P}(\bullet|\bullet)}(\bullet|\bullet)$ of the lower prevision $\underline{P}(\bullet|\bullet)$, in the sense of Definition 13.23$_{259}$, and the natural extension $\underline{E}_{\underline{Q}}$ of the lower prevision $\underline{Q}$, in the sense of Definition 4.8$_{47}$, coincide.*

Proof. As the lower prevision $\underline{Q}$ avoids sure loss, so does the (unconditional) lower prevision $\underline{P}(\bullet|\bullet)$, by Theorem 13.21$_{258}$. We may therefore start from Equation (13.14)$_{260}$ and use an argument similar to the one in the proof of Theorem 13.55$_\frown$ to find that for all bounded gambles f on $\mathscr{X}$, because all $\underline{P}(f_k|\mathscr{X}) = \underline{Q}(f_k) \in \mathbb{R}$,

$$\underline{E}_{\underline{P}(\bullet|\bullet)}(f|\mathscr{X})$$

$$= \sup\left\{\alpha \in \mathbb{R} : f - \alpha \geq \sum_{k=1}^{n} \lambda_k [f_k - \underline{P}(f_k|\mathscr{X})],\, n \in \mathbb{N},\, \lambda_k \geq 0,\, f_k \in \operatorname{dom} \underline{Q}\right\}$$

$$= \sup\left\{\alpha \in \mathbb{R} : f - \alpha \geq \sum_{k=1}^{n} \lambda_k[f_k - \underline{Q}(f_k)],\ n \in \mathbb{N},\ \lambda_k \geq 0,\ f_k \in \operatorname{dom}\underline{Q}\right\}$$

$$= \underline{E}_{\underline{Q}}(f),$$

where the last equality follows from Equation $(4.7)_{47}$. □

13.8 Alternative characterisations for avoiding sure loss, coherence and natural extension

In Section $13.5.4_{264}$ we introduced linear conditional previsions. We denoted the set of all linear conditional previsions on $\mathbb{G} \times \mathscr{P}^\circ$ by CP.

With any conditional lower prevision $\underline{P}(\bullet|\bullet)$, we can associate the set of all linear conditional previsions on $\mathbb{G} \times \mathscr{P}^\circ$ that dominate $\underline{P}(\bullet|\bullet)$:

$$\operatorname{lins}(\underline{P}(\bullet|\bullet)) := \left\{Q(\bullet|\bullet) \in \mathbb{CP} : (\forall (f, A) \in \operatorname{dom}\underline{P}(\bullet|\bullet))\left(Q(f|A) \geq \underline{P}(f|A)\right)\right\},$$

called the **dual model** of $\underline{P}(\bullet|\bullet)$.

The next theorem, due to (Troffaes, 2006, Theorem 6), extends the lower envelope Theorem 4.38_{71}. Williams (1975b, Section 2.1.1, Theorem 2) has already proved a similar result for conditional lower previsions on bounded gambles.

Theorem 13.57 (Lower envelope theorem) *Let $\underline{P}(\bullet|\bullet)$ be any conditional lower prevision. Then the following statements hold:*

(i) *$\underline{P}(\bullet|\bullet)$ avoids sure loss if and only if* $\operatorname{lins}(\underline{P}(\bullet|\bullet)) \neq \emptyset$.

(ii) *$\underline{P}(\bullet|\bullet)$ is coherent if and only if it avoids sure loss and*

$$\underline{P}(f|A) = \min\left\{Q(f|A) : Q(\bullet|\bullet) \in \operatorname{lins}(\underline{P}(\bullet|\bullet))\right\}$$

for all (f, A) in $\operatorname{dom}\underline{P}(\bullet|\bullet)$.

(iii) *If $\underline{P}(\bullet|\bullet)$ avoids sure loss, then its natural extension $\underline{E}_{\underline{P}(\bullet|\bullet)}$ is the lower envelope of the non-empty set* $\operatorname{lins}(\underline{P}(\bullet|\bullet))$, *that is, it satisfies*

$$\underline{E}_{\underline{P}(\bullet|\bullet)}(f|A) = \min\left\{Q(f|A) : Q(f|A) \in \operatorname{lins}(\underline{P}(\bullet|\bullet))\right\}$$

for all (f, A) in $\mathbb{G} \times \mathscr{P}^\circ$.

The proof of this theorem, which we exceptionally omit for the sake of brevity, can be established using Zorn's lemma (a version of the Axiom of Choice) in a straightforward manner. Interestingly, we cannot apply any of the usual versions of the Hahn–Banach theorem because they apply to real-valued functionals only.

13.9 Marginal extension

As a simple yet very important example of natural extension, we consider the so-called **marginal extension** (Walley, 1991, Section 6.7). Consider two random variables X and Y, taking values in $\mathcal{X}$ and $\mathcal{Y}$, respectively. For simplicity, we assume that they are **logically independent**: any combination of values of X and Y is possible, so the set of possible values for the joint random variable (X, Y) is simply $\mathcal{X} \times \mathcal{Y}$.

We are interested in finding the natural extension of a conditional lower prevision $\underline{P}(\bullet|\bullet)$ with domain

$$\left\{(f, \mathcal{X} \times \mathcal{Y}) : f \in \mathcal{K}_{\mathcal{X}}\right\} \cup \left\{(g, \{x\} \times \mathcal{Y}) : g \in \mathcal{K}_{\mathcal{Y}}(x), x \in \mathcal{X}\right\}$$

for certain given subsets $\mathcal{K}_{\mathcal{X}} \subseteq \mathbb{G}(\mathcal{X})$ and $\mathcal{K}_{\mathcal{Y}}(x) \subseteq \mathbb{G}(\mathcal{Y})$, $x \in \mathcal{X}$. Above, and in the remainder of this section, we identify – and use the same notation for – a map f on $\mathcal{X}$ and its **cylindrical extension** f' to a map on $\mathcal{X} \times \mathcal{Y}$, defined by $f'(x, y) := f(x)$ for all $(x, y) \in \mathcal{X} \times \mathcal{Y}$ and, similarly, for maps g on $\mathcal{Y}$.

Because $\{\{x\} \times \mathcal{Y} : x \in \mathcal{X}\}$ is a partition of $\mathcal{X} \times \mathcal{Y}$, this puts us squarely in the context of Section 13.6.4$_{278}$ and in particular Theorem 13.46$_{278}$. It will, therefore, be useful to take a closer look at the *separate* (unconditional) lower previsions that can be associated with $\underline{P}(\bullet|\bullet)$. First of all,

$$\underline{P}_{\mathrm{s}}(f|\mathcal{X} \times \mathcal{Y}) = \underline{P}(f|\mathcal{X} \times \mathcal{Y}) =: \underline{P}_{\mathcal{X}}(f) \text{ for all } f \in \mathcal{K}_{\mathcal{X}},$$

and for all $x \in \mathcal{X}$,

$$\underline{P}_{\mathrm{s}}(h|_{\{x\}\times\mathcal{Y}}|\{x\} \times \mathcal{Y}) = \underline{P}(h(x, \bullet)|\{x\} \times \mathcal{Y}) =: \underline{P}_{\mathcal{Y}}(h(x, \bullet)|x) \text{ for all } h \in \mathcal{K}^{\star}_{\mathcal{Y}}(x),$$

where

$$\mathcal{K}^{\star}_{\mathcal{Y}}(x) := \left\{h \in \mathbb{G}(\mathcal{X} \times \mathcal{Y}) : h(x, \bullet) \in \mathcal{K}_{\mathcal{Y}}(x)\right\}.$$

It is clear that $\underline{P}_{\mathrm{s}}(\bullet|\mathcal{X} \times \mathcal{Y})$ is essentially the (unconditional) lower prevision $\underline{P}_{\mathcal{X}}$ on $\mathcal{K}_{\mathcal{X}}$ and that each lower prevision $\underline{P}_{\mathrm{s}}(\bullet|\{x\} \times \mathcal{Y})$ is essentially the (unconditional) lower prevision $\underline{P}_{\mathcal{Y}}(\bullet|x)$ on $\mathcal{K}_{\mathcal{Y}}(x)$.

So, we consider $\underline{P}_{\mathcal{X}}$ as an (unconditional) lower prevision on the set $\mathcal{K}_{\mathcal{X}}$ of gambles on $\mathcal{X}$, whose natural extension to a(n unconditional) lower prevision on $\mathbb{G}(\mathcal{X})$ we denote by $\underline{E}_{\mathcal{X}}$. In other words, we have that for all $f \in \mathbb{G}(\mathcal{X})$,

$$\underline{E}_{\mathcal{X}}(f) = \sup \left\{ \alpha \in \mathbb{R} : f - \alpha \geq \sum_{k=1}^{n} \lambda_k (f_k - a_k), \right.$$
$$\left. n \in \mathbb{N}, a_k \in \mathbb{R}, a_k < \underline{P}_{\mathcal{X}}(f_k), \lambda_k \geq 0, f_k \in \mathcal{K}_{\mathcal{X}} \right\}. \quad (13.30)$$

We formally extend this formula so that it also applies to maps $f : \mathcal{X} \to \mathbb{R}^*$.

Similarly, for every $x \in \mathcal{X}$, we consider $\underline{P}_{\mathcal{Y}}(\bullet|x)$ as an (unconditional) lower prevision on the set $\mathcal{K}_{\mathcal{Y}}(x)$ of gambles on $\mathcal{Y}$, whose natural extension to $\mathbb{G}(\mathcal{Y})$ we denote by $\underline{E}_{\mathscr{Y}}(\bullet|x)$. In other words, we have that for all $x \in \mathscr{X}$ and all $g \in \mathbb{G}(\mathscr{Y})$,

$$\underline{E}_{\mathscr{Y}}(g|x) = \sup\left\{\beta \in \mathbb{R} : g - \beta \geq \sum_{k=1}^{n} \mu_k(g_k - b_k),\right.$$
$$\left. n \in \mathbb{N}, b_k \in \mathbb{R}, b_k < \underline{P}_{\mathscr{Y}}(g_k|x), \mu_k \geq 0, g_k \in \mathscr{K}_{\mathscr{Y}}(x)\right\}. \tag{13.31}$$

The natural extension we are after is given by the following theorem. We introduce the notation $A \Subset \mathscr{X}$ for 'A is a *finite* subset of $\mathscr{X}$'.

Theorem 13.58 (Marginal Extension Theorem) *$\underline{P}(\bullet|\bullet)$ avoids sure loss if and only if $\underline{P}_{\mathscr{X}}$ avoids sure loss and $\underline{P}_{\mathscr{Y}}(\bullet|x)$ avoids sure loss for all $x \in \mathscr{X}$. In that case, we have that, for all gambles h on $\mathscr{X} \times \mathscr{Y}$ and all $x \in \mathscr{X}$,*

$$\underline{E}_{\underline{P}(\bullet|\bullet)}(h|\{x\} \times \mathscr{Y}) = \underline{E}_{\mathscr{Y}}(h(x, \bullet)|x) \tag{13.32}$$

and

$$\underline{E}_{\underline{P}(\bullet|\bullet)}(h|\mathscr{X} \times \mathscr{Y}) = \sup_{A \Subset \mathscr{X}} \underline{E}_{\mathscr{X}}\left(\sum_{x \in A} I_{\{x\}} \underline{E}_{\mathscr{Y}}(h(x, \bullet)|x) + \sum_{x \in A^c} I_{\{x\}} \inf_{y \in \mathscr{Y}} h(x, y)\right). \tag{13.33}$$

Before we go on to the proof, it will pay to consider in more detail the expression

$$h_A := \sum_{x \in A} I_{\{x\}} \underline{E}_{\mathscr{Y}}(h(x, \bullet)|x) + \sum_{x \in A^c} I_{\{x\}} \inf_{y \in \mathscr{Y}} h(x, y) \tag{13.34}$$

for any gamble h on $\mathscr{X} \times \mathscr{Y}$ and any $A \Subset \mathscr{X}$, which occurs as an argument for $\underline{E}_{\mathscr{X}}$ in Equation (13.33). This h_A is an extended real-valued map on $\mathscr{X}$, so not necessarily a gamble, but we have already mentioned that we can easily extend the domain of $\underline{E}_{\mathscr{X}}$ to include such maps as well. Because the collection of all finite subsets of $\mathscr{X}$ is directed under the inclusion relation, the h_A constitute a net of maps that is increasing, because by coherence, $\underline{E}_{\mathscr{Y}}(h(x, \bullet)|x) \geq \inf_{y \in \mathscr{Y}} h(x, y)$ for all $x \in \mathscr{X}$:

$$A \subseteq B \Rightarrow h_A \leq h_B \text{ for all } A, B \Subset \mathscr{X}.$$

The supremum in Equation (13.33) is therefore actually a limit: $\sup_{A \Subset \mathscr{X}} \underline{E}_{\mathscr{X}}(h_A) = \lim_{A \Subset \mathscr{X}} \underline{E}_{\mathscr{X}}(h_A)$.

Proof. If $\underline{P}(\bullet|\bullet)$ avoids sure loss, then obviously $\underline{P}_{\mathscr{X}}$ and all $\underline{P}_{\mathscr{Y}}(\bullet|x)$ avoid sure loss as well, because they are restrictions of $\underline{P}(\bullet|\bullet)$ (see Theorem 13.45$_{278}$).

Conversely, assume that $\underline{P}_{\mathscr{X}}$ and all $\underline{P}_{\mathscr{Y}}(\bullet|x)$ avoid sure loss. This already implies (use Theorem 13.47$_{280}$) that the (unconditional) lower previsions $\underline{E}_{\mathscr{X}}$ on $\mathbb{G}(\mathscr{X})$ and $\underline{E}_{\mathscr{Y}}(\bullet|x)$ on $\mathbb{G}(\mathscr{Y})$ are coherent, for all $x \in \mathscr{X}$. Consider, for ease of notation, the domain

$$\mathscr{O} := \mathbb{G}(\mathscr{X} \times \mathscr{Y}) \times (\{\mathscr{X} \times \mathscr{Y}\} \cup \{\{x\} \times \mathscr{Y} : x \in \mathscr{X}\})$$

and define the conditional lower prevision $\underline{F}(\bullet|\bullet)$ on $\mathscr{O}$ by letting

$$\underline{F}(h|\{x\} \times \mathscr{Y}) := \underline{E}_{\mathscr{Y}}(h(x, \bullet)|x) \tag{13.35}$$

and

$$\underline{F}(h|\mathscr{X} \times \mathscr{Y}) := \sup_{A \Subset \mathscr{X}} \underline{E}_{\mathscr{X}}\left(\sum_{x \in A} I_{\{x\}} \underline{E}_{\mathscr{Y}}(h(x, \bullet)|x) + \sum_{x \in A^c} I_{\{x\}} \inf_{y \in \mathscr{Y}} h(x, y)\right) \tag{13.36}$$

for all $h \in \mathbb{G}(\mathscr{X} \times \mathscr{Y})$ and all $x \in \mathscr{X}$. We will first invoke Theorem 13.46_{278}, with $\mathscr{B} = \{\{x\} \times \mathscr{Y} : x \in \mathscr{X}\}$, to prove that $\underline{F}(\bullet|\bullet)$ is coherent. We will then use the coherence of $\underline{F}(\bullet|\bullet)$ to show that $\underline{F}(\bullet|\bullet)$ coincides with the natural extension of $\underline{P}(\bullet|\bullet)$.

In order to be able to invoke Theorem 13.46_{278}, we need to know what are the separate (unconditional) lower previsions, associated with $\underline{F}(\bullet|\bullet)$. Proceeding in a manner similarly to what we did earlier for $\underline{P}(\bullet|\bullet)$, we see that

$$\underline{F}_{\mathrm{s}}(h|\mathscr{X} \times \mathscr{Y}) = \underline{F}(h|\mathscr{X} \times \mathscr{Y})$$

and

$$\underline{F}_{\mathrm{s}}(h|_{\{x\}\times\mathscr{Y}}|\{x\} \times \mathscr{Y}) = \underline{F}(h(x, \bullet)|\{x\} \times \mathscr{Y}) = \underline{E}_{\mathscr{Y}}(h(x, \bullet)|x)$$

for all $h \in \mathbb{G}(\mathscr{X} \times \mathscr{Y})$. In order to prove that condition $(\mathrm{i})_{279}$ of Theorem 13.46_{278} is verified, we have to prove that these (unconditional) lower previsions are coherent. As the (unconditional) lower previsions $\underline{E}_{\mathscr{Y}}(\bullet|x)$ on $\mathbb{G}(\mathscr{Y})$ are coherent for all $x \in \mathscr{X}$, we now concentrate on $\underline{F}_{\mathrm{s}}(\bullet|\mathscr{X} \times \mathscr{Y})$ and invoke Theorem 13.34_{271} to prove that it is coherent too.

For condition $(\mathrm{i})_{271}$ of Theorem 13.34_{271}, consider any gamble h on $\mathscr{X} \times \mathscr{Y}$ and observe that by coherence (bounds, Theorem $13.31(\mathrm{i})_{266}$) of each $\underline{E}_{\mathscr{Y}}(\bullet|x)$,

$$\begin{aligned} h_A &= \sum_{x \in A} I_{\{x\}} \underline{E}_{\mathscr{Y}}(h(x, \bullet)|x) + \sum_{x \in A^c} I_{\{x\}} \inf_{y \in \mathscr{Y}} h(x, y) \geq \sum_{x \in \mathscr{X}} I_{\{x\}} \inf_{y \in \mathscr{Y}} h(x, y) \\ &= \inf_{y \in \mathscr{Y}} h(\bullet, y), \end{aligned}$$

and, therefore, indeed, by coherence[9] (again bounds, Theorem $13.31(\mathrm{i})_{266}$) of $\underline{E}_{\mathscr{X}}$,

$$\underline{F}_{\mathrm{s}}(h|\mathscr{X} \times \mathscr{Y}) = \lim_{A \Subset \mathscr{X}} \underline{E}_{\mathscr{X}}(h_A) \geq \underline{E}_{\mathscr{X}}\left(\inf_{y \in \mathscr{Y}} h(\bullet, y)\right) \geq \inf_{x \in \mathscr{X}}\left(\inf_{y \in \mathscr{Y}} h(x, y)\right) \geq \inf h.$$

For condition $(\mathrm{ii})_{271}$ of Theorem 13.34_{271}, consider any gambles h_1, h_2 on $\mathscr{X} \times \mathscr{Y}$ and observe that, again by coherence (super-additivity, Theorem $13.31(\mathrm{iv})_{266}$) of each

[9] An astute reader will observe that the coherence argument only applies for h_A and $\inf_{y \in \mathscr{Y}} h(\bullet, y)$ that are real-valued. If these maps can also assume infinite values, the same result can also be derived directly from the expression for $\underline{E}_{\mathscr{X}}$ in Equation $(13.30)_{288}$. Similar observations can be made for a few more of the following derivations.

$\underline{E}_{\mathscr{Y}}(\bullet|x)$,

$$\begin{aligned}(h_1+h_2)_A &= \sum_{x\in A} I_{\{x\}}\underline{E}_{\mathscr{Y}}(h_1(x,\bullet)+h_2(x,\bullet)|x)+\sum_{x\in A^c} I_{\{x\}}\inf_{y\in\mathscr{Y}}[h_1(x,y)+h_2(x,y)]\\ &\geq \sum_{x\in A} I_{\{x\}}\underline{E}_{\mathscr{Y}}(h_1(x,\bullet)|x)+\sum_{x\in A^c} I_{\{x\}}\inf_{y\in\mathscr{Y}} h_1(x,y)\\ &\quad+\sum_{x\in A} I_{\{x\}}\underline{E}_{\mathscr{Y}}(h_2(x,\bullet)|x)+\sum_{x\in A^c} I_{\{x\}}\inf_{y\in\mathscr{Y}} h_2(x,y)\\ &= (h_1)_A+(h_2)_A,\end{aligned}$$

and, therefore, indeed, by coherence (super-additivity and monotonicity, Theorem 13.31(iv) and (vi)$_{266}$) of $\underline{E}_{\mathscr{X}}$ and the additivity and monotonicity of limits,

$$\begin{aligned}\underline{F}_{\mathrm{s}}(h_1+h_2|\mathscr{X}\times\mathscr{Y}) &= \lim_{A\Subset\mathscr{X}}\underline{E}_{\mathscr{X}}((h_1+h_2)_A)\geq\lim_{A\Subset\mathscr{X}}\underline{E}_{\mathscr{X}}((h_1)_A+(h_2)_A)\\ &\geq\lim_{A\Subset\mathscr{X}}\Big(\underline{E}_{\mathscr{X}}((h_1)_A)+\underline{E}_{\mathscr{X}}((h_2)_A)\Big)\\ &= \lim_{A\Subset\mathscr{X}}\underline{E}_{\mathscr{X}}((h_1)_A)+\lim_{A\Subset\mathscr{X}}\underline{E}_{\mathscr{X}}((h_2)_A))\\ &= \underline{F}_{\mathrm{s}}(h_1|\mathscr{X}\times\mathscr{Y})+\underline{F}_{\mathrm{s}}(h_2|\mathscr{X}\times\mathscr{Y}).\end{aligned}$$

The proof for condition (iii)$_{271}$ of Theorem 13.34$_{271}$ is similar, if somewhat easier, and left to the reader.

So, to prove that $\underline{F}(\bullet|\bullet)$ is coherent, we are now only left to show that $\underline{F}(\bullet|\bullet)$ also satisfies condition (ii)$_{279}$ of Theorem 13.46$_{278}$. In what follows, $I_{\{x\}}$ is a shorthand notation for $I_{\{x\}\times\mathscr{Y}}$ – remember that we use the same notation for a gamble on $\mathbb{G}(\mathscr{X})$ and its cylindrical extension to a gamble in $\mathbb{G}(\mathscr{X}\times\mathscr{Y})$. Consider any $h\in\mathbb{G}(\mathscr{X}\times\mathscr{Y})$, any $z\in\mathscr{X}$ and any $a\in\mathbb{R}$, then we must show that

$$\underline{F}((h-a)I_{\{z\}}|\mathscr{X}\times\mathscr{Y})\begin{cases}\geq 0 \text{ if } a<\underline{F}(h|\{z\}\times\mathscr{Y}),\\ \leq 0 \text{ if } a>\underline{F}(h|\{z\}\times\mathscr{Y}).\end{cases}\tag{13.37}$$

Clearly, by its definition (see Equation (13.35)), $\underline{F}(h|\{z\}\times\mathscr{Y})=\underline{E}_{\mathscr{Y}}(h(z,\bullet)|z)$. What about $\underline{F}((h-a)I_{\{z\}}|\mathscr{X}\times\mathscr{Y})$? Let, for ease of notation, $g:=(h-a)I_{\{z\}}$. If we choose $A\Subset\mathscr{X}$ large enough so as to contain z, then because $g(x,\bullet)=0$ for all $x\in\mathscr{X}\setminus\{z\}$

$$\begin{aligned}g_A &= \sum_{x\in A} I_{\{x\}}\underline{E}_{\mathscr{Y}}(g(x,\bullet)|x)+\sum_{x\in A^c} I_{\{x\}}\inf_{y\in\mathscr{Y}} g(x,y)\\ &= I_{\{z\}}\underline{E}_{\mathscr{Y}}(g(z,\bullet)|z)=I_{\{z\}}[\underline{E}_{\mathscr{Y}}(h(z,\bullet)|z)-a],\end{aligned}$$

where we also used the constant additivity of the coherent $\underline{E}_{\mathscr{Y}}(\bullet|z)$ (Theorem 13.31 (iii)$_{266}$). Therefore, by Equation (13.36),

$$\underline{F}((h-a)I_{\{z\}}|\mathscr{X}\times\mathscr{Y})=\underline{E}_{\mathscr{X}}\left(I_{\{z\}}[\underline{E}_{\mathscr{Y}}(h(z,\bullet)|z)-a]\right).$$

Recall that, to establish coherence of $\underline{F}(\bullet|\bullet)$, we must show that

$$\underline{E}_{\mathcal{X}}\left(I_{\{z\}}[\underline{E}_{\mathcal{Y}}(h(z,\bullet)|z)-a]\right)\begin{cases}\geq 0 \text{ if } a<\underline{E}_{\mathcal{Y}}(h(z,\bullet)|z),\\ \leq 0 \text{ if } a>\underline{E}_{\mathcal{Y}}(h(z,\bullet)|z).\end{cases} \tag{13.38}$$

It is important to observe that $I_{\{z\}}[\underline{E}_{\mathcal{Y}}(h(z,\bullet)|z)-a]$ is not necessarily a real-valued gamble, so we are led to consider three possible cases:

1. When $\underline{E}_{\mathcal{Y}}(h(z,\bullet)|z)\in\mathbb{R}$, then Equation (13.38) follows at once from the monotonicity of the coherent $\underline{E}_{\mathcal{X}}$ on $\mathbb{G}(\mathcal{X})$ (Theorem 13.31(vi)$_{266}$).

2. When $\underline{E}_{\mathcal{Y}}(h(z,\bullet)|z)=-\infty$, then also

$$\underline{E}_{\mathcal{X}}\left(I_{\{z\}}[\underline{E}_{\mathcal{Y}}(h(z,\bullet)|z)-a]\right)=\underline{E}_{\mathcal{X}}\left(-\infty I_{\{z\}}\right)=-\infty, \tag{13.39}$$

where the last equality holds because in the expression for $\underline{E}_{\mathcal{X}}$, that is, Equation (13.30)$_{288}$, we see that $-\infty I_{\{z\}}-\alpha$ cannot dominate any gamble, so the supremum is over the empty set, which yields $-\infty$. So Equation (13.38) holds in this case as well.

3. When $\underline{E}_{\mathcal{Y}}(h(z,\bullet)|z)=+\infty$, then

$$\underline{E}_{\mathcal{X}}\left(I_{\{z\}}[\underline{E}_{\mathcal{Y}}(h(z,\bullet)|z)-a]\right)=\underline{E}_{\mathcal{X}}\left(+\infty I_{\{z\}}\right)\geq 0, \tag{13.40}$$

where the last inequality follows at once from Equation (13.30)$_{288}$. Here too, Equation (13.38) holds.

So condition (ii)$_{279}$ of Theorem 13.46$_{278}$ is indeed satisfied, and therefore, $\underline{F}(\bullet|\bullet)$ is a coherent conditional lower prevision. If we can now show that $\underline{P}(\bullet|\bullet)\leq\underline{F}(\bullet|\bullet)$ on $\operatorname{dom}\underline{P}(\bullet|\bullet)$ and $\underline{F}(\bullet|\bullet)\leq\underline{E}_{\underline{P}}(\bullet|\bullet)$ on $\operatorname{dom}\underline{F}(\bullet|\bullet)=\mathcal{O}$, then the first inequality guarantees that $\underline{P}(\bullet|\bullet)$ avoids sure loss (criterion (C)$_{255}$ with $\underline{Q}(\bullet|\bullet)=\underline{E}_{\underline{F}}(\bullet|\bullet)$) and consequently $\underline{F}(\bullet|\bullet)$ must coincide with $\underline{E}_{\underline{P}}(\bullet|\bullet)$ on $\operatorname{dom}\underline{F}(\bullet|\bullet)$, because $\underline{E}_{\underline{P}(\bullet|\bullet)}(\bullet|\bullet)$ is the point-wise smallest coherent conditional lower prevision that dominates $\underline{P}(\bullet|\bullet)$ (Theorem 13.51(i)$_{282}$ and Corollary 13.52$_{282}$ with $\mathcal{O}=\operatorname{dom}\underline{F}(\bullet|\bullet)$). But if $\underline{F}(\bullet|\bullet)=\underline{E}_{\underline{P}}(\bullet|\bullet)$ on $\operatorname{dom}\underline{F}(\bullet|\bullet)$, then Equations (13.32)$_{289}$ and (13.33)$_{289}$ hold, because of Equations (13.35)$_{290}$ and (13.36)$_{290}$, finishing the proof.

Let us, therefore, first show that $\underline{P}(\bullet|\bullet)\leq\underline{F}(\bullet|\bullet)$ on $\operatorname{dom}\underline{P}(\bullet|\bullet)$. It follows from Equation (13.35)$_{290}$ that for all $f\in\mathcal{K}_{\mathcal{X}}$ and all $x\in\mathcal{X}$, $\underline{F}(f|\{x\}\times\mathcal{Y})=\underline{E}_{\mathcal{Y}}(f(x)|x)=f(x)$, so Equation (13.36)$_{290}$ yields

$$\underline{F}(f|\mathcal{X}\times\mathcal{Y})=\underline{E}_{\mathcal{X}}(f)\geq\underline{P}_{\mathcal{X}}(f)=\underline{P}(f|\mathcal{X}\times\mathcal{Y}),$$

where the inequality follows from Equation (13.30)$_{288}$ ($\underline{E}_{\mathcal{X}}$ dominates $\underline{P}_{\mathcal{X}}$; alternatively, use Theorem 13.47(i)$_{280}$). Similarly, it follows from Equation (13.35)$_{290}$ that for all $x\in\mathcal{X}$ and all $g\in\mathcal{K}_{\mathcal{Y}}(x)$,

$$\underline{F}(g|\{x\}\times\mathcal{Y})=\underline{E}_{\mathcal{Y}}(g|x)\geq\underline{P}_{\mathcal{Y}}(g|x)=\underline{P}(g|\{x\}\times\mathcal{Y}),$$

where the inequality follows from Equation $(13.31)_{289}$ ($\underline{E}_{\mathscr{Y}}(\bullet|x)$ dominates $\underline{P}_{\mathscr{Y}}(\bullet|x)$; alternatively, use Theorem 13.47(i)$_{280}$).

We complete the proof by showing that $\underline{F}(\bullet|\bullet) \leq \underline{E}_{P(\bullet|\bullet)}(\bullet|\bullet)$ on $\operatorname{dom}\underline{F}(\bullet|\bullet) = \mathscr{O}$. Consider any gamble h on $\mathscr{X} \times \mathscr{Y}$. We will only prove that $\underline{E}_{P(\bullet|\bullet)}(h|\mathscr{X} \times \mathscr{Y}) \geq \underline{F}(h|\mathscr{X} \times \mathscr{Y})$, where, by Equation $(13.14)_{260}$, $\underline{E}_{P(\bullet|\bullet)}(h|\mathscr{X} \times \mathscr{Y})$ is equal to

$$\sup\Bigg\{\gamma \in \mathbb{R} : h - \gamma \geq \sum_{k=1}^{n} \lambda_k(f_k - a_k) + \sum_{x\in A} I_{\{x\}} \sum_{k_x=1}^{n_x} \mu_{x,k_x}(g_{x,k_x} - b_{x,k_x}),$$
$$n \in \mathbb{N}, \lambda_k \geq 0, f_k \in \mathscr{K}_{\mathscr{X}}, a_k \in \mathbb{R}, a_k < \underline{P}_{\mathscr{X}}(f_k), A \Subset \mathscr{X},$$
$$n_x \in \mathbb{N}, \mu_{x,k_x} \geq 0, g_{x,k_x} \in \mathscr{K}_{\mathscr{Y}}(x), b_{x,k_x} \in \mathbb{R}, b_{x,k_x} < \underline{P}_{\mathscr{Y}}(g_{x,k_x}|x)\Bigg\}. \tag{13.41}$$

For the other type of inequality $\underline{E}_{P(\bullet|\bullet)}(h|\{x\} \times \mathscr{Y}) \geq \underline{F}(h|\{x\} \times \mathscr{Y})$, the proof is similar but less involved and therefore omitted.

We need only to show that $\underline{E}_{P(\bullet|\bullet)}(h|\mathscr{X} \times \mathscr{Y}) \geq \underline{E}_{\mathscr{X}}(h_A)$ for all $A \Subset \mathscr{X}$ (h_A was defined in Equation $(13.34)_{289}$). Consider any real α and assume that $\alpha < \underline{E}_{\mathscr{X}}(h_A)$. We show that then $\alpha \leq \underline{E}_{P(\bullet|\bullet)}(h|\mathscr{X} \times \mathscr{Y})$. We may clearly assume without loss of generality that $\underline{E}_{\mathscr{X}}(h_A) > -\infty$ (otherwise the desired inequality holds trivially). This then implies that h_A can only assume values in $\mathbb{R} \cup \{+\infty\}$ (because we infer from Equation $(13.30)_{288}$ that otherwise $\underline{E}_{\mathscr{X}}(h_A) = -\infty$). It follows from Equation $(13.30)_{288}$ that there are $n \in \mathbb{N}$, real $\lambda_k \geq 0$, $f_k \in \mathscr{K}_{\mathscr{X}}$ and real $a_k < \underline{P}_{\mathscr{X}}(f_k)$ such that

$$h_A - \alpha \geq \sum_{k=1}^{n} \lambda_k(f_k - a_k). \tag{13.42}$$

Let, for ease of notation, $A_{<+\infty} := \{x \in A : h_A(x) < +\infty\}$, and consider any $\varepsilon > 0$. Then there are three possibilities for any $x \in \mathscr{X}$:

$x \in A_{<+\infty}$. Then $h_A(x) - \varepsilon < \underline{E}_{\mathscr{Y}}(h(x,\bullet)|x)$, so it follows from Equation $(13.31)_{289}$ that there are $n_x \in \mathbb{N}$, real $\mu_{x,k_x} \geq 0$, gambles $g_{x,k_x} \in \mathscr{K}_{\mathscr{Y}}(x)$ and real $b_{x,k_x} < \underline{P}_{\mathscr{Y}}(g_{x,k_x}|x)$ such that $h(x,\bullet) - h_A(x) + \varepsilon \geq \sum_{k_x=1}^{n_x} \mu_{x,k_x}(g_{x,k_x} - b_{x,k_x})$ and, therefore, taking into account Equation (13.42), that

$$h(x,\bullet) - \alpha + \varepsilon \geq \sum_{k=1}^{n} \lambda_k[f_k(x) - a_k] + \sum_{k_x=1}^{n_x} \mu_{x,k_x}(g_{x,k_x} - b_{x,k_x}). \tag{13.43}$$

$x \in A \setminus A_{<+\infty}$. Then $h_A(x) = \underline{E}_{\mathscr{Y}}(h(x,\bullet)|x) = +\infty$, so Equation $(13.31)_{289}$ guarantees that for all real β, and therefore in particular also for $\beta := \alpha - \varepsilon - \sum_{k=1}^{n} \lambda_k[f_k(x) - a_k]$, there are $n_x \in \mathbb{N}$, real $\mu_{x,k_x} \geq 0$, gambles $g_{x,k_x} \in \mathscr{K}_{\mathscr{Y}}(x)$ and real $b_{x,k_x} < \underline{P}_{\mathscr{Y}}(g_{x,k_x}|x)$ such that $h(x,\bullet) - \beta \geq \sum_{k_x=1}^{n_x} \mu_{x,k_x}(g_{x,k_x} - b_{x,k_x})$ and, therefore,

$$h(x,\bullet) - \alpha + \varepsilon \geq \sum_{k=1}^{n} \lambda_k[f_k(x) - a_k] + \sum_{k_x=1}^{n_x} \mu_{x,k_x}(g_{x,k_x} - b_{x,k_x}). \tag{13.44}$$

$x \in A^c$. Then $h_A(x) - \varepsilon = \inf_{y\in\mathcal{Y}} h(x,y) - \varepsilon \leq h(x,\bullet)$ and, therefore, taking into account Equation (13.42)$_{\curvearrowleft}$, we see that

$$h(x,\bullet) - \alpha + \varepsilon \geq \sum_{k=1}^{n} \lambda_k[f_k(x) - a_k]. \tag{13.45}$$

If we compare Equations (13.43)$_{\curvearrowleft}$–(13.45) with Equation (13.41)$_{\curvearrowleft}$, we conclude that $\alpha - \varepsilon \leq \underline{E}_{\underline{P}(\bullet|\bullet)}(h|\mathcal{X}\times\mathcal{Y})$, and because this holds for any $\varepsilon > 0$, indeed also $\alpha \leq \underline{E}_{\underline{P}(\bullet|\bullet)}(h|\mathcal{X}\times\mathcal{Y})$. □

Walley (1991) describes a stronger form of marginal extension, which also invokes *conglomerability* over the conditioning partition. We call this extension, which can be different from the natural extension, the **conglomerable marginal extension**:

$$\underline{F}_c(h|\mathcal{X}\times\mathcal{Y}) := \underline{E}_{\mathcal{X}}\left(\sum_{x\in\mathcal{X}} I_{\{x\}}\underline{E}_{\mathcal{Y}}(h(x,\bullet)|x)\right) \text{ for all } h \in \mathbb{G}(\mathcal{X}\times\mathcal{Y}). \tag{13.46}$$

If $\mathcal{X}$ is a finite set, then the conglomerable marginal extension coincides with the (unconditional) natural extension. It is easy to show that the conglomerable marginal extension, along with $\underline{F}_c(h|\{x\}\times\mathcal{Y}) := \underline{E}_{\mathcal{Y}}(h(x,\bullet)|x)$, is also a coherent extension of $\underline{P}(\bullet|\bullet)$; in fact the proof of Theorem 13.58$_{289}$ applies with $\underline{F}_c(\bullet|\bullet)$ almost unmodified, except of course that one cannot prove that $\underline{F}_c(\bullet|\bullet)$ is a restriction of $\underline{E}_P(\bullet|\bullet)$, as the following counterexample shows.

Example 13.59 Consider the bounded gamble $f\colon (x,y) \mapsto y$ on $\mathcal{X}\times\mathcal{Y} := \mathbb{R}\times[-1,1]$. Let $\underline{P}_{\mathcal{X}}(g) := \inf_{x\in\mathcal{X}} g(x)$ for all $g \in \mathbb{B}(\mathcal{X})$ and $\underline{P}_{\mathcal{Y}}(f(x,\bullet)|x) := 0$ for all $x \in \mathcal{X}$. These (unconditional) lower previsions are coherent (for the inf, see Section 5.4$_{81}$), and therefore, we see in particular that $\underline{E}_{\mathcal{Y}}(f(x,\bullet)|x) = 0$ for all $x \in \mathcal{X}$. Hence, by Theorem 13.58$_{289}$,

$$\begin{aligned}\underline{E}_{\underline{P}(\bullet|\bullet)}(f|\mathcal{X}\times\mathcal{Y}) &= \sup_{A\Subset\mathcal{X}} \underline{E}_{\mathcal{X}}\left(\sum_{x\in A} I_{\{x\}}\underline{E}_{\mathcal{Y}}(f(x,\bullet)|x) + \sum_{x\in A^c} I_{\{x\}} \inf_{y\in\mathcal{Y}} f(x,y)\right)\\ &= \sup_{A\Subset\mathcal{X}} \inf_{x\in\mathcal{X}} [-I_{A^c}] = -1.\end{aligned}$$

However, the conglomerable marginal extension is different:

$$\underline{F}_c(f|\mathcal{X}\times\mathcal{Y}) = \underline{E}_{\mathcal{X}}\left(\sum_{x\in\mathcal{X}} I_{\{x\}}\underline{E}_{\mathcal{Y}}(f(x,\bullet)|x)\right) = \inf_{x\in\mathcal{X}}[0] = 0.$$

We conclude that the marginal – or in other words, natural – extension and the conglomerable marginal extension do not coincide. ♦

13.10 Extending a lower prevision from bounded gambles to conditional gambles

13.10.1 General case

If we have a lower prevision $\underline{Q}$ defined on some set of bounded gambles, we have seen earlier that we can identify it with the (unconditional) lower prevision $\underline{P}(\bullet|\mathcal{X})$, for simplicity also denoted by $\underline{P}$ and defined on $\operatorname{dom}\underline{Q} \times \{\mathcal{X}\}$ by letting $\underline{P}(f|\mathcal{X}) = \underline{P}(f) := \underline{Q}(f)$ for all bounded gambles f in $\operatorname{dom}\underline{Q}$.

Moreover, the notions of avoiding sure loss, coherence and natural extension (to bounded gambles) for the lower prevision $\underline{Q}$ and the (unconditional) lower prevision $\underline{P}$ are in direct correspondence, as we have seen in Theorems 13.21_{258}, 13.27_{264} and 13.56_{286} respectively.

If the lower prevision $\underline{Q}$ avoids sure loss, it therefore makes perfect sense to use the procedure of natural extension to correct and extend it from a lower prevision on bounded gambles to a coherent conditional lower prevision $\underline{E}_{\underline{P}(\bullet|\mathcal{X})}(\bullet|\bullet)$ on $\mathbb{G} \times \mathscr{P}^{\circ}$. As no contradictions can arise, we will from now on denote this natural extension $\underline{E}_{\underline{P}(\bullet|\mathcal{X})}(\bullet|\bullet)$ simply by $\underline{E}_{\underline{Q}}(\bullet|\bullet)$, and its restriction $\underline{E}_{\underline{Q}}(\bullet|\mathcal{X})$ to a(n unconditional) coherent lower prevision simply by $\underline{E}_{\underline{Q}}$.

In this section, we study this natural extension in more detail. This will prepare us for a more detailed discussion on extensions of lower previsions on bounded gambles to (unconditional) lower previsions on unbounded gambles in Chapters 14_{304} and 15_{327}. As 'being the natural extension of' is a transitive relation by Corollaries 4.32_{67} and 13.52_{282}, we may assume without loss of generality that the lower prevision $\underline{Q}$ that is to be extended is actually a coherent lower prevision defined on the set of all bounded gambles $\mathbb{B}$.

Theorem 13.60 *Let $\underline{Q}$ be a coherent lower prevision defined on $\mathbb{B}$. Then its natural extension $\underline{E}_{\underline{Q}}(\bullet|\bullet)$ is given by*

$$\underline{E}_{\underline{Q}}(f|A) = \begin{cases} \sup\left\{\alpha \in \mathbb{R} : (f-\alpha)I_A \geq g,\ g \in \mathbb{B},\ \underline{Q}(g) \geq 0\right\} & \text{if } \underline{Q}(A) > 0 \\ \inf(f|A) & \text{if } \underline{Q}(A) = 0 \end{cases}$$

$$\underline{E}_{\underline{Q}}(f) = \sup\left\{\underline{Q}(g) : g \leq f,\ g \in \mathbb{B}\right\}$$

for all (f,A) in $\mathbb{G} \times \mathscr{P}^{\circ}$. In particular, when $\inf(f|A) = -\infty$, then $\underline{E}_{\underline{Q}}(f|A) = -\infty$.

Proof. We start by constructing a slightly simplified expression for $\underline{E}_{\underline{Q}}(f|A)$, which will be used throughout the proof. The definition of the natural extension in Equation $(13.14)_{260}$ tells us that $\underline{E}_{\underline{Q}}(f|A)$ is equal to

$$\sup\left\{\alpha \in \mathbb{R} : (f-\alpha)I_A \geq \sum_{k=1}^{n} \lambda_k(f_k - a_k),\ n \in \mathbb{N},\ a_k < \underline{Q}(f_k),\ \lambda_k \geq 0,\ f_k \in \mathbb{B}\right\}. \tag{13.47}$$

Assume that there are $n \in \mathbb{N}$, $f_k \in \mathbb{B}$, $a_k < \underline{Q}(f_k)$ and $\lambda_k \geq 0$ such that $(f - \alpha)I_A \geq \sum_{k=1}^{n} \lambda_k(f_k - a_k)$. If we let $g := \sum_{k=1}^{n} \lambda_k(f_k - a_k)$, then it follows from the coherence of the lower prevision $\underline{Q}$ (Theorem 4.13(v), (vi) and (iii)$_{53}$) that either $\underline{Q}(g) \geq \sum_{k=1}^{n} \lambda_k[\underline{Q}(f_k) - a_k] > 0$ (when at least one λ_k is non-zero) or $g = 0$, which tells us that there is some $g \in \mathbb{B}$ with either $\underline{Q}(g) > 0$ or $g = 0$ such that $(f - \alpha)I_A \geq g$. This already implies that

$$\underline{E}_{\underline{Q}}(f|A) = \sup\left\{\alpha \in \mathbb{R} : (f - \alpha)I_A \geq g,\ g \in \mathbb{B},\ \underline{Q}(g) > 0 \text{ or } g = 0\right\} \tag{13.48}$$

and because, for $g = 0$, the supremum equals $\inf(f|A)$, we arrive at

$$= \sup\left\{\alpha \in \mathbb{R} : (f - \alpha)I_A \geq g,\ g \in \mathbb{B},\ \underline{Q}(g) > 0\right\} \vee \inf(f|A). \tag{13.49}$$

It is worth emphasising that the supremum in Equation (13.49) really can become strictly less than $\inf(f|A)$, so we cannot just drop it. For example, if $\underline{Q} = \inf$ (vacuous), f is bounded and A is a proper non-empty subset of $\mathcal{X}$, then there is no g such that $\underline{Q}(g) > 0$ and $(f - \alpha)I_A \geq g$, because $(f - \alpha)I_A$ is zero on A^c and $\underline{Q}(g) > 0$ implies that g is strictly positive everywhere. In this example, the supremum is therefore $-\infty$, whilst $\inf(f|A)$ is finite because of the boundedness of f.

We now turn to the proof of the first equality. Assume that $\underline{Q}(A) > 0$. Then Equation (13.48) immediately implies that

$$\underline{E}_{\underline{Q}}(f|A) \leq \sup\left\{\alpha \in \mathbb{R} : (f - \alpha)I_A \geq g,\ g \in \mathbb{B},\ \underline{Q}(g) \geq 0\right\}.$$

To establish the converse inequality, we fix $\varepsilon > 0$ and rewrite Equation (13.49) in a slightly different manner,

$$\begin{aligned}\underline{E}_{\underline{Q}}(f|A) &= \sup\left\{\alpha \in \mathbb{R} : (f - \alpha)I_A \geq g + \varepsilon I_A,\ g \in \mathbb{B},\ \underline{Q}(g + \varepsilon I_A) > 0\right\} \\ &\qquad \vee \inf(f|A) \\ &\geq \sup\left\{\alpha \in \mathbb{R} : (f - \alpha)I_A \geq g + \varepsilon I_A,\ g \in \mathbb{B},\ \underline{Q}(g + \varepsilon I_A) > 0\right\}\end{aligned}$$

and because $\underline{Q}(g + \varepsilon I_A) \geq \underline{Q}(g) + \varepsilon \underline{Q}(A) > 0$ whenever $\underline{Q}(g) \geq 0$ (due to the super-additivity of the coherent $\underline{Q}$ and our assumption that $\underline{Q}(A) > 0$),

$$\begin{aligned}&\geq \sup\left\{\alpha \in \mathbb{R} : (f - \alpha)I_A \geq g + \varepsilon I_A,\ g \in \mathbb{B},\ \underline{Q}(g) \geq 0\right\} \\ &= \sup\left\{\beta - \varepsilon \in \mathbb{R} : (f - \beta)I_A \geq g,\ g \in \mathbb{B},\ \underline{Q}(g) \geq 0\right\} \\ &= -\varepsilon + \sup\left\{\beta \in \mathbb{R} : (f - \beta)I_A \geq g,\ g \in \mathbb{B},\ \underline{Q}(g) \geq 0\right\},\end{aligned}$$

where we used the substitution $\beta := \alpha + \varepsilon$. As this inequality holds for every $\varepsilon > 0$, it must also hold for $\varepsilon = 0$.

Next, we consider the case that $\underline{Q}(A) = 0$ and prove that then $\underline{E}_Q(f|A) = \inf(f|A)$. It follows from Equation (13.49) that we only need to prove that $\underline{E}_Q(f|A) \leq \inf(f|A)$. If $\inf(f|A) = -\infty$, this follows trivially from Equation (13.49) (there is no $\alpha \in \mathbb{R}$ satisfying the condition, so the supremum there is taken over the empty set). If $\inf(f|A) > -\infty$, then again by Equation (13.49) it suffices to show that for all $\alpha > \inf(f|A)$ and all $g \in \mathbb{B}$,

$$(f - \alpha)I_A \geq g \Rightarrow \underline{Q}(g) \leq 0. \tag{13.50}$$

So fix $\alpha > \inf(f|A)$ and $g \in \mathbb{B}$, and assume that $(f - \alpha)I_A \geq g$. This implies that g is non-positive on A^c and, therefore, $g \leq \sup(g|A)I_A$, whence indeed

$$\underline{Q}(g) \leq \underline{Q}(\sup(g|A)I_A) = \begin{cases} \sup(g|A)\underline{Q}(A) & \text{if } \sup(g|A) \geq 0 \\ \sup(g|A)\overline{Q}(A) & \text{if } \sup(g|A) < 0 \end{cases} \leq 0,$$

where the first inequality follows from the monotonicity (Theorem 4.13(iv)$_{53}$) and the equality from the homogeneity properties (use Theorem 4.13(vi)$_{53}$ and conjugacy) of the coherent lower prevision $\underline{P}$. The last inequality follows from $\underline{Q}(A) = 0$.

To prove the second equality, we start from the first one for $A = \mathcal{X}$,

$$\begin{aligned} \underline{E}_Q(f) = \underline{E}_Q(f|\mathcal{X}) &= \sup\left\{\alpha \in \mathbb{R} : f - \alpha \geq g - a,\ g \in \mathbb{B},\ a \in \mathbb{R},\ a < \underline{Q}(g)\right\} \\ &= \sup\left\{\beta + a \colon \beta \in \mathbb{R}, f \geq g + \beta,\ g \in \mathbb{B},\ a \in \mathbb{R},\ a < \underline{Q}(g)\right\} \\ &= \sup\left\{\beta + \underline{Q}(g) \colon \beta \in \mathbb{R}, f \geq g + \beta,\ g \in \mathbb{B}\right\} \\ &= \sup\left\{\underline{Q}(\beta + g) \colon \beta \in \mathbb{R}, f \geq g + \beta,\ g \in \mathbb{B}\right\} \\ &= \sup\left\{\underline{Q}(h) \colon f \geq h,\ h \in \mathbb{B}\right\}, \end{aligned}$$

which completes the proof. □

An important observation is that $\underline{E}_Q(\bullet|A)$ is vacuous when $\underline{Q}(A) = 0$. This agrees with the existing conditional theory for bounded gambles; see, for instance, Walley (1991, Section 6.8.2(c)).

We now look at a number of very simple examples.

13.10.2 Linear previsions and probability charges

As already discussed in Section 9.4$_{183}$, the natural extension of a probability charge μ to bounded gambles is simply given by its lower S-integral, lower Lebesgue integral or Choquet integral. The following proposition characterises the natural extension to arbitrary gambles.

Proposition 13.61 *Let $\mathcal{F}$ be a field on $\mathcal{X}$, and let μ be a probability charge on $\mathcal{F}$. The natural extension of the prevision P_μ to an (unconditional) lower prevision on $\mathbb{G}$ is given by*

$$\underline{E}_{P_\mu}(f) = \begin{cases} \sup\left\{D\int g \,\mathrm{d}\mu : g \in \mathrm{span}(\mathcal{F}),\ g \leq f\right\} & \textit{if } \inf f \in \mathbb{R} \\ -\infty & \textit{otherwise.} \end{cases} \tag{13.51}$$

Proof. Immediate, by Theorem 13.60$_{295}$, Corollary 13.52$_{282}$ (transitivity of natural extension), Definition 8.27$_{171}$ (definition of the Lebesgue integral in terms of the Dunford integral on simple gambles) and Theorem 8.32$_{174}$ (the Dunford integral on simple gambles equals natural extension). □

When $\mathscr{F}$ is a σ-field, μ is a σ-additive probability charge – a probability measure in the usual sense – and f is $\mathscr{F}$-measurable in the sense that $\{f \geq t\} \in \mathscr{F}$ for all $t \in \mathbb{R}$, then the Lebesgue integral of f with respect to μ can be defined as

$$L\int f \,\mathrm{d}\mu := \sup\left\{D\int g\,\mathrm{d}\mu : g \in \operatorname{span}(\mathscr{F}),\ g \leq f^+\right\} - \sup\left\{D\int g\,\mathrm{d}\mu : g \in \operatorname{span}(\mathscr{F}),\ g \leq f^-\right\},$$

provided the difference on the right-hand side is well defined, where we have used the previously introduced notations $f^+ := 0 \vee f$ and $f^- := -(0 \wedge f)$. In other words, natural extension generalises Lebesgue integration to arbitrary probability charges and arbitrary – but non-negative – gambles. The generalisation actually holds not just for non-negative gambles but also for any gambles that are bounded below, because both natural extension and Lebesgue integration satisfy constant additivity.

However, the natural extension for gambles that are not bounded below is always $-\infty$, which is not always in agreement with the Lebesgue integral. This points to an interesting 'limitation' in the definition of natural extension, which we will overcome in Chapters 14$_{304}$ and 15$_{327}$ and more specifically in Section 15.10.1$_{358}$.

Example 13.62 Let $\mathscr{X} = \mathbb{N}_{>0}$, and let μ be the probability measure on the power set of $\mathscr{X}$ that is uniquely determined by $\mu(\{x\}) := 2^{-x}$ for all $x \in \mathbb{N}_{>0}$. Consider the gamble f defined by $f(x) := -x$ for all $x \in \mathbb{N}_{>0}$. Because, in this case, the Lebesgue integral simply coincides with expectation, we get

$$L\int f\,\mathrm{d}\mu = \sum_{x\in\mathbb{N}_{>0}} (-x)2^{-x} = -2.$$

To see this, let $\alpha := \sum_{n=1}^{+\infty} n2^{-n}$, then

$$2\alpha = \sum_{n=1}^{+\infty} n2^{-n+1} = \sum_{k=0}^{+\infty} (k+1)2^{-k} = \alpha + \sum_{k=0}^{+\infty} 2^{-k} = \alpha + 2.$$

But $\underline{E}_{P_\mu}(f) = -\infty$ because f is not bounded below. In contrast, $L\int -f\,\mathrm{d}\mu = 2 = \underline{E}_{P_\mu}(-f)$. To see why, note that $-f$ can be approximated from below by the sequence of bounded gambles $g_n(x) := \min\{-f(x), n\}$ and use Theorem 13.60$_{295}$. ◆

13.10.3 Vacuous lower previsions

Let us consider the vacuous lower prevision $\underline{P}_B$ relative to some non-empty subset B of $\mathscr{X}$. We know from the discussion in Section 5.4$_{81}$ that $\underline{P}_B$ is a coherent lower prevision on the set $\mathbb{B}$ of all bounded gambles. When we look at the natural extension to all gambles of this vacuous lower prevision $\underline{P}_B$ defined on bounded gambles, we arrive at a first apparent contradiction: it does not entirely correspond to the vacuous conditional lower prevision relative to B discussed in Example 13.10_{242}.

Proposition 13.63 *Consider a non-empty subset B of $\mathscr{X}$. Then the natural extension of $\underline{P}_B$ to a conditional lower prevision on $\mathbb{G} \times \mathscr{P}^\circ$ is given by*

$$\underline{E}_{\underline{P}_B}(f|A) = \begin{cases} \inf(f|B) & \text{if } \inf(f|A) > -\infty \text{ and } B \subseteq A, \\ \inf(f|A) & \text{otherwise.} \end{cases}$$

Proof. We invoke Theorem 13.60_{295} and distinguish between three cases.

First of all, if $\inf(f|A) = -\infty$, then also $\underline{E}_{\underline{P}_B}(f|A) = -\infty$, by Theorem 13.60_{295}. In the remainder of the proof, we therefore assume that f is bounded below on A.

When $\underline{P}_B(A) = 0$ or, equivalently, when $B \not\subseteq A$, then obviously $\underline{E}_{\underline{P}_B}(f|A) = \inf(f|A)$, again by Theorem 13.60_{295}.

Finally, when $\underline{P}_B(A) > 0$ or, equivalently, when $B \subseteq A$, then we infer from Theorem 13.60_{295} that

$$\underline{E}_{\underline{P}_B}(f|A) = \sup_{g \in \mathbb{B}, \underline{P}_B(g) \geq 0} \sup \left\{ \alpha \in \mathbb{R} : (f - \alpha) I_A \geq g \right\}, \tag{13.52}$$

and we have to prove that $\underline{E}_{\underline{P}_B}(f|A) = \inf(f|B)$.

On the one hand, consider the bounded gamble $h := \left[\inf(f|A \setminus B) - \inf(f|B)\right] I_{A \setminus B}$, then $\underline{P}_B(h) \geq 0$ (actually, $\underline{P}_B(h) = 0$) and

$$\left[f - \inf(f|B)\right] I_A = \left[f - \inf(f|B)\right] I_B + \left[f - \inf(f|B)\right] I_{A \setminus B} \geq 0 + h = h,$$

whence $\underline{E}_{\underline{P}_B}(f|A) \geq \inf(f|B)$, using Equation (13.52).

For the converse inequality, it suffices to prove that for any $\alpha > \inf(f|B)$ and $g \in \mathbb{B}$,

$$(f - \alpha) I_A \geq g \Rightarrow \inf(g|B) < 0.$$

Indeed, it follows from $(f - \alpha) I_A \geq g$ and $B \subseteq A$ that

$$\inf(g|B) \leq \inf((f - \alpha) I_A|B) = \inf(f - \alpha|B) = \inf(f|B) - \alpha < 0,$$

so indeed $\underline{E}_{\underline{P}_B}(f|A) \leq \inf(f|B)$. □

The special case $B = \mathscr{X}$ leads us to the coherent conditional lower prevision $\inf(\bullet|\bullet)$ on $\mathbb{G} \times \mathscr{P}^\circ$, which is the natural extension of the coherent lower prevision $\underline{P}_{\mathscr{X}} = \inf$.

We can also look at the natural extension of the lower prevision $\underline{P}_B$ to an (unconditional) lower prevision. This yields, for any $f \in \mathbb{G}$,

$$\underline{E}_{\underline{P}_B}(f) = \begin{cases} \inf(f|B) & \text{if } \inf f \in \mathbb{R} \\ -\infty & \text{if } \inf f = -\infty, \end{cases}$$

in agreement with Example 13.67_{302} further on.

13.10.4 Lower previsions associated with proper filters

Consider the coherent lower prevision $\underline{P}_{\mathscr{F}}$ on $\mathbb{B}$ associated with a proper filter $\mathscr{F}$, defined earlier in Equation $(5.5)_{83}$:

$$\underline{P}_{\mathscr{F}}(f) = \sup_{F\in\mathscr{F}} \inf_{x\in F} f(x) \text{ for any bounded gamble } f \in \mathbb{B}.$$

Proposition 13.64 *Let $\mathscr{F}$ be a proper filter. The natural extension of $\underline{P}_{\mathscr{F}}$ on $\mathbb{B}$ to an (unconditional) lower prevision on $\mathbb{G}$ is given by*

$$\underline{E}_{\underline{P}_{\mathscr{F}}}(f) = \begin{cases} \sup_{F\in\mathscr{F}} \inf_{x\in F} f(x) & \textit{if } \inf f \in \mathbb{R} \\ -\infty & \textit{otherwise.} \end{cases} \tag{13.53}$$

Proof. We invoke Theorem 13.60_{295} to find that

$$\underline{E}_{\underline{P}_{\mathscr{F}}}(f) = \sup\left\{\underline{P}_{\mathscr{F}}(g) \colon g \leq f,\ g \in \mathbb{B}\right\} \text{ for all gamble } f \text{ on } \mathscr{X}. \tag{13.54}$$

If f is not bounded below, then there is no bounded gamble g for which $g \leq f$ and, therefore, the supremum on the right-hand side assumes the value $-\infty$. Assume, therefore, that f is bounded below but not bounded above. For any bounded gamble $g \leq f$, obviously,

$$\underline{P}_{\mathscr{F}}(g) = \sup_{F\in\mathscr{F}} \inf_{x\in F} g(x) \leq \sup_{F\in\mathscr{F}} \inf_{x\in F} f(x),$$

so we infer from Equation (13.54) that $\underline{E}_{\underline{P}_{\mathscr{F}}}(f) \leq \sup_{F\in\mathscr{F}} \inf_{x\in F} f(x)$. To prove the converse inequality, define, for every $F \in \mathscr{F}$, the bounded gamble f_F on $\mathscr{X}$ by letting

$$f_F(x) := \begin{cases} \inf_{x'\in F} f(x') & \text{if } x \in F \\ \inf f & \text{otherwise,} \end{cases}$$

which is guaranteed to be a bounded gamble because f is bounded below. Moreover, it satisfies $f_F \leq f$. Clearly, $\underline{P}_{\mathscr{F}}(f_F) = \inf_{x\in F} f(x)$, and therefore, we infer from Equation (13.54) that $\underline{E}_{\underline{P}_{\mathscr{F}}}(f) \geq \sup\left\{\underline{P}_{\mathscr{F}}(f_F) \colon F \in \mathscr{F}\right\} = \sup_{F\in\mathscr{F}} \inf_{x\in F} f(x)$. □

13.10.5 Limits inferior

As another interesting example, we consider a directed set $\mathscr{A}$ and the coherent lower prevision lim inf on $\mathbb{B}(\mathscr{A})$, introduced in Section 5.3_{80}. We look at its natural extension $\underline{E}_{\liminf}$ to an (unconditional) lower prevision on $\mathbb{G}(\mathscr{A})$.

Proposition 13.65 *Consider a directed set $\mathscr{A}$. The natural extension of the coherent lower prevision* lim inf *to an (unconditional) lower prevision on $\mathbb{G}(\mathscr{A})$ is given by*

$$\underline{E}_{\liminf}(f) = \begin{cases} \liminf f & \textit{if } \inf f \in \mathbb{R} \\ -\infty & \textit{otherwise.} \end{cases}$$

Proof. We invoke Theorem 13.60$_{295}$ to find that

$$\underline{E}_{\liminf}(f) = \sup\{\liminf g : g \leq f, g \in \mathbb{B}(\mathscr{A})\} \text{ for all gambles } f \text{ on } \mathscr{A}. \quad (13.55)$$

If f is not bounded below, then there is no bounded gamble g for which $g \leq f$ and, therefore, the supremum on the right-hand side assumes the value $-\infty$.

Assume, therefore, that f is bounded below but not bounded above. As for any bounded gamble $g \leq f$, it follows from the definition of the lim inf operator that $\liminf g = \sup_\alpha \inf_{\alpha \leq \beta} g(\beta) \leq \sup_\alpha \inf_{\alpha \leq \beta} f(\beta) = \liminf f$, we infer from Equation (13.55) that $\underline{E}_{\liminf}(f) \leq \liminf f$.

To prove the converse inequality, define, for every $R \in \mathbb{R}$, the bounded gamble f_R on $\mathscr{A}$ by letting $f_R(\alpha) := \min\{R, \inf_{\alpha \leq \beta} f(\beta)\}$ for all $\alpha \in \mathscr{A}$ (this is a bounded gamble because f is bounded below). Also,

$$\begin{aligned}\liminf f_R &= \sup_\alpha \inf_{\alpha \leq \beta} f_R(\beta) = \sup_\alpha \inf_{\alpha \leq \beta} \min\{R, \inf_{\beta \leq \gamma} f(\gamma)\} \\ &= \sup_\alpha \min\{R, \inf_{\alpha \leq \beta} \inf_{\beta \leq \gamma} f(\gamma)\} = \sup_\alpha \min\{R, \inf_{\alpha \leq \beta} f(\beta)\} = \sup_\alpha f_R(\alpha). \quad (13.56)\end{aligned}$$

Because clearly $f_R \leq f$ for all $R \in \mathbb{R}$, we infer from Equation (13.55) that

$$\begin{aligned}\underline{E}_{\liminf}(f) &\geq \sup\left\{\liminf f_R : R \in \mathbb{R}\right\} \\ &= \sup_R \sup_\alpha f_R(\alpha) = \sup_\alpha \sup_R f_R(\alpha) = \sup_\alpha \inf_{\alpha \leq \beta} f(\beta) = \liminf f,\end{aligned}$$

where we also used Equation (13.56). □

We conclude that the natural extension to gambles $\underline{E}_{\liminf}$ of the lim inf operator on bounded gambles need not coincide with the lim inf operator on gambles: $\underline{E}_{\liminf}$ and lim inf differ on all gambles (real nets) f that are unbounded below and have $\liminf f > -\infty$. Let us show that this is indeed possible.

Example 13.66 Consider the set $\mathbb{Z}$ of all integers, which is directed by the natural ordering. Then the net (gamble) f, defined by

$$f(k) = \begin{cases} k & \text{if } k < 0 \\ 2^{-k} & \text{if } k \geq 0, \end{cases}$$

is unbounded below ($\inf f = -\infty$) but $\liminf f = 0$. ♦

13.11 The need for infinity?

As always when developing new theories, it is instructive to look back and reflect on what has been achieved, in order to see whether the results make sense. The purpose of this section is to discuss whether or not is reasonable that conditional lower previsions can assume infinite values, which is something we found emerging from the new theory. Avoiding these infinite values seems to be extremely difficult, not just in the theory presented in this chapter but also in many of the theories it aims to generalise, such as that of expectation – linear previsions – and integration.

As we have already indicated previously in this chapter, $\underline{P}(f|A) = -\infty$ is taken to mean that contingent on A, our subject is not willing to buy the gamble f for any real price $t \in \mathbb{R}$. If the subject is to be coherent, such an assessment is only possible for f that are unbounded below on A. It may indeed be eminently reasonable for a subject not to accept the risk of losing an uncertain amount of utility on which there is no bound. It, therefore, seems quite natural and reasonable to admit $-\infty$ as a supremum buying price in such cases. In perfect accordance with this, Example 13.8_{241} has made clear that, even when $\underline{P}(\bullet|\bullet)$ is real-valued, $\underline{E}_{\underline{P}(\bullet|\bullet)}(\bullet|\bullet)$ can take the value $-\infty$. The next example, due to Troffaes (2005, pp. 214), shows that this seems to be the rule, rather than the exception, and provides us with further motivation for including $-\infty$ in the range of conditional lower previsions.

Example 13.67 Let $\underline{Q}$ be any coherent lower prevision on $\mathbb{B}$, and consider its natural extension $\underline{E}_{\underline{Q}}$ to an (unconditional) lower prevision on $\mathbb{G}$, as described in Section $13.10.1_{295}$. Then $\underline{E}_{\underline{Q}}$ will take the value $-\infty$ in any gamble f that is unbounded below. Indeed, we infer from Theorem 13.60_{295} that

$$\underline{E}_{\underline{Q}}(f) = \sup\left\{\underline{Q}(g) : g \leq f,\, g \in \mathbb{B}\right\} = \sup \emptyset = -\infty.$$

Similarly, $\overline{E}_{\underline{Q}}$ will assume the value $+\infty$ in any gamble f that is unbounded above. ♦

The other extreme case, $\underline{P}(f|A) = +\infty$, is taken to mean that contingent on A, the subject is willing to buy the gamble f at any price. Clearly, such behaviour is far from being reasonable. Yet, we have argued in Section $13.3.1_{240}$, and demonstrated in Example 13.9_{241}, that the rationality requirements for acceptability in Axiom 13.1_{236} do not exclude such infinite supremum buying prices for unbounded gambles. The following example, again due to Troffaes (2005, pp. 214–215), provides similar arguments directly in terms of conditional lower previsions.

Example 13.68 Let $\mathscr{X} := \mathbb{R}_{\geq 0}$ and define the sequence of gambles $f_n \in \mathbb{G}$ by

$$f_n(x) := \min\{n, x\} = \begin{cases} x & \text{if } x < n, \\ n & \text{otherwise.} \end{cases}$$

Define the (unconditional) lower prevision $\underline{P}(\bullet|\bullet)$ on $\left\{f_n : n \in \mathbb{N}\right\} \times \{\mathscr{X}\}$ by $\underline{P}(f_n|\mathscr{X}) := \sup f_n = n$ for all $n \in \mathbb{N}$. This $\underline{P}(\bullet|\bullet)$ is coherent as a restriction of the coherent (unconditional) lower prevision defined by $\underline{Q}(g|\mathscr{X}) := \lim_{n\to+\infty} \inf_{x\geq n} g(x)$ for all gambles $g \in \mathbb{G}$ (use Example 13.10_{242} as well as Propositions $13.39(\text{ii})_{273}$ and $13.43(\text{ii})_{276}$). Clearly, $X \geq f_n(X)$ for all $n \in \mathbb{N}$.[10] By coherence of $\underline{P}(\bullet|\bullet)$ and its

[10] Remember that we write X also for the identity gamble.

the natural extension $\underline{E}_{\underline{P}(\bullet|\bullet)}(\bullet|\bullet)$, this implies that

$$\underline{E}_{\underline{P}(\bullet|\bullet)}(X|\mathcal{X}) \geq \underline{E}_{\underline{P}(\bullet|\bullet)}(f_n|\mathcal{X}) = \underline{P}(f_n|\mathcal{X}) = n \text{ for all } n \in \mathbb{N}$$

and therefore $\underline{E}_{\underline{P}(\bullet|\bullet)}(X|\mathcal{X}) = +\infty$. ♦

Seidenfeld et al. (2009) provide many further examples for linear previsions. In particular, they show, amongst other things, that the linear prevision of an unbounded gamble is not uniquely determined through integration with respect to a probability charge.

So, perhaps something is wrong with our axioms of acceptability? Let us turn to a well-known example involving unbounded utility, in order to get more insight into what might be going on here. The Saint Petersburg paradox, introduced by Bernoulli (1713), provides an instance of a gamble, unbounded above, that is possibly worth to be bought at any price – although of course this does not go without any controversy: consider a fair coin, flipped until it comes up tails, and assume you will then receive 2^n utility, where n is the total number of flips. The expected value of this transaction is $\sum_{n=0}^{+\infty} \frac{1}{2^n} 2^n = +\infty$. Common sense might tell us we should not stake all our money on this gamble. Unless the subject has access to unlimited resources, a lower prevision of $+\infty$ for this gamble is unrealistic. The classical solution to this paradox is that our choicc of utility is wrong and that we should, for instance, consider the log of monetary value for its utility, instead of choosing a linear utility function. Of course, even when taking the log, it is easy to come up with another example that admits infinite expectation again. In conclusion, as soon as utility is unbounded, it is always possiblc to come up with gambles that yield infinite expectation.

So, we will also have to allow for the $+\infty$ value, because it seems to be the only way to arrive at a complete theory that does not exclude particular gambles. One pragmatic view could be the following. The rationality axioms are only an approximation of how agents really ought to behave. This is evident even in the bounded case; for example, see Clemen and Reilly (2001, Chapter 14). When utility becomes unbounded, we accept these rationality axioms as a matter of convenience in modelling a particular problem at hand. If things really matter at the infinite end, as with the Saint Petersburg paradox, perhaps one should rethink one's choice of utility.

Having said this, supremum buying prices that are $+\infty$ do not seem to prevail that commonly in practice. In fact, there are coherent conditional lower previsions that do not take the value $+\infty$ at all, such as the vacuous conditional lower previsions relative to subsets, and any convex combination of these.

Given all this, perhaps, starting from the well-understood bounded case, and taking limits, might lead to more practical answers. We will pursue this route in the next few chapters. Using the general framework depicted in the present chapter, we will investigate there how lower previsions on bounded gambles can be extended to gambles using various limit arguments. In doing so, we will also be able to characterise those gambles for which infinity can be avoided and we solve at least partly some of the issues raised by Seidenfeld et al. (2009).

14

Lower previsions for essentially bounded gambles

In Chapter 13, we have explained how to define and generalise notions of avoiding sure loss, coherence and natural extension for conditional lower previsions defined on gambles that need not be bounded.

In this and the next chapter, we investigate further ways of extending coherent lower previsions that are defined on bounded gambles to (unconditional) coherent lower previsions on all gambles. Not to unduly complicate matters, we restrict ourselves to the unconditional case, also because conditional coherent lower previsions can always be determined from unconditional ones through natural extension, as discussed in Chapter 13_{235}.

The initial ideas for these extensions were published in Troffaes and De Cooman (2002a, 2003), and treated in detail in Troffaes (2005, Chapter 5). The present chapter closely follows Troffaes (2005, Section 5.3), and the next one is based on the discussion in Troffaes (2005, Sections 5.4 and 5.5). In this book, we improve on and extend our earlier results in numerous ways. Most notably, in Section 15.7_{352}, we prove equivalence of Choquet integrability and previsibility under 2-monotonicity, and in Sections 14.5_{322} and 15.10_{358}, we provide new practical examples that should be of interest to anyone who seeks to apply the theory. In Section 14.4_{316}, we also link back to our earlier discussion of sets of acceptable gambles (Section 13.2_{236}).

At first sight, the study of alternative techniques for extending coherent lower previsions from bounded gambles to arbitrary gambles might seem like a pointless exercise, because natural extension already provides one way of performing such an extension. However, for practical applications, one quickly runs into the problem that natural extension can be too conservative when applied to unbounded gambles: for instance, as shown in Example 13.67_{302} near the end of Chapter 13, the natural extension of *any* coherent lower prevision on bounded gambles will assume the value $-\infty$

Lower Previsions, First Edition. Matthias C.M. Troffaes and Gert de Cooman.

on *all* gambles that are unbounded below. Such conservative inferences are generally speaking not very useful.

This leaves us with a number of important questions: (i) if we have a coherent lower prevision defined on bounded gambles, can it be extended to a(n unconditional) coherent lower prevision on gambles in such a way that it remedies the perceived shortcomings of natural extension, and if so, (ii) to what set of gambles; and furthermore, (iii) how are such extensions related to natural extension?

A first step towards extending lower previsions to unbounded gambles is based on the observation that a coherent lower prevision may be invariant under a particular change of values of a bounded gamble, as in the following example, taken from Troffaes (2005, p. 225).

Example 14.1 You throw a pebble and when it lands, the distance is measured between its centre and a reference point near you. Let the random variable X be the measured distance (in metres). After the measurement $X = x$ you – the subject – receive a reward of $f(x) := \min\{x, 100\}$ (in units of some linear utility).

Now consider the bounded gamble defined by $f'(x) := f(x)$ for every $x \leq 100$ and $f'(x) := -100$ otherwise. If you are very confident that you cannot throw a pebble further than 100 m, then f' is equivalent with f in that you will accept f if and only if you accept f'. Changing the value of the bounded gamble f in any $x \geq 100$ does not alter the reward you expect from it. ♦

This example shows that, depending on your beliefs, gambles can sometimes be modified at some points without changing your expected reward. There is a similar phenomenon in the theory of integration: for example, changing the value of the integrand in a countable number of points of the real line does not change the value of a Riemann integral. Therefore, it does not matter if the value of integrand is changed in these points, or whether it is bounded on such a countable set of points.

In this chapter and the next, $\underline{P}$ denotes a coherent lower prevision on $\mathbb{B}$ and $\overline{P}$ is its conjugate upper prevision, intended to represent a subject's information about the value that a random variable X assumes in the set $\mathcal{X}$. This is not really a restriction because, if we start out with a lower prevision $\underline{Q}$ that is defined on a smaller set of bounded gambles and that avoids sure loss, then we can always end up with a coherent lower prevision $\underline{P}$ defined on all bounded gambles by taking the natural extension of $\underline{Q}$.

In fact, in the rest of this book, we only rely on the coherence of $\underline{P}$ and nothing else – apart from the fact that we will sometimes investigate if n-monotonicity simplifies matters. We will make extensive use of the properties of coherent lower and upper previsions, listed in Theorem 4.13_{53}.

So in summary, our aim in the rest of this book is to find interesting ways of extending a coherent $\underline{P}$ to a(n unconditional) coherent lower prevision on gambles.

14.1 Null sets and null gambles

Definition 14.2 (Null sets) *A subset A of $\mathcal{X}$ is called **$\underline{P}$-null** if $\overline{P}(A) = 0$. The set of all $\underline{P}$-null sets is denoted by $\mathcal{N}_{\underline{P}}$.*

The following theorem assures us that $\underline{P}$-null sets are just those sets for which our subject is **practically certain** – because prepared to bet at all odds on the fact – that they do not contain the random variable X.

Theorem 14.3 *A set A is a $\underline{P}$-null set if and only if $\underline{P}(-KI_A + \varepsilon I_{A^c}) > 0$ for all $K \geq 0$ and $\varepsilon > 0$.*

Proof. 'if'. If A could contain the actual value of the random variable X, then accepting the bounded gamble $-KI_A + \varepsilon I_{A^c}$ could result in an arbitrary large loss by choosing K sufficiently large and ε sufficiently small. Hence, intuitively, it is clear that the condition yields a sufficient condition for A to be a $\underline{P}$-null set. Mathematically, this follows from the coherence of $\underline{P}$ (see Theorem 4.13(v) and (vi)$_{53}$):

$$0 < \underline{P}(-KI_A + \varepsilon I_{A^c}) \leq K\underline{P}(-I_A) + \varepsilon \overline{P}(A^c),$$

and therefore, by conjugacy, $K\overline{P}(A) < \varepsilon \overline{P}(A^c)$ for all $K \geq 0$ and $\varepsilon > 0$. This can only be satisfied if $\overline{P}(A) = 0$. Indeed, if $\overline{P}(A) \neq 0$, then any $K > \varepsilon \overline{P}(A^c)/\overline{P}(A)$ will violate the inequality.

'only if'. Assume that A is $\underline{P}$-null, then observe that, by coherence and conjugacy, $\underline{P}(-I_A) = -\overline{P}(A) = 0$ and $\underline{P}(A^c) = \underline{P}(1 - I_A) = 1 + \underline{P}(-I_A) = 1$ (see Theorem 4.13(iii)$_{53}$). Using these equalities, we find that indeed

$$\underline{P}(-KI_A + \varepsilon I_{A^c}) \geq K\underline{P}(-I_A) + \varepsilon \underline{P}(A^c) = \varepsilon > 0 \text{ for all } K \geq 0 \text{ and } \varepsilon > 0,$$

where we invoked Theorem 4.13(v) and (vi)$_{53}$. □

The collection $\mathscr{N}_{\underline{P}}$ of null sets has interesting properties.

Proposition 14.4 *The following statements hold:*

(i) *$\emptyset$ is a $\underline{P}$-null set, and $\mathscr{X}$ is not a $\underline{P}$-null set.*

(ii) *If A is a $\underline{P}$-null set and $B \subseteq A$, then B is a $\underline{P}$-null set.*

(iii) *If $A_1, \ldots, A_n$ are $\underline{P}$-null sets, then $\bigcup_{i=1}^n A_i$ is a $\underline{P}$-null set.*

Proof. (i) follows from Theorem 4.13(ii)$_{53}$.
(ii) follows from Theorem 4.13(i) and (iv)$_{53}$.
(iii) follows from Theorem 4.13(i) and (v)$_{53}$. □

Corollary 14.5 *The following statements hold:*

(i) *$\mathscr{N}_{\underline{P}}$ is a proper ideal of subsets of $\mathscr{X}$.*

(ii) *$(\mathscr{N}_{\underline{P}}, \subseteq)$ is a directed set.*

Proof. (i). Simply observe that (i)–(iii) in Proposition 14.4 are the defining properties of a proper ideal of subsets, as defined in Section 1.4$_5$.

(ii). The statement follows from Proposition 14.4(iii). Indeed, if A and $B \in \mathscr{N}_{\underline{P}}$, then $A \subseteq C$ and $B \subseteq C$ for $C = A \cup B \in \mathscr{N}_{\underline{P}}$. □

These results imply that the complements of null sets constitute a proper filter (and in particular also a directed set). As discussed in Section 1.5_7, such directed sets – and in particular proper filters, see Example 1.11_8 – can be used to introduce Moore–Smith limits (Moore and Smith, 1922). We will use this idea in Section 14.3_{311} to define essential infima and suprema as limits with respect to this directed set (or proper filter).

A gamble will be called null if the subject is practically certain – because prepared to bet at all odds, in the sense of Theorem 14.3, on the fact – that its absolute value will not exceed any positive real number $\varepsilon > 0$, however, small.

Definition 14.6 (Null gambles) *We call a gamble f $\underline{P}$-null if $\overline{P}(\{|f| > \varepsilon\}) = 0$ for every $\varepsilon > 0$. A bounded gamble f is called **$\underline{P}$-null** if it is a $\underline{P}$-null gamble. The set of all $\underline{P}$-null gambles is denoted by $\mathbb{G}^0_{\underline{P}}$.*

In this definition, why not use the requirement that the subject should be practically certain that f is zero or, in other words, that

$$\overline{P}(\{|f| > 0\}) = 0 \text{ or, equivalently, } \underline{P}(\{|f| = 0\}) = 1,$$

for a gamble to be null? It is clear that this alternative condition is at least as strong as – implies – the one we use in Definition 14.6, as $\{|f| > \varepsilon\} \subseteq \{|f| \neq 0\}$ for all $\varepsilon > 0$. That it is generally strictly stronger follows from the fact that not all coherent lower previsions satisfy monotone convergence, as we have, for instance, seen in Proposition 5.12_{91}. The following example illustrates that this alternative definition can be too restrictive. It is an adaptation from Bhaskara Rao and Bhaskara Rao (1983, Proposition 4.2.7(ii) and Example 2.3.5(1)) and builds on the arguments set out in Example 5.13_{92}.

Example 14.7 Consider a random variable X that has only natural numbers as its possible values, so $\mathcal{X} = \mathbb{N}_{>0}$. We model a subject's belief that X is infinitely large, that is, larger than any natural number anyone cares to name. The corresponding upper prevision $\overline{P}$, therefore, satisfies $\overline{P}(A) := 0$ if A is finite and $\overline{P}(A) := 1$ otherwise. We have found in Example 5.13_{92} (see also Corollary 5.9_{86}) that necessarily $\underline{P} = \underline{P}_{\mathcal{F}}$, where

$$\mathcal{F} = \left\{A \subseteq \mathbb{N}_{>0} : A^c \text{ is finite}\right\}$$

is the proper filter made up of all co-finite subsets of $\mathbb{N}_{>0}$ and

$$\overline{P}_{\mathcal{F}}(f) = \inf_{n\in\mathbb{N}_{>0}} \sup_{m\geq n} f(m) = \limsup f \text{ for all bounded gambles } f \text{ on } \mathbb{N}_{>0}.$$

Consider the bounded gamble g on $\mathbb{N}_{>0}$ defined by $g(n) := 1/n$, $n \in \mathbb{N}_{>0}$. Then, intuitively, because X is believed to be infinitely large, we would expect $g(X) = 1/X$ to be zero, so we would want g to be $\underline{P}$-null. Now it is clear that $\overline{P}_{\mathcal{F}}(\{|g| > \varepsilon\}) = 0$ for all $\varepsilon > 0$, so g is indeed $\underline{P}$-null according to our definition, but it does not satisfy the stronger alternative condition, because $\{|g| \neq 0\} = \mathbb{N}_{>0}$ and, therefore, $\overline{P}_{\mathcal{F}}(\{|g| \neq 0\}) = 1 > 0$. ♦

We can use null gambles to define almost everywhere equality and almost everywhere dominance.

Definition 14.8 (Almost everywhere equality) *Two gambles f and g are* ***equal almost everywhere [$\underline{P}$]*** *if $f - g$ is a $\underline{P}$-null gamble. In this case, we use the notation $f = g$ a.e. [$\underline{P}$].*

Definition 14.9 (Almost everywhere dominance) *A gamble f is* ***dominated almost everywhere [$\underline{P}$]*** *by a gamble g if there is some $\underline{P}$-null gamble N such that $f \leq g + N$. In this case, we use the notation $f \leq g$ a.e. [$\underline{P}$]. We also define the expressions $f \leq +\infty$ a.e. [$\underline{P}$] and $-\infty \leq f$ a.e. [$\underline{P}$] to be true for every gamble f.*

The following proposition lists some properties of $\underline{P}$-null gambles, almost everywhere dominance and equality.

Proposition 14.10 *Let $A \subseteq \mathcal{X}$. Let f, g, h, f_1, f_2, g_1 and g_2 be gambles. Let a and b be real numbers. Then the following statements hold:*

(i) *A is a $\underline{P}$-null set if and only if I_A is a $\underline{P}$-null gamble.*

(ii) *If A is a $\underline{P}$-null set then fI_A is a $\underline{P}$-null gamble.*

(iii) *If f and g are $\underline{P}$-null, then so are I_Af, $|f|$, $af + bg$, $f \vee g$ and $f \wedge g$.*

(iv) *If $|f| \leq |g|$ a.e. [$\underline{P}$] and g is $\underline{P}$-null, then so is f.*

(v) *If $f \leq g$, then $f \leq g$ a.e. [$\underline{P}$].*

(vi) *If $f \leq g$ a.e. [$\underline{P}$], then $fI_A \leq gI_A$ a.e. [$\underline{P}$].*

(vii) *If $f \leq g$ a.e. [$\underline{P}$] and $g \leq h$ a.e. [$\underline{P}$], then $f \leq h$ a.e. [$\underline{P}$].*

(viii) *$f \leq g$ a.e. [$\underline{P}$] and $g \leq f$ a.e. [$\underline{P}$] if and only if $f = g$ a.e. [$\underline{P}$].*

(ix) *Assume that a and b are positive. If $f_1 \leq f_2$ a.e. [$\underline{P}$] and $g_1 \leq g_2$ a.e. [$\underline{P}$], then $af_1 + bg_1 \leq af_2 + bg_2$ a.e. [$\underline{P}$].*

(x) *If $f = g$ a.e. [$\underline{P}$] and $g = h$ a.e. [$\underline{P}$], then $f = h$ a.e. [$\underline{P}$].*

(xi) *If $f_1 = f_2$ a.e. [$\underline{P}$] and $g_1 = g_2$ a.e. [$\underline{P}$], then $|f_1| = |f_2|$ a.e. [$\underline{P}$], $af_1 + bg_1 = af_2 + bg_2$ a.e. [$\underline{P}$], $f_1 \vee g_1 = f_2 \vee g_2$ a.e. [$\underline{P}$] and $f_1 \wedge g_1 = f_2 \wedge g_2$ a.e. [$\underline{P}$].*

Proof. (i). This follows from Definitions 14.2_{305} and $14.6_{\curvearrowleft}$.

(ii). Let $\varepsilon > 0$. We have that $\left\{|fI_A| > \varepsilon\right\} \subseteq A$. Using the monotonicity of the coherent $\overline{P}$ (see Theorem 4.13(iv)$_{53}$), we find that fI_A is a $\underline{P}$-null gamble.

(iii). Treating sum and scalar multiplication separately and assuming that $a \neq 0$ (the case $a = 0$ is trivial), this follows from

$$\left\{|I_Af| > \varepsilon\right\} \subseteq \{|f| > \varepsilon\}$$
$$\left\{||f|| > \varepsilon\right\} = \{|f| > \varepsilon\},$$
$$\{|af| > \varepsilon\} = \left\{|f| > \frac{\varepsilon}{|a|}\right\},$$

$$\{|f+g|>\varepsilon\}\subseteq\left\{|f|>\frac{\varepsilon}{2}\right\}\cup\left\{|g|>\frac{\varepsilon}{2}\right\},$$

and the monotonicity and sub-additivity of the coherent $\overline{P}$ (Theorem 4.13(iv) and (v)$_{53}$). The maximum and the minimum of f and g can be written as linear combinations of f, g and $|f-g|$:

$$f\vee g=\frac{(f+g+|f-g|)}{2}\text{ and }f\wedge g=\frac{(f+g-|f-g|)}{2},$$

so these cases follow from the previous ones.

(iv). By definition of $|f|\leq|g|$ a.e. [$\underline{P}$], there is some $\underline{P}$-null gamble N such that $|f|\leq|g|+N$ and, therefore, $|f|\leq|g|+|N|$. We find that

$$\{|f|>\varepsilon\}\subseteq\left\{|g|>\frac{\varepsilon}{2}\right\}\cup\left\{|N|>\frac{\varepsilon}{2}\right\}.$$

Now use the monotonicity and sub-additivity of the coherent upper prevision $\overline{P}$ (see Theorem 4.13(iv) and (v)$_{53}$).

(v). Obvious, because $f\leq g$ implies that $f\leq g+0$ and 0 is a $\underline{P}$-null gamble.

(vi). If $f\leq g$ a.e. [$\underline{P}$], then there is some $\underline{P}$-null gamble N such that $f\leq g+N$ and, therefore, $I_Af\leq I_Ag+I_AN$. Now use that I_AN is $\underline{P}$-null too, by (iii).

(vii). By definition, there are two $\underline{P}$-null gambles N and M such that $f\leq g+N$ and $g\leq h+M$. It follows that $f\leq h+(N+M)$. By (iii), $N+M$ is a $\underline{P}$-null gamble. We find that $f\leq h$ a.e. [$\underline{P}$].

(viii). 'if'. Obvious.

'only if'. By definition, there are two $\underline{P}$-null gambles N and M such that $f\leq g+N$ and $g\leq f+M$. It follows that $-|M|\leq f-g\leq|N|$, which implies that $|f-g|\leq|N|\vee|M|$. By (iii), $|N|\vee|M|$ is a $\underline{P}$-null gamble. It follows from (iv) that $f-g$ is also a $\underline{P}$-null gamble. We find that $f=g$ a.e. [$\underline{P}$].

(ix). By definition, there are two $\underline{P}$-null gambles N and M such that $f_1\leq f_2+N$ and $g_1\leq g_2+M$. It follows that $af_1+bg_1\leq af_2+bg_2+(aN+bM)$. By (iii), $aN+bM$ is a $\underline{P}$-null gamble. We find that $af_1+bg_1\leq af_2+bg_2$ a.e. [$\underline{P}$].

(x). By definition, there are two $\underline{P}$-null gambles N and M such that $f=g+N$ and $g=h+M$. It follows that $f=h+(N+M)$. By (iii), $N+M$ is a $\underline{P}$-null gamble. We find that $f=h$ a.e. [$\underline{P}$].

(xi). By definition, there are two $\underline{P}$-null gambles N and M such that $f_1=f_2+N$ and $g_1=g_2+M$. It follows that $|f_2|-|N|\leq|f_1|\leq|f_2|+|N|$. By (iii), $-|N|$ and $|N|$ are $\underline{P}$-null gambles. Invoking (viii), we find that $|f_1|=|f_2|$ a.e. [$\underline{P}$]. It also follows that $af_1+bg_1=af_2+bg_2+(aN+bM)$. By (iii), $aN+bM$ is a $\underline{P}$-null gamble. We find that $af_1+bg_1=af_2+bg_2$ a.e. [$\underline{P}$]. The maximum and the minimum of f_1 and g_1 can be written as linear combinations of f_1, g_1 and $|f_1-g_1|$, and a similar statement holds for the maximum and the minimum of f_2 and g_2, so these cases follow from the previous ones. □

This leads at once to the following conclusions.[1]

[1] See Definition A.1$_{368}$ for preorders.

Corollary 14.11 *The binary relation '$\leq$ a.e. [$\underline{P}$]' is a preorder on the set of gambles with associated equivalence relation '$=$ a.e. [$\underline{P}$]'.*

Corollary 14.12 *The set of $\underline{P}$-null gambles constitutes a linear lattice with respect to the point-wise ordering of gambles.*

14.2 Null bounded gambles

It turns out that null bounded gambles are precisely those bounded gambles whose absolute value has upper prevision zero. Proving this is the subject of this section.

Lemma 14.13 *Let f be any bounded gamble and a be any real number. If $\overline{P}(|f|) = 0$, then $\underline{P}(af) = \overline{P}(af) = 0$.*

Proof. Assume that $\overline{P}(|f|) = 0$. Then we infer from the coherence of $\underline{P}$ (Theorem 4.13(vi)$_{53}$) that also $\overline{P}(|af|) = |a|\overline{P}(|f|) = 0$. Using Theorem 4.13(i) and (viii)$_{53}$, we then get

$$0 = -\overline{P}(|af|) = \underline{P}(-|af|) \leq \underline{P}(af) \leq \overline{P}(af) \leq \overline{P}(|af|) = 0.$$

We may therefore conclude that $\underline{P}(af) = \overline{P}(af) = 0$. □

Proposition 14.14 *Let f and g be bounded gambles and a and b real numbers. The following statements hold:*

(i) *f is $\underline{P}$-null if and only if $\overline{P}(|f|) = 0$.*

(ii) *If f and g are $\underline{P}$-null, then $\underline{P}(af + bg) = \overline{P}(af + bg) = 0$.*

Proof. (i). Define $A_\varepsilon := \{|f| > \varepsilon\}$ for every $\varepsilon > 0$.

We first show that the condition is sufficient. Suppose $\overline{P}(|f|) = 0$. Let $\varepsilon > 0$. Then $|f(x)| > \varepsilon$ for every $x \in A_\varepsilon$ and, therefore, $|f| \geq \varepsilon I_{A_\varepsilon}$. These inequalities and the coherence of $\underline{P}$ (see Theorem 4.13(i), (iv) and (vi)$_{53}$) imply that

$$0 \leq \overline{P}(A_\varepsilon) \leq \overline{P}\left(\frac{|f|}{\varepsilon}\right) = \frac{\overline{P}(|f|)}{\varepsilon} = 0,$$

so $\overline{P}(A_\varepsilon) = 0$ for all $\varepsilon > 0$ and f is indeed $\underline{P}$-null.

Next we show that the condition is necessary. Suppose that f is $\underline{P}$-null, so $\overline{P}(A_\varepsilon) = 0$ for every $\varepsilon > 0$. Consider the sequence $f_n := fI_{A_{1/n}}$, $n \in \mathbb{N}_{>0}$. As $0 \leq |f_n| \leq I_{A_{1/n}} \sup|f|$ for every $n \in \mathbb{N}_{>0}$, we have that $\overline{P}(|f_n|) = 0$, because $\underline{P}$ is coherent (see Theorem 4.13(iv) and (vi)$_{53}$), and because assumption $\overline{P}(A_{1/n}) = 0$ for every $n \in \mathbb{N}_{>0}$. Now it follows from the definition of $A_{1/n}$ and f_n that $\sup\big||f_n| - |f|\big| \leq 1/n$ and, therefore, $\sup\big||f_n| - |f|\big| \to 0$. This implies that $\overline{P}(|f_n|) \to \overline{P}(|f|)$, because $\underline{P}$ is coherent (see Theorem 4.13(i) and (xiv)$_{53}$). As we know that $\overline{P}(|f_n|) = 0$ for all $n \in \mathbb{N}_{>0}$, it follows that $\overline{P}(|f|)$ must be zero too.

(ii). We infer from (i) that $\overline{P}(|f|) = 0$ and $\overline{P}(|g|) = 0$. Lemma 14.13, therefore, implies that $\overline{P}(af) = \underline{P}(af) = 0$ and $\overline{P}(bg) = \underline{P}(bg) = 0$. Using the super-additivity of the coherent $\underline{P}$, the sub-additivity of $\overline{P}$ and $\underline{P}(h) \leq \overline{P}(h)$ for every bounded gamble h (see Theorem 4.13(i) and (v)$_{53}$), we find that

$$0 = \underline{P}(af) + \underline{P}(bg) \leq \underline{P}(af + bg) \leq \overline{P}(af + bg) \leq \overline{P}(af) + \overline{P}(bg) = 0,$$

so indeed $\underline{P}(af + bg) = \overline{P}(af + bg) = 0$. □

14.3 Essentially bounded gambles

We now set out to explain how lower and upper previsions can be extended quite easily to gambles that are bounded on the complement of a $\underline{P}$-null set.

Definition 14.15 (Essential boundedness from above, below) *Let f be a gamble. Then the following conditions are equivalent; if any (and hence all) of them are satisfied, we say that f is **$\underline{P}$-essentially bounded above**:*

(A) *There is a $\underline{P}$-null set $A \subset \mathcal{X}$ such that f is bounded above on A^c.*

(B) *There is a bounded gamble g such that $f \leq g$ a.e. $[\underline{P}]$.*

(C) *There is a bounded gamble h such that $\overline{P}(\{f > h\}) = 0$.*

*We say that f is **$\underline{P}$-essentially bounded below** if $-f$ is $\underline{P}$-essentially bounded above.*

Let us prove that these conditions are indeed equivalent.

Proof. (A)⇒(B). Let $A \subset \mathcal{X}$ be a $\underline{P}$-null set such that f is bounded above on A^c. If we define the *bounded gamble* $g := (fI_{A^c}) \vee 0$ and the gamble $N := fI_A$, then clearly $f \leq g + N$. If we can prove that N is a $\underline{P}$-null gamble, then (B) is established. Let $\varepsilon > 0$. As $|N(x)| > \varepsilon$ implies that $x \in A$, we find that $\{|N| > \varepsilon\} \subseteq A$. Now use the monotonicity of $\overline{P}$ (which follows from the coherence of $\underline{P}$, see Theorem 4.13(iv)$_{53}$).

(B)⇒(C). Let g be a bounded gamble such that $f \leq g$ a.e. $[\underline{P}]$. Then there is a $\underline{P}$-null gamble N such that $f \leq g + N$. We also have that $f \leq g + |N|$. It follows that if $f(x) > g(x) + 1$, then $|N(x)| > 1$. If we define the bounded gamble $h := g + 1$, then this tells us that $\{f > h\} \subseteq \{|N| > 1\}$. Now use the monotonicity of the coherent $\overline{P}$ (Theorem 4.13(iv)$_{53}$) and the fact that N is $\underline{P}$-null.

(C)⇒(A). Let h be a bounded gamble such that $\overline{P}(\{f > h\}) = 0$. If we define $A := \{f > h\}$, then A is a $\underline{P}$-null set. As $fI_{A^c} \leq hI_{A^c}$, it follows that f is bounded above (by $\sup h$) on A^c. This establishes (A). □

Definition 14.16 (Essential boundedness) *Let f be a gamble. Then the following conditions are equivalent; if any (and hence all) of them are satisfied, we say that f is **$\underline{P}$-essentially bounded**.*

(A) *f is $\underline{P}$-essentially bounded above and below.*

(B) *$|f|$ is $\underline{P}$-essentially bounded above.*

(C) *There is a bounded gamble $f_\flat$ such that $f = f_\flat$ a.e. $[\underline{P}]$.*

The set of all $\underline{P}$-essentially bounded gambles is denoted by $\mathbb{G}^\sharp_{\underline{P}}(\mathcal{X})$ or, more simply, by $\mathbb{G}^\sharp_{\underline{P}}$ if no confusion is possible.

We prove that these conditions are indeed equivalent.

Proof. (A)$_\frown$⇒(B). Assume that $g_1 \le f$ a.e. $[\underline{P}]$ and $f \le g_2$ a.e. $[\underline{P}]$ for some bounded gambles g_1 and g_2. This implies that there are $\underline{P}$-null gambles N_1 and N_2 such that $g_1 \le f + N_1$ and $f \le g_2 + N_2$. This in turn leads to $|f| \le |g_1| + |g_2| + |N_1| + |N_2|$ and, therefore, $|f| \le |g_1| + |g_2|$ a.e. $[\underline{P}]$, because $|N_1| + |N_2|$ is a $\underline{P}$-null gamble too, by Proposition 14.10(iii)$_{308}$.

(B)⇒(C). Assume that there is a bounded gamble g such that $\overline{P}(\{|f| > g\}) = 0$. Define $A := \{|f| > g\}$ and $f_\flat := fI_{A^c}$. If we can show that $f - f_\flat$ is a $\underline{P}$-null gamble, then $f = f_\flat$ a.e. $[\underline{P}]$. Let $\varepsilon > 0$. Then $\left\{|f - f_\flat| > \varepsilon\right\} \subseteq A$. Now use the monotonicity of the coherent upper prevision $\overline{P}$ (Theorem 4.13(iv)$_{53}$).

(C)⇒(A)$_\frown$. It suffices to recall that $f = f_\flat$ a.e. $[\underline{P}]$ implies that both $f \le f_\flat$ a.e. $[\underline{P}]$ and $f_\flat \le f$ a.e. $[\underline{P}]$. □

Definition 14.17 (Essential supremum and infimum) *The **$\underline{P}$-essential supremum** of a gamble f is defined by*

$$\operatorname{ess\,sup}_{\underline{P}} f = \inf\left\{\lambda \in \mathbb{R} : f \le \lambda \text{ a.e. } [\underline{P}]\right\} = \inf_{N \in \mathbb{G}^0_{\underline{P}}} \sup(f + N)$$

$$= \lim_{A \in \mathcal{N}_{\underline{P}}} \sup(f|A^c) = \inf_{A \in \mathcal{N}_{\underline{P}}} \sup(f|A^c),$$

*its **$\underline{P}$-essential infimum** by*

$$\operatorname{ess\,inf}_{\underline{P}} f = \sup\left\{\lambda \in \mathbb{R} : \lambda \le f \text{ a.e. } [\underline{P}]\right\} = \sup_{N \in \mathbb{G}^0_{\underline{P}}} \inf(f + N)$$

$$= \lim_{A \in \mathcal{N}_{\underline{P}}} \inf(f|A^c) = \sup_{A \in \mathcal{N}_{\underline{P}}} \inf(f|A^c)$$

*and its **$\underline{P}$-essential supremum norm** by*

$$\|f\|_{\underline{P},\infty} = \operatorname{ess\,sup}_{\underline{P}} |f| = \inf\left\{\lambda \in \mathbb{R} : |f| \le \lambda \text{ a.e. } [\underline{P}]\right\} = \inf_{N \in \mathbb{G}^0_{\underline{P}}} \sup|f + N|$$

$$= \lim_{A \in \mathcal{N}_{\underline{P}}} \sup(|f||A^c) = \inf_{A \in \mathcal{N}_{\underline{P}}} \sup(|f||A^c).$$

In these expressions, $\inf(\bullet|\bullet)$ and $\sup(\bullet|\bullet)$ are the vacuous conditional lower and upper previsions introduced in Example 13.10$_{242}$: $\inf(f|B) = \inf\{f(x) : x \in B\}$ and $\sup(f|B) = \sup\{f(x) : x \in B\}$ for any gamble f and non-empty event B. The limit over $\mathcal{N}_{\underline{P}}$ is understood to be taken toward larger $\underline{P}$-null sets, that is, as a limit associated with the proper filter made up of the complements of the null sets.

Let us prove that the alternative expressions are indeed equivalent.

Proof. Whichever *corresponding* expressions we take for $\operatorname{ess\,inf}_{\underline{P}}$, $\operatorname{ess\,sup}_{\underline{P}}$ and $\|\bullet\|_{\underline{P},\infty}$, we get $\operatorname{ess\,inf}_{\underline{P}} f = -\operatorname{ess\,sup}_{\underline{P}} -f$ and $\|f\|_{\underline{P},\infty} = \operatorname{ess\,sup}_{\underline{P}} |f|$. So if we can prove that all the expressions for the $\underline{P}$-essential supremum are equivalent, then the expressions for the $\underline{P}$-essential infimum and the $\underline{P}$-essential supremum norm will be equivalent too.

As, for any gamble f and any $\lambda \in \mathbb{R}$, $f \leq \lambda$ a.e. $[\underline{P}]$ if and only if there is some $\underline{P}$-null gamble $N \in \mathbb{G}^0_{\underline{P}}$ such that $f + N \leq \lambda$ and therefore $\sup(f + N) \leq \lambda$, we already find that

$$\inf\left\{\lambda \in \mathbb{R} : f \leq \lambda \text{ a.e. } [\underline{P}]\right\} = \inf\left\{\lambda \in \mathbb{R} : (\exists N \in \mathbb{G}^0_{\underline{P}})(\sup(f + N) \leq \lambda)\right\}$$
$$= \inf_{N \in \mathbb{G}^0_{\underline{P}}} \sup(f + N).$$

Next, because, by Corollary 14.5(ii)$_{306}$, $(\mathscr{N}_{\underline{P}}, \subseteq)$ is a directed set, we can take the Moore–Smith limit over $\mathscr{N}_{\underline{P}}$. As the extended real net $A \mapsto \sup(f|A^c)$, $A \in \mathscr{N}_{\underline{P}}$ is non-increasing, its limit exists in $\mathbb{R}^*$ (see Proposition 1.13(ii)$_9$) and it coincides with the infimum of the net. This proves that $\lim_{A \in \mathscr{N}_{\underline{P}}} \sup(f|A^c) = \inf_{A \in \mathscr{N}_{\underline{P}}} \sup(f|A^c)$.

To conclude, we prove that $\inf\left\{\lambda \in \mathbb{R} : f \leq \lambda \text{ a.e. } [\underline{P}]\right\} = \inf_{A \in \mathscr{N}_{\underline{P}}} \sup(f|A^c)$. Let A be any $\underline{P}$-null set. If $\sup(f|A^c)$ is finite, then

$$f \leq \sup(f|A^c) I_{A^c} + f I_A = \sup(f|A^c) + [f - \sup(f|A^c)] I_A$$

so it follows from Proposition 14.10(ii)$_{308}$ that $f \leq \sup(f|A^c)$ a.e. $[\underline{P}]$. Moreover, if $\sup(f|A^c)$ is infinite, it must be equal to $+\infty$, as there can only be equality to $-\infty$ when $A = \mathscr{X}$, which is impossible because $\mathscr{N}_{\underline{P}}$ is a proper ideal. Hence,

$$\left\{\sup(f|A^c) : A \in \mathscr{N}_{\underline{P}}\right\} \subseteq \left\{\lambda \in \mathbb{R} : f \leq \lambda \text{ a.e. } [\underline{P}]\right\} \cup \{+\infty\},$$

and taking the infimum over both sides, this implies that

$$\inf_{A \in \mathscr{N}_{\underline{P}}} \sup(f|A^c) \geq \inf\left\{\lambda \in \mathbb{R} : f \leq \lambda \text{ a.e. } [\underline{P}]\right\}.$$

It only remains to prove the converse inequality. Consider any $\lambda \in \mathbb{R}$, and assume that $f \leq \lambda$ a.e. $[\underline{P}]$. Then there is some $\underline{P}$-null gamble N_λ such that $f \leq \lambda + N_\lambda$. For every $\varepsilon > 0$, define the $\underline{P}$-null set $A_{\lambda,\varepsilon} := \left\{|N_\lambda| > \varepsilon\right\}$. If $x \in A^c_{\lambda,\varepsilon}$, then $|N_\lambda(x)| \leq \varepsilon$, which implies that $f(x) \leq \lambda + \varepsilon$. Therefore,

$$\lambda = \inf_{\varepsilon > 0}(\lambda + \varepsilon) \geq \inf_{\varepsilon > 0} \sup\left\{f(x) : x \in A^c_{\lambda,\varepsilon}\right\} \geq \inf_{A \in \mathscr{N}_{\underline{P}}} \sup(f|A^c),$$

whence indeed

$$\inf_{A \in \mathscr{N}_{\underline{P}}} \sup(f|A^c) \leq \inf\left\{\lambda \in \mathbb{R} : f \leq \lambda \text{ a.e. } [\underline{P}]\right\},$$

so we are done. □

Contrary to the supremum $\sup f$, the essential supremum $\operatorname{ess\,sup}_{\underline{P}} f$ may assume the value $-\infty$ for some gambles f, and similarly, $\operatorname{ess\,inf}_{\underline{P}} f$ may assume the value $+\infty$, as the following example shows.

Example 14.18 Consider the coherent lower prevision lim inf defined on the set of all bounded gambles on $\mathbb{N}_{>0}$. It has already made an appearance in Sections 5.3_{80} and $13.10.5_{300}$, as well as in Examples 5.13_{92} and 14.7_{307}, where we discussed its interpretation. The set $\mathscr{N}_{\liminf}$ of null sets for this lower prevision is the proper ideal consisting of all finite subsets of $\mathbb{N}_{>0}$.

Consider the gamble f defined by $f(n) := -n$ for all $n \in \mathbb{N}_{>0}$. For any $A \in \mathscr{N}_{\liminf}$, we see that $\sup(f|A^c) = -\min(A^c)$, and therefore

$$\operatorname{ess\,sup}_{\liminf} f = \inf_{A \in \mathscr{N}_{\liminf}} \sup(f|A^c) = -\sup_{A \in \mathscr{N}_{\liminf}} \min(A^c) = -\infty,$$

while, of course, $\sup f = -\inf \mathbb{N}_{>0} = -1$. ♦

The next proposition summarises the most important properties of the $\underline{P}$-essential supremum and the $\underline{P}$-essential infimum.

Proposition 14.19 *The following statements hold for any gambles f and g:*

(i) $\inf f \leq \operatorname{ess\,inf}_{\underline{P}} f \leq \operatorname{ess\,sup}_{\underline{P}} f \leq \sup f$.

(ii) *If* $\operatorname{ess\,inf}_{\underline{P}} f < +\infty$, *then* $\operatorname{ess\,inf}_{\underline{P}} f \leq f$ *a.e.* $[\underline{P}]$, *and if* $\operatorname{ess\,sup}_{\underline{P}} f > -\infty$, *then* $f \leq \operatorname{ess\,sup}_{\underline{P}} f$ *a.e.* $[\underline{P}]$.

(iii) *If* $f \leq g$ *a.e.* $[\underline{P}]$, *then* $\operatorname{ess\,sup}_{\underline{P}} f \leq \operatorname{ess\,sup}_{\underline{P}} g$ *and* $\operatorname{ess\,inf}_{\underline{P}} f \leq \operatorname{ess\,inf}_{\underline{P}} g$.

(iv) *If* $f = g$ *a.e.* $[\underline{P}]$, *then* $\operatorname{ess\,sup}_{\underline{P}} f = \operatorname{ess\,sup}_{\underline{P}} g$ *and* $\operatorname{ess\,inf}_{\underline{P}} f = \operatorname{ess\,inf}_{\underline{P}} g$.

(v) *f is $\underline{P}$-null if and only if* $\|f\|_{\underline{P},\infty} = 0$.

(vi) *f is $\underline{P}$-essentially bounded if and only if* $\|f\|_{\underline{P},\infty} < +\infty$.

Proof. (i). This follows from $\sup f \geq \sup(f|A^c) \geq \inf(f|A^c) \geq \inf f$ for every $A \in \mathscr{N}_{\underline{P}}$ and every gamble f, because taking the limit over A in $\mathscr{N}_{\underline{P}}$ preserves the inequalities.

(ii). It suffices to prove the second statement: the first then follows from conjugacy. Assume, therefore, that $\operatorname{ess\,sup}_{\underline{P}} f > -\infty$. We may also assume without loss of generality that $\operatorname{ess\,sup}_{\underline{P}} f < +\infty$, so $\operatorname{ess\,sup}_{\underline{P}} f$ is finite. In that case, clearly, $f \leq \operatorname{ess\,sup}_{\underline{P}} f \vee f = \operatorname{ess\,sup}_{\underline{P}} f + (f - \operatorname{ess\,sup}_{\underline{P}} f) \vee 0$. Let $N := (f - \operatorname{ess\,sup}_{\underline{P}} f) \vee 0$. If we can prove that the gamble N is $\underline{P}$-null, then the claim is established.

Consider $\varepsilon > 0$ and $x \in \mathscr{X}$, and assume that $N(x) > \varepsilon$. It follows that $f(x) > \operatorname{ess\,sup}_{\underline{P}} f + \varepsilon$. As $\operatorname{ess\,sup}_{\underline{P}} f = \inf_{A \in \mathscr{N}_{\underline{P}}} \sup \{f(x') : x' \in A^c\}$, there is some $\underline{P}$-null set A_ε such that $\operatorname{ess\,sup}_{\underline{P}} f + \varepsilon \geq \sup \{f(x') : x' \in A^c_\varepsilon\}$ and, therefore, also $f(x) > \sup \{f(x') : x' \in A^c_\varepsilon\}$. This implies that $x \in A_\varepsilon$. It follows that $\{|N| > \varepsilon\} \subseteq A_\varepsilon$, because N is moreover non-negative. Now use the monotonicity of the coherent upper prevision $\overline{P}$ (Theorem $4.13(\text{iv})_{53}$).

(iii). If $f \leq g$ a.e. [$\underline{P}$], then there is a $\underline{P}$-null gamble N such that $f \leq g + N$. It follows that, because $\mathbb{G}^0_{\underline{P}}$ is a linear space (see Corollary 14.12$_{310}$) and $N \in \mathbb{G}^0_{\underline{P}}$,

$$\operatorname{ess\,sup}_{\underline{P}} f \leq \operatorname{ess\,sup}_{\underline{P}}(g+N) = \inf_{M \in \mathbb{G}^0_{\underline{P}}} \sup(g+N+M)$$

$$= \inf_{N' \in \mathbb{G}^0_{\underline{P}}} \sup(g+N') = \operatorname{ess\,sup}_{\underline{P}} g.$$

That also $\operatorname{ess\,inf}_{\underline{P}} f \leq \operatorname{ess\,inf}_{\underline{P}} g$ now follows from conjugacy: $\operatorname{ess\,sup}_{\underline{P}}(f) = -\operatorname{ess\,inf}_{\underline{P}}(-f)$ (see Definition 14.17$_{312}$).

(iv). Immediate from (iii).

(v). 'if'. Assume that $\|f\|_{\underline{P},\infty} = \inf_{A \in \mathcal{N}_{\underline{P}}} \sup(|f|\,|A^c) = 0$. Let $\varepsilon > 0$. Then there is some $\underline{P}$-null set A_ε such that $\sup(|f|\,|A^c_\varepsilon) < \varepsilon$. This implies that $|f(x)| \leq \varepsilon$ for every $x \in A^c_\varepsilon$, whence $\{|f| > \varepsilon\} \subseteq A_\varepsilon$. Now use the monotonicity of the coherent upper prevision $\overline{P}$ (Theorem 4.13(iv)$_{53}$).

'only if'. Assume that f is a $\underline{P}$-null gamble. Let $\varepsilon > 0$. Then $A_\varepsilon := \{|f| > \varepsilon\}$ is a $\underline{P}$-null set, and indeed

$$\|f\|_{\underline{P},\infty} = \inf_{A \in \mathcal{N}_{\underline{P}}} \sup(|f|\,|A^c) \leq \inf_{\varepsilon > 0} \sup(|f|\,|A^c_\varepsilon) \leq \inf_{\varepsilon > 0} \varepsilon = 0.$$

(vi). 'only if'. Assume that f is $\underline{P}$-essentially bounded. Then there is some bounded gamble g such that $|f| \leq g$ a.e. [$\underline{P}$] and, therefore, also $|f| \leq \sup|g|$ a.e. [$\underline{P}$]. Therefore, by Definition 14.17$_{312}$, $\|f\|_{\underline{P},\infty} \leq \sup|g| < +\infty$.

'if'. Assume that $\|f\|_{\underline{P},\infty} < +\infty$. Then, by Definition 14.17$_{312}$, there must be some constant gamble $\lambda < +\infty$ such that $|f| \leq \lambda$ a.e. [$\underline{P}$]. Hence, f is $\underline{P}$-essentially bounded. □

We conclude this discussion with an interpretation of the $\underline{P}$-essential infimum as a(n unconditional) coherent lower prevision. As a consequence, we can actually use it for extending $\underline{P}$. This will be the approach taken Section 14.4.

Theorem 14.20 $\operatorname{ess\,inf}_{\underline{P}}$ *is a coherent lower prevision on* $\mathbb{G}$ *and* $\operatorname{ess\,sup}_{\underline{P}}$ *is its conjugate upper prevision.*

Proof. As $\mathbb{G}$ is a linear space, we can use Theorem 13.34$_{271}$ to establish coherence. That condition (i)$_{271}$ is satisfied follows from Proposition 14.19(i). Condition (iii)$_{271}$ follows from $\inf(\lambda f|A^c) = \lambda \inf(f|A^c)$ for all $\lambda > 0$, $A \in \mathcal{N}_{\underline{P}}$ and gambles f. Condition (ii)$_{271}$ follows from a similar argument, because $\inf(f+g|A^c) \geq \inf(f|A^c) + \inf(g|A^c)$ for all $A \in \mathcal{N}_{\underline{P}}$ and every pair of gambles f and g. We conclude that $\operatorname{ess\,inf}_{\underline{P}}$ is a coherent lower prevision. Since $\operatorname{ess\,sup}_{\underline{P}}(f) = -\operatorname{ess\,inf}_{\underline{P}}(-f)$ for all gambles f (see Definition 14.17$_{312}$), $\operatorname{ess\,sup}_{\underline{P}}$ is its conjugate upper prevision. □

14.4 Extension of lower and upper previsions to essentially bounded gambles

Consider the set of *bounded gambles* associated with the coherent lower prevision as defined in Section 4.2_{41} (see Equation $(4.2)_{42}$):

$$\mathscr{A}_{\underline{P}} := \left\{f - \mu : f \in \mathbb{B} \text{ and } \mu < \underline{P}(f)\right\}.$$

To see what these acceptability assessments for bounded gambles imply for the acceptability of gambles that are not necessarily bounded, we take its natural extension, as discussed in Section $13.2.2_{238}$ (see Equation $(13.1)_{238}$):

$$\mathscr{E}_{\mathscr{A}_{\underline{P}}} = \mathbb{G}_{\geq 0} + \text{nonneg}(\mathscr{A}_{\underline{P}}).$$

It is now a straightforward exercise to verify that

$$\underline{P}(f) = \text{lpr}(\mathscr{E}_{\mathscr{A}_{\underline{P}}})(f) = \sup\left\{\mu \in \mathbb{R} : f - \mu \in \mathscr{E}_{\mathscr{A}_{\underline{P}}}\right\} \text{ for all } f \in \mathbb{B},$$

where we recalled Equation $(13.2)_{240}$.

But what happens if we use Equation $(13.2)_{240}$ to calculate the (unconditional) lower prevision $\text{lpr}(\mathscr{E}_{\mathscr{A}_{\underline{P}}})(h)$ of a gamble h that is not necessarily bounded?

If h is not bounded below, then neither is any $h - \mu$, meaning that it cannot dominate any element of $\text{nonneg}(\mathscr{A}_{\underline{P}})$, as these are all bounded below. Hence, $\text{lpr}(\mathscr{E}_{\mathscr{A}_{\underline{P}}})(h) = \sup \emptyset = -\infty$. Similarly, $\text{upr}(\mathscr{E}_{\mathscr{A}_{\underline{P}}})(h) = +\infty$ for any gamble h that is not bounded above.

This revisits, using a slightly different argument, the conclusions reached in Section 13.11_{301} and in particular in Example 13.67_{302}: as soon as we leave the ambit of bounded gambles, natural extension of the assessments implicit in the definition of a coherent lower prevision makes the lower or the upper prevision of an unbounded gamble infinite.

One way to remedy this, consists in making an extra assessment involving unbounded gambles, *effectively making all $\underline{P}$-null gambles acceptable too:* we consider the (unconditional) lower prevision $\text{lpr}(\mathscr{E}_{\mathscr{A}_{\underline{P}} \cup \mathbb{G}^0_{\underline{P}}})$ – recall that $\mathbb{G}^0_{\underline{P}}$ denotes the set of all $\underline{P}$-null gambles. It turns out that, formally, this lower prevision on all gambles looks very much like an inner extension, where we replace point-wise dominance with almost everywhere dominance.

Theorem 14.21 *For all gambles f on $\mathcal{X}$,*

$$\text{lpr}(\mathscr{E}_{\mathscr{A}_{\underline{P}} \cup \mathbb{G}^0_{\underline{P}}})(f) = \sup\left\{\underline{P}(g) : g \in \mathbb{B} \textit{ and } g \leq f \textit{ a.e. } [\underline{P}]\right\}.$$

Proof. Indeed, by Equation $(13.2)_{240}$,

$$\text{lpr}(\mathscr{E}_{\mathscr{A}_{\underline{P}} \cup \mathbb{G}^0_{\underline{P}}})(f)$$

$$= \sup\left\{\mu \in \mathbb{R} : f - \mu \in \mathscr{E}_{\mathscr{A}_{\underline{P}} \cup \mathbb{G}^0_{\underline{P}}}\right\}$$

$$= \sup\left\{\mu \in \mathbb{R} : (\exists h \in \mathbb{B})(\exists N \in \mathbb{G}^0_{\underline{P}})(\underline{P}(h) > 0 \text{ and } f - \mu \geq h + N)\right\}$$

and if we use the correspondence $g := h + \mu$ and, therefore $\underline{P}(g) = \underline{P}(h) + \mu$,

$$= \sup\left\{\mu \in \mathbb{R} : (\exists g \in \mathbb{B})(\exists N \in \mathbb{G}^0_{\underline{P}})(\underline{P}(g) > \mu \text{ and } f \geq g + N)\right\}$$

$$= \sup\left\{\underline{P}(g) : g \in \mathbb{B} \text{ and } (\exists N \in \mathbb{G}^0_{\underline{P}}) f \geq g + N\right\}$$

and if we recall that $N \in \mathbb{G}^0_{\underline{P}} \Leftrightarrow -N \in \mathbb{G}^0_{\underline{P}}$,

$$= \sup\left\{\underline{P}(g) : g \in \mathbb{B} \text{ and } g \leq f \text{ a.e. } [\underline{P}]\right\},$$

which completes the proof. □

It turns out that we can achieve the same result by considering as acceptable all gambles whose $\underline{P}$-essential infimum is strictly positive. Consider the set of acceptable gambles corresponding to the lower prevision $\operatorname{ess\,inf}_{\underline{P}}$:

$$\mathscr{A}_{\operatorname{ess\,inf}_{\underline{P}}} = \left\{f - \mu : f \in \mathbb{G} \text{ and } \mu < \operatorname{ess\,inf}_{\underline{P}}(f)\right\}.$$

Theorem 14.22 *For all gambles f on $\mathscr{X}$,*

$$\operatorname{lpr}(\mathscr{E}_{\mathscr{A}_{\underline{P}} \cup \mathscr{A}_{\operatorname{ess\,inf}_{\underline{P}}}})(f) = \sup\left\{\underline{P}(g) : g \in \mathbb{B} \textit{ and } g \leq f \textit{ a.e. } [\underline{P}]\right\}.$$

Proof. Using the coherence of $\underline{P}$ and $\operatorname{ess\,inf}_{\underline{P}}$, and Equation $(13.13)_{259}$, it is easy to see that $\mathscr{E}_{\mathscr{A}_{\underline{P}} \cup \mathscr{A}_{\operatorname{ess\,inf}_{\underline{P}}}}$ equals

$$\left\{g + h_1 + h_2 : g, h_1 \in \mathbb{G}, h_2 \in \mathbb{B}, g \geq 0, \operatorname{ess\,inf}_{\underline{P}}(h_1) > 0 \text{ and } \underline{P}(h_2) > 0\right\},$$

and because, by Definition 14.17_{312}, $\operatorname{ess\,inf}_{\underline{P}}(h_1) > 0$ if and only if there are $\varepsilon > 0$ and $N \in \mathbb{G}^0_{\underline{P}}$ such that $h_1 \geq N + \varepsilon$, this is also equal to

$$\left\{g + N + \varepsilon + h_2 : g \in \mathbb{G}, N \in \mathbb{G}^0_{\underline{P}}, h_2 \in \mathbb{B}, \varepsilon > 0, g \geq 0 \text{ and } \underline{P}(h_2) > 0\right\},$$

and therefore, by Equation $(13.2)_{240}$,

$$\operatorname{lpr}(\mathscr{E}_{\mathscr{A}_{\underline{P}} \cup \mathscr{A}_{\operatorname{ess\,inf}_{\underline{P}}}})(f)$$

$$= \sup\left\{\mu \in \mathbb{R} : f - \mu \in \mathscr{E}_{\mathscr{A}_{\underline{P}} \cup \mathscr{A}_{\operatorname{ess\,inf}_{\underline{P}}}}\right\}$$

$$= \sup \left\{ \mu \in \mathbb{R} : (\exists h_2 \in \mathbb{B})(\exists \varepsilon > 0)(\exists N \in \mathbb{G}^0_{\underline{P}}) \right.$$
$$\left. (\underline{P}(h_2) > 0 \text{ and } f - \mu \geq h_2 + N + \varepsilon) \right\}$$
$$= \sup \left\{ \mu \in \mathbb{R} : (\exists h_2 \in \mathbb{B})(\exists N \in \mathbb{G}^0_{\underline{P}})(\underline{P}(h_2) > 0 \text{ and } f - \mu \geq h_2 + N) \right\}.$$

Now continue as in the proof of Theorem 14.21_{316}. □

These results show that for unbounded gambles h, contrary to $\mathrm{lpr}(\mathscr{E}_{\mathscr{A}_{\underline{P}}})(h)$, $\mathrm{lpr}(\mathscr{E}_{\mathscr{A}_{\underline{P}} \cup \mathbb{G}^0_{\underline{P}}})(h)$ will not be $-\infty$ unless h is not $\underline{P}$-essentially bounded below and, similarly, that $\mathrm{upr}(\mathscr{E}_{\mathscr{A}_{\underline{P}} \cup \mathbb{G}^0_{\underline{P}}})(h)$ will not be $+\infty$ unless h is not $\underline{P}$-essentially bounded above. In other words, the set $\mathbb{G}^\sharp_{\underline{P}}$ of $\underline{P}$-essentially bounded gambles contains exactly those gambles h for which

$$-\infty < \mathrm{lpr}(\mathscr{E}_{\mathscr{A}_{\underline{P}} \cup \mathbb{G}^0_{\underline{P}}})(h) \leq \mathrm{upr}(\mathscr{E}_{\mathscr{A}_{\underline{P}} \cup \mathbb{G}^0_{\underline{P}}})(h) < +\infty.$$

This leads to the following immediate conclusion.

Theorem 14.23 *The functionals* $\underline{P}^\sharp$ *and* $\overline{P}^\sharp$*, defined on* $\mathbb{G}^\sharp_{\underline{P}}$ *by*

$$\underline{P}^\sharp(f) := \mathrm{lpr}(\mathscr{E}_{\mathscr{A}_{\underline{P}} \cup \mathbb{G}^0_{\underline{P}}})(f) = \sup \left\{ \underline{P}(g) : g \in \mathbb{B} \text{ and } g \leq f \text{ a.e. } [\underline{P}] \right\} \tag{14.1}$$

$$\overline{P}^\sharp(f) := \mathrm{upr}(\mathscr{E}_{\mathscr{A}_{\underline{P}} \cup \mathbb{G}^0_{\underline{P}}})(f) = \inf \left\{ \overline{P}(g) : g \in \mathbb{B} \text{ and } g \geq f \text{ a.e. } [\underline{P}] \right\} \tag{14.2}$$

for all $\underline{P}$*-essentially bounded gambles* f*, are real-valued conjugate coherent lower and upper previsions on* $\mathbb{G}^\sharp_{\underline{P}}$.

In the rest of this section, we study these lower previsions in more detail. We begin by taking a closer look at their domain.

Theorem 14.24 $\mathbb{G}^\sharp_{\underline{P}}$ *is a linear lattice with respect to the point-wise ordering of gambles that contains all bounded gambles and all* $\underline{P}$*-null gambles.*

Proof. It is clear that all bounded gambles and all $\underline{P}$-null gambles are also $\underline{P}$-essentially bounded.

Consider any gambles f and g in $\mathbb{G}^\sharp_{\underline{P}}$ and any real numbers a and b. Then because $0 \leq |af + bg| \leq |a||f| + |b||g|$, if follows from the monotonicity, sub-additivity and non-negative homogeneity (Theorem 13.31(vi), (iv) and $(\mathrm{v})_{266}$) of the coherent upper prevision $\operatorname{ess\,sup}_{\underline{P}}$ (Theorem 14.20_{315}) that

$$0 \leq \operatorname{ess\,sup}_{\underline{P}} |af + bg| \leq |a| \operatorname{ess\,sup}_{\underline{P}} |f| + |b| \operatorname{ess\,sup}_{\underline{P}} |g| = 0,$$

which shows that $af + bg \in \mathbb{G}^\sharp_{\underline{P}}$ (see Proposition $14.19(\mathrm{vi})_{314}$), so $\mathbb{G}^\sharp_{\underline{P}}$ is a linear space.

To show that $\mathbb{G}^\sharp_{\underline{P}}$ is a linear lattice, first observe that both $|f \wedge g|$ and $|f \vee g|$ are dominated by $|f| \vee |g|$. As, by Definition 14.16(C)$_{311}$, there are $\underline{P}$-null gambles N and M and bounded gambles $f_\flat$ and $g_\flat$ such that $f = f_\flat + N$ and $g = g_\flat + M$, we find that $|f| \vee |g| \leq |f_\flat| \vee |g_\flat| + |N| \vee |M|$, and because $|f_\flat| \vee |g_\flat|$ is a bounded gamble and $|N| \vee |M|$ is $\underline{P}$-null (see Corollary 14.12$_{310}$), it follows that $|f| \vee |g|$, and therefore also $f \wedge g$ and $f \vee g$, are $\underline{P}$-essentially bounded. □

As for every $\underline{P}$-essentially bounded gamble f, there is some bounded gamble $f_\flat$ such that $f = f_\flat$ a.e. [$\underline{P}$], we could wonder whether $\underline{P}^\sharp(f) = \underline{P}(f_\flat)$ and $\overline{P}^\sharp(f) = \overline{P}(f_\flat)$. Checking that this is indeed the case is the subject of the following propositions and corollary.

Proposition 14.25 *Let f and g be bounded gambles. If $f \leq g$ a.e. [$\underline{P}$], then $\underline{P}(f) \leq \underline{P}(g)$ and $\overline{P}(f) \leq \overline{P}(g)$.*

Proof. Assume that $f \leq g$ a.e. [$\underline{P}$]. By Definition 14.9$_{308}$, there is some $\underline{P}$-null gamble N' such that $f \leq g + N'$. But this means that there is also some $\underline{P}$-null bounded gamble N such that $f \leq g + N$. Indeed, let $N := |N'| \wedge |f - g|$. Then N is $\underline{P}$-null because $N \leq |N'|$, and N is a bounded gamble because $\sup |N| \leq \sup |f - g| < +\infty$. It then follows from the coherence of the lower prevision $\underline{P}$ (see Theorem 4.13(iv) and (v)$_{53}$) that

$$\underline{P}(f) \leq \underline{P}(g + N) \leq \underline{P}(g) + \overline{P}(N),$$
$$\overline{P}(f) \leq \overline{P}(g + N) \leq \overline{P}(g) + \overline{P}(N).$$

By Proposition 14.14(i)$_{310}$, $\overline{P}(N) = 0$, so indeed $\underline{P}(f) \leq \underline{P}(g)$ and $\overline{P}(f) \leq \overline{P}(g)$. □

Corollary 14.26 *Let f and g be bounded gambles. If $f = g$ a.e. [$\underline{P}$], then $\underline{P}(f) = \underline{P}(g)$ and $\overline{P}(f) = \overline{P}(g)$.*

The coherent lower previsions $\underline{P}^\sharp$ and $\overline{P}^\sharp$ also satisfy a generalised monotonicity property.

Proposition 14.27 *Consider any f_1 and f_2 in $\mathbb{G}^\sharp_{\underline{P}}$ such that $f_1 \leq f_2$ a.e. [$\underline{P}$], then $\underline{P}^\sharp(f_1) \leq \underline{P}^\sharp(f_2)$ and $\overline{P}^\sharp(f_1) \leq \overline{P}^\sharp(f_2)$.*

Proof. We only give the proof for $\underline{P}^\sharp$, as the proof for $\overline{P}^\sharp$ is similar. Consider any N_1 in $\mathbb{G}^0_{\underline{P}}$ and g in $\mathbb{B}$ such that $g \leq f_1 + N_1$. By assumption, there is some N_2 in $\mathbb{G}^0_{\underline{P}}$ such that $f_1 \leq f_2 + N_2$ and, therefore, $g \leq f_2 + N_1 + N_2$. This tells us that if $g \leq f_1$ a.e. [$\underline{P}$], then also $g \leq f_2$ a.e. [$\underline{P}$]. Now use Theorem 14.23. □

Proposition 14.28 *$\underline{P}^\sharp$ is a coherent extension of the coherent lower prevision $\underline{P}$ to the set $\mathbb{G}^\sharp_{\underline{P}}$.*

Proof. It follows from Theorem 14.23_{318} that we only have to show that $\underline{P}^{\sharp}$ is an extension of $\underline{P}$ to $\mathbb{G}^{\sharp}_{\underline{P}}$. Let f be a bounded gamble on $\mathcal{X}$. Then because $f \leq f$ a.e. $[\underline{P}]$, it follows from Equation $(14.1)_{318}$ that $\underline{P}^{\sharp}(f) \geq \underline{P}(f)$. On the other hand, if $h \in \mathbb{B}$ and $h \leq f$ a.e. $[\underline{P}]$, then it follows from Proposition $14.25_{\curvearrowleft}$ that $\underline{P}(h) \leq \underline{P}(f)$, and if we take the supremum of all such h, we infer from Equation $(14.1)_{318}$ that $\underline{P}^{\sharp}(f) \leq \underline{P}(f)$. Hence, indeed $\underline{P}^{\sharp}(f) = \underline{P}(f)$. □

Proposition 14.29 *Consider any $f \in \mathbb{G}^{\sharp}_{\underline{P}}$, then for all $g \in \mathbb{B}$ and all $N \in \mathbb{G}^{0}_{\underline{P}}$ such that $f = g + N$, we have that $\underline{P}^{\sharp}(f) = \underline{P}(g)$ and $\overline{P}^{\sharp}(f) = \overline{P}(g)$.*

Proof. Consider any $g \in \mathbb{B}$ and all $N \in \mathbb{G}^{0}_{\underline{P}}$ such that $f = g + N$. Then in particular $g \leq f$ a.e. $[\underline{P}]$ and, therefore, $\underline{P}^{\sharp}(f) \geq \underline{P}(g)$, by Equation $(14.1)_{318}$. On the other hand, $f = g + N$ also implies that $f \leq g$ a.e. $[\underline{P}]$ and, therefore, $\underline{P}^{\sharp}(f) \leq \underline{P}^{\sharp}(g) = \underline{P}(g)$, where the inequality follows from Proposition $14.27_{\curvearrowleft}$ and the equality from Proposition $14.28_{\curvearrowleft}$. □

If a gamble f is $\underline{P}$-essentially bounded, then it follows from Definition $14.16(\mathrm{C})_{311}$ that there are a bounded gamble $f_{\flat}$ and a $\underline{P}$-null gamble N such that $f = f_{\flat} + N$ and it follows from Proposition 14.29 that $\underline{P}^{\sharp}(f) = \underline{P}(f_{\flat})$ and $\overline{P}^{\sharp}(f) = \overline{P}(f_{\flat})$. Any such $f_{\flat}$ is called a **determining bounded gamble** for f with respect to $\underline{P}$, or simply a determining bounded gamble for f. Proposition 14.29 guarantees that the lower and upper previsions of f are independent of the choice of the determining bounded gamble $f_{\flat}$.

We continue with a few more interesting properties of the extensions $\underline{P}^{\sharp}$.

Proposition 14.30 *If $f \in \mathbb{G}^{\sharp}_{\underline{P}}$, then* $\operatorname{ess\,inf}_{\underline{P}} f \leq \underline{P}^{\sharp}(f) \leq \overline{P}^{\sharp}(f) \leq \operatorname{ess\,sup}_{\underline{P}} f$.

Proof. That $f \in \mathbb{G}^{\sharp}_{\underline{P}}$ means that f is a $\underline{P}$-essentially bounded gamble, so $\operatorname{ess\,sup}_{\underline{P}} |f| < +\infty$, by Proposition $14.19(\mathrm{vi})_{314}$. As $\operatorname{ess\,inf}_{\underline{P}}$ and $\operatorname{ess\,sup}_{\underline{P}}$ are conjugate coherent lower and upper previsions by Theorem 14.20_{315}, we infer from $f \leq |f|$ and Theorem 13.31(i) and $(\mathrm{vi})_{266}$ that $\operatorname{ess\,inf}_{\underline{P}} f \leq \operatorname{ess\,sup}_{\underline{P}} f \leq \operatorname{ess\,sup}_{\underline{P}} |f|$ and, therefore, $\operatorname{ess\,inf}_{\underline{P}} f < +\infty$. By conjugacy and the fact that $\mathbb{G}^{\sharp}_{\underline{P}}$ is a linear space (see Theorem 14.24_{318}), this also leads to $\operatorname{ess\,sup}_{\underline{P}} f > -\infty$. Now use Proposition $14.19(\mathrm{ii})_{314}$ to find that $\operatorname{ess\,inf}_{\underline{P}} f \leq f$ a.e. $[\underline{P}]$ and $f \leq \operatorname{ess\,sup}_{\underline{P}} f$ a.e. $[\underline{P}]$. Hence, $\operatorname{ess\,inf}_{\underline{P}} f \leq f_{\flat}$ a.e. $[\underline{P}]$ and $f_{\flat} \leq \operatorname{ess\,sup}_{\underline{P}} f$ a.e. $[\underline{P}]$ (use Corollary 14.11_{310}). Now use Proposition $14.25_{\curvearrowleft}$ and the fact that $\underline{P}(f_{\flat}) \leq \overline{P}(f_{\flat})$. □

The following result generalises Proposition $14.14(\mathrm{i})_{310}$.

Proposition 14.31 *A gamble f is $\underline{P}$-null if and only if it is $\underline{P}$-essentially bounded and $\overline{P}^{\sharp}(|f|) = 0$:*

$$\mathbb{G}^{0}_{\underline{P}} = \left\{ f \in \mathbb{G}^{\sharp}_{\underline{P}} : \overline{P}^{\sharp}(|f|) = 0 \right\}. \tag{14.3}$$

Proof. 'if'. Suppose that f is $\underline{P}$-essentially bounded with $\overline{P}^{\sharp}(|f|) = 0$. Let $\varepsilon > 0$ and $A_\varepsilon := \{|f| > \varepsilon\}$. As $|f| \geq \varepsilon A_\varepsilon \geq 0$, it follows from Theorem 13.31(v) and $(\mathrm{vi})_{266}$ that $0 = \overline{P}^{\sharp}(|f|) \geq \varepsilon \overline{P}(A_\varepsilon) \geq 0$. We conclude that $\overline{P}(A_\varepsilon) = 0$ for every $\varepsilon > 0$.

'only if'. First recall from Theorem 14.24_{318} that $\mathbb{G}^0_{\underline{P}} \subseteq \mathbb{G}^{\sharp}_{\underline{P}}$. So consider any $\underline{P}$-essentially bounded gamble f. Then we infer from Equation $(14.2)_{318}$ that

$$\overline{P}^{\sharp}(|f|) = \inf\left\{\overline{P}(g) : g \in \mathbb{B} \text{ and } g \geq |f| \text{ a.e. } [\underline{P}]\right\}$$
$$\leq \inf\left\{\lambda : \lambda \in \mathbb{R} \text{ and } \lambda \geq |f| \text{ a.e. } [\underline{P}]\right\} = \|f\|_{\underline{P},\infty}$$

where the inequality follows because we are restricting the infimum to run only over constant gambles $\lambda \in \mathbb{R}$ and using $\overline{P}(\lambda) = \lambda$ and the last equality follows from Definition 14.17_{312}. So if f is $\underline{P}$-null, then we infer from Proposition $14.19(\mathrm{v})_{314}$ that $\|f\|_{\underline{P},\infty} = 0$ and, therefore, also $\overline{P}^{\sharp}(|f|) = 0$. □

It turns out to be quite straightforward to come up with a determining bounded gamble constructively, as follows. We begin by associating with a gamble a collection of bounded gambles, leading to a definition that will also prove useful in Chapter 15_{327}:

Definition 14.32 *Let f be a gamble. For any real numbers a and b such that $a \leq b$, the (a, b)-**cut of** f is the bounded gamble f_a^b defined by*

$$f_a^b(x) := \begin{cases} a & \text{if } f(x) < a, \\ b & \text{if } f(x) > b, \\ f(x) & \text{otherwise.} \end{cases}$$

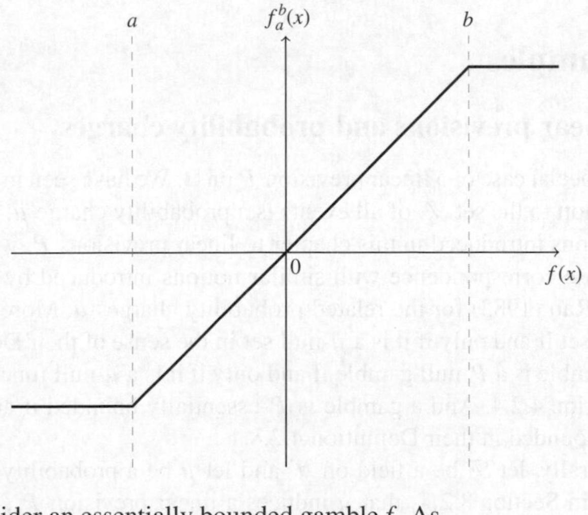

Now, consider an essentially bounded gamble f. As

$$-\infty < -\operatorname{ess\,sup}_{\underline{P}} |f| \leq \operatorname{ess\,inf}_{\underline{P}} f \leq \operatorname{ess\,sup}_{\underline{P}} f \leq \operatorname{ess\,sup}_{\underline{P}} |f| < +\infty,$$

we infer from Proposition 14.19(ii)$_{314}$ that

$$(\text{ess inf}_{\underline{P}} f \leq f \leq \text{ess sup}_{\underline{P}} f) \text{ a.e. } [\underline{P}], \tag{14.4}$$

and therefore we may replace the infimum in the first of the equivalent expressions in Definition 14.17$_{312}$ with a minimum:

$$\text{ess sup}_{\underline{P}} f = \min \left\{ \lambda \in \mathbb{R} : f \leq \lambda \text{ a.e. } [\underline{P}] \right\},$$

and there are similar expressions for the $\underline{P}$-essential infimum and the $\underline{P}$-essential supremum norm. This allows us to prove the validity of the following constructive recipe for finding determining bounded gambles.

Proposition 14.33 *If f is $\underline{P}$-essentially bounded, then the bounded gamble f_a^b, with $a := \text{ess inf}_{\underline{P}} f$ and $b = \text{ess sup}_{\underline{P}} f$, is a determining bounded gamble for f and, therefore, $\underline{P}^\sharp(f) = \underline{P}(f_a^b)$.*

Proof. Fix $\varepsilon > 0$ and infer from

$$\left\{ |f - f_a^b| > \varepsilon \right\} = \left\{ f > \text{ess sup}_{\underline{P}} f + \varepsilon \right\} \cup \left\{ f < \text{ess inf}_{\underline{P}} f - \varepsilon \right\}$$

and the monotonicity and sub-additivity (Theorem 4.13(iv) and (v)$_{53}$) of the coherent $\overline{P}$ that

$$0 \leq \overline{P}(\left\{ |f - f_a^b| > \varepsilon \right\}) \leq \overline{P}\left(\left\{ f > \text{ess sup}_{\underline{P}} f + \varepsilon \right\} \right) + \overline{P}\left(\left\{ f < \text{ess inf}_{\underline{P}} f - \varepsilon \right\} \right).$$

The terms on the right-hand side are zero due to Equation (14.4). So too is, therefore, the term in the middle. □

14.5 Examples

14.5.1 Linear previsions and probability charges

Consider the special case of a linear prevision P on $\mathbb{B}$. We have seen in Section 8.2$_{159}$ that its restriction to the set $\mathscr{P}$ of all events is a probability charge μ. When we specialise the notions introduced in this chapter to linear previsions P, we see that they are in very close correspondence with similar notions introduced by Bhaskara Rao and Bhaskara Rao (1983) for the related probability charges μ. More specifically, a set is a P-null set if and only if it is a μ-null set in the sense of their Definition 4.2.2. Similarly, a gamble is a P-null gamble if and only if it is a μ-null function according to their Definition 4.2.4. And a gamble is P-essentially bounded if and only if it is μ-essentially bounded in their Definition 4.2.8.

More generally, let $\mathscr{F}$ be a field on $\mathscr{X}$ and let μ be a probability charge on $\mathscr{F}$. We have seen in Section 8.2$_{159}$ that μ induces a linear prevision P_μ, whose natural extension to bounded gambles we have denoted by $\underline{E}_\mu$. Then, because for all events $A \subseteq \mathscr{X}$, we have that $\overline{E}_\mu(I_A) = \mu^*(A)$ (see Theorem 8.18$_{163}$), a set is a $\underline{E}_\mu$-null set if and only if it is a μ-null set in the sense of their Definition 4.2.2. Similarly, a gamble is a $\underline{E}_\mu$-null gamble if and only if it is a μ-null function according to their Definition 4.2.4. And a gamble is $\underline{E}_\mu$-essentially bounded if and only if it is μ-essentially bounded in their Definition 4.2.8.

14.5.2 Vacuous lower previsions

Let $\underline{P} = \inf$ be the vacuous lower prevision on $\mathbb{B}$. Then the empty set is the only inf-null set. The only inf-null gamble is 0. The inf-essential supremum, inf-essential infimum and inf-essential supremum norm coincide with the supremum, the infimum and the supremum norm, respectively. We have that $\mathbb{G}^\sharp_{\inf} = \mathbb{B}$ and $\inf^\sharp = \inf$.

More generally, let A be any non-empty subset of $\mathcal{X}$ and let $\underline{P}_A$ be the vacuous lower prevision on $\mathbb{B}$ relative to the event A:

$$\underline{P}_A(f) = \inf_{x \in A} f(x) \text{ for all } f \in \mathbb{B}.$$

We have also previously (see Example 13.10_{242}) denoted by $\underline{P}_A$ the (unconditional) vacuous lower prevision – defined on all gambles – relative to A, so in what follows, whether $\underline{P}_A$ is a lower prevision on bounded gambles or a lower prevision on gambles should be inferred from the context – from the argument it is applied to: no conflict can arise.

As $\mathcal{N}_{\underline{P}_A} = \left\{B \subseteq \mathcal{X} : B \cap A = \emptyset\right\}$, the null sets are the ones included in A^c. The null gambles are the ones that are zero on A:

$$\mathbb{G}^0_{\underline{P}_A} = \left\{f \in \mathbb{G} : I_A f = 0\right\}.$$

f is dominated almost everywhere by g if g dominates f on A and equal almost everywhere to g if equal to it on A. The essentially bounded gambles are the ones that are bounded on A:

$$\mathbb{G}^\sharp_{\underline{P}_A} = \left\{f \in \mathbb{G} : I_A f \in \mathbb{B}\right\}.$$

For any such essentially bounded gamble f, we get $\underline{P}^\sharp_A(f) = \inf_{x \in A} f(x)$, and the following equalities are easily seen to hold:

$$\operatorname{ess\,inf}_{\underline{P}_A} g = \inf_{x \in A} g(x) = \underline{P}_A(g) = \underline{E}_{\underline{P}^\sharp_A}(g) \text{ for any gamble } g.$$

14.5.3 Lower previsions associated with proper filters

Consider the coherent lower prevision $\underline{P}_{\mathcal{F}}$ on $\mathbb{B}$ associated with a proper filter $\mathcal{F}$, defined earlier in Equation $(5.5)_{83}$:

$$\underline{P}_{\mathcal{F}}(f) = \sup_{F \in \mathcal{F}} \inf_{x \in F} f(x).$$

A set A is null if and only if $(\exists F \in \mathcal{F}) A \cap F = \emptyset$, meaning that it eventually becomes *practically impossible* in the corresponding vacuous model.

A gamble f is null if and only if $(\forall\varepsilon > 0)(\exists F_\varepsilon \in \mathscr{F})F_\varepsilon \subseteq \{|f| \leq \varepsilon\}$, meaning that $|f|$ eventually becomes smaller than, or bounded by, any positive real number. Similarly, two gambles are equal almost everywhere if the absolute value of their difference eventually becomes smaller than any positive real number.

A gamble f is essentially bounded if and only if $|f|$ eventually becomes smaller than some bounded gamble h,[2] meaning that $(\exists F \in \mathscr{F})F \subseteq \{|f| \leq h\}$, which is also equivalent to

$$\inf_{F\in\mathscr{F}} \sup_{x\in F} |f|(x) = \inf_{F\in\mathscr{F}} \sup\big(|f||F\big) < +\infty. \tag{14.5}$$

And the essential infimum of a gamble f is, roughly speaking, the infimum of the values it eventually reaches:

$$\operatorname{ess\,inf}_{\underline{P}_{\mathscr{F}}} f = \sup_{F\in\mathscr{F}} \inf(f|F) = \sup_{F\in\mathscr{F}} \inf_{x\in F} f(x),$$

so $\operatorname{ess\,inf}_{\underline{P}_{\mathscr{F}}}$ can also be obtained by *formally extending the defining expression for $\underline{P}_{\mathscr{F}}$ from bounded gambles to gambles.*

To conclude, we show that

$$\underline{P}^\sharp_{\mathscr{F}}(f) = \sup_{F\in\mathscr{F}} \inf_{x\in F} f(x) \text{ for any essentially bounded gamble } f$$

as well, so $\underline{P}^\sharp_{\mathscr{F}}$ and $\operatorname{ess\,inf}_{\underline{P}_{\mathscr{F}}}$ coincide on all essentially bounded gambles. Indeed, we already know from Proposition 14.30$_{320}$ that $\operatorname{ess\,inf}_{\underline{P}_{\mathscr{F}}} f \leq \underline{P}^\sharp_{\mathscr{F}}(f)$. Assume *ex absurdo* that $\operatorname{ess\,inf}_{\underline{P}_{\mathscr{F}}} f < \underline{P}^\sharp_{\mathscr{F}}(f)$. If we let $a := \sup_{F\in\mathscr{F}} \inf_{x\in F} f(x)$ and $b := \inf_{F\in\mathscr{F}} \sup_{x\in F} f(x)$ and use Proposition 14.33$_{322}$, this implies that $a < \sup_{F\in\mathscr{F}} \inf_{x\in F} f_a^b(x)$. So we infer that there are $\varepsilon > 0$ and $F_0 \in \mathscr{F}$ such that $f_a^b(x) > a + \varepsilon$, and therefore also $f(x) > a + \varepsilon$, for all $x \in F_0$. But then $\inf_{x\in F_0} f(x) > a = \sup_{F\in\mathscr{F}} \inf_{x\in F} f(x)$, a contradiction.

14.5.4 Limits inferior

Consider a set $\mathscr{A}$ directed by a binary relation $\leq$. If we recall from Section 5.5.3$_{90}$ that the limit inferior operator lim inf associated with this directed set can be seen as a lower prevision associated with a proper filter of subsets of $\mathscr{A}$, we can derive the following conclusions from the discussion in the previous section.

A subset A of $\mathscr{A}$ is null if and only if $(\exists\alpha \in \mathscr{A})(\forall\beta \geq \alpha)\beta \notin A$, meaning that the real net $I_A(\alpha)$ converges to zero: $I_A \to 0$.

A real net f is null if and only if $(\forall\varepsilon > 0)(\exists\alpha_\varepsilon \in \mathscr{A})(\forall\beta \geq \alpha)|f(\beta)| \leq \varepsilon$, meaning that $|f| \to 0$. Similarly, two real nets f and g are equal almost everywhere if the absolute value of their difference converges to zero: $|f - g| \to 0$.

A real net f is essentially bounded if and only if $|f|$ eventually becomes smaller than some bounded net h,[3] meaning that $(\exists\alpha \in \mathscr{A})(\forall\beta \geq \alpha)|f|(\beta) \leq h(\beta)$ or, equivalently, $\limsup |f| < +\infty$:

$$\mathbb{G}^\sharp_{\liminf}(\mathscr{A}) = \{f \in \mathbb{G}(\mathscr{A}) \colon \limsup |f| < +\infty\}.$$

[2] Or, equivalently, eventually becomes smaller than some real number.

[3] Or, equivalently, eventually becomes smaller than some real number.

The essential infimum of a real net g coincides with its limit inferior,

$$\operatorname{ess\,inf}_{\liminf} g = \sup_{\alpha \in \mathscr{A}} \inf_{\beta \geq \alpha} g(\beta) = \liminf g,$$

and for any essentially bounded gamble f, $\liminf^{\sharp}(f) = \liminf f$ as well, so $\liminf^{\sharp}$ and $\operatorname{ess\,inf}_{\liminf}$ coincide on all essentially bounded real nets.

14.5.5 Belief functions

As another example, we turn to the completely monotone coherent lower prevision $\underline{P}_m$ associated with a basic probability assignment m on a finite set of focal elements $\mathscr{H} \subseteq \mathscr{P} \setminus \{\emptyset\}$, where the space $\mathscr{X}$ is not necessarily finite (see the discussion in Section 7.3_{128}):

$$\underline{P}_m(f) = \sum_{F \in \mathscr{H}} m(F)\underline{P}_F(f) = \sum_{F \in \mathscr{H}} m(F) \inf_{x \in F} f(x) \text{ for all bounded gambles } f \text{ on } \mathscr{X}.$$

An event A is $\underline{P}_m$-null if and only if $A \cap \bigcup \mathscr{H} = \emptyset$, and a gamble f is $\underline{P}_m$-null if and only if it is zero on $\bigcup \mathscr{H}$.

A gamble f dominates a gamble g almost everywhere $[\underline{P}_m]$ if and only if f dominates g on $\bigcup \mathscr{H}$, and $f = g$ a.e. $[\underline{P}_m]$ if and only if f and g coincide on $\bigcup \mathscr{H}$. Also, f is essentially bounded $[\underline{P}_m]$ if and only if f is bounded – coincides with a bounded gamble – on $\bigcup \mathscr{H}$. Moreover,

$$\operatorname{ess\,inf}_{\underline{P}_m} f = \sup_{A \subseteq (\bigcup \mathscr{H})^c} \inf(f|A^c) = \sup_{\bigcup \mathscr{H} \subseteq A^c} \inf(f|A^c) = \inf\left(f \middle| \bigcup \mathscr{H}\right)$$

$$\operatorname{ess\,sup}_{\underline{P}_m} f = \inf_{A \subseteq (\bigcup \mathscr{H})^c} \sup(f|A^c) = \inf_{\bigcup \mathscr{H} \subseteq A^c} \sup(f|A^c) = \sup\left(f \middle| \bigcup \mathscr{H}\right),$$

and for any essentially bounded f, we see that

$$\underline{P}_m^{\sharp}(f) = \sum_{F \in \mathscr{H}} m(F) \inf_{x \in F} f(x) = \sum_{F \in \mathscr{H}} m(F) \inf(f|F),$$

because we can take as a determining bounded gamble for f any bounded gamble that coincides with f on $\bigcup \mathscr{H}$.

14.5.6 Possibility measures

Next in this list of examples, we consider the case of a possibility measure Π on a non-empty set $\mathscr{X}$, with possibility distribution π. We have seen in Section 7.8_{143} that its natural extension to an upper prevision on all bounded gambles f is given by

$$\overline{P}_\pi(f) = R\int_0^1 \sup\{f(x) : \pi(x) > \alpha\}\, \mathrm{d}\alpha.$$

For events, $\overline{P}_\pi(A) = \sup_{x \in \mathscr{X}} \pi(x)$, which is zero if and only if $\pi(x) = 0$ for all $x \in A$. In other words, if we consider the strict level set $\{\pi > 0\} := \{x \in \mathscr{X} : \pi(x) > 0\}$, then an event A is $\underline{P}_\pi$-null if and only if $A \cap \{\pi > 0\} = \emptyset$ or, equivalently, if and only if $A \subseteq \{\pi = 0\}$. Therefore, a gamble f is $\underline{P}_\pi$-null if and only if it is zero on – at least – $\{\pi > 0\}$.

A gamble f dominates a gamble g almost everywhere [$\underline{P}_\pi$] if and only if f dominates g on $\{\pi > 0\}$, and $f = g$ a.e. [$\underline{P}_\pi$] if and only if f and g coincide on $\{\pi > 0\}$. Also, f is essentially bounded [$\underline{P}_\pi$] if and only if f is bounded – coincides with a bounded gamble – on $\{\pi > 0\}$. Moreover,

$$\operatorname{ess\,inf}_{\underline{P}_\pi} f = \sup_{A \subseteq \{\pi>0\}^c} \inf(f|A^c) = \sup_{\{\pi>0\} \subseteq A^c} \inf(f|A^c) = \inf(f|\{\pi > 0\})$$

$$\operatorname{ess\,sup}_{\underline{P}_\pi} f = \inf_{A \subseteq \{\pi>0\}^c} \sup(f|A^c) = \inf_{\{\pi>0\} \subseteq A^c} \sup(f|A^c) = \sup(f|\{\pi > 0\}),$$

and for any essentially bounded f, we see that

$$\overline{P}^{\sharp}_\pi(f) = R\int_0^1 \sup\{f(x) : \pi(x) > \alpha\}\, \mathrm{d}\alpha,$$

because we can take as a determining bounded gamble for f any bounded gamble that coincides with f on $\{\pi > 0\}$.

15

Lower previsions for previsible gambles

In this chapter, we continue the line of reasoning in Chapter 14_{304}: we want to extend a coherent lower prevision $\underline{P}$, defined on the set $\mathbb{B}$ of all bounded gambles, to a larger set of not necessarily bounded gambles. But instead of considering gambles that are essentially equal to some bounded gamble, we now move our attention to those gambles that can be approximated by a sequence of bounded gambles, loosely speaking in the sense that the sequence of bounded gambles converges to the gamble outside a null set. This type of convergence will be called *convergence in probability*, as it is a direct generalisation of its counterpart in probabilistic measure theory. It is introduced and discussed in Section $15.1_{\frown}$.

However, it is usually quite easy to construct different sequences of bounded gambles that converge in probability to the same gamble but for which the corresponding sequences of lower previsions do not converge to the same number. Proposition 15.4_{331} gives a sufficient Cauchy-type condition for such candidate sequences to converge to the same lower prevision. The gambles for which such a construction is possible are called *previsible*; it is for such gambles that our extension method works. Previsibility is introduced and studied in Section 15.2_{331}. We prove that all properties of coherent lower and upper previsions on bounded gambles still hold for the thus extended lower and upper previsions, that is, the extension is coherent.

A very reasonable way of approximating an unbounded gamble is by a sequence of it cuts: bounded gambles that coincide with the given gamble up to certain thresholds. We succeed in proving that every previsible gamble can be approximated by a sequence of cuts in the sense described earlier. Moreover, the converse also holds: if a gamble can be approximated by cuts in the sense described earlier, then it is previsible. To establish this result, we need to prove a version of the Lebesgue dominated convergence theorem for coherent lower previsions. All of this is covered in Sections 15.3_{340}–15.5_{348}.

Lower Previsions, First Edition. Matthias C.M. Troffaes and Gert de Cooman.

It is then not so difficult to come up with a simple sufficient condition for previsibility without referring explicitly to approximating sequences of bounded gambles – but we still need to calculate two limits. This is the subject of Section 15.6_{350}.

In Section 15.7_{352}, we consider the special case of extending 2-monotone lower previsions. We find a necessary and sufficient condition for previsibility, establish equivalence between previsibility and Choquet integrability and prove equality of the previsible extension and the Choquet integral. As a special case, we obtain equivalence between the Dunford integral and the Choquet integral.

In Section 15.8_{355}, we look at previsibility for convex combinations of lower previsions.

In Section 15.9_{355}, we establish a one-to-one correspondence between dominating linear previsions on bounded gambles and dominating linear previsions on the extended domain, and we prove that every extended lower prevision is the lower envelope of its extended dominating linear previsions.

In Section 15.10_{358}, we investigate previsibility in a wide range of examples: linear previsions, probability density functions, vacuous lower previsions, lower previsions associated with proper filters, limits inferior, belief functions and possibility measures. We also take a brief look at the problem of estimation with quadratic loss functions.

The key idea of this chapter can be traced back to Dunford (1935) who defined an integral on gambles by taking Cauchy sequences of simple gambles – we will use the same technique to extend lower previsions from bounded gambles to (unbounded) gambles by taking Cauchy sequences of bounded gambles. Our approach is strongly reminiscent of, and inspired by, the way Bhaskara Rao and Bhaskara Rao (1983, Chapter 4) extend the Dunford integral from simple gambles to gambles: most of the arguments they make can be adapted and extended to work for this more general case as well.[1] This means that in a number of cases, we will be able to follow their argumentation quite closely – we will clearly indicate when and where we do so. But we are also able in some places to simplify and strengthen their results beyond direct generalisation and uncover new connections, most notably by using our cut approximation method. In particular, from Section 15.5_{348} onwards, we believe the material to be entirely new.

As already mentioned in the introduction of Chapter 14_{304}, this chapter closely follows Troffaes (2005, Sections 5.4 and 5.5), with some improvements and extensions. Most notably as far as extensions are concerned, we prove equivalence of Choquet integrability and previsibility under 2-monotonicity in Section 15.7_{352} and provide a range of examples in Section 15.10_{358}.

15.1 Convergence in probability

In the rest of this chapter, $\underline{P}$ is a coherent lower prevision defined on the set $\mathbb{B}$ of all bounded gambles on $\mathcal{X}$. We begin with associating with $\underline{P}$ a special type of

[1] We do not start from simple gambles because that type of extension is already taken care of by using the procedure of natural extension.

convergence of gambles. We have already introduced this type of convergence in our discussion of the Dunford integrability of bounded gambles in Section 8.7$_{172}$.

Definition 15.1 (Convergence in probability) *Let f be a gamble and let f_α be a net of gambles. Then the following conditions are equivalent; if any (and hence all) of them are satisfied, we say that the net f_α* ***converges in probability*** *$[\underline{P}]$ to f, which we denote as $f_\alpha \xrightarrow{\underline{P}} f$.*

(A) *For every $\varepsilon > 0$, $\lim_\alpha \overline{P}\left(\{|f - f_\alpha| > \varepsilon\}\right) = 0$.*

(B) *For every $\varepsilon > 0$, eventually, $\overline{P}\left(\{|f - f_\alpha| > \varepsilon\}\right) < \varepsilon$.*

Let us prove that these conditions are indeed equivalent.

Proof. (A)⇒(B). Consider $\varepsilon > 0$. Then (A) implies that for every $\delta > 0$, there is some $\alpha_{\varepsilon,\delta}$ such that $\overline{P}\left(\{|f - f_\alpha| > \varepsilon\}\right) < \delta$ for all $\alpha \geq \alpha_{\varepsilon,\delta}$. Now take $\delta := \varepsilon$ to see that for all $\varepsilon > 0$, there is some $\alpha_\varepsilon := \alpha_{\varepsilon,\varepsilon}$ such that $\overline{P}\left(\{|f - f_\alpha| > \varepsilon\}\right) < \varepsilon$ for all $\alpha \geq \alpha_\varepsilon$.

(B)⇒(A). Consider $\varepsilon > 0$. Let, for any $\delta > 0$, $\gamma := \min\{\varepsilon, \delta\}$. Then (B) implies that there is some α_δ such that for every $\alpha \geq \alpha_\delta$, $\overline{P}\left(\{|f - f_\alpha| > \gamma\}\right) < \gamma \leq \delta$. As $\{|f - f_\alpha| > \varepsilon\} \subseteq \{|f - f_\alpha| > \gamma\}$ and therefore also, by the coherence of the lower prevision $\underline{P}$ (Theorem 4.13(iv)$_{53}$), $\overline{P}\left(\{|f - f_\alpha| > \varepsilon\}\right) \leq \overline{P}\left(\{|f - f_\alpha| > \gamma\}\right)$, we find that $\overline{P}\left(\{|f - f_\alpha| > \varepsilon\}\right) < \delta$ for every $\alpha \geq \alpha_\delta$, so indeed $\overline{P}\left(\{|f - f_\alpha| > \varepsilon\}\right) \to 0$. □

When $\underline{P}$ is the natural extension $\underline{E}_\mu$ to all bounded gambles of the probability charge μ or, to be more precise, of the probability P_μ, our definition coincides with Bhaskara Rao and Bhaskara Rao's (1983, Section 4.3) notion of *hazy convergence*, which requires that $\lim_\alpha \mu^*\left(\{|f - f_\alpha| > \varepsilon\}\right) = 0$. Indeed, by Theorem 8.18$_{163}$, $\overline{E}_\mu$ coincides on events with the outer set function μ^* of μ.

As a direct consequence, when μ is in particular a probability measure, that is, σ-additive, this reduces to the classical notion of convergence in probability (see, for instance, Kallenberg, 2002, Chapter 4, p. 63).

Finally, when $\underline{P}$ is the vacuous lower prevision inf, this type of convergence reduces to uniform convergence.

The topology on $\mathbb{G}$ associated with convergence in probability is not generally Hausdorff, because limits are only unique up to almost everywhere equality, as we now show.

Proposition 15.2 *Let f and g be gambles, and let f_α be a net of gambles. Then the following statements hold:*

(i) *If $f_\alpha \xrightarrow{\underline{P}} f$ and $f = g$ a.e. $[\underline{P}]$, then $f_\alpha \xrightarrow{\underline{P}} g$.*

(ii) *Conversely, if $f_\alpha \xrightarrow{\underline{P}} f$ and $f_\alpha \xrightarrow{\underline{P}} g$, then $f = g$ a.e. $[\underline{P}]$.*

Proof. The proof follows from

$$\{|f_\alpha - g| > \varepsilon\} \subseteq \left\{|f_\alpha - f| > \frac{\varepsilon}{2}\right\} \cup \left\{|f - g| > \frac{\varepsilon}{2}\right\}$$
$$\{|f - g| > \varepsilon\} \subseteq \left\{|f - f_\alpha| > \frac{\varepsilon}{2}\right\} \cup \left\{|f_\alpha - g| > \frac{\varepsilon}{2}\right\}$$

and the monotonicity and sub-additivity of the coherent upper prevision $\overline{P}$ (see Theorem 4.13(iv) and (v)$_{53}$). □

We shall need the following convenient properties of convergence in probability a number of times.

Proposition 15.3 *Let f and g be two gambles, and let f_α and g_α be two nets of gambles. Assume that $f_\alpha \xrightarrow{P} f$ and $g_\alpha \xrightarrow{P} g$. Let a and b be real numbers, and let A be any subset of $\mathcal{X}$. Then $I_A f_\alpha$, $af_\alpha + bg_\alpha$, $|f_\alpha|$, $f_\alpha \vee g_\alpha$ and $f_\alpha \wedge g_\alpha$ converge in probability [$\underline{P}$] to $I_A f$, $af + bg$, $|f|$, $f \vee g$ and $f \wedge g$, respectively.*

Proof. For the combination with the indicator, we have $\{|I_A f - I_A f_\alpha| > \varepsilon\} \subseteq \{|f - f_\alpha| > \varepsilon\}$. Use the monotonicity of the coherent upper prevision $\overline{P}$ (see Theorem 4.13(iv)$_{53}$) to find that

$$\lim_\alpha \overline{P}\left(\{|I_A f - I_A f_\alpha| > \varepsilon\}\right) \leq \lim_\alpha \overline{P}\left(\{|f - f_\alpha| > \varepsilon\}\right) = 0,$$

whence indeed $I_A f_\alpha \xrightarrow{P} I_A f$.

For the linear combination, we treat addition and scalar multiplication separately. For the addition, we have that

$$\{|f + g - f_\alpha - g_\alpha| > \varepsilon\} \subseteq \left\{|f - f_\alpha| > \frac{\varepsilon}{2}\right\} \cup \left\{|g - g_\alpha| > \frac{\varepsilon}{2}\right\},$$

so we find, using the monotonicity and sub-additivity of the coherent upper prevision $\overline{P}$ (see Theorem 4.13(iv) and (v)$_{53}$), that

$$\lim_\alpha \overline{P}\left(\{|f + g - f_\alpha - g_\alpha| > \varepsilon\}\right)$$
$$\leq \lim_\alpha \overline{P}\left(\left\{|f - f_\alpha| > \frac{\varepsilon}{2}\right\}\right) + \lim_\alpha \overline{P}\left(\left\{|g - g_\alpha| > \frac{\varepsilon}{2}\right\}\right) = 0 + 0,$$

whence indeed $f_\alpha + g_\alpha \xrightarrow{P} f + g$. For scalar multiplication, we may assume that the scalar a is non-zero (the case $a = 0$ is trivial). Then

$$\lim_{n \to +\infty} \overline{P}\left(\{|af - af_\alpha| > \varepsilon\}\right) = \lim_{n \to +\infty} \overline{P}\left(\left\{|f - f_\alpha| > \frac{\varepsilon}{|a|}\right\}\right) = 0,$$

whence indeed $af_\alpha \xrightarrow{P} af$.

For the absolute value, we have $\{||f| - |f_\alpha|| \geq \varepsilon\} \subseteq \{|f - f_\alpha| > \varepsilon\}$. Again use the monotonicity of the coherent upper prevision $\overline{P}$ (see Theorem 4.13(iv)$_{53}$).

The maximum and the minimum of f and g can be written as linear combinations of f, g and $|f-g|$, because $f \vee g = (f+g+|f-g|)/2$ and $f \wedge g = (f+g-|f-g|)/2$, so this case follows from the previous cases. □

15.2 Previsibility

Now suppose we have a sequence of bounded gambles f_n that converges in probability $[\underline{P}]$ to a gamble f. It then seems obvious to try and extend $\underline{P}$ to f by considering the limit of the corresponding sequence of lower previsions $\underline{P}(f_n)$. But at this point, we have no guarantee that this sequence converges to some real number p or that this extension method leads to a unique value for p: if we have another sequence of bounded gambles $g_n \xrightarrow{\underline{P}} f$, then there is no guarantee that the sequence $\underline{P}(g_n)$ converges to the same limit p. In the next proposition, we establish a sufficient condition for the suggested extension method to work. It is a generalisation from Dunford integrals to lower and upper previsions of Proposition 4.4.10 in Bhaskara Rao and Bhaskara Rao (1983). Our proof follows theirs quite closely.

Proposition 15.4 *Let f_n and g_n be two sequences of bounded gambles converging in probability $[\underline{P}]$ to the same gamble h. Suppose that both*[2]

$$\lim_{n,m\to+\infty} \overline{P}\left(|f_n - f_m|\right) = 0 \text{ and } \lim_{n,m\to+\infty} \overline{P}\left(|g_n - g_m|\right) = 0.$$

Then the limits $\lim_{n\to+\infty} \overline{P}(f_n)$ and $\lim_{n\to+\infty} \overline{P}(g_n)$ exist, are finite and coincide. Also the limits $\lim_{n\to+\infty} \underline{P}(f_n)$ and $\lim_{n\to+\infty} \underline{P}(g_n)$ exist, are finite and coincide.

Proof. First, we prove that the limits exist and are finite. This follows from the following inequalities (guaranteed by the coherence of the lower prevision $\underline{P}$, see Theorem 4.13(x)$_{53}$):

$$\left|\overline{P}(f_n) - \overline{P}(f_m)\right| \le \overline{P}\left(|f_n - f_m|\right) \qquad \left|\underline{P}(f_n) - \underline{P}(f_m)\right| \le \overline{P}\left(|f_n - f_m|\right)$$
$$\left|\overline{P}(g_n) - \overline{P}(g_m)\right| \le \overline{P}\left(|g_n - g_m|\right) \qquad \left|\underline{P}(g_n) - \underline{P}(g_m)\right| \le \overline{P}\left(|g_n - g_m|\right).$$

As the right-hand sides converge to zero, the left-hand sides must converge to zero too. This means that the real sequences $\overline{P}(f_n)$, $\underline{P}(f_n)$, $\overline{P}(g_n)$ and $\underline{P}(g_n)$ are Cauchy. By the completeness of $\mathbb{R}$, their limits therefore exist and are finite.

It remains to prove that

$$\lim_{n\to+\infty} \overline{P}(f_n) = \lim_{n\to+\infty} \overline{P}(g_n) \text{ and } \lim_{n\to+\infty} \underline{P}(f_n) = \lim_{n\to+\infty} \underline{P}(g_n).$$

Define the sequence of bounded gambles $N_n := |f_n - g_n|$. Again using Theorem 4.13(x)$_{54}$, we find that $\left|\underline{P}(f_n) - \underline{P}(g_n)\right| \le \overline{P}(N_n)$ and $\left|\overline{P}(f_n) - \overline{P}(g_n)\right| \le \overline{P}(N_n)$. The

[2] These limits of real nets are introduced in Section 1.5$_7$ and, in particular, Example 1.10$_8$.

proof will, therefore, be complete if we can show that the real sequence $\overline{P}(N_n)$ converges to zero. This is what we now set out to do.

For every $n \in \mathbb{N}$ and every $A \subseteq \mathcal{X}$, define $a_n(A) := \overline{P}(N_n I_A)$. We must prove that the real sequence $a_n(\mathcal{X})$ converges to zero.

Every a_n is an element of the function space $\mathbb{R}^{\mathscr{P}}$. Equip this function space with the topology of uniform convergence on $\mathscr{P}$. From the completeness of $\mathbb{R}$, it follows by a well-established argument (see, for instance, Schechter, 1997, Section 19.12) that $\mathbb{R}^{\mathscr{P}}$ is complete with respect to the topology of uniform convergence on $\mathscr{P}$.

We first claim that a_n converges with respect to the topology of uniform convergence on $\mathscr{P}$. Indeed, consider $A \subseteq \mathcal{X}$, then, using the inequalities $\left|\overline{P}(f) - \overline{P}(g)\right| \leq \overline{P}(|f-g|)$, $||f| - |g|| \leq |f-g| \leq |f| + |g|$ and $|f| I_A \leq |f|$ for all bounded gambles f and g, as well as the monotonicity and sub-linearity of the coherent upper prevision $\overline{P}$ (see Theorem 4.13(iv) and (v)$_{53}$), we find that

$$\begin{aligned}\left|a_n(A) - a_m(A)\right| &= \left|\overline{P}\big(|f_n - g_n| I_A\big) - \overline{P}\big(|f_m - g_m| I_A\big)\right| \\ &\leq \overline{P}\Big(\big||f_n - g_n| - |f_m - g_m|\big| I_A\Big) \leq \overline{P}\Big(\big|(f_n - g_n) - (f_m - g_m)\big| I_A\Big) \\ &\leq \overline{P}\Big(\big|(f_n - f_m) - (g_n - g_m)\big|\Big) \leq \overline{P}\big(|f_n - f_m|\big) + \overline{P}\big(|g_n - g_m|\big).\end{aligned}$$

As the right-hand side converges to zero independently of A, it follows that a_n is Cauchy with respect to the topology of uniform convergence on $\mathscr{P}$. By the completeness of $\mathbb{R}^{\mathscr{P}}$ with respect to this topology, we find that a_n indeed converges uniformly on $\mathscr{P}$.

Uniform convergence implies point-wise convergence, so, for every $A \subseteq \mathcal{X}$, we can define $a(A) := \lim_{n \to +\infty} a_n(A)$. Again, we must prove that $a(\mathcal{X}) = 0$.

Let $\varepsilon > 0$. By the convergence of a_n with respect to the topology of uniform convergence on $\mathscr{P}$, there is some $M_\varepsilon \in \mathbb{N}$ such that

$$\left|a_n(A) - a(A)\right| < \varepsilon \text{ for all } A \subseteq \mathcal{X} \text{ and all } n \geq M_\varepsilon. \tag{15.1}$$

By Lemma 15.5 further on, there is some $\delta_\varepsilon > 0$ such that for every $A \subseteq \mathcal{X}$, we have that if $\overline{P}(A) < \delta_\varepsilon$, then $a_{M_\varepsilon}(A) = \overline{P}(N_{M_\varepsilon} I_A) < \varepsilon$. As $a(A) \leq \left|a(A) - a_{M_\varepsilon}(A)\right| + a_{M_\varepsilon}(A)$, it follows from the inequality (15.1) that

$$\overline{P}(A) < \delta_\varepsilon \Rightarrow a(A) < 2\varepsilon \text{ for all } A \subseteq \mathcal{X}. \tag{15.2}$$

Define $B := \left\{N_{M_\varepsilon} \neq 0\right\}$, then $N_{M_\varepsilon} I_{B^c} = 0$. This implies that $a_{M_\varepsilon}(B^c) = \overline{P}(N_{M_\varepsilon} I_{B^c}) = 0$. From the inequality (15.1), it follows that $a(B^c) < \varepsilon$.

Next we prove that $a(\mathcal{X}) < 5\varepsilon$. As this will hold for any $\varepsilon > 0$, we may then conclude that indeed $a(\mathcal{X}) = 0$. There are two possible cases.

1. Consider the case $\overline{P}(B) = 0$. Then $a(B) = \lim_{n \to +\infty} \overline{P}(N_n I_B) = 0$ because $0 \leq \overline{P}(N_n I_B) \leq \overline{P}(B) \sup N_n = 0$ for every $n \in \mathbb{N}$. By the sub-additivity of the coherent upper prevision $\overline{P}$ (see Theorem 4.13(v)$_{53}$), it follows that $a(\mathcal{X}) \leq a(B) + a(B^c) < 0 + \varepsilon < 5\varepsilon$.

2. Next, consider the other case $\overline{P}(B) > 0$. As f_n and g_n converge in probability [$\underline{P}$] to h, it follows from Proposition 15.3$_{330}$ that $N_n = |f_n - g_n|$ converges in probability [$\underline{P}$] to 0. Consequently, for all $\beta > 0$ and $\gamma > 0$, there is some $K_{\beta,\gamma} \in \mathbb{N}$ such that $\overline{P}\left(\{N_n > \beta\}\right) < \gamma$ for all $n \geq K_{\beta,\gamma}$. In particular, for the ε, M_ε and δ_ε constructed above and letting $\beta := \varepsilon/\overline{P}(B)$ and $\gamma := \delta_\varepsilon$, there is some $K_\varepsilon := \max\{K_{\varepsilon/\overline{P}(B),\delta_\varepsilon}, M_\varepsilon\} \geq M_\varepsilon$ such that

$$\overline{P}\left(\left\{N_n > \frac{\varepsilon}{\overline{P}(B)}\right\}\right) < \delta_\varepsilon \text{ for all } n \geq K_\varepsilon. \tag{15.3}$$

Let $C := \left\{N_{K_\varepsilon} \leq \varepsilon/\overline{P}(B)\right\}$. By the sub-additivity of the coherent upper prevision $\overline{P}$ (see Theorem 4.13(v)$_{53}$), we have that $a(\mathcal{X}) \leq a(B \cap C) + a(B \cap C^c) + a(B^c)$. We now investigate each term of this sum. From the inequality (15.1), we infer that $a(B \cap C) < a_{K_\varepsilon}(B \cap C) + \varepsilon$, because $K_\varepsilon \geq M_\varepsilon$. As $N_{K_\varepsilon}(x) \leq \varepsilon/\overline{P}(B)$ for all $x \in C$ and $\overline{P}(B \cap C) \leq \overline{P}(B)$, we find that $a_{K_\varepsilon}(B \cap C) = \overline{P}(N_{K_\varepsilon} I_{B\cap C}) \leq \varepsilon\overline{P}(B \cap C)/\overline{P}(B) \leq \varepsilon$. Hence, $a(B \cap C) < 2\varepsilon$. From the inequality (15.3), it follows that $\overline{P}(B \cap C^c) \leq \overline{P}(C^c) < \delta_\varepsilon$. But then (15.2) tells us that $a(B \cap C^c) < 2\varepsilon$. We have already proved that $a(B^c) < \varepsilon$. Hence, $a(\mathcal{X}) \leq a(B \cap C) + a(B \cap C^c) + a(B^c) < 2\varepsilon + 2\varepsilon + \varepsilon = 5\varepsilon$. □

Lemma 15.5 *Let f be a non-negative bounded gamble. Then for all $\varepsilon > 0$, there is some $\delta_\varepsilon > 0$ such that for all $A \subseteq \mathcal{X}$: if $\overline{P}(A) < \delta_\varepsilon$, then $\overline{P}(fI_A) < \varepsilon$.*

Proof. Let $\varepsilon > 0$. If $\sup f = 0$, then $f = 0$ and therefore $\overline{P}(fI_A) = 0$, independently of $A \subseteq \mathcal{X}$, whence trivially $\overline{P}(fI_A) < \varepsilon$. Hence, we may assume that $\sup f > 0$. Define $\delta_\varepsilon := \varepsilon/\sup f$. If $\overline{P}(A) < \delta_\varepsilon$, then $\overline{P}(fI_A) \leq \overline{P}(A)\sup f < \delta_\varepsilon \sup f = \varepsilon$. □

This leads us to the following definition. It is a generalisation of Bhaskara Rao and Bhaskara Rao's (1983, Definition 4.4.11) approach to the Dunford integral and Dunford integrability; see also Definition 8.29$_{172}$.

Definition 15.6 (Previsibility) *A gamble f is called **$\underline{P}$-previsible** if there is some sequence f_n of bounded gambles such that*

(i) *f_n converges in probability [$\underline{P}$] to f,*

(ii) $\lim_{n,m\to+\infty} \overline{P}\left(|f_n - f_m|\right) = 0$.

*If f is $\underline{P}$-previsible, then the **extended lower and upper prevision** of f are defined by $\underline{P}^{\mathbf{x}}(f) := \lim_{n\to+\infty} \underline{P}(f_n)$ and $\overline{P}^{\mathbf{x}}(f) := \lim_{n\to+\infty} \overline{P}(f_n)$, respectively. The sequence f_n is called a **determining sequence** of bounded gambles for f with respect to $\underline{P}$ or, simply, a determining sequence for f. The set of all $\underline{P}$-previsible gambles is denoted by $\mathbb{G}^{\mathbf{x}}_{\underline{P}}(\mathcal{X})$ or, more simply, by $\mathbb{G}^{\mathbf{x}}_{\underline{P}}$ if no confusion can arise.*

If a gamble f is $\underline{P}$-previsible, then it follows from Proposition 15.4$_{331}$ that the extended lower prevision $\underline{P}^{\mathbf{x}}(f)$ and upper prevision $\overline{P}^{\mathbf{x}}(f)$ of f exist, are finite and independent of the choice of the determining sequence for f.

Condition (ii) for previsibility is called the **Cauchy condition**, because it means that f_n is a Cauchy sequence with respect to the seminorm $\|\bullet\|_{\underline{P}} = \overline{P}(|\bullet|)$, introduced in Definition 4.25$_{64}$.

Every bounded gamble f is $\underline{P}$-previsible, because the constant sequence $f_n := f$ is a determining sequence for f. It follows that $\mathbb{B} \subseteq \mathbb{G}^{\mathrm{x}}_{\underline{P}}$. The following results take this a step further. We begin by showing that the extension method introduced here subsumes the one in Chapter 14.

Theorem 15.7 $\mathbb{B} \subseteq \mathbb{G}^{\sharp}_{\underline{P}} \subseteq \mathbb{G}^{\mathrm{x}}_{\underline{P}}$, *and* $\underline{P}^{\mathrm{x}}$ *and* $\overline{P}^{\mathrm{x}}$ *are extensions of* $\underline{P}^{\sharp}$ *and* $\overline{P}^{\sharp}$, *and therefore of* $\underline{P}$ *and* $\overline{P}$, *respectively.*

Proof. Let $f \in \mathbb{G}^{\sharp}_{\underline{P}}$. Let $f_\flat$ be a determining bounded gamble for f. Define $f_n := f_\flat$ for every $n \in \mathbb{N}$. Then f_n is a determining sequence for f: $\overline{P}\left(|f_m - f_n|\right) = \overline{P}(0) = 0$ for all $n, m \in \mathbb{N}$ and $\overline{P}\left(\{|f_n - f| > \varepsilon\}\right) = \overline{P}\left(\{|f_\flat - f| > \varepsilon\}\right) = 0$ for all $n \in \mathbb{N}$ and $\varepsilon > 0$, because $f_\flat$ is determining bounded gamble for f. It follows from Definition 15.6$_{\curvearrowleft}$ that $\overline{P}^{\mathrm{x}}(f) = \overline{P}(f_\flat) = \overline{P}^{\sharp}(f)$ and $\underline{P}^{\mathrm{x}}(f) = \underline{P}(f_\flat) = \underline{P}^{\sharp}(f)$. □

Theorem 15.8 *Let f and g be $\underline{P}$-previsible gambles, with respective determining sequences f_n and g_n. Let a and b be real numbers, and let A be any subset of $\mathcal{X}$. Then $I_A f_n$, $af_n + bg_n$, $|f_n|$, $f_n \vee g_n$ and $f_n \wedge g_n$ are determining sequences for $I_A f$, $af + bg$, $|f|$, $f \vee g$ and $f \wedge g$, respectively. Hence, $I_A f$, $af + bg$, $|f|$, $f \vee g$ and $f \wedge g$ are $\underline{P}$-previsible.*

Proof. The convergence in probability $[\underline{P}]$ follows from Proposition 15.3$_{330}$. To check the Cauchy condition, use the inequalities

$$\overline{P}\left(|I_A f_n - I_A f_m|\right) \leq \overline{P}\left(|f_n - f_m|\right)$$

$$\overline{P}\left(|(af_n + bg_n) - (af_m + bg_m)|\right) \leq |a|\overline{P}\left(|f_n - f_m|\right) + |b|\overline{P}\left(|g_n - g_m|\right)$$

$$\overline{P}\left(||f_n| - |f_m||\right) \leq \overline{P}\left(|f_n - f_m|\right),$$

which follow from the coherence of $\underline{P}$ (use Theorem 4.13(v)–(vi)$_{53}$) and use the fact that the maximum and the minimum of f and g can be written as linear combinations of f, g and $|f - g|$. □

Corollary 15.9 $\mathbb{G}^{\mathrm{x}}_{\underline{P}}$ *is a linear lattice.*

Corollary 15.10 *Let f be any gamble, then the following statements are equivalent:*

(i) *f is $\underline{P}$-previsible;*

(ii) *f^+ and f^- are $\underline{P}$-previsible.*

Proof. Immediately from Theorem 15.8 if we recall that $f^+ = 0 \vee f$, $f^- = -(0 \wedge f)$ and $f = f^+ - f^-$. □

Proposition 15.11 *$\underline{P}^{\mathbf{x}}$ is a coherent lower prevision on $\mathbb{G}^{\mathbf{x}}_{\underline{P}}$.*

Proof. We use the coherence of $\underline{P}$ on $\mathbb{B}$. Indeed, let f_n be a determining sequence for f, and let g_n be a determining sequence for g. We know that $\underline{P}(f_n) \geq \inf f_n$, $\underline{P}(\lambda f_n) \geq \lambda \underline{P}(f_n)$ and $\underline{P}(f_n + g_n) \geq \underline{P}(f_n) + \underline{P}(g_n)$ for all f_n and g_n and real $\lambda \geq 0$. As inequalities are preserved when taking the limit over n (if the limits exist on both sides), we find that the extended lower prevision $\underline{P}^{\mathbf{x}}$, defined on the linear space $\mathbb{G}^{\mathbf{x}}_{\underline{P}}$, satisfies all the properties required for coherence in Theorem 13.34_{271}. We have also used Theorem 15.8 in order to be assured that λf_n is a determining sequence for λf and that $f_n + g_n$ is a determining sequence for $f + g$, whenever f_n and g_n are determining sequences for f and g, respectively. □

Definition 15.12 (Previsible seminorm) *The **$\underline{P}$-previsible seminorm** of a $\underline{P}$-previsible gamblef is defined by $\|f\|^{\mathbf{x}}_{\underline{P}} := \overline{P}^{\mathbf{x}}(|f|)$. A net f_α of $\underline{P}$-previsible gambles is said to **converge in $\underline{P}$-previsible seminorm** to a $\underline{P}$-previsible gamble f if the net of real numbers $\|f - f_\alpha\|^{\mathbf{x}}_{\underline{P}}$ converges to zero. We then use the notation*

$$f_\alpha \xrightarrow{\|\bullet\|^{\mathbf{x}}_{\underline{P}}} f.$$

It will follow from Theorem 15.14(ii), (vii) and (xi) further on that $\|\bullet\|^{\mathbf{x}}_{\underline{P}}$ is indeed a seminorm on $\mathbb{G}^{\mathbf{x}}_{\underline{P}}$.

We now turn to a number of important properties for the coherent extended lower prevision $\underline{P}^{\mathbf{x}}$. We start with a useful lemma, of some interest by itself.

Lemma 15.13 *Let f be a $\underline{P}$-previsible gamble. If $0 \leq f$ a.e. $[\underline{P}]$, then $\underline{P}^{\mathbf{x}}(f) \geq 0$ and $\overline{P}^{\mathbf{x}}(f) \geq 0$.*

Proof. If $0 \leq f$ a.e. $[\underline{P}]$, then there is some $\underline{P}$-null gamble N such that $0 \leq f + N$. Let f_n be a determining sequence for f. As trivially $f = f + N$ a.e. $[\underline{P}]$, it follows from Proposition 15.2_{329} that f_n converges in probability $[\underline{P}]$ to $f + N$, so f_n is also a determining sequence for $f + N$. By Theorem 15.8, $(f_n)^+ := f_n \vee 0$ is also a determining sequence for $(f + N)^+ = f + N$. Hence, $\underline{P}^{\mathbf{x}}(f) = \lim_{n\to+\infty} \underline{P}(f_n) = \underline{P}^{\mathbf{x}}(f + N) = \lim_{n\to+\infty} \underline{P}((f_n)^+) \geq 0$ and $\overline{P}^{\mathbf{x}}(f) = \lim_{n\to+\infty} \overline{P}(f_n) = \overline{P}^{\mathbf{x}}(f + N) = \lim_{n\to+\infty} \overline{P}((f_n)^+) \geq 0$. □

As $\underline{P}^{\mathbf{x}}$ is a coherent (unconditional) lower prevision, most of the following properties follow from Theorem 13.31_{266}.

Theorem 15.14 *Let f and g be $\underline{P}$-previsible gambles. Let a be a constant gamble. Let λ and κ be real numbers with $\lambda \geq 0$ and $0 \leq \kappa \leq 1$. Then the following statements hold:*

(i) $\overline{P}^{\mathbf{x}}(f) = -\underline{P}^{\mathbf{x}}(-f)$. **(conjugacy)**

(ii) $\operatorname{ess\,inf}_{\underline{P}} f \leq \underline{P}^{\mathbf{x}}(f) \leq \overline{P}^{\mathbf{x}}(f) \leq \operatorname{ess\,sup}_{\underline{P}} f$. **(bounds)**

(iii) $\underline{P}^{\mathbf{x}}(a) = \overline{P}^{\mathbf{x}}(a) = a$. **(normality)**

(iv) $\underline{P}^{\mathbf{x}}(f + a) = \underline{P}^{\mathbf{x}}(f) + a$ and $\overline{P}^{\mathbf{x}}(f + a) = \overline{P}^{\mathbf{x}}(f) + a$. **(constant additivity)**

(v) If $f \leq g + a$ a.e. $[\underline{P}]$, then $\underline{P}^{\mathbf{x}}(f) \leq \underline{P}^{\mathbf{x}}(g) + a$ and $\overline{P}^{\mathbf{x}}(f) \leq \overline{P}^{\mathbf{x}}(g) + a$. **(monotonicity)**

(vi) $\underline{P}^{\mathbf{x}}(f) + \underline{P}^{\mathbf{x}}(g) \leq \underline{P}^{\mathbf{x}}(f + g) \leq \underline{P}^{\mathbf{x}}(f) + \overline{P}^{\mathbf{x}}(g) \leq \overline{P}^{\mathbf{x}}(f + g) \leq \overline{P}^{\mathbf{x}}(f) + \overline{P}^{\mathbf{x}}(g)$. **(mixed super-/sub-additivity)**

(vii) $\underline{P}^{\mathbf{x}}(\lambda f) = \lambda \underline{P}^{\mathbf{x}}(f)$ and $\overline{P}^{\mathbf{x}}(k\lambda f) = \lambda \overline{P}^{\mathbf{x}}(f)$. **(non-negative homogeneity)**

(viii) $\kappa \underline{P}^{\mathbf{x}}(f) + (1 - \kappa)\underline{P}^{\mathbf{x}}(g) \leq \underline{P}^{\mathbf{x}}(\kappa f + (1 - \kappa)g) \leq \kappa \underline{P}^{\mathbf{x}}(f) + (1 - \kappa)\overline{P}^{\mathbf{x}}(g) \leq \overline{P}^{\mathbf{x}}(\kappa f + (1 - \kappa)g) \leq \kappa \overline{P}^{\mathbf{x}}(f) + (1 - \kappa)\overline{P}^{\mathbf{x}}(g)$. **(mixed convexity/concavity)**

(ix) $\underline{P}^{\mathbf{x}}(|f|) \geq \underline{P}^{\mathbf{x}}(f)$ and $\overline{P}^{\mathbf{x}}(|f|) \geq \overline{P}^{\mathbf{x}}(f)$.

(x) $\left|\underline{P}^{\mathbf{x}}(f) - \underline{P}^{\mathbf{x}}(g)\right| \leq \overline{P}^{\mathbf{x}}(|f - g|)$ and $\left|\overline{P}^{\mathbf{x}}(f) - \overline{P}^{\mathbf{x}}(g)\right| \leq \overline{P}^{\mathbf{x}}(|f - g|)$.

(xi) $\underline{P}^{\mathbf{x}}(|f + g|) \leq \underline{P}^{\mathbf{x}}(|f|) + \overline{P}^{\mathbf{x}}(|g|)$ and $\overline{P}^{\mathbf{x}}(|f + g|) \leq \overline{P}^{\mathbf{x}}(|f|) + \overline{P}^{\mathbf{x}}(|g|)$. **(mixed Cauchy–Schwartz inequalities)**

(xii) $\underline{P}^{\mathbf{x}}(f \vee g) + \underline{P}^{\mathbf{x}}(f \wedge g) \leq \underline{P}^{\mathbf{x}}(f) + \overline{P}^{\mathbf{x}}(g) \leq \overline{P}^{\mathbf{x}}(f \vee g) + \overline{P}^{\mathbf{x}}(f \wedge g)$, $\underline{P}^{\mathbf{x}}(f) + \underline{P}^{\mathbf{x}}(g) \leq \underline{P}^{\mathbf{x}}(f \vee g) + \overline{P}^{\mathbf{x}}(f \wedge g) \leq \overline{P}^{\mathbf{x}}(f) + \overline{P}^{\mathbf{x}}(g)$ and $\underline{P}^{\mathbf{x}}(f) + \underline{P}^{\mathbf{x}}(g) \leq \overline{P}^{\mathbf{x}}(f \vee g) + \underline{P}^{\mathbf{x}}(f \wedge g) \leq \overline{P}^{\mathbf{x}}(f) + \overline{P}^{\mathbf{x}}(g)$.

(xiii) If $f_\alpha \xrightarrow{\|\bullet\|_{\underline{P}}^{\mathbf{x}}} f$, then $\underline{P}^{\mathbf{x}}(f_\alpha) \to \underline{P}^{\mathbf{x}}(f)$ and $\overline{P}^{\mathbf{x}}(f_\alpha) \to \overline{P}^{\mathbf{x}}(f)$. **(continuity)**

However, most of these properties can also be proved directly, using the determining sequences and the coherence of the lower prevision $\underline{P}$.

Proof. Indeed, let f_n be a determining sequence for f, and let g_n be a determining sequence for g. Statements (i)↶–(iv) and (vi)–(xii) follow immediately from Theorem 4.13_{53} applied to every term of some determining sequence and from the fact that limits preserve an (in)equality if the limits exist on both sides of that (in)equality. We give the proof of statement (ii)↶ as an example. By Theorem $4.13(\mathrm{i})_{53}$, we have that $\underline{P}(f_n) \leq \overline{P}(f_n)$ for all n. It follows that $\lim_{n\to+\infty} \underline{P}(f_n) \leq \lim_{n\to+\infty} \overline{P}(f_n)$ or, in other words, $\underline{P}^{\mathbf{x}}(f) \leq \overline{P}^{\mathbf{x}}(f)$.

There are only a few statements left to prove.

(v). If $f \leq g + a$ a.e. $[\underline{P}]$, then we infer from Lemma 15.13↶ that $\underline{P}^{\mathbf{x}}(g + a - f) \geq 0$. Using (iv) and (vi), we find that indeed

$$\underline{P}^{\mathbf{x}}(g) + a - \underline{P}^{\mathbf{x}}(f) = \underline{P}^{\mathbf{x}}(g) + a + \overline{P}^{\mathbf{x}}(-f) \geq \underline{P}^{\mathbf{x}}(g + a - f) \geq 0$$

and, similarly,

$$\overline{P}^{\mathbf{x}}(g) + a - \overline{P}^{\mathbf{x}}(f) = \overline{P}^{\mathbf{x}}(g) + a + \underline{P}^{\mathbf{x}}(-f) \geq \underline{P}^{\mathbf{x}}(g + a - f) \geq 0.$$

(ii)$_{335}$. It still remains to prove that $\operatorname{ess\,inf}_{\underline{P}} f \leq \underline{P}^{\mathrm{x}}(f)$ and $\overline{P}^{\mathrm{x}}(f) \leq \operatorname{ess\,sup}_{\overline{P}} f$. We infer from Proposition 14.19(ii)$_{314}$ that $\operatorname{ess\,inf}_{\underline{P}} f \leq f$ a.e. $[\underline{P}]$ and $f \leq \operatorname{ess\,sup}_{\overline{P}} f$ a.e. $[\underline{P}]$. Now use (v).

(xiii). This follows from (x) and $\|f\|_{\underline{P}}^{\mathrm{x}} = \overline{P}^{\mathrm{x}}(|f|)$. □

We already know from Proposition 14.19(v)$_{314}$ that a gamble is $\underline{P}$-null if and only if its $\underline{P}$-essential supremum norm $\|f\|_{\underline{P},\infty} = \operatorname{ess\,sup}_{\overline{P}} |f| = 0$. If f is $\underline{P}$-previsible, its being null can also be checked using the $\underline{P}$-previsible seminorm. The following result generalises Proposition 14.31$_{320}$.

Proposition 15.15 *A gamble f is $\underline{P}$-null if and only if it is $\underline{P}$-previsible and $\|f\|_{\underline{P}}^{\mathrm{x}} = 0$:*

$$\mathbb{G}_{\underline{P}}^{0} = \left\{ f \in \mathbb{G}_{\underline{P}}^{\mathrm{x}} : \overline{P}^{\mathrm{x}}(|f|) = 0 \right\}. \tag{15.4}$$

Proof. 'if'. Suppose that f is $\underline{P}$-previsible with $\|f\|_{\underline{P}}^{\mathrm{x}} = 0$. Let $\varepsilon > 0$ and $A_\varepsilon := \{|f| > \varepsilon\}$. As $|f| \geq \varepsilon A_\varepsilon \geq 0$, it follows from Theorem 13.31(v) and (vi)$_{266}$ that $0 = \|f\|_{\underline{P}}^{\mathrm{x}} \geq \varepsilon \overline{P}(A_\varepsilon) \geq 0$. We conclude that $\overline{P}(A_\varepsilon) = 0$ for every $\varepsilon > 0$.

'only if'. Suppose that f is $\underline{P}$-null, then it follows from Proposition 14.31$_{320}$ that f is $\underline{P}$-essentially bounded and that $\overline{P}^{\sharp}(|f|) = 0$. Now use Theorem 15.7$_{334}$. □

The two requirements for $\underline{P}$-previsibility in Definition 15.6$_{333}$ can be extended from bounded gambles to $\underline{P}$-previsible gambles, in the sense of the next theorem. It is a generalisation from Dunford integrals to lower and upper previsions of Theorem 4.4.20 in Bhaskara Rao and Bhaskara Rao (1983). Our proof is able to follow theirs quite closely.

Theorem 15.16 *Let f_n be a sequence of $\underline{P}$-previsible gambles. Consider a gamble f and suppose that*

(i) *f_n converges in probability $[\underline{P}]$ to f,*

(ii) $\lim_{n,m\to+\infty} \overline{P}^{\mathrm{x}}\left(|f_n - f_m|\right) = 0$.

Then f is $\underline{P}$-previsible, $\underline{P}^{\mathrm{x}}(f) = \lim_{n\to+\infty} \underline{P}^{\mathrm{x}}(f_n)$ and $\overline{P}^{\mathrm{x}}(f) = \lim_{n\to+\infty} \overline{P}^{\mathrm{x}}(f_n)$.

Proof. Consider any $n \in \mathbb{N}_{>0}$. As the gamble f_n is $\underline{P}$-previsible, we infer from the definition of $\underline{P}$-previsibility and Lemma 15.17$_\curvearrowright$ that there is some bounded gamble g_n for which

$$\overline{P}^{\mathrm{x}}\left(|f_n - g_n|\right) < \frac{1}{n} \text{ and } \overline{P}\left(\left\{|f_n - g_n| > \frac{1}{n}\right\}\right) < \frac{1}{n}. \tag{15.5}$$

In this way, we find a sequence of bounded gambles g_n and prove that it is a determining sequence for f.

First, we show that g_n converges in probability $[\underline{P}]$ to f. Consider any $\varepsilon > 0$. As f_n converges in probability $[\underline{P}]$ to f, we now that for all $\alpha > 0$ and $\beta > 0$, there is some

$K_{\alpha,\beta}$ such that $\overline{P}\left(\{|f_n - f| > \alpha\}\right) < \beta$ for all $n \geq K_{\alpha,\beta}$. If we let $N_\varepsilon := K_{\varepsilon/2,\varepsilon/3}$, this implies in particular that

$$\overline{P}\left(\left\{|f_n - f| > \frac{\varepsilon}{2}\right\}\right) < \frac{\varepsilon}{3} \text{ for all } n \geq N_\varepsilon. \tag{15.6}$$

Now let $M_\varepsilon := \max\{N_\varepsilon, \lceil 2/\varepsilon \rceil\}$, where $\lceil 2/\varepsilon \rceil$ is the smallest natural number greater than or equal to $2/\varepsilon$. Then for every $n \geq M_\varepsilon$,

$$\begin{aligned}\{|g_n - f| > \varepsilon\} &\subseteq \left\{|g_n - f_n| > \frac{\varepsilon}{2}\right\} \cup \left\{|f_n - f| > \frac{\varepsilon}{2}\right\}\\ &\subseteq \left\{|g_n - f_n| > \frac{1}{M_\varepsilon}\right\} \cup \left\{|f_n - f| > \frac{\varepsilon}{2}\right\}\\ &\subseteq \left\{|g_n - f_n| > \frac{1}{n}\right\} \cup \left\{|f_n - f| > \frac{\varepsilon}{2}\right\}.\end{aligned}$$

It follows from Equations $(15.5)_{\curvearrowleft}$ and (15.6) and the monotonicity and sub-additivity of the coherent upper prevision $\overline{P}$ (see Theorem 4.13(iv) and $(\mathrm{v})_{53}$) that

$$\overline{P}\left(\{|g_n - f| > \varepsilon\}\right) \leq \frac{1}{n} + \frac{\varepsilon}{3} < \frac{\varepsilon}{2} + \frac{\varepsilon}{2} = \varepsilon \text{ for all } n \geq M_\varepsilon,$$

so g_n converges in probability $[\underline{P}]$ to f.

Next, we have that

$$\overline{P}\left(|g_n - g_m|\right) = \|g_n - g_m\|^{\mathrm{x}}_{\underline{P}} \leq \|g_n - f_n\|^{\mathrm{x}}_{\underline{P}} + \|f_n - f_m\|^{\mathrm{x}}_{\underline{P}} + \|f_m - g_m\|^{\mathrm{x}}_{\underline{P}},$$

where the equality holds because $\underline{P}$ and $\underline{P}^{\mathrm{x}}$ coincide on bounded gambles (Theorem 15.7_{334}) and the inequality follows from the sub-additivity of $\underline{P}^{\mathrm{x}}$ (see Theorem $15.14(\mathrm{vi})_{335}$ or, alternatively, use the sub-additivity of seminorms). As all terms in the sum on the right-hand side of the inequality converge to zero, it follows that $\overline{P}\left(|g_n - g_m|\right)$ converges to zero. We conclude that g_n is a determining sequence for f, and it follows that f is $\underline{P}$-previsible.

Finally, we have that (again, use the sub-additivity of seminorms)

$$\|f_n - f\|^{\mathrm{x}}_{\underline{P}} \leq \|f_n - g_n\|^{\mathrm{x}}_{\underline{P}} + \|g_n - f\|^{\mathrm{x}}_{\underline{P}}.$$

As both terms in the sum on the right-hand side of the inequality converge to zero (see Equation $(15.5)_{\curvearrowleft}$ and Lemma 15.17, respectively), it follows that $\|f_n - f\|^{\mathrm{x}}_{\underline{P}}$ converges to zero. Now use Theorem $15.14(\mathrm{xiii})_{335}$. □

Lemma 15.17 *Let f be a $\underline{P}$-previsible gamble, and let the bounded gambles f_n constitute a determining sequence for f. Then f_n converges to f in $\underline{P}$-previsible seminorm.*

Proof. First of all, for any $n \in \mathbb{N}$, we infer from Theorem 15.8_{334} that $|f_m - f_n|$ is a determining sequence for $|f - f_n|$. This tells us a number of useful things. First of all, that $|f - f_n|$ is $\underline{P}$-previsible, implying that $h(n) := \overline{P}^{\mathrm{x}}\left(|f - f_n|\right)$ is a real number.

Secondly, we infer that if we define the real-valued map g on $\mathbb{N}^2$ by letting $g(n,m) := \overline{P}\left(|f_m - f_n|\right)$ for all $n, m \in \mathbb{N}$, then

$$h(n) = \overline{P}^{\mathbf{x}}\left(|f - f_n|\right) = \lim_{m\to+\infty} \overline{P}^{\mathbf{x}}\left(|f_m - f_n|\right) = \lim_{m\to+\infty} g(n,m) \in \mathbb{R}, \tag{15.7}$$

where we used that $\overline{P}$ and $\overline{P}^{\mathbf{x}}$ coincide on bounded gambles (Theorem 15.7$_{334}$). Thirdly, we infer that

$$0 = \lim_{n,m\to+\infty} \overline{P}\left(|f_m - f_n|\right) = \lim_{n,m\to+\infty} g(n,m). \tag{15.8}$$

We need to show that

$$0 = \lim_{n\to+\infty} \overline{P}^{\mathbf{x}}\left(|f - f_n|\right) = \lim_{n\to+\infty} h(n) = \lim_{n\to+\infty}\lim_{m\to+\infty} g(n,m),$$

where the last two equalities follow from Equation (15.7). So Equation (15.8) tells us that our goal is established if we can show that

$$\lim_{n\to+\infty}\lim_{m\to+\infty} g(n,m) = \lim_{n,m\to+\infty} g(n,m).$$

We show that this is indeed the case, simply using the fact that $\lim_{n,m\to+\infty} g(n,m)$ and $\lim_{m\to+\infty} g(n,m) = h(n)$ exist. Indeed, because both limits exist, we see that

$$\begin{aligned}\lim_{n,m\to+\infty} g(n,m) &= \liminf_{n,m\to+\infty} g(n,m) = \sup_{n,m} \inf_{\substack{N\geq n\\ M\geq m}} g(N,M) = \sup_n \sup_m \inf_{N\geq n} \inf_{M\geq m} g(N,M)\\ &\leq \sup_n \inf_{N\geq n} \sup_m \inf_{M\geq m} g(N,M) = \liminf_{n\to+\infty} \liminf_{m\to+\infty} g(n,m) = \liminf_{n\to+\infty} h(n)\end{aligned}$$

and, similarly,

$$\begin{aligned}\lim_{n,m\to+\infty} g(n,m) &= \limsup_{n,m\to+\infty} g(n,m) = \inf_{n,m} \sup_{\substack{N\geq n\\ M\geq m}} g(N,M) = \inf_n \inf_m \sup_{N\geq n} \sup_{M\geq m} g(N,M)\\ &\geq \inf_n \sup_{N\geq n} \inf_m \sup_{M\geq m} g(N,M) = \limsup_{n\to+\infty} \limsup_{m\to+\infty} g(n,m) = \limsup_{n\to+\infty} h(n),\end{aligned}$$

where the inequalities follow from Lemma 15.18. Hence,

$$\lim_{n,m\to+\infty} g(n,m) \leq \liminf_{n\to+\infty} h(n) \leq \limsup_{n\to+\infty} h(n) \leq \lim_{n,m\to+\infty} g(n,m).$$

This proves that $\lim_{n\to+\infty} h(n) = \lim_{n\to+\infty} \lim_{m\to+\infty} g(n,m)$ exists and is indeed equal to $\lim_{n,m\to+\infty} g(n,m)$. □

Lemma 15.18 *For any non-empty sets A and B and any real-valued function h defined on (at least) $A \times B$, we have that*

$$\inf_{y\in B} \sup_{x\in A} h(x,y) \geq \sup_{x\in A} \inf_{y\in B} h(x,y). \tag{15.9}$$

Proof. Consider any $x \in A$ and $y \in B$, then $h(x,y) \geq \inf_{v\in B} h(x,v)$, and hence also, taking the supremum over $x \in A$ on both sides: $\sup_{x\in A} h(x,y) \geq \sup_{x\in A} \inf_{v\in B} h(x,v)$. Hence, indeed, $\inf_{y\in B} \sup_{x\in A} h(x,y) \geq \sup_{x\in A} \inf_{v\in B} h(x,v)$. □

15.3 Measurability

So far, we have managed to extend lower previsions from bounded gambles to previsible gambles. Our definition of previsibility involves the existence of a *determining sequence* of bounded gambles that is Cauchy and converges in probability to the gamble. But we have not given any *systematic way* of *constructing* such a determining sequence and, hence, determining the value of the extended lower prevision. We deal with this problem in Section 15.5_{348}. In this and the next section, we lay the theoretical foundations that make this possible.

Our main purpose here is to associate a notion of *measurability* with a coherent lower prevision. In Section 1.8_{12}, we have called a bounded gamble measurable if it is a uniform limit of simple gambles. In a similar way, we will now define a gamble to be $\underline{P}$-measurable if there is a sequence of simple gambles that converges to it in probability [$\underline{P}$]. As convergence in probability [$\underline{P}$] reduces to uniform convergence when $\underline{P} = \inf$, this new definition is actually a generalisation of the classical measurability definition. We also want to mention that previsibility will turn out to be a special form of measurability (see Theorem 15.24_{345} further on).

We call two sequences $a_n \leq b_n$ of real numbers such that a_n converges to $-\infty$ and b_n converges to $+\infty$, a pair of **cut sequences**. A similar definition can be given for a pair of **cut nets** $a_\alpha \leq b_\alpha$.

Definition 15.19 *Let f be a gamble. Then the following conditions are equivalent; if any (and hence all) of them are satisfied, we say that f is* ***$\underline{P}$-measurable****.*

(A) *There is a sequence of simple gambles f_n converging in probability [$\underline{P}$] to f.*

(B) *For every $\varepsilon > 0$, there is a partition $\{F_0, F_1, \dots, F_n\}$ of $\mathcal{X}$ such that $\overline{P}(F_0) < \varepsilon$ and $|f(x) - f(x')| < \varepsilon$ for every $x, x' \in F_i$, $i = 1, \dots, n$.*

(C) *There is a pair of cut sequences $a_n \leq b_n$ such that the sequence of bounded gambles $f_{a_n}^{b_n}$ converges in probability [$\underline{P}$] to f.*

(D) *For any pair of cut sequences $a_n \leq b_n$, the sequence of bounded gambles $f_{a_n}^{b_n}$ converges in probability [$\underline{P}$] to f.*

(E) $\lim_{t\to+\infty} \overline{P}(\{|f| \geq t\}) = 0$.

The bounded gambles $f_{a_n}^{b_n}$ in this definition are the (a_n, b_n)-*cuts* of f, introduced in Definition 14.32_{321}.

For the special case that $\underline{P}$ is a linear prevision P associated with a probability charge μ, Bhaskara Rao and Bhaskara Rao call the notion defined by (A) **T_1-measurability**, the one defined by (B) **T_2-measurability** and the one defined by (E) **smoothness** (see Bhaskara Rao and Bhaskara Rao, 1983, Definitions 4.2.14, 4.4.5, and 4.5.6). They prove equivalence between T_1-measurability and T_2-measurability (Bhaskara Rao and Bhaskara Rao, 1983, Theorem 4.4.7) and prove that these measurability conditions imply smoothness (Bhaskara Rao and Bhaskara Rao, 1983, Corollary 4.4.8). So the theorem above generalises these results to arbitrary coherent lower previsions, adds an extra implication[3] and adds cuts to the picture of measurability.

In the remainder of this section, we prove that these conditions are indeed equivalent. We first concentrate on the equivalence of (A) and (B), a generalisation from charges to lower and upper previsions of Theorem 4.4.7 in Bhaskara Rao and Bhaskara Rao (1983). Our proof is able to follow theirs quite closely.

Proof. (A)⇒(B). Consider any $\varepsilon > 0$. As the sequence of simple gambles f_n converges in probability $[\underline{P}]$ to f, there is some m_ε in $\mathbb{N}$ such that

$$\overline{P}\left(\left\{\left|f - f_{m_\varepsilon}\right| > \frac{\varepsilon}{3}\right\}\right) < \varepsilon.$$

Let $g := f_{m_\varepsilon} = \sum_{i=1}^{n} c_i I_{E_i}$, where $g(\mathscr{X}) =: \{c_1, \dots, c_n\}$ and $E_i := g^{-1}(\{c_i\})$ for $i = 1$, …, n. Let $F_0 := \{|f - g| > \varepsilon/3\}$, so $\overline{P}(F_0) < \varepsilon$ and $F_i := E_i \cap F_0^c$, $i = 1$, …, n. As $\{E_1, \dots, E_n\}$ is a partition of $\mathscr{X}$, it follows that $\{F_0, F_1, \dots, F_n\}$ is also a partition[4] of $\mathscr{X}$. For $i = 1, \dots, n$, we have for every $x, x' \in F_i$ that $g(x) = g(x') = c_i$, $|f(x) - g(x)| \leq \varepsilon/3$ and $|f(x') - g(x')| \leq \varepsilon/3$. Hence,

$$\left|f(x) - f(x')\right| = \left|f(x) - g(x) + g(x') - f(x')\right| \leq \frac{\varepsilon}{3} + \frac{\varepsilon}{3} < \varepsilon.$$

(B)⇒(A). Consider any $n \in \mathbb{N}_{>0}$, then there is some partition

$$\{F_{n,0}, F_{n,1}, \dots, F_{n,k_n}\}$$

of $\mathscr{X}$ such that $\overline{P}(F_{n,0}) < 1/n$ and $\left|f(x) - f(x')\right| < 1/n$ for every $x, x' \in F_{n,i}$, $i = 1$, …, k_n. Fix some $x_{n,i}$ in $F_{n,i}$ for each $i = 1, \dots, k_n$, and use these to define the simple gamble $f_n := \sum_{i=1}^{k_n} f(x_{n,i}) I_{F_{n,i}}$. In this way, we construct a sequence of simple gambles f_n for which we now prove that it converges in probability $[\underline{P}]$ to f. Let $\varepsilon > 0$ and $M_\varepsilon := \lceil 1/\varepsilon \rceil$, the smallest natural number that is still greater than or equal to $1/\varepsilon$. It follows from the construction of f_n that $\left\{|f - f_n| > \varepsilon\right\} \subseteq \left\{|f - f_n| > 1/n\right\} \subseteq F_{n,0}$ for all $n \geq M_\varepsilon$ and, therefore, $\overline{P}\left(\left\{|f - f_n| > \varepsilon\right\}\right) \leq \overline{P}(F_{n,0}) < 1/n$ for all $n \geq M_\varepsilon$. □

To prove the equivalence of conditions (B), (C), (D) and (E) in Definition 15.19, we begin by proving a number of lemmas. Their proofs rely on conditions (A) or (B) to characterise $\underline{P}$-measurability.

Lemma 15.20 *Every bounded gamble is $\underline{P}$-measurable.*

Proof. Consider any bounded gamble f and any $\varepsilon > 0$. As f is bounded, there is some sequence f_n of simple gambles converging uniformly to f. This implies

---

[3] Bhaskara Rao and Bhaskara Rao (1983, Section 4.4.8) claim, apparently without proof, that smoothness does not imply T_1-measurability. This claim seems contradicted by our proof that the conditions (A) and (E) are in fact equivalent. We suspect, but cannot be certain, that the apparent contradiction derives form the fact that in their definition, not all events are assumed to be measurable: their definitions are all relative to some field of subsets.

[4] We are allowing ourselves some leeway here in calling this a partition, as some of its elements might be empty.

that there is some $N_\varepsilon \in \mathbb{N}$ such that $|f(x) - f_n(x)| < \varepsilon$ for all $n \geq N_\varepsilon$ and all $x \in \mathscr{X}$. Hence, $\overline{P}(\{|f - f_n| > \varepsilon\}) = \overline{P}(\emptyset) = 0$ for all $n \geq N_\varepsilon$ and, therefore, f_n converges in probability $[\underline{P}]$ to f. We conclude that f is $\underline{P}$-measurable, using condition $(A)_{340}$ of Definition 15.19_{340}. □

The following lemma generalises Proposition 4.6.13 in Bhaskara Rao and Bhaskara Rao (1983) from charges to lower and upper previsions.

Lemma 15.21 *Let f be a gamble, and let f_α be a net of $\underline{P}$-measurable gambles. If f_α converges in probability $[\underline{P}]$ to f, then f is $\underline{P}$-measurable.*

Proof. We verify that f satisfies condition $(B)_{340}$ of Definition 15.19_{340}. Consider any $\varepsilon > 0$. As f_α converges in probability $[\underline{P}]$ to f, there is some α_ε such that

$$\overline{P}\left(\left\{\left|f - f_{\alpha_\varepsilon}\right| > \frac{\varepsilon}{4}\right\}\right) < \frac{\varepsilon}{2}.$$

Let $A := \left\{\left|f - f_{\alpha_\varepsilon}\right| > \varepsilon/4\right\}$, then $\overline{P}(A) < \varepsilon/2$ and $\left|f(x) - f_{\alpha_\varepsilon}(x)\right| \leq \varepsilon/4$ for all $x \in A^c$. As f_{α_ε} is $\underline{P}$-measurable, there is some partition $\{F_0, F_1, \dots, F_n\}$ of $\mathscr{X}$ such that $\overline{P}(F_0) < \varepsilon/2$ and $\left|f_{\alpha_\varepsilon}(x) - f_{\alpha_\varepsilon}(x')\right| < \varepsilon/2$ for all $x, x' \in F_i$, $i = 1, \dots, n$. Define $E_0 := A \cup F_0$ and $E_i := (A^c) \cap F_i$, $i = 1, \dots, n$. As $\{F_0, F_1, \dots, F_n\}$ is a partition of $\mathscr{X}$, so is $\{E_0, E_1, \dots, E_n\}$.[5] It follows that (see Theorem $4.13(\mathrm{v})_{53}$)

$$\overline{P}(E_0) = \overline{P}(A \cup F_0) \leq \overline{P}(A) + \overline{P}(F_0) < \frac{\varepsilon}{2} + \frac{\varepsilon}{2} = \varepsilon$$

and

$$\begin{aligned}
|f(x) - f(x')| &= \left|f(x) - f_{\alpha_\varepsilon}(x) + f_{\alpha_\varepsilon}(x) - f_{\alpha_\varepsilon}(x') + f_{\alpha_\varepsilon}(x') - f(x')\right| \\
&\leq \left|f(x) - f_{\alpha_\varepsilon}(x)\right| + \left|f_{\alpha_\varepsilon}(x) - f_{\alpha_\varepsilon}(x')\right| + \left|f_{\alpha_\varepsilon}(x') - f(x')\right| \\
&< \frac{\varepsilon}{4} + \frac{\varepsilon}{2} + \frac{\varepsilon}{4} = \varepsilon
\end{aligned}$$

for all $x, x' \in (A^c) \cap F_i = E_i$ This establishes the claim. □

We are now ready to prove the equivalence of conditions $(B)_{340}$, $(C)_{340}$, $(D)_{340}$ and $(E)_{340}$ in Definition 15.19_{340}. Only the proof of the first implication, which does not involve cuts, generalises an argument by Bhaskara Rao and Bhaskara Rao (1983, Corollary 4.4.8).

Proof. $(B)_{340} \Rightarrow (E)_{340}$. Let f be any $\underline{P}$-measurable gamble, and consider any $\varepsilon > 0$. By condition $(B)_{340}$ of Definition 15.19_{340}, there is some partition $\{F_0, F_1, \dots, F_n\}$ of $\mathscr{X}$ such that $\overline{P}(F_0) < \varepsilon$ and $|f(x) - f(x')| < \varepsilon$ for every $x, x' \in F_i$, $i = 1, \dots, n$. Fix any x_i in F_i for all $i \in \{1, \dots, n\}$. Then $|f(x)| \leq |f(x_i)| + \varepsilon$ for all $x \in F_i$, so f is bounded on all F_i, $i = 1, \dots, n$. If we let $t_\varepsilon := \max_{i=1}^n \sup\left\{|f(x)| : x \in F_i\right\}$ and consider any $t \geq t_\varepsilon$, then $|f(x)| > t$ only if $x \in F_0$. By the monotonicity of the coherent

[5] We are allowing ourselves some leeway here in calling the latter a partition, as some of its elements might be empty.

upper prevision $\overline{P}$ (see Theorem 4.13(iv)$_{53}$), we find that $\overline{P}(\{|f| > t\}) \leq \overline{P}(F_0) < \varepsilon$, which establishes the claim.

(E)$_{340}$⇒(D)$_{340}$. We are given that, for every $\delta > 0$, there is some $t_\delta > 0$ such that

$$\overline{P}\left(\{|f| > t_\delta\}\right) < \delta. \tag{15.10}$$

Let $a_n \leq b_n$ be any pair of cut sequences. Then there is some N_δ in $\mathbb{N}$ such that $a_n \leq -t_\delta < t_\delta \leq b_n$ for all $n \geq N_\delta$. Consider any $\varepsilon > 0$. For every $n \geq N_\delta$, we find that

$$\{|f_{a_n}^{b_n} - f| > \varepsilon\} \subseteq \{|f| > \varepsilon + \min\{-a_n, b_n\}\} \subseteq \{|f| > \varepsilon + t_\delta\} \subseteq \{|f| > t_\delta\}$$

Now use the monotonicity of the coherent upper prevision $\overline{P}$ (Theorem 4.13(iv)$_{53}$) and the inequality (15.10) to conclude that

$$\overline{P}\Big(\{|f_{a_n}^{b_n} - f| > \varepsilon\}\Big) < \delta,$$

and therefore, $f_{a_n}^{b_n}$ converges in probability [$\underline{P}$] to f.

(D)$_{340}$⇒(C)$_{340}$. Trivial.

(C)$_{340}$⇒(B)$_{340}$. Assume that there is some pair of cut sequences $a_n \leq b_n$ such that the sequence of bounded gambles $f_{a_n}^{b_n}$ converges in probability [$\underline{P}$] to f. As each $f_{a_n}^{b_n}$ is bounded, it follows from Lemma 15.20$_{341}$ that $f_{a_n}^{b_n}$ is $\underline{P}$-measurable for every $n \in \mathbb{N}$. Now use Lemma 15.21. □

15.4 Lebesgue's dominated convergence theorem

As a second step in establishing a practical criterion for determining whether a gamble is $\underline{P}$-previsible, and at the same time constructing a determining sequence for it, we now consider a result of some interest by itself: we generalise Lebesgue's celebrated dominated convergence theorem (see, for instance, Schechter, 1997, Section 22.29) from measures to extended coherent lower previsions.

We begin by putting in some preparatory work, in the form of lemmas. The first one is a generalisation from charges to lower and upper previsions of Lemma 4.4.17 in Bhaskara Rao and Bhaskara Rao (1983). Our proof is similar to theirs but corrects a minor glitch near the end.

Lemma 15.22 *Let f_n be a sequence of bounded gambles converging in probability [$\underline{P}$] to some gamble f. Then there is a sequence of bounded gambles g_n converging in probability [$\underline{P}$] to f such that $|g_n| \leq 2|f|$ for all $n \in \mathbb{N}_{>0}$.*

Proof. As f_n converges in probability [$\underline{P}$] to f, there is some subsequence f_{n_k} such that $\overline{P}\left(\left\{\left|f - f_{n_k}\right| > 1/k\right\}\right) < 1/k$ for all $k \in \mathbb{N}_{>0}$. Let $A_k := \left\{\left|f - f_{n_k}\right| > 1/k\right\}$, then we have

$$\overline{P}(A_k) < \frac{1}{k} \text{ and } \left|f(x) - f_{n_k}(x)\right| \leq \frac{1}{k} \text{ for all } x \in A_k^c \text{ and all } k \in \mathbb{N}_{>0}. \tag{15.11}$$

Define, for all $k \in \mathbb{N}_{>0}$,

$$g_k(x) := \begin{cases} f_{n_k}(x) & \text{if } x \in A_k^c \text{ and } \left|f_{n_k}(x)\right| > \frac{2}{k} \\ 0 & \text{otherwise.} \end{cases} \tag{15.12}$$

As f_{n_k} is bounded, so is g_k. We show that the sequence g_k satisfies the conditions of the theorem.

First, we show that $|g_k| \leq 2|f|$ for all $k \in \mathbb{N}_{>0}$. Consider any x in $\mathcal{X}$, then there are two possibilities. Either $x \in A_k$ or $\left|f_{n_k}(x)\right| \leq 2/k$. In this case, $|g_k(x)| = 0$ and the inequality $|g_k(x)| \leq 2|f(x)|$ is satisfied trivially. Or $x \in A_k^c$ and $\left|f_{n_k}(x)\right| > 2/k$. Then we infer from Equation (15.11)$_\curvearrowleft$ that

$$|g_k(x)| = \left|f_{n_k}(x)\right| \leq \left|f_{n_k}(x) - f(x)\right| + |f(x)| \leq \frac{1}{k} + |f(x)| \tag{15.13}$$

and, therefore, that $1/k \geq \left|f_{n_k}(x)\right| - |f(x)| > 2/k - |f(x)|$ or, equivalently,

$$|f(x)| > \frac{1}{k}. \tag{15.14}$$

From Equations (15.13) and (15.14), we deduce that $|g_k(x)| \leq 2|f(x)|$ here too.

It remains to prove that the sequence g_k converges in probability $[\underline{P}]$ to f. Consider any $\varepsilon > 0$, then

$$\{|f - g_k| > \varepsilon\} = \{x \in A_k : |f(x) - g_k(x)| > \varepsilon\} \cup \{x \in A_k^c : |f(x) - g_k(x)| > \varepsilon\}.$$

Using the monotonicity and sub-linearity of the coherent upper prevision $\overline{P}$ (see Theorem 4.13(iv) and (v)$_{53}$), we find that for every $k \geq 3/\varepsilon$,

$$\overline{P}(\{|f - g_k| > \varepsilon\}) \leq \overline{P}(A_k) + \overline{P}(\emptyset) < \frac{1}{k} + 0,$$

proving the lemma, if we can show that $\{x \in A_k^c : |f(x) - g_k(x)| > \varepsilon\} = \emptyset$ when $k \geq 3/\varepsilon$.

Indeed, for $x \in A_k^c$, by Equation (15.12),

$$|f(x) - g_k(x)| = \begin{cases} \left|f(x) - f_{n_k}(x)\right| & \text{if } \left|f_{n_k}(x)\right| > \frac{2}{k}, \\ |f(x)| & \text{otherwise,} \end{cases}$$

so, using Equation (15.11)$_\curvearrowleft$ in both cases,

$$\begin{aligned} |f(x) - g_k(x)| &\leq \begin{cases} \frac{1}{k} & \text{if } \left|f_{n_k}(x)\right| > \frac{2}{k}, \\ \left|f_{n_k}(x)\right| + \frac{1}{k} & \text{otherwise} \end{cases} \\ &\leq \frac{3}{k}, \end{aligned}$$

whence indeed $\{x \in A_k^c : |f(x) - g_k(x)| > \varepsilon\} = \emptyset$ when $k \geq 3/\varepsilon$. □

Next comes a result that extends Lemma 15.5_{333} for bounded gambles to $\underline{P}$-previsible gambles.

Lemma 15.23 *Let f be a non-negative $\underline{P}$-previsible gamble. Then for all $\varepsilon > 0$, there is some $\delta_\varepsilon > 0$ such that for all $A \subseteq \mathscr{X}$: $\overline{P}(A) < \delta_\varepsilon \Rightarrow \overline{P}^{\mathbf{x}}(fI_A) < \varepsilon$.*

Proof. First of all, any $I_A f$ is guaranteed to be $\underline{P}$-previsible by Theorem 15.8_{334}. Fix any $\varepsilon > 0$. As f is $\underline{P}$-previsible, it has a determining sequence of bounded gambles, which, by Lemma 15.17_{338}, converges to f with respect to the $\underline{P}$-previsible seminorm $\|\bullet\|^{\mathbf{x}}_{\underline{P}}$. This implies that there is some bounded gamble g for which $\overline{P}^{\mathbf{x}}(|f-g|) < \varepsilon/2$. It is clear that we can assume without loss of generality that $g \neq 0$. Let $\delta_\varepsilon := \varepsilon/(2\sup|g|)$, and consider any $A \subseteq \mathscr{X}$ such that $\overline{P}^{\mathbf{x}}(A) = \overline{P}(A) < \delta_\varepsilon$. Taking into account Theorems 15.14_{335} and 4.13_{53}, we find that indeed

$$\begin{aligned}\overline{P}^{\mathbf{x}}(fI_A) &\leq \overline{P}^{\mathbf{x}}(|f-g|I_A) + \overline{P}^{\mathbf{x}}(|g|I_A) = \overline{P}^{\mathbf{x}}(|f-g|I_A) + \overline{P}(|g|I_A)\\ &\leq \overline{P}^{\mathbf{x}}(|f-g|) + \overline{P}(A)\sup|g| < \varepsilon/2 + \varepsilon/2 = \varepsilon,\end{aligned}$$

where we again used Theorem 15.8_{334} to guarantee the $\underline{P}$-previsibility of intervening gambles. □

The following theorem establishes a close relationship between $\underline{P}$-previsibility and $\underline{P}$-measurability. It is a generalisation from Dunford integrals to lower and upper previsions of Theorem 4.4.18 in Bhaskara Rao and Bhaskara Rao (1983), with a proof that resembles theirs quite closely.

Theorem 15.24 *Let f be a $\underline{P}$-previsible gamble, and let g be a gamble. Suppose that $|g| \leq |f|$ a.e. $[\underline{P}]$. Then g is $\underline{P}$-previsible if and only if g is $\underline{P}$-measurable.*

Proof. 'only if'. Suppose g is $\underline{P}$-previsible, then there is some determining sequence of bounded gambles g_n that converges in probability $[\underline{P}]$ to g. Each bounded gamble g_n is $\underline{P}$-measurable by Lemma 15.20_{341}, and therefore the limit g is $\underline{P}$-measurable as well, by Lemma 15.21_{342}.

'if'. Assume that g is $\underline{P}$-measurable. By Lemma 15.22_{343}, there is a sequence g_n of bounded gambles converging in probability $[\underline{P}]$ to g such that $g_n \leq 2|g|$ for every $n \in \mathbb{N}$.

Consider any $A \subseteq \mathscr{X}$. We infer from $g_n \leq 2|g|$ that $|g_n - g_m|I_A \leq 4|g|I_A$, and because $|g| \leq |f|$ a.e. $[\underline{P}]$, this implies that $|g_n - g_m|I_A \leq 4|f|I_A$ a.e. $[\underline{P}]$ for all $n, m \in \mathbb{N}$ (use Proposition 14.10(v)–(vii) and $(\text{ix})_{308}$). We find using the monotonicity of $\overline{P}$ (Theorem $15.14(\text{v})_{335}$) that

$$\overline{P}\left(|g_n - g_m|I_A\right) \leq \overline{P}^{\mathbf{x}}(4|f|I_A) \text{ for all } A \subseteq \mathscr{X} \text{ and all } n, m \in \mathbb{N}. \tag{15.15}$$

Now we show that $\overline{P}\left(|g_n - g_m|\right)$ converges to zero. Let $\varepsilon > 0$. By Lemma 15.23, there is some $\delta_\varepsilon > 0$ such that for every $A \subseteq \mathscr{X}$,

$$\overline{P}^{\mathbf{x}}(A) < \delta_\varepsilon \Rightarrow \overline{P}^{\mathbf{x}}(4|f|I_A) < \varepsilon. \tag{15.16}$$

As g_n converges in probability $[\underline{P}]$ to g, there is, for this ε and δ_ε, some N_ε in $\mathbb{N}$ such that for all $n \geq N_\varepsilon$, $\overline{P}\left(\{|g_n - g| > \varepsilon/2\}\right) < \delta_\varepsilon/2$. Therefore, for all $n, m \geq N_\varepsilon$, we have that

$$\overline{P}\left(\{|g_n - g_m| > \varepsilon\}\right) \leq \overline{P}\left(\left\{|g_n - g| > \frac{\varepsilon}{2}\right\}\right) + \overline{P}\left(\left\{|g - g_m| > \frac{\varepsilon}{2}\right\}\right) < \frac{\delta_\varepsilon}{2} + \frac{\delta_\varepsilon}{2} = \delta_\varepsilon,$$

where we have used the sub-additivity of the coherent $\overline{P}$ (Theorem 4.13(v)$_{53}$). Let $B_{nm} := \{|g_n - g_m| > \varepsilon\})$ then for all $n, m \geq N_\varepsilon$, we have $\overline{P}^{\mathbf{x}}(B_{nm}) < \delta_\varepsilon$. From (15.16)$_\frown$, it follows that $\overline{P}^{\mathbf{x}}(4|f|I_{B_{nm}}) < \varepsilon$ for all $n, m \geq N_\varepsilon$. We also have that $|g_n(x) - g_m(x)| \leq \varepsilon$ for all x in B^c_{nm}. Using the sub-additivity of the coherent $\overline{P}^{\mathbf{x}}$ (Theorem 15.14(vi)$_{335}$) and Equation (15.15)$_\frown$, it follows that

$$\overline{P}\left(|g_n - g_m|\right) \leq \overline{P}\left(|g_n - g_m|I_{B_{nm}}\right) + \overline{P}\left(|g_n - g_m|I_{B^c_{nm}}\right) \leq \overline{P}^{\mathbf{x}}(4|f|I_{B_{nm}}) + \varepsilon\overline{P}^{\mathbf{x}}(B^c_{nm}) < \varepsilon + \varepsilon = 2\varepsilon$$

for all $n, m \geq N_\varepsilon$. We conclude that g_n is a determining sequence for g, so g is indeed $\underline{P}$-previsible. □

The next theorem – a generalisation from Dunford integrals to lower and upper previsions of Theorem 4.6.14 in Bhaskara Rao and Bhaskara Rao (1983) – is the result we are after: it can be seen as a form of Lebesgue's dominated convergence theorem for our extension $\underline{P}^{\mathbf{x}}$.

Theorem 15.25 *Let g be a $\underline{P}$-previsible gamble, and let f_n be a sequence of $\underline{P}$-measurable gambles such that $|f_n| \leq |g|$ a.e. $[\underline{P}]$ for all $n \in \mathbb{N}$. Let f be any gamble. Then the following statements are equivalent:*

(i) *f_n converges in probability $[\underline{P}]$ to f;*

(ii) *f is $\underline{P}$-previsible and $\|f - f_n\|^{\mathbf{x}}_{\underline{P}}$ converges to zero and, hence, $\underline{P}^{\mathbf{x}}(f_n) \to \underline{P}^{\mathbf{x}}(f)$ and $\overline{P}^{\mathbf{x}}(f_n) \to \overline{P}^{\mathbf{x}}(f)$.*

Proof. By Theorem 15.24$_\frown$, f_n is $\underline{P}$-previsible for all $n \in \mathbb{N}$.

(i)⇒(ii). Assume that f_n converges in probability $[\underline{P}]$ to f. Then (ii) will follow from Theorem 15.16$_{337}$ if we can show that $\lim_{n,m\to+\infty} \|f_n - f_m\|^{\mathbf{x}}_{\underline{P}} = 0$.

Let $\varepsilon > 0$. As g is $\underline{P}$-previsible, so is $|g|$, by Theorem 15.8$_{334}$. Invoking Lemma 15.23$_\frown$, this implies that there is some $\delta_\varepsilon > 0$ such that for every $A \subseteq \mathscr{X}$, if $\overline{P}(A) < \delta_\varepsilon$, then $\overline{P}^{\mathbf{x}}(|g|I_A) < \varepsilon$. As $|f_n| \leq |g|$ a.e. $[\underline{P}]$ and therefore also $|f_n|I_A \leq |g|I_A$ a.e. $[\underline{P}]$ (use Proposition 14.10(vi)$_{308}$) for all $n \in \mathbb{N}$, it also follows from Theorems 15.8$_{334}$ and 15.14$_{335}$that

$$\overline{P}^{\mathbf{x}}(A) < \delta_\varepsilon \Rightarrow \overline{P}^{\mathbf{x}}\left(|f_n|I_A\right) < \varepsilon \text{ for all } n \in \mathbb{N}. \tag{15.17}$$

As f_n converges in probability $[\underline{P}]$ to f, there is some N_ε in $\mathbb{N}$ such that $\overline{P}(\{|f-f_n|>\varepsilon/2\})<\delta_\varepsilon/2$ for all $n\geq N_\varepsilon$. Hence, $\overline{P}(\{|f_n-f_m|>\varepsilon\})<\delta_\varepsilon$ for all $n,m\geq N_\varepsilon$. Let $E_{nm}:=\{|f_n-f_m|>\varepsilon\}$, then both $\overline{P}^{\mathbf{x}}(E_{nm})<\delta_\varepsilon$ and $|f_n-f_m|I_{E_{nm}^c}\leq\varepsilon$ for all $n,m\geq N_\varepsilon$. By Equation (15.17), we also have that $\overline{P}^{\mathbf{x}}\left(|f_n|I_{E_{nm}}\right)<\varepsilon$ and $\overline{P}^{\mathbf{x}}\left(|f_m|I_{E_{nm}}\right)<\varepsilon$ for all $n,m\geq N_\varepsilon$. Using the coherence of $\overline{P}^{\mathbf{x}}$, and Theorem 15.14$_{335}$ in particular, we find that for all $n,m\geq N_\varepsilon$,

$$\begin{aligned}\overline{P}^{\mathbf{x}}(|f_n-f_m|)&\leq\overline{P}^{\mathbf{x}}\left(|f_n-f_m|I_{E_{nm}}\right)+\overline{P}^{\mathbf{x}}\left(|f_n-f_m|I_{E_{nm}^c}\right)\\&\leq\overline{P}^{\mathbf{x}}\left(|f_n|I_{E_{nm}}\right)+\overline{P}^{\mathbf{x}}\left(|f_m|I_{E_{nm}}\right)+\varepsilon\\&<\varepsilon+\varepsilon+\varepsilon=3\varepsilon,\end{aligned}$$

which completes this part of the proof.

(ii)⇒(i). Assume that f is $\underline{P}$-previsible and $\|f-f_n\|_{\underline{P}}^{\mathbf{x}}$ converges to zero. We need to prove that f_n converges in probability $[\underline{P}]$ to f or, in other words, that for all $\varepsilon_1,\varepsilon_2>0$, there is some $N_{\varepsilon_1,\varepsilon_2}\in\mathbb{N}$ such that $\overline{P}(\{|f-f_n|>\varepsilon_1\})<\varepsilon_2$ for all $n\geq N_{\varepsilon_1,\varepsilon_2}$. This will be established if, for every $\varepsilon_1,\varepsilon_2>0$, we can prove the existence of a sequence of sets E_n (which may depend on ε_1 and ε_2) such that $\overline{P}(E_n)<\varepsilon_2$ and $|f-f_n|I_{E_n^c}\leq\varepsilon_1$ for all $n\geq N_{\varepsilon_1,\varepsilon_2}$. Indeed, suppose that the second inequality holds, then $E_n^c\subseteq\{|f-f_n|\leq\varepsilon_1\}$, which implies that $E_n\supseteq\{|f-f_n|>\varepsilon_1\}$, and now the first inequality and the monotonicity of the coherent $\overline{P}$ lead to $\overline{P}(\{|f-f_n|>\varepsilon_1\})\leq\overline{P}(E_n)<\varepsilon_2$.

So let $\varepsilon_1,\varepsilon_2>0$ and define $r_{\varepsilon_1,\varepsilon_2}:=\varepsilon_1/3>0$ and $\varepsilon_{\varepsilon_1,\varepsilon_2}:=\varepsilon_1\varepsilon_2/10>0$. Consider the sequence $g_n:=|f-f_n|$. As $\|f-f_n\|_{\underline{P}}^{\mathbf{x}}$ converges to zero by assumption, there is some $N_{\varepsilon_1,\varepsilon_2}\in\mathbb{N}$ such that

$$\overline{P}^{\mathbf{x}}(g_n)<\varepsilon_{\varepsilon_1,\varepsilon_2}\text{ for all }n\geq N_{\varepsilon_1,\varepsilon_2}.\tag{15.18}$$

Consider a fixed n, then g_n is $\underline{P}$-previsible by Theorem 15.8$_{334}$, so, by Definition 15.6$_{333}$, it has a determining sequence of bounded gambles h_m^n, $m\in\mathbb{N}$ such that h_m^n converges to h^n in probability $[\underline{P}]$ and such that $\overline{P}(h_m^n)\to\overline{P}^{\mathbf{x}}(g_n)$. As g_n is non-negative, we may assume (use Theorem 15.8$_{334}$) that all $h_m^n\geq 0$. So we know that there is some $m\in\mathbb{N}$ (depending on ε_1 and ε_2) such that, with $h_{n,\varepsilon_1,\varepsilon_2}:=h_m^n\geq 0$,

$$\left|\overline{P}^{\mathbf{x}}(g_n)-\overline{P}(h_{n,\varepsilon_1,\varepsilon_2})\right|<\varepsilon_{\varepsilon_1,\varepsilon_2}\text{ and }\overline{P}\left(\left\{\left|g_n-h_{n,\varepsilon_1,\varepsilon_2}\right|>r_{\varepsilon_1,\varepsilon_2}\right\}\right)<\frac{\varepsilon_{\varepsilon_1,\varepsilon_2}}{r_{\varepsilon_1,\varepsilon_2}}.\tag{15.19}$$

Let $A_{n,\varepsilon_1,\varepsilon_2}:=\left\{\left|g_n-h_{n,\varepsilon_1,\varepsilon_2}\right|>r_{\varepsilon_1,\varepsilon_2}\right\}$, then this leads to

$$\overline{P}(A_{n,\varepsilon_1,\varepsilon_2})<\frac{\varepsilon_{\varepsilon_1,\varepsilon_2}}{r_{\varepsilon_1,\varepsilon_2}}\text{ and }\left|g_n-h_{n,\varepsilon_1,\varepsilon_2}\right|I_{A_{n,\varepsilon_1,\varepsilon_2}^c}\leq r_{\varepsilon_1,\varepsilon_2}\text{ for all }n\geq N_{\varepsilon_1,\varepsilon_2}.\tag{15.20}$$

Also let $B_{n,\varepsilon_1,\varepsilon_2} := \{h_{n,\varepsilon_1,\varepsilon_2} > r_{\varepsilon_1,\varepsilon_2}\}$, then (recall that $h_{n,\varepsilon_1,\varepsilon_2} \geq 0$, so $I_{B_{n,\varepsilon_1,\varepsilon_2}} \leq h_{n,\varepsilon_1,\varepsilon_2}/r_{\varepsilon_1,\varepsilon_2}$)

$$\overline{P}(B_{n,\varepsilon_1,\varepsilon_2}) \leq \frac{\overline{P}(h_{n,\varepsilon_1,\varepsilon_2})}{r_{\varepsilon_1,\varepsilon_2}} \text{ and } h_{n,\varepsilon_1,\varepsilon_2} I_{B^c_{n,\varepsilon_1,\varepsilon_2}} \leq r_{\varepsilon_1,\varepsilon_2} \text{ for all } n \in \mathbb{N}. \tag{15.21}$$

Finally, let $E_{n,\varepsilon_1,\varepsilon_2} := A_{n,\varepsilon_1,\varepsilon_2} \cup B_{n,\varepsilon_1,\varepsilon_2}$ then we have for all $n \geq N_{\varepsilon_1,\varepsilon_2}$ that

$$\overline{P}(E_{n,\varepsilon_1,\varepsilon_2}) \leq \overline{P}(A_{n,\varepsilon_1,\varepsilon_2}) + \overline{P}(B_{n,\varepsilon_1,\varepsilon_2}) < \frac{\varepsilon_{\varepsilon_1,\varepsilon_2}}{r_{\varepsilon_1,\varepsilon_2}} + \frac{\overline{P}(h_{n,\varepsilon_1,\varepsilon_2})}{r_{\varepsilon_1,\varepsilon_2}}$$

$$< \frac{\varepsilon_{\varepsilon_1,\varepsilon_2}}{r_{\varepsilon_1,\varepsilon_2}} + \frac{\left(\overline{P}^{\mathbf{x}}(g_n) + \varepsilon_{\varepsilon_1,\varepsilon_2}\right)}{r_{\varepsilon_1,\varepsilon_2}} < \frac{3\varepsilon_{\varepsilon_1,\varepsilon_2}}{r_{\varepsilon_1,\varepsilon_2}} = \frac{9\varepsilon_2}{10} < \varepsilon_2,$$

where we used the inequalities $(15.20)_\frown$, (15.21), $(15.19)_\frown$ and $(15.18)_\frown$, in that order. We also have that

$$|f - f_n| I_{E^c_{n,\varepsilon_1,\varepsilon_2}} = g_n I_{E^c_{n,\varepsilon_1,\varepsilon_2}} \leq (h_{n,\varepsilon_1,\varepsilon_2} + r_{\varepsilon_1,\varepsilon_2}) I_{E^c_{n,\varepsilon_1,\varepsilon_2}} \leq r_{\varepsilon_1,\varepsilon_2} + r_{\varepsilon_1,\varepsilon_2}$$

$$= \frac{2\varepsilon_1}{3} < \varepsilon_1,$$

where we used the inequalities $(15.20)_\frown$ and (15.21), in that order. □

15.5 Previsibility by cuts

We can now combine the arguments in the previous sections to derive a workable criterion for determining whether a gamble is $\underline{P}$-previsible.

Criterion 15.26 *Let $a_n \leq b_n$ be any pair of cut sequences. A gamble f is $\underline{P}$-previsible if and only if $f_{a_n}^{b_n}$ is a determining sequence for f, and in that case,*

$$\underline{P}^{\mathbf{x}}(f) = \lim_{n\to+\infty} \underline{P}(f_{a_n}^{b_n}) = \lim_{(a,b)\to(-\infty,+\infty)} \underline{P}(f_a^b) \tag{15.22}$$

Proof. If $f_{a_n}^{b_n}$ is a determining sequence for f, then f is $\underline{P}$-previsible by Definition 15.6_{333}. This proves sufficiency.

To prove necessity, suppose that f is $\underline{P}$-previsible, then, by Theorem 15.24_{345}, f is $\underline{P}$-measurable. It follows that the sequence $f_{a_n}^{b_n}$ converges in probability $[\underline{P}]$ to f (use Definition $15.19(\mathrm{D})_{340}$). Also the bounded gamble $f_{a_n}^{b_n}$ is $\underline{P}$-measurable by Lemma 15.20_{341}. Moreover, $\left|f_{a_n}^{b_n}\right| \leq |f|$. So, it follows from Theorem 15.25_{346} that

$$\lim_{n,m\to+\infty} \|f_{a_n}^{b_n} - f_{a_m}^{b_m}\|_{\underline{P}}^{\mathbf{x}} \leq \lim_{n\to+\infty} \|f_{a_n}^{b_n} - f\|_{\underline{P}}^{\mathbf{x}} + \lim_{m\to+\infty} \|f - f_{a_m}^{b_m}\|_{\underline{P}}^{\mathbf{x}} = 0 + 0 = 0.$$

We find that $f_{a_n}^{b_n}$ is a determining sequence for f.

The first equality in Equation (15.22) follows from the definition of $\underline{P}^{\mathbf{x}}$. The second equality follows from Lemma 15.27. □

In the above proof, we have used the following lemma, which links convergence of a net over $\mathbb{R}^2$ to convergence of a net over $\mathbb{N}$; see Section 1.5_7 and in particular Example 1.10_8 for these types of convergence.

Lemma 15.27 *Let $f : \mathbb{R}^2 \to \mathbb{R}$ and $f^* \in \mathbb{R}$. Then $\lim_{(a,b)\to(-\infty,+\infty)} f(a,b) = f^*$ if and only if $\lim_{n\to+\infty} f(a_n, b_n) = f^*$ for all pairs of cut sequences $a_n \leq b_n$.*

Proof. 'if'. Assume that $f(a,b)$ does not converge to f^* as $(a,b) \to (-\infty,+\infty)$. This means that there is some $\varepsilon > 0$ such that for all $R > 0$, there are $-a_R, b_R \geq R$ such that $|f(a_R, b_R) - f^*| \geq \varepsilon$. This holds in particular for all $R = n$, $n \in \mathbb{N}$. Hence, there are two sequences a_n and b_n such that $-a_n, b_n \geq n$ and $|f(a_n, b_n) - f^*| \geq \varepsilon$. Consequently, there is a pair of cut sequences $a_n \leq b_n$ for which $f(a_n, b_n)$ does not converge to f^* as $n \to +\infty$.

'only if'. Let $a_n \leq b_n$ be any pair of cut sequences: for all $R > 0$, there is some $N_R \in \mathbb{N}$ such that $n \geq N_R$ implies that $-a_n, b_n \geq R$. As $\lim_{(a,b)\to(-\infty,+\infty)} f(a,b) = f^*$, for all $\varepsilon > 0$, there is some $R_\varepsilon > 0$ such that $-a, b \geq R_\varepsilon$ implies that $|f(a,b) - f^*| < \varepsilon$. Hence, it holds in particular that $|f(a_n, b_n) - f^*| < \varepsilon$ for all $n \geq N_{R_\varepsilon}$ or, equivalently, $\lim_{n\to+\infty} f(a_n, b_n) = f^*$. □

We can use this previsibility criterion in terms of cuts to add an extra equivalent statement to the formulation of Corollary 15.10_{334}.

Corollary 15.28 *Let f be any gamble, then the following statements are equivalent:*

(i) *f is $\underline{P}$-previsible;*

(ii) *$|f|$ is $\underline{P}$-previsible.*

The implication (i)⇒(ii) has already been established earlier in Theorem 15.8_{334}. The converse implication is new.

Proof. We use Criterion 15.26 with the pair of cut sequences $(-n, n)$. Simply observe that for any $n \in \mathbb{N}$ and any $x \in \mathcal{X}$,

$$|f_{-n}^n(x)| = \begin{cases} n & \text{if } f(x) \leq -n \\ |f(x)| & \text{if } -n \leq f(x) \leq n \\ n & \text{if } f(x) \geq n \end{cases} = \begin{cases} |f(x)| & \text{if } |f(x)| \leq n \\ n & \text{if } |f(x)| \geq n \end{cases}$$

and therefore $|f_{-n}^n| = |f|_{-n}^n$; similarly,

$$|f(x) - f_{-n}^n(x)| = \begin{cases} -n - f(x) & \text{if } f(x) \leq -n \\ 0 & \text{if } -n \leq f(x) \leq n \\ f(x) - n & \text{if } n \leq f(x) \end{cases} = \begin{cases} 0 & \text{if } |f(x)| \leq n \\ |f(x)| - n & \text{if } |f(x)| \geq n \end{cases}$$

and therefore $\left|f - f_{-n}^{n}\right| = \left||f| - |f|_{-n}^{n}\right|$, and also, with $m \leq n \in \mathbb{N}$,

$$\left|f_{-m}^{m}(x) - f_{-n}^{n}(x)\right| = \begin{cases} n - m & \text{if } f(x) \leq -n \\ -f(x) - m & \text{if } -n \leq f(x) \leq -m \\ 0 & \text{if } -m \leq f(x) \leq m \\ f(x) - m & \text{if } m \leq f(x) \leq n \\ n - m & \text{if } f(x) \geq n \end{cases}$$

$$= \begin{cases} 0 & \text{if } |f(x)| \leq m \\ |f(x)| - m & \text{if } m \leq |f(x)| \leq n \\ n - m & \text{if } |f(x)| \geq n \end{cases}$$

and therefore $\left|f_{-m}^{m} - f_{-n}^{n}\right| = \left||f|_{-m}^{m} - |f|_{-n}^{n}\right|$. This shows that f_{-n}^{n} is a determining sequence for f if and only if $|f|_{-n}^{n}$ is a determining sequence for $|f|$. □

15.6 A sufficient condition for previsibility

We will now formulate a simpler *sufficient* condition for $\underline{P}$-previsibility. First, we need some lemmas. The convergence for the nets on $\mathbb{N}^2$ and $\mathbb{R}^2$ in this discussion was introduced in Section 1.5_7 and in particular in Example 1.10_8.

Lemma 15.29 *Let f be any map from non-negative real numbers to non-negative real numbers. If $\lim_{t\to+\infty} tf(t) = 0$, then $\lim_{t\to+\infty} f(t) = 0$.*

Proof. Assume that $\lim_{t\to+\infty} tf(t) = 0$. Let $\varepsilon > 0$, then this means that there is some $T_\varepsilon \geq 0$ such that $tf(t) < \varepsilon$ for every $t \geq T_\varepsilon$. Define $U_\varepsilon := \max\{1, T_\varepsilon\}$, then $0 \leq f(t) \leq tf(t) < \varepsilon$ for every $t \geq U_\varepsilon$. It follows that $\lim_{t\to+\infty} f(t) = 0$. □

Lemma 15.30 *Let s_α and t_α, $\alpha \in \mathscr{A}$ be nets of real numbers converging to $+\infty$, and let $f : \mathbb{R}^2 \to \mathbb{R}$. Then $\lim_{s,t\to+\infty} f(s,t) = \lim_\alpha f(s_\alpha, t_\alpha)$ if $\lim_{s,t\to+\infty} f(s,t)$ exists.*

Proof. Assume $y := \lim_{s,t\to+\infty} f(s,t)$ exists. We give a proof for $y \in \mathbb{R}$. The cases $y = \pm\infty$ are proved in an analogous manner. Let $\varepsilon > 0$, then there is some R_ε in $\mathbb{R}$ such that $|f(s,t) - y| < \varepsilon$ for all $s, t \geq R_\varepsilon$. As s_α converges to infinity, there is some $\alpha_{R_\varepsilon} \in A$ such that $s_\alpha \geq R_\varepsilon$ for all $\alpha \geq \alpha_{R_\varepsilon}$. As t_α converges to infinity, there is some $\beta_{R_\varepsilon} \in A$ such that $t_\alpha \geq R_\varepsilon$ for all $\alpha \geq \beta_{R_\varepsilon}$. This implies that $\left|f(s_\alpha, t_\alpha) - y\right| < \varepsilon$ for all $\alpha \geq \max\{\alpha_{R_\varepsilon}, \beta_{R_\varepsilon}\}$ or, equivalently, $\lim_\alpha f(s_\alpha, t_\alpha) = y$. □

Condition 15.31 *A gamble f is $\underline{P}$-previsible if the following conditions are satisfied:*

(i) $\lim\limits_{a\to+\infty} a\overline{P}(\{|f| \geq a\}) = 0$;

(ii) $\lim\limits_{a,b\to+\infty} (b - a)\overline{P}(\{a \leq |f| \leq b\}) = 0$.

Proof. Let $a_n \leq b_n$ be a pair of cut sequences. We prove that if the conditions are satisfied, then $f_{a_n}^{b_n}$ is a determining sequence for f. By Criterion 15.26348, it will then follow that f is $\underline{P}$-previsible. As eventually $a_n \leq 0$ and $b_n \geq 0$, we may assume without loss of generality that $a_n \leq 0 \leq b_n$ for all natural numbers n.

First, we prove convergence in probability [$\underline{P}$]. Fix any $\varepsilon > 0$. Define $t_n := \min\{-a_n, b_n\} + \varepsilon$, then t_n converges to $+\infty$ because $-a_n$ and b_n do. From (i) and Lemmas 15.29 and 15.30 (for a map f depending only on a single variable), it follows that

$$\lim_{n \to +\infty} \overline{P}\left(\{|f| \geq t_n\}\right) = 0. \tag{15.23}$$

As $\left|f(x) - f_{a_n}^{b_n}(x)\right| > \varepsilon$ implies that $|f(x)| \geq \min\{-a_n, b_n\} + \varepsilon = t_n$, we also find that $\left\{\left|f - f_{a_n}^{b_n}\right| > \varepsilon\right\} \subseteq \{|f| \geq t_n\}$. Now use Equation (15.23) and the monotonicity of the coherent $\overline{P}$ to find that $\lim_{n\to+\infty} \overline{P}\left(\left\{\left|f - f_{a_n}^{b_n}\right| > \varepsilon\right\}\right) = 0$, so the sequence of bounded gambles $f_{a_n}^{b_n}$ indeed converges in probability [$\underline{P}$] to f.

Next, we prove that the Cauchy condition $\lim_{n,m\to+\infty} \overline{P}\left(\left|f_{a_n}^{b_n} - f_{a_m}^{b_m}\right|\right) = 0$ holds. Define the nets $c_{n,m} := \min\{a_n, a_m\} \leq 0$ and $d_{n,m} := \max\{a_n, a_m\} \leq 0$ and the nets $e_{n,m} := \min\{b_n, b_m\} \geq 0$ and $g_{n,m} := \max\{b_n, b_m\} \geq 0$. Then $-c_{n,m}$, $-d_{n,m}$, $e_{n,m}$ and $g_{n,m}$ converge to $+\infty$. Observe that (see Figure 15.1↷)

$$\begin{aligned}
|f_{a_n}^{b_n} - f_{a_m}^{b_m}| &\leq (g_{n,m} - e_{n,m})\left[I_{\{e_{n,m} \leq f \leq g_{n,m}\}} + I_{\{f \geq g_{n,m}\}}\right] \\
&\quad + (d_{n,m} - c_{n,m})\left[I_{\{c_{n,m} \leq f \leq d_{n,m}\}} + I_{\{f \leq c_{n,m}\}}\right] \\
&\leq g_{n,m} I_{\{f \geq g_{n,m}\}} + (g_{n,m} - e_{n,m}) I_{\{e_{n,m} \leq f \leq g_{n,m}\}} \\
&\quad + (-c_{n,m}) I_{\{f \leq c_{n,m}\}} + (d_{n,m} - c_{n,m}) I_{\{c_{n,m} \leq f \leq d_{n,m}\}} \\
&\leq g_{n,m} I_{\{|f| \geq g_{n,m}\}} + (g_{n,m} - e_{n,m}) I_{\{e_{n,m} \leq |f| \leq g_{n,m}\}} \\
&\quad + (-c_{n,m}) I_{\{|f| \geq -c_{n,m}\}} + (d_{n,m} - c_{n,m}) I_{\{-d_{n,m} \leq |f| \leq -c_{n,m}\}}.
\end{aligned}$$

If we now use the sub-additivity, monotonicity and non-negative homogeneity of the coherent $\overline{P}$, we find that

$$\begin{aligned}
\overline{P}\left(\left|f_{a_n}^{b_n} - f_{a_m}^{b_m}\right|\right) &\leq g_{n,m} \overline{P}\left(\{|f| \geq g_{n,m}\}\right) + (-c_{n,m}) \overline{P}\left(\{|f| \geq -c_{n,m}\}\right) \\
&\quad + (g_{n,m} - e_{n,m}) \overline{P}\left(\{e_{n,m} \leq |f| \leq g_{n,m}\}\right) \\
&\quad + (d_{n,m} - c_{n,m}) \overline{P}\left(\{-d_{n,m} \leq |f| \leq -c_{n,m}\}\right),
\end{aligned}$$

and the assumptions and Lemma 15.30 guarantee that the right-hand side converges to zero. Therefore, the left-hand side converges to zero as well. □

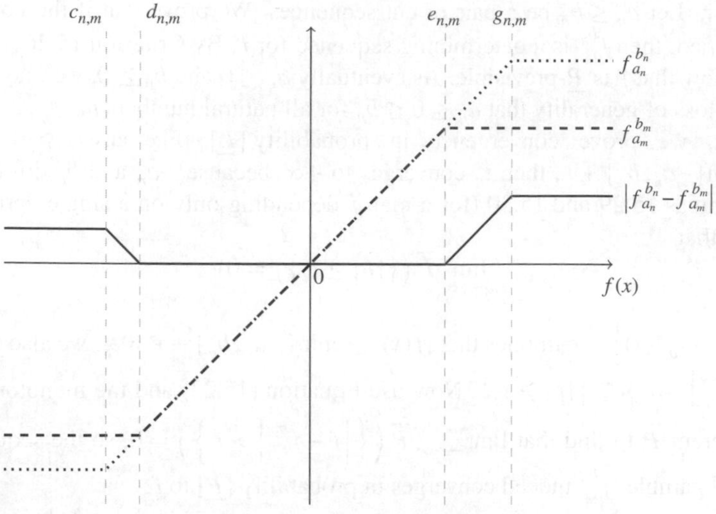

Figure 15.1 *Bounding the cut differences* $|f_{a_n}^{b_n} - f_{a_m}^{b_m}|$.

15.7 Previsibility for 2-monotone lower previsions

We now turn to the special case that the lower prevision $\underline{P}$ is 2-monotone. It is an immediate consequence of Corollary 6.23_{118} and Theorem 4.26(ii)_{65} that $\underline{P}$ coincides on all bounded gambles with the Choquet functional associated with its restriction to events:

$$\underline{P}(f) = C_{\underline{P}}(f) = C\int f \, \mathrm{d}\underline{P} \text{ for all } f \in \mathbb{B}. \tag{15.24}$$

We will provide a simple characterisation of previsibility for 2-monotone lower previsions and show that Equation (15.24) can be extended to $\mathbb{G}^{\mathbf{x}}_{\underline{P}}$. More details on Choquet integration of bounded and unbounded gambles can be found in Appendix C_{376}.

Theorem 15.32 *Let $\underline{P}$ be a 2-monotone coherent lower prevision on $\mathbb{B}$. A gamble f is $\underline{P}$-previsible if and only if*

$$\lim_{s\to+\infty} R\int_s^{+\infty} \overline{P}(\{|f| \geq t\}) \, \mathrm{d}t = 0, \tag{15.25}$$

or, equivalently, if and only if

$$\lim_{s\to+\infty} C\int (|f| - s)^+ \, \mathrm{d}\overline{P} = 0. \tag{15.26}$$

Proof. We establish that f_{-n}^n is a determining sequence for f with respect to $\underline{P}$ if and only if Equation (15.25) is satisfied.

Equation (15.25) obviously implies that $\lim_{t\to+\infty} \overline{P}(\{|f| \geq t\}) = 0$, and therefore, by Definition 15.19_{340} ((D)_{340} and (E)_{340} in particular), f_{-n}^n converges in probability $[\underline{P}]$ to f. So it suffices to prove that the Cauchy condition is equivalent to Equation (15.25). To see that such is indeed the case, consider that for any natural numbers m and n with $m \leq n$,

$$|f_{-n}^{n}(x) - f_{-m}^{m}(x)| = \begin{cases} 0 & \text{if } |f(x)| \leq m \\ |f(x)| - m & \text{if } m \leq |f(x)| \leq n \\ n - m & \text{if } |f(x)| \geq n, \end{cases} \tag{15.27}$$

and therefore, with $g_{m,n} := |f_{-n}^{n} - f_{-m}^{m}|$ and assuming that f is unbounded (the bounded case is trivially satisfied),

$$\begin{aligned} \overline{P}\left(\left|f_{-n}^{n} - f_{-m}^{m}\right|\right) &= \inf g_{m,n} + R\int_{\inf g_{m,n}}^{\sup g_{m,n}} \overline{P}\left(\left\{g_{m,n}(x) \geq t\right\}\right) \mathrm{d}t \\ &= R\int_{0}^{n-m} \overline{P}(\{|f(x)| - m \geq t\}) \,\mathrm{d}t \\ &= R\int_{m}^{n} \overline{P}(\{|f(x)| \geq t\}) \,\mathrm{d}t, \end{aligned}$$

so the Cauchy condition for previsibility turns into

$$\lim_{m,n\to+\infty} R\int_{\min\{m,n\}}^{\max\{m,n\}} \overline{P}(\{|f(x)| \geq t\}) \,\mathrm{d}t = 0.$$

We show that this is simply another way of writing Equation (15.25). On the one hand, the limit above being zero means that for all $\varepsilon > 0$, we can find m and $n \in \mathbb{N}$ such that the Riemann integral from $\min\{m, n\}$ to $\max\{m, n\}$ is strictly smaller than ε. On the other hand, Equation (15.25) means that for all $\varepsilon > 0$, we can find some $s \in \mathbb{R}$ such that the Riemann integral from s to $+\infty$ is strictly smaller than ε, which is in turn equivalent to saying that for all $\varepsilon > 0$, we can find s and u in $\mathbb{R}$ such that the Riemann integral from $\min\{s, u\}$ to $\max\{s, u\}$ is strictly smaller than ε. Now apply a trivial variation of Lemma 15.27_{349}.

Equation (15.26) follows immediately from Equation (15.25), the definition of the Choquet integral (Equation $(\text{C.5})_{379}$) and $(|f| - s)^{+} \geq t \Leftrightarrow |f| \geq s + t$ for non-negative real s and t. □

Theorem 15.33 *Let $\underline{P}$ be a 2-monotone coherent lower prevision on $\mathbb{B}$. A gamble f is $\underline{P}$-previsible if and only if f is Choquet integrable with respect to $\underline{P}$, and in that case,*

$$\underline{P}^{\mathrm{x}}(f) = C\int f \,\mathrm{d}\underline{P}.$$

Proof. 'if'. First, we consider the case that f is non-negative. Assume that f is Choquet integrable, meaning that $C\int f \,\mathrm{d}\underline{P}$ is a (finite) real number. Then it follows from Equations $(\text{C.5})_{379}$ and $(\text{C.4})_{377}$ that for all $s \in \mathbb{R}_{>0}$,

$$C\int f \,\mathrm{d}\underline{P} = R\int_{0}^{+\infty} \underline{P}(\{f \geq t\}) \,\mathrm{d}t = R\int_{0}^{s} \underline{P}(\{f \geq t\}) \,\mathrm{d}t + R\int_{s}^{+\infty} \underline{P}(\{f \geq t\}) \,\mathrm{d}t.$$

Taking the limit for $s \to +\infty$ on both sides of the inequality, we get, also recalling Equation (C.3),

$$C\int f\,\mathrm{d}\underline{P} = \lim_{s\to+\infty} R\int_0^s \underline{P}(\{f \geq t\})\,\mathrm{d}t + \lim_{s\to+\infty} R\int_s^{+\infty} \underline{P}(\{f \geq t\})\,\mathrm{d}t$$
$$= C\int f\,\mathrm{d}\underline{P} + \lim_{s\to+\infty} R\int_s^{+\infty} \underline{P}(\{f \geq t\})\,\mathrm{d}t.$$

Because $C\int f\,\mathrm{d}\underline{P}$ is finite, the limit on the right-hand side must be zero and, therefore, f is $\underline{P}$-previsible by Theorem 15.32_{352}.

Now let f be an arbitrary gamble. If f is Choquet integrable, then clearly so are the non-negative gambles f^+ and f^-, and by what we have just shown, both f^+ and f^- are therefore $\underline{P}$-previsible, implying that $f^+ - f^- = f$ is $\underline{P}$-previsible too (use Corollary 15.10_{334}).

'only if'. Assume that f is $\underline{P}$-previsible, so we know from Corollary 15.10_{334} that f^+ and f^- are $\underline{P}$-previsible as well. With the notations established in Appendix C_{376} and taking into account Proposition C.3(i) and $(\mathrm{ii})_{379}$,[6]

$$C\int f^+\,\mathrm{d}\underline{P} = R\int_0^{+\infty} \underline{P}\left(\left\{f^+ \geq t\right\}\right)\mathrm{d}t = \lim_{b\to+\infty} R\int_0^b \underline{P}\left(\left\{f^+ \geq t\right\}\right)\mathrm{d}t$$
$$= \lim_{(a,b)\to(-\infty,+\infty)} R\int_0^b \underline{P}\left(\left\{(f^+)_a^b \geq t\right\}\right)\mathrm{d}t$$
$$= \lim_{(a,b)\to(-\infty,+\infty)} C\int (f^+)_a^b\,\mathrm{d}\underline{P}.$$

If we now use Equation $(15.24)_{352}$ and the $\underline{P}$-previsibility of f^+ (in particular, Equation $(15.22)_{348}$), we see that

$$C\int f^+\,\mathrm{d}\underline{P} = \lim_{(a,b)\to(-\infty,+\infty)} C\int (f^+)_a^b\,\mathrm{d}\underline{P} = \lim_{(a,b)\to(-\infty,+\infty)} \underline{P}\left((f^+)_a^b\right) = \underline{P}^{\mathbf{x}}(f^+)$$

is a real number and so, for similar reasons, is

$$C\int f^-\,\mathrm{d}\overline{P} = \lim_{(a,b)\to(-\infty,+\infty)} C\int (f^-)_a^b\,\mathrm{d}\overline{P} = \lim_{(a,b)\to(-\infty,+\infty)} \overline{P}\left((f^-)_a^b\right) = \overline{P}^{\mathbf{x}}(f^-).$$

This implies that f is Choquet integrable and that, moreover,

$$C\int f\,\mathrm{d}\underline{P} = C\int f^+\,\mathrm{d}\underline{P} - C\int f^-\,\mathrm{d}\overline{P}$$
$$= \lim_{(a,b)\to(-\infty,+\infty)} C\int (f^+)_a^b\,\mathrm{d}\underline{P} - \lim_{(a,b)\to(-\infty,+\infty)} C\int (f^-)_a^b\,\mathrm{d}\overline{P}$$
$$= \lim_{(a,b)\to(-\infty,+\infty)} C\int (f_a^b)^+\,\mathrm{d}\underline{P} - \lim_{(a,b)\to(-\infty,+\infty)} C\int (f_{-b}^{-a})^-\,\mathrm{d}\overline{P}$$

[6] In considering the limits below, it should be remembered that, *eventually*, $a \leq 0 \leq b$. The argument only acquires validity when (as soon as) we pass into in this region.

$$= \lim_{(a,b)\to(-\infty,+\infty)} C\int (f_a^b)^+ \,\mathrm{d}\underline{P} - \lim_{(a,b)\to(-\infty,+\infty)} C\int (f_a^b)^- \,\mathrm{d}\overline{P}$$

$$= \lim_{(a,b)\to(-\infty,+\infty)} \left(C\int (f_a^b)^+ \,\mathrm{d}\underline{P} - C\int (f_a^b)^- \,\mathrm{d}\overline{P} \right)$$

$$= \lim_{(a,b)\to(-\infty,+\infty)} C\int f_a^b \,\mathrm{d}\underline{P} = \lim_{(a,b)\to(-\infty,+\infty)} \underline{P}(f_a^b) = \underline{P}^{\mathbf{x}}(f),$$

where we have used Proposition C.3(i)$_{379}$, the fact that, *eventually*, $a \le 0$ and $b \ge 0$ and therefore $(f^+)_a^b = (f_a^b)^+$ and also $(f^-)_a^b = [(-f)^+]_a^b = [(-f)_a^b]^+ = (-f_{-b}^{-a})^+ = (f_{-b}^{-a})^-$, Equation (15.24)$_{352}$ and the $\underline{P}$-previsibility of f. □

15.8 Convex combinations

Another way to characterise previsibility is to express the lower prevision $\underline{P}$ as a convex combination of lower previsions for which we already know the previsible gambles and the previsible extensions.

Proposition 15.34 *Suppose $\Gamma = \{\underline{P}_1, \underline{P}_2, \dots, \underline{P}_p\}$ is a finite collection of coherent lower previsions defined on $\mathbb{B}$. Let $\underline{Q}$ be a convex combination of the elements of Γ, meaning that*

$$\underline{Q}(f) = \sum_{i=1}^{p} \lambda_i \underline{P}_i(f) \text{ for all } f \in \mathbb{B},$$

where $\lambda_1, \dots, \lambda_p \in \mathbb{R}_{>0}$ and $\sum_{i=1}^{p} \lambda_i = 1$. Then

$$\mathbb{G}_{\underline{Q}}^{\mathbf{x}} = \bigcap_{i=1}^{p} \mathbb{G}_{\underline{P}_i}^{\mathbf{x}} \text{ and } \underline{Q}^{\mathbf{x}}(f) = \sum_{i=1}^{p} \lambda_i \underline{P}_i^{\mathbf{x}}(f) \text{ for all } f \in \mathbb{G}_{\underline{Q}}^{\mathbf{x}}.$$

Proof. To see that an gamble is $\underline{Q}$-previsible if and only if it is $\underline{P}_i$-previsible for all $i \in \{1, \dots, p\}$, simply use Definitions 15.6$_{333}$ and 15.1$_{329}$ and recall that the limit of a convex combination is the convex combination of the limits.

Using a similar argument,

$$\underline{Q}^{\mathbf{x}}(f) = \lim_{n\to+\infty} \underline{Q}(f_n) = \lim_{n\to+\infty} \sum_{i=1}^{p} \lambda_i \underline{P}_i(f_n) = \sum_{i=1}^{p} \lambda_i \lim_{n\to+\infty} \underline{P}_i(f_n) = \sum_{i=1}^{p} \lambda_i \underline{P}_i^{\mathbf{x}}(f_n),$$

where f_n is any determining sequence for f with respect to $\underline{Q}$ and, therefore, also with respect to all $\underline{P}_i$. □

15.9 Lower envelope theorem

To polish off the theoretical part of this chapter, we turn to considerations of duality and prove that every extended lower prevision $\underline{P}^{\mathbf{x}}$ is the lower envelope of the set of linear previsions on $\mathbb{G}^{\mathrm{x}}_{\underline{P}}$ that are obtained through extension of a linear prevision in $\mathrm{lins}(\underline{P})$.

Our discussion hinges on the following three basic propositions. We can roughly summarise the first by stating that the more precise – or the less committal – a coherent lower prevision is, the more previsible gambles it has.

Proposition 15.35 *Consider any coherent lower prevision $\underline{Q}$ on $\mathbb{B}$ that dominates $\underline{P}$. Then the following statements hold:*

(i) *If f_n is a determining sequence for gamble f with respect to $\underline{P}$, then f_n is also a determining sequence for f with respect to $\underline{Q}$.*

(ii) $\mathbb{G}^{\mathrm{x}}_{\underline{Q}} \supseteq \mathbb{G}^{\mathrm{x}}_{\underline{P}}$.

(iii) *$\underline{Q}^{\mathrm{x}}$ dominates $\underline{P}^{\mathrm{x}}$ on $\mathbb{G}^{\mathrm{x}}_{\underline{P}}$.*

Proof. (i). As $\underline{Q}$ dominates $\underline{P}$, we also have that $\overline{Q}(h) \leq \overline{P}(h)$ for all bounded gambles h, which implies that for all $\varepsilon > 0$ and all $n, m \in \mathbb{N}$,

$$\overline{Q}\left(\{|f - f_n| > \varepsilon\}\right) \leq \overline{P}\left(\{|f - f_n| > \varepsilon\}\right) \text{ and } \overline{Q}\left(|f_n - f_m|\right) \leq \overline{P}\left(|f_n - f_m|\right).$$

As f_n is a determining sequence for f with respect to $\underline{P}$, the right-hand sides of both inequalities converge to zero (see Definition 15.6_{333}) and so do, therefore, the left-hand sides. This shows that f_n is a determining sequence for f with respect to $\underline{Q}$ as well.

(ii). Immediate consequence of (i).

(iii). Consider any $\underline{P}$-previsible gamble f, and let f_n be any determining sequence for f with respect to $\underline{P}$. We have already proved in (i) that f_n is also a determining sequence for f with respect to $\underline{Q}$. As $\underline{Q}(f_n) \geq \underline{P}(f_n)$ for all $n \in \mathbb{N}$, we find that $\lim_{n\to+\infty} \underline{Q}(f_n) \geq \lim_{n\to+\infty} \underline{P}(f_n)$, because the limits on both sides are guaranteed to exist. Hence, $\underline{Q}^{\mathrm{x}}(f) \geq \underline{P}^{\mathrm{x}}(f)$, and therefore $\underline{Q}^{\mathrm{x}}$ dominates $\underline{P}^{\mathrm{x}}$ on the set of $\underline{P}$-previsible gambles. □

The second proposition combines the first with the discussion in Section $15.10.1_{358}$ to identify previsible extension of dominating linear previsions with Dunford integration.

Proposition 15.36 *Let $Q \in \mathrm{lins}(\underline{P})$, and let f be a $\underline{P}$-previsible gamble. Then the following statements hold:*

(i) *f is Q-previsible.*

(ii) $Q^{\mathrm{x}}(f) = D\int f \, \mathrm{d}Q$.

Proof. (i). The Q-previsibility of f follows from Proposition 15.35(ii).

(ii). This follows immediately from the discussion of the previsible extension of linear previsions in Section $15.10.1_{358}$ further on. □

The third proposition identifies the (unconditional) linear previsions that dominate a previsible extension with previsible extensions of linear previsions.

Proposition 15.37 *Let Λ be any linear prevision on $\mathbb{G}^{\mathbf{x}}_{\underline{P}}$ that dominates $\underline{P}^{\mathbf{x}}$: $\Lambda(f) \geq \underline{P}^{\mathbf{x}}(f)$ for all $f \in \mathbb{G}^{\mathbf{x}}_{\underline{P}}$. Let Q be the restriction of Λ to the set $\mathbb{B}$ of all bounded gambles. Then*

(i) $Q \in \mathrm{lins}(\underline{P})$;

(ii) $\Lambda(f) = Q^{\mathbf{x}}(f)$ *for all* $f \in \mathbb{G}^{\mathbf{x}}_{\underline{P}}$.

Proof. (i). First, observe that Q is a linear prevision because it is the restriction of a linear prevision Λ (compare the linearity characterisations $(\mathrm{D})_{52}$ and $(\mathrm{D})_{265}$). As Q dominates $\underline{P}$, we see that indeed $Q \in \mathrm{lins}(\underline{P})$.

(ii). First, we establish that $|\Lambda(g)| \leq \Lambda(|g|)$ for all $g \in \mathbb{G}^{\mathbf{x}}_{\underline{P}}$. Indeed,

$$|g| - g \geq 0 \Rightarrow \Lambda(|g| - g) \geq \underline{P}^{\mathbf{x}}(|g| - g) \geq 0 \Rightarrow \Lambda(|g|) \geq \Lambda(g),$$
$$|g| + g \geq 0 \Rightarrow \Lambda(|g| + g) \geq \underline{P}^{\mathbf{x}}(|g| + g) \geq 0 \Rightarrow \Lambda(|g|) \geq -\Lambda(g).$$

Let f be a $\underline{P}$-previsible gamble, and let f_n be a determining sequence for f with respect to $\underline{P}$. Then it follows from Proposition 15.35(i) that

$$Q^{\mathbf{x}}(f) = \lim_{n \to +\infty} Q(f_n) = \lim_{n \to +\infty} \Lambda(f_n),$$

and because

$$\lim_{n \to +\infty} |\Lambda(f) - \Lambda(f_n)| = \lim_{n \to +\infty} |\Lambda(f - f_n)| \leq \lim_{n \to +\infty} \Lambda\left(|f - f_n|\right) \leq \lim_{n \to +\infty} \overline{P}^{\mathbf{x}}\left(|f - f_n|\right) = 0,$$

we see that $\lim_{n \to +\infty} \Lambda(f_n) = \Lambda(f)$, so indeed $Q^{\mathbf{x}}(f) = \Lambda(f)$. □

Theorem 15.38 *For all $\underline{P}$-previsible gambles f,*

$$\underline{P}^{\mathbf{x}}(f) = \min_{Q \in \mathrm{lins}(\underline{P})} Q^{\mathbf{x}}(f) = \min_{Q \in \mathrm{lins}(\underline{P})} D\int f \,\mathrm{d}Q.$$

Proof. Given Proposition 15.37, the proof here is almost identical to the one we have given for the original Lower Envelope Theorem $4.38(\mathrm{iii})_{71}$ for bounded gambles.

Fix any f in $\mathbb{G}^{\mathbf{x}}_{\underline{P}}$. By Proposition 15.35(iii), we see that

$$\underline{P}^{\mathbf{x}}(f) \leq \inf\left\{Q^{\mathbf{x}}(f) : Q \in \mathrm{lins}(\underline{P})\right\},$$

so we are left to show that $\underline{P}^{\mathbf{x}}(f) = Q^{\mathbf{x}}(f)$ for at least one $Q \in \mathrm{lins}(\underline{P})$: the desired result will then follow immediately from Proposition 15.36(ii).

Define the real functional Ψ on $\mathbb{G}^{\mathrm{x}}_{\underline{P}}$ by letting $\Psi(g) := \underline{P}^{\mathrm{x}}(g+f) - \underline{P}^{\mathrm{x}}(f)$ for all $g \in \mathbb{G}^{\mathrm{x}}_{\underline{P}}$. For all $\kappa \in [0,1]$ and all g and $h \in \mathbb{G}^{\mathrm{x}}_{\underline{P}}$,

$$\begin{aligned}\Psi(\kappa g + (1-\kappa)h) &= \underline{P}^{\mathrm{x}}(\kappa g + (1-\kappa)h + f) - \underline{P}^{\mathrm{x}}(f)\\ &\geq \kappa \underline{P}^{\mathrm{x}}(g+f) + (1-\kappa)\underline{P}^{\mathrm{x}}(h+f) - \underline{P}^{\mathrm{x}}(f)\\ &= \kappa\Psi(g) + (1-\kappa)\Psi(h),\end{aligned}$$

where the inequality follows from the coherence (concavity, Theorem 15.14(viii)$_{335}$) of $\underline{P}^{\mathrm{x}}$. So Ψ is concave and $\Psi(0) = 0$. It, therefore, follows from the Hahn–Banach Theorem A.12$_{370}$ (with $V = \mathbb{G}^{\mathrm{x}}_{\underline{P}}$ and $V' = \{0\}$) that there is a linear functional Λ on $\mathbb{G}^{\mathrm{x}}_{\underline{P}}$ that dominates Ψ on $\mathbb{G}^{\mathrm{x}}_{\underline{P}}$. Now for any $g \in \mathbb{G}^{\mathrm{x}}_{\underline{P}}$,

$$\Lambda(g) \geq \Psi(g) = \underline{P}^{\mathrm{x}}(g+f) - \underline{P}^{\mathrm{x}}(f) \geq \underline{P}^{\mathrm{x}}(g),$$

where the last inequality follows from the super-additivity (Theorem 15.14(vi)$_{335}$) of $\underline{P}^{\mathrm{x}}$. The converse inequality holds for $g = f$ as well, because

$$-\Lambda(f) = \Lambda(-f) \geq \Psi(-f) = \underline{P}^{\mathrm{x}}(-f+f) - \underline{P}^{\mathrm{x}}(f) = -\underline{P}^{\mathrm{x}}(f),$$

which guarantees that $\underline{P}^{\mathrm{x}}(f) = \Lambda(f)$. Now apply Proposition 15.37$_{\curvearrowleft}$ to see that there is indeed a $Q \in \mathrm{lins}(\underline{P})$ such that $\underline{P}^{\mathrm{x}}(f) = Q^{\mathrm{x}}(f)$. □

15.10 Examples

15.10.1 Linear previsions and probability charges

Consider a linear prevision P on $\mathbb{B}(\mathcal{X})$. Let f be P-previsible gamble, and let f_n be a determining sequence for f.

By the correspondence between linear previsions on $\mathbb{B}(\mathcal{X})$ and probability charges on $\mathscr{P}(\mathcal{X})$ (Corollary 8.23$_{167}$) and the correspondence between Dunford integration and natural extension (Theorem 8.32$_{174}$), we have that

$$P(f_n) = D\int f_n \,\mathrm{d}P.$$

By Theorem 4.4.20 in Bhaskara Rao and Bhaskara Rao (1983), P-previsibility is equivalent with Dunford integrability (see Section 8.7$_{172}$) with respect to P. It also follows from Theorem 4.4.20 in Bhaskara Rao and Bhaskara Rao (1983) that $D\int |f - f_n| \,\mathrm{d}P$ converges to zero. By Theorem 4.4.13(iii) in Bhaskara Rao and Bhaskara Rao (1983), we conclude that $D\int f_n \,\mathrm{d}P$ converges to $D\int f \,\mathrm{d}P$. Summarising,

$$P^{\mathrm{x}}(f) = \lim_{n\to+\infty} P(f_n) = \lim_{n\to+\infty} D\int f_n \,\mathrm{d}P = D\int f \,\mathrm{d}P.$$

So, for linear previsions, previsibility effectively is the same thing as Dunford integrability and previsible extension coincides with Dunford integration.

Moreover, because linear previsions are 2-monotone, we have for any P-previsible gamble $f : \mathcal{X} \to \mathbb{R}$ that

$$P^{\mathrm{x}}(f) = C\int f\,\mathrm{d}P = R\int_0^{+\infty} P\left(\{f^+ \geq t\}\right)\,\mathrm{d}t - R\int_0^{+\infty} P\left(\{f^- \geq t\}\right)\,\mathrm{d}t, \quad (15.28)$$

by Theorem 15.33_{353} and the definition of the Choquet integral (Equation $(\mathrm{C}.5)_{379}$). Also by Theorem 15.33_{353}, f is P-previsible if and only if f is Choquet integrable with respect to P, that is, if and only if both integrals on the right-hand side of Equation (15.28) are finite. This provides the proof of Theorem 8.33_{177} near the end of Section 8.7_{172}, postponed until now.

15.10.2 Probability density functions: The normal density

Let us consider a normal density function,

$$\mathrm{Nm}(x|\mu,\sigma^2) = \frac{1}{\sqrt{2\pi\sigma^2}}\exp\left(-\frac{(x-\mu)^2}{2\sigma^2}\right) \text{ for all } x \in \mathbb{R},$$

with mean μ and variance $\sigma^2 > 0$. It can be used to define a linear prevision $P(\bullet|\mu,\sigma^2)$ on all continuous bounded gambles f by letting

$$P(f|\mu,\sigma^2) := R\int f(x)\,\mathrm{Nm}(x|\mu,\sigma^2)\,\mathrm{d}x.$$

We will denote the natural extension of this linear prevision to all bounded gambles by $\underline{P}(\bullet|\mu,\sigma^2)$.

As a simple demonstration of the theory, we show that the gamble X – or, in other words, the identity map on $\mathbb{R}$ – is $\underline{P}(\bullet|\mu,\sigma^2)$-previsible, and we calculate its extended lower and upper prevision. As the linear prevision $P(\bullet|\mu,\sigma^2)$ is completely monotone by Theorem 6.6_{105}, so is its natural extension $\underline{P}(\bullet|\mu,\sigma^2)$, by Theorem 6.19_{114}. Consequently, it suffices to verify the previsibility condition of Theorem 15.32_{352} for the gamble X, for which we will use the well-known tail bound for the standard normal distribution function Φ:

$$1 - \Phi(u) \leq \frac{e^{-u^2/2}}{u\sqrt{2\pi}} \text{ for } u \geq 0. \quad (15.29)$$

Indeed,

$$\begin{aligned}\lim_{s\to+\infty} R\int_s^{+\infty} \overline{P}\left(\{X \geq t\}\,|\mu,\sigma^2\right)\,\mathrm{d}t &= \lim_{s\to+\infty} R\int_s^{+\infty} \overline{P}\left(\left\{X \geq \tfrac{t-\mu}{\sigma}\right\}|0,1\right)\,\mathrm{d}t\\ &= \sigma \lim_{s\to+\infty} R\int_{\frac{s-\mu}{\sigma}}^{+\infty} \overline{P}(\{X \geq u\}\,|0,1)\,\mathrm{d}u = \sigma \lim_{v\to+\infty} R\int_v^{+\infty} [1-\Phi(u)]\,\mathrm{d}u\end{aligned}$$

$$\leq \sigma \lim_{v \to +\infty} R \int_{v}^{+\infty} \frac{e^{-u^2/2}}{u\sqrt{2\pi}} \, du$$

$$\leq \sigma \lim_{v \to +\infty} R \int_{v}^{+\infty} \frac{e^{-u^2/2}}{\sqrt{2\pi}} \, du = \sigma \lim_{v \to +\infty} [1 - \Phi(v)] = 0,$$

where the first inequality follows from Equation $(15.29)_{\curvearrowleft}$. A very similar argument shows that

$$\lim_{s \to +\infty} R \int_{s}^{+\infty} \overline{P}(X \leq -t|\mu, \sigma^2) \, \mathrm{d}t = 0,$$

which allows us to infer that X is indeed $\underline{P}(\bullet|\mu, \sigma^2)$-previsible. Alternatively, we could have verified the Choquet integrability of X and applied Theorem 15.33_{353}. We leave this as an exercise to the reader.

Finally, because X is $\underline{P}(\bullet|\mu, \sigma^2)$-previsible and all cuts X^t_{-t} are continuous, we infer from Criterion 15.26_{348} that

$$\underline{P}^{\mathrm{x}}(X|\mu, \sigma^2) = \overline{P}^{\mathrm{x}}(X|\mu, \sigma^2) = \lim_{t \to +\infty} P(X^t_{-t}|\mu, \sigma^2)$$

$$= \lim_{t \to +\infty} (-t) R \int_{-\infty}^{-t} \mathrm{Nm}(x|\mu, \sigma^2) \, \mathrm{d}x + \lim_{t \to +\infty} R \int_{-t}^{t} x \, \mathrm{Nm}(x|\mu, \sigma^2) \, \mathrm{d}x$$

$$+ \lim_{t \to +\infty} t R \int_{t}^{+\infty} \mathrm{Nm}(x|\mu, \sigma^2) \, \mathrm{d}x$$

$$= \lim_{t \to +\infty} (-t) \Phi\left(\frac{-t - \mu}{\sigma}\right) + R \int_{-\infty}^{+\infty} x \, \mathrm{Nm}(x|\mu, \sigma^2) \, \mathrm{d}x + \lim_{t \to +\infty} t \Phi\left(\frac{t - \mu}{\sigma}\right)$$

$$= 0 + \mu + 0 = \mu,$$

where, in the last step, the limits are zero by the tail bound of Equation $(15.29)_{\curvearrowleft}$.

Interestingly, simple natural extension of $\underline{P}(\bullet|\mu, \sigma^2)$ to arbitrary gambles is really useless here:

$$\underline{E}_{\underline{P}(\bullet|\mu, \sigma^2)}(X) = -\infty \text{ and } \overline{E}_{\underline{P}(\bullet|\mu, \sigma^2)}(X) = +\infty.$$

These equalities follow immediately from Example 13.67_{302}.

15.10.3 Vacuous lower previsions

We can also go back to the other extreme end from linear previsions and look at the very special case of vacuous lower previsions relative to some subset: let A be any non-empty subset of $\mathcal{X}$ and let $\underline{P}_A$ be the vacuous lower prevision on $\mathbb{B}$ relative to the event A. We have already discussed its fairly trivial extension to essentially bounded gambles in Section $14.5.2_{323}$. Its extension to previsible gambles is no less straightforward.

A net of gambles f_α converges in probability $[\underline{P}_A]$ to a gamble f if its restriction to A converges uniformly to the restriction of f to A: convergence in probability $[\underline{P}_A]$ is uniform convergence on A. As a sequence of bounded gambles can only converge uniformly to a bounded gamble, this implies that the $\mathbb{G}^{\mathbf{x}}_{\underline{P}_A}$ set of $\underline{P}_A$-previsible gambles is just the set $\mathbb{G}^{\sharp}_{\underline{P}_A} = \{f \in \mathbb{G} : I_A f \in \mathbb{B}\}$ of all gambles bounded on A: the domain of a vacuous lower prevision cannot be extended further using arguments of previsibility.

15.10.4 Lower previsions associated with proper filters

We can take this one step further by considering a proper filter $\mathscr{F}$ of subsets of $\mathscr{X}$ and the associated lower prevision $\underline{P}_{\mathscr{F}}$ defined by

$$\underline{P}_{\mathscr{F}}(f) = \sup_{F \in \mathscr{F}} \inf_{x \in F} f(x) \text{ for all bounded gambles } f \text{ on } \mathscr{X},$$

see the discussion in Sections 7.4_{129} and $14.5.3_{323}$.

Proposition 15.39 *A gamble f is $\underline{P}_{\mathscr{F}}$-previsible if and only if it is $\underline{P}_{\mathscr{F}}$-essentially bounded. Consequently,*

$$\mathbb{G}^{\mathbf{x}}_{\underline{P}_{\mathscr{F}}} = \mathbb{G}^{\sharp}_{\underline{P}_{\mathscr{F}}} = \left\{ f \in \mathbb{G} : \inf_{F \in \mathscr{F}} \sup\big(|f|\big|F\big) < +\infty \right\}$$

and

$$\underline{P}^{\mathbf{x}}_{\mathscr{F}}(f) = \underline{P}^{\sharp}_{\mathscr{F}}(f) = \sup_{F \in \mathscr{F}} \inf(f|F) \text{ for all } f \in \mathbb{G}^{\mathbf{x}}_{\underline{P}_{\mathscr{F}}}.$$

Proof. Owing to Theorem 15.7_{334}, it suffices to prove that $\mathbb{G}^{\mathbf{x}}_{\underline{P}_{\mathscr{F}}} \subseteq \mathbb{G}^{\sharp}_{\underline{P}_{\mathscr{F}}}$. Assume, therefore, that $f \in \mathbb{G}^{\mathbf{x}}_{\underline{P}_{\mathscr{F}}}$, which implies that f is $\underline{P}_{\mathscr{F}}$-measurable: there is a sequence of bounded gambles ($\underline{P}_{\mathscr{F}}$-measurable by Lemma 15.20_{341}) that converges in probability $[\underline{P}_{\mathscr{F}}]$ to f, and therefore, the limit f is $\underline{P}_{\mathscr{F}}$-measurable too (by Lemma 15.21_{342}). So we infer from Definition 15.19(E)$_{340}$ that $\lim_{t\to+\infty} \overline{P}_{\mathscr{F}}(\{|f| \geq t\}) = 0$. As $\overline{P}_{\mathscr{F}}$ is $\{0,1\}$-valued on events, it follows that $\overline{P}_{\mathscr{F}}(\{|f| \geq t\}) = 0$ for any sufficiently large t. Fix any such t, then the equality just obtained means that there is some $F \in \mathscr{F}$ such that $|f(x)| < t$ for all $x \in F$, which in turn implies that $\inf_{F \in \mathscr{F}} \sup\big(|f|\big|F\big) \leq t < +\infty$, implying that indeed $f \in \mathbb{G}^{\sharp}_{\underline{P}_{\mathscr{F}}}$. □

15.10.5 Limits inferior

We can now continue the discussion in Section $14.5.4_{324}$ of the limit inferior operator associated with bounded gambles on a set $\mathscr{A}$ directed by a relation $\leq$. Recall again from Section $5.5.3_{90}$ that lim inf can be written as a lower prevision associated with a proper filter. As a consequence, we can derive the following special case from Proposition 15.39.

Proposition 15.40 *A gamble f is* lim inf*-previsible if and only if it is* lim inf*-essentially bounded. Consequently,*

$$\mathbb{G}^{\mathbf{x}}_{\liminf}(\mathscr{A}) = \mathbb{G}^{\sharp}_{\liminf}(\mathscr{A}) = \{f \in \mathbb{G}(\mathscr{A}) : \limsup |f| < +\infty\}$$

and

$$\liminf{}^{\mathbf{x}} f = \liminf{}^{\sharp} f = \liminf f \text{ for all } f \in \mathbb{G}^{\mathbf{x}}_{\liminf}(\mathscr{A}).$$

15.10.6 Belief functions

Next, we continue the discussion of Section 14.5.5$_{325}$ and consider the completely monotone coherent lower prevision $\underline{P}_m$ associated with a basic probability assignment m on a finite set of focal elements $\mathscr{H} \subseteq \mathscr{P} \setminus \{\emptyset\}$, where the space $\mathscr{X}$ is not necessarily finite:

$$\underline{P}_m(f) = \sum_{F \in \mathscr{H}} m(F)\underline{P}_F(f) = \sum_{F \in \mathscr{H}} m(F) \inf_{x \in F} f(x) \text{ for all bounded gambles } f \text{ on } \mathscr{X}$$

Proposition 15.41 *A gamble f is $\underline{P}_m$-previsible if and only if it is $\underline{P}_m$-essentially bounded. Consequently,*

$$\mathbb{G}^{\mathbf{x}}_{\underline{P}_m} = \mathbb{G}^{\sharp}_{\underline{P}_m} = \left\{ f \in \mathbb{G} : \sup\left(|f| \middle| \bigcup \mathscr{H}\right) < +\infty \right\}$$

and

$$\underline{P}^{\mathbf{x}}_m(f) = \underline{P}^{\sharp}_m(f) = \sum_{F \in \mathscr{H}} m(F) \inf(f|F) \textit{ for all } f \in \mathbb{G}^{\mathbf{x}}_{\underline{P}_m}.$$

Proof. Immediate consequence of Proposition 15.34$_{355}$ and the results in Section 15.10.3$_{360}$. □

15.10.7 Possibility measures

We now turn to a possibility measure Π on a non-empty set $\mathscr{X}$, with possibility distribution π, continuing the discussion of Section 14.5.6$_{325}$.

By Theorem 15.32$_{352}$, a gamble f is $\underline{P}_\pi$-previsible if and only if

$$\lim_{s \to +\infty} R \int_s^{+\infty} \sup\{\pi(x) : |f(x)| \geq t\} \, dt = 0. \tag{15.30}$$

It turns out that we can calculate $\overline{P}^{\mathbf{x}}_\pi(f)$ for previsible f using an expression that is formally similar to the one for bounded gambles given in Equation (7.31)$_{144}$:

Proposition 15.42 *For any $\underline{P}_\pi$-previsible gamble f,*

$$\begin{aligned}\overline{P}^{\mathbf{x}}_\pi(f) &= R \int_{0\leftarrow}^{\rightarrow 1} \sup\{f(x) : \pi(x) > \alpha\} \, d\alpha \\ &:= R \int_{0\leftarrow}^{1} \sup\left\{f^+(x) : \pi(x) > \alpha\right\} d\alpha - R \int_{0\leftarrow}^{1} \inf\left\{f^-(x) : \pi(x) > 1 - \alpha\right\} d\alpha.\end{aligned}$$

The integrals above are improper Riemann integrals, compare with the definition in Section C.1.1$_{376}$. For the first one, we need the given explicit decomposition in terms of f^+ and f^- because in Section C.1.1$_{376}$, we only define improper Riemann integrals for *non-negative* and non-increasing integrands.

Proof. As $(f^n_{-n})^+ = \min\{n, f^+\}$ and $(f^n_{-n})^- = \min\{n, f^-\}$, we infer from Theorems 7.14$_{144}$ and C.3(i)$_{379}$ that

$$\begin{aligned}\overline{P}_\pi(f^n_{-n}) &= \overline{P}_\pi((f^n_{-n})^+) - \underline{P}_\pi((f^n_{-n})^-)\\ &= R\int_0^1 \sup\left\{\min\{n, f^+(x)\} : \pi(x) > \alpha\right\} \mathrm{d}\alpha\\ &\qquad - R\int_0^1 \inf\left\{\min\{n, f^-(x)\} : \pi(x) > \alpha\right\} \mathrm{d}\alpha.\end{aligned}$$

As we infer from Corollary 15.10$_{334}$ that the non-negative gambles f^+ and f^- are $\underline{P}_\pi$-previsible too, we may invoke Lemma 15.43 to complete the proof. □

Lemma 15.43 *For any $\underline{P}_\pi$-previsible non-negative gamble f and all $\alpha > 0$,*[7] *$\sup\{f(x) : \pi(x) > \alpha\}$ and $\inf\{f(x) : \pi(x) > 1 - \alpha\}$ are (finite) real numbers, and*

$$\lim_{n\to+\infty} R\int_0^1 \sup\{\min\{n, f(x)\} : \pi(x) > \alpha\}\,\mathrm{d}\alpha = R\int_{0\leftarrow}^1 \sup\{f(x) : \pi(x) > \alpha\}\,\mathrm{d}\alpha$$

$$\lim_{n\to+\infty} R\int_0^1 \inf\{\min\{n, f(x)\} : \pi(x) > \alpha\}\,\mathrm{d}\alpha = R\int_{0\leftarrow}^1 \inf\{f(x) : \pi(x) > 1 - \alpha\}\,\mathrm{d}\alpha.$$

The integrals on the right-hand sides are again improper Riemann integrals (see Section C.1.1$_{376}$).

Proof. We begin with the first equality. If there is some n_0 such that $\{\pi > \alpha\} \cap \{f > n_0\} = \emptyset$ for all $\alpha \in (0, 1)$, then $\sup\{f(x) : \pi(x) > \alpha\} \leq n_0$ for all $\alpha \in (0, 1)$ and the desired equality holds trivially. So we may assume that for all $n \in \mathbb{N}$ the down-set[8] $A_f(n) := \left\{\alpha \in (0, 1) : \{\pi > \alpha\} \cap \{f > n\} \neq \emptyset\right\}$ is non-empty and, therefore, has a non-zero supremum:

$$\alpha_f(n) := \sup A_f(n) = \sup\{\pi(x) : f(x) > n\} = \overline{P}_\pi(\{f > n\}) \in (0, 1].$$

This defines a non-increasing map $\alpha_f : \mathbb{N} \to (0, 1]$. Clearly,

$$\begin{aligned}R\int_0^1 &\sup\{\min\{n, f(x)\} : \pi(x) > \alpha\}\,\mathrm{d}\alpha\\ &= R\int_0^{\alpha_f(n)} n\,\mathrm{d}\alpha + R\int_{\alpha_f(n)}^1 \sup\{f(x) : \pi(x) > \alpha\}\,\mathrm{d}\alpha\\ &= n\alpha_f(n) + R\int_{\alpha_f(n)}^1 \sup\{f(x) : \pi(x) > \alpha\}\,\mathrm{d}\alpha,\end{aligned}$$

[7] As is usual, in these expressions, we let the supremum over the empty set $\emptyset$ be equal to the smallest element 0 of the set $\mathbb{R}_{\geq 0}$ the supremum operator is taken to be working on.

[8] See the definition of a down-set in Appendix A$_{368}$.

so the equality is proved if we can show that $\lim_{n\to+\infty} n\alpha_f(n) = 0$ (because then necessarily also $\lim_{n\to+\infty} \alpha_f(n) = 0$). On the one hand, $\overline{P}_\pi(|f_{-n}^n - f_{-n/2}^{n/2}|) \to 0$, because f is $\underline{P}_\pi$-previsible. On the other hand, Equation $(15.27)_{353}$ for $m = n/2$ tells us that $|f_{-n}^n - f_{-n/2}^{n/2}| \geq \frac{n}{2} I_{\{f>n\}}$ and, therefore, $\overline{P}_\pi(|f_{-n}^n - f_{-n/2}^{n/2}|) \geq \frac{n}{2}\overline{P}_\pi(\{f > n\}) = \frac{n}{2}\alpha_f(n)$. (An alternative argument would start from the $\underline{P}_\pi$-previsibility of f and rely on Condition $15.31(\text{i})_{350}$.)

The proof of the second equality is similar. If there is some n_0 such that $\{\pi > 1 - \alpha\} \cap \{f < n_0\} \neq \emptyset$ for all $\alpha \in (0, 1)$, then $\inf\{f(x) : \pi(x) > 1 - \alpha\} < n_0$ for all $\alpha \in (0, 1)$ and the desired equality holds trivially. So we may assume that for all $n \in \mathbb{N}$, the down-set $B_f(n) := \{\alpha \in (0, 1) : \{\pi > 1 - \alpha\} \subseteq \{f \geq n\}\}$ is non-empty and, therefore, has a non-zero supremum:

$$\begin{aligned}\beta_f(n) &:= \sup B_f(n) = \inf\left\{\alpha \in (0,1) : \alpha \notin B_f(n)\right\} = 1 - \sup\{\pi(x) : f(x) < n\}\\ &= \underline{P}_\pi(\{f \geq n\}).\end{aligned}$$

This defines a non-increasing map $\beta_f : \mathbb{N} \to (0, 1]$. Clearly,

$$\begin{aligned}&R\int_0^1 \inf\{\min\{n, f(x)\} : \pi(x) > 1 - \alpha\}\,\mathrm{d}\alpha\\ &\qquad = R\int_0^{\beta_f(n)} n\,\mathrm{d}\alpha + R\int_{\beta_f(n)}^1 \inf\{f(x) : \pi(x) > 1 - \alpha\}\,\mathrm{d}\alpha,\end{aligned}$$

so the equality is proved if we can show that $\lim_{n\to+\infty} n\beta_f(n) = 0$ (because then necessarily also $\lim_{n\to+\infty} \beta_f(n) = 0$). On the one hand, $\overline{P}_\pi(|f_{-n}^n - f_{-n/2}^{n/2}|) \to 0$, since f is previsible. On the other hand, Equation $(15.27)_{353}$ for $m = n/2$ tells us that $|f_{-n}^n - f_{-n/2}^{n/2}| \geq \frac{n}{2} I_{\{f\geq n\}}$ and, therefore, $\overline{P}_\pi(|f_{-n}^n - f_{-n/2}^{n/2}|) \geq \underline{P}_\pi(|f_{-n}^n - f_{-n/2}^{n/2}|) \geq \frac{n}{2}\underline{P}_\pi(\{f \geq n\}) = \frac{n}{2}\beta_f(n)$.

The rest of the proof is now obvious. □

The reader might be led to think that the proposition above can also be deduced by a straightforward application of Proposition C.8(ii)_{388}, following an argument similar to that in the proof of Proposition 7.14_{144}. To see that such is not the case, recall that in that proof, use was made of the set function – or vacuous upper probability – ϕ defined by $\phi(A) := 1$ for all non-empty subsets A of $\mathcal{X}$. Interestingly, only *bounded* gambles are Choquet integrable with respect to this ϕ (recall the discussion in Section $15.10.3_{360}$ and the relationship between Choquet integrability and previsibility we uncovered in Section 15.7_{352}). Nevertheless, we have reasons to believe that it is possible to extend Proposition C.8(ii)_{388} in such a way that it could be applied here as well, under a notion of Choquet integration of non-negative gambles with respect to set functions that can also assume infinite values, allowing the condition of Choquet integrability to be altogether dropped from Proposition C.8(ii)_{388}.

We end this section with an example to show that here previsibility may indeed allow us to extend $\overline{P}_\pi$ beyond the essentially bounded gambles. Consider the case where $\mathscr{X} := \mathbb{R}$, $\pi(x) := \exp(-x^2)$ and $f(x) := x^2$. As $\{\pi > 0\} = \mathbb{R}$, we see that $\sup(|f|\,|\{\pi > 0\}) = \sup f = +\infty$, so f is not essentially bounded. But for any $t > 0$, we see that $|f|(x) > t \Leftrightarrow |x| > \sqrt{t}$ and, therefore,

$$\overline{P}_\pi(\{|f| > t\}) = \sup\left\{\exp(-x^2) : x^2 > t\right\} = e^{-t}.$$

This guarantees that f is $\underline{P}_\pi$-previsible: Condition $(15.30)_{362}$ holds because

$$R\int_s^{+\infty} \sup\{\pi(x) : |f(x)| \geq t\}\,\mathrm{d}t = R\int_s^{+\infty} e^{-t}\,\mathrm{d}t = e^{-s} \to 0.$$

If we now use Proposition 15.42_{362}, we find that

$$\begin{aligned}\overline{P}^{\mathbf{x}}_\pi(f) &= R\int_{0\leftarrow}^{1} \sup\left\{x^2 :\ \exp(-x^2) > \alpha\right\}\,\mathrm{d}\alpha = R\int_{0\leftarrow}^{1} \sup\left\{x^2 : x^2 < -\ln\alpha\right\}\,\mathrm{d}\alpha \\ &= R\int_{0\leftarrow}^{1} (-\ln\alpha)\,\mathrm{d}\alpha = \lim_{\varepsilon\to 0+}\left((-\alpha\ln\alpha)\Big|_\varepsilon^1 + R\int_\varepsilon^1 \mathrm{d}\alpha\right) = 1,\end{aligned}$$

where the next to last equality follows from partial integration.

15.10.8 Estimation

Suppose we have a real random variable X and a coherent lower prevision $\underline{P}$ modelling beliefs about the value of X, defined on the set $\mathbb{B}(\mathbb{R})$ of bounded gambles on $\mathbb{R}$. We want an *optimal estimate* $\hat{X}$ for X when, with each possible estimate θ for X, there is associated a **quadratic gain function**

$$g_\theta(X) := -(X-\theta)^2.$$

These g_θ are unbounded gambles on X, so we will need to consider the extended lower prevision $\underline{P}^{\mathbf{x}}$, defined on $\underline{P}$-previsible gambles.

We are now faced with a *decision problem*: we need to find an optimal value of θ. In order to deal with it, we will extend two approaches from the literature on decision making with coherent lower previsions for bounded gambles (for an overview, see, for instance, Troffaes, 2007).

The first and, as it will turn out, simplest one is based on the notion of (Walley–Sen) **maximality** (Walley, 1991, Section 3.9). It consists in making a pairwise comparison of any two possible estimates θ and ϑ based on their associated gain functions. We start by defining a strict partial order[9] $\sqsupset$ on the set $\mathbb{R}$ of all estimates. Specifically, we say that θ **is strictly preferred to** ϑ and write $\theta \sqsupset \vartheta$, whenever

$$\underline{P}^{\mathbf{x}}(g_\theta - g_\vartheta) > 0. \tag{15.31}$$

[9] See Definition A.2_{368} for strict partial orders.

Of course, for this to work, the gamble

$$g_\theta(X) - g_\vartheta(X) = -(X-\theta)^2 + (X-\vartheta)^2 = (\theta-\vartheta)(2X-\theta-\vartheta) \tag{15.32}$$

must be $\underline{P}$-previsible for all values of θ and ϑ, which clearly amounts to requiring that X should be $\underline{P}$-previsible. We will assume that this is indeed the case.

A *behavioural* interpretation of $\theta \sqsupset \vartheta$ is that our subject is prepared to pay some positive price in order to exchange the uncertain reward $g_\vartheta(X)$ associated with using ϑ as an estimate, for the uncertain reward $g_\theta(X)$ associated with using θ as an estimate. A more sensitivity analysis (or *robustness*) oriented interpretation is based on Theorem 15.38_{357}, which allows us to write

$$\theta \sqsupset \vartheta \Leftrightarrow (\forall Q \in \operatorname{lins}(\underline{P}))Q^{\mathbf{x}}(g_\theta) > Q^{\mathbf{x}}(g_\vartheta),$$

meaning that for each of the compatible linear models $Q \in \operatorname{lins}(\underline{P})$, the expected gain $Q^{\mathbf{x}}(g_\theta)$ associated with the estimate θ is strictly higher than the expected gain associated with the estimate $Q^{\mathbf{x}}(g_\vartheta)$.

The 'best' estimates θ are now the ones that are undominated, or maximal, in this strict partial order: θ is **maximal** when

$$\underline{P}^{\mathbf{x}}(g_\vartheta - g_\theta) \leq 0 \text{ for all } \vartheta \in \mathbb{R}. \tag{15.33}$$

Now it follows from Equation (15.32) and the coherence properties of $\underline{P}^{\mathbf{x}}$ that

$$\underline{P}^{\mathbf{x}}(g_\vartheta(X) - g_\theta(X)) = \begin{cases} (\vartheta-\theta)\,[2\underline{P}^{\mathbf{x}}(X) - \vartheta - \theta] & \text{if } \vartheta \geq \theta \\ (\vartheta-\theta)\,[2\overline{P}^{\mathbf{x}}(X) - \vartheta - \theta] & \text{if } \vartheta \leq \theta, \end{cases}$$

so the condition (15.33) for maximality of the estimate θ can be rewritten as

$$\underline{P}^{\mathbf{x}}(X) \leq \theta \leq \overline{P}^{\mathbf{x}}(X),$$

meaning that any real value between $\underline{P}^{\mathbf{x}}(X)$ and $\overline{P}^{\mathbf{x}}(X)$ is a maximal (undominated) estimate $\hat{X}$ for the value of X under a quadratic gain function.

The second approach, called **Γ-maximin**, consists in finding the value of θ for which the lower prevision $\underline{P}^{\mathbf{x}}(g_\theta)$ of the gain is highest:

$$\hat{X} \in \operatorname{argmax}_{\theta\in\mathbb{R}} \underline{P}^{\mathbf{x}}(g_\theta) = \operatorname{argmax}_{\theta\in\mathbb{R}} \underline{P}^{\mathbf{x}}(-[X-\theta]^2). \tag{15.34}$$

Of course, for this to work, the gamble $g_\theta(X)$ must be $\underline{P}$-previsible for all values of θ, which clearly amounts to requiring that both X and X^2 should be $\underline{P}$-previsible.

When $\underline{P}^{\mathbf{x}}$ is, for instance, a linear prevision $Q^{\mathbf{x}}$, we get that $f(\theta) = Q^{\mathbf{x}}(-[X-\theta]^2) = -\theta^2 + 2\theta Q^{\mathbf{x}}(X) - Q^{\mathbf{x}}(X^2)$. This is a differentiable function of θ, with $Df(\theta) = -2\theta + 2Q^{\mathbf{x}}(X)$, so there is a unique maximum for $\theta = Q^{\mathbf{x}}(X)$. In this case, the Γ-maximin solution $\hat{X}$ and unique maximal solution coincide.

There appears to be no simple way to solve Equation (15.34) generically, as we have done for maximality. Finding Γ-maximin values for θ will therefore have to rely on explicit optimisation, either numerical or analytical. Usually, the Γ-maximin solution will be unique. We can guarantee that this solution must lie between $\underline{P}^{\mathbf{x}}(X)$ and $\overline{P}^{\mathbf{x}}(X)$, because any Γ-maximin solution $\hat{X}$ must also be maximal. Indeed,[10]

$$\underline{P}^{\mathbf{x}}(g_{\vartheta} - g_{\hat{X}}) \leq \underline{P}^{\mathbf{x}}(g_{\vartheta}) - \underline{P}^{\mathbf{x}}(g_{\hat{X}}) \leq 0 \text{ for all } \vartheta \in \mathbb{R},$$

where the first inequality follows from coherence (Theorem 15.14(vi)$_{335}$) and the second from Equation (15.34). Now use Equation (15.33) to find that $\hat{X}$ is indeed maximal.

[10] This argument is essentially due to Walley (1991, Section 3.9.7).

Appendix A

Linear spaces, linear lattices and convexity

In this appendix, we have gathered, for easy reference, a number of basic definitions and results concerning partially ordered sets and lattices, linear spaces and linear lattices. More information about partially ordered sets and lattices can be found in the books by Davey and Priestley (1990) and Birkhoff (1995). For a discussion of linear spaces and lattices, we can refer, amongst many others, to the book by Schechter (1997).

We begin with the definition of preorder and partial order.

Definition A.1 (Preorder and partial order) *A binary relation $\leq$ on a non-empty set V is called a* ***preorder*** *(relation) if it satisfies the following properties:*

PO1. $(\forall x \in V) x \leq x$ **(reflexivity)**

PO2. $(\forall x, y, z \in V)(x \leq y \text{ and } y \leq z \Rightarrow x \leq z)$. **(transitivity)**

The set V is then called a ***preordered set****. If additionally also*

PO3. $(\forall x, y \in V)(x \leq y \text{ and } y \leq x \Rightarrow x = y)$, **(antisymmetry)**

then the relation $\leq$ is called a ***partial order*** *and the set V is called a* ***partially ordered set****, or* ***poset****.*

The following definition is also useful.

Definition A.2 (Strict partial order) *A binary relation $<$ on a non-empty set V is called a* ***strict partial order*** *(relation) if it satisfies the following properties:*

SPO1. $(\forall x \in V)\neg(x < x)$ **(irreflexivity)**

SPO2. $(\forall x, y, z \in V)(x < y \text{ and } y < z \Rightarrow x < z)$. **(transitivity)**

It is then automatically also antisymmetric.

Lower Previsions, First Edition. Matthias C.M. Troffaes and Gert de Cooman.

A subset W of a *poset* V is a **down-set** if it is decreasing in the sense that $(\forall x \in W)(\forall y \in V)(y \leq x \Rightarrow y \in W)$. A similar definition can be given for an up-set.

An element x is called an **upper bound** of a subset W of a poset V if $y \leq x$ for all $y \in W$ and a **lower bound** if $x \leq y$ for all $y \in W$. An upper bound for V, if it exists, is unique and called its **top**. Similarly, if V has a lower bound, it is unique and called the **bottom** of V. The **supremum** of a set W is a smallest upper bound: an upper bound that is smaller than or equal to – in the sense of $\leq$ – all upper bounds. If it exists, it is unique and denoted by $\sup W$. Similarly, for the **infimum** $\inf W$, which is the greatest lower bound of W, if it exists.

Definition A.3 (Lattice) *If the infimum $x \wedge y := \inf\{x, y\}$ exists for any two elements x and y of a partially ordered set V, then V is called a* **$\wedge$-semilattice**; *if the supremum $x \vee y := \sup\{x, y\}$ exists for any two elements x and y of V, then V is called a* **$\vee$-semilattice**. *V is a* **lattice** *if the infimum and the supremum of any two (or equivalently, any finite number of) elements of V exist, and a* **complete lattice** *if the infimum and supremum of any of its subsets exist.*

Next, we turn to linear spaces.

Definition A.4 (Linear space) *A* ***(real) linear space*** *is a set V equipped with two operations, addition $+ : V \times V \to V$ and scalar multiplication $\cdot : \mathbb{R} \times V \to V$, satisfying the following conditions:*

LS1. $x + (y + z) = (x + y) + z$ *for all* $x, y, z \in V$

LS2. $x + y = y + x$ *for all* $x, y \in V$

LS3. *there is an element* $0 \in V$ *such that* $x + 0 = x$ *for all* $x \in V$

LS4. *for every* $x \in V$, *there is an element* $y \in V$, *such that* $x + y = 0$

LS5. *for all* $\lambda \in \mathbb{R}$ *and all* $x, y \in V$, *we have* $\lambda \cdot (x + y) = \lambda \cdot x + \lambda \cdot y$

LS6. *for all* $\lambda, \mu \in \mathbb{R}$ *and all* $x \in V$, *we have* $(\lambda + \mu) \cdot x = \lambda \cdot x + \mu \cdot x$

LS7. *for all* $\lambda, \mu \in \mathbb{R}$ *and all* $x \in V$ *we have* $\lambda \cdot (\mu \cdot x) = (\lambda\mu) \cdot x$

LS8. *for all* $x \in V$, *we have* $1 \cdot f = f$.

A subset V' of V is called a **linear subspace** *of V if V' is a linear space when equipped with the restriction of the operations of V or, in other words, if it is closed under these operations.*

We also use the simpler notation λx for $\lambda \cdot x$. Elements of $\mathbb{R}$ are called **scalars**, and elements of V are called **vectors**.

Linear spaces can also be defined in a completely similar way over other fields than $\mathbb{R}$, such as the field $\mathbb{Q}$ of all rationals or the field of all complex numbers and so on.

Definition A.5 (Ordered linear space) *Let $(V, +, \cdot)$ be a linear space, and let $\leq$ be a partial ordering on V. The tuple $(V, +, \cdot, \leq)$ is called an* **ordered linear space**, *if for every triple of vectors x, y and z and every non-negative real number λ,*

(i) $x \leq y \Rightarrow x + z \leq y + z$

(ii) $x \leq y \Rightarrow \lambda \cdot x \leq \lambda \cdot y$.

Definition A.6 (Linear lattice) *An ordered linear space $(L, +, \cdot, \leq)$ is called a* ***linear lattice*** *if every pair of vectors (x, y) has a supremum $x \vee y$ and an infimum $x \wedge y$ with respect to the partial ordering $\leq$. In that case, for every vector x, we define the vectors $x^+ := x \vee 0$, $x^- := (-x) \vee 0$ and $|x| := x^+ + x^-$. A subset W of L is called a* ***linear sublattice*** *of L if W is a linear lattice when equipped with the restriction of the operations of L or, in other words, if it is a linear subspace that is closed under the operations $\vee$ and $\wedge$.*

Example A.7 Let $\mathscr{X}$ be a set. The set of real-valued maps on $\mathscr{X}$ is a linear lattice when equipped with the point-wise addition, the point-wise scalar multiplication and the point-wise ordering. The set of bounded real-valued maps on $\mathscr{X}$ is a linear sublattice of this linear lattice. ♦

Notation A.8 *Let $(V, +, \cdot)$ be a linear space. Let $n_1, n_2 \in \mathbb{N}$, and let $x_i \in V$ and $\lambda_i \in \mathbb{R}$ for every $i \in \mathbb{N}$, $n_1 \leq i \leq n_2$. Then*

$$\sum_{i=n_1}^{n_2} \lambda_i x_i := \lambda_{n_1} x_{n_1} + \lambda_{n_1+1} x_{n_1+1} + \cdots + \lambda_{n_2} x_{n_2},$$

if $n_1 \leq n_2$ and $\sum_{n_1}^{n_2} \lambda_i x_i := 0$ otherwise.

Definition A.9 (Convex combination) *For any vectors x_1, ..., x_n and non-negative scalars λ_i, ..., λ_n such that $\sum_{i=1}^{n} \lambda_i = 1$, we call $\sum_{i=1}^{n} \lambda_i x_i$ a* ***convex combination*** *of x_1, ..., x_n.*

Definition A.10 (Convex set) *A subset A of a linear space is called* ***convex*** *if it contains all convex combinations of any finite number of its elements, that is, whenever for all finite subsets $\{x_1, \ldots, x_n\}$ of A, where $n \in \mathbb{N}_{>0}$, and all non-negative scalars λ_i, ..., λ_n such that $\sum_{i=1}^{n} \lambda_i = 1$, we have that $\sum_{i=1}^{n} \lambda_i x_i \in A$.*

Definition A.11 (Extreme point) *Let A be a convex subset of a linear space. A vector in A is said to be an* ***extreme point*** *of A if it cannot be written as a convex combination of any other elements in A.*

The following result (see, for instance, Schechter, 1997, Section 12.31 (HB3)) is of central importance in this book.

Theorem A.12 (Hahn–Banach theorem) *Let V be a linear space and V' a linear subspace of V. Let Ψ be a concave real functional on V, dominated by a real linear functional Λ' defined on V': $(\forall x \in V')\Psi(x) \leq \Lambda'(x)$. Then Λ' can be extended to a real linear functional Λ on V that dominates Ψ: $(\forall x \in V)(\Psi(x) \leq \Lambda(x))$.*

Appendix B

Notions and results from topology

In this appendix, we have gathered, for convenient reference, a few definitions and results from topology. We make no effort at being complete, and we refer to, say, the book by Willard (1970) for an extensive discussion.

B.1 Basic definitions

Definition B.1 (Topology) *A **topology** on a non-empty set V is a collection $\mathscr{T}$ of the so-called **open** subsets of V satisfying*

TO1. *$\mathscr{T}$ is closed under arbitrary unions*

TO2. *$\mathscr{T}$ is closed under finite intersections*

TO3. *$\mathscr{T}$ contains both $\emptyset$ and V.*

*The complement O^c of an open set $O \in \mathscr{T}$ is called **closed**.*

The **closure** of a set $A \subseteq V$ is the smallest closed set that includes A – we denote it by $\mathrm{cl}(A)$. The **interior** of a set $A \subseteq V$ is the largest open set that is included in A – we denote it by $\mathrm{int}(A)$. A set A is **dense** in a set B when the closure of A is B.

A **neighbourhood** N of an element x of V is a set that includes an open set $O \in \mathscr{Y}$ that contains x. We denote the set of all neighbourhoods of x by $\mathscr{N}_x$:

$$\mathscr{N}_x := \{N \subseteq V : (\exists O \in \mathscr{T})(x \in O \text{ and } O \subseteq N)\}.$$

$\mathscr{N}_x$ is a *filter*: it is closed under finite intersections and increasing. It is a proper filter because $V \in \mathscr{N}_x$ and $\emptyset \notin \mathscr{N}_x$. Filter bases and sub-bases for $\mathscr{N}_x$ are called

Lower Previsions, First Edition. Matthias C.M. Troffaes and Gert de Cooman.

neighbourhood bases and **sub-bases**, respectively. In particular, we infer from Definition 1.4_6 that a neighbourhood base for $\mathcal{N}_x$ is a set of events such that a set is a neighbourhood of x if and only if it includes some element of it.

Interestingly, a set is open if and only if it includes a neighbourhood of any of its points:

$$O \in \mathcal{T} \Leftrightarrow (\forall x \in O)(\exists N_x \in \mathcal{N}_x) N_x \subseteq O.$$

A **topological space** V is a set provided with a topology $\mathcal{T}$. It is **compact** if every open cover of V has a finite sub-cover: for any subset $\mathcal{O} \subseteq \mathcal{T}$ such that $\bigcup \mathcal{O} = V$, there is a finite subset $\mathcal{O}' \subseteq \mathcal{O}$ such that $\bigcup \mathcal{O}' = V$. A subset A of V is **compact** if it is a compact topological space when provided with the so-called **relative topology** $\mathcal{T} \cap A := \{O \cap A \colon O \in \mathcal{T}\}$. A topological space is **locally compact** if every $x \in V$ has a compact neighbourhood.

A topological space is **Hausdorff** if any two distinct elements have disjoint (open) neighbourhoods:

$$x \neq y \Rightarrow (\exists N_x \in \mathcal{N}_x)(\exists N_y \in \mathcal{N}_y) N_x \cap N_y = \emptyset.$$

Every closed subset of a compact set is automatically also compact, but in a Hausdorff space, every compact subset is also closed. So in a Hausdorff space, for subsets of a compact set, compactness and closedness are equivalent. Also, a topological space V is Hausdorff if and only if limits of nets in V are *unique*: for any net x_α in V and all $x, y \in V$, if $x_\alpha \to x$ and $x_\alpha \to y$, then $x = y$.[1]

B.2 Metric spaces

Definition B.2 (Pseudometric and metric) *Consider an arbitrary set V. A **pseudometric** is a map $d \colon V \times V \to \mathbb{R}$ that satisfies*

M1. $d(x, y) \geq 0$ *for all x and $y \in V$*

M2. $d(x, x) = 0$ *for all $x \in V$*

M3. $d(x, y) = d(y, x)$ *for all x and $y \in V$*

M4. $d(x, z) \leq d(x, y) + d(y, z)$ *for all x, y, and $z \in V$.*

*We call such a map d a **metric** if, moreover,*

N5. $d(x, y) = 0 \Rightarrow x = y$ *for all x and $y \in V$.*

Given a (pseudo)metric d, we can define a topology of open sets by taking all the ε-**balls** about x

$$B(x, \varepsilon) := \{y \in V \colon d(x, y) < \varepsilon\}, \quad \varepsilon > 0$$

as a neighbourhood base for the filter of neighbourhoods of x, meaning that a set O is open if, for each of its elements $x \in O$, there is some $\varepsilon > 0$ such that $B(x, \varepsilon) \subseteq O$.

[1] For convergence of nets, see Section 1.5_7.

A topological space is (pseudo)**metrisable** if there is some (pseudo)metric d that generates its open sets in the way just described. Clearly then, in a metrisable space,

$$\{x\} = \bigcap_{n \in \mathbb{N}_{>0}} B\left(x, \frac{1}{n}\right) \text{ for all } x \in V,$$

so any singleton is the intersection of a countable number of open neighbourhoods.

Example B.3 The set $\mathbb{R}$ of all real numbers is usually provided with the Euclidean topology. A neighbourhood base at x is, for instance, the set of open intervals containing x, so a set is open if and only if it includes an open interval around each of its points. This topology is metrisable, and the metric $d(x, y) = |x - y|$ does the job. ◆

Example B.4 The set $\mathbb{B}(\mathcal{X})$ of all bounded gambles on a set $\mathcal{X}$, with the topology of point-wise convergence, is not metrisable unless $\mathcal{X}$ is finite. ◆

Consider a (pseudo)metric d on V. It is easy to see that a net x_α in V **converges**[2] to x in V if and only if the real net $d(x, x_\alpha)$ converges to zero in the sense of Section 1.5_7. A net x_α in V is **Cauchy** if, for every $\varepsilon > 0$, there is some γ such that, for all $\alpha, \beta \geq \gamma$, we have that $d(x_\alpha, x_\beta) \leq \varepsilon$. We say that V is **complete** if every Cauchy net has at least one limit in V.

B.3 Continuity

Consider two topological spaces $(V, \mathcal{T}_V)$ and $(W, \mathcal{T}_W)$. Then a map $f : V \to W$ is **continuous** (with respect to these topologies) at $x \in V$ if, for every neighbourhood $N_{f(x)}$ of $f(x)$, there is some neighbourhood N_x of x such that $f(N_x) \subseteq N_{f(x)}$; and f is **continuous** if it is continuous at all $x \in V$, which is also equivalent to

$$(\forall O \in \mathcal{T}_W) f^{-1}(O) \in \mathcal{T}_V.$$

If W is the set of reals $\mathbb{R}$ provided with the usual Euclidean topology, then it is easy to see that the real-valued map f on V is continuous if and only if $f^{-1}((a, b)) \in \mathcal{T}_V$ for all $a < b$ in $\mathbb{R}$. This leads us to the definition of the following, weaker, notions:

(a) f is **lower semi-continuous** if $\{f > a\} = f^{-1}((a, +\infty)) \in \mathcal{T}_V$ for all $a \in \mathbb{R}$;

(b) f is **upper semi-continuous** if $\{f < b\} = f^{-1}((-\infty, b)) \in \mathcal{T}_V$ for all $b \in \mathbb{R}$.

Clearly, a map is continuous if and only if it is both lower and upper semi-continuous.

We will need the following theorem (see, for instance, Schechter, 1997, Section 17.7) a number of times.

Theorem B.5 (Dini's Monotone Convergence Theorem) *Consider a net g_α of upper semi-continuous real functions on a compact topological space V. If, for each $x \in V$, the real net $g_\alpha(x)$ is non-increasing and converges to 0, then the net g_α converges uniformly to 0: $\lim_\alpha \sup g_\alpha = 0$.*

[2] Again, for convergence of nets, see Section 1.5_7.

B.4 Topological linear spaces

A linear space V provided with a topology of open sets $\mathscr{Y}$ is called a **topological linear space** if the operations '+' and '·' are continuous. It is **locally convex** if, for every $x \in V$, $\mathscr{N}_x$ has a neighbourhood base consisting of convex sets, or equivalently, if $\mathscr{N}_0$ – the neighbourhood filter at the origin of V – has a neighbourhood base consisting of convex sets.

A special way of defining a topology on a linear space goes as follows.

Definition B.6 (Seminorm and norm) *Consider a real linear space V. We call* ***seminorm*** *any map $\rho : V \to \mathbb{R}_{\geq 0}$ that satisfies*

N1. $\rho(\lambda x) = |\lambda| \rho(x)$ *for all $x \in V$ and all real λ*

N2. $\rho(x + y) \leq \rho(x) + \rho(y)$ *for all $x, y \in V$.*

We call such a map ρ a ***norm*** *if, moreover,*

N3. $\rho(x) = 0 \Rightarrow x = 0$ *for all $x \in V$.*

A (semi)norm ρ induces a (pseudo)metric $d(x, y) := \rho(x - y)$ and, therefore, a topology. In fact, every seminormed linear space is a locally convex topological linear space.

Example B.7 The set $\mathbb{B}(\mathscr{X})$ of all bounded gambles on a set $\mathscr{X}$ is a linear space, and the supremum norm $\|\bullet\|_{\inf}$ given by

$$\|f\|_{\inf} := \sup |f| \text{ for all bounded gambles } f \text{ on } \mathscr{X}$$

has all the properties of a norm. Convergence in norm $\|\bullet\|_{\inf}$ is also called **uniform convergence**, and the topology that is generated by this norm is also called the **topology of uniform convergence**. ◆

B.5 Extreme points

The following theorems tell us something about convex sets and their extreme points in locally convex topological linear spaces. The first can be found in the book by Holmes (1975, p. 74) on convexity and topological linear spaces.

Theorem B.8 (Krein–Milman Theorem) *Let V be a locally convex topological linear space, and let W be a non-empty convex and compact subset of V. Then the following statements hold:*

(i) $\operatorname{ext}(W) \neq \emptyset$.

(ii) *W is the smallest convex and compact subset of V that includes* $\operatorname{ext}(W)$.

(iii) *Every continuous linear functional on V attains its minimum (and maximum) on W at an extreme point of W.*

The second theorem is a generalisation of this result and is due to Bishop and De Leeuw (1959), but see also Phelps (2001) and Maaß (2002) for more recent formulations and discussions. It is also a generalisation of Choquet's representation theorem (Choquet, 1953–1954) from metric spaces to locally convex Hausdorff spaces. The reader may wish to refer to the definition of the Baire σ-field given in Definition 1.8_6.

Theorem B.9 (Bishop–De Leeuw Theorem) *Let V be a locally convex Hausdorff topological linear space, and let W be a non-empty convex and compact subset of V. Denote by $A(W)$ the linear space of all continuous real maps a on W that are affine, meaning that $a(\lambda v + (1-\lambda)w) = \lambda a(v) + (1-\lambda)a(w)$ for all $v, w \in W$ and all $0 \leq \lambda \leq 1$, and denote by $\mathscr{B}_0(W)$ the Baire σ-field on W. Then for every $w \in W$, there is some probability measure μ_w on the σ-field $\{B \cap \operatorname{ext}(W) : B \in \mathscr{B}_0(W)\}$ such that*

$$a(w) = L\int a|_{\operatorname{ext}(W)}\,\mathrm{d}\mu_w \text{ for all } a \in A(W).$$

It is instructive to analyse the assumptions of the above theorem in more detail. First, note that a is a bounded gamble on W because a is continuous on W, and W is compact, so the Lebesgue integral as introduced in Definition 8.27_{171} applies.

That a is measurable with respect to $\mathscr{B}_0(W)$ is easy to see. Indeed, a is continuous and W is compact; therefore, a has compact support. Now recall that $\mathscr{B}_0(W)$ is the smallest σ-field that makes all continuous bounded gambles with compact support measurable (Theorem 1.9_7).

Consequently, the restriction of a to $\operatorname{ext}(W)$, denoted by $a|_{\operatorname{ext}(W)}$, is measurable with respect to the restriction of $\mathscr{B}_0(W)$ to $\operatorname{ext}(W)$, which is precisely the domain of μ_w. Integrability of $a|_{\operatorname{ext}(W)}$ with respect to μ_w then simply follows from Proposition 8.17_{163} all measurable bounded gambles are integrable and Theorem 8.32_{174} equivalence of Lebesgue integration and natural extension.

Appendix C

The Choquet integral

The Choquet integral allows us to associate a special real-valued functional with a set function. It was first introduced in an influential and seminal paper by Choquet (1953–1954). As with so many other things in this paper, the definition of the integral is hidden amongst a plethora of other results, comments and examples. The integral is introduced in Section 48.1 of Choquet (1953–1954) as a 'functional' on the convex cone of non-negative functions on locally compact space, associated with a set function defined on the class of compact sets of this space, and only serves as an auxiliary tool in the construction of a useful topology on these set functions. The definition is easily generalised towards much more general contexts, as has, for instance, been done by Greco (1982), Denneberg (1994) and König (1997). In the interest of completeness, we include a basic discussion of this integral, limited to those aspects that are relevant to, and necessary for, the developments in this book: essentially in Chapters 6_{101} and 15_{327}. We only give the definition of the Choquet integral for set functions that are defined on all events. We prove a number of its properties, as well as a special version of Greco's (1982) representation theorem. We only prove these properties at the level of generality that is required for the developments in Chapter 6_{101}. For a more detailed and general treatment, we refer to the books by Denneberg (1994) and König (1997). Our present discussion is based on their treatment, and all results presented here can be assumed to derive from their work, unless explicitly stated otherwise.

C.1 Preliminaries

C.1.1 The improper Riemann integral of a non-increasing function

First, we consider any map F from a real interval $[a, b]$ to $\mathbb{R}_{\geq 0}$ that is non-increasing, so $F(x) \geq F(y)$ for all $a \leq x \leq y \leq b$. Then the (proper) **Riemann integral** of F over

Lower Previsions, First Edition. Matthias C.M. Troffaes and Gert de Cooman.

the finite interval $[a, b]$ is always defined and is given by

$$R\int_a^b F(t)\,\mathrm{d}t = \sup_{D\in\mathfrak{S}([a,b])} \sum_{(x,y)\in D} (y-x)F(y), \tag{C.1}$$

where $\mathfrak{S}([a, b])$ is the set of finite **subdivisions** of $[a, b]$:

$$\mathfrak{S}([a,b]) := \left\{(a,d_1),(d_1,d_2),\ldots,(d_n,b) : n\in\mathbb{N}, d_k\in(a,b), d_k<d_{k+1}\right\}.$$

Now if a and b are *extended* real numbers with $a < b$, and F is defined, non-negative and non-increasing on the open interval (a, b), then we can consider the **improper Riemann integral** (König, 1997, p. 112):

$$R\int_{a\leftarrow}^{\rightarrow b} F(t)\,\mathrm{d}t := \sup\left\{R\int_c^d F(t)\,\mathrm{d}t : a<c<d<b\right\}. \tag{C.2}$$

It will, for the purposes of this book, be enough to define improper Riemann integrals for such non-negative and non-increasing maps F. This definition covers the case that a or b are infinite or that F is not defined in a or b. $R\int_a^{\rightarrow b} F(t)\,\mathrm{d}t$ and $R\int_{a\leftarrow}^{b} F(t)\,\mathrm{d}t$ are defined similarly, when F is also defined in a and b, respectively. Following the usual conventions, we write $R\int_a^{+\infty} F(t)\,\mathrm{d}t := R\int_a^{\rightarrow+\infty} F(t)\,\mathrm{d}t$ and $R\int_{-\infty}^{b} F(t)\,\mathrm{d}t := R\int_{-\infty\leftarrow}^{b} F(t)\,\mathrm{d}t$, and so on. In particular, if the non-increasing and non-negative F is defined on $\mathbb{R}_{\geq 0}$, then

$$R\int_0^{+\infty} F(t)\,\mathrm{d}t = \sup\left\{R\int_0^c F(t)\,\mathrm{d}t : 0<c<+\infty\right\} = \lim_{c\to+\infty} R\int_0^c F(t)\,\mathrm{d}t \tag{C.3}$$

is either a finite real number or $+\infty$. It follows readily that for all $s \in \mathbb{R}_{>0}$,

$$R\int_0^{+\infty} F(t)\,\mathrm{d}t = R\int_0^s F(t)\,\mathrm{d}t + R\int_s^{+\infty} F(t)\,\mathrm{d}t. \tag{C.4}$$

This improper Riemann integral has a simple but interesting convergence property.

Proposition C.1 *Let a and b be extended real numbers with $a < b$. Let F_n be a non-decreasing sequence of non-negative and non-increasing real functions defined on the open interval (a, b), converging point-wise to the non-negative and non-increasing real function $F := \lim_{n\to+\infty} F_n = \sup_{n\in\mathbb{N}} F_n$ defined on (a, b), then*

$$\lim_{n\to+\infty} R\int_{a\leftarrow}^{\rightarrow b} F_n(t)\,\mathrm{d}t = R\int_{a\leftarrow}^{\rightarrow b} F(t)\,\mathrm{d}t.$$

Proof. The idea for this simple proof is due to Denneberg (1994, Proposition 1.1). Consider the following chain of equalities:

$$\lim_{n\to+\infty} R\int_{a\leftarrow}^{\rightarrow b} F_n(t)\,\mathrm{d}t = \sup_{n\in\mathbb{N}} R\int_{a\leftarrow}^{\rightarrow b} F_n(t)\,\mathrm{d}t$$

$$= \sup_{n\in\mathbb{N}} \sup_{a<c<d<b} \sup_{D\in\mathfrak{S}([c,d])} \sum_{(x,y)\in D} (y-x)F_n(y)$$

$$= \sup_{a<c<d<b} \sup_{D\in\mathfrak{S}([c,d])} \sup_{n\in\mathbb{N}} \sum_{(x,y)\in D} (y-x)F_n(y)$$

$$= \sup_{a<c<d<b} \sup_{D\in\mathfrak{S}([c,d])} \sum_{(x,y)\in D} (y-x)F(y)$$

$$= R\int_{a\leftarrow}^{\rightarrow b} F(t)\,\mathrm{d}t.$$

The first equality follows from the non-decreasing character of the sequence F_n and the monotonicity of the (proper and improper) Riemann integral, the second and fifth from Equations (C.1)$_\curvearrowleft$ and (C.2)$_\curvearrowleft$, the third from the associativity of supremum and the fourth from the fact that the finite sum is taken over subdivisions. □

C.1.2 Comonotonicity

Next, we turn to the notion of comonotonicity, introduced by Dellacherie (1971)[1] and coined by Schmeidler (1986).

Definition C.2 (Comonotonicity) *Two gambles f and g on $\mathcal{X}$ are called* ***comonotone*** *if any (and hence all) of the following equivalent conditions are satisfied:*

(A) *There are no x_1 and x_2 in $\mathcal{X}$ such that $f(x_1) < f(x_2)$ and $g(x_1) > g(x_2)$.*

(B) *The collection of level sets $\{\{f \geq s\} : s \in \mathbb{R}\} \cup \{\{g \geq t\} : t \in \mathbb{R}\}$ is a* ***chain of sets****, that is, totally ordered by set inclusion $\subseteq$.*

Let us prove that these two conditions are indeed equivalent.

Proof. (A)⇒(B). It suffices to prove that for any real s and t, the sets $\{f \geq s\}$ and $\{g \geq t\}$ are totally ordered by set inclusion $\subseteq$. Assume *ex absurdo* that $\{g \geq t\} \nsubseteq \{f \geq s\}$ and $\{f \geq s\} \nsubseteq \{g \geq t\}$. This means that there are x_1 and x_2 in $\mathcal{X}$ such that $g(x_1) \geq t$ and $f(x_1) < s$ and $f(x_2) \geq s$ and $g(x_2) < t$. Hence, $f(x_1) < f(x_2)$ and $g(x_1) > g(x_2)$, which contradicts (A).

(B)⇒(A). Assume *ex absurdo* that there are x_1 and x_2 in $\mathcal{X}$ such that $f(x_1) < f(x_2)$ and $g(x_1) > g(x_2)$. This implies that there are real s and t such that $f(x_1) < s \leq f(x_2)$ and $g(x_2) < t \leq g(x_1)$. Hence, $x_1 \in \{g \geq t\} \setminus \{f \geq s\}$ and $x_2 \in \{f \geq s\} \setminus \{g \geq t\}$, which contradicts (B). □

C.2 Definition of the Choquet integral

Next, we consider a set function $\phi\colon \mathscr{P} \to \mathbb{R}_{\geq 0}$. In other words (see Section 1.7$_{10}$), (i) $\phi(\emptyset) = 0$ and (ii) ϕ is monotone, meaning that $A \subseteq B$ implies $\phi(A) \leq \phi(B)$ for all subsets A and B of $\mathcal{X}$.

[1] Dellacherie (1971) used the phrase 'f et g ont le même tableau de variation' for 'f and g are comonotone'.

It follows that $0 \leq \phi(A) \leq \phi(\mathscr{X})$, and therefore the map $\overline{\phi}$, defined by

$$\overline{\phi}(A) := \phi(\mathscr{X}) - \phi(A^c) \text{ for all } A \subseteq \mathscr{X},$$

is again a set function, called the **conjugate set function** of ϕ.

With this ϕ and its conjugate $\overline{\phi}$, and for each real-valued function f on $\mathscr{X}$, we associate the extended real number[2]

$$C\int f \, \mathrm{d}\phi = R\int_0^{+\infty} \phi(\{f^+ \geq t\}) \, \mathrm{d}t - R\int_0^{+\infty} \overline{\phi}(\{f^- \geq t\}) \, \mathrm{d}t, \tag{C.5}$$

provided the difference on the right-hand side is well defined, where $f^+ := 0 \vee f$ and $f^- := -(0 \wedge f)$. In that case, we say that the **Choquet integral** $C\int f \, \mathrm{d}\phi$ of f with respect to ϕ **exists**. When both terms are finite or, in other words, when $C\int f \, \mathrm{d}\phi$ is a real number, then f is called **Choquet integrable** with respect to ϕ. So f is Choquet integrable if and only if both f^+ and f^- are.

C.3 Basic properties of the Choquet integral

We now list a number of properties of the Choquet integral.

Proposition C.3 (Alternative expressions) *The following statements hold:*

(i) *Suppose the Choquet integral of f with respect to ϕ exists. Then*

$$\begin{aligned} C\int f \, \mathrm{d}\phi &= C\int f^+ \, \mathrm{d}\phi - C\int f^- \, \mathrm{d}\overline{\phi} \\ &= R\int_0^{+\infty} \phi(\{f \geq t\}) \, \mathrm{d}t - R\int_{-\infty}^0 [\phi(\mathscr{X}) - \phi(\{f \geq t\})] \, \mathrm{d}t, \end{aligned}$$

and in this expression, we can replace any '$\geq$' by '$>$'.

(ii) *If f is bounded, then f is Choquet integrable with respect to ϕ, and*

$$C\int f \, \mathrm{d}\phi = \phi(\mathscr{X}) \inf f + R\int_{\inf f}^{\sup f} \phi(\{f \geq t\}) \, \mathrm{d}t \tag{C.6}$$

$$= \phi(\mathscr{X}) \inf f + R\int_{\inf f}^{\sup f} \phi(\{f > t\}) \, \mathrm{d}t. \tag{C.7}$$

(iii) *Suppose there are $n > 0$, real α_1 and non-negative real α_k, $k = 2, \ldots, n$ and subsets $F_n \subseteq F_{n-1} \subseteq \cdots \subseteq F_2 \subseteq F_1 := \mathscr{X}$ of $\mathscr{X}$ such that $f = \sum_{k=1}^n \alpha_k I_{F_k}$. Then*

$$C\int f \, \mathrm{d}\phi = \sum_{k=1}^n \alpha_k \phi(F_k). \tag{C.8}$$

[2] When in second term on the right-hand side, we replace $\overline{\phi}$ by ϕ, we get what Denneberg (1994, Chapter 7) calls the 'symmetric integral'. This is also the type of integral studied by König (1997, Section 13, p. 133). We will not consider this alternative definition here.

Proof. (i)$_\frown$ Observe that $(f^+)^- = (f^-)^- = 0$, $(f^+)^+ = f^+$, and $(f^-)^+ = f^-$, and, therefore, using Equation (C.5)$_\frown$, that

$$C\int f^+ \,\mathrm{d}\phi = R\int_0^{+\infty} \phi(\{f^+ \geq t\})\,\mathrm{d}t \text{ and } C\int f^- \,\mathrm{d}\overline{\phi} = R\int_0^{+\infty} \overline{\phi}(\{f^- \geq t\})\,\mathrm{d}t,$$

which, again using Equation (C.5)$_\frown$, leads to the first equality. The second equality now follows at once from

$$R\int_0^{+\infty} \phi(\{f^+ \geq t\})\,\mathrm{d}t = R\int_0^{+\infty} \phi(\{f \geq t\})\,\mathrm{d}t$$

and

$$R\int_0^{+\infty} \overline{\phi}(\{f^- \geq t\})\,\mathrm{d}t = R\int_0^{+\infty} \overline{\phi}(\{-f \geq t\})\,\mathrm{d}t = R\int_{-\infty}^{0} [\phi(\mathcal{X}) - \phi(\{f > t\}]\,\mathrm{d}t.$$

We are left to show that we can replace any inequalities by strict inequalities. To this end, it suffices to show that, for any non-negative real-valued function g on $\mathcal{X}$,

$$C\int g\,\mathrm{d}\phi := R\int_0^{+\infty} \phi(\{g \geq t\})\,\mathrm{d}t = R\int_0^{+\infty} \phi(\{g > t\})\,\mathrm{d}t. \tag{C.9}$$

The desired result then follows by applying the equality above for $g := f^+$ and $g := f^-$.

Indeed, first of all, notice that for any $t \geq 0$ and $\varepsilon > 0$, we have that $\{f \geq t + \varepsilon\} \subseteq \{f > t\} \subseteq \{f \geq t\}$. Because the set function ϕ is monotone, it follows that

$$R\int_0^{+\infty} \phi(\{g \geq t + \varepsilon\})\,\mathrm{d}t \leq R\int_0^{+\infty} \phi(\{g > t\})\,\mathrm{d}t \leq R\int_0^{+\infty} \phi(\{g \geq t\})\,\mathrm{d}t. \tag{C.10}$$

The first term of this sequence of inequalities can be rewritten as

$$\begin{aligned} R\int_0^{+\infty} \phi(\{g \geq t + \varepsilon\})\,\mathrm{d}t &= R\int_\varepsilon^{+\infty} \phi(\{g \geq t\})\,\mathrm{d}t \\ &= -R\int_0^{\varepsilon} \phi(\{g \geq t\})\,\mathrm{d}t + R\int_0^{+\infty} \phi(\{g \geq t\})\,\mathrm{d}t. \end{aligned}$$

Clearly, $R\int_0^\varepsilon \phi(\{g \geq t\})\,\mathrm{d}t \leq \varepsilon\phi(\mathcal{X})$, so the sequence of inequalities (C.10) leads to

$$-\varepsilon\phi(\mathcal{X}) + R\int_0^{+\infty} \phi(\{g \geq t\})\,\mathrm{d}t \leq R\int_0^{+\infty} \phi(\{g > t\})\,\mathrm{d}t \leq R\int_0^{+\infty} \phi(\{g \geq t\})\,\mathrm{d}t.$$

Because this holds for any $\varepsilon > 0$, Equation (C.9) follows.

(ii)$_\frown$ We prove the first equality; the proof for the second is completely similar. We consider the case that $\inf f < 0$ and $\sup f > 0$. The other cases can be treated similarly. For any real $a > \sup f$, we have that $\{f \geq t\} = \emptyset$, and therefore $R\int_0^a \phi(\{f \geq t\})\,\mathrm{d}t =$

$R\int_0^{\sup f} \phi(\{f \geq t\})\,\mathrm{d}t$. Using the definition (C.2)$_{377}$ of the improper Riemann integral, this leads to

$$R\int_0^{+\infty} \phi(\{f \geq t\})\,\mathrm{d}t = \sup\left\{R\int_0^a \phi(\{f \geq t\})\colon 0 < a\right\} = R\int_0^{\sup f} \phi(\{f \geq t\})\,\mathrm{d}t.$$

Similarly, for any real $b < \inf f$, we have that $\{f \geq t\} = \mathcal{X}$ and, therefore, $R\int_b^0 [\phi(\mathcal{X}) - \phi(\{f \geq t\}]\,\mathrm{d}t = R\int_{\inf f}^0 [\phi(\mathcal{X}) - \phi(\{f \geq t\}]\,\mathrm{d}t$. Again using the definition (C.2)$_{377}$ of the improper Riemann integral, now with non-increasing integrand $-[\phi(\mathcal{X}) - \phi(\{f \geq t\})]$, this leads to

$$\begin{aligned}
-R\int_{-\infty}^0 [\phi(\mathcal{X}) - \phi(\{f \geq t\})]\,\mathrm{d}t &= \sup\left\{-R\int_b^0 [\phi(\mathcal{X}) - \phi(\{f \geq t\})]\,\mathrm{d}t\colon b < 0\right\}\\
&= -R\int_{\inf f}^0 [\phi(\mathcal{X}) - \phi(\{f \geq t\}]\,\mathrm{d}t\\
&= \phi(\mathcal{X})\inf f + R\int_{\inf f}^0 \phi(\{f \geq t\})\,\mathrm{d}t.
\end{aligned}$$

Now use (i)$_{379}$.

(iii)$_{379}$ Let, for ease of notation, $f_k \coloneqq \sum_{\ell=1}^k \alpha_k$ for $k = 1, \ldots, n$. Then it follows from the assumptions that $f_1 \leq f_2 \leq \cdots \leq f_n$. We now use Equation (C.6)$_{379}$ and take into account that $\inf f = \alpha_1 = f_1$ and $\sup f = \sum_{k=1}^n \alpha_k = f_n$ to find that

$$\begin{aligned}
C\int f\,\mathrm{d}\phi &= f_1\phi(\mathcal{X}) + R\int_{f_1}^{f_n} \phi(\{f \geq t\})\,\mathrm{d}t = f_1\phi(F_1) + \sum_{k=2}^n R\int_{f_{k-1}}^{f_k} \phi(\{f \geq t\})\,\mathrm{d}t\\
&= f_1\phi(F_1) + \sum_{k=2}^n R\int_{f_{k-1}}^{f_k} \phi(F_k)\,\mathrm{d}t = f_1\phi(F_1) + \sum_{k=2}^n (f_k - f_{k-1})\phi(F_k)\\
&= \alpha_1\phi(F_1) + \sum_{k=2}^n \alpha_k\phi(F_k) = \sum_{k=1}^n \alpha_k\phi(F_k),
\end{aligned}$$

which completes the proof. □

Corollary C.4 *If f is simple, that is, has a finite number $n > 0$ of values $f_1 < f_2 < \cdots < f_n$, then*

$$f = f_1 + \sum_{k=2}^n (f_k - f_{k-1}) I_{\{f \geq f_k\}} \tag{C.11}$$

and

$$C\int f\,\mathrm{d}\phi = \phi(\mathcal{X})f_1 + \sum_{k=2}^n (f_k - f_{k-1})\phi(\{f \geq f_k\}). \tag{C.12}$$

Proof. To prove Equation (C.11), observe that if $f(x) = f_\ell$, then the value of the right-hand side in x is $f_1 + \sum_{k=1}^{\ell} (f_k - f_{k-1}) = f_1 + (f_\ell - f_1) = f_\ell$. To prove Equation (C.12), apply Proposition C.3(iii)$_{379}$ with $F_k \coloneqq \{f \geq f_k\}$, $\alpha_1 \coloneqq f_1$ and $\alpha_k \coloneqq f_k - f_{k-1}$ for $f = 2, \ldots, n$. □

For the following properties, we will restrict ourselves to the Choquet integral of bounded real-valued maps, or *bounded gambles*. As already mentioned earlier, these properties can still be extended to larger domains of real-valued maps, but we do not need such extended properties in the context of this book.

It follows from the considerations above that with a set function ϕ on $\mathscr{P}$, we can associate the so-called **Choquet functional** C_ϕ on the set $\mathbb{B}$ or all bounded gambles on $\mathscr{X}$, which is the real-valued functional defined by

$$C_\phi(f) := C\int f\,\mathrm{d}\phi \text{ for all } f \text{ in } \mathbb{B}.$$

Proposition C.5 *Let ϕ be any set function on $\mathscr{P}$, then the following statements hold:*

(i) C_ϕ *is bounded:* $\phi(\mathscr{X})\inf f \leq C\int f\,\mathrm{d}\phi \leq \phi(\mathscr{X})\sup f$ *for all bounded gambles f on $\mathscr{X}$.*

(ii) $C_\phi(I_A) = \phi(A)$ *for all $A \subseteq \mathscr{X}$.*

(iii) C_ϕ *is non-negatively homogeneous:* $C\int \lambda f\,\mathrm{d}\phi = \lambda C\int f\,\mathrm{d}\phi$ *for all non-negative real λ and all bounded gambles f on $\mathscr{X}$.*

(iv) $C\int -f\,\mathrm{d}\phi = -C\int f\,\mathrm{d}\overline{\phi}$ *for all bounded gambles f on $\mathscr{X}$.*

(v) C_ϕ *is constant additive:* $C\int f+\alpha\,\mathrm{d}\phi = C\int f\,\mathrm{d}\phi + \alpha\phi(\mathscr{X})$ *for all bounded gambles f on $\mathscr{X}$ and all real α.*

(vi) C_ϕ *is monotone:* $C\int f\,\mathrm{d}\phi \leq C\int g\,\mathrm{d}\phi$ *for all bounded gambles $f \leq g$ on $\mathscr{X}$.*

(vii) C_ϕ *is comonotone additive:* $C\int f+g\,\mathrm{d}\phi = C\int f\,\mathrm{d}\phi + C\int g\,\mathrm{d}\phi$ *for all comonotone bounded gambles f and g on $\mathscr{X}$.*

(viii) $C\int \alpha\,\mathrm{d}\phi = \alpha\phi(\mathscr{X})$ *for all real α.*

(ix) C_ϕ *is continuous with respect to the topology of uniform convergence: given that* $\sup|f_n - f| \to 0$, *it follows that* $\lim_{n\to\infty} C\int f_n\,\mathrm{d}\phi = C\int f\,\mathrm{d}\phi$.

Proof. (i) For all $t \in \mathbb{R}$, $0 \leq \phi(\{f \geq t\}) \leq \phi(\mathscr{X})$, because ϕ is a set function. Hence, we derive from Proposition C.3(ii)$_{379}$ that indeed

$$\phi(\mathscr{X})\inf f + R\int_{\inf f}^{\sup f} 0\,\mathrm{d}t \leq C\int f\,\mathrm{d}\phi \leq \phi(\mathscr{X})\inf f + R\int_{\inf f}^{\sup f} \phi(\mathscr{X})\,\mathrm{d}t.$$

(viii) This is an immediate consequence of (i).

(ii) Because of (viii), we may assume without loss of generality that A is a proper subset of $\mathscr{X}$. Hence, $\inf I_A = 0$, $\sup I_A = 1$ and $\{I_A \geq t\} = A$ for all $t \in (0,1]$, so we derive from Proposition C.3(ii)$_{379}$ that indeed $C\int I_A\,\mathrm{d}\phi = 0 + R\int_0^1 \phi(A)\,\mathrm{d}t = \phi(A)$.

(iii) Because of (viii), we may assume without loss of generality that $\lambda > 0$. But then $\inf(\lambda f) = \lambda\inf f$, $\sup(\lambda f) = \lambda\sup f$ and $\{\lambda f \geq t\} = \{f \geq t/\lambda\}$, so Proposition C.3(ii)$_{379}$ tells us that indeed

$$C_\phi(\lambda f) = \lambda\inf f + R\int_{\lambda\inf f}^{\lambda\sup f} \phi\left(\left\{f \geq \frac{t}{\lambda}\right\}\right)\mathrm{d}t = \lambda C_\phi(f).$$

(iv) As $\sup(-f) = -\inf f$, $\inf(-f) = -\sup f$ and $\{-f \geq t\} = \{f \leq -t\}$, we derive from Proposition C.3(ii)$_{379}$ that

$$
\begin{aligned}
C_\phi(-f) &= \phi(\mathscr{X})\inf(-f) + R\int_{\inf(-f)}^{\sup(-f)} \phi(\{-f \geq t\})\,\mathrm{d}t \\
&= -\phi(\mathscr{X})\sup f + R\int_{-\sup f}^{-\inf f} \phi(\{f \leq -t\})\,\mathrm{d}t \\
&= -\phi(\mathscr{X})\sup f + R\int_{\inf f}^{\sup f} \phi(\{f \leq t\})\,\mathrm{d}t \\
&= -\phi(\mathscr{X})\sup f + R\int_{\inf f}^{\sup f} [\phi(\mathscr{X}) - \overline{\phi}(\{f > t\})\,\mathrm{d}t \\
&= -\phi(\mathscr{X})\inf f - R\int_{\inf f}^{\sup f} \overline{\phi}(\{f > t\})\,\mathrm{d}t = -C_{\overline{\phi}}(f).
\end{aligned}
$$

(vi) If $f \leq g$, then $\{f \geq t\} \subseteq \{g \geq t\}$ and, therefore, also $\phi(\{f \geq t\}) \leq \phi(\{g \geq t\})$ for all $t \in \mathbb{R}$. Now use Proposition C.3(i)$_{379}$ and the monotonicity of the Riemann integral.

(ix) Assume that $\sup|f_n - f| \to 0$ and consider any $\varepsilon > 0$. Then there is some N_ε such that for all $n \geq N_\varepsilon, f_n - \varepsilon \leq f \leq f_n + \varepsilon$, and therefore, using (vi) and (v),

$$
C\int f_n\,\mathrm{d}\phi - \varepsilon\phi(\mathscr{X}) \leq C\int f\,\mathrm{d}\phi \leq C\int f_n\,\mathrm{d}\phi + \varepsilon\phi(\mathscr{X}).
$$

Hence, indeed, $C\int f_n\,\mathrm{d}\phi \to C\int f\,\mathrm{d}\phi$.

(vii) We first assume that the comonotone f and g are simple: they assume the values $f_1 < \cdots < f_n$ and $g_1 < \cdots < g_m$, respectively. We can always label the elements F_k in the finite chain of r subsets

$$
\begin{aligned}
&\{\{f \geq s\} : s \in \mathbb{R}\} \cup \{\{g \geq t\} : t \in \mathbb{R}\} \\
&\qquad = \left\{\{f \geq f_k\} : k \in \{1, \ldots, n\}\right\} \cup \left\{\{g \geq g_\ell\} : \ell \in \{1, \ldots, m\}\right\},
\end{aligned}
$$

in such a way that $F_r \subseteq F_{r-1} \subseteq \cdots \subseteq F_2 \subseteq F_1 = \mathscr{X}$, and we know that there are real λ_1 and α_1, and non-negative real λ_k and α_k, $k = 2, \ldots, r$, such that $f = \sum_{k=1}^r \lambda_k I_{F_k}$ and $g = \sum_{k=1}^r \alpha_k I_{F_k}$ and, therefore, also $f + g = \sum_{k=1}^r (\lambda_k + \alpha_k) I_{F_k}$. Hence, indeed,

$$
\begin{aligned}
C\int (f+g)\,\mathrm{d}\phi &= \sum_{k=1}^r (\lambda_k + \alpha_k)\phi(F_k) \\
&= \sum_{k=1}^r \lambda_k\phi(F_k) + \sum_{k=1}^r \alpha_k\phi(F_k) = C\int f\,\mathrm{d}\phi + C\int g\,\mathrm{d}\phi,
\end{aligned}
$$

where we have made repeated use of Proposition C.3(iii)$_{379}$.

Next, let f and g be general bounded gambles that are comonotone, and let $h := f + g$, then we have to show that $C_\phi(h) = C_\phi(f) + C_\phi(g)$. We know from Lemma 1.21_{16} that we can uniformly approximate f, g and h by the respective sequences of simple gambles $f_n := u_n \circ f$, $g_n := u_n \circ g$ and $h_n := u_n \circ h$. Now it follows from Lemma 1.21(iii)$_{16}$ that $f_n + g_n - 2/n \leq h_n \leq f_n + g_n$, so

$$\begin{aligned} C_\phi(f_n) + C_\phi(g_n) - \frac{2}{n}\phi(\mathcal{X}) &= C_\phi\left(f_n + g_n - \frac{2}{n}\right) \\ &\leq C_\phi(h_n) \leq C_\phi(f_n + g_n) = C_\phi(f_n) + C_\phi(g_n), \end{aligned}$$

where the equalities follows from (v)$_{382}$ and the fact that f_n and g_n are comonotone (use Lemma 1.21(iv)$_{16}$), and the inequalities follow from (vi)$_{382}$. Taking the limit for $n \to +\infty$ now yields the desired equality, if we realise that C_ϕ is continuous with respect to the topology of uniform convergence (use (ix)$_{382}$). □

The first part – for simple gambles – of the proof of Proposition C.5(vii)$_{382}$ is due to Dellacherie (1971, p. 81).[3]

We will need the following lemma, whose proof is due to Walley (1981, Lemma 6.2, p. 54).

Lemma C.6 *Let ϕ be any 2-monotone set function on $\mathscr{P}$, meaning that $\phi(A \cup B) + \phi(A \cap B) \geq \phi(A) + \phi(B)$ for all subsets A and B of $\mathcal{X}$. Then, for all $m \in \mathbb{N}_{>0}$, $A_1 \subseteq A_2 \subseteq \cdots \subseteq A_m \subseteq \mathcal{X}$ and $\mathcal{X} \supseteq B_1 \supseteq B_2 \supseteq \cdots \supseteq B_m$*

$$\phi\left(\bigcup_{k=1}^{m}[A_k \cap B_k]\right) \geq \sum_{k=1}^{m}\phi(A_k \cap B_k) - \sum_{k=1}^{m-1}\phi(A_k \cap B_{k+1}).$$

Proof. We proceed by induction on m. Clearly, the inequality holds trivially for $m = 1$. Suppose it holds for $m = n$. Does it also hold for $m = n + 1$? By 2-monotonicity, we get

$$\begin{aligned} &\phi\left(\bigcup_{k=1}^{n}[A_k \cap B_k]\right) + \phi(A_{n+1} \cap B_{n+1}) \\ &\leq \phi\left(\left[\bigcup_{k=1}^{n}(A_k \cap B_k)\right] \cup [A_{n+1} \cap B_{n+1}]\right) + \phi\left(\left[\bigcup_{k=1}^{n}(A_k \cap B_k)\right] \cap [A_{n+1} \cap B_{n+1}]\right) \\ &= \phi\left(\bigcup_{k=1}^{n+1}[A_k \cap B_k]\right) + \phi(A_n \cap B_{n+1}), \end{aligned}$$

[3] After correcting a simple mistake in his proof, which is not more than an oversight: he states that the sets F_k should be mutually disjoint, rather than nested.

once it is noted that

$$\left[\bigcup_{k=1}^{n}(A_k \cap B_k)\right] \cap [A_{n+1} \cap B_{n+1}]$$
$$= \bigcup_{k=1}^{n}[(A_k \cap B_k) \cap (A_{n+1} \cap B_{n+1})] = \bigcup_{k=1}^{n}(A_k \cap B_{n+1}) = A_n \cap B_{n+1}.$$

After rearranging the terms, and applying the induction hypothesis, we find

$$\phi\left(\bigcup_{k=1}^{n+1}[A_k \cap B_k]\right)$$
$$\geq \phi\left(\bigcup_{k=1}^{n}[A_k \cap B_k]\right) + \phi(A_{n+1} \cap B_{n+1}) - \phi(A_n \cap B_{n+1})$$
$$\geq \sum_{k=1}^{n}\phi(A_k \cap B_k) - \sum_{k=1}^{n-1}\phi(A_k \cap B_{k+1}) + \phi(A_{n+1} \cap B_{n+1}) - \phi(A_n \cap B_{n+1})$$
$$= \sum_{k=1}^{n+1}\phi(A_k \cap B_k) - \sum_{k=1}^{n}\phi(A_k \cap B_{k+1}),$$

which completes the proof. □

Now we can prove the following important result, due to Choquet (1953–1954, p. 289). We essentially follow the proof given by Walley (1981, pp. 54–55) – he proves it for 2-monotone lower probabilities, whereas we prove it more generally for arbitrary 2-monotone set functions: the difference is mostly cosmetic.

Proposition C.7 *Let ϕ be any set function on $\mathscr{P}$, then the following statements are equivalent:*

(i) *ϕ is 2-monotone: $\phi(A \cup B) + \phi(A \cap B) \geq \phi(A) + \phi(B)$ for all subsets A and B of $\mathcal{X}$;*

(ii) *C_ϕ is super-additive: $C\int(f+g)\,\mathrm{d}\phi \geq C\int f\,\mathrm{d}\phi + C\int g\,\mathrm{d}\phi$ for all bounded gambles f and g on $\mathcal{X}$.*

Proof. (ii)⇒(i). As $I_A + I_B = I_{A\cup B} + I_{A\cap B}$ and $A \cap B \subseteq A \cup B$, we derive from Proposition C.3(iii)$_{379}$ that $C\int(I_A + I_B)\,\mathrm{d}\phi = \phi(A \cup B) + \phi(A \cap B)$. By Proposition C.5(ii)$_{382}$, also $C\int I_A\,\mathrm{d}\phi = \phi(A)$ and $C\int I_B\,\mathrm{d}\phi = \phi(B)$.

(i)⇒(ii). We first prove super-additivity of C_ϕ on non-negative simple gambles. Let f and g be non-negative simple gambles assuming the non-negative values $f_1 < \cdots < f_n$ and $g_1 > \cdots > g_m$, respectively. Define $A_k := \{g \geq g_k\}$ and $B_k(t) := \{f \geq t - g_k\}$ for $k \in \{1, \ldots, m\}$ and $t \in \mathbb{R}$. Clearly, $A_1 \subseteq A_2 \subseteq \cdots \subseteq A_m$ and $B_1(t) \supseteq B_2(t) \supseteq \cdots \supseteq B_m(t)$ for all $t \in \mathbb{R}$. Moreover,

$$\{f + g \geq t\} = \bigcup_{k=1}^{m} [\{g \geq g_k\} \cap \{f \geq t - g_k\}] = \bigcup_{k=1}^{m} [A_k \cap B_k(t)] \text{ for all } t \in \mathbb{R},$$

so, by Lemma C.6$_{384}$ and the linearity of the Riemann integral, we find that

$$\begin{aligned} C_\phi(f + g) &= R\int_0^{+\infty} \phi(\{f + g \geq t\})\,\mathrm{d}t \\ &\geq \sum_{k=1}^{m} R\int_0^{+\infty} \phi(A_k \cap B_k(t))\,\mathrm{d}t - \sum_{k=1}^{m-1} R\int_0^{+\infty} \phi(A_k \cap B_{k+1}(t))\,\mathrm{d}t \\ &= \sum_{k=1}^{m-1} \left[R\int_0^{+\infty} \phi(A_k \cap B_k(t))\,\mathrm{d}t - R\int_0^{+\infty} \phi(A_k \cap B_{k+1}(t))\,\mathrm{d}t \right] \\ &\qquad + R\int_0^{+\infty} \phi(A_m \cap B_m(t))\,\mathrm{d}t. \end{aligned}$$

Now, note that for all $t \in \mathbb{R}$,

$$B_{k+1}(t) = \{f \geq t - g_{k+1}\} = \{f \geq t - g_{k+1} + g_k - g_k\} = B_k(t - g_{k+1} + g_k),$$

and therefore, for any $k \in \{1, \ldots, m - 1\}$,

$$\begin{aligned} R\int_0^{+\infty} &\phi(A_k \cap B_k(t))\,\mathrm{d}t - R\int_0^{+\infty} \phi(A_k \cap B_{k+1}(t))\,\mathrm{d}t \\ &= R\int_0^{+\infty} \phi(A_k \cap B_k(t))\,\mathrm{d}t - R\int_{g_k - g_{k+1}}^{+\infty} \phi(A_k \cap B_k(t))\,\mathrm{d}t \\ &= R\int_0^{g_k - g_{k+1}} \phi(A_k \cap B_k(t))\,\mathrm{d}t = (g_k - g_{k+1})\phi(A_k), \end{aligned}$$

because for $t \leq g_k$, it holds that $B_k(t) = \{f \geq t - g_k\} = \mathcal{X}$ (recall that f is assumed to be non-negative). Concluding, also taking into account that $A_m = \mathcal{X}$ and $A_m \cap B_m(t) = \{f \geq t - g_m\}$:

$$\begin{aligned} C_\phi(f + g) &\geq \sum_{k=1}^{m-1} (g_k - g_{k+1})\phi(A_k) + R\int_0^{+\infty} \phi(A_m \cap B_m(t))\,\mathrm{d}t \\ &= \sum_{k=1}^{m-1} (g_k - g_{k+1})\phi(A_k) + R\int_0^{+\infty} \phi(\{f \geq t - g_m\})\,\mathrm{d}t \end{aligned}$$

$$= \sum_{k=1}^{m-1}(g_k - g_{k+1})\phi(A_k) + R\int_{-g_m}^{0}\phi(\{f \geq t\})\,\mathrm{d}t + R\int_{0}^{+\infty}\phi(\{f \geq t\})\,\mathrm{d}t$$

$$= \sum_{k=1}^{m-1}(g_k - g_{k+1})\phi(A_k) + g_m\phi(A_m) + R\int_{0}^{+\infty}\phi(\{f \geq t\})\,\mathrm{d}t$$

$$= C_\phi(g) + C_\phi(f),$$

where the last equality follows from Proposition C.3_{379}. This tells us that C_ϕ is super-additive on non-negative simple gambles. Using the constant additivity of the Choquet functional (Proposition C.5(v)$_{382}$), it readily follows that C_ϕ is super-additive on all simple gambles.

Next, we turn to the super-additivity of C_ϕ on all bounded gambles. Consider any bounded gambles f and g on $\mathcal{X}$, and let $h := f + g$. Then we have to show that $C_\phi(h) \geq C_\phi(f) + C_\phi(g)$. We consider the step functions u_n, $n \in \mathbb{N}$ introduced in Equation (1.6)$_{15}$, and use these to construct the sequences of simple gambles (Lemma 1.21(ii)$_{16}$) $f_n := u_n \circ f$, $g_n := u_n \circ g$ and $h_n := u_n \circ h$. We know from Lemma 1.21_{16} that these sequences converge uniformly to f, g and h, respectively, and so the continuity of the Choquet functional with respect to the topology of uniform convergence (Proposition C.5(ix)$_{382}$) lets us conclude that $C_\phi(f) = \lim_{n\to+\infty} C_\phi(f_n)$, $C_\phi(g) = \lim_{n\to+\infty} C_\phi(g_n)$ and $C_\phi(h) = \lim_{n\to+\infty} C_\phi(h_n)$. By Lemma 1.21(iii)$_{16}$, $h_n + 2/n \geq f_n + g_n$ and, therefore,

$$C_\phi(h_n) + \frac{2}{n}\phi(\mathcal{X}) = C_\phi\left(h_n + \frac{2}{n}\right) \geq C_\phi(f_n + g_n) \geq C_\phi(f_n) + C_\phi(g_n),$$

where the equality follows from Proposition C.5(v)$_{382}$, the first inequality from Proposition C.5(vi)$_{382}$ and the second one from the super-additivity of C_ϕ for simple gambles proved earlier. Taking the limit for $n \to +\infty$ now yields the desired inequality. □

As already mentioned, the proof of (i)$_{385}$ and the first part (concerning simple gambles) of the proof of (ii)$_{385}$ are due to Walley (1981, Lemma 6.3, p. 54). The second part of the proof of (ii)$_{385}$ is actually due to Denneberg (1994, proof of Theorem 6.3(b), p. 76) as Walley only provides a hint.

C.4 A simple but useful equality

We now turn to a symmetry result that is inspired by the treatment of Radon–Nikodym derivatives in measure theory (see, for instance, Halmos, 1974). It is used in Section 7.8$_{143}$ to derive alternative expressions for the Choquet functional associated with a possibility measure.

Consider a set function ϕ on $\mathscr{P}$ and any subset A of $\mathcal{X}$. Then we can define a new set function ϕ_A on $\mathscr{P}$, by letting

$$\phi_A(B) := \phi(A \cap B) \text{ for all } B \subseteq \mathcal{X}.$$

We can use this new set function to generalise the definition of the Choquet integral:

$$C\int_A f\,\mathrm{d}\phi := C\int f\,\mathrm{d}\phi_A$$

for all gambles f on $\mathcal{X}$ whose Choquet integral with respect to ϕ_A exists.

Similarly, with any non-negative gamble h on $\mathcal{X}$, we can now associate yet another set function ϕ_h in the following way, provided that h is Choquet integrable with respect to ϕ:[4]

$$\phi_h(B) := C\int_B h\,\mathrm{d}\phi = R\int_0^{+\infty} \phi(B \cap \{h \geq t\})\,\mathrm{d}t \text{ for all } B \subseteq \mathcal{X}.$$

We need h to be non-negative and Choquet integrable with respect to ϕ in order for ϕ_h to be a set function. Observe that $\phi_A = \phi_{I_A}$.

And as a final step, we use the set function ϕ_h to define an extended real functional, also denoted by ϕ_h:

$$\phi_h(g) := C_{\phi_h}(g) = C\int g\,\mathrm{d}\phi_h$$

for all gambles g on $\mathcal{X}$ whose Choquet integral with respect to the set function ϕ_h exists. For non-negative g, this Choquet integral always exists, and we can write

$$\phi_h(g) = R\int_0^{+\infty}\left(R\int_0^{+\infty} \phi(\{g \geq t\} \cap \{h \geq s\})\,\mathrm{d}s\right)\mathrm{d}t. \tag{C.13}$$

Proposition C.8 *Let $n \in \mathbb{N}^*_{>0}$, $n \geq 2$ and consider a set function ϕ on $\mathcal{P}$ and arbitrary non-negative gambles g and h on $\mathcal{X}$ that are Choquet integrable with respect to ϕ. Then the following statements hold:*

(i) *If ϕ is n-monotone, then the real functional ϕ_h is n-monotone as well;*

(ii) $\phi_h(g) = \phi_g(h)$.

Proof. (i). The proof of this statement proceeds along exactly the same lines as the proof of Theorem 6.15_{111}.

(ii). Immediate if we look at Equation (C.13) and realise that we can change the order of integration in the double Riemann integral. □

Although we believe this result to be new, the idea behind it was first used by De Cooman (2000, Section 5) in a discussion about integration of possibility measures, in a context related to Section 7.8_{143}. De Cooman (2001, Proof of Theorem 10) also mentions its validity for bounded gambles without an explicit proof.

[4] Because h is non-negative, h is Choquet integrable with respect to ϕ if and only if $C\int h\,\mathrm{d}\phi < +\infty$. In that case, h is also Choquet integrable with respect to ϕ_B for all $B \subseteq \mathcal{X}$, as $C\int h\,\mathrm{d}\phi_B \leq C\int h\,\mathrm{d}\phi$ (because h is non-negative and $\phi_B \leq \phi$).

C.5 A simplified version of Greco's representation theorem

Greco's (1982) representation theorem states that a real functional Γ that satisfies a number of conditions can be represented as a Choquet functional C_ϕ associated with some set function ϕ. Amongst these conditions, those that stand out are monotonicity and comonotone additivity. A similar but weaker result was proved somewhat later – and apparently independently – by Schmeidler (1986). We refer to Denneberg (1994, Chapter 13) for a detailed discussion of this important theorem and its many special cases, amongst which the classical Daniell–Stone representation theorem (see also König, 1997, Chapter V).

Here, we give a direct proof for a simplified version that is more limited in scope but eminently suited for the developments in Chapter 6_{101}. This proof is a simplification of the one given by Denneberg (1994, Theorem 13.2) for the full-blown version of Greco's representation theorem.

Theorem C.9 (Greco Light) *Consider a lower prevision $\underline{P}$ defined on a linear lattice of bounded gambles that contains all constant gambles. Assume that $\underline{P}$ is coherent and comonotone additive, and consider the set functions α and β defined on $\mathscr{P}$ by*

$$\left.\begin{aligned}\alpha(A) &:= \sup\left\{\underline{P}(f) : f \le I_A \text{ and } f \in \operatorname{dom}\underline{P}\right\} = \underline{P}_*(I_A)\\ \beta(A) &:= \inf\left\{\underline{P}(f) : I_A \le f \text{ and } f \in \operatorname{dom}\underline{P}\right\} = \underline{P}^*(I_A)\end{aligned}\right\}\ \text{for all } A \subseteq \mathscr{X}.$$

Then $\underline{P}(f) = C_\phi(f)$ for all bounded gambles f in $\operatorname{dom}\underline{P}$ and any set function ϕ on $\mathscr{P}(\mathscr{X})$ with $\alpha \le \phi \le \beta$.

Proof. As $\underline{P}$ is monotone (because coherent, see Theorem 4.13(iv)$_{53}$) and $\operatorname{dom}\underline{P}$ contains all constant gambles, we see that $\alpha(\emptyset) = \beta(\emptyset) = \underline{P}(0) = 0$ and $\alpha(\mathscr{X}) = \beta(\mathscr{X}) = \underline{P}(1) = 1$, where the last equalities follow from the coherence of $\underline{P}$ (Theorem 4.13(ii)$_{53}$). Hence, also $\phi(\emptyset) = 0$ and $\phi(\mathscr{X}) = 1$.

Consider any bounded gamble f, then $f = h + \inf f$ with $h = f - \inf f \ge 0$. Then it follows from the constant additivity of both $\underline{P}$ (Theorem 4.13(iii)$_{53}$) and C_ϕ (Proposition C.5(v)$_{382}$) that $\underline{P}(f) = \underline{P}(h) + \inf f$ and $C_\phi(f) = C_\phi(h) + \phi(\mathscr{X}) \inf f = C_\phi(h) + \inf f$. We may therefore assume without loss of generality that $f \ge 0$.

We now consider the sequence of bounded gambles $f_n := u_n \circ f$, $n \in \mathbb{N}_{>0}$, where the step functions u_n were introduced in Equation $(1.6)_{15}$. We begin by observing that $f_n \ge 0$ and, therefore,

$$\begin{aligned}C_\phi(f_n) &= R\int_0^{+\infty} \phi(\{u_n \circ f > t\})\,\mathrm{d}t = \sum_{k=0}^{+\infty} R\int_{k/n}^{k+1/n} \phi(\{u_n \circ f > t\})\,\mathrm{d}t\\ &= \sum_{k=0}^{+\infty} R\int_{k/n}^{k+1/n} \phi\left(\left\{f > \frac{k}{n}\right\}\right)\mathrm{d}t = \frac{1}{n}\sum_{k=0}^{n_f} \phi\left(\left\{f > \frac{k}{n}\right\}\right),\end{aligned} \tag{C.14}$$

where n_f is the greatest integer that is still strictly smaller than $n\sup f$. The third equality holds because for all $t \in (k/n, k + 1/n)$, $u_n(s) > t \Leftrightarrow k/n < s$, and the fourth because $\{f > k/n\} = \emptyset \Leftrightarrow k \geq n\sup f$.

As a next step, we observe that for all $k \in \mathbb{Z}$ and all $x \in \mathbb{R}$,

$$n\left(\min\left\{x, \frac{k+1}{n}\right\} - \min\left\{x, \frac{k}{n}\right\}\right) \leq I_{\left\{x>\frac{k}{n}\right\}}$$
$$\leq n\left(\min\left\{x, \frac{k}{/}n\right\} - \min\left\{x, \frac{k-1}{n}\right\}\right),$$

where $I_{\{x>k/n\}} = 1$ for $x > k/n$ and zero otherwise. This implies that

$$n\left(f \wedge k + 1/n - f \wedge \frac{k}{n}\right) \leq I_{\left\{f>\frac{k}{n}\right\}} \leq n\left(f \wedge \frac{k}{n} - f \wedge \frac{k-1}{n}\right).$$

Hence, using the definitions of α and β, the non-negative homogeneity of $\underline{P}$ (which follows from coherence, see Theorem 4.13(vi)$_{53}$) and the fact that the bounded gambles on the left and the right in the inequalities above belong to dom $\underline{P}$, we get

$$n\underline{P}\left(f \wedge \frac{k+1}{n} - f \wedge \frac{k}{n}\right) \leq \alpha(\left\{f > \frac{k}{n}\right\})$$
$$\leq \phi\left(\left\{f > \frac{k}{n}\right\}\right)$$
$$\leq \beta\left(\left\{f > \frac{k}{n}\right\}\right) \leq n\underline{P}\left(f \wedge \frac{k}{n} - f \wedge \frac{k-1}{n}\right).$$

After summing all these inequalities for $k = 0, 1 \ldots, n_f$, and realising that for any real $a_1 \leq a_2 \leq a_3$, the bounded gambles $f \wedge a_3 - f \wedge a_2$ and $f \wedge a_2 - f \wedge a_1$ are comonotone, we may invoke the comonotone additivity of $\underline{P}$ to find that

$$\underline{P}\left(\sum_{k=0}^{n_f}\left[f \wedge \frac{k+1}{n} - f \wedge \frac{k}{n}\right]\right) \leq \frac{1}{n}\sum_{k=0}^{n_f}\phi\left(\left\{f > \frac{k}{n}\right\}\right)$$
$$\leq \underline{P}\left(\sum_{k=0}^{n_f}\left[f \wedge \frac{k}{n} - f \wedge \frac{k-1}{n}\right]\right),$$

or also, using Equation (C.14)$_{\curvearrowleft}$,

$$\underline{P}\left(f \wedge \frac{n_f+1}{n}\right) \leq C_\phi(f_n) \leq \underline{P}\left(f \wedge \frac{n_f}{n} + \frac{1}{n}\right) = \underline{P}\left(f \wedge \frac{n_f}{n}\right) + \frac{1}{n} \leq \underline{P}(f) + \frac{1}{n}.$$

Now, because by definition, $n_f + 1 \geq n\sup f$, we see that this leads to $\underline{P}(f) \leq C_\phi(f_n) \leq \underline{P}(f) + 1/n$, and therefore, taking the limit, $\underline{P}(f) = \lim_{n\to+\infty} C_\phi(f_n) = C_\phi(f)$. The last equality follows from the fact that the sequence f_n converges uniformly to f (Lemma 1.21$_{16}$) and from the continuity of C_ϕ with respect to the topology of uniform convergence (Proposition C.5(ix)$_{382}$). □

Appendix D

The extended real calculus

In this appendix, we recall the definition and elementary properties of the extended real calculus. For the sake of completeness, the (very easy!) proofs are given too: apparently, the properties of the extended real number system we rely on in this work are rather hard to find in the literature.

D.1 Definitions

The set $\mathbb{R}^*$ of **extended real numbers** is $\mathbb{R} \cup \{-\infty, +\infty\}$.

Definition D.1 *The addition '+' on $\mathbb{R}$ is extended to $\mathbb{R}^*$ as follows:*

$$-\infty + (-\infty) = -\infty, \qquad +\infty + (+\infty) = +\infty,$$

$$a + (-\infty) = -\infty + a = -\infty, \quad a + (+\infty) = +\infty + a = +\infty, \quad \text{if } a \in \mathbb{R}.$$

We call a sum of extended real numbers **well defined** if it cannot be reduced to $+\infty + (-\infty)$ or $-\infty + (+\infty)$.

As usual, '$a + (+\infty)$' is abbreviated to '$a + \infty$' and '$a + (-\infty)$' is abbreviated to '$a - \infty$'. We also write, for instance, '$\left\{\sum_{i=1}^n a_n \text{ w.d.} : a_1 \in A_1, \dots, a_n \in A_n\right\}$' as an abbreviation for the set '$\left\{\sum_{i=1}^n a_n : \sum_{i=1}^n a_n \text{ well defined, and } a_1 \in A_1, \dots, a_n \in A_n\right\}$', where A_1, …, A_n are subsets of $\mathbb{R}^*$.

The **well-defined sum** of two subsets A and B of $\mathbb{R}^*$ is defined by

$$A + B := \{a + b \text{ w.d.} : a \in A \text{ and } b \in B\}.$$

Definition D.2 *The multiplication '×' on $\mathbb{R}$ is extended to $\mathbb{R}^*$ as follows:*

$$-\infty \times -\infty = +\infty \times +\infty = +\infty, \quad -\infty \times +\infty = +\infty \times -\infty = -\infty,$$

$$a \times -\infty = -\infty \times a = -\infty, \qquad a \times +\infty = +\infty \times a = +\infty, \quad \text{if } a > 0,$$

Lower Previsions, First Edition. Matthias C.M. Troffaes and Gert de Cooman.

$$a \times -\infty = -\infty \times a = +\infty, \qquad a \times +\infty = +\infty \times a = -\infty, \quad \textit{if } a < 0,$$

$$a \times -\infty = -\infty \times a = 0, \qquad a \times +\infty = +\infty \times a = 0, \quad \textit{if } a = 0.$$

Definition D.3 *The ordering* $\leq$ *on* $\mathbb{R}$ *is extended to* $\mathbb{R}^*$ *by defining* $-\infty \leq a$ *and* $a \leq +\infty$ *for any* $a \in \mathbb{R}^*$.

Definition D.4 *The equality relation* $=$ *on* $\mathbb{R}$ *is extended to* $\mathbb{R}^*$ *by defining* $-\infty = -\infty$ *and* $+\infty = +\infty$.

We write '$x = \pm\infty$*' as an abbreviation of '*$x = -\infty$ *or* $x = +\infty$*', where* x *denotes any extended real number.*

D.2 Properties

Proposition D.5 *The addition '+' on* $\mathbb{R}^*$ *is commutative and associative, the multiplication '×' on* $\mathbb{R}^*$ *is commutative and associative.*

Proof. Immediate. □

Lemma D.6 *For any non-zero real number* λ*, and any extended real numbers* $a_1, \ldots, a_n$*,* $\sum_{i=1}^n a_i$ *is well defined if and only if* $\sum_{i=1}^n \lambda a_i$ *is well defined, and in that case,*

$$\lambda \sum_{i=1}^{n} a_i = \sum_{i=1}^{n} \lambda a_i.$$

Proof. Immediate. □

Lemma D.7 *For any extended real numbers* $a_1, \ldots, a_n, b_1, \ldots, b_n$ *such that* $a_1 \geq b_1, \ldots, a_n \geq b_n$*, we have that* $\sum_{i=1}^n a_i \geq \sum_{i=1}^n b_i$ *whenever both sums are well defined.*

Proof. Assume that $a_1 \geq b_1, \ldots, a_n \geq b_n$ and that both $\sum_{i=1}^n a_i$ and $\sum_{i=1}^n b_i$ are well defined. The inequality $\sum_{i=1}^n a_i \geq \sum_{i=1}^n b_i$ obviously holds if all a_i and b_i are real valued.

If $a_j = +\infty$ for some j, then $\sum_{i=1}^n a_i = +\infty$ and the inequality is trivially satisfied. Similarly, if $b_j = -\infty$ for some j, then $\sum_{i=1}^n b_i = -\infty$ and the inequality is again trivially satisfied.

If $a_j = -\infty$ for some j, then b_j must be $-\infty$ as well because $a_j \geq b_j$, and we have already considered this case earlier. If $b_j = +\infty$ for some j, then a_j must be $+\infty$ as well because $a_j \geq b_j$, and again, we have already considered this case earlier.

In summary, the inequality $\sum_{i=1}^n a_i \geq \sum_{i=1}^n b_i$ holds in all possible cases. □

For the following lemma, we refer to the material in the beginning of Section 1.5_7 for a discussion of nets of extended real numbers and their convergence.

Lemma D.8 *Let* $(\mathscr{A}, \leq)$ *be a directed set. Let* x_α *be a net of extended real numbers converging to an extended real number* x*. Let* y *be an extended real number.*

(i) *If* $x > y$, *then there is some* $\alpha^* \in \mathscr{A}$ *such that* $x_\alpha > y$ *for all* $\alpha \in \mathscr{A}$ *such that* $\alpha \geq \alpha^*$.

(ii) *If* $x < y$, *then there is some* $\alpha^* \in \mathscr{A}$ *such that* $x_\alpha < y$ *for all* $\alpha \in \mathscr{A}$ *such that* $\alpha \geq \alpha^*$.

Proof. (i). If x is a real number, then the convergence of x_α to x means that, for every real number $\varepsilon > 0$, there is some α_ε such that $|x - x_\alpha| < \varepsilon$ for all $\alpha \geq \alpha_\varepsilon$. Take any ε satisfying $0 < \varepsilon < x - y$, then, for this choice of ε, it holds that $x - x_\alpha < \varepsilon < x - y$ and, hence, $x_\alpha > y$, for all $\alpha \geq \alpha_\varepsilon$.

If $x = +\infty$, then the convergence of x_α to x means that, for every real number $R > 0$, there is an α_R such that $x_\alpha > R$ for all $\alpha \geq \alpha_R$. Take any real number R satisfying $R > y$ [there is such an R because $y \neq +\infty$ since $x > y$ by assumption], then $x_\alpha > y$ for all $\alpha \geq \alpha_R$.

If $x = -\infty$, then there can be no extended real number y such that $x > y$, so there is nothing to prove in this case.

(ii). As $-x > -y$ and $-x_\alpha \to -x$, it follows from (i) that there is some α^* such that $-x_\alpha > -y$ and, hence, $x_\alpha < y$ for all $\alpha \geq \alpha^*$. □

Lemma D.9 *Let a, b, c and d be sums of extended real numbers. Let A and B be subsets of* $\mathbb{R}^*$. *The following statements hold:*

(i) *If* $a + b$ *is well defined, then a and b are well defined. Conversely, if a or b are not well defined , then* $a + b$ *is also not well defined.*

(ii) *If a and b are well defined and* $a - b$ *is not well defined, then it can only be that* $a = b = \pm\infty$.

(iii) *'*$a + b - a$ *well defined' implies '*$a \in \mathbb{R}$*, b well defined and* $a + b - a = b$*'. Conversely, '*$a \in \mathbb{R}$ *and b well defined' implies '*$a + b - a$ *well defined and* $a + b - a = b$*'.*

(iv) *'*$a \geq b + c$ *whenever a and* $b + c$ *are well defined' is equivalent to '*$a - b \geq c$ *whenever* $a - b$ *and c are well defined'. Hence, also '*$a + b \geq c$ *whenever* $a + b$ *and c are well defined' is equivalent to '*$a \geq c - b$ *whenever a and* $c - b$ *are well defined'.*

(v) *If '*$c \geq d$ *whenever c and d are well defined', then '*$a \geq b + c$ *whenever a and* $b + c$ *are well defined' implies that '*$a \geq b + d$ *whenever a and* $b + d$ *are well defined'.*

(vi) *'*$a = b + c$ *whenever a and* $b + c$ *are well defined' is equivalent to '*$a - b = c$ *whenever* $a - b$ *and c are well defined'.*

(vii) $\sup A = \sup(A \setminus \{-\infty\})$ *and* $\inf A = \inf(A \setminus \{+\infty\})$.

(viii) $\sup(A + B) = \sup A + \sup B$ *whenever the right-hand side is well defined.*

Proof. (i) and (ii). Immediate from the definition of being well defined.

(iii). If $a + b - a$ is well defined, then by (i) a and b are well defined. Also, a must be a real number because $a + b - a$ would reduce to $+\infty - \infty$ otherwise. In all three possible cases, namely, that $b = \pm\infty$, or b real, the equality follows. The other implication is proved in a similar way.

(iv)$_\frown$. If a, b or c are not well defined, then the equivalence is trivial by (i)$_\frown$. Therefore, we can assume without loss of generality that a, b and c are well defined.

Assume that $a - b$ is well defined. We show that under the assumption '$b + c$ well defined implies $a \geq b + c$', $a - b \geq c$ holds.

If $b + c$ is not well defined, then there are the following possible cases:

(Ia) $b = -c = -\infty$. In this case, $a - b = +\infty$ and $c = +\infty$ too, so $a - b \geq c$ holds.

(Ib) $b = -c = +\infty$. In this case, $a - b = -\infty$ and $c = -\infty$ too, so $a - b \geq c$ holds.

If, on the other hand, $b + c$ is well defined, then we know that $a \geq b + c$. We now consider the following possible cases:

(IIa) $(b + c) - b$ not well defined, $b + c = b = +\infty$. We have that $a \geq b + c = +\infty$, which implies that also $a = +\infty$, so $a - b$ is not well defined; we have reached a contradiction, which means that this case cannot occur.

(IIb) $(b + c) - b$ not well defined, $b + c = b = -\infty$. As $a - b$ is well defined by assumption, $a > -\infty$, and therefore, $a - b = +\infty$ and $a - b \geq c$ holds.

(IIc) $(b + c) - b$ well defined. Then $b \in \mathbb{R}$. If $c = -\infty$, then the inequality $a - b \geq c$ is obvious. If $c = +\infty$, then $a \geq b + c = +\infty$, so $a = +\infty$ and the inequality $a - b \geq c$ holds. If $c \in \mathbb{R}$ and $a = -\infty$, then we have a contradiction because $a \geq b + c$, so this case cannot occur. If $c \in \mathbb{R}$ and $a \in \mathbb{R}$, then the inequality $a - b \geq c$ follows from the usual real calculus. Finally, if $c \in \mathbb{R}$ and $a = +\infty$, then the inequality $a - b \geq c$ is obvious.

Conversely, assume that $b + c$ is well defined. We show that under the assumption '$a - b$ well defined implies $a - b \geq c$', $a \geq b + c$ holds.

If $a - b$ is not well defined, then there are the following possible cases:

(Ia) $a = b = +\infty$. In this case, $b + c = +\infty$ and $a = +\infty$ too, so $a \geq b + c$ holds.

(Ib) $a = b = -\infty$. In this case, $b + c = -\infty$ and $a = -\infty$ too, so $a \geq b + c$ holds.

If, on the other hand, $a - b$ is well defined, then we know that $a - b \geq c$. We now consider the following possible cases:

(IIa) $(a - b) + b$ not well defined, $a - b = -b = -\infty$. We have that $-\infty = a - b \geq c$, which implies that also $c = -\infty$, so $b + c$ is not well defined; we have reached a contradiction, which means that this case cannot occur.

(IIb) $(a - b) + b$ not well defined, $a - b = -b = +\infty$. As $b + c$ is well defined by assumption, $c < +\infty$ and, therefore, $b + c = -\infty$ and $a \geq b + c$ holds.

(IIc) $(a - b) + b$ well defined. Then $b \in \mathbb{R}$. If $a = +\infty$, then the inequality $a \geq b + c$ is obvious. If $a = -\infty$,then $-\infty = a - b \geq c$, so $c = -\infty$, hence $b + c = -\infty$, and the inequality $a \geq b + c$ holds. If $a \in \mathbb{R}$ and $c = +\infty$, then we have a contradiction because $a - b \geq c$ so this cannot occur. If $a \in \mathbb{R}$ and $c \in \mathbb{R}$, then the inequality $a \geq b + c$ follows from the usual real calculus. Finally, if $a \in \mathbb{R}$ and $c = -\infty$, then the inequality $a \geq b + c$ is obvious.

The second equivalence follows simply by replacing b by $-b$ in the first equivalence and using commutativity of the addition on $\mathbb{R}^*$.

(v)$_{393}$. Assume that c and d are well defined. If a or b is not well defined then the equivalence is trivial by (i)$_{393}$. Therefore, we can assume without loss of generality that a and b are well defined.

If $b = -\infty$ or $c = -\infty$, then $b = -\infty$ or $d = -\infty$, and the statement trivially holds. Without loss of generality, we may thus assume that b and c are strictly larger than $-\infty$. In particular, we only need to consider cases in which $b + c$ is well defined.

If $b = +\infty$ or $c = +\infty$, then $a = +\infty$ (whenever $b + c$ is well defined, which is the case) and the statement holds trivially. If both b and c are real, again the statement trivially holds, even though d may be $-\infty$.

(vi)$_{393}$. If a, b or c are not well defined, then the equivalence is trivial by (i)$_{393}$. Therefore, we can assume without loss of generality that a, b and c are well defined.

The equivalence follows from (iv)$_{393}$. Indeed,

$$
\begin{aligned}
&(b + c \text{ well defined} \Rightarrow a = b + c)\\
\Leftrightarrow&(b + c \text{ well defined} \Rightarrow (a \geq b + c \text{ and } b + c \geq a))\\
\Leftrightarrow&(b + c \text{ well defined} \Rightarrow a \geq b + c) \text{ and } (b + c \text{ well defined} \Rightarrow b + c \geq a)\\
\Leftrightarrow&(a - b \text{ well defined} \Rightarrow a - b \geq c) \text{ and } (a - b \text{ well defined} \Rightarrow c \geq a - b)\\
\Leftrightarrow&(a - b \text{ well defined} \Rightarrow (a - b \geq c \text{ and } c \geq a - b))\\
\Leftrightarrow&(a - b \text{ well defined} \Rightarrow a - b = c).
\end{aligned}
$$

(vii)$_{393}$. This follows from the fact that $\sup \emptyset = -\infty$ and $\inf \emptyset = +\infty$.

(viii)$_{393}$. If $\sup A = -\infty$, then (a) $A = \emptyset$, in which case $A + B = \emptyset$, or (b) $A = \{-\infty\}$, in which case $A + B = \emptyset$ or $A + B = \{-\infty\}$. So the proposition holds if $\sup A = -\infty$ or $\sup B = -\infty$ (by commutativity of the addition and symmetry).

If $\sup A = +\infty$, then B must contain extended real numbers strictly larger than $-\infty$ (otherwise $\sup A + \sup B$ would not be well defined). For any such number $b \in B$, $b > -\infty$, we have that $\sup A + b = +\infty$, whence $\sup \{a + b : a \in A, b \in B, a, b > -\infty\} = +\infty$ and, consequently, $\sup \{a + b : a\} \in A,\ b \in B, a + b$ well defined $= \sup(A + B) = +\infty$. So the proposition holds if $\sup A = +\infty$ or $\sup B = +\infty$ (by commutativity of the addition and symmetry).

If $\sup A$ and $\sup B$ are real numbers, then the property follows from the continuity of the addition. □

Appendix E

Symbols and notation

Symbol	Meaning	Where Defined
X	Random variable	Section 3.1_{26}
$\mathscr{X}$	Set of values of X	Section 3.1_{26}
x	An element of $\mathscr{X}$	Section 3.1_{26}
f, g, h	gamble (mapping from $\mathscr{X}$ to $\mathbb{R}$)	Section 1.2_{2}
f_a^b	(a, b)-cut of a gamble f	Definition 14.32_{321}
$\mathbb{B}$	Set of bounded gambles on $\mathscr{X}$	Section 1.2_{2}
$\mathbb{B}_{\geq 0}$	Set of non-negative bounded gambles	Section 1.2_{2}
$\mathbb{G}$	Set of gambles on $\mathscr{X}$	Section 1.2_{2}
$\mathbb{G}_{\geq 0}$	Set of non-negative gambles	Section 1.2_{2}
$\operatorname{nonneg}(\mathscr{K})$	Convex cone spanned by $\mathscr{K}$	Definition 1.1_{4}
$\operatorname{span}(\mathscr{K})$	Linear span of $\mathscr{K}$	Definition 1.1_{4}
$A, B, \dots$	Event (subset of $\mathscr{X}$) or indicator	
I_A	Indicator of A	Section 1.3_{5}
A^c	Complement of A	
$\mathscr{P}$	Power set of $\mathscr{X}$	
$A \subseteq B$	A is a subset of B	
$A \Subset B$	A is a finite subset of B	
$\mathscr{F}$	Field or σ-field	Definition 1.6_{6}
$\mathscr{B}$	Borel σ-field of a topological space $\mathscr{X}$	Definition 1.7_{6}
$\mathscr{B}_0$	Baire σ-field of a topological space $\mathscr{X}$	Definition 1.8_{6}
$\mathscr{F}_\mu^C$	Carathéodory field of a probability charge μ	Definition 8.19_{164}
$\mathscr{F}_\mu^J$	Jordan field of a probability charge μ	Definition 8.19_{164}
$\mathbb{N}$	$\{0, 1, 2, 3, \dots\}$ (set of natural numbers)	Section 1.1_{1}
$\mathbb{N}_{>0}$	$\mathbb{N} \setminus \{0\}$	Section 1.1_{1}
$\mathbb{Z}$	Set of integer numbers	Section 1.1_{1}
$\mathbb{Q}$	Set of rational numbers	Section 1.1_{1}

Lower Previsions, First Edition. Matthias C.M. Troffaes and Gert de Cooman.

Symbol	Meaning	Where Defined
$\mathbb{R}$	Set of real numbers	Section 1.1_1
$\mathbb{R}_{\geq 0}$	Set of non-negative real numbers	Section 1.1_1
$\mathbb{R}_{>0}$	Set of positive real numbers	Section 1.1_1
$\mathbb{R}^*$	$\mathbb{R} \cup \{-\infty, +\infty\}$ (set of extended real numbers)	Section 1.1_1
$(\mathscr{A}, \leq)$	Directed set	Section 1.5_7
x_n	Sequence (map from $(\mathbb{N}, \leq)$ to a set)	Section 1.5_7
x_α	Net (map from $(\mathscr{A}, \leq)$ to a set)	Section 1.5_7
$\mathscr{D}$	Set of acceptable bounded gambles	Section 3.4_{29}
	Set of acceptable gambles	Section 13.2_{236}
$\mathbb{D}_b$	Set of coherent sets of acceptable bounded gambles	Definition 3.2_{31}
$\mathbb{D}$	Set of coherent sets of acceptable gambles	Definition 13.2_{237}
$\mathrm{Cl}_{\mathbb{D}_b}$	Natural extension for sets of bounded gambles	Section $3.4.2_{32}$
$\mathrm{Cl}_{\mathbb{D}}$	Natural extension for sets of gambles	Section $13.2.2_{238}$
lpr, upr	Lower/upper prevision operator	Section $4.1.1_{38}$
		Section $13.3.1_{240}$
	Duality map (lower/upper envelope)	Definition 4.36_{71}
$\underline{P}$, $\overline{P}$, P	Lower/upper prevision, prevision	Section $4.1.2_{40}$
		Section $13.3.2_{252}$
lins	Duality map (set of linear previsions)	Definition 4.36_{71}
$\underline{E}_{\underline{P}}$	Natural extension of $\underline{P}$	Definition 4.8_{47}
$E_{\underline{P}}$	Linear extension of $\underline{P}$	Definition 8.1_{153}
$\mathscr{I}_{\underline{P}}$	Set of $\underline{P}$-integrable bounded gambles	Definition 8.1_{153}
$\mathscr{N}_{\underline{P}}$	Set of $\underline{P}$-null sets	Definition 14.2_{305}
$\mathbb{G}^0_{\underline{P}}$	Set of $\underline{P}$-null gambles	Definition 14.6_{307}
$\mathbb{G}^\sharp_{\underline{P}}$	Set of $\underline{P}$-essentially bounded gambles	Definition 14.16_{311}
$\mathrm{ess\,sup}_{\underline{P}}$	Essential supremum	Definition 14.17_{312}
$\mathrm{ess\,inf}_{\underline{P}}$	Essential infimum	Definition 14.17_{312}
$\lVert\bullet\rVert_{\underline{P},\infty}$	Essential supremum norm	Definition 14.17_{312}
$f_\alpha \xrightarrow{\underline{P}} f$	Convergence in probability	Definition 15.1_{329}
$\mathbb{G}^{\mathrm{x}}_{\underline{P}}$	Set of $\underline{P}$-previsible gambles	Definition 15.6_{333}
$\lVert\bullet\rVert^{\mathrm{x}}_{\underline{P}}$	$\underline{P}$-previsible seminorm	Definition 15.12_{335}

References

Agnew, R.P. and Morse, A.P. (1938) Extensions of linear functionals, with applications to limits, integrals, measures and densities. *Annals of Mathematics*, **39**, 20–30.

Aigner, M. (1997) *Combinatorial Theory*, Springer, Berlin. Reprint of the 1979 edition. Originally published as Vol. 234 of the 'Grundlehren der mathematischen Wissenschaften'.

Alaoglu, L. and Birkhoff, G. (1940) General ergodic theorems. *Annals of Mathematics*, **41**, 293–309.

Anscombe, F.J. and Aumann, R.J. (1963) A definition of subjective probability. *Annals of Mathematical Statistics*, **34** (1), 199–205.

Bernoulli, J. (1713) *Ars Conjectandi*, Thurnisiorum, Basel.

Bhaskara Rao, K. and Bhaskara Rao, M. (1983) *Theory of Charges: A Study of Finitely Additive Measures*, Academic Press, London.

Birkhoff, G. (1995) *Lattice Theory*, A.M.S. Colloquium Publications, Providence, RI. Eight printing with corrections of the third edition, first published in 1967.

Bishop, E. and De Leeuw, K. (1959) The representations of linear functionals by measures on sets of extreme points. *Annales de l'Institut Fourier*, **9**, 305–331.

Bollobás, B. (1999) *Linear Analysis: An Introductory Course*, 2nd edn, Cambridge University Press, Cambridge.

Boole, G. (1854) *An Investigation of the Laws of Thought on Which are Founded the Mathematical Theories of Logic and Probabilities*, Walton and Maberly, London.

Boyd, S. and Vandenberghe, L. (2004) *Convex Optimization*, Cambridge University Press, Cambridge.

Chevé, M. and Congar, R. (2000) Optimal pollution control under imprecise environmental risk and irreversibility. *Risk Decision and Policy*, **5**, 151–164.

Choquet, G. (1953–1954) Theory of capacities. *Annales de l'Institut Fourier*, **5**, 131–295.

Clemen, R.T. and Reilly, T. (2001) *Making Hard Decisions*, Duxbury.

Couso, I. and Moral, S. (2009) Sets of desirable gambles and credal sets, in *ISIPTA '09: Proceedings of the 6th International Symposium on Imprecise Probabilities: Theories and Applications* (eds T. Augustin, F.P.A. Coolen, S. Moral and M.C.M. Troffaes), SIPTA, Durham, pp. 99–108.

Couso, I. and Moral, S. (2011) Sets of desirable gambles: conditioning, representation, and precise probabilities. *International Journal of Approximate Reasoning*, **52** (7), 1034–1055.

Lower Previsions, First Edition. Matthias C.M. Troffaes and Gert de Cooman.

Couso, I., Moral, S. and Walley, P. (2000) Examples of independence for imprecise probabilities. *Risk Decision and Policy*, **5**, 165–181.

Crisma, L., Gigante, P. and Millossovich, P. (1997a) A notion of coherent prevision for arbitrary random numbers. *Quaderni del Dipartimento di Matematica Applicata alle Scienze Economiche Statistiche e Attuariali 'Bruno de Finetti'*, **6**, 233–243.

Crisma, L., Gigante, P. and Millossovich, P. (1997b) A notion of coherent prevision for arbitrary random quantities. *Journal of the Italian Statistical Society*, **6** (3), 233–243.

Darboux, G. (1875) Mémoire sur les fonctions discontinues. *Annales Scientifiques de l'École Normale Supérieure (2e Série)*, **IV**, 57–112.

Davey, B.A. and Priestley, H.A. (1990) *Introduction to Lattices and Order*, Cambridge University Press, Cambridge.

Day, M.M. (1942) Ergodic theorems for Abelian semi-groups. *Transactions of the American Mathematical Society*, **51**, 399–412.

De Bock, J. and De Cooman, G. (2013) Extreme lower previsions and Minkowski indecomposability, in *Symbolic and Quantitative Approaches to Reasoning with Uncertainty*, Lecture Notes in Computer Science (ed. L.C. vander Gaag), Springer, pp. 157–168.

de Campos, L.M., Huete, J.F. and Moral, S. (1994) Probability intervals: a tool for uncertain reasoning. *International Journal of Uncertainty Fuzziness and Knowledge-Based Systems*, **2**, 167–196.

De Cooman, G. (1997) Possibility theory I–III. *International Journal of General Systems*, **25**, 291–371.

De Cooman, G. (2000) Integration in possibility theory, in *Fuzzy Measures and Integrals – Theory and Applications* (eds M. Grabisch, T. Murofushi and M. Sugeno), Physica-Verlag (Springer), Heidelberg, pp. 124–160.

De Cooman, G. (2001) Integration and conditioning in numerical possibility theory. *Annals of Mathematics and Artificial Intelligence*, **32**, 87–123.

De Cooman, G. (2005a) A behavioural model for vague probability assessments. *Fuzzy Sets and Systems*, **154**, 305–358. With discussion.

De Cooman, G. (2005b) Belief models: an order-theoretic investigation. *Annals of Mathematics and Artificial Intelligence*, **45**, 5–34.

De Cooman, G. and Aeyels, D. (1999) Supremum preserving upper probabilities. *Information Sciences*, **118**, 173–212.

De Cooman, G. and Miranda, E. (2007) Symmetry of models versus models of symmetry, in *Probability and Inference: Essays in Honor of Henry E. Kyburg, Jr.* (eds W.L. Harper and G.R. Wheeler), King's College Publications, pp. 67–149.

De Cooman, G. and Miranda, E. (2008) The F. Riesz Representation Theorem and finite additivity, in *Soft Methods for Handling Variability and Imprecision (Proceedings of SMPS 2008)* (eds D. Dubois, M.A. Lubiano, H. Prade, M.A. Gil, P. Grzegorzewski and O. Hryniewicz), Springer, pp. 243–252.

De Cooman, G. and Miranda, E. (2009) Forward irrelevance. *Journal of Statistical Planning and Inference*, **139** (2), 256–276.

De Cooman, G. and Quaeghebeur, E. (2009) Exchangeability for sets of desirable gambles, in *ISIPTA '09 – Proceedings of the 6th International Symposium on Imprecise Probability: Theories and Applications* (eds T. Augustin, F.P.A. Coolen, S. Moral and M.C.M. Troffaes), SIPTA, Durham, pp. 159–168.

De Cooman, G. and Quaeghebeur, E. (2012) Exchangeability and sets of desirable gambles. *International Journal of Approximate Reasoning*, **53**, 363–395. Special issue in honour of Henry E. Kyburg, Jr.

De Cooman, G. and Troffaes, M.C.M. (2003) Dynamic programming for discrete-time systems with uncertain gain, in *ISIPTA '03 – Proceedings of the 3rd International Symposium on Imprecise Probabilities and Their Applications* (eds J.M. Bernard, T. Seidenfeld and M. Zaffalon), Carleton Scientific, pp. 162–176.

De Cooman, G. and Troffaes, M.C.M. (2004) Coherent lower previsions in systems modelling: products and aggregation rules. *Reliability Engineering & System Safety*, **85**, 113–134.

De Cooman, G., Miranda, E. and Zaffalon, M. (2011) Independent natural extension. *Artificial Intelligence*, **175**, 1911–1950.

De Cooman, G., Quaeghebeur, E. and Miranda, E. (2009) Exchangeable lower previsions. *Bernoulli*, **15** (3), 721–735.

De Cooman, G., Troffaes, M.C.M. and Miranda, E. (2005a) *n*-Monotone lower previsions. *Journal of Intelligent & Fuzzy Systems*, **16** (4), 253–263.

De Cooman, G., Troffaes, M.C.M. and Miranda, E. (2005b) *n*-Monotone lower previsions and lower integrals, in *Proceedings of the 4th International Symposium on Imprecise Probabilities and Their Applications* (eds F.G. Cozman, R. Nau and T. Seidenfeld), SIPTA, pp. 145–154.

De Cooman, G., Troffaes, M.C.M. and Miranda, E. (2008a) *n*-Monotone exact functionals. *Journal of Mathematical Analysis and Applications*, **347**, 143–156.

De Cooman, G., Troffaes, M.C.M. and Miranda, E. (2008b) A unifying approach to integration for bounded positive charges. *Journal of Mathematical Analysis and Applications*, **340** (2), 982–999.

de Finetti, B. (1937) La prévision: ses lois logiques, ses sources subjectives. *Annales de l'Institut Henri Poincaré*, **7**, 1–68.

de Finetti, B. (1970) *Teoria delle Probabilità*, Einaudi, Turin.

de Finetti, B. (1974–1975) *Theory of Probability: A Critical Introductory Treatment*, John Wiley & Sons, Inc., New York. English translation of de Finetti (1970), two volumes.

DeGroot, M.H. and Schervisch, M.J. 2011 *Probability and Statistics*, 4th edn. Pearson.

Dellacherie, C. (1971) Quelques commentaires sur les prolongements de capacités. *Séminaire de probabilités (Strasbourg)*, **5**, 77–81.

Dempster, A.P. (1967a) Upper and lower probabilities generated by a random closed interval. *Annals of Mathematical Statistics*, **39**, 957–966.

Dempster, A.P. (1967b) Upper and lower probabilities induced by a multivalued mapping. *Annals of Mathematical Statistics*, **38**, 325–339.

Denneberg, D. (1994) *Non-Additive Measure and Integral*, Kluwer, Dordrecht.

Destercke, S. and Dubois, D. (2006) A unified view of some representations of imprecise probabilities, in *Soft Methods in Probability for Integrated Uncertainty Modelling*, Advances in Soft Computing (eds J. Lawry, E. Miranda, A. Bugarin, S. Li, M.Á. Gil, P. Grzegorzewski and O. Hryniewicz), Springer, pp. 249–257.

Destercke, S., Dubois, D. and Chojnacki, E. (2008) Unifying practical uncertainty representations: I. Generalized p-boxes. *International Journal of Approximate Reasoning*, **49** (3), 649–663.

Dubois, D. and Prade, H. (1988) *Possibility Theory – An Approach to Computerized Processing of Uncertainty*, Plenum Press, New York.

Dubois, D. and Prade, H. (1992) When upper probabilities are possibility measures. *Fuzzy Sets and Systems*, **49**, 65–74.

Dunford, N. (1935) Integration in general analysis. *Transactions of the American Mathematical Society*, **37** (3), 441–453.

Dunford, N. and Schwartz, J.T. (1957) *Linear Operators*, John Wiley & Sons, Inc., New York.

Ferson, S., Kreinovich, V., Ginzburg, L., Myers, D.S. and Sentz, K. (2003) Constructing probability boxes and Dempster-Shafer structures. Technical Report SAND2002–4015, Sandia National Laboratories.

Ferson, S. and Tucker, W. (2006) Sensitivity analysis using probability bounding. *Reliability Engineering & System Safety*, **91** (10-11), 1435–1442.

Good, I.J. (1950) *Probability and the Weighing of Evidence*, Griffin, London.

Gould, G.G. (1965) Integration over vector-valued measures. *Proceedings of the London Mathematical Society*, **15** (3), 193–225.

Greco, G.H. (1981) Sur la mesurabilité d'une fonction numérique par rapport à une famille d'ensembles. *Rendiconti del Seminario Matematico della Università di Padova*, **65**, 163–176.

Greco, G.H. (1982) Sulla rappresentazione di funzionali mediante integrali. *Rendiconti del Seminario Matematico della Università di Padova*, **66**, 21–42.

Hacking, I. (1975) *The Emergence of Probability: A Philosophical Study of Early Ideas about Probability, Induction and Statistical Inference*, Cambridge University Press.

Halmos, P. (1974) *Measure Theory*, Springer, New York.

Hermans, F. (2012) An operational approach to graphical uncertainty modelling. PhD thesis, Faculty of Engineering and Architecture.

Herstein, I.N. and Milnor, J. (1953) An axiomatic approach to measurable utility. *Econometrica*, **21** (2), 291–297.

Hewitt, E. and Savage, L.J. (1955) Symmetric measures on Cartesian products. *Transactions of the American Mathematical Society*, **80**, 470–501.

Hildebrandt, T.H. (1934) On bounded functional operations. *Transactions of the American Mathematical Society*, **36** (4), 868–875.

Hildebrandt, T.H. (1963) *Introduction to the Theory of Integration*, Academic Press, London.

Holmes, R. (1975) *Geometric Functional Analysis and Its Applications*, Springer, New York.

Huygens, C. (1657) De ratiociniis in ludo aleæ, in *Exercitationum Mathematicarum Libri Quinque: Quibus accedit Christiani Hugenii Tractatus de Ratiociniis in Aleæ Ludo* (ed. Fá. Schooten), Ex officina Johannis Elsevirii Lugd, Batav, pp. 517–524.

Janssen, H. (1999) Een ordetheoretische en behavioristische studie van possibilistische processen. PhD thesis, Ghent University.

Jaynes, E.T. (2003) *Probability Theory: The Logic of Science*, Cambridge University Press.

Johnson, N.L., Kotz, S. and Balakrishnan, N. (1997) *Discrete Multivariate Distributions*, Wiley Series in Probability and Statistics, John Wiley & Sons, Inc., New York.

Kahneman, D. and Tversky, A. (1979) Prospect theory: an analysis of decision under risk. *Econometrica*, **47** (2), 263–291.

Kallenberg, O. (2002) *Foundations of Modern Probability (Probability and Its Applications)*, 2nd edn. Springer.

Kallenberg, O. (2005) *Probabilistic Symmetries and Invariance Principles*, Springer, New York.

Keynes, J.M. (1921) *A Treatise on Probability*, Macmillan, London.

Kolmogorov, A. (1930) Untersuchungen über den Integralbegriff. *Mathematische Annalen*, **103**, 654–696.

König, H. (1997) *Measure and Integration: An Advanced Course in Basic Procedures and Applications*, Springer, Berlin.

Kriegler, E. and Held, H. (2005) Utilizing belief functions for the estimation of future climate change. *International Journal of Approximate Reasoning*, **39**, 185–209.

Krylov, N. and Bogolioubov, N. (1937) La théorie générale de la mesure dans son application à l'étude des systèmes dynamiques de la mécanique non linéaire. *Annals of Mathematics (2)*, **38** (1), 65–113.

Laplace, P.S. (1951) *Philosophical Essay on Probabilities*, Dover Publications. English translation of Laplace (1986).

Lebesgue, H. (1904) *Leçons sur l'intégration et la recherche des fonctions primitives*, Gauthier-Villars, Paris.

Levi, I. (1983) *The Enterprise of Knowledge: An Essay on Knowledge, Credal Probability, and Chance*, MIT Press, Cambridge.

Lindley, D.V. (1982) Scoring rules and the inevitability of probability. *International Statistical Review*, **50**, 1–26. With discussion.

Lindley, D.V. (1987) The probability approach to the treatment of uncertainty in artificial intelligence and expert systems. *Statistical Science*, **2**, 3–44. With discussion.

Luxemburg, W.A.J. (1962) Two applications of the method of construction by ultrapowers to analysis. *Bulletin of the American Mathematical Society*, **68**, 416–419.

Maaß, S. (2002) Exact functionals and their core. *Statistical Papers*, **43**, 75–93.

Maaß, S. (2003) Exact functionals, functionals preserving linear inequalities, Lévy's metric. PhD thesis, University of Bremen.

Marinacci, M. and Montrucchio, L. (2008) On concavity and supermodularity. *Journal of Mathematical Analysis and Applications*, **344** (2), 642–654.

Miranda, E. and De Cooman, G. (2007) Marginal extension in the theory of coherent lower previsions. *International Journal of Approximate Reasoning*, **46** (1), 188–225.

Miranda, E., Couso, I. and Gil, P. (2010) Approximation of upper and lower probabilities by measurable selections. *Information Sciences*, **180** (8), 1407–1417.

Miranda, E., De Cooman, G. and Couso, I. (2005) Lower previsions induced by multi-valued mappings. *Journal of Statistical Planning and Inference*, **133** (1), 173–197.

Miranda, E., De Cooman, G. and Quaeghebeur, E. (2007) The Hausdorff moment problem under finite additivity. *Journal of Theoretical Probability*, **20** (3), 663–693.

Miranda, E., De Cooman, G. and Quaeghebeur, E. (2008a) Finitely additive extensions of distribution functions and moment sequences: the coherent lower prevision approach. *International Journal of Approximate Reasoning*, **48**, 132–155.

Miranda, E., Troffaes, M.C.M. and Destercke, S. (2008b) Generalised p-boxes on totally ordered spaces, in *Soft Methods for Handling Variability and Imprecision*, Advances in Soft Computing (eds D. Dubois, M.A. Lubiano, H. Prade, M.A. Gil, P. Grzegorzewski and O. Hryniewicz), Springer, pp. 235–242.

Moore, E.H. and Smith, H.L. (1922) A general theory of limits. *American Journal of Mathematics*, **44** (2), 102–121.

Moral, S. (2000) Epistemic irrelevance on sets of desirable gambles, in *ISIPTA '01 – Proceedings of the 2nd International Symposium on Imprecise Probabilities and Their Applications* (eds G. DeCooman, T.L. Fine and T. Seidenfeld), Shaker Publishing Maastricht, pp. 247–254.

Moral, S. (2005) Epistemic irrelevance on sets of desirable gambles. *Annals of Mathematics and Artificial Intelligence*, **45**, 197–214.

Nelson, E. (1987) *Radically Elementary Probability Theory*, Annals of Mathematical Studies, Princeton University Press, Princeton, NJ.

Neumaier, A. (2004) Clouds, fuzzy sets and probability intervals. *Reliable Computing*, **10**, 249–272.

Nguyen, H.T. (2006) *An Introduction to Random sets*, Chapman & Hall/CRC, Boca Raton, FL.

Nguyen, H.T., Nguyen, N.T. and Wang, T. (1997) On capacity functionals in interval probabilities. *International Journal of Uncertainty Fuzziness and Knowledge-Based Systems*, **5** (3), 359–377.

Pascal, B. (2001) *Pensées*, Maxi-Livres, Paris, Unfinished work, published posthumously from collected fragments. First incomplete edition: Port-Royal, 1670. First complete reproduction: Michaut, Basle, 1896.

Phelps, R.R. (2001) *Lectures on Choquet's Theorem*, Lecture Notes in Mathematics, 2nd edn, Springer.

Pincus, D. (1972) Independence of the prime ideal theorem from the Hahn Banach theorem. *Bulletin of the American Mathematical Society*, **78** (5), 766–770.

Pincus, D. 1974 The strength of the Hahn-Banach theorem, in *Victoria Symposium on Nonstandard Analysis*, Lecture Notes in Mathematics, Vol. **369** (eds A. Hurd and P. Loeb), Springer, pp. 203–248.

Quaeghebeur, E. (2010) Characterizing the set of coherent lower previsions with a finite number of constraints or vertices, in *Proceedings of the 26th Conference on Uncertainty in Artificial Intelligence* (eds P. Grunwald and P. Spirtes), AUAI Press, pp. 466–473.

Quaeghebeur, E. (2013) Desirability To appear as a chapter of the book 'Introduction to Imprecise Probabilities'.

Quaeghebeur, E. and De Cooman, G. (2008) Extreme lower probabilities. *Fuzzy Sets and Systems*, **159** (16), 2163–2175.

Quaeghebeur, E., De Cooman, G. and Hermans, F. (2012) Accept & reject statement-based uncertainty models, Submitted.

Ramsey, F.P. (1931) Truth and probability, in *Foundations of Mathematics and other Logical Essays*, Chapter VII (ed. R.B. Braithwaite), Routledge and Kegan Paul, London, pp. 156–198, pp. Published posthumously.

Riemann, B. (1868) Ueber die Darstellbarkeit einer Function durch eine trigonometrische Reihe. *Abhandlungen der Königlichen Gesellschaft der Wissenschaften zu Göttingen*, **13**, 87–131. Originally delivered in 1854 by Riemann as part of his habilitation at the University of Göttingen.

Rockafellar, R.T. (1970) *Convex Analysis*, Princeton University Press, Princeton, NJ.

Savage, L.J. (1972) *The Foundations of Statistics*, Dover, New York. Second revised edition.

Schechter, E. (1997) *Handbook of Analysis and Its Foundations*, Academic Press, San Diego, CA.

Schmeidler, D. (1986) Integral representation without additivity. *Proceedings of the American Mathematical Society*, **97**, 255–261.

Seidenfeld, T., Schervish, M.J. and Kadane, J.B. (1995) A representation of partially ordered preferences. *Annals of Statistics*, **23**, 2168–2217.

Seidenfeld, T., Schervish, M.J. and Kadane, J.B. (2009) Preference for equivalent random variables: a price for unbounded utilities. *Journal of Mathematical Economics*, **45**, 329–340.

Shackle, G.L.S. (1961) *Decision, Order and Time in Human Affairs*, Cambridge University Press, Cambridge, MA.

Shafer, G. (1976) *A Mathematical Theory of Evidence*, Princeton University Press, Princeton, NJ.

Shafer, G. (1979) Allocations of probability. *Annals of Probability*, **7** (5), 827–839.

Shapley, L.S. (1971) Cores of convex games. *International Journal of Game Theory*, **1**, 11–26.

Smith, C.A.B. (1961) Consistency in statistical inference and decision. *Journal of the Royal Statistical Society, Series B*, **23**, 1–37.

Straßen, V. (1964) Meßfehler und Information. *Zeitschrift für Wahrscheinlichkeitstheorie und Verwandte Gebiete*, **2**, 273–305.

Suppes, P. (1974) The measurement of belief. *Journal of the Royal Statistical Society, Series B-Methodological*, **36** (2), 160–191.

Troffaes, M.C.M. (2005) Optimality, uncertainty, and dynamic programming with lower previsions. PhD thesis, Universiteit Gent Onderzoeksgroep SYSTeMS, Technologiepark Zwijnaarde 914, B-9052 Zwijnaarde, België.

Troffaes, M.C.M. (2006) Conditional lower previsions for unbounded random quantities, in *Soft Methods in Probability for Integrated Uncertainty Modelling*, Advances in Soft Computing (eds J. Lawry, E. Miranda, A. Bugarin, S. Li, M.Á. Gil, P. Grzegorzewski and O. Hryniewicz), Springer, pp. 201–209.

Troffaes, M.C.M. (2007) Decision making under uncertainty using imprecise probabilities. *International Journal Of Approximate Reasoning*, **45** (1), 17–29.

Troffaes, M.C.M. and De Cooman, G. (2002a) Extension of coherent lower previsions to unbounded random variables, in *Proceedings of the 9th International Conference IPMU 2002 (Information Processing and Management of Uncertainty in Knowledge-Based Systems)*, ESIA – Université de Savoie, Annecy, France, pp. 735–42.

Troffaes, M.C.M. and De Cooman, G. (2002b) Lower previsions for unbounded random variables, in *Soft Methods in Probability, Statistics and Data Analysis*, Advances in Soft Computing (eds P. Grzegorzewski, O. Hryniewicz and M.Á. Gil), Physica-Verlag, New York, pp. 146–155.

Troffaes, M.C.M. and De Cooman, G. (2003) Extension of coherent lower previsions to unbounded random variables, in *Intelligent Systems for Information Processing: From Representation to Applications* (eds B. Bouchon-Meunier, L. Foulloy and R.R. Yager), North-Holland, pp. 277–288.

Troffaes, M.C.M. and Destercke, S. (2011) Probability boxes on totally preordered spaces for multivariate modelling. *International Journal of Approximate Reasoning*, **52** (6), 767–791.

Utkin, L.V. and Gurov, S.V. (1999) Imprecise reliability models for the general lifetime distribution classes, In *ISIPTA '99 – Proceedings of the 1st International Symposium on Imprecise Probabilities and Their Applications* (eds G. DeCooman, F.G. Cozman, S. Moral and P. Walley), SIPTA, pp. 333–342.

Vitali, G. (1905) Sul problema della misura dei gruppi di punti di una retta.

von Neumann, J. and Morgenstern, O. (1944) *Theory of Games and Economic Behavior*, Princeton University Press, Princeton, NJ.

von Winterfeldt, D. and Edwards, W. (1986) *Decision Analysis and Behavioral Research*, Cambridge University Press, Cambridge, MA.

Walley, P. (1981) Coherent lower (and upper) probabilities. Technical Report Statistics Research Report 22, University of Warwick, Coventry.

Walley, P. (1991) *Statistical Reasoning with Imprecise Probabilities*, Chapman and Hall, London.

Walley, P. (1997) Statistical inferences based on a second-order possibility distribution. *International Journal of General Systems*, **26**, 337–383.

Walley, P. (2000) Towards a unified theory of imprecise probability. *International Journal of Approximate Reasoning*, **24**, 125–148.

Walley, P. and De Cooman, G. (2001) A behavioral model for linguistic uncertainty. *Information Sciences*, **134**, 1–37.

Wasserman, L.A. (1990a) Belief functions and statistical inference. *Canadian Journal of Statistics-La Revue Canadienne de Statistique*, **18** (3), 183–196.

Wasserman, L.A. (1990b) Prior envelopes based on belief functions. *Annals Of Statistics*, **18** (1), 454–464.

Whittle, P. (1992) *Probability via Expectation*, 3rd edn, Springer, New York. First edition, Penguin, 1970.

Willard, S. (1970) *General Topology*, Addison-Wesley, Reading, MA.

Williams, P.M. (1975a) Coherence, strict coherence and zero probabilities. *Proceedings of the Fifth International Congress on Logic, Methodology and Philosophy of Science, vol. VI Reidel Dordrecht*, pp. 29–33. Proceedings of a 1974 conference held in Warsaw.

Williams, P.M. (1975b) Notes on conditional previsions. Technical report, School of Mathematical and Physical Science, University of Sussex. Revised journal version: Williams (2007).

Williams, P.M. (1976) Indeterminate probabilities, in *Formal Methods in the Methodology of Empirical Sciences* (eds M. Przelecki, K. Szaniawski and R. Wojcicki), Reidel Dordrecht, pp. 229–246. In Proceedings of a 1974 conference held in Warsaw.

Wong, Y.C. and Ng, K.F. (1973) *Partially Ordered Topological Vector Spaces*, Oxford Mathematical Monographs, Clarendon Press, Oxford.

Young, W.H. (1905) A general theory of integration. *Philosophical Transactions of the Royal Society of London A*, **204**, 211–252.

Zadeh, L.A. (1978) Fuzzy sets as a basis for a theory of possibility. *Fuzzy Sets and Systems*, **1**, 3–28.

Index

WILEY SERIES IN PROBABILITY AND STATISTICS

ESTABLISHED BY WALTER A. SHEWHART AND SAMUEL S. WILKS

Editors: *David J. Balding, Noel A. C. Cressie, Garrett M. Fitzmaurice, Geof H. Givens, Harvey Goldstein, Geert Molenberghs, David W. Scott, Adrian F. M. Smith, Ruey S. Tsay, Sanford Weisberg*
Editors Emeriti: *J. Stuart Hunter, Iain M. Johnstone, J. B. Kadane, Jozef L. Teugels*

The ***Wiley Series in Probability and Statistics*** is well established and authoritative. It covers many topics of current research interest in both pure and applied statistics and probability theory. Written by leading statisticians and institutions, the titles span both state-of-the-art developments in the field and classical methods.

Reflecting the wide range of current research in statistics, the series encompasses applied, methodological and theoretical statistics, ranging from applications and new techniques made possible by advances in computerized practice to rigorous treatment of theoretical approaches.

This series provides essential and invaluable reading for all statisticians, whether in academia, industry, government, or research.

† ABRAHAM and LEDOLTER · Statistical Methods for Forecasting
AGRESTI · Analysis of Ordinal Categorical Data, *Second Edition*
AGRESTI · An Introduction to Categorical Data Analysis, *Second Edition*
AGRESTI · Categorical Data Analysis, *Third Edition*
ALSTON, MENGERSEN and PETTITT (editors) · Case Studies in Bayesian Statistical Modelling and Analysis
ALTMAN, GILL, and McDONALD · Numerical Issues in Statistical Computing for the Social Scientist
AMARATUNGA and CABRERA · Exploration and Analysis of DNA Microarray and Protein Array Data
ANDĚL · Mathematics of Chance
ANDERSON · An Introduction to Multivariate Statistical Analysis, *Third Edition*
* ANDERSON · The Statistical Analysis of Time Series
ANDERSON, AUQUIER, HAUCK, OAKES, VANDAELE, and WEISBERG · Statistical Methods for Comparative Studies
ANDERSON and LOYNES · The Teaching of Practical Statistics
ARMITAGE and DAVID (editors) · Advances in Biometry
ARNOLD, BALAKRISHNAN, and NAGARAJA · Records
* ARTHANARI and DODGE · Mathematical Programming in Statistics
AUGUSTIN, COOLEN, DE COOMAN and TROFFAES · Introduction to Imprecise Probabilities
* BAILEY · The Elements of Stochastic Processes with Applications to the Natural Sciences
BAJORSKI · Statistics for Imaging, Optics, and Photonics

*Now available in a lower priced paperback Edition in the Wiley Classics Library.
†Now available in a lower priced paperback Edition in the Wiley–Interscience Paperback Series.

BALAKRISHNAN and KOUTRAS · Runs and Scans with Applications
BALAKRISHNAN and NG · Precedence-Type Tests and Applications
BARNETT · Comparative Statistical Inference, *Third Edition*
BARNETT · Environmental Statistics
BARNETT and LEWIS · Outliers in Statistical Data, *Third Edition*
BARTHOLOMEW, KNOTT, and MOUSTAKI · Latent Variable Models and Factor Analysis: A Unified Approach, *Third Edition*
BARTOSZYNSKI and NIEWIADOMSKA-BUGAJ · Probability and Statistical Inference, *Second Edition*
BASILEVSKY · Statistical Factor Analysis and Related Methods: Theory and Applications
BATES and WATTS · Nonlinear Regression Analysis and Its Applications
BECHHOFER, SANTNER, and GOLDSMAN · Design and Analysis of Experiments for Statistical Selection, Screening, and Multiple Comparisons
BEIRLANT, GOEGEBEUR, SEGERS, TEUGELS, and DE WAAL · Statistics of Extremes: Theory and Applications
BELSLEY · Conditioning Diagnostics: Collinearity and Weak Data in Regression
† BELSLEY, KUH, and WELSCH · Regression Diagnostics: Identifying Influential Data and Sources of Collinearity
BENDAT and PIERSOL · Random Data: Analysis and Measurement Procedures, *Fourth Edition*
BERNARDO and SMITH · Bayesian Theory
BHAT and MILLER · Elements of Applied Stochastic Processes, *Third Edition*
BHATTACHARYA and WAYMIRE · Stochastic Processes with Applications
BIEMER, GROVES, LYBERG, MATHIOWETZ, and SUDMAN · Measurement Errors in Surveys
BILLINGSLEY · Convergence of Probability Measures, *Second Edition*
BILLINGSLEY · Probability and Measure, *Anniversary Edition*
BIRKES and DODGE · Alternative Methods of Regression
BISGAARD and KULAHCI · Time Series Analysis and Forecasting by Example
BISWAS, DATTA, FINE, and SEGAL · Statistical Advances in the Biomedical Sciences: Clinical Trials, Epidemiology, Survival Analysis, and Bioinformatics
BLISCHKE and MURTHY (editors) · Case Studies in Reliability and Maintenance
BLISCHKE and MURTHY · Reliability: Modeling, Prediction, and Optimization
BLOOMFIELD · Fourier Analysis of Time Series: An Introduction, *Second Edition*

*Now available in a lower priced paperback Edition in the Wiley Classics Library.
†Now available in a lower priced paperback Edition in the Wiley–Interscience Paperback Series.

BOLLEN · Structural Equations with Latent Variables

BOLLEN and CURRAN · Latent Curve Models: A Structural Equation Perspective

BOROVKOV · Ergodicity and Stability of Stochastic Processes

BOSQ and BLANKE · Inference and Prediction in Large Dimensions

BOULEAU · Numerical Methods for Stochastic Processes

* BOX and TIAO · Bayesian Inference in Statistical Analysis

BOX · Improving Almost Anything, *Revised Edition*

* BOX and DRAPER · Evolutionary Operation: A Statistical Method for Process Improvement

BOX and DRAPER · Response Surfaces, Mixtures, and Ridge Analyses, *Second Edition*

BOX, HUNTER, and HUNTER · Statistics for Experimenters: Design, Innovation, and Discovery, *Second Editon*

BOX, JENKINS, and REINSEL · Time Series Analysis: Forcasting and Control, *Fourth Edition*

BOX, LUCEÑO, and PANIAGUA-QUIÑONES · Statistical Control by Monitoring and Adjustment, *Second Edition*

* BROWN and HOLLANDER · Statistics: A Biomedical Introduction

CAIROLI and DALANG · Sequential Stochastic Optimization

CASTILLO, HADI, BALAKRISHNAN, and SARABIA · Extreme Value and Related

Models with Applications in Engineering and Science

CHAN · Time Series: Applications to Finance with R and S-Plus®, *Second Edition*

CHARALAMBIDES · Combinatorial Methods in Discrete Distributions

CHATTERJEE and HADI · Regression Analysis by Example, *Fourth Edition*

CHATTERJEE and HADI · Sensitivity Analysis in Linear Regression

CHERNICK · Bootstrap Methods: A Guide for Practitioners and Researchers, *Second Edition*

CHERNICK and FRIIS · Introductory Biostatistics for the Health Sciences

CHILÈS and DELFINER · Geostatistics: Modeling Spatial Uncertainty, *Second Edition*

CHIU, STOYAN, KENDALL and MECKE · Stochastic Geometry and its Applications, *Third Edition*

CHOW and LIU · Design and Analysis of Clinical Trials: Concepts and Methodologies, *Third Edition*

CLARKE · Linear Models: The Theory and Application of Analysis of Variance

CLARKE and DISNEY · Probability and Random Processes: A First Course with Applications, *Second Edition*

* COCHRAN and COX · Experimental Designs, *Second Edition*

*Now available in a lower priced paperback Edition in the Wiley Classics Library.

†Now available in a lower priced paperback Edition in the Wiley–Interscience Paperback Series.

COLLINS and LANZA · Latent Class and Latent Transition Analysis: With Applications in the Social, Behavioral, and Health Sciences
CONGDON · Applied Bayesian Modelling
CONGDON · Bayesian Models for Categorical Data
CONGDON · Bayesian Statistical Modelling, *Second Edition*
CONOVER · Practical Nonparametric Statistics, *Third Edition*
COOK · Regression Graphics
COOK and WEISBERG · An Introduction to Regression Graphics
COOK and WEISBERG · Applied Regression Including Computing and Graphics
CORNELL · A Primer on Experiments with Mixtures
CORNELL · Experiments with Mixtures, Designs, Models, and the Analysis of Mixture Data, *Third Edition*
COX · A Handbook of Introductory Statistical Methods
CRESSIE · Statistics for Spatial Data, *Revised Edition*
CRESSIE and WIKLE · Statistics for Spatio-Temporal Data
CSÖRGÖ and HORVÁTH · Limit Theorems in Change Point Analysis
DAGPUNAR · Simulation and Monte Carlo: With Applications in Finance and MCMC
DANIEL · Applications of Statistics to Industrial Experimentation
DANIEL · Biostatistics: A Foundation for Analysis in the Health Sciences, *Eighth Edition*
* DANIEL · Fitting Equations to Data: Computer Analysis of Multifactor Data, *Second Edition*
DASU and JOHNSON · Exploratory Data Mining and Data Cleaning
DAVID and NAGARAJA · Order Statistics, *Third Edition*
DAVINO, FURNO and VISTOCCO · Quantile Regression: Theory and Applications
* DEGROOT, FIENBERG, and KADANE · Statistics and the Law
DEL CASTILLO · Statistical Process Adjustment for Quality Control
DEMARIS · Regression with Social Data: Modeling Continuous and Limited Response Variables
DEMIDENKO · Mixed Models: Theory and Applications
DENISON, HOLMES, MALLICK and SMITH · Bayesian Methods for Nonlinear Classification and Regression
DETTE and STUDDEN · The Theory of Canonical Moments with Applications in Statistics, Probability, and Analysis
DEY and MUKERJEE · Fractional Factorial Plans
DE ROCQUIGNY · Modelling Under Risk and Uncertainty: An Introduction to Statistical, Phenomenological and Computational Models
DILLON and GOLDSTEIN · Multivariate Analysis: Methods and Applications

*Now available in a lower priced paperback Edition in the Wiley Classics Library.
†Now available in a lower priced paperback Edition in the Wiley–Interscience Paperback Series.

* DODGE and ROMIG · Sampling Inspection Tables, *Second Edition*
* DOOB · Stochastic Processes
DOWDY, WEARDEN, and CHILKO · Statistics for Research, *Third Edition*
DRAPER and SMITH · Applied Regression Analysis, *Third Edition*
DRYDEN and MARDIA · Statistical Shape Analysis
DUDEWICZ and MISHRA · Modern Mathematical Statistics
DUNN and CLARK · Basic Statistics: A Primer for the Biomedical Sciences, *Fourth Edition*
DUPUIS and ELLIS · A Weak Convergence Approach to the Theory of Large Deviations
EDLER and KITSOS · Recent Advances in Quantitative Methods in Cancer and Human Health Risk Assessment
* ELANDT-JOHNSON and JOHNSON · Survival Models and Data Analysis
ENDERS · Applied Econometric Time Series, *Third Edition*
† ETHIER and KURTZ · Markov Processes: Characterization and Convergence
EVANS, HASTINGS, and PEACOCK · Statistical Distributions, *Third Edition*
EVERITT, LANDAU, LEESE, and STAHL · Cluster Analysis, *Fifth Edition*
FEDERER and KING · Variations on Split Plot and Split Block Experiment Designs
FELLER · An Introduction to Probability Theory and Its Applications, Volume I, *Third Edition*, Revised; Volume II, *Second Edition*
FITZMAURICE, LAIRD, and WARE · Applied Longitudinal Analysis, *Second Edition*
* FLEISS · The Design and Analysis of Clinical Experiments
FLEISS · Statistical Methods for Rates and Proportions, *Third Edition*
† FLEMING and HARRINGTON · Counting Processes and Survival Analysis
FUJIKOSHI, ULYANOV, and SHIMIZU · Multivariate Statistics: High-Dimensional and Large-Sample Approximations
FULLER · Introduction to Statistical Time Series, *Second Edition*
† FULLER · Measurement Error Models
GALLANT · Nonlinear Statistical Models
GEISSER · Modes of Parametric Statistical Inference
GELMAN and MENG · Applied Bayesian Modeling and Causal Inference from Incomplete-Data Perspectives
GEWEKE · Contemporary Bayesian Econometrics and Statistics
GHOSH, MUKHOPADHYAY, and SEN · Sequential Estimation
GIESBRECHT and GUMPERTZ · Planning, Construction, and Statistical Analysis of Comparative Experiments
GIFI · Nonlinear Multivariate Analysis
GIVENS and HOETING · Computational Statistics

*Now available in a lower priced paperback Edition in the Wiley Classics Library.

†Now available in a lower priced paperback Edition in the Wiley–Interscience Paperback Series.

GLASSERMAN and YAO · Monotone Structure in Discrete-Event Systems

GNANADESIKAN · Methods for Statistical Data Analysis of Multivariate Observations, *Second Edition*

GOLDSTEIN · Multilevel Statistical Models, *Fourth Edition*

GOLDSTEIN and LEWIS · Assessment: Problems, Development, and Statistical Issues

GOLDSTEIN and WOOFF · Bayes Linear Statistics

GRAHAM · Markov Chains: Analytic and Monte Carlo Computations

GREENWOOD and NIKULIN · A Guide to Chi-Squared Testing

GROSS, SHORTLE, THOMPSON, and HARRIS · Fundamentals of Queueing Theory, *Fourth Edition*

GROSS, SHORTLE, THOMPSON, and HARRIS · Solutions Manual to Accompany Fundamentals of Queueing Theory, *Fourth Edition*

* HAHN and SHAPIRO · Statistical Models in Engineering

HAHN and MEEKER · Statistical Intervals: A Guide for Practitioners

HALD · A History of Probability and Statistics and their Applications Before 1750

† HAMPEL · Robust Statistics: The Approach Based on Influence Functions

HARTUNG, KNAPP, and SINHA · Statistical Meta-Analysis with Applications

HEIBERGER · Computation for the Analysis of Designed Experiments

HEDAYAT and SINHA · Design and Inference in Finite Population Sampling

HEDEKER and GIBBONS · Longitudinal Data Analysis

HELLER · MACSYMA for Statisticians

HERITIER, CANTONI, COPT, and VICTORIA-FESER · Robust Methods in Biostatistics

HINKELMANN and KEMPTHORNE · Design and Analysis of Experiments, Volume 1: Introduction to Experimental Design, *Second Edition*

HINKELMANN and KEMPTHORNE · Design and Analysis of Experiments, Volume 2: Advanced Experimental Design

HINKELMANN (editor) · Design and Analysis of Experiments, Volume 3: Special Designs and Applications

* HOAGLIN, MOSTELLER, and TUKEY · Fundamentals of Exploratory Analysis of Variance

* HOAGLIN, MOSTELLER, and TUKEY · Exploring Data Tables, Trends and Shapes

* HOAGLIN, MOSTELLER, and TUKEY · Understanding Robust and Exploratory Data Analysis

HOCHBERG and TAMHANE · Multiple Comparison Procedures

HOCKING · Methods and Applications of Linear Models: Regression and the Analysis of Variance, *Third Edition*

HOEL · Introduction to Mathematical Statistics, *Fifth Edition*

*Now available in a lower priced paperback Edition in the Wiley Classics Library.

†Now available in a lower priced paperback Edition in the Wiley–Interscience Paperback Series.

HOGG and KLUGMAN · Loss Distributions

HOLLANDER and WOLFE · Nonparametric Statistical Methods, *Second Edition*

HOSMER and LEMESHOW · Applied Logistic Regression, *Second Edition*

HOSMER, LEMESHOW, and MAY · Applied Survival Analysis: Regression Modeling of Time-to-Event Data, *Second Edition*

HUBER · Data Analysis: What Can Be Learned From the Past 50 Years

HUBER · Robust Statistics

† HUBER and RONCHETTI · Robust Statistics, *Second Edition*

HUBERTY · Applied Discriminant Analysis, *Second Edition*

HUBERTY and OLEJNIK · Applied MANOVA and Discriminant Analysis, *Second Edition*

HUITEMA · The Analysis of Covariance and Alternatives: Statistical Methods for Experiments, Quasi-Experiments, and Single-Case Studies, *Second Edition*

HUNT and KENNEDY · Financial Derivatives in Theory and Practice, *Revised Edition*

HURD and MIAMEE · Periodically Correlated Random Sequences: Spectral Theory and Practice

HUSKOVA, BERAN, and DUPAC · Collected Works of Jaroslav Hajek – with Commentary

HUZURBAZAR · Flowgraph Models for Multistate Time-to-Event Data

INSUA, RUGGERI and WIPER · Bayesian Analysis of Stochastic Process Models

JACKMAN · Bayesian Analysis for the Social Sciences

† JACKSON · A User's Guide to Principle Components

JOHN · Statistical Methods in Engineering and Quality Assurance

JOHNSON · Multivariate Statistical Simulation

JOHNSON and BALAKRISHNAN · Advances in the Theory and Practice of Statistics:
A Volume in Honor of Samuel Kotz

JOHNSON, KEMP, and KOTZ · Univariate Discrete Distributions, *Third Edition*

JOHNSON and KOTZ (editors) · Leading Personalities in Statistical Sciences: From the Seventeenth Century to the Present

JOHNSON, KOTZ, and BALAKRISHNAN · Continuous Univariate Distributions, Volume 1, *Second Edition*

JOHNSON, KOTZ, and BALAKRISHNAN · Continuous Univariate Distributions, Volume 2, *Second Edition*

JOHNSON, KOTZ, and BALAKRISHNAN · Discrete Multivariate Distributions

JUDGE, GRIFFITHS, HILL, LÜTKEPOHL, and LEE · The Theory and Practice of Econometrics, *Second Edition*

*Now available in a lower priced paperback Edition in the Wiley Classics Library.

†Now available in a lower priced paperback Edition in the Wiley–Interscience Paperback Series.

JUREK and MASON · Operator-Limit Distributions in Probability Theory
KADANE · Bayesian Methods and Ethics in a Clinical Trial Design
KADANE AND SCHUM · A Probabilistic Analysis of the Sacco and Vanzetti Evidence
KALBFLEISCH and PRENTICE · The Statistical Analysis of Failure Time Data, *Second Edition*
KARIYA and KURATA · Generalized Least Squares
KASS and VOS · Geometrical Foundations of Asymptotic Inference
† KAUFMAN and ROUSSEEUW · Finding Groups in Data: An Introduction to Cluster Analysis
KEDEM and FOKIANOS · Regression Models for Time Series Analysis
KENDALL, BARDEN, CARNE, and LE · Shape and Shape Theory
KHURI · Advanced Calculus with Applications in Statistics, *Second Edition*
KHURI, MATHEW, and SINHA · Statistical Tests for Mixed Linear Models
* KISH · Statistical Design for Research
KLEIBER and KOTZ · Statistical Size Distributions in Economics and Actuarial Sciences
KLEMELÄ · Smoothing of Multivariate Data: Density Estimation and Visualization
KLUGMAN, PANJER, and WILLMOT · Loss Models: From Data to Decisions, *Third Edition*
KLUGMAN, PANJER, and WILLMOT · Solutions Manual to Accompany Loss Models: From Data to Decisions, *Third Edition*
KOSKI and NOBLE · Bayesian Networks: An Introduction
KOTZ, BALAKRISHNAN, and JOHNSON · Continuous Multivariate Distributions, Volume 1, *Second Edition*
KOTZ and JOHNSON (editors) · Encyclopedia of Statistical Sciences: Volumes 1 to 9 with Index
KOTZ and JOHNSON (editors) · Encyclopedia of Statistical Sciences: Supplement Volume
KOTZ, READ, and BANKS (editors) · Encyclopedia of Statistical Sciences: Update Volume 1
KOTZ, READ, and BANKS (editors) · Encyclopedia of Statistical Sciences: Update Volume 2
KOWALSKI and TU · Modern Applied U-Statistics
KRISHNAMOORTHY and MATHEW · Statistical Tolerance Regions: Theory, Applications, and Computation
KROESE, TAIMRE, and BOTEV · Handbook of Monte Carlo Methods
KROONENBERG · Applied Multiway Data Analysis
KULINSKAYA, MORGENTHALER, and STAUDTE · Meta Analysis: A Guide to Calibrating and Combining Statistical Evidence
KULKARNI and HARMAN · An Elementary Introduction to Statistical Learning Theory

*Now available in a lower priced paperback Edition in the Wiley Classics Library.
†Now available in a lower priced paperback Edition in the Wiley–Interscience Paperback Series.

KUROWICKA and COOKE · Uncertainty Analysis with High Dimensional Dependence Modelling

KVAM and VIDAKOVIC · Nonparametric Statistics with Applications to Science and Engineering

LACHIN · Biostatistical Methods: The Assessment of Relative Risks, *Second Edition*

LAD · Operational Subjective Statistical Methods: A Mathematical, Philosophical, and Historical Introduction

LAMPERTI · Probability: A Survey of the Mathematical Theory, *Second Edition*

LAWLESS · Statistical Models and Methods for Lifetime Data, *Second Edition*

LAWSON · Statistical Methods in Spatial Epidemiology, *Second Edition*

LE · Applied Categorical Data Analysis, *Second Edition*

LE · Applied Survival Analysis

LEE · Structural Equation Modeling: A Bayesian Approach

LEE and WANG · Statistical Methods for Survival Data Analysis, *Third Edition*

LEPAGE and BILLARD · Exploring the Limits of Bootstrap

LESSLER and KALSBEEK · Nonsampling Errors in Surveys

LEYLAND and GOLDSTEIN (editors) · Multilevel Modelling of Health Statistics

LIAO · Statistical Group Comparison

LIN · Introductory Stochastic Analysis for Finance and Insurance

LITTLE and RUBIN · Statistical Analysis with Missing Data, *Second Edition*

LLOYD · The Statistical Analysis of Categorical Data

LOWEN and TEICH · Fractal-Based Point Processes

MAGNUS and NEUDECKER · Matrix Differential Calculus with Applications in Statistics and Econometrics, *Revised Edition*

MALLER and ZHOU · Survival Analysis with Long Term Survivors

MARCHETTE · Random Graphs for Statistical Pattern Recognition

MARDIA and JUPP · Directional Statistics

MARKOVICH · Nonparametric Analysis of Univariate Heavy-Tailed Data: Research and Practice

MARONNA, MARTIN and YOHAI · Robust Statistics: Theory and Methods

MASON, GUNST, and HESS · Statistical Design and Analysis of Experiments with Applications to Engineering and Science, *Second Edition*

McCULLOCH, SEARLE, and NEUHAUS · Generalized, Linear, and Mixed Models, *Second Edition*

McFADDEN · Management of Data in Clinical Trials, *Second Edition*

* McLACHLAN · Discriminant Analysis and Statistical Pattern Recognition

McLACHLAN, DO, and AMBROISE · Analyzing Microarray Gene Expression Data

*Now available in a lower priced paperback Edition in the Wiley Classics Library.

†Now available in a lower priced paperback Edition in the Wiley–Interscience Paperback Series.

McLACHLAN and KRISHNAN · The EM Algorithm and Extensions, *Second Edition*

McLACHLAN and PEEL · Finite Mixture Models

McNEIL · Epidemiological Research Methods

MEEKER and ESCOBAR · Statistical Methods for Reliability Data

MEERSCHAERT and SCHEFFLER · Limit Distributions for Sums of Independent Random Vectors: Heavy Tails in Theory and Practice

MENGERSEN, ROBERT, and TITTERINGTON · Mixtures: Estimation and Applications

MICKEY, DUNN, and CLARK · Applied Statistics: Analysis of Variance and Regression, *Third Edition*

* MILLER · Survival Analysis, *Second Edition*

MONTGOMERY, JENNINGS, and KULAHCI · Introduction to Time Series Analysis and Forecasting

MONTGOMERY, PECK, and VINING · Introduction to Linear Regression Analysis, *Fifth Edition*

MORGENTHALER and TUKEY · Configural Polysampling: A Route to Practical Robustness

MUIRHEAD · Aspects of Multivariate Statistical Theory

MULLER and STOYAN · Comparison Methods for Stochastic Models and Risks

MURTHY, XIE, and JIANG · Weibull Models

MYERS, MONTGOMERY, and ANDERSON-COOK · Response Surface Methodology: Process and Product Optimization Using Designed Experiments, *Third Edition*

MYERS, MONTGOMERY, VINING, and ROBINSON · Generalized Linear Models. With Applications in Engineering and the Sciences, *Second Edition*

NATVIG · Multistate Systems Reliability Theory With Applications

† NELSON · Accelerated Testing, Statistical Models, Test Plans, and Data Analyses

† NELSON · Applied Life Data Analysis

NEWMAN · Biostatistical Methods in Epidemiology

NG, TAIN, and TANG · Dirichlet Theory: Theory, Methods and Applications

OKABE, BOOTS, SUGIHARA, and CHIU · Spatial Tesselations: Concepts and Applications of Voronoi Diagrams, *Second Edition*

OLIVER and SMITH · Influence Diagrams, Belief Nets and Decision Analysis

PALTA · Quantitative Methods in Population Health: Extensions of Ordinary Regressions

PANJER · Operational Risk: Modeling and Analytics

PANKRATZ · Forecasting with Dynamic Regression Models

*Now available in a lower priced paperback Edition in the Wiley Classics Library.

†Now available in a lower priced paperback Edition in the Wiley–Interscience Paperback Series.

PANKRATZ · Forecasting with Univariate Box-Jenkins Models: Concepts and Cases

PARDOUX · Markov Processes and Applications: Algorithms, Networks, Genome and Finance

PARMIGIANI and INOUE · Decision Theory: Principles and Approaches

* PARZEN · Modern Probability Theory and Its Applications

PEÑA, TIAO, and TSAY · A Course in Time Series Analysis

PESARIN and SALMASO · Permutation Tests for Complex Data: Applications and Software

PIANTADOSI · Clinical Trials: A Methodologic Perspective, *Second Edition*

POURAHMADI · Foundations of Time Series Analysis and Prediction Theory

POURAHMADI · High-Dimensional Covariance Estimation

POWELL · Approximate Dynamic Programming: Solving the Curses of Dimensionality, *Second Edition*

POWELL and RYZHOV · Optimal Learning

PRESS · Subjective and Objective Bayesian Statistics, *Second Edition*

PRESS and TANUR · The Subjectivity of Scientists and the Bayesian Approach

PURI, VILAPLANA, and WERTZ · New Perspectives in Theoretical and Applied Statistics

† PUTERMAN · Markov Decision Processes: Discrete Stochastic Dynamic Programming

QIU · Image Processing and Jump Regression Analysis

* RAO · Linear Statistical Inference and Its Applications, *Second Edition*

RAO · Statistical Inference for Fractional Diffusion Processes

RAUSAND and HØYLAND · System Reliability Theory: Models, Statistical Methods, and Applications, *Second Edition*

RAYNER, THAS, and BEST · Smooth Tests of Goodnes of Fit: Using R, *Second Edition*

RENCHER and SCHAALJE · Linear Models in Statistics, *Second Edition*

RENCHER and CHRISTENSEN · Methods of Multivariate Analysis, *Third Edition*

RENCHER · Multivariate Statistical Inference with Applications

RIGDON and BASU · Statistical Methods for the Reliability of Repairable Systems

* RIPLEY · Spatial Statistics

* RIPLEY · Stochastic Simulation

ROHATGI and SALEH · An Introduction to Probability and Statistics, *Second Edition*

ROLSKI, SCHMIDLI, SCHMIDT, and TEUGELS · Stochastic Processes for Insurance and Finance

*Now available in a lower priced paperback Edition in the Wiley Classics Library.

†Now available in a lower priced paperback Edition in the Wiley–Interscience Paperback Series.

ROSENBERGER and LACHIN · Randomization in Clinical Trials: Theory and Practice

ROSSI, ALLENBY, and McCULLOCH · Bayesian Statistics and Marketing

† ROUSSEEUW and LEROY · Robust Regression and Outlier Detection

ROYSTON and SAUERBREI · Multivariate Model Building: A Pragmatic Approach to Regression Analysis Based on Fractional Polynomials for Modeling Continuous Variables

* RUBIN · Multiple Imputation for Nonresponse in Surveys

RUBINSTEIN and KROESE · Simulation and the Monte Carlo Method, *Second Edition*

RUBINSTEIN and MELAMED · Modern Simulation and Modeling

RUBINSTEIN, RIDDER, and VAISMAN · Fast Sequential Monte Carlo Methods for Counting and Optimization

RYAN · Modern Engineering Statistics

RYAN · Modern Experimental Design

RYAN · Modern Regression Methods, *Second Edition*

RYAN · Sample Size Determination and Power

RYAN · Statistical Methods for Quality Improvement, *Third Edition*

SALEH · Theory of Preliminary Test and Stein-Type Estimation with Applications

SALTELLI, CHAN, and SCOTT (editors) · Sensitivity Analysis

SCHERER · Batch Effects and Noise in Microarray Experiments: Sources and Solutions

* SCHEFFE · The Analysis of Variance

SCHIMEK · Smoothing and Regression: Approaches, Computation, and Application

SCHOTT · Matrix Analysis for Statistics, *Second Edition*

SCHOUTENS · Levy Processes in Finance: Pricing Financial Derivatives

SCOTT · Multivariate Density Estimation: Theory, Practice, and Visualization

* SEARLE · Linear Models

† SEARLE · Linear Models for Unbalanced Data

† SEARLE · Matrix Algebra Useful for Statistics

† SEARLE, CASELLA, and McCULLOCH · Variance Components

SEARLE and WILLETT · Matrix Algebra for Applied Economics

SEBER · A Matrix Handbook For Statisticians

† SEBER · Multivariate Observations

SEBER and LEE · Linear Regression Analysis, *Second Edition*

† SEBER and WILD · Nonlinear Regression

SENNOTT · Stochastic Dynamic Programming and the Control of Queueing Systems

*Now available in a lower priced paperback Edition in the Wiley Classics Library.

†Now available in a lower priced paperback Edition in the Wiley–Interscience Paperback Series.

* SERFLING · Approximation Theorems of Mathematical Statistics

SHAFER and VOVK · Probability and Finance: It's Only a Game!

SHERMAN · Spatial Statistics and Spatio-Temporal Data: Covariance Functions and Directional Properties

SILVAPULLE and SEN · Constrained Statistical Inference: Inequality, Order, and Shape Restrictions

SINGPURWALLA · Reliability and Risk: A Bayesian Perspective

SMALL and McLEISH · Hilbert Space Methods in Probability and Statistical Inference

SRIVASTAVA · Methods of Multivariate Statistics

STAPLETON · Linear Statistical Models, *Second Edition*

STAPLETON · Models for Probability and Statistical Inference: Theory and Applications

STAUDTE and SHEATHER · Robust Estimation and Testing

STOYAN · Counterexamples in Probability, *Second Edition*

STOYAN, KENDALL, and MECKE · Stochastic Geometry and Its Applications, *Second Edition*

STOYAN and STOYAN · Fractals, Random Shapes and Point Fields: Methods of Geometrical Statistics

STREET and BURGESS · The Construction of Optimal Stated Choice Experiments: Theory and Methods

STYAN · The Collected Papers of T. W. Anderson: 1943–1985

SUTTON, ABRAMS, JONES, SHELDON, and SONG · Methods for Meta-Analysis in Medical Research

TAKEZAWA · Introduction to Nonparametric Regression

TAMHANE · Statistical Analysis of Designed Experiments: Theory and Applications

TANAKA · Time Series Analysis: Nonstationary and Noninvertible Distribution Theory

THOMPSON · Empirical Model Building: Data, Models, and Reality, *Second Edition*

THOMPSON · Sampling, *Third Edition*

THOMPSON · Simulation: A Modeler's Approach

THOMPSON and SEBER · Adaptive Sampling

THOMPSON, WILLIAMS, and FINDLAY · Models for Investors in Real World Markets

TIERNEY · LISP-STAT: An Object-Oriented Environment for Statistical Computing and Dynamic Graphics

TROFFAES and DE COOMAN · Lower Previsions

TSAY · Analysis of Financial Time Series, *Third Edition*

TSAY · An Introduction to Analysis of Financial Data with R

*Now available in a lower priced paperback Edition in the Wiley Classics Library.

†Now available in a lower priced paperback Edition in the Wiley–Interscience Paperback Series.

TSAY · Multivariate Time Series Analysis: With R and Financial Applications

UPTON and FINGLETON · Spatial Data Analysis by Example, Volume II: Categorical and Directional Data

† VAN BELLE · Statistical Rules of Thumb, *Second Edition*

VAN BELLE, FISHER, HEAGERTY, and LUMLEY · Biostatistics: A Methodology for the Health Sciences, *Second Edition*

VESTRUP · The Theory of Measures and Integration

VIDAKOVIC · Statistical Modeling by Wavelets

VIERTL · Statistical Methods for Fuzzy Data

VINOD and REAGLE · Preparing for the Worst: Incorporating Downside Risk in Stock Market Investments

WALLER and GOTWAY · Applied Spatial Statistics for Public Health Data

WANG and WANG · Structural Equation Modeling: Applications Using M*plus*

WEISBERG · Applied Linear Regression, *Fourth Edition*

WEISBERG · Bias and Causation: Models and Judgment for Valid Comparisons

WELSH · Aspects of Statistical Inference

WESTFALL and YOUNG · Resampling-Based Multiple Testing: Examples and Methods for *p*-Value Adjustment

* WHITTAKER · Graphical Models in Applied Multivariate Statistics

WINKER · Optimization Heuristics in Economics: Applications of Threshold Accepting

WOODWORTH · Biostatistics: A Bayesian Introduction

WOOLSON and CLARKE · Statistical Methods for the Analysis of Biomedical Data, *Second Edition*

WU and HAMADA · Experiments: Planning, Analysis, and Parameter Design Optimization, *Second Edition*

WU and ZHANG · Nonparametric Regression Methods for Longitudinal Data Analysis

YAKIR · Extremes in Random Fields

YIN · Clinical Trial Design: Bayesian and Frequentist Adaptive Methods

YOUNG, VALERO-MORA, and FRIENDLY · Visual Statistics: Seeing Data with Dynamic Interactive Graphics

ZACKS · Stage-Wise Adaptive Designs

* ZELLNER · An Introduction to Bayesian Inference in Econometrics

ZELTERMAN · Discrete Distributions – Applications in the Health Sciences

ZHOU, OBUCHOWSKI, and MCCLISH · Statistical Methods in Diagnostic Medicine, *Second Edition*

[*]Now available in a lower priced paperback Edition in the Wiley Classics Library.

[†]Now available in a lower priced paperback Edition in the Wiley–Interscience Paperback Series.